Lecture Notes in Computer Science

Commenced Publication in 1973
Founding and Former Series Editors:
Gerhard Goos, Juris Hartmanis, and Jan van Leeuwen

Beniamino Murgante Osvaldo Gervasi
Andrés Iglesias David Taniar
Bernady O. Apduhan (Eds.)

Computational Science and Its Applications - ICCSA 2011

International Conference
Santander, Spain, June 20-23, 2011
Proceedings, Part I

Volume Editors

Beniamino Murgante
Basilicata University Potenza, Italy
E-mail: beniamino.murgante@unibas.it

Osvaldo Gervasi
University of Perugia, Italy
E-mail: osvaldo@unipg.it

Andrés Iglesias
University of Cantabria, Santander, Spain
E-mail: iglesias@unican.es

David Taniar
Monash University, Clayton, VIC, Australia
E-mail: david.taniar@infotech.monash.edu.au

Bernady O. Apduhan
Kyushu Sangyo University
Fukuoka, Japan
E-mail: bob@is.kyusan-u.ac.jp

ISSN 0302-9743
e-ISSN 1611-3349
ISBN 978-3-642-21927-6
e-ISBN 978-3-642-21928-3
DOI 10.1007/978-3-642-21928-3
Springer Heidelberg Dordrecht London New York

Library of Congress Control Number: 2011929636

CR Subject Classification (1998): C.2, H.4, F.2, H.3, D.2, C.2.4, F.1, H.5

LNCS Sublibrary: SL 1 – Theoretical Computer Science and General Issues

Typesetting: Camera-ready by author, data conversion by Scientific Publishing Services, Chennai, India

Printed on acid-free paper

Springer is part of Springer Science+Business Media (www.springer.com)

Preface

These multiple volumes (LNCS volumes 6782, 6783, 6784, 6785 and 6786) consist of the peer-reviewed papers from the 2011 International Conference on Computational Science and Its Applications (ICCSA 2011) held in Santander, Spain during June 20-23, 2011. ICCSA 2011 was a successful event in the International Conferences on Computational Science and Its Applications (ICCSA) conference series, previously held in Fukuoka, Japan (2010), Suwon, South Korea (2009), Perugia, Italy (2008), Kuala Lumpur, Malaysia (2007), Glasgow, UK (2006), Singapore (2005), Assisi, Italy (2004), Montreal, Canada (2003), and (as ICCS) Amsterdam, The Netherlands (2002) and San Francisco, USA (2001).

Computational science is a main pillar of most of the present research, as well as industrial and commercial activities and plays a unique role in exploiting ICT innovative technologies. The ICCSA conferences have been providing a venue to researchers and industry practitioners to discuss new ideas, to share complex problems and their solutions, and to shape new trends in computational science.

Apart from the general tracks, ICCSA 2011 also included 31 special sessions and workshops, in various areas of computational science, ranging from computational science technologies to specific areas of computational science, such as computer graphics and virtual reality. We accepted 52 papers for the general track, and 210 in special sessions and workshops. These represent an acceptance rate of 29.7%. We would like to show our appreciations to the Workshop and Special Session Chairs and co-Chairs.

The success of the ICCSA conference series, in general, and ICCSA 2011, in particular, is due to the support of many people: authors, presenters, participants, keynote speakers, Session Chairs, Organizing Committee members, student volunteers, Program Committee members, International Liaison Chairs, and people in other various roles. We would like to thank them all. We would also like to thank Springer for their continuous support in publishing ICCSA conference proceedings.

June 2011

Osvaldo Gervasi
David Taniar

Message from the ICCSA 2011 General Chairs

These five volumes contain an outstanding collection of refereed papers selected for the 11th International Conference on Computational Science and Its Applications, ICCSA 2011, held in Santander (Spain), June 20-23, 2011. We cordially invite you to visit the ICCSA website `http://www.iccsa.org` where you can find all relevant information about this interesting and exciting event.

ICCSA 2011 marked the beginning of the second decade of this conference series. Previous editions in this series of highly successful International Conferences on Computational Science and Its Applications (ICCSA) were held in Fukuoka, Japan (2010), Suwon, Korea (2009), Perugia, Italy (2008), Kuala Lumpur, Malaysia (2007), Glasgow, UK (2006), Singapore (2005), Assisi, Italy (2004), Montreal, Canada (2003), and (as ICCS) Amsterdam, The Netherlands (2002) and San Francisco, USA (2001).

As we enter the second decade of ICCSA, we realize the profound changes and spectacular advances in the world of computational science. This discipline plays a unique role in fostering new technologies and knowledge, and is crucial for most of the present research, and industrial and commercial activities. We believe that ICCSA has contributed to this change by offering a real opportunity to explore innovative approaches and techniques to solve complex problems. Reciprocally, the computational science community has enthusiastically embraced the successive editions of ICCSA, thus contributing to making ICCSA a focal meeting point for those interested in innovative, cutting-edge research about the latest and most exciting developments in the field. We are grateful to all those who have contributed to the current success of ICCSA with their continued support over the past ten years.

ICCSA 2011 would not have been made possible without the valuable contribution from many people. We would like to thank all session organizers for their diligent work, which further enhanced the conference levels and all reviewers for their expertise and generous effort which led to a very high quality event with excellent papers and presentations. We especially recognize the contribution of the Program Committee and Local Organizing Committee members for their tremendous support and for making this congress a very sucessful event.

We would like to sincerely thank our keynote speakers, who willingly accepted our invitation and shared their expertise through illuminating talks, helping us to fully meet the conference objectives.

We highly appreciate the University of Cantabria for their enthusiastic acceptance to host the conference on its main campus, their logistic assistance and additional financial support. The conference was held in the Faculty of Sciences of the University of Cantabria. We thank the Dean of the Faculty of Sciences, Ernesto Anabitarte, for his support before and during the congress, and for providing the venue of the conference and the use of all needed facilities.

ICCSA 2011 was jointly organized by the Department of Applied Mathematics and Computational Sciences and the Department of Mathematics, Statistics and Computation of the University of Cantabria, Spain. We thank both departments for their encouranging support of this conference from the very beginning. We would like to express our gratitude to the Local Organizing Committee for their persistent and enthusiastic work towards the success of this conference.

We owe special thanks to all our sponsors: the Faculty of Sciences, the University of Cantabria, the Municipality of Santander, the Regional Government of Cantabria and the Spanish Ministry of Science and Innovation, for their continuous support without which this conference would not be possible. We also thank our publisher, Springer, for their acceptance to publish the proceedings and for their kind assistance and cooperation during the editing process.

Finally, we thank all authors for their submissions and all conference attendants for making ICCSA 2011 truly an excellent forum on computational science, facilitating exchange of ideas, fostering new collaborations and shaping the future of this exciting field. Last, but certainly not least, we wish to thank our readers for their interest in these proceedings. We really hope you find in these pages interesting material and fruitful ideas for your future work.

June 2011

Andrés Iglesias
Bernady O. Apduhan

The Wisdom of Ancient Masters

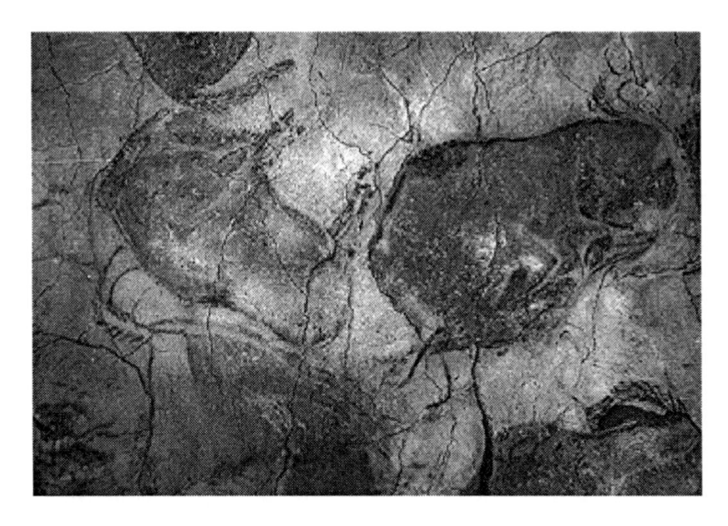

In 1879, Marcelino Sanz de Sautuola and his young daugther María incidentally noticed that the ceiling of the Altamira cave was covered by images of bisons and other animals, some as old as between 25,000 and 35,000 years. They had discovered what came to be called the Sistine Chapel of Paleolithic Art. When the discovery was first made public in 1880, many experts rejected it under the belief that prehistoric man was unable to produce such beautiful and ellaborated paintings. Once their authenticity was later confirmed, it changed forever our perception of prehistoric human beings.

Today, the cave of Altamira and its paintings are a symbol of the wisdom and ability of our ancient ancestors. They remind us that our current technological development is mostly based on the work, genius and efforts of our predecessors over many generations.

The cave of Altamira (UNESCO World Heritage Site) is located in the region of Cantabria, near the city of Santander (ICCSA 2011 conference venue). The original cave is closed to the public for preservation, but conference attendees visited the "Neocave", an exact reproduction of the original space with all its cracks and textures and the permanent exhibition "The Times of Altamira", which introduces visitors to the prehistory of the peninsula and rupestrian art.

"After Altamira, all is decadence" (Pablo Picasso, famous Spanish painter)

ICCSA 2011 Welcome Message

Welcome to the proceedings of the 11th International Conference on Computational Science and Its Applications, ICCSA 2011, held in Santander, Spain.

The city of Santander is located in the self-governed region of Cantabria, on the northern coast of Spain between Asturias and the Basque Country. This beautiful region of half a million inhabitants is on the shores of the Cantabrian Sea and is crossed by a mountain range. The shores and inland valleys offer a wide variety of landscapes as a consequence of the mild, moist climate of so-called Green Spain. The coastal landscape of beaches, bays and cliffs blends together with valleys and highland areas. All along the coast there are typical traditional fishing ports and innumerable diverse beaches of soft white sand.

However, Cantabria's attractions are not limited to its natural treasures. History has provided a rich artistic and cultural heritage found in towns and villages that are outstanding in their own right. The archaeological remains and historic buildings bear the mark of a unique history starting with the world-famous Altamira cave paintings, a veritable shrine to the prehistoric age. In addition, there are remarkable remains from the Romans, the Mozarabic presence and the beginnings of the Reconquest of Spain, along with an artistic heritage of Romanesque, Gothic and Baroque styles. Examples include the Prehistoric Era (the Altamira and Puente Viesgo Caves), Roman ruins such as those of Julióbriga, medieval settlements, such as Santillana del Mar, and several examples of the civil and religious architecture of the nineteenth and twentieth centuries.

The surrounding natural landscape and the historical importance of many of its villages and buildings make this region very appealing for tourism, especially during the spring and summer seasons, when the humid, mild weather gives the region a rich and varied nature of woods and meadows. At the time of the conference, attendees enjoyed the gentle climate (with temperatures averaging 18-20 degrees Celsius) and the longest days of the year. They found themselves waiting for sunset at the beach at about 11 pm!

Capital of the autonomous region of Cantabria, the city of Santander is also a very popular destination for tourism. Based around one of the most beautiful bays in the world, this modern city is famous for its sparkling beaches of yellow sand and clean water, the hospitality of its people and the high reputation of its celebrated gastronomy, mostly based on fish and shellfish. With a population of about 200,000 inhabitants, Santander is a very safe city, with a vibrant tourist scene filled with entertainment and a high quality of life, matching the best standards in the world. The coastal side of the city boasts a long string of top–quality beaches and recreational areas, such as the Magdalena Peninsula, the Sardinero and Mataleñas Park. There are several beaches and harbors limiting the city on the northern side, toward the southern part there is the old city center and a bit further on the green mountains. We proudly say that Santander is between the blue and the green.

The University of Cantabria (in Spanish, *the Universidad de Cantabria, UC*) is the only public university in Cantabria, Spain. It was founded in 1972 and is organized in 12 faculties and schools. With about 13,000 students and 1,000 academic staff, the University of Cantabria is one of the most reputed universities in the country, ranking in the highest positions of Spanish universities in relation to its size. Not surprisingly, it was selected as a Campus of International Excellence by the Spanish Government in 2009.

Besides the technical sessions and presentations, ICCSA 2011 provided an interesting, must-attend social program. It started with a Welcome Reception at the Royal Palace of the Magdalena (Sunday 19), the most emblematic building of Santander and also the most visited place in the city. The royal family used the palace during the period 1913–1930 as a base for numerous recreational and sporting activities, and the king sometimes also held government meetings at the property. Conference delegates had the wonderful opportunity to visit this splendid palace, enjoy the magnificent views and see some rooms where royalty lived. The Gala Dinner (Tuesday 21) took place at the Grand Casino, in the "Sardinero" area, a regal, 1920's building with large windows and spacious terraces offering superb views of the Sardinero beach. The Casino was King Alfonso XIII and Queen Victoria Eugenia's main place of entertainment during their summer holidays in the city between 1913 and 1930. The gala also included some cultural and musical events. Finally, a half-day conference tour (Wednesday 22) covered the "live museum" of the Middle Ages, Santillana del Mar (a medieval town with cobbled streets, declared "Site of Artistic and Historical Importance" and one of the best-known cultural and tourist centers in Cantabria) and the Altamira Neocave, an exact reproduction of the original Altamira cave (now closed to the public for preservation) with all its cracks and textures and the permanent exhibition "The Times of Altamira", which introduces visitors to the prehistory of the peninsula and rupestrian art.

To close the conference, attendees could join the people of Santander for St. John's day, celebrated in the night between June 23 and 24 to commemorate the summer solstice with bonfires on the beach.

We believe that all these attractions made the conference an unforgettable experience.

On behalf of the Local Organizing Committee members, I thank all attendees fot their visit.

June 2011 Andrés Iglesias

Message from the Chairs of the Session: 6th International Workshop on "Geographical Analysis, Urban Modeling, Spatial Statistics" (GEOG-AN-MOD 2011)

During the past few decades the main problem in geographical analysis was the lack of spatial data availability. Nowadays the wide diffusion of electronic devices containing geo-referenced information generates a great production of spatial data. Volunteered geographic information activities (e.g., Wikimapia, OpenStreetMap), public initiatives (e.g., spatial data infrastructures, geo-portals) and private projects (e.g., Google Earth, Microsoft Virtual Earth, etc.) produced an overabundance of spatial data, which, in many cases, do not help the efficiency of decision processes. The increase of geographical data availability has not been fully coupled by an increase of knowledge to support spatial decisions.

The inclusion of spatial simulation techniques in recent GIS software favored the diffusion of these methods, but in several cases led to mechanisms based on which buttons have to be pressed without having geography or processes in mind. Spatial modeling, analytical techniques and geographical analyses are therefore required in order to analyze data and to facilitate the decision process at all levels, with a clear identification of the geographical information needed and reference scale to adopt. Old geographical issues can find an answer thanks to new methods and instruments, while new issues are developing, challenging researchers for new solutions. This workshop aims at contributing to the development of new techniques and methods to improve the process of knowledge acquisition.

Conference themes include:

Geostatistics and spatial simulation
Agent-based spatial modeling
Cellular automata spatial modeling
Spatial statistical models
Space-temporal modeling
Environmental modeling
Geovisual analytics, geovisualization, visual exploratory data analysis
Visualization and modeling of track data
Spatial optimization
Interaction simulation models
Data mining, spatial data mining
Spatial data warehouse and spatial OLAP
Integration of spatial OLAP and spatial data mining
Spatial decision support systems

Spatial multicriteria decision analysis
Spatial rough set
Spatial extension of fuzzy set theory
Ontologies for spatial analysis
Urban modeling
Applied geography
Spatial data analysis
Dynamic modeling
Simulation, space-time dynamics, visualization and virtual reality.

Giuseppe Borruso
Beniamino Murgante
Stefania Bertazzon

Message from the Chairs of the Session: "Cities, Technologies and Planning" (CTP 2011)

'Share' term has turned into a key issue of many successful initiatives in recent times. Following the advent of Web 2.0, positive experiences based on mass collaboration generated "Wikinomics" hnd ave become 'Socialnomics", where 'Citizens are voluntary sensors'.

During the past few decades, the main issue in GIS implementation has been the availability of sound spatial information. Nowadays, the wide diffusion of electronic devices providing geo-referenced information resulted in the production of extensive spatial information datasets. This trend has led to "GIS wikification", where mass collaboration plays a key role in the main components of spatial information frameworks (hardware, software, data, and people).

Some authors (Goodchild, 2007) talk about 'volunteered geographic information' (VGI), as the harnessing of tools to create, assemble, and disseminate geographic information provided by individuals voluntarily creating their own contents by marking the locations of occurred events or by labeling certain existing features not already shown on a map. The term "neogeography" is often adopted to describe peoples activities when using and creating their own maps, geo-tagging pictures, movies, websites, etc. It could be defined as a new bottom up approach to geography prompted by users, therefore introducing changes in the roles of traditional' geographers and consumers' of geographical contents themselves. The volunteered approach has been adopted by important American organizations, such as US Geological Survey, US Census Bureau, etc. While technologies (e.g. GPS, remote sensing, etc.) can be useful in producing new spatial data, volunteered activities are the only way to update and describe such data. If spatial data have been produced in various ways, remote sensing, sensor networks and other electronic devices generate a great flow of relevant spatial information concerning several aspects of human activities or of environmental phenomena monitoring. This 'information-explosion era' is characterized by a large amount of information produced both by human activities and by automated systems; the capturing and the manipulation of this information leads to 'urban computing' and represents a sort of bridge between computers and the real world, accounting for the social dimension of human environments. This technological evolution produced a new paradigm of urban development, called 'u-City'. Such phenomena offer new challenges to scholars (geographers, engineers, planners, economists, sociologists, etc.) as well as to spatial planners in addressing spatial issues and a wealth of brand-new, updated data, generally created by people who are interested in geographically related phenomena. As attention is to date dedicated to visualization and content creation, little has been done from the spatial analytical point of view and in involving users as citizens in participatory geographical activities.

Conference themes include:

SDI and planning
Planning 2.0, participation 2.0
Urban social networks, urban sensing
E-democracy, e-participation, participatory GIS
Technologies for e-participation, policy modeling, simulation and visualization
Second Life and participatory games
Ubiquitous computing environment; urban computing; ubiquitous-city
Neogeography
Collaborative mapping
Geotagging
Volunteered geographic information
Crowdsourcing
Ontologies for urban planning
City Gml
Geo-applications for mobile phones
Web 2.0, Web 3.0
Wikinomics, socialnomics
WikiCities
Maps mash up
Tangible maps and planning
Augmented reality,
Complexity assessment and mapping

Giuseppe Borruso
Beniamino Murgante

Message from the Chairs of the Session: 11th Annual International Workshop on "Computational Geometry and Applications" (CGA 2011)

The 11th International Workshop on Computational Geometry and Applications CGA 2011, held in conjunction with the International Conference on Computational Science and Applications, took place in Santander, Spain. The workshop has run annually since it was founded in 2001, and is intended as an international forum for researchers in computational geometry and related areas, with the goal of advancing the state of research in computational geometry and related disciplines. This year, the workshop was chaired for 11th year by CGA workshop series Founding Chair Marina Gavrilova, University of Calgary, joined by co-Chair Ovidiu Daescu, University of Texas at Dallas. Selected papers from the previous CGA Workshops have appeared in special issues in the following highly regarded journals: *International Journal of Computational Geometry and Applications*, Springer (three special issues), *International Journal of Computational Science and Engineering* (IJCSE), Journal of CAD/CAM , *Transactions on Computational Sciences*, Springer. A special issue comprising best papers presented at CGA 2011 is currently being planned.

The workshop attracts international attention and receives papers presenting high-quality original research in the following tracks:

- Theoretical computational geometry
- Applied computational geometry
- Optimization and performance issues in geometric algorithms implementation Workshop topics of interest include:
- Design and analysis of geometric algorithms
- Geometric algorithms in path planning and robotics
- Computational geometry in biometrics
- Intelligent geometric computing
- Geometric algorithms in computer graphics and computer vision
- Voronoi diagrams and their generalizations
- 3D Geometric modeling
- Geometric algorithms in geographical information systems
- Algebraic geometry
- Discrete and combinatorial geometry
- Implementation issues and numerical precision
- Applications in computational biology, physics, chemistry, geography, medicine, education
- Visualization of geometric algorithms

CGA 2011 was located in beautiful Santander, Cantabria, Spain. Santander, the capital city of Cantabria, is located on the northern coast of Spain, between Asturias and the Basque Country overlooking the Cantabrian Sea, and is surrounded by beaches. The conference preceded the Spanish Meeting on Computational Geometry, which took place in Madrid, facilitating interested researchers to attend both events. The 14 articles presented in this Springer LNCS proceeding volume represent papers selected fromi a large number of submissions to this year's workshop. We would like to express our sincere gratitude to the following International Program Committee members who performed their duties diligently and provided constructive feedback for authors to improve on their presentation:

Tetsuo Asano (Japan Advanced Institute of Science and Technology, Japan)
Sergei Bereg (University of Texas at Dallas, USA)
Karoly Bezdek (University of Calgary, Canada)
Ovidiu Daescu (University of Texas at Dallas, USA)
Mirela Damian (Villanova University, USA)
Tamal Dey (Ohio State University, USA)
Marina L. Gavrilova (University of Calgary, Canada)
Christopher Gold (University of Glamorgan, UK)
Hisamoto Hiyoshi (Gunma University, Japan)
Andrés Iglesias (University of Cantabria, Spain)
Anastasia Kurdia (Smith College, USA)
Deok-Soo Kim (Hanyang University, Korea)
Ivana Kolingerova (Unversity of West Bohemia, Czech Republic)
Nikolai Medvedev (Novosibirsk Russian Academy of Science, Russia)
Asish Mukhopadhyay (University of Windsor, Canada)
Dimitri Plemenos (Université de Limoges, France)
Val Pinciu (Southern Connecticut State University, USA)
Jon Rokne (University of Calgary, Canada)
Carlos Seara (Universitat Politecnica de Catalunya, Spain)
Kokichi Sugihara (University of Tokyo, Japan)
Vaclav Skala (University of West Bohemia, Czech Republic)
Muhammad Sarfraz (KFUPM, Saudi Arabia)
Alexei Sourin (Nanyang Technological University, Singapore)
Ryuhei Uehara (Japan Advanced Institute of Science and Technology, Japan)
Chee Yap (New York University, USA)
Kira Vyatkina (Sanct Petersburg State University, Russia)

We also would like to acknowledge the independent referees, ICCSA 2011 organizers, sponsors, volunteers, and Springer for their continuing collaboration and support.

Marina C. Gavrilova
Ovidiu Daescu

Message from the Chair of the Session: 3rd International Workshop on "Software Engineering Processes and Applications" (SEPA 2011)

The Third International Workshop on Software Engineering Processes and Applications (SEPA 2011) covered the latest developments in processes and applications of software engineering. SEPA includes process models, agile development, software engineering practices, requirements, system and design engineering including architectural design, component level design, formal methods, software modeling, testing strategies and tactics, process and product metrics, Web engineering, project management, risk management, and configuration management and all those areas which are related to the processes and any type of software applications. This workshop attracted papers from leading researchers in the field of software engineering and its application areas. Seven regular research papers were accepted as follows.

Sanjay Misra, Ibrahim Akman and Ferid Cafer presented a paper on "A Multi-Paradigm Complexity Metric(MCM)" The authors argued that there are not metrics in the literature for multi-paradigm. MCM is developed by using function points and procedural and object–oriented language's features. In this view, MCM involves most of the factors which are responsible for the complexity of any multi-paradigm language. MCM can be used for most programming paradigms, including both procedural and object–oriented languages.

Mohamed A. El-Zawawy's paper entitled 'Flow Sensitive-Insensitive Pointer Analysis Based Memory Safety for Multithreaded Programs' presented approaches for the pointer analysis and memory safety of multithreaded programs as simply structured type systems. The author explained that in order to balance accuracy and scalability, the proposed type system for pointer analysis of multithreaded programs is flow-sensitive, which invokes another flow-insensitive type system for parallel constructs.

Cesar Pardo, Francisco Pino, Felix Garcia, Francisco Romero, Mario Piattini, and Maria Teresa Baldassarre presented their paper entitled 'HProcessTOOL: A Support Tool in the Harmonization of Multiple Reference Models'. The authors have developed the tool HProcessTOOL, which guides harmonization projects by supporting specific techniques, and supports their management by controlling and monitoring the resulting harmonization projects. The validation of the tool is performed by two case studies.

Wasi Haider Butt, Sameera Amjad and Farooque Azam presented a paper on 'Requirement Conflicts Resolution: Using Requirement Filtering and Analysis'. The authors presented a systematic approach toward resolving software requirements spanning from requirement elicitation to the requirement analysis activity of the requirement engineering process. The authors developed a model 'conflict resolution strategy' (CRS) which employs a requirement filter and an analysis strategy for resolving any conflict arising during software development. They also implemented their model on a real project.

Rajesh Prasad, Suneeta Agarwal, Anuj Kumar Sharma, Alok Singh and Sanjay Misra presented a paper on 'Efficient Algorithm for Detecting Parameterized Multiple Clones in a Large Software System'. In this paper the authors have tried to solve the word length problem in a bit-parallel parameterized matching by extending the BLIM algorithm of exact string matching. The authors further argued that the extended algorithm is also suitable for searching multiple patterns simultaneously. The authors presented a comparison in support of their algorithm.

Takahiro Uchiya and Tetsuo Kinoshita presented the paper entitled 'Behavior Analyzer for Developing Multiagent Systems on Repository-Based Multiagent Framework'. In this paper the authors proposed an interactive design environment of agent system (IDEA) founded on an agent-repository-based multiagent framework. They focused on the function of the behavior analyzer for developing multiagent systems and showed the effectiveness of the function.

Jose Alfonso Aguilar, Irene Garrigos, and Jose-Norberto Mazon presented a paper on 'Impact Analysis of Goal-Oriented Requirements in Web Engineering'. This paper argues that Web developers need to know dependencies among requirements to ensure that Web applications finally satisfy the audience. The authors developed an algorithm to deal with dependencies among functional and non-functional requirements so as to understand the impact of making changes when developing a Web application.

Sanjay Misra

Message from the Chair of the Session: 2nd International Workshop on "Software Quality" (SQ 2011)

Following the success of SQ 2009, the Second International Workshop on "Software Quality" (SQ 2011) was organized in conjunction with ICCSA 2011. This workshop extends the discussion on software quality issues in the modern software development processes. It covers all the aspects of process and product quality, quality assurance and standards, quality planning, quality control and software quality challenges. It also covers the frontier issues and trends for achieving the quality objectives. In fact this workshop covers all areas, that are concerned with the quality issue of software product and process. In this workshop, we featured nine articles devoted to different aspects of software quality.

Roberto Espinosa, Jose Zubcoff, and Jose-Norberto Mazon's paper entitled "A Set of Experiments to Consider Data Quality Criteria in Classification Techniques for Data Mining" analyzed data–mining techniques to know the behavior of different data quality criteria from the sources. The authors have conducted a set of experiments to assess three data quality criteria: completeness, correlation and balance of data.

In their paper, Ivaylo Spassov, Valentin Pavlov, Dessislava Petrova-Antonova, and Sylvia Ilieva's have developed a tool "DDAT: Data Dependency Analysis Tool for Web Service Business Processes". The authors have implemented and shown experimental results from the execution of the DDAT over BPEL processes.

Filip Radulovic and Raul Garca-Castro presented a paper on "Towards a Quality Model for Semantic Technologies". The authors presented some well-known software quality models, after which a quality model for semantic technologies is designed by extending the ISO 9126 quality model.

Luis Fernandez-Sanz and Sanjay Misra authored the paper "Influence of Human Factors in Software Quality and Productivity". The authors first analyzed the existing contributions in the area and then presented empirical data from specific initiatives to know more about real practices and situations in software organizations.

Eudisley Anjos, and Mario Zenha-Rela presented a paper on "A Framework for Classifying and Comparing Software Architecture Tools for Quality Evaluation". This framework identifies the most relevant features for categorizing different architecture evaluation tools according to six different dimensions. The authors reported that the attributes that a comprehensive tool should support include: the ability to handle multiple modeling approaches, integration with the industry standard UML or specific ADL, support for trade–off analysis of competing quality attributes and the reuse of knowledge through the build-up of new architectural patterns.

Hendrik Decker presented a paper on "Causes of the Violation of Integrity Constraints for Supporting the Quality of Databases". He presented a quality metric with the potential of more accuracy by measuring the causes. He further argued that such measures also serve for controlling quality impairment across updates.

Csaba Nagy, Laszlo Vidacs , Rudolf Ferenc, Tibor Gyimothy Ferenc Kocsis, and Istvan Kovacs's presented a paper on "Complexity measures in a 4GL environment". The authors discussed the challenges in adopting the metrics from 3GL environments. Based on this, they presented a complexity measure in 4GL environments. They performed the experimentations and demonstrated the results.

Lukasz Radlinski's paper on "A Framework for Integrated Software Quality Prediction Using Bayesian Nets" developed a framework for integrated software quality prediction. His framework is developed and formulated using a Bayesian net, a technique that has already been used in various software engineering studies. The author argues that his model may be used in decision support for software analysts and managers.

Seunghun Park, Sangyoon Min, and Doohwan Bae authored the paper entitled "Process Instance Management Facilities Based on the Meta-Process Models". Based on the metar-process models, the authors proposed a process model and two types of process instance models: the structural instance model and the behavioral instance model. The authors' approach enables a project manager to analyze structural and behavioral properties of a process instance and allows a project manager to make use of the formalism for management facilities without knowledge of the formalism.

Sanjay Misra

Message from the Chairs of the Session: "Remote sensing Data Analysis, Modeling, Interpretation and Applications: From a Global View to a Local Analysis" (RS 2011)

Remotely sensed data provide temporal and spatial consistent measurements useful for deriving information on the dynamic nature of Earth surface processes (sea, ice, land, atmosphere), detecting and identifying land changes, discovering cultural resources, studying the dynamics of urban expansions. Thanks to the establishment and maintenance of long-term observation programs, presently a huge amount of multiscale and multifrequency remotely sensed data are available.

To fully exploit such data source for various fields of application (environmental, cultural heritage, urban analysis, disaster management) effective and reliable data processing, modeling and interpretation are required. This session brought together scientists and managers from the fields of remote sensing, ICT, geospatial analysis and modeling, to share information on the latest advances in remote sensing data analysis, product development, validation and data assimilation.

Main topics included:

Remotely sensed data – Multispectral satellite : from medium to very high spatial resolution; airborne and spaceborne Hypespectral data; open data source (Modis, Vegetation, etc..); airborne Laser Scanning; airborne and spaceborne Radar imaging; thermal imaging; declassified Intelligence Satellite Photographs (Corona, KVR); ground remote sensing

Methods and procedures – change detection; classification Data fusion / Data integration; data mining; geostatistics and Spatial statistics; image processing; image interpretation; linear and on linear statistical analysis; segmentation Pattern recognition and edge detection; time space modeling

Fields of application and products – archaeological site discovery; cultural Heritage management; disaster management; environmental sciences; mapping Landscape and digital elevation models; land cover analysis; open source softwares; palaeoenvironmental studies; time series

Nicola Masini
Rosa Lasaponara

Message from the Chairs of the Session: "Approximation, Optimization and Applications" (AOA 2011)

The objective of the session Approximation, Optimization and Applications during the 11th International Conference on Computational Science and Its Applications was to bring together scientists working in the areas of Approximation Theory and Numerical Optimization, including their applications in science and engineering.

Hypercomplex function theory, renamed Clifford analysis in the 1980s, studies functions with values in a non-commutative Clifford algebra. It has its roots in quaternionic analysis, developed as another generalization of the classic theory of functions of one complex variable compared with the theory of functions of several complex variables. It is well known that the use of quaternions and their applications in sciences and engineering is increasing, due to their advantages for fast calculations in 3D and for modeling mathematical problems. In particular, quasi-conformal 3D-mappings can be realized by regular (monogenic) quaternionic functions. In recent years the generalization of classical polynomials of a real or complex variable by using hypercomplex function theoretic tools has been the focus of increased attention leading to new and interesting problems. All these aspects led to the emergence of new software tools in the context of quaternionic or, more generally, Clifford analysis.

Irene Falcão
Ana Maria A.C. Rocha

Message from the Chair of the Session: "Symbolic Computing for Dynamic Geometry" (SCDG 2011)

The papers comprising in the Symbolic Computing for Dynamic Geometry technical session correspond to talks delivered at the conference. After the evaluation process, six papers were accepted for oral presentation, according to the recommendations of the reviewers. Two papers, "Equal bisectors at a vertex of a triangle" and "On Equivalence of Conditions for a Quadrilateral to Be Cyclica", study geometric problem by means of symbolic approaches.

Another contributions deal with teaching ("Teaching geometry with TutorMates" and "Using Free Open Source Software for Intelligent Geometric Computing"), while the remaining ones propose a framework for the symbolic treatment of dynamic geometry ("On the Parametric Representation of Dynamic Geometry Constructions") and a formal library for plane geometry ("A Coq-based Library for Interactive and Automated Theorem Proving in Plane Geometry").

Francisco Botana

Message from the Chairs of the Session: "Computational Design for Technology Enhanced Learning" (CD4TEL 2011)

Providing computational design support for orchestration of activities, roles, resources, and systems in technology-enhanced learning (TEL) is a complex task. It requires integrated thinking and interweaving of state-of-the-art knowledge in computer science, human–computer interaction, pedagogy, instructional design and curricular subject domains. Consequently, even where examples of successful practice or even standards and specifications like IMS learning design exist, it is often hard to apply and (re)use these efficiently and systematically. This interdisciplinary technical session brought together practitioners and researchers from diverse backgrounds such as computer science, education, and cognitive sciences to share their proposals and findings related to the computational design of activities, resources and systems for TEL applications.

The call for papers attracted 16 high-quality submissions. Each submission was reviewed by three experts. Eventually, five papers were accepted for presentation. These contributions demonstrate different perspectives of research in the CD4TEL area, dealing with standardization in the design of game-based learning; the integration of individual and collaborative electronic portfolios; the provision of an editing environment for different actors designing professional training; a simplified graphical notation for modeling the flow of activities in IMS learning design units of learning; and a pattern ontology-based model to support the selection of good-practice scripts for designing computer–supported collaborative learning.

Michael Derntl
Manuel Caeiro-Rodríguez
Davinia Hernández-Leo

Message from the Chair of the Session: "Chemistry and Materials Sciences and Technologies" (CMST 2011)

The CMST workshop is a typical example of how chemistry and computer science benefit from mutual interaction when operating within a grid e-science environment. The scientific contributions to the workshop, in fact, contain clear examples of chemical problems solved by exploiting the extra power offered by the grid infrastructure to the computational needs of molecular scientists when trying to push ahead the frontier of research and innovation.

Ideal examples of this are the papers on the coulomb potential decomposition in the multiconfiguration time–dependent Hartree method, on the extension of the grid–empowered simulator GEMS to the a priori evaluation of the crossed beam measurements and on the evaluation of the concentration of pollutants when using a box model version of the Community Multiscale Air Quality Modeling System 4.7. Another example of such progress in computational molecular science is offered by the paper illustrating the utilization of a fault–tolerant workflow for the DL-POLY package for molecular dynamics studies.

At the same time molecular science studies are an excellent opportunity for investigating the use of new (single or clustered) GPU chips as in the case of the papers related to their use for computationally demanding quantum calculations of atom diatom reactive scattering. In addition, of particular interest are the efforts spent to develop tools for evaluating user and service quality to the end of promoting collaborative work within virtual organizations and research communities through the awarding and the redeeming of credits.

Antonio Laganà

Message from the Chairs of the Session: "Cloud for High Performance Computing" (C4HPC 2011)

On behalf of the Program Committee, it is a pleasure for us to introduce the proceedings of this First International Workshop on Cloud for High–Performance Computing held in Santander (Spain) in 2011 during the 11th International Conference on Computational Science and Its Applications. The conference joined high quality researchers around the world to present the latest results in the usage of cloud computing for high–performance computing.

High–performance computing, or HPC, is a great tool for the advancement of science, technology and industry. It intensively uses computing resources, both CPU and storage, to solve technical or scientific problems in minimum time. It also uses the most advanced techniques to achieve this objective and evolves along with computing technology as fast as possible. During the last few years we have seen the introduction of new hardware isuch as multi-core and GPU representing a formidable challenge for the scientific and technical developers that need time to absorb these additional characteristics. At the same time, scientists and technicians have learnt to make faster and more accurate measurements, accumulating a large set of data which need more processing capacity. While these new paradigms were entering the field of HPC, virtualization was suddenly introduced in the market, generating a new model for provisioning computing capacity: the cloud. Although conceptually the cloud is not completely new, because it follows the old dream of computing as a utility, it has introduced new characteristics such as elasticity, but at the cost of losing some performance.

Consequently, HPC has a new challenge: how to tackle or solve this reduction in performance while adapting methods to the elasticity of the new platform. The initial results show the feasibility of using cloud infrastructures to execute HPC applications. However, there is also some consensus that the cloud is not the solution for grand challenges, which will still require dedicated supercomputers. Although recently a cluster of more than 4000 CPUs has been deployed, there are still many technical barriers to allow technicians to use it frequently. This is the reason for this workshop which we had the pleasure of introducing.

This First International Workshop on Cloud for High–Performance Computing was an original idea of Osvaldo Gervasi. We were working on the proposal of a COST action devoted to the cloud for HPC which would link the main researchers in Europe. He realized that the technical challenges HPC has to solve in the next few years to use the Cloud efficiently, need the collaboration of as many scientists and technicians as possible as well as to rethink the way the applications are executed.

This first workshop, which deserves in the next ICCSA conferences, joined together experts in the field that presented high quality research results in the area. They include the first approximations of topology methods such as cellular data system to cloud to be used to process data. Data are also the main issue for the TimeCloud front end, an interface for time series analysis based on Hadop and Hbase, designed to work with massive datasets. In fact, cloud can generate such a large amount of data when accurate information about its infrastructure and executing applications is needed. This is the topic of the third paper which introduce LISA algorithm to tackle the problem of information retrieval in cloud environment where the methods must adapt to the elasticity, scalability and possibility of failure. In fact, to understand Cloud infrastructures, researchers and technicians will need these series of data as well as the usage of tools that allow better knowledge to be gained. In this sense, iCanCloud, a novel simulator of cloud infrastructures, is introduced presenting its results for the most used and cited service: Amazon.

We strongly believe that the reader will enjoy the selected papers, which represent only a minimal, but important, part of the effervescent activity in Cloud for HPC. This selection was only possible thanks to the members of the Program Committee, all of them supporting actively the initiative. We appreciate their commitment to the workshop. Also, we want to thank all of the reviewers who kindly participated in the review of the papers and, finally, to all the scientists who submitted a paper, even if it was not accepted. We hope that they will have the opportunity to join us in the next editions.

Andrés Gomez
Osvaldo Gervasi

ICCSA 2011 Invited Speakers

Ajith Abraham
Machine Intelligence Research Labs, USA

Marina L. Gavrilova
University of Calgary, Canada

Yee Leung
The Chinese University of Hong Kong, China

Evolving Future Information Systems: Challenges, Perspectives and Applications

Ajith Abraham

Machine Intelligence Research Labs, USA
ajith.abraham@ieee.org

Abstract

We are blessed with the sophisticated technological artifacts that are enriching our daily lives and society. It is believed that the future Internet is going to provide us with the framework to integrate, control or operate virtually any device, appliance, monitoring systems, infrastructures etc. The challenge is to design intelligent machines and networks that could communicate and adapt according to the environment. In this talk, we first present the concept of a digital ecosystem and various research challenges from several application perspectives. Finally, we present some real–world applications.

Biography

Ajith Abraham received a PhD degree in Computer Science from Monash University, Melbourne, Australia. He is currently the Director of Machine Intelligence Research Labs (MIR Labs), Scientific Network for Innovation and Research Excellence, USA, which has members from more than 75 countries. He serves/has served the editorial board of over 50 international journals and has also guest edited 40 special issues on various topics. He has authored/co-authored more than 700 publications, and some of the works have also won best paper awards at international conferences. His research and development experience includes more than 20 years in industry and academia. He works in a multidisciplinary environment involving machine intelligence, network security, various aspects of networks, e-commerce, Web intelligence, Web services, computational grids, data mining, and their applications to various real-world problems. He has given more than 50 plenary lectures and conference tutorials in these areas.

Dr. Abraham is the Chair of IEEE Systems Man and Cybernetics Society Technical Committee on Soft Computing. He is a Senior Member of the IEEE, the IEEE Computer Society, the Institution of Engineering and Technology (UK) and the Institution of Engineers Australia (Australia). He is actively involved in the Hybrid Intelligent Systems (HIS), Intelligent Systems Design and Applications (ISDA), Information Assurance and Security (IAS), and Next–Generation Web Services Practices (NWeSP) series of international conferences, in addition to other conferences. More information can be found at: `http://www.softcomputing.net`.

Recent Advances and Trends in Biometric

Marina L. Gavrilova

Department of Computer Science, University of Calgary
marina@cpsc.ucalgary.ca

Extended Abstract

The area of biometric, without a doubt, is one of the most dynamic areas of interest, which recently has displayed a gamut of broader links to other fields of sciences. Among those are visualization, robotics, multi-dimensional data analysis, artificial intelligence, computational geometry, computer graphics, e-learning, data fusion and data synthesis. The theme of this keynote is reviewing the state of the art in multi-modal data fusion, fuzzy logic and neural networks and its recent connections to advanced biometric research.

Over the past decade, multimodal biometric systems emerged as a feasible and practical solution to counterweight the numerous disadvantages of single biometric systems. Active research intoi the design of a multimodal biometric system has started, mainly centered around: types of biometrics, types of data acquisition and decision-making processes. Many challenges originating from non-uniformity of biometric sources and biometric acquisition devices result in significant differences on which information is extracted, how is it correlated, the degree of allowable error, cost implications, ease of data manipulation and management, and also reliability of the decisions being made. With the additional demand of computational power and compact storage, more emphasis is shifted toward database design and computational algorithms.

One of the actively researched areas in multimodal biometric systems is information fusion. Which information needs to be fused and what level is needed to obtain the maximum recognition performance is the main focus of current research. In this talk I concentrate on an overview of the current trends in recent multimodal biometric fusion research and illustrate in detail one fusion strategy: rank level fusion. From the discussion, it is seen that rank level fusion often outperforms other methods, especially combined with powerful decision models such as Markov chain or fuzzy logic.

Another aspect of multi-modal biometric system development based on neural networks is discussed further. Neural networks have the capacity to simulate learning processes of a human brain and to analyze and compare complex patters, which can originate from either single or multiple biometric sources, with amazing precision. Speed and complexity have been the downsides of neural networks, however, recent advancements in the area, especially in chaotic neural networks, allow these barriers to be overcome.

The final part of the presentation concentrates on emerging areas utilizing the above developments, such as decision making in visualization, graphics, e-learning, navigation, robotics, and security of web-based and virtual worlds. The extent to which biometric advancements have an impact on these emerging areas makes a compelling case for the bright future of this area.

References

1. Ross, A., Nandakumar, K., and Jain, A.K., Handbook of multibiometrics, New York, Springer (2006).
2. Jain, A.K., Ross, A., Prabhakar, S., An introduction to biometric recognition, IEEE Trans. on Circuits and Systems for Video Technology, Special Issue on Image- and Video-Based Biometrics, 14 (1): 420 (2004)
3. Nandakumar, K., Jain, A.K., Ross, A., Fusion in multibiometric identification systems: What about the missing data?, in LNCS 5558: 743752, Springer (2009).
4. Monwar, M. M., and Gavrilova, M.L., A multimodal biometric system using rank level fusion approach, IEEE Trans. SMC - B: Cybernetics, 39(4): 867-878 (2009).
5. Monwar, M. M., and Gavrilova, M.L., Secured access control through Markov chain based rank level fusion method, in proc. of 5th Int. Conf. on Computer Vision Theory and Applications (VISAPP), 458-463, Angres, France (2010).
6. Monwar, M. M., and Gavrilova, M.L., FES: A system of combining face, ear and signature biometrics using rank level fusion, in proc. 5th IEEE Int. Conf. IT: New Generations, pp 922-927, (2008).
7. Wang, C., Gavrilova, M.L., Delaunay Triangulation Algorithm for Fingerprint Matching. ISVD'2006. pp.208 216
8. Wecker, L., Samavati, F.F., Gavrilova, M.L., Iris synthesis: a reverse subdivision application. GRAPHITE'2005. pp.121 125
9. Anikeenko, A.V., Gavrilova, M.L., Medvedev, N.N., A Novel Delaunay Simplex Technique for Detection of Crystalline Nuclei in Dense Packings of Spheres. ICCSA (1)'2005. pp.816 826
10. Luchnikov, V.A., Gavrilova, M.L., Medvedev, N.N., Voloshin, V. P., The Voronoi-Delaunay approach for the free volume analysis of a packing of balls in a cylindrical container. Future Generation Comp. Syst., 2002: 673 679
11. Frischholz, R., and Dieckmann, U., BioID: A multimodal biometric identification system, IEEE Computer, 33 (2): 64-68 (2000).
12. Latifi, S., Solayappan, N. A survey of unimodal biometric methods, in proc. of Int. Conf. on Security & Management, 57-63, Las Vegas, USA (2006).
13. Dunstone, T., and Yager, N., Biometric system and data analysis: Design, evaluation, and data mining. Springer, New York (2009).
14. Ho, T.K., Hull, J.J., and Srihari, S.N., Decision combination in multiple classifier systems, IEEE Trans. on Pattern Analysis and Machine Intelligence, 16 (1): 66-75 (1994)

Biography

Marina L. Gavrilova is an Associate Professor in the Department of Computer Science, University of Calgary. Prof. Gavrilova's research interests lie in the area of computational geometry, image processing, optimization, spatial and biometric modeling. Prof. Gavrilova is founder and co-director of two innovative research laboratories: the Biometric Technologies Laboratory: Modeling and Simulation and the SPARCS Laboratory for Spatial Analysis in Computational Sciences. Prof. Gavrilova publication list includes over 120 journal and conference papers, edited special issues, books and book chapters, including World Scientific Bestseller of the Month (2007) *Image Pattern Recognition: Synthesis and Analysis in Biometric* and the Springer book Computational Intelligence: A Geometry-Based Approach. Together with Dr. Kenneth Tan, Prof. Gavrilova founded the ICCSA series of successful international events in 2001. She founded and chaired the International Workshop on Computational Geometry and Applications for over ten years, was co-Chair of the International Workshop on Biometric Technologies BT 2004, Calgary, served as Overall Chair of the Third International Conference on Voronoi Diagrams in Science and Engineering (ISVD) in 2006, was Organizing Chair of WADS 2009 (Banff), and general chair of the International Conference on Cyberworlds CW2011 (October 4-6, Banff, Canada). Prof. Gavrilova is an Editor-in-Chief of the successful LNCS Transactions on Computational Science Journal, Springer-Verlag since 2007 and serves on the Editorial Board of the International Journal of Computational Sciences and Engineering, CAD/CAM Journal and Journal of Biometrics. She has been honored with awards and designations for her achievements and was profiled in numerous newspaper and TV interviews, most recently being chosen together with other outstanding Canadian scientists to be featured in the National Museum of Civilization and National Film Canada production.

Theories and Applications of Spatial-Temporal Data Mining and Knowledge Discovery

Yee Leung

The Chinese University of Hong Kong, China
yeeleung@cuhk.edu.hk

Abstract

Basic theories of knowledge discovery in spatial and temporal data are examined in this talk. Fundamental issues in the discovery of spatial structures and processes will first be discussed. Real-life spatial data mining problems are employed as the background on which concepts, theories and methods are scrutinized. The unraveling of land covers, seismic activities, air pollution episodes, rainfall regimes, epidemics, patterns and concepts hidden in spatial and temporal data are employed as examples to illustrate the theoretical arguments and algorithms performances. To round up the discussion, directions for future research are outlined.

Biography

Yee Leung is currently Professor of Geography and Resource Management at The Chinese University of Hong Kong. He is also the Associate Academic Director of the Institute of Space and Earth Information Science of The Chinese University of Hong Kong. He is adjunct professor of several universities in P.R. China. Professor Leung had also served on public bodies including the Town Planning Board and the Environmental Pollution Advisory Committee of Hong Kong SAR. He is now Chair of The Commission on Modeling Geographical Systems, International Geographical Union, and Chair of The Commission on Quantitative and Computational Geography of The Chinese Geographical Society. He serves on the editorial board of several international journals such as *Annals of Association of American Geographers*, *Geographical Analysis*, *GeoInformatica*, *Journal of Geographical Systems*, *Acta Geographica Sinica*, *Review of Urban and Regional Development Studies*, etc. Professor Leung is also Council member of The Society of Chinese Geographers.

Professor Leung carried out pioneer and influential research in imprecision/uncertainty analysis in geography, intelligent spatial decision support systems, geocomputation (particularly on fuzzy sets, rough sets, spatial statistics, fractal analysis, neural networks and genetic algorithms), and knowledge discovery and data mining. He has obtained more than 30 research grants, authored and co-authored six books and over 160 papers in international journals and book chapters on geography, computer science, and information engineering. His landmark books are: *Spatial Analysis and Planning under Imprecision* (Elsevier, 1988), *Intelligent Spatial Decision Support Systems* (Springer-Verlag, 1997), and *Knowledge Discovery in Spatial Data* (Springer-Verlag, 2010).

Organization

ICCSA 2011 was organized by the University of Cantabria (Spain), Kyushu Sangyo University (Japan), the University of Perugia (Italy), Monash University (Australia) and the University of Basilicata (Italy).

Honorary General Chairs

Antonio Laganà	University of Perugia, Italy
Norio Shiratori	Tohoku University, Japan
Kenneth C.J. Tan	Qontix, UK

General Chairs

Bernady O. Apduhan	Kyushu Sangyo University, Japan
Andrés Iglesias	University of Cantabria, Spain

Program Committee Chairs

Osvaldo Gervasi	University of Perugia, Italy
David Taniar	Monash University, Australia

Local Arrangements Chairs

Andrés Iglesias	University of Cantabria, Spain (Chair)
Akemi Gálvez	University of Cantabria, Spain
Jaime Puig-Pey	University of Cantabria, Spain
Angel Cobo	University of Cantabria, Spain
José L. Montaña	University of Cantabria, Spain
César Otero	University of Cantabria, Spain
Marta Zorrilla	University of Cantabria, Spain
Ernesto Anabitarte	University of Cantabria, Spain
Unal Ufuktepe	Izmir University of Economics, Turkey

Workshop and Session Organizing Chair

Beniamino Murgante	University of Basilicata, Italy

International Liaison Chairs

Jemal Abawajy	Deakin University, Australia
Marina L. Gavrilova	University of Calgary, Canada
Robert C.H. Hsu	Chung Hua University,Taiwan
Tai-Hoon Kim	Hannam University, Korea
Takashi Naka	Kyushu Sangyo University, Japan

Awards Chairs

Wenny Rahayu	LaTrobe University, Australia
Kai Cheng	Kyushu Sangyo University, Japan

Workshop Organizers

Approaches or Methods of Security Engineering (AMSE 2011)

Tai-hoon Kim	Hannam University, Korea

Approximation, Optimization and Applications (AOA 2011)

Ana Maria A.C. Rocha	University of Minho, Portugal
Maria Irene Falcao	University of Minho, Portugal

Advances in Web–Based Learning (AWBL 2011)

Mustafa Murat Inceoglu	Ege University (Turkey)

Computational Aspects and Methods in Renewable Energies (CAMRE 2011)

Maurizio Carlini	University of Tuscia, Italy
Sonia Castellucci	University of Tuscia, Italy
Andrea Tucci	University of Tuscia, Italy

Computer–Aided Modeling, Simulation, and Analysis (CAMSA 2011)

Jie Shen	University of Michigan, USA

Computer Algebra Systems and Applications (CASA 2011)

Andrés Iglesias	University of Cantabria (Spain)
Akemi Gálvez	University of Cantabria (Spain)

Computational Design for Technology–Enhanced Learning: Methods, Languages, Applications and Tools (CD4TEL 2011)

Michael Derntl	University of Vienna, Austria
Manuel Caeiro-Rodriguez	University of Vigo, Spain
Davinia Hernandez-Leo	Universitat Pompeu Fabra, Spain

Computational Geometry and Applications (CGA 2011)

Marina L. Gavrilova	University of Calgary, Canada

Computer Graphics and Virtual Reality (CGVR 2011)

Osvaldo Gervasi	University of Perugia, Italy
Andrés Iglesias	University of Cantabria, Spain

Chemistry and Materials Sciences and Technologies (CMST 2011)

Antonio Laganà	University of Perugia, Italy

Consulting Methodology and Decision Making for Security Systems (CMDMSS 2011)

Sangkyun Kim	Kangwon National University, Korea

Cities, Technologies and Planning (CTP 2011)

Giuseppe Borruso	University of Trieste, Italy
Beniamino Murgante	University of Basilicata, Italy

Cloud for High–Performance Computing (C4HPC 2011)

Andrés Gomez	CESGA, Santiago de Compostela, Spain
Osvaldo Gervasi	University of Perugia, Italy

Future Information System Technologies and Applications (FISTA 2011)

Bernady O. Apduhan	Kyushu Sangyo University, Japan
Jianhua Ma	Hosei University, Japan
Qun Jin	Waseda University, Japan

Geographical Analysis, Urban Modeling, Spatial Statistics (GEOG-AN-MOD 2011)

Stefania Bertazzon	University of Calgary, Canada
Giuseppe Borruso	University of Trieste, Italy
Beniamino Murgante	University of Basilicata, Italy

International Workshop on Biomathematics, Bioinformatics and Biostatistics (IBBB 2011)

Unal Ufuktepe	Izmir University of Economics, Turkey
Andrés Iglesias	University of Cantabria, Spain

International Workshop on Collective Evolutionary Systems (IWCES 2011)

Alfredo Milani	University of Perugia, Italy
Clement Leung	Hong Kong Baptist University, China

Mobile Communications (MC 2011)

Hyunseung Choo	Sungkyunkwan University, Korea

Mobile Sensor and Its Applications (MSA 2011)

Moonseong Kim	Korean Intellectual Property Office, Korea

Mobile Systems and Applications (MoSA 2011)

Younseung Ryu	Myongji University, Korea
Karlis Kaugars	Western Michigan University, USA

Logical, Scientific and Computational Aspects of Pulse Phenomena in Transitions (PULSES 2011)

Carlo Cattani	University of Salerno, Italy
Cristian Toma	Corner Soft Technologies, Romania
Ming Li	East China Normal University, China

Resource Management and Scheduling for Future–Generation Computing Systems (RMS 2011)

Jemal H. Abawajy	Deakin University, Australia

Remote Sensing Data Analysis, Modeling, Interpretation and Applications: From a Global View to a Local Analysis (RS 2011)

Rosa Lasaponara	IRMMA, CNR, Italy
Nicola Masini	IBAM, CNR, Italy

Symbolic Computing for Dynamic Geometry (SCDG 2011)

Francisco Botana	Vigo University, Spain

Software Engineering Processes and Applications (SEPA 2011)

Sanjay Misra   Atilim University, Turkey

Software Quality (SQ 2011)

Sanjay Misra   Atilim University, Turkey

Tools and Techniques in Software Development Processes (TTSDP 2011)

Sanjay Misra   Atilim University, Turkey

Virtual Reality in Medicine and Surgery (VRMS 2011)

Giovanni Aloisio   University of Salento, Italy
Lucio T. De Paolis   University of Salento, Italy

Wireless and Ad-Hoc Networking (WADNet 2011)

Jongchan Lee   Kunsan National University, Korea
Sangjoon Park   Kunsan National University, Korea

WEB 2.0 and Social Networks (Web2.0 2011)

Vidyasagar Potdar   Curtin University of Technology, Australia

Workshop on Internet Communication Security (WICS 2011)

Josè Maria Sierra Camara   University of Madrid, Spain

Wireless Multimedia Sensor Networks (WMSN 2011)

Vidyasagar Potdar   Curtin University of Technology, Australia
Yan Yang   Seikei University, Japan

Program Committee

Jemal Abawajy   Deakin University, Australia
Kenneth Adamson   Ulster University, UK
Michela Bertolotto   University College Dublin, Ireland
Sandro Bimonte   CEMAGREF, TSCF, France
Rod Blais   University of Calgary, Canada
Ivan Blecic   University of Sassari, Italy
Giuseppe Borruso   Università degli Studi di Trieste, Italy
Martin Buecker   Aachen University, Germany
Alfredo Buttari   CNRS-IRIT, France
Yves Caniou   Lyon University, France
Carlo Cattani   University of Salerno, Italy
Mete Celik   Erciyes University, Turkey

Nicola Masini	National Research Council, Italy
Nirvana Meratnia	University of Twente, The Netherlands
Alfredo Milani	University of Perugia, Italy
Sanjay Misra	Atilim University, Turkey
Josè Luis Montaña	University of Cantabria, Spain
Beniamino Murgante	University of Basilicata, Italy
Jiri Nedoma	Academy of Sciences of the Czech Republic, Czech Republic
Laszlo Neumann	University of Girona, Spain
Kok-Leong Ong	Deakin University, Australia
Belen Palop	Universidad de Valladolid, Spain
Marcin Paprzycki	Polish Academy of Sciences, Poland
Eric Pardede	La Trobe University, Australia
Kwangjin Park	Wonkwang University, Korea
Vidyasagar Potdar	Curtin University of Technology, Australia
David C. Prosperi	Florida Atlantic University, USA
Wenny Rahayu	La Trobe University, Australia
Jerzy Respondek	Silesian University of Technology Poland
Alexey Rodionov	Institute of Computational Mathematics and Mathematical Geophysics, Russia
Jon Rokne	University of Calgary, Canada
Octavio Roncero	CSIC, Spain
Maytham Safar	Kuwait University, Kuwait
Haiduke Sarafian	The Pennsylvania State University, USA
Qi Shi	Liverpool John Moores University, UK
Dale Shires	U.S. Army Research Laboratory, USA
Carmelo Torre	Polytechnic of Bari, Italy
Giuseppe A. Trunfio	University of Sassari, Italy
Unal Ufuktepe	Izmir University of Economics, Turkey
Mario Valle	Swiss National Supercomputing Centre, Switzerland
Piero Giorgio Verdini	INFN Pisa and CERN, Italy
Andrea Vittadini	University of Padova, Italy
Koichi Wada	University of Tsukuba, Japan
Krzysztof Walkowiak	Wroclaw University of Technology, Poland
Robert Weibel	University of Zurich, Switzerland
Roland Wismüller	Universität Siegen, Germany
Markus Wolff	University of Potsdam, Germany
Mudasser Wyne	National University, USA
Chung-Huang Yang	National Kaohsiung Normal University, Taiwan
Xin-She Yang	National Physical Laboratory, UK
Salim Zabir	France Telecom Japan Co., Japan
Albert Y. Zomaya	University of Sydney, Australia

Sponsoring Organizations

ICCSA 2011 would not have been possible without tremendous support of many organizations and institutions, for which all organizers and participants of ICCSA 2011 express their sincere gratitude:

- The Department of Applied Mathematics and Computational Sciences, University of Cantabria, Spain
- The Department of Mathematics, Statistics and Computation, University of Cantabria, Spain
- The Faculty of Sciences, University of Cantabria, Spain
- The Vicerrector of Research and Knowledge Transfer, University of Cantabria, Spain
- The University of Cantabria, Spain
- The University of Perugia, Italy
- Kyushu Sangyo University, Japan
- Monash University, Australia
- The University of Basilicata, Italy
- Cantabria Campus Internacional, Spain
- The Municipality of Santander, Spain
- The Regional Government of Cantabria, Spain
- The Spanish Ministry of Science and Innovation, Spain
- GeoConnexion (`http://www.geoconnexion.com/`)
- Vector1 Media (`http://www.vector1media.com/`)

Table of Contents – Part I

Workshop on Geographical Analysis, Urban Modeling, Spatial Statistics (GEO-AN-MOD 2011)

Towards a Spatio-Temporal Information System for Moving Objects 1
Maribel Yasmina Santos, José Mendes, Adriano Moreira, and Monica Wachowicz

Analyzing Demographic and Economic Simulation Model Results: A Semi-automatic Spatial OLAP Approach 17
Hadj Mahboubi, Sandro Bimonte, and Guillaume Deffuant

Linking SLEUTH Urban Growth Modeling to Multi Criteria Evaluation for a Dynamic Allocation of Sites to Landfill 32
Abdolrassoul Salman Mahiny and Mehdi Gholamalifard

Statistical Evaluation of Spatial Interpolation Methods for Small-Sampled Region: A Case Study of Temperature Change Phenomenon in Bangladesh 44
Avit Kumar Bhowmik and Pedro Cabral

A High Resolution Land Use/Cover Modelling Framework for Europe: Introducing the EU-ClueScanner100 Model 60
Carlo Lavalle, Claudia Baranzelli, Filipe Batista e Silva, Sarah Mubareka, Carla Rocha Gomes, Eric Koomen, and Maarten Hilferink

An Adaptive Neural Network-Based Method for Tile Replacement in a Web Map Cache 76
Ricardo García, Juan Pablo de Castro, María Jesús Verdú, Elena Verdú, Luisa María Regueras, and Pablo López

Mapping the Anthropic Backfill of the Historical Center of Rome (Italy) by Using Intrinsic Random Functions of Order k (IRF-k) 92
Giancarlo Ciotoli, Francesco Stigliano, Fabrizio Marconi, Massimiliano Moscatelli, Marco Mancini, and Gian Paolo Cavinato

Spatio-Temporal Analysis Using Urban-Rural Gradient Modelling and Landscape Metrics 103
Marco Vizzari

Urban Development Scenarios and Probability Mapping for Greater Dublin Region: The MOLAND Model Applications 119
Harutyun Shahumyan, Roger White, Laura Petrov, Brendan Williams, Sheila Convery, and Michael Brennan

Hierarchical Clustering through Spatial Interaction Data: The Case of Commuting Flows in South-Eastern France 135
Giovanni Fusco and Matteo Caglioni

Analyzing and Testing Strategies to Guarantee the Topological Relationships and Quality Requirements between Networks Inside BTA and BTU Data Models 152
Francisco Ruiz-Lopez, Eloina Coll, and Jose Martinez-Llario

An Assessment-Based Process for Modifying the Built Fabric of Historic Centres: The Case of Como in Lombardy 162
Pier Luigi Paolillo, Alberto Benedetti, Umberto Baresi, Luca Terlizzi, and Giorgio Graj

XPath for Querying GML-Based Representation of Urban Maps 177
Jesús M. Almendros-Jiménez, Antonio Becerra-Terón, and Francisco García-García

Accounting for Spatial Heterogeneity in Educational Outcomes and International Migration in Mexico 192
Edith Gutiérrez, Landy Sánchez, and Silvia Giorguli

Accessibility Analysis and Modelling in Public Transport Networks – A Raster Based Approach 207
Morten Fuglsang, Henning Sten Hansen, and Bernd Münier

MyTravel: A Geo-referenced Social-Oriented Web 2.0 Application 225
Gabriele Cestra, Gianluca Liguori, and Eliseo Clementini

GIS and Remote Sensing to Study Urban-Rural Transformation during a Fifty-Year Period 237
Carmelo Riccardo Fichera, Giuseppe Modica, and Maurizio Pollino

Spatial Clustering to Uncluttering Map Visualization in SOLAP 253
Ricardo Silva, João Moura-Pires, and Maribel Yasmina Santos

Mapping the Quality of Life Experience in Alfama: A Case Study in Lisbon, Portugal 269
Pearl May dela Cruz, Pedro Cabral, and Jorge Mateu

Evolution Trends of Land Use/Land Cover in a Mediterranean Forest Landscape in Italy 284
Salvatore Di Fazio, Giuseppe Modica, and Paolo Zoccali

Application of System Dynamics, GIS and 3D Visualization in a Study of Residential Sustainability 300
Zhao Xu and Volker Coors

The State's Geopolitics versus the Local's Concern, a Case Study on the Land Border between Indonesia and Timor-Leste in West Sector 315
Sri Handoyo

CartoService: A Web Service Framework for Quality On-Demand Geovisualisation 329
Rita Engemaier and Hartmut Asche

An Analysis of Poverty in Italy through a Fuzzy Regression Model 342
Silvestro Montrone, Francesco Campobasso, Paola Perchinunno, and Annarita Fanizzi

Fuzzy Logic Approach for Spatially Variable Nitrogen Fertilization of Corn Based on Soil, Crop and Precipitation Information 356
Yacine Bouroubi, Nicolas Tremblay, Philippe Vigneault, Carl Bélec, Bernard Panneton, and Serge Guillaume

The Characterisation of "Living" Landscapes: The Role of Mixed Descriptors and Volunteering Geographic Information 369
Ernesto Marcheggiani, Andrea Galli, and Hubert Gulinck

Conceptual Approach to Measure the Potential of Urban Heat Islands from Landuse Datasets and Landuse Projections 381
Christian Daneke, Benjamin Bechtel, Jürgen Böhner, Thomas Langkamp, and Jürgen Oßenbrügge

Integration of Temporal and Semantic Components into the Geographic Information through Mark-Up Languages. Part I: Definition 394
Willington Siabato and Miguel-Angel Manso-Callejo

Resilient City and Seismic Risk: A Spatial Multicriteria Approach 410
Lucia Tilio, Beniamino Murgante, Francesco Di Trani, Marco Vona, and Angelo Masi

Towards a Planning Decision Support System for Low-Carbon Urban Development 423
Ivan Blecic, Arnaldo Cecchini, Matthias Falk, Serena Marras, David R. Pyles, Donatella Spano, and Giuseppe A. Trunfio

Quantitative Analysis of Pollutant Emissions in the Context of Demand Responsive Transport 439
Julie Prud'homme, Didier Josselin, and Jagannath Aryal

Individual Movements and Geographical Data Mining: Clustering Algorithms for Highlighting Hotspots in Personal Navigation Routes 454
Gabriella Schoier and Giuseppe Borruso

How Measure Estate Value in the Big Apple? Walking along Districts of Manhattan 466
Carmelo M. Torre

Modelling Proximal Space in Urban Cellular Automata 477
Ivan Blecic, Arnaldo Cecchini, and Giuseppe A. Trunfio

Improvement of Spatial Data Quality Using the Data Conflation 492
Silvija Stankutė and Hartmut Asche

Territories of Digital Communities: Representing the Social Landscape of Web Relationships 501
Letizia Bollini

General Tracks

Decentralized Distributed Computing System for Privacy-Preserving Combined Classifiers – Modeling and Optimization 512
Krzysztof Walkowiak, Szymon Sztajer, and Michał Woźniak

A New Adaptive Framework for Classifier Ensemble in Multiclass Large Data 526
Hamid Parvin, Behrouz Minaei, and Hosein Alizadeh

An Efficient Hash-Based Load Balancing Scheme to Support Parallel NIDS 537
Nam-Uk Kim, Sung-Min Jung, and Tai-Myoung Chung

Penalty Functions for Genetic Programming Algorithms 550
José L. Montaña, César L. Alonso, Cruz E. Borges, and Javier de la Dehesa

Generation of Pseudorandom Binary Sequences with Controllable Cryptographic Parameters 563
Amparo Fúster-Sabater

Mobility Adaptive CSMA/CA MAC for Wireless Sensor Networks 573
Bilal Muhammad Khan and Falah H. Ali

A Rough Set Approach Aim to Space Weather and Solar Storms Prediction 588
Reza Mahini, Caro Lucas, Masoud Mirmomeni, and Hassan Rezazadeh

A Discrete Flow Simulation Model for Urban Road Networks, with Application to Combined Car and Single-File Bicycle Traffic........... 602
Jelena Vasic and Heather J. Ruskin

A GPU-Based Implementation for Range Queries on Spaghettis Data Structure 615
Roberto Uribe-Paredes, Pedro Valero-Lara, Enrique Arias, José L. Sánchez, and Diego Cazorla

A Concurrent Object-Oriented Approach to the Eigenproblem Treatment in Shared Memory Multicore Environments 630
Alfonso Niño, Camelia Muñoz-Caro, and Sebastián Reyes

Geospatial Orchestration Framework for Resolving Complex User Query 643
Sudeep Singh Walia, Arindam Dasgupta, and S.K. Ghosh

A Bio-Inspired Approach for Risk Analysis of ICT Systems 652
Aurelio La Corte, Marialisa Scatá, and Evelina Giacchi

Maximization of Network Survivability Considering Degree of Disconnectivity 667
Frank Yeong-Sung Lin, Hong-Hsu Yen, and Pei-Yu Chen

Application of the GFDM for Dynamic Analysis of Plates 677
Francisco Ureña, Luis Gavete, Juan José Benito, and Eduardo Salete

A Viewer-Dependent Tensor Field Visualization Using Particle Tracing 690
Gildo de Almeida Leonel, João Paulo Peçanha, and Marcelo Bernardes Vieira

Statistical Behaviour of Discrete-Time Rössler System with Time Varying Delay 706
Madalin Frunzete, Adrian Luca, Adriana Vlad, and Jean-Pierre Barbot

Author Index 721

Table of Contents – Part II

Cities, Technologies and Planning (CTP 2011)

Games and Serious Games in Urban Planning: Study Cases 1
Alenka Poplin

Spatial Analysis with a Tool GIS via Systems of Fuzzy Relation Equations 15
Ferdinando Di Martino and Salvatore Sessa

Web-GIS Based Green Landscape and Urban Planning System Development 31
Junghoon Ki and Kunesook Hur

Building a Crowd-Sourcing Tool for the Validation of Urban Extent and Gridded Population 39
Steffen Fritz, Linda See, Ian McCallum, Christian Schill, Christoph Perger, and Michael Obersteiner

View- and Scale-Based Progressive Transmission of Vector Data 51
Padraig Corcoran, Peter Mooney, Michela Bertolotto, and Adam Winstanley

Mapping Invisibles - Acquiring GIS for Urban Planner Workshop 63
Małgorzata Hanzl, Ewa Stankiewicz, Agata Wierzbicka, Tomasz Kujawski, Karol Dzik, Paulina Kowalczyk, Krystian Kwiecinski, Maciej Burdalski, Anna Śliwka, Mateusz Wójcicki, Michał Miszkurka, Semir Poturak, and Katarzyna Westrych

On How to Build SDI Using Social Networking Principles in the Scope of Spatial Planning and Vocational Education 78
Karel Janecka, Raitis Berzins, Karel Charvat, and Andris Dzerve

Smarter Urban Planning: Match Land Use with Citizen Needs and Financial Constraints 93
Maria-Lluïsa Marsal-Llacuna, Ying Tat Leung, and Guang-Jie Ren

Seismic Vulnerability Assessment Using Field Survey and Remote Sensing Techniques 109
Paolo Ricci, Gerardo Mario Verderame, Gaetano Manfredi, Maurizio Pollino, Flavio Borfecchia, Luigi De Cecco, Sandro Martini, Carmine Pascale, Elisabetta Ristoratore, and Valentina James

Connecting Geodata Initiatives to Contribute to a European Spatial Data Infrastructure - The CRISOLA Case for Malta and the Project Plan4all 125
Saviour Formosa, Vincent Magri, Julia Neuschmid, and Manfred Schrenk

Network Based Kernel Density Estimation for Cycling Facilities Optimal Location Applied to Ljubljana 136
Nicolas Lachance-Bernard, Timothée Produit, Biba Tominc, Matej Nikšič, and Barbara Goličnik Marušić

Monitoring Temporary Populations through Cellular Core Network Data 151
Fabio Manfredini, Paolo Tagliolato, and Carmelo Di Rosa

Towards 'Resilient Cities' – Harmonisation of Spatial Planning Information as One Step along the Way 162
Manfred Schrenk, Julia Neuschmid, and Daniela Patti

The Participation Loop: Helping Citizens to Get in 172
Jorge Gustavo Rocha

An OWL2 Land Use Ontology: LBCS 185
Nuno Montenegro, Jorge Gomes, Paulo Urbano, and José Duarte

E-Democracy in Collaborative Planning: A Critical Review 199
Francesco Rotondo and Francesco Selicato

Web Based Integrated Models for Participatory Planning 210
Grammatikogiannis Elias and Maria Giaoutzi

A Service Quality Model for Web-Services Evaluation in Cultural Heritage Management 227
Aline Chiabai, Lorena Rocca, and Livio Chiarullo

Onto-Planning: Innovation for Regional Development Planning within EU Convergence Framework 243
Francesco Scorza, Giuseppe Las Casas, and Angelo Carlucci

Ontology and Spatial Planning 255
Beniamino Murgante and Francesco Scorza

Crowd-Cloud Tourism, New Approaches to Territorial Marketing 265
Beniamino Murgante, Lucia Tilio, Francesco Scorza, and Viviana Lanza

Constructing Strategies in Strategic Urban Planning: A Case Study of a Decision Support and Evaluation Model 277
Ivan Blecic, Arnaldo Cecchini, and Alessandro Plaisant

Remote Sensing Data Analysis, Modeling, Interpretation and Applications: From a Global View to a Local Analysis (RS 2011)

Speed-Up of GIS Processing Using Multicore Architectures 293
Iulian Nita, Teodor Costachioiu, and Vasile Lazarescu

From the Point Cloud to Virtual and Augmented Reality: Digital Accessibility for Disabled People in San Martin's Church (Segovia) and Its Surroundings 303
Juan Mancera-Taboada, Pablo Rodríguez-Gonzálvez, Diego González-Aguilera, Javier Finat, Jesús San José, Juan J. Fernández, José Martínez, and Rubén Martínez

Comparative Algorithms for Oil Spill Detection from Multi Mode RADARSAT-1 SAR Satellite Data 318
Maged Marghany and Mazlan Hashim

Pre and Post Fire Vegetation Behavioral Trends from Satellite MODIS/NDVI Time Series in Semi-natural Areas 330
Tiziana Montesano, Antonio Lanorte, Fortunato De Santis, and Rosa Lasaponara

On the Use of Satellite Remote Sensing Data to Characterize and Map Fuel Types 344
Antonio Lanorte and Rosa Lasaponara

A Computationally Efficient Method for Sequential MAP-MRF Cloud Detection 354
Paolo Addesso, Roberto Conte, Maurizio Longo, Rocco Restaino, and Gemine Vivone

Assessment for Remote Sensing Data: Accuracy of Interactive Data Quality Interpretation 366
Erik Borg, Bernd Fichtelmann, and Hartmut Asche

An Open Source GIS System for Earthquake Early Warning and Post-Event Emergency Management 376
Maurizio Pollino, Grazia Fattoruso, Antonio Bruno Della Rocca, Luigi La Porta, Sergio Lo Curzio, Agnese Arolchi, Valentina James, and Carmine Pascale

On the Processing of Aerial LiDAR Data for Supporting Enhancement, Interpretation and Mapping of Archaeological Features 392
Rosa Lasaponara and Nicola Masini

Satellite Based Observations of the Dynamic Expansion of Urban Areas in Southern Italy Using Geospatial Analysis 407
Gabriele Nolè and Rosa Lasaponara

General Tracks

Design and Results of a Statistical Survey in a School 422
Francesca Pierri

MathATESAT: A Symbolic-Numeric Environment in Astrodynamics and Celestial Mechanics 436
Juan Félix San-Juan, Luis María López, and Rosario López

Hybrid Analytical-Statistical Models 450
Juan Félix San-Juan, Montserrat San-Martín, and David Ortigosa

Thermal-Aware Floorplan Schemes for Reliable 3D Multi-Core Processors 463
Dong Oh Son, Young Jin Park, Jin Woo Ahn, Jae Hyung Park, Jong Myon Kim, and Cheol Hong Kim

Self-Adaptive Deployment of Parametric Sweep Applications through a Complex Networks Perspective 475
María Botón-Fernández, Francisco Prieto Castrillo, and Miguel A. Vega-Rodríguez

An Adaptive N-Variant Software Architecture for Multi-Core Platforms: Models and Performance Analysis 490
Li Tan and Axel Krings

Towards an Architecture for the Notification of Events from Tools of Third-Party Providers 506
Manuel Caeiro-Rodriguez and Jorge Fontenla-Gonzalez

Visual Perception Substitution by the Auditory Sense 522
Brian David Cano Martínez, Osslan Osiris Vergara Villegas, Vianey Guadalupe Cruz Sánchez, Humberto de Jesús Ochoa Domínguez, and Leticia Ortega Maynez

Data Driven Bandwidth for Medoid Shift Algorithm 534
Syed Zulqarnain Ahmad Gilani and Naveed Iqbal Rao

Optimization of Sparse Matrix-Vector Multiplication by Auto Selecting Storage Schemes on GPU 547
Yuji Kubota and Daisuke Takahashi

Summarizing Cluster Evolution in Dynamic Environments 562
Irene Ntoutsi, Myra Spiliopoulou, and Yannis Theodoridis

A Speed-Up Technique for Distributed Shortest Paths Computation 578
Gianlorenzo D'Angelo, Mattia D'Emidio, Daniele Frigioni, and Vinicio Maurizio

MulO-AntMiner: A New Ant Colony Algorithm for the Multi-objective Classification Problem 594
Nesrine Said, Moez Hammami, and Khaled Ghedira

Software Fault Localization via Mining Execution Graphs 610
Saeed Parsa, Somaye Arabi Naree, and Neda Ebrahimi Koopaei

Removal of Surface Artifacts of Material Volume Data with Defects 624
Jie Shen, Vela Diego, and David Yoon

Two-Stage Fish Disease Diagnosis System Based on Clinical Signs and Microscopic Images 635
Chang-Min Han, Sang-Woong Lee, Soonhee Han, and Jeong-Seon Park

Proteins Separation in Distributed Environment Computation 648
Ming Chau, Thierry Garcia, and Pierre Spiteri

Synchronous and Asynchronous Distributed Computing for Financial Option Pricing 664
Thierry Garcia, Ming Chau, and Pierre Spiteri

A Set of Experiments to Consider Data Quality Criteria in Classification Techniques for Data Mining 680
Roberto Espinosa, José Zubcoff, and Jose-Norberto Mazón

A Numerical Algorithm for the Solution of Simultaneous Nonlinear Equations to Simulate Instability in Nuclear Reactor and Its Analysis 695
Goutam Dutta and Jagdeep B. Doshi

Author Index 711

Table of Contents – Part III

Workshop on Computational Geometry and Applications (CGA 2011)

Optimizing the Layout of Proportional Symbol Maps 1
Guilherme Kunigami, Pedro J. de Rezende, Cid C. de Souza, and Tallys Yunes

An Optimal Hidden-Surface Algorithm and Its Parallelization 17
F. Dévai

Construction of Pseudo-triangulation by Incremental Insertion 30
Ivana Kolingerová, Jan Trčka, and Ladislav Hobza

Non-uniform Geometric Matchings 44
Christian Knauer, Klaus Kriegel, and Fabian Stehn

Multi-robot Visual Coverage Path Planning: Geometrical Metamorphosis of the Workspace through Raster Graphics Based Approaches 58
João Valente, Antonio Barrientos, Jaime del Cerro, Claudio Rossi, Julian Colorado, David Sanz, and Mario Garzón

A Practical Solution for Aligning and Simplifying Pairs of Protein Backbones under the Discrete Fréchet Distance 74
Tim Wylie, Jun Luo, and Binhai Zhu

k-Enclosing Axis-Parallel Square 84
Priya Ranjan Sinha Mahapatra, Arindam Karmakar, Sandip Das, and Partha P. Goswami

Tree Transformation through Vertex Contraction with Application to Skeletons 94
Arseny Smirnov and Kira Vyatkina

Topology Construction for Rural Wireless Mesh Networks – A Geometric Approach 107
Sachin Garg and Gaurav Kanade

An Adapted Version of the Bentley-Ottmann Algorithm for Invariants of Plane Curves Singularities 121
Mădălina Hodorog, Bernard Mourrain, and Josef Schicho

A Heuristic Homotopic Path Simplification Algorithm 132
Shervin Daneshpajouh and Mohammad Ghodsi

An Improved Approximation Algorithm for the Terminal Steiner Tree Problem 141
Yen Hung Chen

Min-Density Stripe Covering and Applications in Sensor Networks 152
Adil I. Erzin and Sergey N. Astrakov

Power Diagrams and Intersection Detection 163
Michal Zemek and Ivana Kolingerová

Workshop on Approximation, Optimization and Applications (AOA 2011)

Heuristic Pattern Search for Bound Constrained Minimax Problems 174
Isabel A.C.P. Espírito Santo and Edite M.G.P. Fernandes

Novel Fish Swarm Heuristics for Bound Constrained Global Optimization Problems 185
Ana Maria A.C. Rocha, Edite M.G.P. Fernandes, and Tiago F.M.C. Martins

Quaternions: A Mathematica Package for Quaternionic Analysis 200
M.I. Falcão and Fernando Miranda

Influence of Sampling in Radiation Therapy Treatment Design 215
Humberto Rocha, Joana M. Dias, Brigida C. Ferreira, and Maria do Carmo Lopes

On Minimizing Objective and KKT Error in a Filter Line Search Strategy for an Interior Point Method 231
M. Fernanda P. Costa and Edite M.G.P. Fernandes

Modified Differential Evolution Based on Global Competitive Ranking for Engineering Design Optimization Problems 245
Md. Abul Kalam Azad and Edite M.G.P. Fernandes

Laguerre Polynomials in Several Hypercomplex Variables and Their Matrix Representation 261
H.R. Malonek and G. Tomaz

On Generalized Hypercomplex Laguerre-Type Exponentials and Applications 271
I. Cação, M.I. Falcão, and H.R. Malonek

Branch-and-Bound Reduction Type Method for Semi-Infinite Programming 287
Ana I. Pereira and Edite M.G.P. Fernandes

On Multiparametric Analysis in Generalized Transportation Problems 300
Sanjeet Singh, Pankaj Gupta, and Milan Vlach

On an Hypercomplex Generalization of Gould-Hopper and Related Chebyshev Polynomials 316
I. Cação and H.R. Malonek

Nonlinear Optimization for Human-Like Movements of a High Degree of Freedom Robotics Arm-Hand System 327
Eliana Costa e Silva, Fernanda Costa, Estela Bicho, and Wolfram Erlhagen

Applying an Elitist Electromagnetism-Like Algorithm to Head Robot Stabilization 343
Miguel Oliveira, Cristina P. Santos, Ana Maria A.C. Rocha, Lino Costa, and Manuel Ferreira

3D Mappings by Generalized Joukowski Transformations 358
Carla Cruz, M.I. Falcão, and H.R. Malonek

Workshop on Chemistry and Materials Sciences and Technologies (CMST 2011)

Evaluation of SOA Formation Using a Box Model Version of CMAQ and Chamber Experimental Data 374
Manuel Santiago, Ariel F. Stein, Fantine Ngan, and Marta G. Vivanco

A Fault Tolerant Workflow for CPU Demanding Calculations 387
A. Costantini, O. Gervasi, and A. Laganà

A Grid Credit System Empowering Virtual Research Communities Sustainability 397
C. Manuali and A. Laganà

A Parallel Code for Time Independent Quantum Reactive Scattering on CPU-GPU Platforms 412
Ranieri Baraglia, Malko Bravi, Gabriele Capannini, Antonio Laganà, and Edoardo Zambonini

Time Dependent Quantum Reactive Scattering on GPU 428
Leonardo Pacifici, Danilo Nalli, Dimitris Skouteris, and Antonio Laganà

Potential Decomposition in the Multiconfiguration Time-Dependent Hartree Study of the Confined H Atom 442
Dimitrios Skouteris and Antonio Laganà

An Extension of the Molecular Simulator GEMS to Calculate the Signal of Crossed Beam Experiments 453
Antonio Laganà, Nadia Balucani, Stefano Crocchianti, Piergiorgio Casavecchia, Ernesto Garcia, and Amaia Saracibar

Federation of Distributed and Collaborative Repositories and Its Application on Science Learning Objects 466
Sergio Tasso, Simonetta Pallottelli, Riccardo Bastianini, and Antonio Laganà

Workshop on Mobile Systems and Applications (MoSA 2011)

HTAF: Hybrid Testing Automation Framework to Leverage Local and Global Computing Resources 479
Keun Soo Yim, David Hreczany, and Ravishankar K. Iyer

Page Coloring Synchronization for Improving Cache Performance in Virtualization Environment 495
Junghoon Kim, Jeehong Kim, Deukhyeon Ahn, and Young Ik Eom

Security Enhancement of Smart Phones for Enterprises by Applying Mobile VPN Technologies 506
Young-Ran Hong and Dongsoo Kim

An Efficient Mapping Table Management in NAND Flash-Based Mobile Computers 518
Soo-Hyeon Yang and Yeonseung Ryu

Performance Improvement of I/O Subsystems Exploiting the Characteristics of Solid State Drives 528
Byeungkeun Ko, Youngjoo Kim, and Taeseok Kim

A Node Placement Heuristic to Encourage Resource Sharing in Mobile Computing 540
Davide Vega, Esunly Medina, Roc Messeguer, Dolors Royo, and Felix Freitag

Session on Cloud for High Performance Computing

Examples of WWW Business Application System Development Using a Numerical Value Identifier 556
Toshio Kodama, Tosiyasu L. Kunii, and Yoichi Seki

Building a Front End for a Sensor Data Cloud 566
Ian Rolewicz, Michele Catasta, Hoyoung Jeung, Zoltán Miklós, and Karl Aberer

Design of a New Cloud Computing Simulation Platform 582
A. Nuñez, J.L. Vázquez-Poletti, A.C. Caminero, J. Carretero, and I.M. Llorente

General Tracks

System Structure for Dependable Software Systems 594
Vincenzo De Florio and Chris Blondia

Robust Attributes-Based Authenticated Key Agreement Protocol Using Smart Cards over Home Network 608
Xin-Yi Chen and Hyun-Sung Kim

$AUTH_{HOTP}$ - HOTP Based Authentication Scheme over Home Network Environment 622
Hyun Jung Kim and Hyun Sung Kim

FRINGE: A New Approach to the Detection of Overlapping Communities in Graphs 638
Camilo Palazuelos and Marta Zorrilla

Parallel Implementation of the Heisenberg Model Using Monte Carlo on GPGPU 654
Alessandra M. Campos, João Paulo Peçanha, Patrícia Pampanelli, Rafael B. de Almeida, Marcelo Lobosco, Marcelo B. Vieira, and Sócrates de O. Dantas

Lecture Notes in Computer Science: Multiple DNA Sequence Alignment Using Joint Weight Matrix 668
Jian-Jun Shu, Kian Yan Yong, and Weng Kong Chan

Seismic Wave Propagation and Perfectly Matched Layers Using a GFDM 676
Francisco Ureña, Juan José Benito, Eduardo Salete, and Luis Gavete

Author Index 693

Table of Contents – Part IV

Workshop on Computer Aided Modeling, Simulation, and Analysis (CAMSA 2011)

On the Stability of Fully-Explicit Finite-Difference Scheme for Two-Dimensional Parabolic Equation with Nonlocal Conditions 1
Svajūnas Sajavičius

Numerical Solution of Multi-scale Electromagnetic Boundary Value Problems by Utilizing Transformation-Based Metamaterials 11
Ozlem Ozgun and Mustafa Kuzuoglu

Coupled Finite Element - Scaled Boundary Finite Element Method for Transient Analysis of Dam-Reservoir Interaction 26
Shangming Li

A Comparison of Different Advective Solvers in the CHIMERE Air Quality Model 35
Pedro Molina, Luis Gavete, Marta García Vivanco, Inmaculada Palomino, M. Lucía Gavete, Francisco Ureña, and Juan José Benito

Chinese Chess Recognition Based on Log-Polar Transform and FFT 50
Shi Lei, Pan Hailang, Cao Guo, and Li Chengrong

Adaptive Discontinuous Galerkin B-Splines on Parametric Geometries 59
Maharavo Randrianarivony

Development of a Didactic Model of the Hydrologic Cycle Using the TerraME Graphical Interface for Modeling and Simulation 75
Tiago Lima, Sergio Faria, and Tiago Carneiro

Visual Quality Control of Planar Working Pieces: A Curve Based Approach Using Prototype Fitting 91
Georg Maier and Andreas Schindler

New Approaches for Model Generation and Analysis for Wire Rope 103
Cengiz Erdönmez and Cevat Erdem İmrak

High-Quality Real-Time Simulation of a Turbulent Flame 112
Piotr Opioła

Workshop on Mobile Sensor and Its Applications (MSA 2011)

Security Improvement on a Group Key Exchange Protocol for Mobile Networks 123
Junghyun Nam, Kwangwoo Lee, Juryon Paik, Woojin Paik, and Dongho Won

Energy and Path Aware Clustering Algorithm (EPAC) for Mobile Ad Hoc Networks 133
Waqar Asif and Saad Qaisar

Employing Energy-Efficient Patterns for Coverage Problem to Extend the Network Lifetime 148
Manh Thuong Quan Dao, Ngoc Duy Nguyen, Vyacheslaw Zalyubovskiy, and Hyunseung Choo

Cooperative Communication for Energy Efficiency in Mobile Wireless Sensor Networks 159
Mehwish Nasim and Saad Qaisar

Towards Fast and Energy-Efficient Dissemination via Opportunistic Broadcasting in Wireless Sensor Networks 173
Minjoon Ahn, Hao Wang, Mihui Kim, and Hyunseung Choo

A Dynamic Multiagent-Based Local Update Strategy for Mobile Sinks in Wireless Sensor Networks 185
Jinkeun Yu, Euihoon Jeong, Gwangil Jeon, Dae-Young Seo, and Kwangjin Park

Multipath-Based Reliable Routing Protocol for Periodic Messages on Wireless Sensor Networks 197
Hoai Phong Ngo and Myung-Kyun Kim

A Multi-hop Based Media Access Control Protocol Using Magnetic Fields in Wireless Sensor Networks 209
EuiHoon Jeong, YunJae Won, SunHee Kim, SeungOk Lim, and Young-Cheol Bang

A Study on Hierarchical Policy Model for Managing Heterogeneous Security Systems 225
DongYoung Lee, Sung-Soo Ahn, and Minsoo Kim

Hashing-Based Lookup Service with Multiple Anchor Cluster Distribution System in MANETs 235
Jongpil Jeong

Performance Analysis of MIMO System Utilizing the Detection Algorithm 248
Sungsoo Ahn and Dongyoung Lee

Workshop on Computational Aspects and Methods in Renewable Energies (CAMRE 2011)

Overview and Comparison of Global Concentrating Solar Power Incentives Schemes by Means of Computational Models ... 258
M. Villarini, M. Limiti, and R. Impero Abenavoli

Economical Analysis of SOFC System for Power Production ... 270
Andrea Colantoni, Menghini Giuseppina, Marco Buccarella, Sirio Cividino, and Michela Vello

Modelling the Vertical Heat Exchanger in Thermal Basin ... 277
Maurizio Carlini and Sonia Castellucci

Optical Modelling of Square Solar Concentrator ... 287
Maurizio Carlini, Carlo Cattani, and Andrea O.M. Tucci

Plant for the Production of Chips and Pellet: Technical and Economic Aspects of an Case Study in the Central Italy ... 296
Danilo Monarca, Massimo Cecchini, and Andrea Colantoni

Feasibility of the Electric Energy Production through Gasification Processes of Biomass: Technical and Economic Aspects ... 307
Danilo Monarca, Massimo Cecchini, Andrea Colantoni, and Alvaro Marucci

Soot Emission Modelization of a Diesel Engine from Experimental Data ... 316
Enrico Bocci and Lorenzo Rambaldi

Workshop on Symbolic Computing for Dynamic Geometry (SCDG 2011)

Equal Bisectors at a Vertex of a Triangle ... 328
R. Losada, T. Recio, and J.L. Valcarce

On the Parametric Representation of Dynamic Geometry Constructions ... 342
Francisco Botana

Using Free Open Source Software for Intelligent Geometric Computing ... 353
Miguel A. Abánades, Francisco Botana, Jesús Escribano, and José L. Valcarce

A Coq-Based Library for Interactive and Automated Theorem Proving in Plane Geometry ... 368
Tuan-Minh Pham, Yves Bertot, and Julien Narboux

Teaching Geometry with TutorMates 384
María José González, Julio Rubio, Tomás Recio, Laureano González-Vega, and Abel Pascual

On Equivalence of Conditions for a Quadrilateral to Be Cyclic 399
Pavel Pech

Workshop on Wireless and Ad Hoc Networking (WADNet 2011)

An Algorithm for Prediction of Overhead Messages in Client-Server Based Wireless Networks 412
Azeem Irshad, Muddesar Iqbal, Amjad Ali, and Muhammad Shafiq

TCP Hybla+ : Making TCP More Robust against Packet Loss in Satellite Networks 424
ManKyu Park, MinSu Shin, DeockGil Oh, ByungChul Kim, and JaeYong Lee

A Secure Privacy Preserved Data Aggregation Scheme in Non Hierarchical Networks 436
Arijit Ukil and Jaydip Sen

An OWL-Based Context Model for U-Agricultural Environments 452
Yongyun Cho, Sangjoon Park, Jongchan Lee, and Jongbae Moon

A Point-Based Inventive System to Prevent Free-Riding on P2P Network Environments 462
Jongbae Moon and Yongyun Cho

A Probability Density Function for Energy-Balanced Lifetime-Enhancing Node Deployment in WSN 472
Subir Halder, Amrita Ghosal, Amartya Chaudhuri, and Sipra DasBit

Session on Computational Design for Technology Enhanced Learning: Methods, Languages, Applications and Tools (CD4TEL 2011)

Towards Combining Individual and Collaborative Work Spaces under a Unified E-Portfolio 488
Hugo A. Parada G., Abelardo Pardo, and Carlos Delgado Kloos

A Scenario Editing Environment for Professional Online Training Systems 502
José Luis Aguirre-Cervantes and Jean-Philippe Pernin

Standardization of Game Based Learning Design 518
Sebastian Kelle, Roland Klemke, Marion Gruber, and Marcus Specht

Simplified Workflow Representation of IMS Learning Design 533
Juan C. Vidal, Manuel Lama, and Alberto Bugarín

From a Pattern Language to a Pattern Ontology Approach for CSCL Script Design 547
Jonathan Chacón, Davinia Hernández-Leo, and Josep Blat

Session on Virtual Reality in Medicine and Surgery (VRMS 2011)

Advanced Interface for the Pre-operative Planning and the Simulation of the Abdominal Access in Pediatric Laparoscopy 562
Lucio Tommaso De Paolis and Giovanni Aloisio

An Augmented Reality Application for the Radio Frequency Ablation of the Liver Tumors 572
Lucio T. De Paolis, Francesco Ricciardi, Aldo F. Dragoni, and Giovanni Aloisio

Virtual Reality and Hybrid Technology for Neurorehabilitations 582
Alessandro De Mauro, Aitor Ardanza, Chao Chen, Eduardo Carrasco, David Oyarzun, Diego Torricelli, Shabs Rajasekharan, Josè Luis Pons, Ángel Gil-Agudo, and Julián Flórez Esnal

Virtual Angioscopy Based on Implicit Vasculatures 592
Qingqi Hong, Qingde Li, and Jie Tian

Session on Logical, Scientific and Computational Aspects of Pulse Phenomena in Transitions (PULSES 2011)

Improvement of Security and Feasibility for Chaos-Based Multimedia Cryptosystem 604
Jianyong Chen and Junwei Zhou

Recursion Formulas in Determining Isochronicity of a Cubic Reversible System 619
Zhiwu Liao, Shaoxiang Hu, and Xianling Hou

Skeletonization of Low-Quality Characters Based on Point Cloud Model 633
X.L. Hou, Z.W. Liao, and S.X. Hu

Family of Curves Based on Riemann Zeta Function 644
Carlo Cattani

Author Index 659

Table of Contents – Part V

Workshop on Mobile Communications (MC 2011)

Selective Placement of High-Reliability Operation in Grid-Style Wireless Control Networks 1
Junghoon Lee, Gyung-Leen Park, and Ho-Young Kwak

Adaptive Clustering Method for Femtocells Based on Soft Frequency Reuse 11
Young Min Kwon, Bum-Gon Choi, Sueng Jae Bae, and Min Young Chung

A Measurement Based Method for Estimating the Number of Contending Stations in IEEE 802.11 WLAN under Erroneous Channel Condition 22
Jun Suk Kim, Bum-Gon Choi, and Min Young Chung

Implementation of WLAN Connection Management Schemes in Heterogeneous Network Environments with WLAN and Cellular Networks 32
Hyung Wook Cho, Sueng Jae Bae, Hyunseung Choo, and Min Young Chung

Wireless Sensor Network's Lifetime Maximization Problem in Case of Given Set of Covers 44
Adil I. Erzin and Roman V. Plotnikov

An Implementation of SVM-Based Gaze Recognition System Using Advanced Eye Region Detection 58
Kue-Bum Lee, Dong-Ju Kim, and Kwang-Seok Hong

Anti Jamming – Based Medium Access Control Using Adaptive Rapid Channel Hopping in 802.11: AJ-MAC 70
Jaemin Jeung, Seungmyeong Jeong, and Jaesung Lim

Mobile Based HIGHT Encryption for Secure Biometric Information Transfer of USN Remote Patient Monitoring System 83
Young-Hyuk Kim, Il-Kown Lim, and Jae-Kwang Lee

Human Neck's Posture Measurement Using a 3-Axis Accelerometer Sensor 96
Soonmook Jeong, Taehoun Song, Hyungmin Kim, Miyoung Kang, Keyho Kwon, and Jae Wook Jeon

A System Consisting of Off-Chip Based Microprocessor and FPGA Interface for Human-Robot Interaction Applications 110
Tae Houn Song, Soon Mook Jeong, Seung Hun Jin, Dong Kyun Kim, Key Ho Kwon, and Jae Wook Jeon

Multi-hop Cooperative Communications Using Multi-relays in IEEE 802.11 WLANs 120
Sook-Hyoun Lee, Chang-Yeong Oh, and Tae-Jin Lee

Adaptive Packing Strategy to Reduce Packing Loss in MF-TDMA Satellite Networks 133
MinWoo Lee, Jae-Joon Lee, Jung-Bin Kim, JiNyoung Jang, GyooPil Chung, and JaeSung Lim

MyUT: Design and Implementation of Efficient User-Level Thread Management for Improving Cache Utilization 147
Inhyuk Kim, Eunhwan Shin, Junghan Kim, and Young Ik Eom

Relay Selection with Limited Feedback for Multiple UE Relays 157
Kyungrok Oh and Dong In Kim

Enhanced Multi-homing Support for Proxy Mobile IPv6 in Heterogeneous Networks 167
Dae Sun Kim, Yoshinori Kitatsuji, and Hidetoshi Yokota

Fast Handover Scheme Based on Mobility Management of Head MAG in PMIPv6 181
NamYeong Kwon, Hongsuk Kim, Seung-Tak Oh, and Hyunseung Choo

A Hole Detour Scheme Using Virtual Position Based on Residual Energy for Wireless Sensor Networks 193
Zeehan Son, Myungsu Cha, Min Han Shon, Moonseong Kim, Mihui Kim, and Hyunseung Choo

An Improved MPLS-MOB for Reducing Packet Loss in NGN 205
Myoung Ju Yu, Seong Gon Choi, and Sung Won Sohn

Strata: Wait-Free Synchronization with Efficient Memory Reclamation by Using Chronological Memory Allocation......................... 217
Eunhwan Shin, Inhyuk Kim, Junghan Kim, and Young Ik Eom

Workshop on Software Quality (SQ 2011)

DDAT: Data Dependency Analysis Tool for Web Service Business Processes 232
Ivaylo Spassov, Valentin Pavlov, Dessislava Petrova-Antonova, and Sylvia Ilieva

Towards a Quality Model for Semantic Technologies 244
Filip Radulovic and Raúl García-Castro

Influence of Human Factors in Software Quality and Productivity 257
Luis Fernández-Sanz and Sanjay Misra

A Framework for Classifying and Comparing Software Architecture Tools for Quality Evaluation 270
Eudisley Anjos and Mário Zenha-Rela

Causes of the Violation of Integrity Constraints for Supporting the Quality of Databases 283
Hendrik Decker

Complexity Measures in 4GL Environment 293
Csaba Nagy, László Vidács, Rudolf Ferenc, Tibor Gyimóthy, Ferenc Kocsis, and István Kovács

A Framework for Integrated Software Quality Prediction Using Bayesian Nets 310
Łukasz Radliński

Process Instance Management Facilities Based on the Meta Process Models 326
Seunghun Park, Sangyoon Min, and Doohwan Bae

Workshop on Software Engineering Processes and Applications (SEPA 2011)

A Multi-paradigm Complexity Metric (MCM) 342
Sanjay Misra, Ibrahim Akman, and Ferid Cafer

Flow Sensitive-Insensitive Pointer Analysis Based Memory Safety for Multithreaded Programs 355
Mohamed A. El-Zawawy

HProcessTOOL: A Support Tool in the Harmonization of Multiple Reference Models 370
César Pardo, Francisco Pino, Félix García, Francisco Romero Romero, Mario Piattini, and Maria Teresa Baldassarre

Requirement Conflicts Resolution: Using Requirement Filtering and Analysis 383
Wasi Haider Butt, Sameera Amjad, and Farooque Azam

Efficient Algorithm for Detecting Parameterized Multiple Clones in a Large Software System 398
Rajesh Prasad, Suneeta Agarwal, Anuj Kumar Sharma, Alok Singh, and Sanjay Misra

Behavior Analyzer for Developing Multiagent System on Repository-Based Multiagent Framework 410
Takahiro Uchiya and Tetsuo Kinoshita

Impact Analysis of Goal-Oriented Requirements in Web Engineering ... 421
José Alfonso Aguilar, Irene Garrigós, and Jose-Norberto Mazón

Session on Approaches or Methods of Security Engineering (AMSE 2011)

An Improved Protocol for Server-Aided Authenticated Group Key Establishment 437
Junghyun Nam, Juryon Paik, Byunghee Lee, Kwangwoo Lee, and Dongho Won

Presence Based Secure Instant Messaging Mechanism for IP Multimedia Subsystem 447
Zeeshan Shafi Khan, Muhammad Sher, and Khalid Rashid

Two-Directional Two-Dimensional Random Projection and Its Variations for Face and Palmprint Recognition 458
Lu Leng, Jiashu Zhang, Gao Chen, Muhammad Khurram Khan, and Khaled Alghathbar

Session on Advances in Web Based Learning (AWBL 2011)

A Framework for Intelligent Tutoring in Collaborative Learning Systems Using Service-Oriented Architecture 471
Fang-Fang Chua and Chien-Sing Lee

Probability Modelling of Accesses to the Course Activities in the Web-Based Educational System 485
Michal Munk, Martin Drlik, and Marta Vrábelová

Integration of ePortfolios in Learning Management Systems 500
Ricardo Queirós, Lino Oliveira, José Paulo Leal, and Fernando Moreira

Session on International Workshop on Biomathematics, Bioinformatics and Biostatistics (IBBB 2011)

Monitoring and Control in a Spatially Structured Population Model 511
Manuel Gámez, Inmaculada López, József Garay, and Zoltán Varga

Limitations of Using Mass Action Kinetics Method in Modeling Biochemical Systems: Illustration for a Second Order Reaction 521
Cigdem Sayikli and Elife Zerrin Bagci

The Impact of Isolation of Identified Active Tuberculosis Cases on the Number of Latently Infected Individuals 527
Schehrazad Selmane

General Tracks

Performance Analysis of an Algorithm for Computation of Betweenness Centrality 537
Shivam Bhardwaj, Rajdeep Niyogi, and Alfredo Milani

Approaches to Preference Elicitation for Group Recommendation 547
Inma Garcia, Laura Sebastia, Sergio Pajares, and Eva Onaindia

Quantifying Downloading Performance of Locality-Aware BitTorrent Protocols 562
Lidong Yu, Ming Chen, and Changyou Xing

Practical and Effective Domain-Specific Function Unit Design for CGRA.. 577
Ming Yan, Ziyu Yang, Liu Yang, Lei Liu, and Sikun Li

Secure Hash-Based Password Authentication Protocol Using Smartcards.. 593
Hyunhee Jung and Hyun Sung Kim

Towards a Better Integration of Patterns in Secure Component-Based Systems Design 607
Rahma Bouaziz, Brahim Hamid, and Nicolas Desnos

Enriching Dynamically Detected Invariants in the Case of Arrays 622
Mohammadhani Fouladgar, Behrouz Minaei-Bidgoli, and Hamid Parvin

On the Problem of Numerical Modeling of Dangerous Convective Phenomena: Possibilities of Real-Time Forecast with the Help of Multi-core Processors 633
N.O. Raba and E.N. Stankova

Efficient Model Order Reduction of Large-Scale Systems on Multi-core Platforms 643
P. Ezzatti, E.S. Quintana-Ortí, and A. Remón

Author Index 655

Towards a Spatio-Temporal Information System for Moving Objects

Maribel Yasmina Santos[1], José Mendes[1], Adriano Moreira[1], and Monica Wachowicz[2]

[1] Information Systems Department,
University of Minho, Portugal
{maribel,adriano}@dsi.uminho.pt, pg12678@alunos.uminho.pt
[2] Geodesy and Geomatics Engineering,
University of New Brunswick, Canada
monicaw@unb.ca

Abstract. The analysis of movement in the geographic space requires the development of systems and data models supporting storage and querying functions useful for the analysis of how movement evolves and how it is constrained by events or obstacles that emerge in space. In the past, the design of information systems has been often based on application driven approaches and restricted to specific types of positioning technologies, leading to a proliferation of a wide range of data models, database models and functionalities. To overcome this proliferation, this paper proposes a spatio-temporal information system that is independent of the application domain and aiming to abstract the positioning technology used for gathering data. Here we describe the initial design of the basic infrastructure of such a system, integrating a data model and its fundamental functionalities for the collection, storage, retrieval and visualization of positioning data. The results achieved so far are promising to demonstrate how the system is able to store positioning data with different formats, and in applying its functionalities to the loaded data.

Keywords: moving objects, spatial data, spatio-temporal information system.

1 Introduction

Data about movement of car fleets, trains, aircrafts, ships, individuals, animals, or even viruses, are being collected in growing amounts by means of current sensing and tracking technologies [1]. A mobile object can be defined as an entity that changes its position over time. Nowadays, databases of moving objects are being developed using a particular data model which is, in turn, related to a specific application domain [2]. Therefore, the proliferation of a wide range of specific data models is the first consequence of this application-driven development approach [3]. Furthermore, databases with modeling and querying mechanisms are also available [4]. These database systems can be filled with implementations of different Database Management Systems (DBMS) models, such as the relational, object-oriented, and

B. Murgante et al. (Eds.): ICCSA 2011, Part I, LNCS 6782, pp. 1–16, 2011.

object-relational models, having to accommodate specific abstract data types, such as spatial data types. More specific data models and applications need to be implemented as those systems do not include interfaces that can be used by end-users to analyze their data sets.

This research work proposes a new approach to address these problems, by focusing on the development of an integrated data model that can be used in different application domains and by different end-users. The main aim is the design of a Spatio-temporal Information System (STAR) for moving objects using the Unified Modeling Language (UML). In the specification of the basic infrastructure of this system, Use Case Diagrams are used to define the system functionalities, and Class Diagrams are used to model the structure of the system's database. The implementation of the primary functionalities of the proposed system is illustrated through the analysis of movement data related with two different application scenarios. However, it is important to point out that several other applications may benefit from this development in the domains of for example, location-based services, tourism services, mobile electronic commerce, transportation and air traffic control, emergency response, mobile resource management, among many others [5].

This paper is organized as follows. Section 2 presents the previous work strongly related to the management of movement data. Section 3 presents some of the Use Case Diagrams specifying the system functionalities. Section 4 presents the Class Diagram of the database that adds persistence and querying capabilities to the proposed STAR system. Section 5 presents the current status of the system implementation, built on top of the SQL Server DBMS and the ArcGIS geographic information system. Section 6 concludes with the discussion about the undertaken work and our guidelines for future design work.

2 Related Work

Movement data can be obtained through location technologies (as add-ons to GSM and UMTS infrastructures), satellite based position technologies (GPS and GLONASS), and indoor positioning systems (Wi-Fi and Bluetooth based RTLS, UltraWideBand (UWB) technologies), among others. Having this variety of technologies for collecting movement data, a wide range of applications have been developed to analyze data associated to moving objects [3, 6, 7, 8], and different database models were designed to support those applications [3, 9-12].

Weng et al. [8] designed and implemented a spatio-temporal data model for a vehicle monitoring system. The model integrates two major concepts: cube cell and trajectory. A cube cell is a three-dimensional space defined by one time and two space dimensions. Given a certain time value, the corresponding time slice can be obtained as well as the vehicle trajectory. In this system, a trajectory represents the minimum querying unit. The sizes or the slices considered in each cube depend on the geographical context, as more positioning data is necessary to describe a trajectory in a dense urban area and fewer positions are needed in lower density areas. Recent work on the impact of data quality in the context of pedestrian movement analysis points out at the same direction, mentioning the importance of the sampling rate on data accuracy [13].

Another work related with data models for mobile objects is presented in [3] by Nóbrega et al. The authors present a model of spatio-temporal classes having mobility aspects. The model focuses on the updating problem associated with positioning. In this case, the authors seek to balance the number of updates with the consistency of the database. The system also integrates some prediction capabilities, based on movement profiles created from history movements. These profiles can store the different speeds that objects normally go on as well as the routes that are usually taken by them.

Lee and Ryu [7] designed a vehicle information management system that is able to manage and retrieve vehicle locations in mobile environments. The design of this system was motivated by the observation that conventional databases are not suitable for retrieving the location of vehicles, as these databases do not take into consideration the property of movement in which an object continuously change its position over time. The proposed system integrates a vehicle information collector, a vehicle information management server, and mobile clients. The system can process spatio-temporal queries related to locations of moving vehicles. It also provides moving vehicles' locations, which are not stored in the system, using a location predictor that estimates these locations.

Erwig et al. [10] proposed the treatment of moving points and moving regions as 3D objects (2D space + time). The structure of these entities is captured by modeling them as abstract data types. The objective is to make these entities available as data types for their integration in relational, object-relational or other DBMS data models. The authors provide the notions of data types for moving points and moving regions as well as a set of operations on such entities. In [14], additional specific spatio-temporal data types are defined, enabling their integration into a query language for querying positioning data.

The DOMINO system [9] introduces a set of capabilities to store and query movement objects that were implemented on top of an existing DBMS (Informix DBMS). The system enables the DBMS to predict the future location of a moving object. Knowing the speed and the route of a moving object, the system computes its location without any additional update to the object's position. The notion of dynamic attribute, an attribute whose value changes continuously as time progresses, is used.

Taking into consideration the advances in sensing and tracking technologies and emerging mobile applications, Praing and Schneider [12] focused their attention on future movements of a mobile object by proposing a universal abstract data model that includes both moving objects models and prediction models. The FuMMO abstract model uses the infinite point-set concept of point-set theory to define future movements of moving objects such as moving points, moving lines and moving regions.

Two other research efforts that need to be highlighted are the SECONDO and HERMES systems. Güting et al. [4] implemented, in the SECONDO system, the moving objects data model previously proposed in [14]. SECONDO is based on an extensible architecture, consisting of three major components: the kernel, the optimizer and the GUI. The kernel can be supported by different DBMS and is extensible by data models. The optimizer assumes an object-relational data model and supports a SQL-like language. SECONDO includes spatio-temporal indexing and was designed to support the development of spatial and spatio-temporal algorithms and applications.

Pelekis et al. [15] implemented the HERMES system for querying trajectory data. Instead of using a database of moving objects for maintaining the objects' current position, HERMES provides a trajectory database. HERMES is developed as an extension of the Oracle DBMS and supports advanced querying on trajectories, including coordinated-based queries (e.g. range, nearest neighbor queries), trajectory-based queries (e.g. topological, navigational, similarity-based queries) or a combination of both.

The previous research in this field has been mainly focused on the storage and querying of movement data. It has been the result of an application-driven strategy in which the analysis of a new dataset demands the development of a new data model, a new database and the implementation of the appropriate querying and analysis mechanisms. In cases where spatio-temporal database operations and functions are available, such as in SECONDO, end-users need to write specific queries (sometimes with a complex syntax) that use these specific operations and functions for data analysis.

In contrast, our approach aims at overcoming the constraints of using a specific type of location technology in order to avoid the proliferation of data models, databases and applications. The basic infrastructure specification of this system is presented in this paper, defining the key functionalities of a data model for a spatio-temporal information system for handling moving objects.

3 System Functionalities

The system functionalities designed for the STAR system include: i) multiuser support and user management; ii) the collection and storage of movement data; and iii) the analysis and visualization of movement data. Three different actors interact with the basic infrastructure of the proposed system – the system Administrator, the User and the Mobile Device – each one of them having access to different functionalities.

The Administrator functionalities include the management of the system database and the user access. The main functionalities for the User actor include: i) collecting data with the assistance of the Mobile Device actor; ii) managing their data records; and iii) managing their folders and files. There are also functionalities for requesting a system account by registering into the system, and for password recovery. The Mobile Device actor is designed to provide real-time movement data feeding into the system database (this functionality is currently under development; therefore its implementation is not described further in this paper).

Fig. 1 depicts an overview of the Use Cases Diagram[1] for these three actors, Administrator, User and Mobile Device.

[1] Detailed descriptions and figures of all the use case diagrams are not provided due to space limitations. However, the systems' main functionalities are always briefly described.

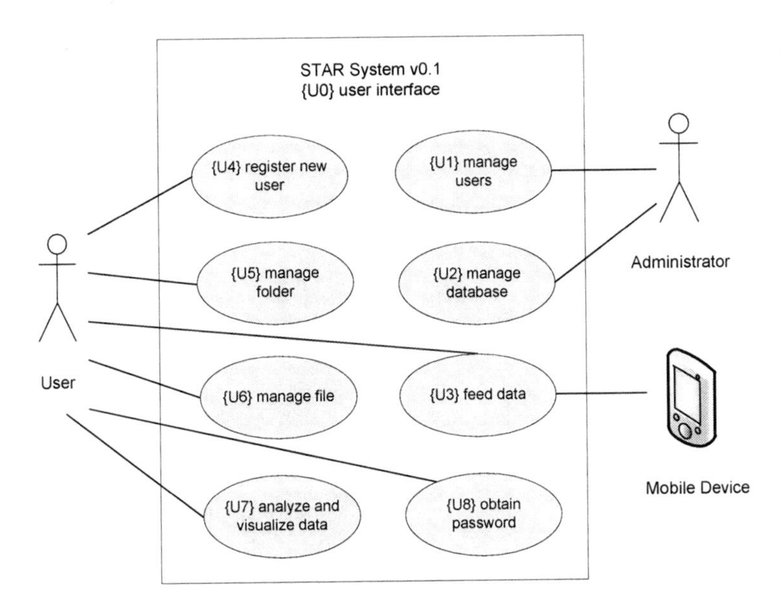

Fig. 1. Overview of the main system functionalities of STAR first specification

As illustrated in Fig. 1, different users can register to use the system. Multiuser support is fundamental in the STAR system, as we aim at designing a system to deal with huge amounts of data, eventually provided by different administrative entities working together in large data analysis projects. After the registration process, and also after the User validation by the Administrator, the User can manage folders, files and data. For each one of these use cases, the User can access several functionalities like creating a new folder, creating a new file, loading data into the system, among other functionalities which will be described latter in this section. The User can also collect real-time movement data, a functionality that is represented by the feed data use case. For the Administrator, and besides all the functionalities that are associated to the management of the users, it is necessary to manage the database that supports the STAR system in order to ensure its integrity. By providing means to create folders and files, the user can organize the several files that are loaded into the STAR environment. This way, the STAR system provides an integrated work environment where different data sets can be combined and analyzed.

Associated with data collection, the User has the possibility to schedule a data collection task, through the Mobile Device actor. In the case of data collection in real-time, the system obtains the readings from the Mobile Device and stores such positions in the database. The User can also list all the scheduled data collection tasks. Both use cases, schedule data collection and get real-time data, make use of the functionalities of the transfer data use case, the interface mechanism between the system and the Mobile Device (Fig. 2). This functionality is being implemented through a web service that is able to receive the positioning readings and check the quality of the gathered data (for example, detecting and cleaning duplicate readings).

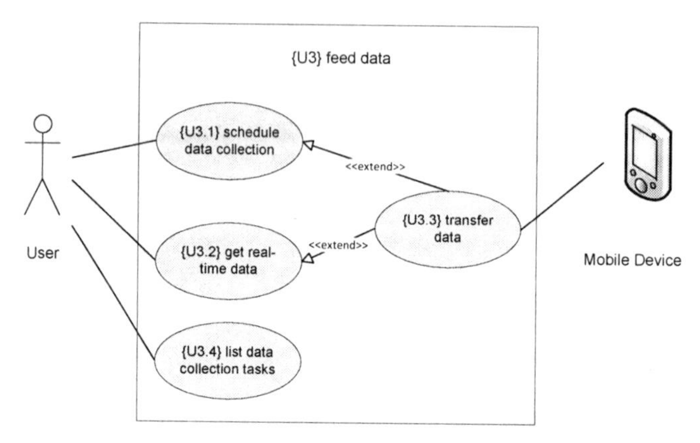

Fig. 2. Feed data use case: detailed view

In each folder, a User can maintain several files or other folders. The manage file use case allows the execution of several operations over files. Among them, the User can create, load, open, delete, rename, classify, or search for a file. The User can also search for specific data records in owned and public files. The same is valid for listing these files.

One of the most relevant functionalities of the manage file use case is related with the uploading of new data into the system, which is done through the load file use case. This use case can make use of the check data quality use case, in which several operations on data can be performed (Fig. 3). Among these operations, the User can: i) verify the structure of the available data; ii) check for missing data; iii) find duplicate records; iv) look for outliers; and, v) look for errors in data.

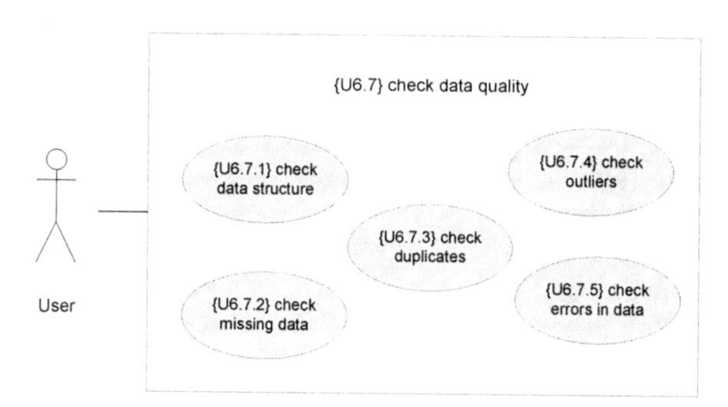

Fig. 3. Check data quality use case: detailed view

After the collection and storage of movement data, the main objective of the proposed system is to make available an integrated environment for data analysis and data visualization, functionalities integrated into the analyze and visualize data use case.

For data analysis, the system makes available a set of functionalities that can be used to compute several variables associated to a specific position, such as its velocity and direction, or associated to several positions from one or more moving objects, such as distances and areas. The system also allows the verification of the proximity between moving objects and also if they are overlapping (Fig. 4). The new operations will be added as the system specification and implementation evolves. The last presented use case diagram is associated with the visualization of data (Fig. 5).

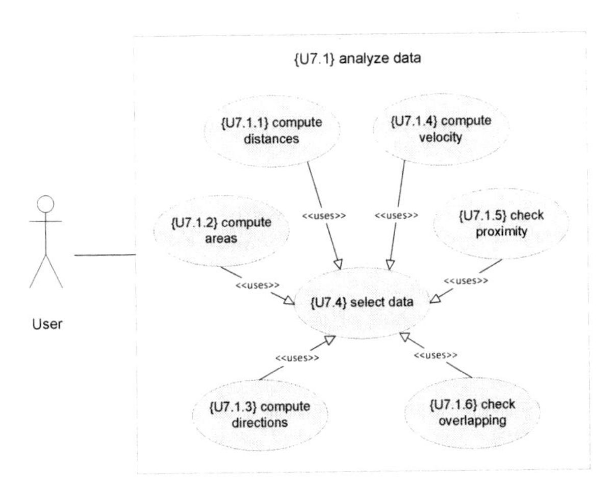

Fig. 4. Analyze data use case: detailed view

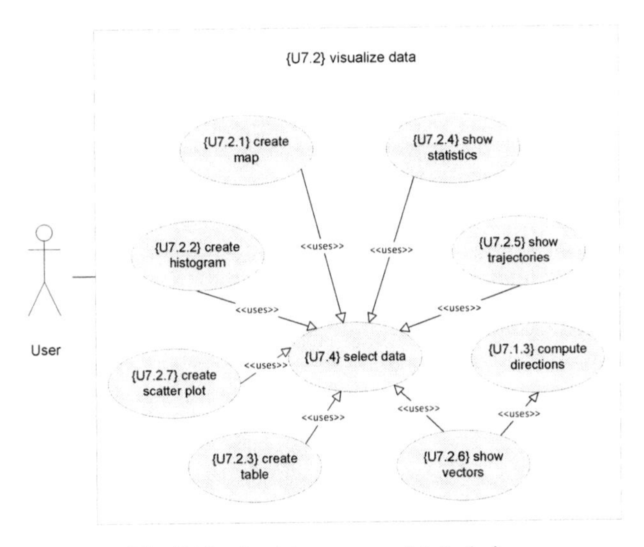

Fig. 5. Visualize data use case: detailed view

As illustrated in Fig. 5, data associated with one or more moving objects can be visualized on a map, table, histogram or scatter plot. Moreover, several statistics associated with the analyzed data can also be visualized. The system also allows the visualization of one or more moving objects as trajectories or as vectors. A trajectory of a moving object can be described as a line, composed by several segments that connect the points of the same trajectory. A vector represents a point in space and time, and the direction of movement. More details about these representations can be found in [13-14].

Although many of the functionalities previously described are common functionalities in a GIS, the STAR system provides an integrated environment with high level specialized functions with which the end-user can load, explore, analyze and visualize data related to movement without having to resort to the generic tools provided by current GISs. These generic systems provide a wide range of functionalities that are not needed or that are not specific for the analysis of movement data. This overload of functionalities leads to a complex system interface that is not suitable for a wide range of end-users.

Besides the fact of being a specific-designed system for the analysis of movement data related to moving objects, the STAR system provides the possibility to capture movement. This functionality is not easily accessible in other systems with spatial and temporal capabilities without any specific integration effort. In [10], Erwig et al. introduced a classification for spatio-temporal applications. In this classification, applications related to moving objects (represented as moving points) or moving objects with an extent (represented by moving regions) will be poorly supported by temporal databases with spatial data types, or by spatial databases, as it is necessary to support queries that refer to the evolution of an object over time, for example, "Find travelers that move frequently" or "What was the largest extent of a country ever?".

To extend spatial data models to capture time, Erwig et al. [10] proposed two basic data types: *mpoint* and *mregion*. These types are defined as mappings from time into space: $mpoint = time \rightarrow point$, $mregion = time \rightarrow region$. In the case of moving points, trajectories can be described as lines that results from applying moving constructor to a point [14]: $trajectory = moving(point) \rightarrow line$. The trajectory of a moving point stored in the database refers to the past. The capability to analyze space over time with an integrated and easy to use environment is the key issue in the proposal of the STAR system. It aims to close the gap currently existing between data collection systems and domain specific data analysis systems.

4 System Database

The main functionalities specified for the first version of the STAR system were described in detail in section 3 using Use Case Diagrams. To add persistence and querying capabilities to the system, this section presents the database model that supports these functionalities.

Worboys et al. [16] proposes the use of an object-oriented data modeling approach to design GIS. Following this line of work, this section presents the design of the supporting system database using Class Diagrams, which are the most frequently used component among the UML diagrams [17].

The proposed Class Diagram adds persistence to the information associated with the users (either system users or system administrators), the databases, the folders, and the files with the collected movement data. Data can be analyzed or visualized through the use of the functionalities presented in the previous section. These functionalities are implemented through the corresponding methods that were included in the several classes (Fig. 6). Using a Class Diagram, the methods and actors of the STAR system, as well as the data that will be stored in the spatial database, are specified.

In the Class Diagram, the Space Geometry package includes classes for dealing with points, lines and/or regions geometries (depending on the application case) of the geographical space, or raster data. Its main purpose is to manage background data used to contextualize the movement data. Good examples are satellite images or road networks. Since they are not directly related to the previously defined functionalities, the classes that belong to this package are not further described.

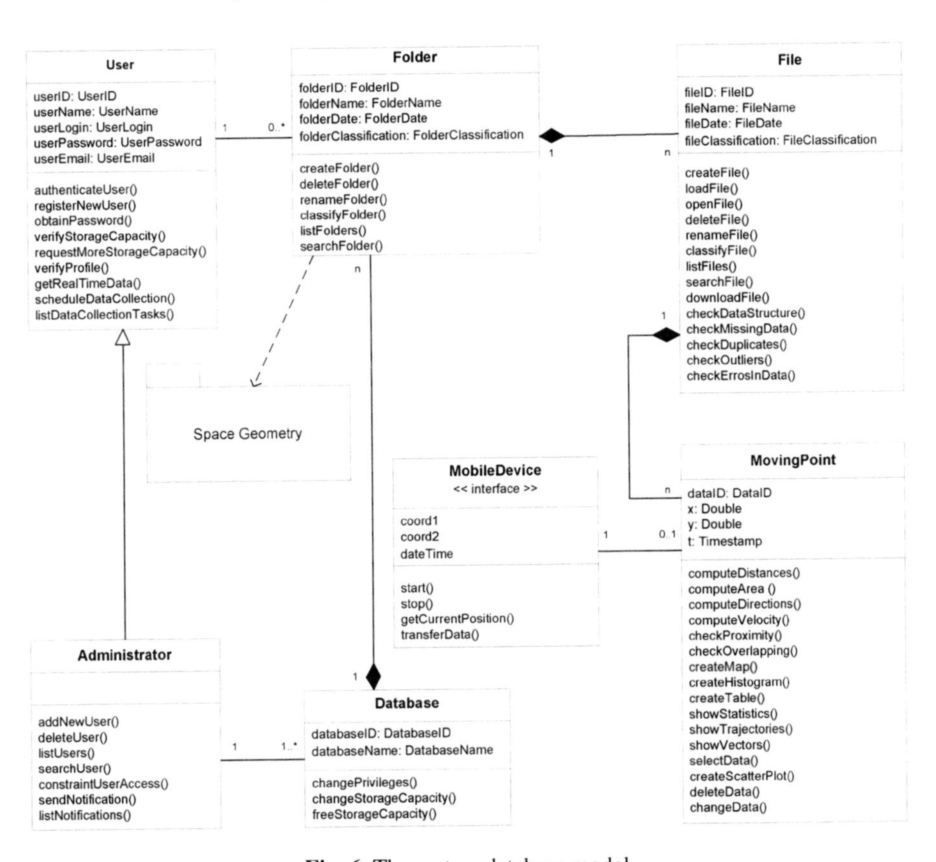

Fig. 6. The system database model

This diagram reveals how the previously defined functionalities are associated with the actors and data structures. For example, the functionalities related to the collection of data through mobile devices are associated with the User class (getRealTimeData(), scheduleDataCollection()), and with the MobileDevice class (getCurrentPosition(), transferData()). The functionalities related to the quality of the loaded data (check data quality use case) are associated with the File class. The functionalities related to the analysis and visualization are associated with the MovingPoint class. This last example emphasizes the need for specific tools for each type of moving object (in this case a moving point).

5 System Implementation

This first specification of the STAR system has been implemented using the Microsoft SQL Server 2005 Database System for the storage of the data loaded and generated by the users, as well as the information about the users and their folders and files; the Geographic Information System ArcView 9.3, which objects' libraries were used to integrate in the STAR system some typical functionalities of a GIS environment; and, the Microsoft Visual Studio 2008 Environment to set up the application using the C# programming language. This application integrates the database component, the objects' libraries and the specified functionalities in the STAR system. Besides the functionalities already presented in previous sections, new functionalities include the integration of data mining algorithms for the identification of movement flows, places of suspension of movement or abnormal behaviors that are present in movement data.

The use of this technological infrastructure (SQL Server, ArcView and C#) was guided by the knowledge that the authors of this paper have with the mentioned technologies. This option allowed a fast development of an early prototype, acting as a proof of concept for the proposed system.

This section focuses on the functionalities specified for the data analysis {U7.1} and data visualization {U7.2} components. Two movement data sets are used to illustrate some of the system functionalities.

The first movement data set is named Recreation Park data set, and consists of the movement of visitors of the Dwingelderveld National Park in The Netherlands. This data set includes GPS recordings collected in seven different days from May to August 2006. More than 140,000 records are available forming the trajectories of more than 370 visitors. In this paper, we have analyzed part of this data set.

The second data set was collected in the city of Amsterdam (The Netherlands) during an outdoor mobile game. Data was collected during ten different days, along June of 2007, totalizing 63,470 records for 419 players. In this paper, this data set is referred to as the Mobile Outdoor Game data set.

Recreation Park Data Set

Fig. 7 shows a small part of this data set which includes attributes such as a user id (R_ID), the point coordinates (Cord_X and Cord_Y), the date and time associated to the registered movement (DATETIME_), a record id (TRACKP), the original coordinates (ORIG_X and ORIG_Y), the calculated bearing (BEARING), the calculated speed (SPEED), among many other attributes that were available in the original text file.

The current implementation of the STAR system supports files in Excel and text formats, namely .xls, .xlsx, .txt, and .csv (comma separated values), which are the usual types of files used by end-users to store the raw data collected from their tracking experiments. Each one of the input files must include a header row indicating the name of each field in the data records. Upon loading, the end-user must identify those fields that represent the point coordinates, once they vary from file to file – (x and y) or (lat, long), for instance.

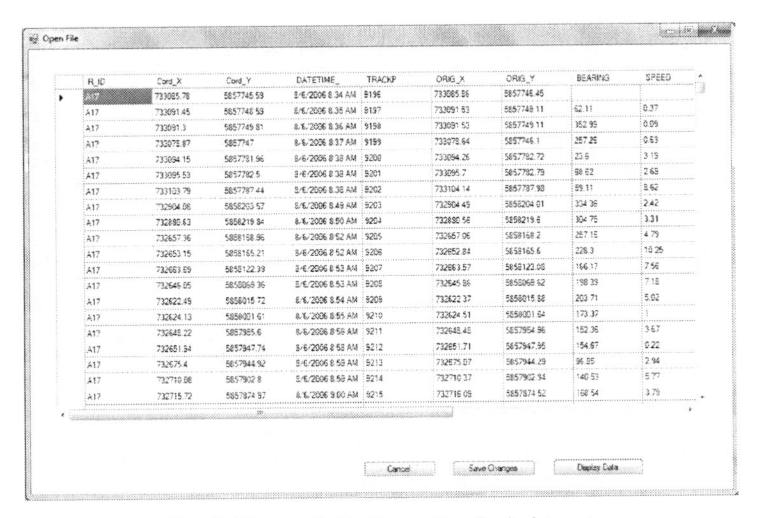

Fig. 7. The available Recreation Park data set

For the above described data set, the user can visualize the available data by displaying the records as points. Points can also be rendered over a map of the region for contextualization purposes. These maps can be loaded from files in shapefile format. Any shapefile can be loaded into the STAR system using the Space Geometry component depicted in Fig. 6. Since the STAR system has been implemented using the objects' library of the ArcGIS platform, maps for supporting different layers of visualization can be loaded into the STAR system. By combining a data selection with a data visualization task, parts of a large data set can be easily visualized. Fig. 8 shows the selection of the records associated with two visitors of the Dwingelderveld National Park, in this case users R127 and R031. As a result, any combination of the available attributes can be used in the design of a query to be processed by the database.

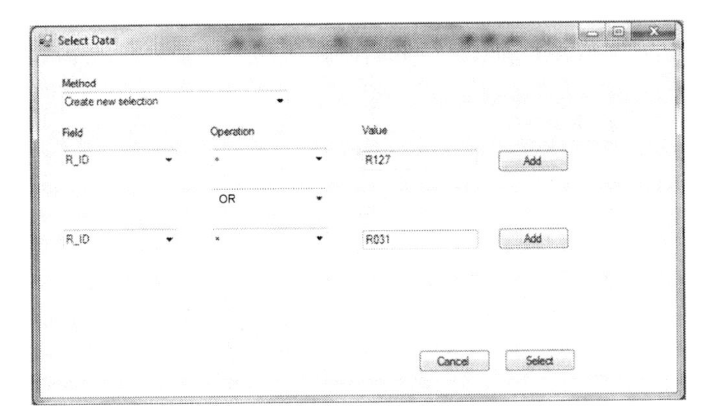

Fig. 8. Selection of the data records associated with two users

Fig. 9 shows the visualization of the query results: the selected points are highlighted with a different color for each user (R127 and R031) over the entire available data.

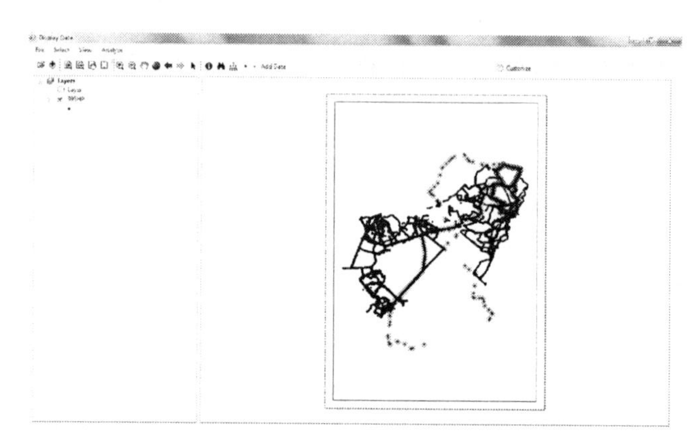

Fig. 9. Display of all the available points and the highlighting of selected data (for two pedestrians)

Mobile Outdoor Game Data Set

Fig. 10 presents the corresponding file loaded into the STAR system for the mobile outdoor gaming data set. It is worth noticing that this data set integrates different columns. All system functionalities can be applied to this data set, as data files are not dependent of any particular data structure. The users can previously define the structure of their files, specifying the name and type of the several attributes to be loaded. Otherwise the structure is inherited from the loaded files, as the first row indicates the name of each field in the data records.

Working with Different Data Sets

One of the advantages of the proposed STAR system is with the support for handling files that can have different structures such as different data attributes (and any number of columns/attributes). Although both the Recreation Park data set and the Mobile Outdoor Game data set are related with pedestrian movement, the analysis of other moving objects within different application areas is also possible. After loading the data, the STAR system adapts all its functionalities to the data set under analysis, integrating the attributes that are specific to each data set. In addition to the ability to deal with different data sets in terms of content and application areas, the analyzed data can also be associated with different movement technologies, as the user has the possibility to choose the attributes that will be used to represent the positions. The position of a moving object can be represented by its Geographic Coordinates (Lat, Lon), e.g. as provided by a GPS receiver, or by its Cartesian Coordinates (X, Y), e.g. as provided by a RTLS (Real Time Location System) indoor positioning system.

Open File

ID_POINT	Cord_X	Cord_Y	TEAM	ZONENAME	LABEL	LATITUDE	LONGITUDE	LONGITUDE_
[illegible]	629393.94	5804421.81	red	zone2	bagoq	52.37	4.9	4.9
43303	629413.85	5804436.64	red	zone2	bagoq	52.37	4.9	4.9
43305	629405.66	5804450.92	red	zone2	bagoq	52.37	4.9	4.9
43308	629406.34	5804463.15	red	zone2	bagoq	52.37	4.9	4.9
43311	629413.51	5804475.58	red	zone2	bagoq	52.37	4.9	4.9
43314	629413.16	5804488.93	red	zone2	bagoq	52.37	4.9	4.9
43317	629408.3	5804492.14	red	zone2	bagoq	52.37	4.9	4.9
43320	629401.92	5804501.98	red	zone2	bagoq	52.37	4.9	4.9
43324	629401.62	5804513.11	red	zone2	bagoq	52.37	4.9	4.9
43328	629395.88	5804524.09	red	zone2	bagoq	52.37	4.9	4.9
43333	629383.93	5804538.24	red	zone2	bagoq	52.37	4.9	4.9
43337	629379.82	5804539.25	red	zone2	bagoq	52.37	4.9	4.9
43341	629384.92	5804552.74	red	zone2	bagoq	52.37	4.9	4.9
43345	629375.59	5804544.7	red	zone2	bagoq	52.37	4.9	4.9
43348	629377.04	5804541.4	red	zone2	bagoq	52.37	4.9	4.9
43352	629379.11	5804540.34	red	zone2	bagoq	52.37	4.9	4.9
43356	629368.15	5804541.11	red	zone2	bagoq	52.37	4.9	4.9
43360	629351.62	5804549.64	red	zone2	bagoq	52.37	4.9	4.9
43364	629345.15	5804562.82	red	zone2	bagoq	52.37	4.9	4.9
43368	629338.08	5804572.65	red	zone2	bagoq	52.37	4.9	4.9

Cancel | Save Changes | Display Data

Fig. 10. Example of the available Mobile Outdoor Game data set

Included in the data visualization functionalities, the user has the possibility to create a scatter plot or a histogram. In both cases, the user only needs to specify the attribute, for a histogram, or the attributes, for a scatter plot, to be included. As an example, Fig. 11a shows the specification of the attributes, in this case it integrates the SPEED and the BEARING attributes, which will be displayed on a scatter plot (Fig. 11b).

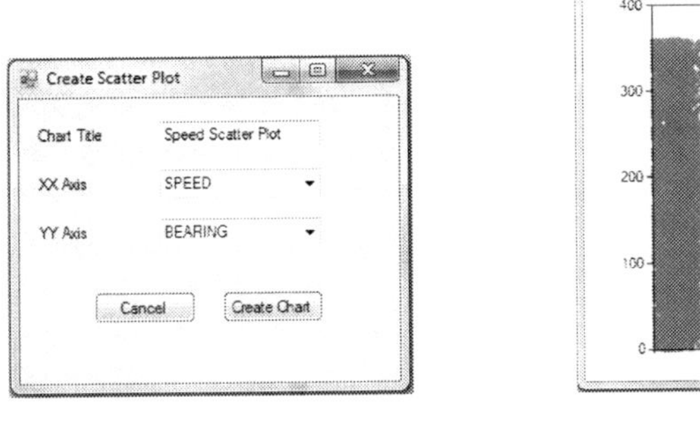

a) Attributes to be used

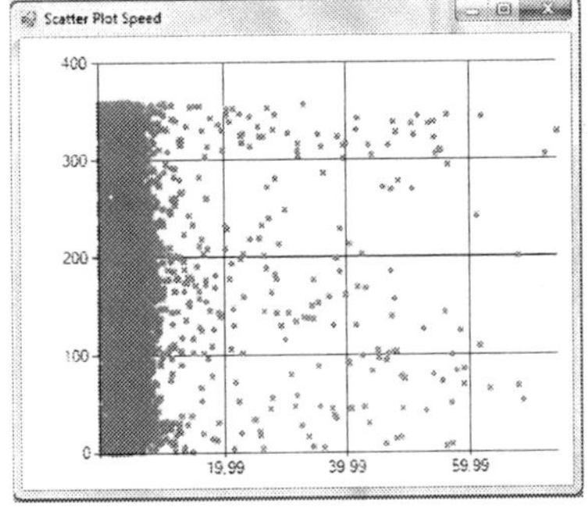

b) Scatter Plot

Fig. 11. Scatter Plot combining Speed and Bearing

In the case of a histogram, the user only needs to specify the attribute to be analyzed, and to define a title for the histogram and the number of intervals (or bins) to be included (Fig. 12).

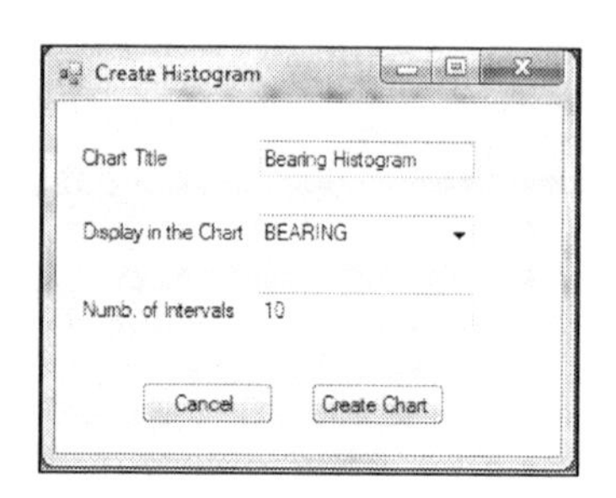

a) Values to be considered in the histogram

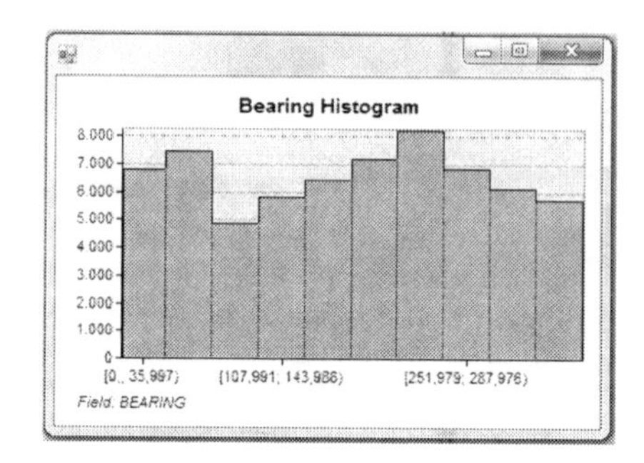

b) Histogram

Fig. 12. Histogram for the Bearing attribute

Furthermore, we proceed with the visualization of trajectories, defined as a sequence of lines connecting consecutive points in time associated with a specific moving object: in our case, a pedestrian trajectory. For this process, the user only needs to specify the information that will be selected from the data set. Finally, the movement of the user R127 is displayed as a trajectory (Fig. 13).

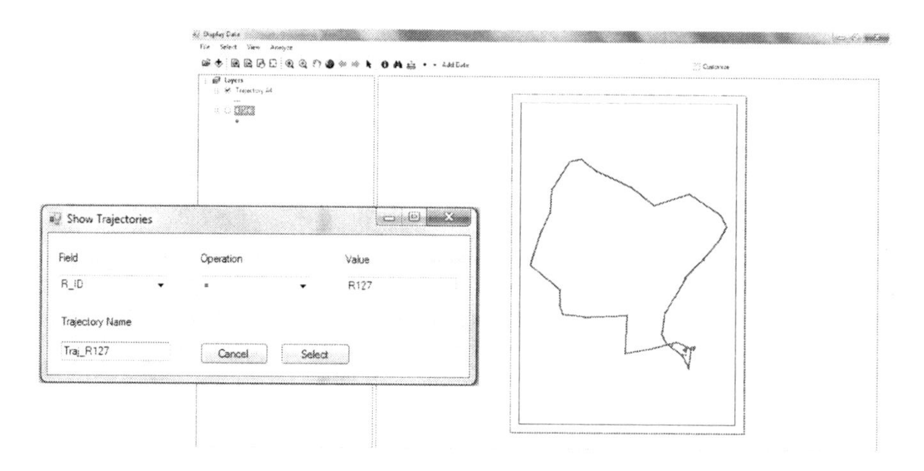

Fig. 13. Trajectory of pedestrian R127

For application scenarios in which the visualization of movement is more appropriate as vectors and not as trajectories, movement can be visualized as a set of arrows representing the direction of the movement.

Fig. 14 shows the interface for selecting the part of a data set to be visualized as vectors, in this example associated with the movement of pedestrian R127, and also shows the corresponding vectors. The {U7.2.6} show vectors use case makes use of the {U7.1.3} compute directions use case functionality, which calculates the bearing between two consecutive points in the data set. The computed bearing values are used to assign a direction to the several arrows.

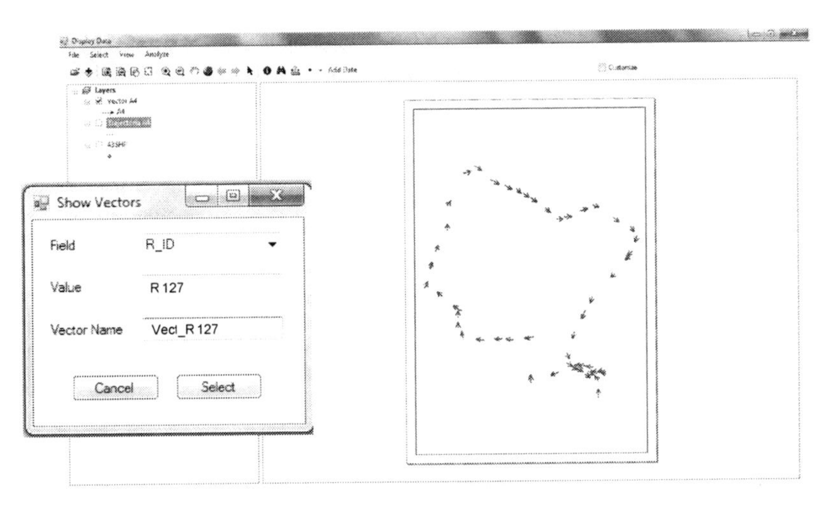

Fig. 14. Vectors of pedestrian R127

6 Conclusions

This paper described the design of a spatio-temporal information system for moving objects. This system includes fundamental functionalities associated with the collection, the storage, the analysis and the visualization of movement data. The motivations for undertaking this direction of research are rooted in the consideration that movement data sets are, and will continue to be, growing rapidly, due to, in particular, the collection of privacy-sensitive telecommunication data from mobile phones and other location-aware devices, as well as the daily collection of transaction data through database systems, network traffic controllers, web servers, and sensors.

The STAR system is intended to be of general use, supporting the functionalities to store and analyze data gathered from different moving objects and associated with different movement technologies. The work developed so far allows the end-user to easily load different data sets and to analyze the corresponding data independently of its initial structure. Two different data sets were loaded and described, demonstrating the functionalities of the STAR system, focusing on the data independency characteristic of the system.

We would like to point out the functionalities that are currently under development to finish the overall implementation of the STAR system: i) the verify data quality use case; ii) the computation of areas and velocities; iii) the verification of proximity and overlapping; and, iv) the collection and loading of real-time data. Also, an implementation using open source technologies is ongoing, in order to make it a system that can be unrestrictedly used by different users.

Further steps in the design and implementation of the STAR system will include the incorporation of functionalities such as reporting, making available the possibility to create, print, delete and store reports with the results of data analysis, data visualization, or both. In terms of advanced data analysis mechanisms, the inclusion of a density-based clustering algorithm for the identification of patterns, trends or flows in data is planned. In terms of moving objects, the inclusion of moving objects that need to be represented as regions will be also considered in the next version of the STAR system.

References

1. Abdelzaher, T.: Mobiscopes for Human Spaces. IEEE Computing – Mobile and Ubiquitous Systems 6(2), 20–29 (2007)
2. Li, B., Cai, G.: A General Object-Oriented Spatial Temporal Data Model. In: The Symposium on Geospatial Theory, Processing and Applications, Ottawa (2002)
3. Nóbrega, E., Rolim, J.T., Times, V.: Representing uncertainty, profile and movement history in mobile objects databases. In: GeoInfo 2004, Brazil (2004)
4. Güting, R., Almeida, V., Ansorge, D., Behr, T., Ding, Z., Hose, T., Hoffmann, F., Spiekermann, M., Telle, U.: Secondo: An extensible DBMS platform for research prototyping and teaching. In: Proceedings of the 21st International Conference on Data Engineering (ICDE 2005), pp. 1115–1116. IEEE, Los Alamitos (2005)
5. Wolfson, O.: Moving Objects Information Management: The Database Challenge. In: The 5th Int. Workshop on Next Generation Information Technologies and Systems (2002)
6. Wachowicz, M., Ligtenberg, A., Renso, C., Gürses, S.: Characterising the Next Generation of Mobile Applications Through a Privacy-Aware Geographic Knowledge Discovery Process. In: Giannotti, F., Pedreschi, D. (eds.) Mobility, Data Mining and Privacy, Springer-Verlag, Berlin, pp. 39–72. Springer, Heidelberg (2008)
7. Lee, E.J., Ryu, K.H.: Design of Vehicle Information Management System for Effective Retrieving of Vehicle Location. In: Int. Conf. on Computational Science and its Applications, Springer, Heidelberg (2005)
8. Weng, J., Wang, W., Fan, K., Huang, J.: Design and Implementation of Spatial-temporal Data Model in Vehicle Monitor System. In: The 8th Int. Conf. on GeoComputation, University of Michigan (2005)
9. Wolfson, O., Sistla, P., Xu, B., Zhou, J., Chamberliam, S.: DOMINO: Databases for Moving Objects Tracking. In: SIGMOD 1999, Phildelphia PA (1999)
10. Erwig, M., Güting, R.H., Schneider, M., Vazirgiannis, M.: Spatio-Temporal Data Types: An Approach to Modeling and Querying Moving Objects in Databases. GeoInformatica 3(3), 269–296 (1999)
11. Forlizzi, L., Güting, R.H., Nardelli, E., Schneider, M.: A Data Model and Data Structures for Moving Objects Databases. In: ACM SIGMOD 2000, Dallas, USA (2000)
12. Praing, R., Schneider, M.: A Universal Abstract Model for Future Movements of Movement Objects. In: Fabrikant, S.I., Wachowicz, M. (eds.) The European Information Society: Leading the way with geo-information, Springer, Heidelberg (2007)
13. Moreira, A., Santos, M.Y., Wachowicz, M., Orellana, D.: The impact of data quality in the context of pedestrian movement analysis. In: Painho, M., Santos, M.Y., Pundt, H. (eds.) Geospatial Thinking, Springer, Heidelberg (2010)
14. Güting, R.H., Böhlen, M.H., Erwig, M., Jensen, C.S., Lorentzos, N.A., Schneider, M., Vazirgiannis, M.: A foundation for representing and querying moving objects. ACM Transactions on Database Systems 25(1), 1–42 (2000)
15. Pelekis, N., Frentzos, E., Giatrakos, N., Theodoridis, Y.: HERMES: aggregative LBS via a trajectory DB engine. In: Proceedings of the 2008 ACM SIGMOD International Conference on Management of Data, Vancouver, Canada, pp. 1255–1258 (2008)
16. Worboys, M.F., Hearnshaw, H.M., Maguire, D.J.: Object-Oriented Modelling for Spatial Databases. Int. J. of Geographical Information Systems 4(4), 369–383 (1990)
17. Dobing, B., Parsons, J.: How UML is used. Communications of the ACM 49, 109–113 (2006)

Analyzing Demographic and Economic Simulation Model Results: A Semi-automatic Spatial OLAP Approach

Hadj Mahboubi, Sandro Bimonte, and Guillaume Deffuant

Cemagref, Campus des Cézeaux,
63173 Aubière, France
{name.surname}@cemagref.fr

Abstract. In this paper, we present a semi-automatic SOLAP approach specially dedicated to the analysis of spatial model simulation results. We illustrate it on demographic and economic data of rural municipalities resulting from a model developed in the context of the European project PRIMA.

Keywords: Spatial simulation, Spatial OLAP, Spatial Data warehouses.

1 Introduction

Data warehousing combined with OLAP (On Line Analytical Processing) technologies provides an innovative support for business intelligence and knowledge discovery [9]. It has now become a leading topic in the commercial world as well as in the research community. The main motivation is to benefit from the enormous amount of data available in distributed and heterogeneous databases in order to enhance data analysis and decision making.

A very important type of data is spatial information, which is present in 80% of the cases. It is obvious that this meaningful information is worth being integrated into the decision making process as a first class knowledge, leading to the concept of *Spatial Online Analytical Processing* (Spatial OLAP or SOLAP). Spatial OLAP has been initially defined by Yvan Bédard as "*a visual platform built especially to support rapid and easy spatiotemporal analysis and exploration of data following a multidimensional approach comprised of aggregation levels available in cartographic displays as well as in tabular and diagram displays*" [2]. SOLAP systems combine OLAP and Geographic Information Systems (GIS) functionalities in a unique coherent framework. It allows decision-makers to analyze huge volumes of spatial data according several axes (dimensions) to produce indicators (measures) that are aggregated using classical functions such as sum, min, max, etc.

Spatial data used by SOLAP systems are from different and heterogeneous sources and are transformed and cleaned using specific ETL tools [19]. Application domains of SOLAP are various such as health, agriculture, risk management, marketing, etc [14].

Another type of data which becomes more and more important is the one generated by simulation models. Indeed, models and simulations are more and more widely

B. Murgante et al. (Eds.): ICCSA 2011, Part I, LNCS 6782, pp. 17–31, 2011.

used to study complex social dynamics and policy scenarios in various contexts such as urbanization, risk management, etc [4]. An example of such a model that we shall consider in this paper, is developed in the PRIMA European project, which aims to "*develop a method for scaling down the analysis of policy impacts on multifunctional land uses and on the economic activities*" [7]. This method relies on micro-simulation and multi-agents models designed and validated at municipality level, using input from stakeholder [5]. The main goal of the model is to analyze the evolution of the populations according to the structural policies of the municipalities.

In this paper we develop a new approach using Spatial OLAP tools for the multidimensional analysis of results issued from the PRIMA simulation model. More precisely, we aim at providing decision-makers (stakeholder) with dedicated tools that allow them exploring model results through cartographic and tabular displays. By this way, massive volumes of data coming from models are easily exploitable by stakeholders and modelers using combined GIS and OLAP functionalities. Consequently, as shown in this paper, modelers and decision makers can share an efficient and user-friendly visual framework for the analysis and exploration of data resulting from models, encompassing the limitations of existing model exploration tools: scalability, performance and support for statistical operators and visual analytical interfaces. Moreover we present a tool, called *SimOLAP* that allows modelers to automatically define and feed their spatial-multidimensional applications from simulation results, minimizing/avoiding work of data warehouse designers and architects.

The remainder of this paper is organized as follows. We first provide details on data warehouses and (Spatial) OLAP technologies, and simulation modeling (Section 2). Then in Section 3, we present requirements for an effective analysis of simulation results. Next in Section 4, we present our tool and its usage for the semi-automatic implementation of a spatial multidimensional application for the analysis of demographic simulation model results. Finally, we present conclusions and future work (Section 5).

2 Background Concepts

2.1 Data Warehouse, OLAP and Spatial OLAP

Data warehousing combined with On Line Analytical Processing (OLAP) are technologies intended to support business intelligence. A data warehouse is defined as "*a subject-oriented, integrated, non-volatile and time-variant collection of data stored in a single site repository and collected from multiple sources*" [8]. Warehoused data are organized according the multidimensional schema (conceptual representation) that defines analysis axes (dimensions) and subjects (facts). In this schema, dimensions are organized into hierarchies (ordered and linked levels), which allow analyzing measures at different granularities. Facts are characterized by measures, whose values are summarized using classical SQL functions, such as min, max, average, etc., when associated to coarser hierarchies' levels. A retail example application concerns the analysis of the sales of some stores. The multidimensional schema for this application presents a fact "Sales" and two measures "profit" and "amount". Usual dimensions are "Products", "Location" and "Time". These dimensions can be organized following hierarchical schemas, for example the hierarchy of the "Products" dimension defines "Product" and "Category" levels. The instance of a level is a set of alphanumeric values, called members. These members are connected by hierarchical links following the hierarchical schema of their dimension. As an example, the "Category" level is composed of the members "Games" and "Hi-fi", and the "Product" level is composed of the members "PlayStations", "SoniA4", etc. Using this schema, decision-makers can easily explore huge volume of data to produce on-line statistical reports on measures according to several dimensions. For example, some typical queries on this retail application are: "*What is the total amount of sales per year in Paris?*" or "*What is the profit average per store and per year?*".

Thus, the decisional process consists of exploring the multidimensional schema instance (or data cube) using OLAP operators. Usual OLAP operators are: Drill-Up (or Roll-Up), Drill-Down, Slice and Dice. The Drill-Up operator permits to climb up a dimension, i.e. move to a higher hierarchy level meanwhile aggregating measures, Slice and Dice select and project pieces of the data cube, and the Drill-Down operator is the reverse of the Drill-Up [8]. Other more complex operators like split, nest, pull, push, etc. have been also defined [15].

Usually OLAP tools are based on the so-called ROLAP architecture, a three-tier architecture composed of (Figure 1):

1) the data warehouse tier where data is stored. It is relational database management system (Relational DBMS; i.e. Oracle, PostgreSQL, etc.). Indeed, it is widely recognized that RDBMS provide an efficient support to store and query data,
2) the OLAP Server tier that implements the OLAP operators providing statistical computation functionalities, and
3) the OLAP client tier that allows visualizing data using tabular and graphic displays. It allows also triggering OLAP operators by the simple interaction with its visual components by hiding complexity of OLAP operators to decision-makers.

One type of important information which is very often embedded in corporate data is geo-referenced information and in despite of the significance and the complexity of this information, standard multidimensional models (and OLAP tools) treat it as traditional (textual) data.

On the other hand, it is widely recognized that cartographic representations allows enhancing decision making in many ways: reveal unknown and hidden spatial relationships and trends, help user to focus on particular subset of data, etc. [10] Therefore, it is important that geographic information is properly represented and integrated within decision-making support systems and particularly in OLAP systems. This integration is not limited to cartographic visualization, but in order to achieve an effective spatial decision support framework, GIS analysis functionalities should be also be provided to the users [1].

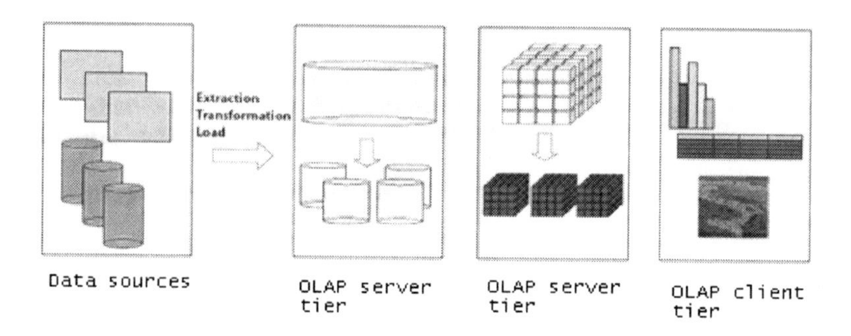

Fig. 1. ROLAP architecture

Then, a new kind of OLAP systems has been developed: Spatial OLAP (SOLAP). These systems integrate OLAP tools, which provide multidimensional behavior, and GIS functionalities for storing, analyzing and visualizing spatial information [1]. An example of SOLAP client is provided in Figure 2. Here measures values are displayed using an interactive thematic map.

SOLAP redefines main multidimensional model's concepts in order to integrate spatial information into OLAP analysis. Thus, SOLAP defines spatial measures and spatial dimensions introducing spatial data as subject and analysis axes respectively [13].

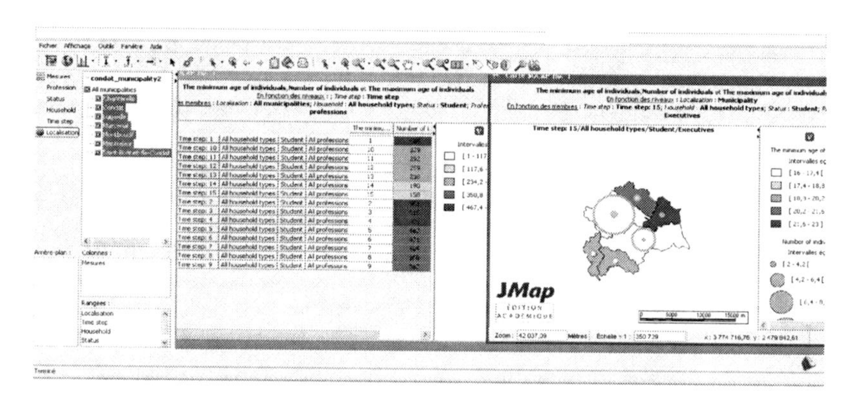

Fig. 2. A SOLAP tool (Map4Decision[1])

Nowadays, Spatial OLAP tools have been reached a good maturity and some commercial and open-source solutions such as SpagoBi, Map4Decision, etc. have been developed. In the same way, several are application domains of SOLAP rising from environmental to marketing, etc., which confirm the importance of this technology that allows spatial decision-makers to on-line analyzing huge volume of geographic data by means of statistical operators supported by interactive maps and pivot tables, and surpass spatial analytical limits of GIS systems [1].

[1] http://www.intelli3.com/

2.2 Modeling Complex Spatial Dynamics

Modeling is more and more widely used for studying complex phenomena and scenarios, such as urbanization, risk management, etc. Indeed, the increasing computing power makes it possible to develop more and more complex models, coupling sub-models of different types. Understanding the behavior of such models through an analytical analysis of the equations has become unfeasible because of their complexity. In general, a lot of simulation runs are necessary to analyze such models (models with hundreds of parameters are common), especially to make sensitivity analyses or to explore critical regimes. To perform these sets of simulations, scientists have developed specific tools, which aim at [6,17,18]:

- providing a programming environment devoted to the development of simulation experiments via a dedicated user interface that is compatible with the largest set of possible models,
- supporting easy access to well-established libraries for experiment designs and data treatments (R, Scilab integration, etc) and,
- defining an information system in a client/server configuration that provides the possibility of sharing or exchanging components of experiments and tracing them.

In any case, performing the simulation experiments often produces huge quantities of data, especially when the models have refined spatial and time resolutions. This is important to design appropriate tools to analyze such data.

3 Requirements for an Effective SOLAP Tool for Simulation Result Analysis

As mentioned earlier, simulation models produce huge amounts of simulation result data, on which simulation modelers need to extract certain summarized data, such as typical regularities or synthetic indicators. To manage these data they need tools that allow them to extract and to construct indicators as well as cartographic visualizations.

In this purpose, some simulation tools (cf. Section 2.2) have been developed. The main limitations of these tools are the lack of relational storage support to grant scalability, structured data representation and efficient querying [17]. On the other hand, current GIS are efficient for cartographic data visualization, but these systems are not especially designed for the analysis of huge volumes of data and most of them do not support geo-visualization techniques for interactive exploration and analysis of multidimensional spatial data [1].

SOLAP technologies provide the means to overcome these limits, as already outlined in [12]. Indeed, such technologies aim at providing decision-makers (and modelers in our case) with dedicated tools that allow them exploring models results through cartographic and tabular displays.

However, the development of a SOLAP architecture generally requires to design and implement multidimensional schemas, where analysis is carried. However, these design and implementation tasks are time-consuming, requiring expert knowledge of logical data warehouse designers [3]. In addition, modelers, as for instance in the case of the PRIMA project, do not always have data warehouse experts with competence to define and validate data cubes. Moreover this process is iterative, because several attempts are often necessary before getting the right one.

Thus, our main objective is to develop new flexible systems facilitating the design and the implementation of multidimensional schemas, and making them easily doable by non data warehouse experts. In this way, such systems allow modelers to implement by their own SOLAP applications for analyzing model results.

Recent theoretical frameworks have been developed to automatically derive data cubes from transactional sources (database tables) [16]. However, these efforts do not seem adequate in our context for two reasons: (1) simulation results are semi-structured data generally stored in text files, meanwhile proposed frameworks are designed to deal with relational data, and (2) no user-friendly interfaces are provided to help decision-makers (and modelers) to define their own spatial data cubes for the analysis of the simulation model dynamic.

To summarize, our aim is to develop tools that:

1. Build spatial data cubes on which SOLAP operations can be performed. This process should avoid experts' intervention and reduce the design task complexity. Thus these tools should provide facilities allowing modelers to specify their analysis needs by themselves and deriving the adequate spatial data cubes accordingly.
2. Exploit methods that handle non classical data to build appropriate data cubes. Indeed, simulation results can be considered as semi-structured data usually stored in text files.

In [11], we have proposed a semi-automatic method for designing and generating spatial multidimensional schema from on results of simulation models.

Our method supposes that the simulation results are organized in a tree like structure [11], which can be used to access the data. This hypothesis is not very constraining, since most data structures can be mapped into such a tree. This data model, as defined by [4], defines three node types:, *ComplexNode*, *SequenceNode*, *AttributeNodepe*. An *AttributeNodepe* node defines the attribute of a object, and is a leaf of the tree. An *AttributeNodepe* includes a single child which can be specfied as *ComplexNode*, *AttributeNode* or *SequenceNode*, describing the type of the element of the list. Finally, a *ComplexNode* includes a given number of children which can be specified as *ComplexNode*, *AttributeNode* or *SequenceNode*. An example of such structure is highlighted in the section 4.1.

Thus, based on this data model we present in the next section a tool that efficiently handle simulation data type and build appropriate data cubes.

4 Spatial-multidimensional Analysis of Demographic and Economic Model Results

In this section, we present a tool for semi-automatically build spatial data cubes to analyze simulation results. We illustrate the usage of our tool to analyze the dynamic of demographic results issued from PRIMA simulation model.

4.1 PRIMA Simulation Model

The simulation model, developed in the PRIMA project is a micro-simulation model that describes the economic and social evolution. Each individual of the population changes of state according to some probabilities derived from real data. This approach shows some advantages for analyzing policies' impacts at various levels [7]. In particular, the objective is to model the evolution of rural municipalities in terms of structure of population, activities, land use, housing, endogenously defined services, under different assumptions about policy and context evolutions. It particularly helps in comparing the evolution of the different types of municipalities according to their characteristics (especially the commuting network) and their policy choices [5]. More precisely, the model should help in understanding better the dynamics leading to the development or, on the contrary, to the decrease and maybe the disappearance of municipalities and settlements.

In the PRIMA project, the model takes as input an artificial population, where individuals are described by their age and status, and they are gathered in households respecting a variety of statistical constraints [5]. Then, the dynamics of the model rule the temporal evolution of the characteristics of the individuals and of the households, with demographic and economic events. The results are made of such populations at different states characterized by time steps. The simulations should be replicated several times because the dynamics are stochastic. Considering the results over these sets of replications allows the modelers to evaluate their variability due to the random aspects of the model.

An example of simulation result structure (tree representation [11]) and its corresponding data instance is shown on Figures 3(a) and 3(b).

It represents simulation results obtained at a set of replications, the *Replications* node. A *Replications* node is composed of the States node, which represents the set of simulation states obtained at each replication. This *SequenceNode* node is composed of a set of states, each one represented by the *State* node. A *State* node is characterized by a *TimeStep* node, and it composed of the *Municipalities* node.

The latter represents a list of municipalities on which changes are observed. It consists on a *Municipality* node that represents a single municipality, characterized by a land use, *Landuse* node, and groups a set of households, the *Households* node. A Household *SequenceNode* node is composed of the *Household* node that defines a single household. This node is characterized of a *Type* node and is composed of a set of individuals, the *Individual* node. This node defines the list of individuals in a household and is composed of an *Individual* node. An *Individual* node is composed by an *Age* node, and a *Status* node, that respectively represent the age and status of an individual.

Figure 3(b) illustrates two output files (population and activity) obtained by simulation and for a specific municipality, state and replication. The population file describes households (file lines). In this file, each household is defined by a type (e.g. 3 for a couple with children) and ages of its individuals. In the activity file, each line corresponds to an individual. It defines the individual status and activity sector (e.g. an employed, value 2, individual having a farmer occupation, value 0).

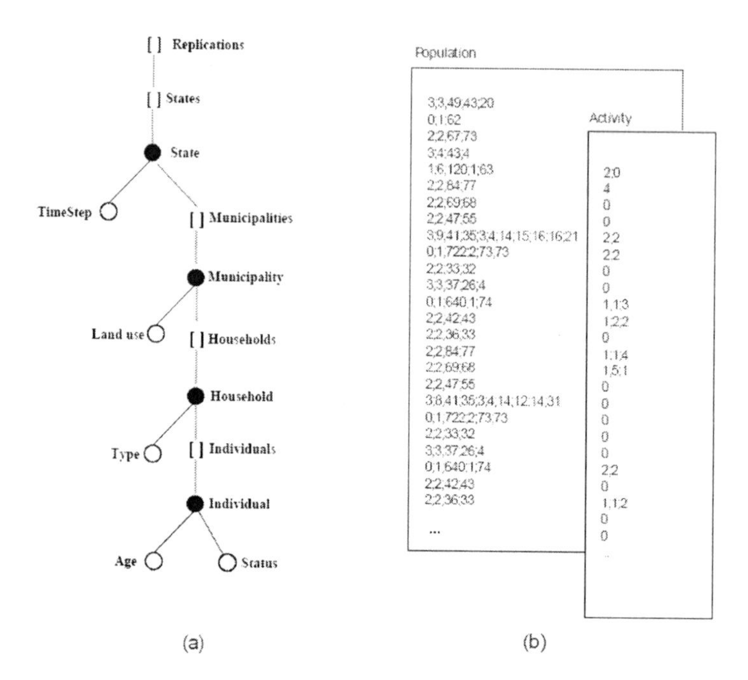

Fig. 3. An example of Prima model: a) result structure, b) data

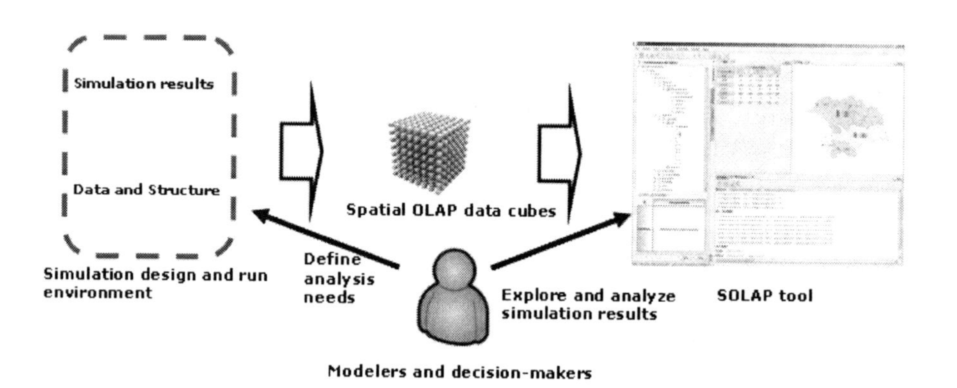

Fig. 4. SimOLAP functionalities

4.2 Spatial OLAP for Simulation Result Models: SimOLAP Tool

To fulfill our requirements presented in Section 3, we define a tool, called *SimOLAP*, which provides the following new features (Figure 4):

(1) ***Capture the simulation modeler analysis needs.*** This is performed using the simulation result structure captured during the simulation runs. It is built based on the tree result structure, presented in the previous section, and is then presented to the users by means of an interactive visual interface that allows users to select an item to analyze (or in other terms to use as fact; for example, the household evolution).
(2) ***Derive the multidimensional schema.*** *SimOLAP* generates a multidimensional schema using the selected item as fact and automatically deriving dimensions and measures from the simulation result structure.
(3) ***Built the appropriate spatial data cube in a SOLAP tool.*** *SimOLAP* also automatically implements the spatial data cube for the previous generated multidimensional schema, and feeds it with simulation results. Finally, this cube can be explored using a SOLAP tool.

4.2.2 The Architecture of SimOLAP

SimOLAP is a Java-based application that guides simulation modelers to define their SOLAP application, and then automatically implements it. The *SimOLAP* architecture is depicted in the Figure 5.

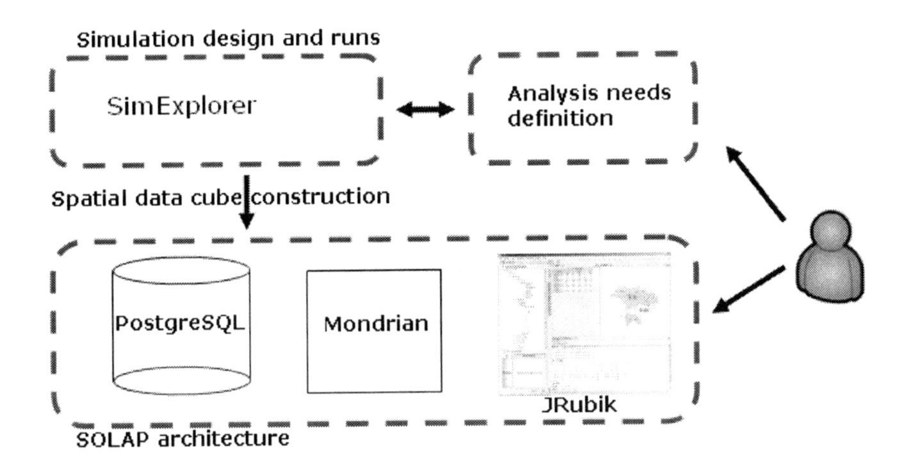

Fig. 5. SimOLAP architecture

Its main components are: (1) Simulation design and runs component: *SimExplorer*; (2) User analysis needs definition component and (3) SOLAP component.

In particular, SimExplorer[2] is a tool dedicated to facilitate the design and the management of simulation runs on any simulation model [4]. The *SimOLAP User analysis needs definition component* analyzes and visualizes the simulation results, defined by the simulation modelers and represented as trees on SimExplorer

[2] http://www.simexplorer.org/

(Figure 6). This user interface is interactive allowing users to choice a result tree element that can be used as analysis subject (fact).

An example is shown on Figure 6 where simulation modeler specifies to analyze individual data among PRIMA results. Then, once dimensions and measures are automatically derived, the multidimensional schema is implemented in the SOLAP component.

The SOLAP component is defined by the following open source technologies: PostgreSQL[3] as Data Base Management System (DBMS), Mondrian as OLAP sever[4] and JRubik as OLAP client tool[5].

Mondrian is an open and extensible framework, on top of a relational database. This mapping is provided via a XML description of the multidimensional schema elements and their associated relational elements in order to guarantee the greatest flexibility. Mondrian includes an OLAP engine layer that validates and executes MDX (Multidimensional Expressions) queries (pseudo-standard for OLAP servers), and an aggregation layer that controls data in memory and request data that is not cached. Finally, JRubik is an OLAP client developed in Java and based on JPivot project components.

The client connects to Mondrian OLAP data sources (i.e. Relational DBMS). OLAP queries could be defined using MDX language or by means of the interaction with the tabular and cartographic displays. Thus, the implementation of the spatio-multidimensional application in this ROLAP architecture consists of: (1) creation and feeding of the logical schema in PostgreSQL, and (2) creation of the XML Mondrian file[6].

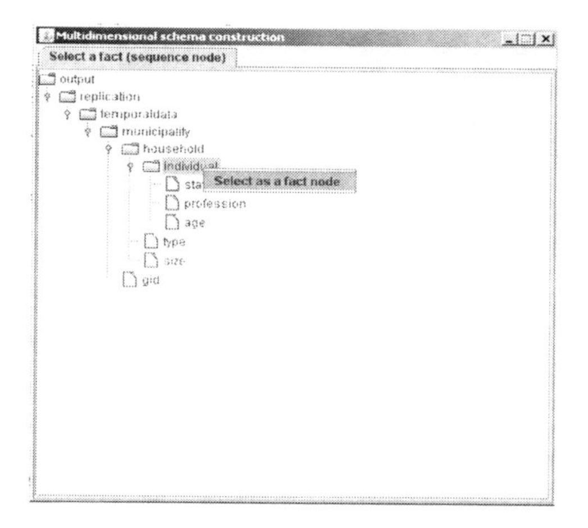

Fig. 6. The interactive graphic user interface of the "*Analysis needs definition component*"

[3] http://www.postgresql.org/
[4] http://mondrian.pentaho.com/
[5] http://rubik.sourceforge.net/
[6] http://mondrian.pentaho.com/documentation/schema.php

4.3 The Use of SimOLAP

Let us now show the usage of *SimOLAP* for the analysis of results obtained from runs of the simulation model developed under PRIMA project on the Auvergne French region (cf. Sec. 3.1).

Let consider that the user chooses *Individuals* as fact (Figure 6), in order to analyze (aggregating with min, max, and avg) the ages of individuals. *SimOLAP* generates and implements hence the multidimensional schema depicted in Figure 7. This schema presents *Individuals* as fact, and defines *Replications*, *States*, *Municipalities*, *Status*, *Age ranges* and *Households* as dimension levels.

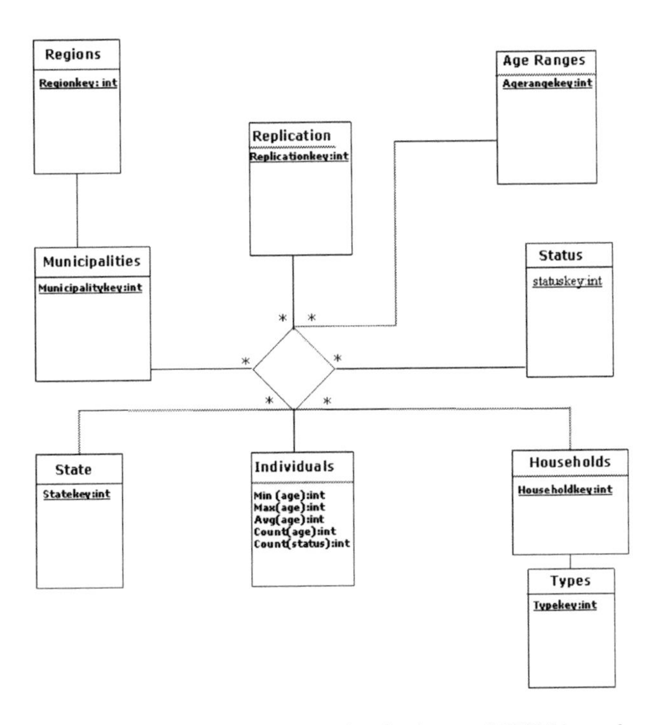

Fig. 7. Conceptual multidimensional schema of PRIMA results

In this multidimensional schema, *States* is a temporal dimension and allows exploring results for various time steps and hence observing the demographic dynamic over time. *Municipalities* is a spatial dimension and consists of a hierarchy describing the municipalities administrative organization (municipalities < departments < regions). This dimension allows cartographic visualization of measure in JRubik. *Replications*, *States*, *Municipalities*, *Status*, *Age ranges* and *Households* are thematic dimensions. *Replications* dimension allows analysis of results over different simulation replications, and hence the analysis of the model results dynamic. Finally, facts (*Individuals*) are described by *min(age)*, *max(age)*, *avg(age)*, *count(age)* and *count(status)* measures.

By using this multidimensional model, to analyze the model results users can answer queries like these:

- *What is the number of individuals (count(age)) per state and age range?*
- *What is the number of individuals (count(status)) per state, age range and status?*
- *What is the maximum age of individuals per replication and municipality?*
- *What is the average age of individuals per state, status and household type?*
- *What is the minimum age of individuals per municipality and age range?*
- ...

In the same way, modelers can analyze the model dynamics:

- *What is the size of the population of Condat (French municipality) for each replication?*
- *What is the number of individuals of a couples with children (household type) for each replication?*

This schema is then implemented in Mondrian (Spatial OLAP Server tier) by means of the XML file shown on Figure 8. In this schema, the spatial dimension is represented by the dimension element having "municipality" as attribute name. Measures are represented by measure elements.

```
  <?xml version="1.0" encoding="UTF-8" ?>
- <Schema name="individual">
  - <Cube name="individual">
      <Table name="individual" />
    - <Dimension name="replication" foreignKey="replicationkey">
      - <Hierarchy hasAll="true" primaryKey="replicationkey" primaryKeyTable="replication">
          <Table name="replication" />
          <Level column="replicationkey" name="replication" type="Numeric" uniqueMembers="true" />
        </Hierarchy>
      </Dimension>
    - <Dimension name="temporaldata" foreignKey="temporaldatakey">
      - <Hierarchy hasAll="true" primaryKey="temporaldatakey" primaryKeyTable="temporaldata">
          <Table name="temporaldata" />
          <Level column="temporaldatakey" name="temporaldata" type="Numeric" uniqueMembers="true" />
        </Hierarchy>
      </Dimension>
    - <Dimension name="municipality" foreignKey="municipalitykey">
      - <Hierarchy name="municipality" hasAll="true" primaryKey="municipalitykey">
          <Table name="municipality" />
        - <Level name="municipality" table="municipality" column="municipalitykey" type="Integer" nameColumn="sgid" uniqueMembers="true">
            <Parameter name="municipality" value="condat.svg" />
          </Level>
        </Hierarchy>
      </Dimension>
    - <Dimension name="household" foreignKey="householdkey">
      - <Hierarchy hasAll="true" primaryKey="householdkey" primaryKeyTable="household">
          <Table name="household" />
          <Level column="type" name="type" uniqueMembers="true" />
        - <Level column="householdkey" name="household" type="Numeric" uniqueMembers="true">
            <Property name="size" column="size" />
          </Level>
        </Hierarchy>
      </Dimension>
    + <Dimension name="status" foreignKey="statuskey">
    + <Dimension name="profession" foreignKey="professionkey">
      <Measure name="status(count)" column="status" datatype="Numeric" aggregator="count" />
      <Measure name="profession(count)" column="profession" datatype="Numeric" aggregator="count" />
      <Measure name="age(max)" column="age" datatype="Numeric" aggregator="max" />
      <Measure name="age(min)" column="age" datatype="Numeric" aggregator="min" />
      <Measure name="age(count)" column="age" datatype="Numeric" aggregator="count" />
      <Measure name="age(avg)" column="age" datatype="Numeric" aggregator="avg" />
    </Cube>
  </Schema>
```

Fig. 8. Mondrian XML file

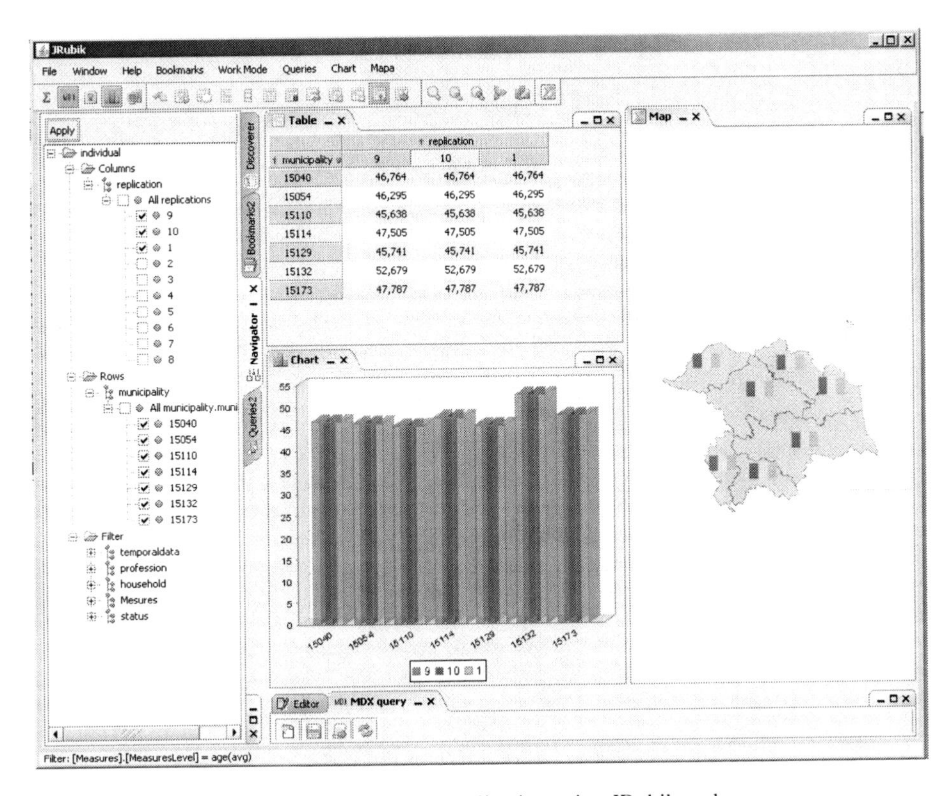

Fig. 9. Example of visualization using JRubik tool

According to [**13**], SimOLAP implements the conceptual spatial-multidimensional schema in PostgreSQL (spatial data warehouse tier) using the star schema approach that denormalizes dimension tables to speedup join queries. Dimension tables are linked to the fact table where measures are stored. In our case study, we have one numerical measure the age, and the other measure is the foreign key of the status dimension (called degenerated dimension [**9**]). It is very important note that spatial data is stored into SVG files.

Finally, the spatial data cube is explored and analyzed using JRubik. An example of SOLAP analysis is presented in Figure 9. It shows the temporal evolution of the maximum of individual ages calculated by the PRIMA model. This visualization corresponds to a specific query on the data cube: "*display min age measure per municipality and for the first and the two last replications*". It is available in tabular and cartographic displays. Other visualizations can hence be carried, like minimum, average of individual ages by time steps and municipalities or the number of individuals having a specific status, by time steps, replications and municipalities.

5 Conclusions and Future Work

With the growing development of simulation models, new emerging needs appear for managing and analyzing simulation results. To address these needs, efficient methods and techniques for explorative data analysis, like SOLAP technologies, are attractive solutions.

We have presented a tool that semi-automatically implements spatial data cubes for exploring the results of simulation models. Our tool helps users to easily specify their analysis needs and the automatic implementation of the corresponding spatial data cube. Then, the analysis and visualization of the spatial data cubes are carried out using appropriate SOLAP client. We illustrate the use of our tool for the analysis of results from a demographic simulation model.

In the next future, we envisage to develop specific data structures for input, calibration, scenarios and experimental design factors to derive the multidimensional schema and data cubes. Indeed, modelers often need to compare results issued from different simulation runs and produced by different input and scenarios sets with observation data. This raises issues that include envisaging factor element in the result tree structure to derive a multidimensional schema. Moreover, we envisage improving geo-visualization interface in SOLAP tools, in order to enable a better perception and visualization of the evolution and dynamics on the results. For this purpose, we plan to exploit existing geo-visualization techniques [20] and adapt them to SOLAP tools.

References

1. Bédard, Y., Merrett, T., Han, J.: Fundaments of Spatial Data Warehousing for Geographic Knowledge Discovery. In: Geographic Data Mining and Knowledge Discovery, pp. 53–73. Taylor & Francis, UK (2001)
2. Bédard, Y.: Spatial OLAP. In 2me Forum Annuel sur R-D, Gomaique VI: Un monde accessible, Montral 13(14) (1997)
3. Carmè, A., Mazón, J.N., Rizzi, S.: A Model-Driven Heuristic Approach for Detecting Multidimensional Facts in Relational Data Sources. In: Bach Pedersen, T., Mohania, M.K., Tjoa, A.M. (eds.) DAWAK 2010. LNCS, vol. 6263, pp. 13–24. Springer, Heidelberg (2010)
4. Chuffart, F., Dumoulin, N., Faure, T., Deffuant, G.: SimExplorer: Programming Experimental Designs on Models and Managing Quality of Modelling Process. International Journal of Agricultural and Environmental Information Systems (IJAEIS) 1(1), 55–68 (2010)
5. Gargiulo, F., Ternes, S., Huet, S., Deffuant, G.: An Iterative Approach for Generating Statistically Realistic Populations of Households. PLoS ONE, 5 (2010)
6. Helton, J.C., Johnson, J.D., Salaberry, C.J., Storlie, C.B.: Survey of sampling based methods for uncertainty and sensitivity analysis. Reliability Engineering and System Safety 91, 1175–1209 (2006)
7. Huet, S., Deffuant, G.: Common Framework for Micro-Simulation Model in PRIMA Project. Cemagref Lisc, 12 (2010)
8. Inmon, W.H.: Building the data warehouse. John Wiley & Sons, Chichester (2005)
9. Kimball, R.: The Data Warehouse Toolkit: Practical Techniques for Building Dimensional Data Warehouses. John Wiley & Sons, Chichester (1996)
10. MacEachren, A., Gahegan, M., Pike, W.: Geovisualization for Knowledge Construction and Decision Support. IEEE Computer Graphics and Applications 24(1), 13–17 (2004)

11. Mahboubi, H., Bimonte, S., Deffuant, G., Pinet, F., Chanet, J.-P.: Semi-automatic Design of Spatial Data Cubes from Structurally Generic Simulation Model Results. Technical report, Cemagref (2011)
12. Mahboubi, H., Faure, T., Bimonte, S., Deffuant, G., Chanet, J.-P., Pinet, F.: A Multidimensional Model for Data Warehouses of Simulation Results. International Journal of Agricultural and Environmental Information Systems, IGI Global 1(2), 1–19 (2010)
13. Malinowski, E., Zimányi, E.: Logical Representation of a Conceptual Model for Spatial Data Warehouses. Geoinformatica 11(4), 431–457 (2007)
14. Nilakantaa, S., Scheibea, K., Raib, A.: Dimensional issues in agricultural data warehouse designs. Computers and electronics in agriculture 60(3), 263–278 (2008)
15. Rafanelli, R.: Operators for Multidimensional Aggregate Data. Multidimensional Databases: Problems and Solutions, pp. 116–165. Idea Group (2003)
16. Romero, O., Abelló, A.: A Survey of Multidimensional Modeling Methodologies. International Journal of Data Warehousing and Mining 5, 1–23 (2009)
17. Sacks, J., Welch, W.J., Mitchell, T.J., Wynn, H.P.: Design and analysis of computer experiments. Statistical Science 4, 409–435 (1989)
18. Saltelli, A., Ratto, M., Andres, T., Campolongo, F., Cariboni, J., Gatelli, D., Saisana, M., Tarantola, S.: Global Sensitivity Analysis. The Primer. John Wiley & Sons, Chichester (2008)
19. Stefanovic, N., Han, J., Koperski, K.: Object-Based Selective Materialization for Efficient Implementation of Spatial Data Cubes. IEEE Transactions on Knowledge and Data Engineering 12, 938–958 (2000)
20. Wood Member, J., Dykes, J., Slingsby, A., Clarke, K.: Interactive Visual Exploration of a Large Spatio-Temporal Dataset: Reflections on a Geovisualization Mashup. IEEE Transactions on Visualization and Computer Graphics 6(13), 1176–1183 (2007)

Linking SLEUTH Urban Growth Modeling to Multi Criteria Evaluation for a Dynamic Allocation of Sites to Landfill

Abdolrassoul Salman Mahiny[1] and Mehdi Gholamalifard[2]

[1] Gorgan University of Agricultural Sciences and Natural Resources
Beheshti St. Gorgan, Iran
a_mahini@yahoo.com
[2] PhD Candiadate of Environmental Sciences, Tarbiat Modares University, Noor, Iran
mehdi_gholamalifard@yahoo.com

Abstract. Taking timely measures for management of the natural resources requires knowledge of the dynamic environment and land use practices in the rapidly changing post- industrial world. We used the SLUETH urban growth modeling and a multi-criteria evaluation (MCE) technique to predict and allocate land available to landfill as affected by the dynamics of the urban growth. The city is Gorgan, the capital of the Golestan Province of Iran. Landsat TM and ETM+ data were used to derive past changes that had occurred in the city extent. Then we employed slope, exclusion zones, urban areas, transportation network and hillshade layer of the study area in the SLEUTH modeling method to predict town sprawl up to the year 2050. We applied weighted linear combination technique of the MCE to define areas suitable for landfill. Linking the results from the two modeling methods yielded necessary information on the available land and the corresponding location for landfill given two different scenarios of town expansion up to the year 2050. These included two scenarios for city expansion and three scenarios for waste disposal. The study proved the applicability of the modeling methods and the feasibility of linking their results. Also, we showed the usefulness of the approach to decision makers in proactively taking measures in managing the likely environment change and possibly directing it towards more sustainable outcomes. This also provided a basis for dynamic land use allocation with regards to the past, present and likely future changes.

Keywords: SLEUTH, MCE, Landfill, Land Use Planning, Gorgan.

1 Introduction

Urbanization is one of the most evident global changes. Small and isolated population centers of the past have become large and complex features, interconnected, economically, physically and environmentally [1]. One hundred years ago, approximately 15% of the world's population was living in urban areas. Today, the

B. Murgante et al. (Eds.): ICCSA 2011, Part I, LNCS 6782, pp. 32–43, 2011.

percentage is nearly 50%. In the last 200 years, while the world population has increased six times, the urban population has multiplied 100 times [1]. Urban settlements and their connectivity will be the dominant driver of global change during the twenty-first century.

Understanding land use change in urban areas is a key aspect of planning for sustainable development. It also helps in designing plans to counter the negative effects of such changes. According to Clarke et al., [2], simulation of future spatial urban patterns can provide insight into how our cities can develop under varying social, economic, and environmental conditions. Since the late 1980s, applications of computers in urban planning have changed dramatically and concepts such as cellular automata have been included in the computer programs. Cellular automata (CA) are discrete dynamic systems whose behavior is completely specified in terms of a local relation. They are composed of four elements: cells, states, neighborhood and transition rules. Cells are objects in any dimensional space that manifest some adjacency or proximity to one another. Each cell can take on only one state at any one time from a set of states that define the attributes of the system. The state of any cell depends on the states of other cells in the neighborhood of that cell, the neighborhood being the immediately adjacent set of cells that are 'next' to the cell in question. Finally, there are transition rules that drive changes of state in each cell as some function of what exists or is happening in the neighborhood of the cell [3].

According to Dietzel and Clarke [4], of all the CA models available, SLEUTH may be the most appropriate because it is a hybrid of the two schools in CA modeling—it has the ability to model urban growth and incorporate detailed land use data. The name SLEUTH has been derived from the simple image input requirements of the model: Slope, Land cover, Exclusion, Urbanization, Transportation, and Hillshade. Reasons attributed to choosing this model are: (1) the shareware availability means that any researcher could perform a similar application or experiment at no cost given they have the data; (2) the model is portable so that it can be applied to any geographic system at any extent or spatial resolution; (3) the presence of a well-established internet discussion board to support any problems and provide insight into the model's application; (4) a well documented history in geographic modeling literature that documents both theory and application of the model; and (5) the ability of the model to project urban growth based on historical trends with urban/non-urban data.

The SLEUTH incorporates two models: The urban growth model (UGM) and the land cover deltatron model (DLM). In order to run the model, one usually prepares the data required, verifies the model functions, calibrates the model, predicts the change and builds the products. The user can implement SLEUTH modeling in different modes. In running the model, five coefficients including diffusion, breed, spread, slope-resistance and road gravity are calculated that are governed by estimation of four growth rules consisting of spontaneous growth, new spreading centre growth, edge growth and road-influenced growth. These are achieved in growth cycles each equal to one year or other appropriate time unit. The coefficients thus acquired are then refined in a self-modification mode. The results are then passed through coarse, fine and final modes during which the growth coefficients are refined and final growth rules are set and a growth rate is calculated. Figure 1 below depicts a growth cycle in the SLEUTH.

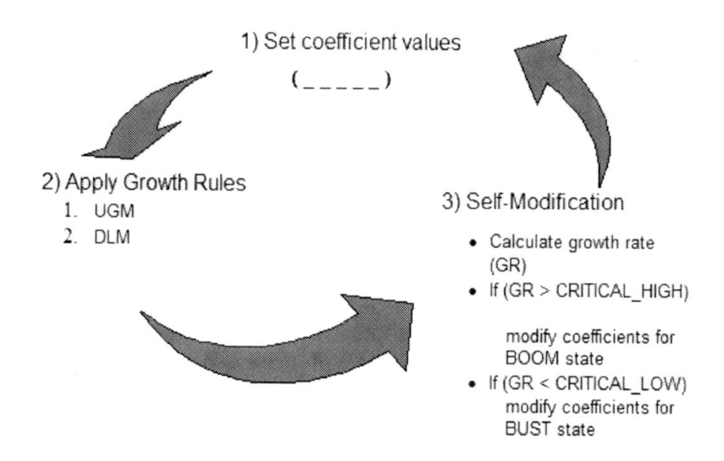
1) Set coefficient values
(_ _ _ _ _)
2) Apply Growth Rules
1. UGM
2. DLM
3) Self-Modification
• Calculate growth rate (GR)
• If (GR > CRITICAL_HIGH)
modify coefficients for BOOM state
• If (GR < CRITICAL_LOW)
modify coefficients for BUST state

Fig. 1. A growth cycle in the SLEUTH

When the best growth rate is achieved and the growth coefficients are calculated, these parameters are then applied to the data layers in the model and those pixels most likely to become urban in the next time periods are determined. This is also applied to other land use/cover types through the land cover deltatron model and in the end model balances change in urban areas and other land use/cover types.

Predicting urban growth can help environmental managers in taking timely measures to counteract or offset possible negative effects. It also helps in locating the affected areas and dynamically planning land use based on the capability and availability of the land under investigation. Besides, it is possible to graphically explore the results of different growth and land use planning scenarios. One of the pressing land use items nowadays is landfill which itself is directly related to the built-up areas and their expansion over time.

Landfill is an essential part of any waste management system which is composed of waste minimisation, reuse of products, recovery of materials and energy from the waste and placing the remaining material in landfills [5]. Even if a combination of the above or other management techniques is utilized and policies of waste reduction and reuse are applied, the existence of a sanitary landfill is necessary to a municipal solid waste management system [6]. In spite of the fact that landfill has been taken to the bottom of the hierarchy of options for waste disposal it has been the most used method for urban solid waste disposal. Landfill has become more difficult to implement because of its increasing cost, community opposition to landfill siting, and more restrictive regulations regarding the siting and operation of landfills. Land is a finite and scarce resource that needs to be used wisely. According to Lane and McDonald [7] a successful landfill site allocation process involves evaluating the basic suitability of all available land for sanitary landfills as an aid in the selection of a limited number of sites for more detailed evaluation. Appropriate allocation of landfills involves the selection of areas that are suitable for waste disposal. With regards to waste management, site selection studies reported in the literature cover the allocation of urban solid waste landfills ([8], [9], [10],[11], [12], [13]), hazardous solid waste centers ([14],[15]), and recycling operation facilities ([16]).

In the present study, we first detected the change in the extent of the Gorgan city using classification of the Landsat TM and ETM+ data. We then modeled the change in the city extent using two different scenarios through the application of the SLEUTH method. SLEUTH with its self-modification rule extracting approach to land use/cover change is deemed an intelligent method of exploring possible future scenarios. Then, we applied the weighted linear combination technique as a multi-criteria evaluation (MCE) method to define areas suitable for landfill. Linking the results of the SLEUTH modeling and the MCE showed the areas available to landfill under scenarios of urban sprawl and waste production and management. We also determined the suitability of land available under each scenario. Our literature review showed no other studies for Iran that contained any attempt to link the results of the two methods for dynamic site selection of the landfill areas.

2 Materials and Methods

Gorgan is the capital city of the Golestan Province in the north east of Iran. The economic growth in the area in the recent past has led to a large increase in population, causing dramatic urban expansion and land use change. We used the SLEUTH modeling method to simulate and project the change in the area of the city. SLEUTH requires an input of five types of digital raster files (six if land use is being analyzed). For all layers, zero is a nonexistent or null value, while values greater than zero and less than 255 represent a live cell. We used a digital elevation (DEM) layer of the area with a 20 meter resolution to derive slope layer. Landsat TM and ETM+ scenes of the Gorgan City covering around 1316 Km^2 were selected for this study. The scenes which dated July 1987, September1988, July 2000 and 2001 were imported into Idrisi 32 software [17], co-registered with other layers and re-sampled to 20 meters resolution. Then, the scenes were classified using knowledge from the area and Maximum Likelihood classifier in supervised classification method with purified training samples [18]. We identified seven classes: water, agriculture, fallow lands, built-up areas, dense broad-leaved forest, thin forest, pastures and needle-leaved woodlands. Total accuracy was 96% and the user's and producer's accuracy for urban class was 98.84% and 99.33% respectively. The urban extent was derived through reclassification of these detailed land cover classifications into a binary urban/non-urban map (Fig. 2).

For deriving the excluded layers and transportation, we used visual image interpretation and on-screen digitizing to generate individual vector layers that were transformed into raster layers with 20 meters resolution. We ensured that all data layers followed the naming protocol for SLEUTH, were in grayscale GIF format and had the same projection, map extent, and resolution. The hillshade map was also generated using the same DEM layer.

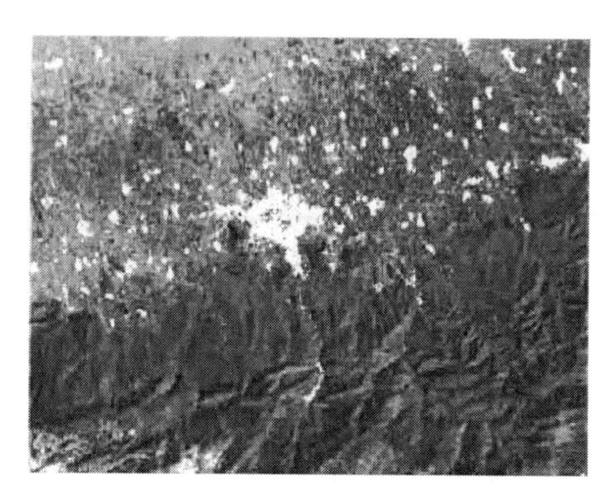

Fig. 2. Grey scale color composite image of the study area, bands 2, 3, and 4 of ETM+ sensor of Landsat satellite, 30th July 2001, with lighter spots showing residential areas

Model calibration was conducted in three phases: coarse, fine and final calibration. The algorithm for narrowing the many runs for calibration is an area of continuous discussion among users, and so far no definitive "right" way has been agreed upon. At the end of each calibration step, several fit metrics are produced which can be used as indicators of modeling success. Examples of the general approaches is use include: sorting on all metrics equally, weighting some metrics more heavily than others, and sorting only on one metric. More recently, the OSM (optimized SLEUTH metric) as the product of 7 metrics including "Compare", "Population", "Edge", "Clusters", "Slope", "Xmean", and "Ymean" [19] has been introduced. In this investigation, the last method, namely sorting on one metric, was applied. Simulations were scored on their performance for the spatial match, using Lee-Sallee metric which was around 0.4 showing the success of the modeling.

Adopting the procedure used by Leao et al., ([12], [13]) and Mahiny [20], we devised two different urban growth scenarios for model prediction. One scenario described the city as growing following historical trends, according to the parameters calibrated based on historical data. The second scenario described a more compact growth as a response to hypothetical policies and the shortage of land to harness urban spreading. Inspection of developing areas in Gorgan showed that at the moment both historical and compact scenarios of urban growth are underway. To apply these, we manipulated the value of some of the calibrated growth parameters. In the historical growth scenario, when the final calibration process was completed, the best selected parameters were run through the historical data many times and their finishing values were averaged considering the self-modification approach towards the included parameters. In the simulation for a compact city, the spread and road-gravity coefficients were reduced to half of the calibrated and the averaged best values were derived in the process.

The resulting forecast of future urban growth was produced as a probabilistic map. In the map, each grid cell will be urbanized at some future date, assuming the same unique ''urban growth signature'' is still in effect as it was in the past, while allowing some system feedbacks termed self-modification. For both the back-cast and projected urban layers, a probability over 70% (given 100 Monte Carlo simulations) was used to consider a grid cell as likely to become urbanized. This was derived through several trial and error attempts and comparison with real maps of the area. The final results of the model application were annual layers of urban extent for the historical time frame (1987–2001) and projected future urban growth (2002–2050).

Multi criteria evaluation (MCE) is most commonly achieved by one of three procedures [17]. The first involves Boolean overlay whereby all criteria are reduced to logical statements of suitability and then combined by means of one or more logical operators such as intersection and union. The second is known as weighted linear combination wherein continuous criteria (factors) are standardized to a common numeric range, and then combined by means of a weighted average. The third option for multi-criteria evaluation is known as the ordered weighted average (OWA) [21]. According to Hopkins [22] the most prevalent procedure for integrating multi- criteria evaluation and multi-objective evaluation (MOE) in GIS for land suitability analysis is using a weighted linear combination approach. The WLC procedure allows full tradeoff among all factors and offers much more flexibility than the Boolean approach.

We applied the weighted linear combination technique to locate areas suitable for landfill. We also employed the zonal land suitability to prioritize land based on the suitability of the pixels comprising zones of suitable land. This was the first application of the MCE for landfill in the area of study. As such, the MCE was faced with shortage of data layers explaining suitability for landfill site selection. Three different scenarios were considered for waste management and disposal. The Maximum Scenario meant that all the waste produced would go to landfill for disposal. In Optimum Scenario, 3% of the waste would be recycled, 11% would be composted and the rest would go to landfill. In the Minimum Scenario, waste production would decrease by 5%, the same amount would be recycled, 27% percent would be composted and the remaining amount would go to landfill. These scenarios were constructed by investigation of the trend in the population size, waste production habits and other social and technical factors involved in waste management currently seen in the area of study.

We used six factors including slope, water permeability, depth of the underground water table, distance from residential areas, distance from roads and wind orientation for the MCE. The factors were all standardized to a range of 0-255 using fuzzy membership functions. Then, weights were derived for the factors using the analytical hierarchy process (AHP) [23] and asking from a range of specialists [24]. We put all the factors on the same level and computed the relative weights through the pairwise comparison technique. Using the standardized factors, their weights and the constraints in the form of Boolean layers, we demonstrated areas suitable for landfill for the three scenarios. We then used the predicted urban sprawl for the Gorgan city as a constraint that limited our choice for the suitable landfill and as a factor affecting some of the other parameters that had been used in the MCE procedure. The result was a dynamic land allocation to landfill based on two scenarios of urban sprawl and three scenarios of waste production and management. In each case, the suitable land for waste disposal was determined and ranked relative to other available areas.

3 Results and Discussion

The three calibration steps and the predictive growth coefficients in the SLEUTH modeling were developed based on the rules that are depicted in the Table 1 below. Most of the statistics for best fit parameters of the simulation results of Gorgan through SLEUTH present high values of fit, indicating the ability of the model to reliably replicate past growth. This suggests that future growth predictions can also be used with confidence.

Table 1. Figures used for calibration and derivation of predictive coefficients in SLEUTH modeling

	Coarse Calibration		Fine Calibration		Final Calibration		Predictive Coefficients	
	Range	Increase	Range	Increase	Range	Increase	Range	Increase
Diffusion Coefficient	0-100	25	0-20	5	1-,5	1	1-,1	1
Breed Coefficient	0-100	25	0-25	5	10-25	5	15-15	1
Spread Coefficient	0-100	25	25-50	5	22-27	1	22-22	1
Slope Coefficient	0-100	25	0-25	5	0-20	5	1,1	1
Road Coefficient	0-100	25	50-100	10	60-80	5	75-75	1
Monte Carlo Simulations	5		8		10		100	
Total Runs	3124		6479		2999		---	

For the simulation of Gorgan city expansion, the final averaged parameters that were used in the prediction phase are presented in Figure 3.

Each parameter in Figure 3 reflects a type of spatial growth. For Gorgan City, the diffusion coefficient is very low, which reflects a low likelihood of dispersive growth. The value for the breed coefficient shows that it is somehow possible to witness growth of new detached urban settlements. The spread coefficient being larger than breed demonstrates the growth outwards of existing and consolidated urban areas is more likely. The high value of the road gravity coefficient denotes that the growth is also highly influenced by the transportation network, occurring along the main roads. Slope resistance shows the slight influence of slope to urbanization. In Gorgan area, topography was shown to have a very small effect in controlling the urban development, where even the hilly areas are likely to urbanize (Fig. 3). Inspection of the newly developed areas in the Gorgan City proved this to be true.

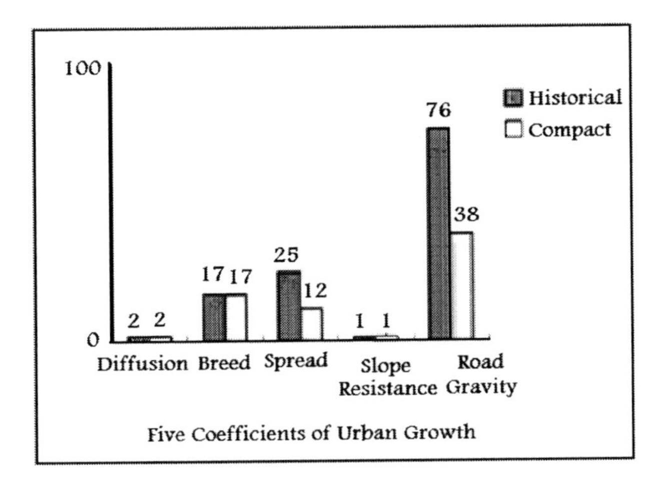

Fig. 3. Best fit parameters for modeling Gorgan city using SLEUTH

Figure 4 illustrates the future urban form and extent of Gorgan City area according to the model simulation using the historical scenario.

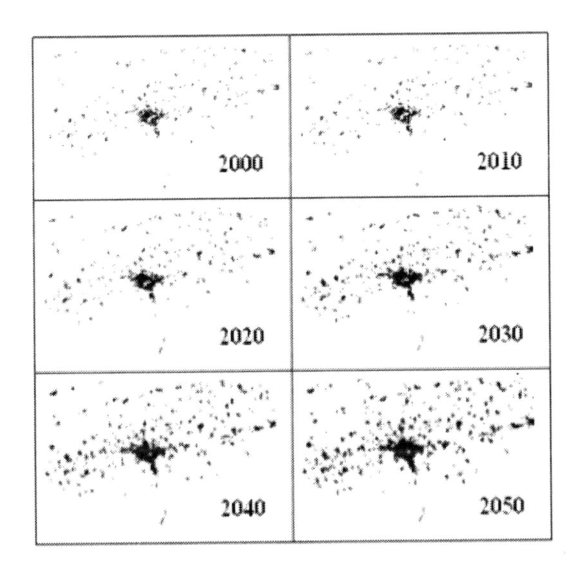

Fig. 4. Simulated Urban Growth in Historical Scenario

Looking at Figure 4, managers and decision makers can easily find the locations and the corresponding intensity of the areas where the city may increase. This information is of great importance, as it gives the managers an upper hand in controlling the unwanted expansion of the built-up areas from happening. It also helps land use planners in optimizing land allocation exercises given the dynamic nature of

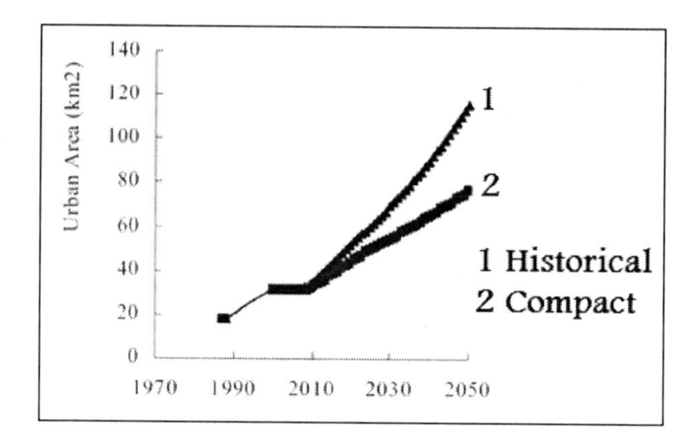

Fig. 5. Gorgan city expansion for the two scenarios

possible changes. Figure 5 shows the extent of urban development over time for the two growth scenarios.

Table 2. Zonal land suitability for 18 suitable land fill sites

Number of zones	Minimum suitability	Maximum suitability	Total suitability	Area (ha.)	Average (zonal) land suitability
1	142	162	90458	23.28	155.42
2	136	169	123346	31.96	154.37
3	127	165	105105	28.28	148.66
4	128	155	140077	38.49	144.26
5	92	169	361351	102.18	141.48
6	89	168	190362	54.65	139.35
7	114	161	255207	73.65	138.62
8	111	162	338971	101.02	134.24
9	116	154	75293	22.52	133.73
10	93	168	675239	207.68	130.07
11	95	161	396163	121.90	130.01
12	104	157	80403	24.80	129.68
13	87	145	80095	27.64	115.91
14	74	147	57110	20.24	112.86
15	65	155	271272	108.34	100.17
16	24	152	477351	204.40	93.43
17	42	129	79193	45.73	69.28
18	29	92	154535	96.38	64.14

Quite expectedly, the compact city scenario predicts a smaller increase for the future as compared to the historical scenario. However, the choices are open to the users to construct different scenarios and immediately assess their effects on the fate of the city. Modification of the driving parameters of city change, as defined in this study, can help in defining the best method for preventive measures in terms of feasibility and economy. Urban change control, cumulative effects assessment of land use/cover changes and land use planning and land allocation optimization are among other applications of the basic research conducted here.

The application of the MCE in the Gorgan city followed by zonal land suitability assessment indicated that initially there are 18 zones for landfill sites. The zonal land suitability of these sites varied from 155.42 to 64.14 (Table 2) and (Fig. 6). The analysis of the level of suitability of the zones selected and the allocation process shows the little available land suitable for landfill. This situation indicates that the areas to be used for landfill are going to become progressively less accessible. This has consequences on the costs of the waste disposal system, as well as on the risks for the environment and the community.

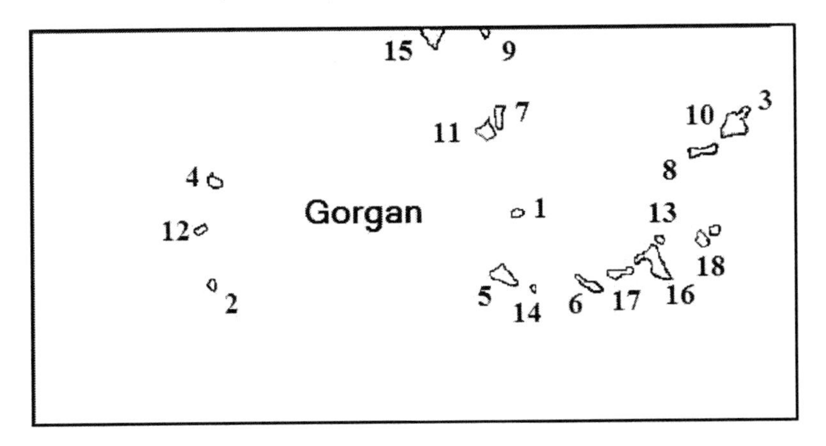

Fig. 6. Suitable zones for landfill and their rank in terms of zonal suitability

4 Conclusions

Planning and management are based on generic problem solving. They begin with problem definition and description, and then turn to various forms of analysis, which might include simulation and modeling, and finally move to prediction and thence to prescription or design, which often involves the evaluation of alternative solutions to the problem [25]. According to Rubenstein-Montano and Zandi [26], modeling tools form the majority of approaches developed to assist decision-makers with planning activities. The method described in this paper combines the power of SLEUTH urban expansion prediction with that of the MCE for landfill site selection. The evaluation abilities of MCE method and the analytical tools of GIS show the use of GIS as a decision support system (DSS). The first step of the process reveals possible areas of urban expansion under two different scenarios. The second step, assesses the availability of land for waste disposal by combining the relevant criteria (constraints and factors) for landfill plus the minimum area requirement constraint (20 ha) under three waste management scenarios. The relative importance weights of factors are estimated using the analytical hierarchy process (AHP). Initially, the land evaluation is performed on a cell by cell basis. The suitability of each cell for landfill is calculated by means of weighted linear combination (WLC) of multiple criteria in raster GIS.

The WLC approach results in a continuous suitability map that requires the user to decide what locations should be chosen from the set of all locations, each of which has some degree of suitability. This was addressed by adding the post-aggregation constraint that suitable sites must be at least 20 hectares in size. The model then calculated the suitability for landfill for each zone. This zonal suitability is obtained by calculating the average of the suitability of all cells belonging to each zone. In the final step, zones were ranked in descending order by the value of their zonal land suitability. Maps of suitable zones under each scenario were produced that provided the decision makers with several options for waste management. From among the zones, managers of the city can choose the best in terms of availability, price and social considerations and so forth such that a sustainable environmental management is achieved.

Results can be useful for policy and decision makers in Gorgan city. It must be noted that the presented method is only a tool to aid decision makers; it is not the decision itself. We successfully modeled the change in the extent of the Gorgan City using the SLEUTH method for the first time in Iran. The process was shown to be feasible, considering the time, facilities and the background knowledge it requires. The results, although not tested thoroughly, were found very useful in terms of providing insight into the process of city change for the managers and decision makers. Using this information, the authorities can take preventive measures for controlling negative effects of the predicted change. They can also use the information for preparing the infrastructure required for waste management in near future and mitigate the unwanted changes through possible means. One such measure can be control of the development of transportation network, as this was shown to have high effect on causing urban sprawl in the area. Also, focusing on consolidated urban areas and minimizing their expansion can be regarded as a measure towards harnessing the unwanted urban growth in Gorgan. This is also the case with the development of new detached urban areas, as shown by the breed coefficient. Using a combination of the past, present and future city sizes and their impact on the surrounding land use and land cover which ultimately affect land suitability, information can be also compiled for a proper, dynamic and timely land allocation for landfill sites in the area.

Acknowledgements. Authors are grateful to authorities of the Gorgan University for supporting development and presentation of this paper at GEOG-AN-MOD 2011.

References

1. Acevedo, W., Foresman, T.W., Buchanan, J.T.: Origins and Philosophy of Building a Temporal Database to Examine Human Transformation Processes. In: Proceedings, ASPRS/ACSM Annual Convention and Exhibition, Baltimore, MD, vol. I, pp. 148–161 (1996)
2. Clarke, K., Gaydos, L.: Loose-coupling a Cellular Automaton Model and GIS: Long Term Urban Growth Prediction for San Francisco and Washington/ Baltimore. Int. J. Geo. Inform. Sci. 12(7), 699–714 (1998)
3. Batty, M., Xie, Y.: Possible Urban Automata. Environ. Plan., B 24, 175–192 (1997)
4. Dietzel, C., Clarke, K.: The Effect of Disaggregating Land Use Categories in Cellular Automata During Model Calibration and Forecasting. Computers, Environ. Urban Sys. 30, 78–101 (2006)
5. Leao, S., Bishop, I., Evans, D.: Spatial- Temporal Model for Demand and Allocation of Waste Landfills in Growing Urban Region. Computers, Environ. Urban Sys. 28, 353–385 (2004)

6. Tchobanoglous, G., Theisen, H., Vigil, S.A.: Integrated Solid Waste Management: Engineering Principles and Management Issues. McGrow-Hill, New York (1993)
7. Lane, W.N., McDonald, R.R.: Land suitability analysis: landfill siting. J. Urban Plan. Develop. 109(1), 50–61 (1983)
8. Chang, N.B., Wang, S.F.: A Locational Model for the Site Selection of Solid Waste Management Facilities with Traffic Congestion Constraints. J. Civil Eng. Sys. 11, 287–306 (1993)
9. Lober, D.J.: Resolving the Siting Impasse: Modelling Social and Environmental Locational Criteria with a Geographic Information System. J. Am. Plan. Assoc. 61(4), 482–495 (1995)
10. Siddiqui, M.Z., Everett, J.W., Vieux, B.E.: Landfill Siting Using Geographic Information Systems: a Demonstration. J. Environ. Eng. 122(6), 515–523 (1996)
11. Kao, J.J., Lin, H.Y., Chen, W.Y.: Network Geographic Information System for Landfill Siting. Waste Manag. Res. 15, 239–253 (1997)
12. Leao, S., Bishop, I., Evans, D.: Assessing the Demand of Solid Waste Disposal in Urban Region by Urban Dynamics Modeling in a GIS Environment. Resources, Conserv. Recyc. 33, 289–313 (2001)
13. Leao, S., Bishop, I., Evans, D.: Spatial-Temporal Model for Demand and Allocation of Waste Landfills in Growing Urban Region. Computers, Environ. Urban Sys. 28, 353–385 (2004)
14. Canter, L.W.: Environmental Impact Assessment for Hazardous Waste Landfills. J. Urban Plan. Develop. 117(2), 59–76 (1991)
15. Koo, J.K., Shin, H.S., Yoo, H.C.: Multi-objective Siting Planning for a Regional Hazardous Waste Treatment Centre. Waste Manag. Res. 9, 205–218 (1991)
16. Hokkanen, J., Salminen, P.: Locating a Waste Treatment Facility by Multi-Criteria Analysis. J. M. Crit. Ana. 6, 175–184 (1997)
17. Eastman, R.J.: Idrisi 32, Release 2, p. 237. Clark University, USA (2001)
18. Mahiny, A.S.: Purifying training site in supervised classification of remote sensing data: A case study in Gorgan City and its environs. The Environment. J. of the Environ. Res. Ins. 1(5), 25–32 (2010)
19. Dietzel, C., Clarke, K.C.: Toward Optimal Calibration of the SLEUTH Land Use Change Model. Transactions in GIS 11(1), 29–45 (2007)
20. Mahiny, A.S.: A Modeling Approach to Cumulative Effects Assessment for Rehabilitation of RemnantVegetation, PhD Thesis, SRES, ANU, Australia (2003)
21. Eastman, J.R., Jiang, H.: Fuzzy Measures in Multi- Criteria Evaluation. In: Proceedings, 2nd. International Symposium on Spatial Accuracy Assessment in Natural Resources and Environmental Studies, Fort Collins, Colorado, May 21-23, pp. 527–534 (1996)
22. Hopkins, L.D.: Methods for Generating Land Suitability Maps: a Comparative Evaluation. J. Am. Inst. Plan. 43(4), 386–400 (1977)
23. Saaty, T.L.: A Scaling Method for Priorities in Hierarchical Structures. J. Math. Psycho. 15, 234–281 (1977)
24. Mahini, A.S., Gholamalifard, A.: Siting MSW landfills with a weighted linear combination methodology in a GIS environment. Int. J. of Environ. Sci.& Tech. 3(4), 435–445 (2006)
25. Batty, M., Densham, P.J.: Decision support, GIS, and Urban Planning (1996), http://www.geog.ucl.ac.uk/~pdensham/SDSS/s_t_paper.html
26. Rubenstein-Montano, B., Zandi, I.: An Evaluative Tool for Solid Waste Management. J. Urban Plan. Develop. 126(3), 119–135 (2000)

Statistical Evaluation of Spatial Interpolation Methods for Small-Sampled Region: A Case Study of Temperature Change Phenomenon in Bangladesh

Avit Kumar Bhowmik and Pedro Cabral

Instituto Superior de Estatística e Gestão de Informação, ISEGI,
Universidade Nova de Lisboa, 1070-312 LISBOA, Portugal
{m2010161,pcabral}@isegi.unl.pt

Abstract. This study compares three interpolation methods to create continuous surfaces that describe temperature trends in Bangladesh between years 1948 and 2007. The reviewed techniques include Spline, Inverse Distance Weighting (IDW) and Kriging. A statistical assessment based on univariate statistics of the resulting continuous surfaces indicates that there is little difference in the predictive power of these techniques making hard the decision of selecting the best interpolation method. A Willmott statistical evaluation has been applied to minimize this uncertainty. Results show that IDW performs better for average and minimum temperature trends and Ordinary Kriging for maximum temperature trends. Results further indicate that temperature has an increasing trend all over Bangladesh noticably in the northern and coastal southern parts of the country. The temperature follows an overall increasing trend of 1.06°C per 100 years.

Keywords: Spatial Interpolation, Spline, Inverse Distance Weighting, Ordinary Kriging, Univariate Statistics, Willmott Statistics.

1 Introduction

The temperature change over the past 30–50 years is unlikely to be entirely due to internal climate variability and has been attributed to changes in the concentrations of greenhouse gases and sulphate aerosols due to human activity [1]. Although global distribution of climate response to many global climate catalysts is reasonably congruent in climate models, suggesting that the global metric is surprisingly useful; climate effects are felt locally and they are region-specific [2]. Spatial interpolation methods have been used to quantify region-specific changes of temperature based on historical data [3]. There is no single preferred method for data interpolation being selection criteria based on the data, the required level of accuracy and the time and/or computer resources available. Geostatistics, is based on the theory of regionalized variables [4, 5 & 6] and allows to capitalize on the spatial correlation between neighboring observations to predict attribute values at unsampled locations. Geostatistical spatial interpolation prediction techniques such as Spline, IDW,

B. Murgante et al. (Eds.): ICCSA 2011, Part I, LNCS 6782, pp. 44–59, 2011.

Kriging, etc. provide better estimates of temperature than conventional methods [7 & 8]. Results strongly depend on the sampling density and, for high-resolution networks, the kriging method does not show significantly greater predictive power than simpler techniques, such as the inverse square distance method [9].

This study compares three spatial interpolators - Spline, IDW, and Kriging – with the goal of determining which one creates the best representation of reality for measured temperatures between years 1948 and 2007 in Bangladesh. Specifically this study aims to describe the temperature change phenomenon in the region by: (1) describing the overall and station specific Average, Maximum and Minimum temperature using trend analysis of the historical dataset; (2) interpolating the trend values obtained from trend analysis; and (3) evaluating the interpolation results using Univariate and Willmott Statistical methods, thus identifying the most appropriate interpolation method. Additionally, the benefits and limitations of these commonly used interpolation methods for small-sampled areas are discussed. This assessment is important because much of geographic research includes the creation of data for spatial analysis. Selecting an appropriate spatial interpolation method is key to surface analysis since different methods of interpolation result in different surfaces.

2 Study Area

Bangladesh is one of the countries most likely to suffer adverse impacts from anthropogenic climate change [2]. Threats include sea level rise (approximately one fifth of the country consists of low-lying coastal zones within 1 meter of the high water mark), droughts, floods, and seasonal shifts. The total area of the country is 147,570 sq.km. with only thirty-four meteorological stations to measure rainfall and temperature all over the country by the Bangladesh Meteorological Department [11] (Fig. 1). A number of studies carried out on trend of climate change in climatic parameters over Bangladesh have pointed out that the mean annual temperature has increased during the period of 1895-1980 at 0.31^{0}C over the past two decades and that the annual mean maximum temperature will increase to 0.4^{0}C and 0.73^{0}C by the year of 2050 and 2100 respectively [1, 2, 12, 13 & 14]. In this context, it is essential to quantify region-specific changes of temperature in Bangladesh in recent years based on historical data.

3 Data and Methods

3.1 Data

Daily temperature data from 1948 to 2007 collected from 34 fixed meteorological stations of the Bangladesh Meteorological Department were used in this study. Average, maximum and minimum daily, monthly and yearly temperature have been derived from this dataset and have been used for further trend analysis. Microsoft Excel 2007 has been used for trend analysis and ArcGIS 9.3.1 and GeoMS [15] have been used for the spatial interpolation of the trend values.

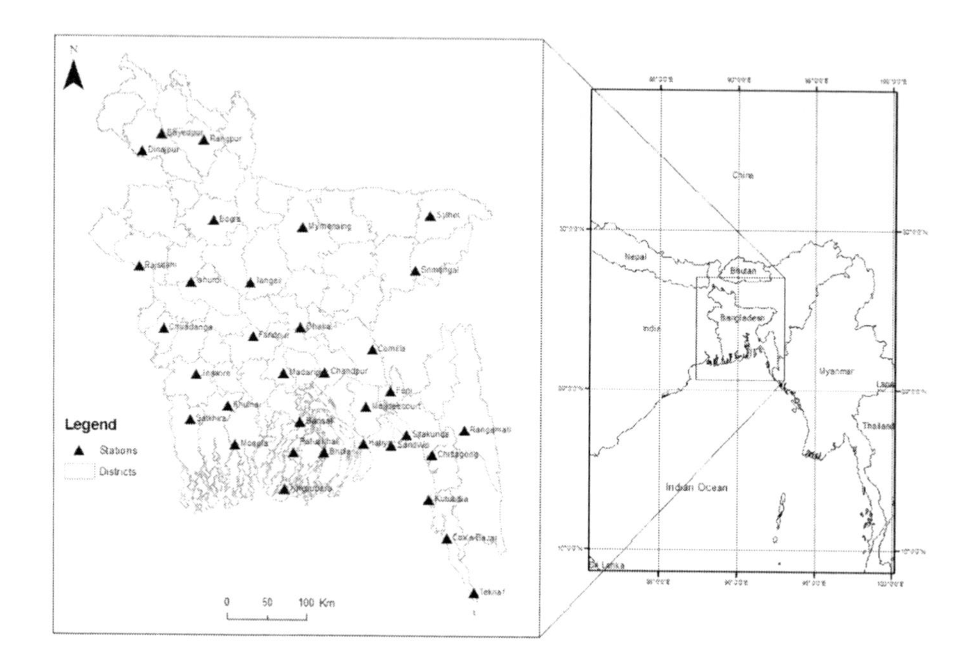

Fig. 1. Study area-Bangladesh with the location of thirty four meteorological stations

3.2 Trend Analysis

Trend analysis is the most commonly used process to describe the temperature change phenomenon of a region using historic data [1]. This is the simplest form of regression, linear regression, and uses the formula of a straight line (1).

$$y = a + bx \tag{1}$$

The equation determines the appropriate values for a and b to predict the value of y based upon a given value of x. Linear regression assumes that an intercept term is to be included and takes two parameters: the independent variables (a matrix whose columns represent the independent variables) and the dependent variable (in a column vector). Trend analysis by linear regression is less affected by large errors than least squares regression [16]. For the region-specific analysis, trend values of average, maximum and minimum temperature change have been calculated for every stations using the formula (2).

$$b = \frac{\sum (x_i - \bar{x})(y_i - \bar{y})}{\sum (x_i - \bar{x})^2} \tag{2}$$

Where, xi is the independent variable, x is the average of the independent variable, y_i is the dependable variable and y is the average of dependable variable. If the value of b is positive then the dataset shows an increasing trend. If it is negative, the dataset shows a decreasing trend. The higher the value of b the higher is the trend of change. One way of testing significance of trends of temperature is calculating the Coefficient of Determination, R^2 of the trend (3). Values of R^2 vary between 0 and 1.

$$R^2 = \frac{\left[\sum (x_i - \bar{x})(y_i - \bar{y})\right]^2}{\sum (x_i - \bar{x})^2 \sum (y_i - \bar{y})^2} \tag{3}$$

Highest correlation of the dataset can be found at 1 and it gradually reduces towards zero. Value less than 0.5 has been considered as less significant correlation.

3.3 Spatial Interpolation

The idea and mechanism of spatial interpolation for the study was generated from [3]. Interpolation is a method or mathematical function that estimates the values at locations where no measured values are available. It can be as simple as a number line; however, most geographic information science research involves spatial data. Spatial interpolation assumes that attribute data are continuous over space. This allows for the estimation of the attribute at any location within the data boundary. Another assumption is that the attribute is spatially dependent, indicating the values closer together are more likely to be similar than the values farther apart. These assumptions allow for the spatial interpolation methods to be formulated [3].

Spatial interpolation is widely used for creating continuous data from data collected at discrete locations, i.e. points. These point data are displayed as interpolated surfaces for qualitative interpretation. In addition to qualitative research, these interpolated surfaces can also be used in quantitative research from climate change to anthropological studies of human locational responses to landscape [3]. However, when an interpolated surface is used as part of larger research project [8] both the method and accuracy of the interpolation technique are important. The goal of spatial interpolation is to create a surface that is intended to best represent empirical reality thus the method selected must be assessed for accuracy.

The techniques assessed in this study include the deterministic interpolation methods of Spline [8] and IDW [8] and the stochastic method of Kriging [8] in an effort to retain actual temperature trend measurement in a final surface. Each selected method requires that the exact trend values for the sample points are included in the final output surface.

The Spline method can be thought of as fitting a rubber-sheeted surface through the known points using a mathematical function. In ArcGIS, the spline interpolation is a Radial Basis Function (RBF). The equation for k-order B-spline with n+1 control points (P0 , P1 , ... , Pn) is (4):

$$P(t) = \sum_{i=0,n} N_{i,k}(t)\, P_i , \quad t_{k-1} <= t <= t_{n+1} . \tag{4}$$

Advantages of Spline functions are that they can generate sufficiently accurate surfaces from only a few sampled points and they retain small features. A disadvantage is that they may have different minimum and maximum values than the data set and the functions are sensitive to outliers due to the inclusion of the original data values at the sample points. This is true for all exact interpolators, which are commonly used in GIS, but can present more serious problems for Spline since it operates best for gently varying surfaces, i.e. those having a low variance.

The input parameters for the Spline function are the input sampled trend values, the interpolation attribute (i.e., temperature), the type (regularized or tension), the weight, number of points to consider for each new value and the output cell size. The settings have been regularized by trial and error for function to obtain minimum mean bias and root mean square error with the default weight of 0.1. The selection of tension parameter was made to keep the range within the actual data range and the weight was increased to 1.0. When using the tension Spline, the higher the weight, the more the values conform to the range of sample data. In addition, for all of the interpolation methods the number of sample points to be used in the analysis of new locations was set to eight points. This was due to the limited number of observed data points, the default value of twelve meant that about half the points were contributing to the area regardless of how far they were from the location being estimated. Reducing the number of points used, assures the use of the closest values in the calculation process.

IDW is based on the assumption that nearby values contribute more to the interpolated values than distant observations. In other words, for this method the influence of a known data point is inversely related to the distance from the unknown location that is being estimated. This interpolation works best with evenly distributed points. Similar to the spline functions, IDW is sensitive to outliers. Furthermore, unevenly distributed data clusters results in introduced errors. To avoid unrealistic patchy maps, the temperature trend z can be estimated as a linear combination of several surrounding observations, with the weights being inversely proportional to the square distance between observations and u:

$$z_{\mathrm{Inv}}(u) = \frac{1}{\sum_{i=1}^{n(u)} \lambda_i(u)} \sum_{i=1}^{n(u)} \binom{n}{k} \lambda_i(u)\, z(u_i) \text{ with: } \lambda_i(u) = \frac{1}{|u-u_i|^2} \quad (5)$$

Similar to IDW, kriging uses a weighting which assigns more influence to the nearest data points in the interpolation of values for unknown locations. Kriging, however, is not deterministic but extends the proximity weighting approach of IDW to include random components where exact point location is not known by the function. Kriging depends on spatial and statistical relationships to calculate the surface. The two-step process of kriging begins with semivariance estimations and then performs the interpolation. Some advantages of this method are the incorporation of variable interdependence and the available error surface output. A disadvantage is that it requires substantially more computing and modeling time as well as more input from the user. Ordinary kriging with known varying means OK (u) derived from the secondary information [4]:

$$z_{OK}(u) = \sum_{i=1}^{n(u)} \lambda_i^{OK}(u)\ \ z(u_i) \quad \text{with: } \sum_{i=1}^{n(u)} \lambda_i^{OK}(u) = 1 \tag{6}$$

3.4 Univariate and Willmott Statistics

In general, the evaluation of the interpolation methods follows the univariate statistical method which calculates error statistics on the control stations with the recorded temperatures as the observed data and the interpolated temperatures as the predicted values. Summary univariate measures include the mean of the observed (Obar), mean of the predicted (Pbar), and their standard deviations (so, sp). It calculates the Mean Bias Errors (MBEs) and Root Mean Square Errors (RMSEs) of different interpolation methods and identifies the method which produces least MBE and RMSE as the best [7]. Another indicator of the models potential is how closely sp approaches so such that the closer to the standard deviation the observed (so) is to the predicted standard deviation (sp) the better the method is at reproducing the observed variance. Equations from univariate statistics can be found in [7].

Some critical limitations of univariate statistics led analysts to go for Willmott Statistics which cautions that these statistical measures should not be over analyzed. The ranking of the interpolation methods is impossible by univariate statistical evaluation when it obtains least value of one of the errors for one interpolation method but least values of other errors for another interpolation method. Nevertheless, it is also possible to obtain least MBE for Spline, least RMSE for IDW and least difference between so and sp in the same univariate evaluation. In such cases, it is sometimes recommended to perform ranking based on the mean bias error and to ignore root mean square error and standard deviations [7]. However, root mean square error and the approaching of the standard deviations cannot be ignored since the distribution of the predicted values based on the measured values and their distances from the ideal predicted values are of utmost importance to evaluate and thus identify appropriate spatial interpolation method.

Willmott statistics use five difference measures that are useful in evaluating the performance of the interpolation methods. These measures are: 1) Mean Absolute Error (MAE), 2) Root Mean Square Errors (RMSE), 3) Systematic Root Mean Square Errors (RMSEs), 4) Unsystematic Root Mean Square Errors (RMSEu), and 5) the Index of Agreement (d). Equations from Willmott statistics are in [10]. The ordinary least-squares (OLS) simple linear regression coefficients of a and b are used to compute the difference measures systematic and unsystematic root mean square errors (RMSEs, RMSEu). MAE is sometimes preferred over the RMSE as an evaluator because it is less sensitive to extreme values; however, RMSE is the error measure commonly computed in geographic applications. The systematic RMSEs assesses whether the model errors are predictable, whereas the unsystematic RMSEu identifies those errors that are not predictable mathematically. The final error measure, d, varies between 0.0 and 1.0. Therefore, the closer d is to 1.0 the better the agreement between O and P with 1.0 conveying perfect agreement and 0.0 complete disagreements.

4 Results

4.1 Trend Analysis

The trend analysis has been performed using temperature dataset from 1948 to 2007 of 34 meteorological stations of Bangladesh for average, maximum and minimum temperature. The analysis resulted in increasing trend of average, maximum and minimum temperature all over the country. Maximum temperature has shown the higher trend of increase (Fig 2).

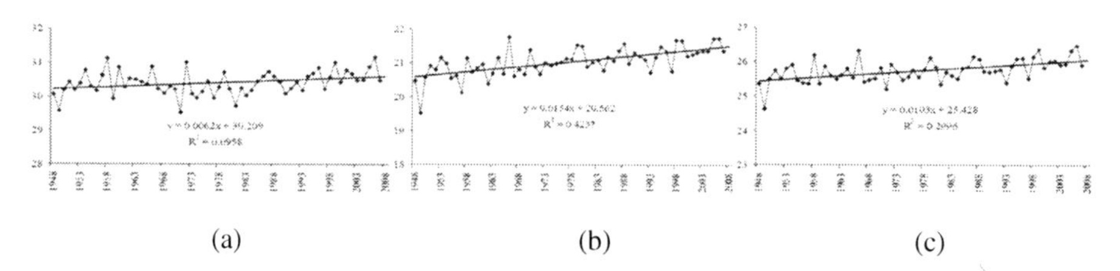

(a) (b) (c)

Fig. 2. Daily (a) maximum, (b) minimum and (c) mean temperature between 1948-2007)

Trends for 100 years of daily maximum, minimum temperature have been analyzed for each station for simplification of capturing a long term scenario of temperature change (Table 1). Since data from 1948 was not available for each station, beginning year of each data set are presented for each station. Trends of average temperature vary from -1.05 to 3.27^{0}C per 100 year. At Kutubdia, trend of average temperature is the highest among all the stations with a value of 3.27^{0}C per 100 year. Variation of maximum temperature trends is higher than the variation of average or minimum temperature trends. Maximum temperature trends vary from -2.59 to 5.8^{0}C per 100 year. Maximum value of this trend has been found at Sitakunda which is 5.8^{0}C per 100 year. On the other hand, trends of daily minimum temperature vary from -2.34 to 4.04^{0}C per 100 year. Station Bogra exhibits the highest value of maximum trend among all the stations with a value of 4.04^{0}C per 100 year. Coefficient of determination, R^2 of the trend analysis of average and maximum temperature varies from 0.01 to 0.66, and minimum temperature from 0 to 0.59. R^2 value less than 0.5 was found in many stations which represents poor statistical significance of the trend. Trends are found more than 2^{0}C per 100 years for stations with R^2 value more than 0.5.

Table 1. 100 years (a) Average (b) Maximum and (c) Minimum Temperature Trends of 34 Meteorological Stations of Bangladesh with Coefficients of Determination

Station	Starting Year	Average Temperature		Maximum Temperature		Minimum Temperature	
		Trend	R2	Trend	R2	Trend	R2
Barisal	1949	-0.47	0.01	0.78	0.06	-1.63	0.06
Bhola	1966	2.07	0.43	1.71	0.26	2.07	0.36
Bogra	1948	2.56	0.1	1.17	0.03	4.04	0.15
Chandpur	1964	1.63	0.09	1.64	0.09	1.62	0.06
Chittagong	1949	1.58	0.41	2.24	0.53	0.9	0.15
Chuadanga	1989	0.9	0.03	-0.38	0.01	2.17	0.14
Comilla	1948	0.23	0.02	0.5	0.03	-0.1	0.01
Cox's Bazar	1948	2.59	0.66	2.95	0.52	2.2	0.59
Dhaka	1953	1.72	0.33	1.19	0.12	2.25	0.4
Dinajpur	1948	-0.27	0.01	-2.13	0.25	1.51	0.09
Faridpur	1948	1.22	0.23	2.75	0.44	1.3	0.21
Feni	1973	2.42	0.11	1.74	0.07	3.31	0.15
Hatiya	1966	0.31	0.01	2.61	0.38	-2.02	0.06
Ishurdi	1961	0.26	0.01	0.32	0.01	0.37	0.01
Jessore	1948	1.39	0.34	1.47	0.2	1.13	0.17
Khepupara	1975	1.57	0.14	2.67	0.37	0.38	0.01
Khulna	1948	-0.16	0.01	0.37	0.01	-0.53	0.01
Kutubdia	1985	3.27	0.41	4.16	0.5	2.48	0.13
Madaripur	1977	1.74	0.17	0.47	0.01	1.6	0.12
Maijdeecourt	1951	2.03	0.36	1.93	0.32	2.04	0.27
Mongla	1989	2.7	0.42	4.4	0.55	1.05	0.09
Mymensing	1948	0.01	0.01	-0.86	0.11	0.86	0.06
Patuakhali	1973	2.18	0.17	3.29	0.46	2.7	0.15
Rajshahi	1964	0.63	0.01	1.0	0.04	0.26	0.01
Rangamati	1957	-1.05	0.09	-0.39	0.01	-1.78	0.09
Rangpur	1957	0.16	0.01	-2.59	0.19	2.81	0.13
Sandwip	1966	-0.52	0.01	0.73	0.01	-1.81	0.09
Satkhira	1948	0.88	0.11	0.65	0.04	1.07	0.09
Sitakunda	1977	1.84	0.2	5.8	0.66	-2.11	0.15
Srimongal	1948	1.39	0.19	0.3	0.01	2.37	0.26
Sayedpur	1991	2.11	0.19	2.66	0.12	1.08	0.09
Sylhet	1956	1.01	0.17	1.08	0.19	0.57	0.01
Tangail	1987	-0.31	0.01	1.74	0.07	-2.34	0.07
Teknaf	1977	2.24	0.32	2.41	0.32	2.36	0.28

4.2 Spatial Interpolation

These spatial interpolation methods have various parameters. The descriptions below include the options and values used in the different modules of ArcGIS. For display purposes, all images are grouped to 10 classes. The images show the decreasing trend values in lighter colors with the increasing trend values in the darkest color. The selected techniques, Spline, IDW and Kriging, are not all of the interpolation methods, nor are they a comprehensive review of the ArcGIS Geostatistical extension. Equivalent geostatistical techniques are available in other software products as well [17].

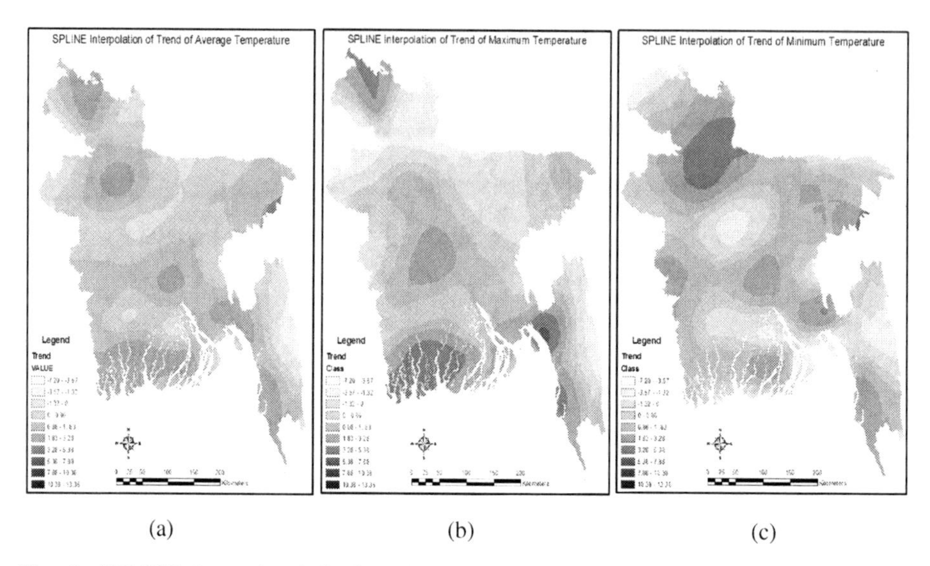

Fig. 3. SPLINE Interpolated Surface for (a) Average (b) Maximum and (c) Minimum Temperature Trends

The results of a tension Spline with weight set at 1.0, number of points at 8 and cell size of 300 are in Fig. 3. The smooth surface created by the Spline function provides the general temperature trends for the area. The regularized option resulted in a range of temperature trends from –7.20 to +13.34^{O}C, this range has been used for other interpolation methods with similar class distributions to maintain conformity of analysis. It correctly shows an increasing trend of average and maximum temperature in the south of Bangladesh which is basically a coastal area of Bay of Bengal. An increasing trend of average and maximum temperature is also found in the critical North of Bangladesh which is the warmest region in summer and coolest region in winter with extreme temperature. Increasing trend of minimum temperature is found on the south-eastern region Bangladesh and in the neck of the northern region. And in all over the study region there are some point regions with increasing temperature surrounded by gradually decreasing temperature area. The decreasing trend of average, maximum and minimum temperature was mostly found in the part of southern eastern region of Bangladesh which is known as Hill Tracts area.

The distance-decay principle is shown by IDW surface in Fig. 4. This surface shows less diversity in the central area than the tension Spline but is far smoother, which is one of the general characteristics of a IDW surface. It may be inappropriate, however, to use the smoothing Spline functions for a highly heterogeneous area since it provides an unrealistic view of reality by reducing spatial variance. The input points are the same for all three methods. As are the use of 8 points and a cell size of 300 m^2. The power was set to the most commonly used value of 2. When using a power of 2, it is known as inverse distance squared weighted interpolation. The search radius was set to variable due to the sparse and irregularly spaced sampled locations.

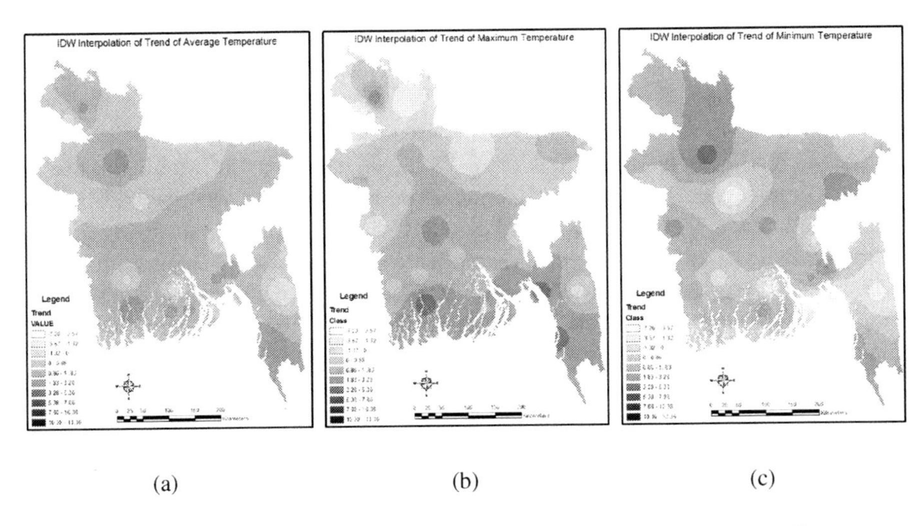

Fig. 4. IDW Interpolated Surface for (a) Average (b) Maximum and (c) Minimum Temperature Trends

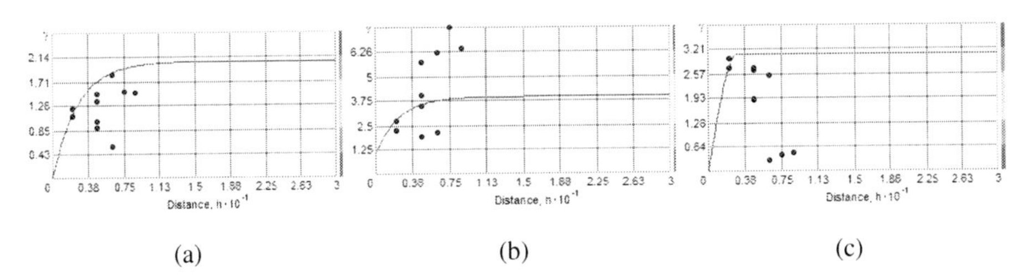

Fig. 5. Semivariograms for (a) Average (b) Maximum and (c) Minimum Temperature Trends

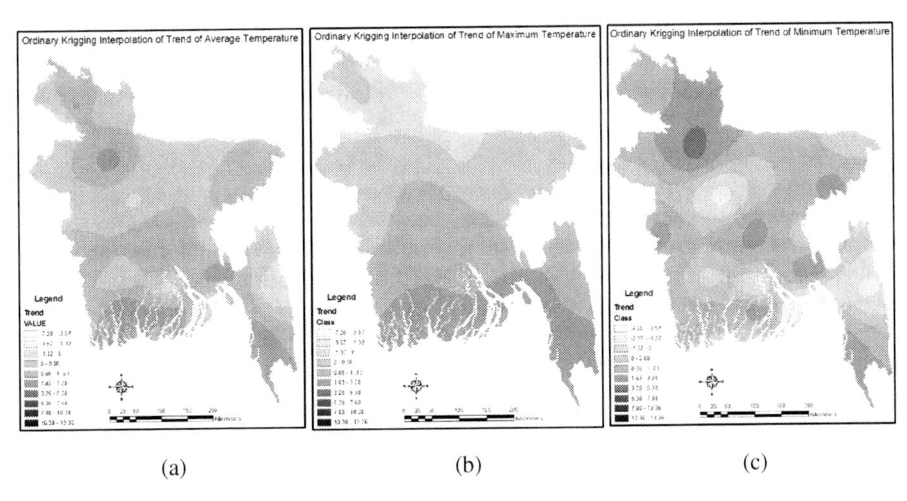

Fig. 6. Ordinary Kriging surface for (a) Average (b) Maximum and (c) Minimum Temperature Trends

In a krigged surface, there is less variation within the central region than with either the Spline or IDW surface (Fig. 6). This surface also has cooler values on the west / southwestern side which are more consistent with reality than the other modeled surfaces. There are several parameters to be set when using the kriging option for creating an interpolated surface. First, is the selection of whether to use the ordinary or universal method. In this study, ordinary kriging was selected because there is no known overriding trend in the data. Ordinary analysis is most widely used in kriging [5]. For this data set, the assumption that the constant means are unknown is true. The exponential models were selected for the average and maximum temperature trend semivariogram and spherical model for minimum temperature trend semivariogram with a variable radius were selected due to the sparse and irregular location of the meteorological stations. 10 lags with lag size of 3 were used for all models and major range used for average, maximum and minimum temperature trends were 8, 7 and 3 respectively. For this surface, the advanced options were not used. The output of the selection criteria has been described in Fig. 5.

4.3 Univariate Statistical Evaluation

The summary statistics Obar and Pbar have shown that, on average, all surfaces under predict the average, maximum and minimum temperature. But most importantly this statistics cannot profoundly establish the decision for the best interpolation method. As such, for average temperature phenomenon lower MBE and RMSE have been found for IDW method but sp more closely approach to so for Spline method (Table 2(a)). Similar confusion occurred for describing the maximum temperature change phenomenon The lower MBE and RMSE have been found for ordinary Kriging method but again sp more closely approach to so for Spline method (Table 2(b)). The worst uncertainty has occurred when the interpolation methods for describing minimum temperature change have been evaluated. The lower MBE has been found for ordinary Kriging, lower RMSE has been found for IDW method and sp more closely approach to so for SPLINE method (Table 2(c)). So the conclusion is that all of the methods are the same which is apparently impossible since different interpolation methods have shown completely different results (Fig. 3, 4 & 6). And the lowest difference between so and sp has been found for SPLINE method for all temperature change trends interpolation. These suggest that the other methods than SPLINE are less able to reproduce the observed variance.

The common suggestion to solve this uncertainty is to go for the MBE. It is commonly recommended to perform ranking on the basis of MBE ignoring the RMSE and approaching of sp to so. This approach has identified IDW as the best interpolation method to describe average temperature change phenomenon and ordinary Kriging as the best to describe maximum and minimum temperature change. But the scattergram of observed temperature trends and estimated temperature trends by different interpolation methods to describe average, maximum and minimum temperature trend; has totally refused to ignore RMSE and standard deviations effects.

Table 2. Univariate Statistical Evaluation Results of SPLINE, IDW and Kriging Methods for Describing (a) Average (b) Maximum and (c) Minimum Temperature Change

(a)

Summary Univariate Measures for Average Temperature Change							
Method	Obar	Pbar	MBE	So	Sp	RMSE	N
SPLINE	1.17	1.32	0.15	1.11	1.11	1.75	34
IDW	1.17	1.18	0.009	1.11	0.44	1.26	34
Kriging	1.17	1.21	0.05	1.11	0.68	1.41	34

(b)

Summary Univariate Measures for Maximum Temperature Change							
Method	Obar	Pbar	MBE	So	Sp	RMSE	N
SPLINE	1.42	1.56	0.14	1.72	1.95	2.81	34
IDW	1.42	1.57	0.15	1.72	1.05	1.927	34
Kriging	1.42	1.46	0.03	1.72	0.98	1.778	34

(c)

Summary Univariate Measures for Minimum Temperature Change							
Method	Obar	Pbar	MBE	SDo	SDe	RMSE	N
SPLINE	0.95	1.03	0.09	1.67	1.37	2.19	34
IDW	0.95	0.87	-0.08	1.67	0.66	1.68	34
Kriging	0.95	0.96	0.02	1.67	0.99	1.92	34

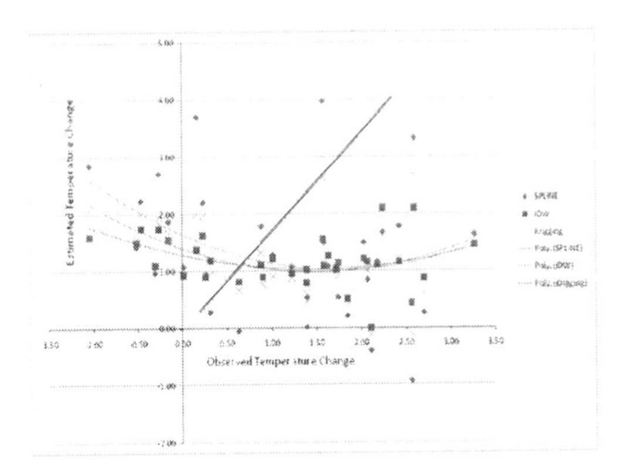

Fig. 7. Scattergram showing the Ideal Prediction and the Distribution of Observed and Estimated Temperature Trends by different Spatial Interpolation Methods to describe Average Temperature Change

The green straight line shows what should be the ideal prediction by the methods. Though none of the methods has shown very good closeness to ideal prediction, the distribution of observed and estimated trends have clearly depicted that in general, the values are under predicted. The estimated temperature trends by Kriging has shown more closeness to the ideal prediction line than IDW for the average temperature and eventually the kriging should be the best method to describe average temperature change (Fig. 7). Similar results have appeared for maximum and minimum temperature trends where IDW has shown more closeness to ideal prediction in case of maximum and minimum temperature trends . This is because of the distribution of observed and estimated temperature trends by different interpolation methods; therefore the RMSE and the standard deviation closeness cannot be just ignored [18] has also discovered the similar idea for the evaluation of spatial interpolation techniques to describe the precipitation change effect The proper way to get rid of this uncertainty problem is to find an approach which calculates MBE, RMSE and the difference of the standard deviations together [17]. This approach will evaluate the method's performance to calculate the errors as well as to take into account the distribution of the estimated value. In other words the approach will look for the error within the errors calculated by a particular method. The evaluation method of spatial interpolation techniques described by [10] is such kind of approach which describes the agreement of the measured trends and estimated trends by a spatial interpolation technique. Thus it takes into account the effects of distribution of the values as well as their quantitative values.

4.4 Willmott Statistical Evaluation

[10] suggests that these statistical measures should not be over analyzed. The evaluation using Willmott equations for the five difference measures have put an end to the limitations of the univariate statistics. The MAE and RMSE measures disagree with the univariate statistics on the potentially better interpolation methods with no uncertainty in ranking, having both a lower MAE and RMSE. The RMSEs and RMSEu measures have a similar response as the MAE and the RMSE in that the best interpolation method has the lowest RMSEu and RMSEs. The final error difference measure, d, also indicates that the interpolation method with lower d is a bit better than other interpolation methods. Therefore, the lower MAE, RMSE, RMSEs and RMSEu were found for IDW method in case of interpolating average temperature trend. The higher index of agreement (d=0.89) was also found for IDW and thus IDW was found to be the best interpolation method to describe the average temperature change phenomenon without further uncertainty (Table 3(a)). And in case of interpolating maximum temperature trend the lower MAE, RMSE, RMSEs and RMSEu were found for ordinary Kriging method along with the higher index of agreement (d=0.88) (Table 3(b)). The lower MAE, RMSE, RMSEs and RMSEu were found for IDW method and the higher index of agreement (d=0.78) was also found for IDW and thus IDW was found to be the best interpolation method to describe the minimum temperature change phenomenon and the uncertainty of univariate statistics has come into end (Table 3(c)).

Table 3. Willmott statistical evaluation results of Spline, IDW and Kriging methods for describing (a) Average (b) Maximum and (c) Minimum temperature change

(a)

	Simple Linear OLS coefficients			Difference Measures				
Method	n	a	b	MAE	RMSE	RMSEs	RMSEu	D
SPLINE	34	1.65	-0.03	1.40	1.75	1.41	1.05	0.25
IDW	34	1.29	-0.09	1.03	1.26	1.19	0.42	0.89
Kriging	34	1.41	-0.16	1.16	1.42	1.26	0.64	0.60

(b)

	Simple Linear OLS coefficients			Difference Measures				
Method	n	a	b	MAE	RMSE	RMSEs	RMSEu	D
SPLINE	34	1.89	-0.23	2.26	2.81	2.09	1.88	0.50
IDW	34	1.51	0.04	1.44	1.93	1.63	1.03	0.62
Kriging	34	1.31	0.12	1.34	1.78	1.51	0.95	0.88

(c)

	Simple Linear OLS coefficients			Difference Measures				
Method	n	a	b	MAE	RMSE	RMSEs	RMSEu	D
SPLINE	34	1.08	-0.05	1.78	2.19	1.73	1.35	0.13
IDW	34	0.81	0.06	1.40	1.68	1.55	0.64	0.78
Kriging	34	0.97	-0.007	1.57	1.92	1.66	0.98	0.43

5 Discussion

The five Willmott difference measures consistently identified Kriging as the best method for interpolating surfaces for maximum temperature trend and IDW for average and minimum temperature trend (Table 4). Nevertheless, as illustrated in Fig. 8, the fact lies in the average standard error of prediction [10]. Since the variation in temperature trend with only 34 sample points is too small to assess temperature trend and the prediction of the average and minimum temperature trend have produced lower standard errors than the average temperature trend prediction (Fig. 8), IDW has performed better for average and minimum temperature trend analysis. It is also true that only having 34 points to calculate interpolation error across that area is also a significant limitation because there are little differences in temperature trends between the meteorological stations. It is impossible, however, to increase the sample size because of the number of existing climate stations.

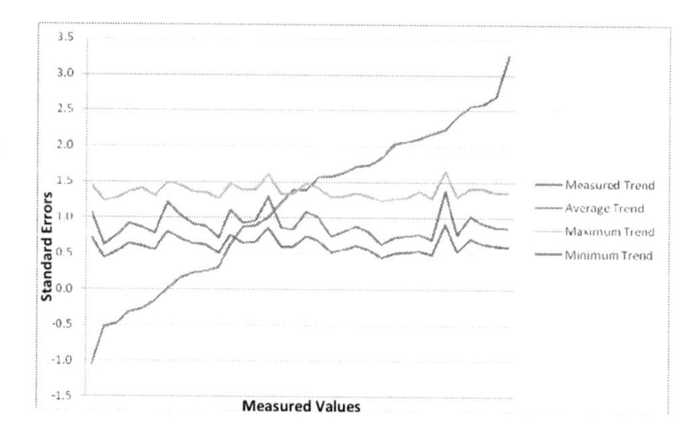

Fig. 8. Standard errors of prediction by different spatial interpolation methods to describe Average, Maximum and Minimum temperature change

Table 4. Willmott Statistical Evaluation Results of SPLINE, IDW and Kriging Methods for Describing (a) Average (b) Maximum and (c) Minimum Temperature Change

Temperature Change Phenomenon	Best Spatial Interpolation Method
Average Temperature	Inverse Distance Weighting
Maximum Temperature	Ordinary Kriging
Minimum Temperature	Inverse Distance Weighting

6 Conclusions

This study has shown that IDW is most likely to produce the best estimation of a temperature surface in a small-sampled region, which has depicted that temperature has an increasing trend all over Bangladesh. Northern and coastal southern parts of the country are experiencing the significant increase in temperature with an increasing trend of 5.38°C per 100 years. Nevertheless, regardless of the approach taken these interpolation methods do not adequately address the temperature variability inherent in a regional setting when the sample size is too small. Thus, it is critical that additional factors specific to the regional environment are incorporated into the spatial interpolation methods. Consequently, an estimated surface of temperature based on IDW can be used as input into a more complex spatio-temporal integration model to generate a better representation of temperature. Results confirm that for a small sampled region, the evaluation of geostatistical interpolation techniques using univariate statistics outperforms, ignoring the pattern of spatial dependence that is usually observed for temperature data. Willmott statistics for evaluation of the interpolation techniques proved to be an effective approach for the identification of best spatial interpolation techniques for further analysis.

Acknowledgements

The work has been supported by the European Commission, Erasmus Mundus Programme, M.Sc. in Geospatial Technologies, project NO. 2007-0064.

References

1. Divya, Mehrotra, R.: Climate Change and hydrology with emphasis on the Indian subcontinent. Hydrologic Sciences Journal 40, 231–241 (1995)
2. Chowdhury, M.H.K., Debsharma, S.K.: Climate change in Bangladesh – A statistical review. Report on IOC-UNEP Workshop on Impacts of Sea Level Rise due to Global Warming, NOAMI, November 16-19, 1992, Bangladesh (1992)
3. Chou, Y.-H.: Exploring Spatial Analysis in Geographic Information Systems, p. 474. OnWord Press, Santa Fe (1997)
4. Goovaerts, P.: Geostatistics for Natural Resources Evaluation. Oxford University Press, New York (1997)
5. Goovaerts, P.: Geostatistical tools for characterizing the spatial variability of microbiological and physico-chemical soil properties. Biol. Fertil. Soils 27(4), 315–334 (1998a)
6. Journel, A.G., Huijbregts, C.J.: Mining Geostatistics. Academic Press, New York (1978)
7. Phillips, D.L., Dolph, J., Marks, D.: A comparison of geostatistical procedures for spatial analysis of precipitations in mountainous terrain. Agric. and Forest Meteor. 58, 119–141 (1992)
8. Tabios, G.Q., Salas, J.D.: A comparative analysis of techniques for spatial interpolation of precipitation. Water Resources Bulletin 21(3), 365–380 (1985)
9. Dirks, K.N., Hay, J.E., Stow, C.D., Harris, D.: High-resolution studies of rainfall on Norfolk Island Part II: interpolation of rainfall data. J. Hydrol. 208(3-4), 187–193 (1998)
10. Willmott, C.J.: On the evaluation of model performance in physical geography. In: Gaile, G.L., Willmott, C.J. (eds.) Spatial Statistics and Models, pp. 443–460 (1984)
11. Statistical Pocket Book Bangladesh, Bangladesh Bureau of Statistics (BBS). Retrieved 2009-10-10 (2008)
12. Karmakar, S., Shrestha, M.L.: Recent climate change in Bangladesh. SMRC No.4, SMRC, Dhaka (2000)
13. Mia, N.M.: Variations of temperature of Bangladesh. In: Proceedings of SAARC Seminars on Climate Variability In the South Asian Region and its Impacts, SMRC, Dhaka (2003)
14. Parthasarathy, B., Sontake, N.A., Monot, A.A., Kothawale, D.R.: Drought-flood in the summer monsoon season over different meteorological subdivisions of India for the period 1871-1984. Journal of Climatology 7, 57–70 (1987)
15. Centro de Modelização de Reservatórios Petrolíferos (Decemeber 01, 2008) Centro de Modelização de Reservatórios Petrolíferos, `http://cmrp.ist.utl.pt/index.php?lg=2&cont=19` (retrieved January 31, 2011)
16. Birkes, Dodge: Alternative Methods of Regression. Word Press (2009)
17. Deutsch, C.V., Journel, A.G.: GSLIB: Geostatistical Software Library and User's Guide, 2nd edn. Oxford University Press, New York (1998)
18. Goovaerts, P.: Ordinary cokriging revisited. Math. Geol. 30(1), 21–42 (1998b)
19. Isaaks, E.H., Srivastava, R.M.: An Introduction to Applied Geostatistics. Oxford University Press, New York (1989)

A High Resolution Land Use/Cover Modelling Framework for Europe: Introducing the EU-ClueScanner100 Model

Carlo Lavalle[1], Claudia Baranzelli[1], Filipe Batista e Silva[1], Sarah Mubareka[1], Carla Rocha Gomes[1], Eric Koomen[2], and Maarten Hilferink[3]

[1] European Commission, Joint Research Centre,
Institute for Environment and Sustainability,
Land Management and Natural Hazards Unit, Via E. Fermi, 2749,
21027 Ispra (VA), Italy
{carlo.lavalle,claudia.baranzelli,filipe.batista,sarah.mubareka,
carla.rocha-gomes}@jrc.ec.europa.eu
[2] Faculty of Economics and Business Administration,
VU University, De Boelelaan 1105,
1081 HV Amsterdam, the Netherlands
e.koomen@vu.nl
[3] Object Vision BV p/a Vrije Universiteit, De Boelelaan 1085,
1081 HV Amsterdam, The Netherlands

Abstract. In this paper we introduce the new configuration of the EU-ClueScanner model (EUCS100) that is designed for evaluating the impact of policy alternatives on the European territory at the high spatial resolution of 100 meters. The high resolution in combination with the vast extent of the model called for considerable reprogramming to optimize processing speed. In addition, the calibration of the model was revised to account for the fact that different spatial processes may be prominent at this more detailed resolution. This new configuration of EU-ClueScanner also differs from its predecessors in that it has increased functionalities which allow the modeller more flexibility. It is now possible to work with irregular regions of interest, composed of any configuration of NUTS 2 regions. The structure of the land allocation model allows it to act as a bridge for different sector and indicator models and has the capacity to connect Global and European scale to the local level of environmental impacts. The EUCS100 model is at the core of a European Land Use Modelling Platform that aims to produce policy-relevant information related to land use/cover dynamics.

Keywords: Land use/cover; Modelling; Europe; Land demand; Factor data.

1 Introduction and Background

A land-use/cover model is an "*interpretive framework*" [1] of the interactions between different non-linear systems (bio-physical and human) that influence the dynamics of

B. Murgante et al. (Eds.): ICCSA 2011, Part I, LNCS 6782, pp. 60–75, 2011.

Land Use/Cover Change (LUCC). Land-use/cover depends on natural systems, but also, as a product of a society, is constrained and driven by demographic, economic, cultural, political, and technological changes [1]. Besides, land-use/cover is among one of the most important determinants of landscape, local climate conditions, biodiversity, and is relevant for biophysical systems and processes and for most land functions [47] and ecosystem services (de Groot *et al.*. 2002, de Groot *et al.*. 2009). However, the results of these interactions are not fully predictable since "*complex systems generate a dynamic which enables their elements to transform in ways that are surprising, through adaptation, mutation, transformation*" [3]. Based on assumptions regarding the future (what-if scenarios), LUCC simulations are relevant elements for structuring discussion and debate in a decision-making process, but do not present the future itself.

Therefore, and considering that land is a finite resource where competing claims interact, LUCC models emerge as an advanced tool to assess the *ex-ante* consequences of different policy options with spatial relevance and thus can be considered as a spatial decision support system. In 2002, the introduction by the European Commission of the Impact Assessment procedures [7] contributed to consolidate the importance of this tool and to multiply the examples of its application in providing scientific support to policy makers, at a European scale. This is due to the fact that the aim of the Impact Assessment procedures is to provide "evidence for political decision-makers on the advantages and disadvantages of possible policy options by assessing their potential impacts" [37].

Several models which can be linked to the impacts of policy alternatives are available today. The European Environmental Agency provides an overview of the models available to simulate environmental change at European scale [13], focusing on those that can support "*forward-looking environmental assessments and outlooks, as well as models that provide outlook indicators of environmental trends*". Nearly a hundred modelling tools are inventoried according to thematic focus (agriculture, biodiversity, climate, land-use, etc.), geographical scale (global, European, regional) and analytical technique (equilibrium model, empirical-statistical models, dynamic system models, and interactive models). This comprehensive list is a sign of the increasing importance of supporting decision making in such a cross-sector field as environmental issues, and more specifically land use management.

Regarding LUCC, many European projects integrate land-use/cover modelling tools to support policy-making. Among others, the EURURALIS project [47] appears as one of these examples, where the Dyna-CLUE model [43] was used to assess the policy impact of CAP measures post-2013 [21]. Another example is the PRELUDE (PRospective Environmental analysis of Land Use Development in Europe) project [12] that used the Louvain-la-Neuve model to assess the environmental implications of LUCC in Europe, by providing quantitative land use/cover change modelling analyses of the scenarios that were defined through a participatory exercise. A modelling approach was also at the core of integrated projects funded by the EU Framework Programme 6, such as SENSOR [22], which aimed to develop *ex-ante* Sustainability Assessment Tools (SIAT) to support policy making for land use in Europe [2].

In the scope of the EU FP6, some specific targeted research projects were developed with the aim of giving scientific support to policy-making. An example is the CAPRI-Dynaspat project, which aims to develop tools and methods to perform ex-ante policy assessment of the Common Agricultural Policy (CAP) based on the CAPRI model. This model is operational at DG-AGRI. In addition, it is also worth to mention FARO-EU (Foresight Analysis of Rural areas Of Europe), which had the objective of providing guidance for future rural development policies in Europe. In order to accomplish this main goal, and to "assess medium term future of rural areas in terms of economic, environmental social and institutional developments by considering sustainability conceptual framework and under alternative policy scenarios" (http://www.faro-eu.org) a chain of models was used (ESIM-CAPRI-LEITAP/IMAGE-CLUE-s).

Since 1998, the Joint Research Centre has been involved in the development of land use modelling. A first step was taken with the MURBANDY (Monitoring Urban Dynamics) project, whose aim was to monitor the dynamics of urban areas and to identify spatial trends at a European scale [28]. In 2004, collaboration between RIKS and JRC originated the follow-up project MOLAND (Monitoring Land Use Changes (http://moland.jrc.ec.europa.eu) [27], whose aim was to "provide a spatial planning tool for monitoring, modelling and assessing the development of urban and regional system" and also to contribute to mitigation of natural hazards [16].

In 2008, DG Environment identified the need to perform integrated assessments including spatial perspectives, and initiated the *Land use modelling – implementation* project which consisted in defining "an integrated land use modelling framework that can support policy needs of different DGs of the European Commission, such as *ex-ante* assessments and more specific impact assessments" [30]. In order to accomplish this goal, it was decided to use available well-established models. The land-use/cover modelling platform chosen, EU-ClueScanner 1km (EUCS1k), is an evolution of the land use allocation model Dyna-CLUE [46], [44] and [43], which in turn is a version of the CLUE model [39]. EUCS1k brought its predecessors to an open source programming environment, the Geo Data and Model Server (GeoDMS), and was built upon a multi-scale approach that integrates sector models whose role is to translate global and regional LUCC drivers into demand for land. The outcome of the land-use/cover simulation results are assessed through a set of indicators that can be either embedded in the model structure or processed *a posteriori*. EUCS1k is currently operated by the JRC.

In parallel, the JRC initiated the development of an integrated framework in "support to the conception, development, implementation and monitoring of EU policies" (http://ec.europa.eu/dgs/jrc/index.cfm?id=1370). Such a framework, defined as the Land Use Modelling Platform (LUMP), aims to integrate explicit land use models with other modelling activities in specific thematic fields, such as hydrology, agriculture, forestry, etc., and to support the *ex-ante* assessment of different policies.

Our land use modelling platform is currently based on three complementary LUCC models so to be able to cope with applications requiring high spatial resolution such as the simulation of urban areas; and pan-European applications. While the first purpose is served by the MOLAND model, the second makes use of EUCS1k.

As operational work began with land use modelling, it became evident that a new model was required to fill certain gaps. The evolution of the EUCS100 model is based on the structure of the EUCS1k model described above. EUCS100 and EUCS1k share several algorithms and processes, but EUCS100 has extended and improved capabilities which we will describe in detail throughout this paper. To summarise, EUCS100 is able to treat demand sets at NUTS 2 regional level and has a simplified and lighter configuration than its parent model EUCS1k. EUCS100 allows for multiple land use typologies and multiple regional divisions, as well as multiple simulation periods. A significant improvement is the benefit of modelling at the same scale as the Corine Land Cover input. Another important evolution in the EUCS100 version with respect to its 1km parent is the tiling of the images, saving valuable processing time and physical disk space. Furthermore, EUCS100 now converts maps to the ETRS 1989 reference system on the fly, and thus complies with the Inspire Directive (http://inspire.jrc.ec.europa.eu). These advantages translate to a higher degree of flexibility in terms of applications of the land use model.

The possibility to apply the most appropriate combination of explicit LUCC models amongst the three available (EUCS100, EUCS1k and MOLAND) represents a unique and powerful opportunity for the evaluation of impact of new or revised policies in Europe. This paper aims to give an overview of the land use/cover model EUCS100. Presently, the model is still under development, and some of its features are being improved. In section 2 its main characteristics are described, followed by an explanation about the structure of the model, in section 3. Lastly, section 4 provides an insight of the work undertaken to calibrate and validate EUCS100. We conclude with final remarks on the challenges to be addressed in future developments.

2 General Overview of the Model and Novelties

EUCS100 is a land allocation model whose main purpose is to support decision making by assessing *ex-ante* policy options in different sectors for the extent of the European Union. The main outputs of EUCS100 are projected land use/cover maps for a specified time in the future. Therefore, impact assessment of policy alternatives can be made specifically in the domain of land use/cover change and its direct or related implications in the context of different applications and studies.

To project land-use/cover into the future, EUCS100 uses known information on the drivers and determinants of LUCC. This information is mainly extracted from the analysis of observed LUCC. In this modelling environment, land-use/cover is depicted in a raster environment. EU-ClueScanner is implemented in GeoDMS (Geo Data and Model Server), which is a programming language optimized for raster operations and was developed and made freely available by Object Vision (http://www.objectvision.nl) as a stand-alone software optimized for desktop computers, provided with a graphical user interface. GeoDMS is a high level programming language that allows 1) transparency regarding the algorithms; 2) continuous improvement and debugging; 3) the management of large raster datasets; and 4) is fit for modular development of extra features and linkages with other models. The GeoDMS also underlies the much-applied Land Use Scanner model (e.g. [23], [25] and [26]).

The characteristics of the model are summarized as follows. It is modular because it is organized in three main components: land demand module, land allocation module and indicator module. It is sector integrative, mainly because land allocations are primarily driven by forecasts in the socio-economic sector at national or regional level. Simulations are yearly based, since there is a projected land-use/cover map for every year from the starting until the ending year of the simulation. It is recursive because simulation results for year *n* are used for the simulation of the year *n+1*. It is statistical because drivers for LUCC are inferred statistically from historic land-use/cover patterns, their neighbourhood relationships and their dependence on physical or socio-economic factors and it is European-wide because it provides simulations for continental Europe. Configured to run with 1 hectare cells (100 m x 100 m), it is considered a very high spatial resolution model. The latter two characteristics constitute together the major improvement of the model compared to previous land-use/cover modelling frameworks.

In the current configuration of the model, the land-use/cover legend is derived from Corine Land Cover (CLC) encompassing a total number of 44 different land-use/cover classes. EU-ClueScanner is capable of dealing with different legend configurations of CLC, each configuration being a different combination of land-use/cover class aggregations. This is a useful feature in order to adapt to different assessment requests. Furthermore, since the model can technically deal with any land-use/cover class for which processes and suitability factors are known, it is not strictly dependent on CLC as source of land-use/cover input.

Each class used in the model falls into one of the three following categories: active classes, passive classes and fixed classes. Active classes are those for which land demand can be exogenously determined using external sector models or projections. Examples of these classes are the built-up areas (residential, industrial or commercial), agriculture and forest. Passive classes are governed primarily by the dynamics of the active classes, as no specific demand has to be defined. Semi-natural land use types typically fall into this category. Finally, the fixed classes are those that remain spatially fixed during the simulation. Fixed classes are those for which processes, demands and suitability criteria are not known or cannot be modelled with current knowledge/data. Examples of fixed classes are water bodies, wetlands, burnt areas and some natural areas like bare rocks, beaches, dunes and glaciers. This categorization of land-use/cover classes is similar to the ones adopted in other modelling frameworks, such as in Moland [27] or Dyna-Clue [43].

3 Structure of the Model

EUCS100 has three main modules (Fig. 1). The first module integrates land claim data derived from external models. These external models take global factors into consideration and are subject to change according to the scenario modelled. The second module consists of the core of EUCS100 and is a dynamic simulation procedure that is the same as in the dyna-Clue model [43]. Starting from the current land use/cover map, the algorithm is designed so that in a specific location, a certain transition from one land use/cover to another actually takes place only if the overall suitability assigned to the latter is higher than the ones assigned to the other land use/covers which are competing for that location. The overall suitability is built up of several elements, namely factor data, neighbourhood effect, location specific drivers, conversion costs. All those elements are designed in order to allow the definition and implementation of EU policy alternatives.

The third module translates the projected land use/cover maps generated at annual time steps to formats usable for impact analysis. As shown in Figure 1, land use/cover based or thematic indicators are computed by built-in algorithms.

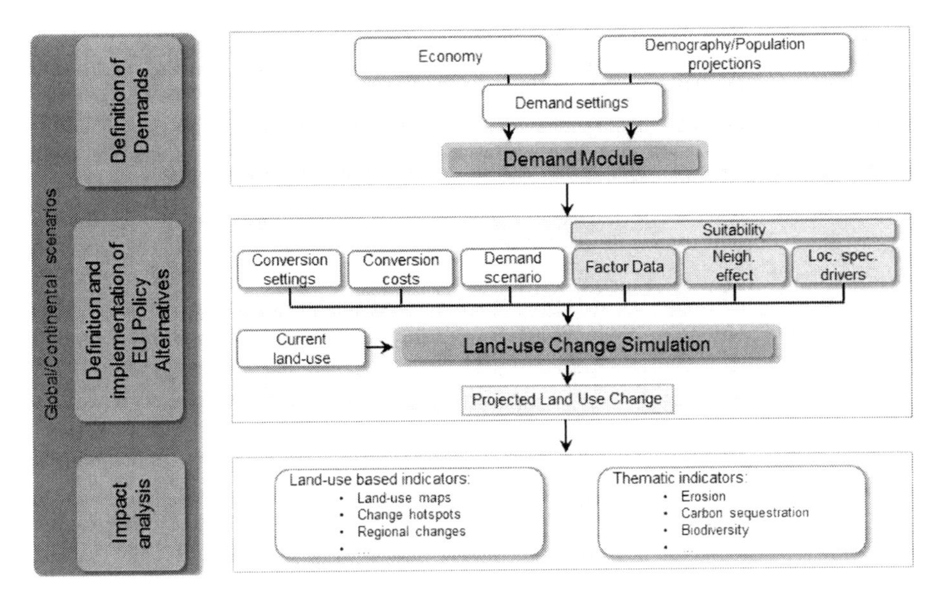

Fig. 1. Overall workflow of EUCS100 highlighting the three main modules of the model

3.1 Demand Module

As shown in Fig. 1, the demand module is the first stage of the modelling workflow. Its purpose is to provide an input for the land allocation module specifying the amount of land per each active land-use/cover class per region. These regions can be any combination of NUTS levels depending on availability of land demand data. Ideally, the total amount of land available is equivalent to the total amount of land required by all sectors. The sources of the demand sets themselves differ substantially depending on the sector. As an example, demographic and employment dynamics, as well as housing market trends are important drivers of residential land developments. A different angle is taken for arable land demand sets. Market conditions, demand for agricultural commodities and public policies are among a number of other relevant factors. Once demands are set for the required land-use/cover classes, the land allocation module allocates the extra land requirements at the pixel level.

We emphasize that 'land claims' are different from 'land demand'. The former is derived from the latter, and is modified to account for regional areal constraints. Land claims are the actual figures to be allocated.

Demand for Artificial Land-Uses/Covers

Since natural growth has been declining in Europe, the modest population growth rate is mainly attributable to immigration. The demographic profile is changing: population is aging and while the number of family nuclei is increasing, their mean size is decreasing. These changes affect future demand for housing. It has also been signalled that the increase of unstable jobs and the foreseen loan restrictions may temper some of the dynamics [35]. In another study, different approaches for calculating future residential land use demand in the context of land-use/cover modelling are reviewed [24]. Three main categories were identified: 1) trend extrapolations, 2) regression models and 3) density measures.

The third approach makes use of density measures, such as population density, to compute urban land demand. Historical measures of density can be kept constant or be dynamically modified during simulation. Then, these measures, together with demographic projections, may be used to estimate marginal urban land demands.

The density approach is taken in EUCS100 to derive residential land demand. Firstly the historical trends in density measures (inhabitants / built-up area) for each region are computed using time series for population (Eurostat) and for urban land (CLC). In a second step, density measures are extrapolated up to 2030 using past trends. Lastly, population projection figures[1] are used together with the densities to estimate future urban land use demands.

In EUCS100, industrial or commercial units are modelled separately from the predominantly residential urban fabric class. The modelling of this class poses a number of challenges. The first is related to the heterogeneity of the class itself as defined in CLC. This class in CLC includes two very different components: industrial and commercial units, which makes isolating the appropriate LUCC driving factors very difficult. The second challenge is related to identifying a source for the demand for these areas. As seen previously, it is possible to conceive a relationship between demographic dynamics, population density trends and requirements for new urban land. But for the case of land demand for industry and commercial areas, the relationship involves more variables and is less straightforward.

It is noteworthy that demand for non-residential built-up areas will be more difficult to model for the next decades due to the transformation of the nature of work and employment patterns that are "creating unstable demands for land use" [35]. While Pratt [35] provides extensive and pertinent considerations on the social and economic drivers of LUCC from a very disaggregated perspective – the firms and the households –, our modelling framework, however, must deal with aggregated trends at regional level, usually expressed by macro-economic indicators, from which demand must be inferred. Consequently, aggregated economic sector perspectives at regional level must be used in an adequate regression model to estimate overall demand. Such a work is currently under development.

Demand for Agriculture and Forest

The agricultural land demand is given by the Common Agricultural Policy Regionalised Impact analysis model (CAPRI). The CAPRI economic model was developed in the late nineties with the scope of modelling the agricultural sector on an EU scale [4] and [5]. CAPRI relies on several datasets (Eurostat, FAOSTAT and OECD) and compiles them into a single complete and consistent database (CoCo). It

---

[1] EUROPOP2008 population projection from Eurostat will be used, as it is considered to be, at the moment, the best available guess regarding future population development. It estimates population by sex and age by NUTS2 region, up to 2031 in annual time steps.

has two major modules: the supply and the market modules. The first module is able to capture the premiums paid under the Common Agricultural Policy (CAP), taking the feeding activities, nutrient requirements of animals and set-aside land obligations, milk quotas and the sugar quota. The market module feeds the supply model with prices through an iterative process.

The data used from CAPRI are Supply details for farms for 1990-2005 (several other indicators are available in the database from 1984-2005). Data is available for all countries in the EU plus Norway, Western Balkans and Turkey, with some data holes where data could not be inferred. In addition to this historical data which we will use for validating the methods described later in this paper, the results for the baseline scenario run are available for 2020 for all NUTS 2 regions for the regions mentioned above.

For forecasts in the agricultural sector, we rely on the CAPRI baseline scenario output for 2020 for agriculture. This baseline scenario is generated in harmony with DG AGRI's outlook projections and can be described as the most probable future development for the agricultural sector in Europe.

Finally, a dedicated module is currently being developed for the evaluation of forest land demand. The module firstly quantifies the demand for services and products provided by the forest in response to various drivers such as market (for wood and non-wood products), policy (e.g. CAP, renewable energy policies), environment (e.g. climate, extreme events, demand for ecosystem-based services) and socio-economy (e.g. population growth, GDP). Relevant data come from existing and well acknowledged sources such as EUROSTAT, FAO, EU funded projects, the UN-ECE EFSOS Committee. These also provide sector outlooks and foresight studies. The second step is the actual computation of forest area demand, which could be either positive or negative, at the most appropriate geographical aggregation level.

3.2 Land Allocation Module

EUCS100 is able to process land demand data for any combination of NUTS1 and NUTS2 regions, and not only at national level, as its predecessor EUCS1k did. Another improvement in this version of the model is that in order to deal with the uncertainty inevitably brought into our approach by the linkages with other models delivering demands of land, it is possible to translate those demands into ranges of claims, i.e. minimum and maximum claims, for each modelled land use/cover class. This approach emphasizes the role of the complex of spatial interactions carried out by the allocation module. The mechanism is implemented in a way that, if the total amount of available land (i.e. the area of the region minus the area of the fixed classes) is below the sum of minimum demands, the minimum claims are adjusted so to take into account that threshold, while the maximum claims hold steady. On the contrary, having more land use/cover at its disposal as land prone to be potentially changed than the actual total amount of maximum demands allows the model larger maximum claims for each land use/cover. To take matters further, implementation of a spill-over effect it is currently being considered. This implies that when the demand is not met within the region for which it has been defined, the excess quantity of LUCC can be transferred to neighbouring regions. The rules about the definition of these neighbourhoods can be either policy-related or taking account for known local/regional economic dynamics.

Suitability

Land use and land cover models deal with the interactions and competitions among different land use/cover classes, thus facing a complex system in which several cross-scale dynamics take place [45]. All the processes involved may relate to variables (either exogenous or endogenous with respect to the land use/cover system), commonly referred to as driving forces (e.g. demographic, economic dynamics, bio-physical characteristics). Thus, according to our approach, factor data represent exogenous factors responsible for the land use/cover change dynamics.

In order to quantify the relation between driving forces and land use/cover, an inductive approach is implemented, thus meaning that the suitability of a location for a certain land use/cover is determined in an empirical way by means of a regression analysis [29]. This approach applies for both the core elements which define the overall suitability (factor data and neighbourhood effect). Nevertheless, it is noteworthy that the inclusion of economic or social factors is often reduced due to a lack of spatially defined data and methodological issues regarding the knowledge of specific cause-effect relations. Indeed, it often happens that proxy variables in place of underlying driving forces are included in the modelling framework [40].

The current configuration of EUCS100 enables to process large datasets, as they are meant to be factor data relevant at a fine resolution (100m) and available for a wide territory (the entire European territory).

As for socio-economic factors, accessibility patterns are considered a major driver of land use/cover change at regional and local level due to their importance as location factor for both households and firms [17]. Accessibility is defined depending on the type of both origins and destinations it is computed for, and the considered transportation system. Several definitions and respective measures can be found in a rich bibliography (see [8], [15], [17], [18]).

In the context of LUMP, accessibility depends on travel origins, travel destinations and characteristics of the transportation network (e.g. geometry, speed limit). Accessibility is used in EUCS100 mainly as a factor for the allocation of residential and industrial land use classes. Several location-based accessibility maps have been thus produced, each taking into account a different set of destinations, i.e. main nodes of transport (ports and airports), entry points of high-speed road networks (freeway and highway entries/exits), major European cities and towns of regional importance. The set of origins is invariant and has been implemented to mimic real travel origins (nearly 200.000 points across Europe). The network used in the analysis was obtained from the Transtools project, completed, in some regions, by TeleAtlas 2008 network. An origin-destination matrix was finally computed whereby each origin is assigned the travel cost to the closest destination. The final maps in raster format are then obtained from the spatial interpolation of travel cost assigned to origins.

Proximity to roads of regional/local importance is strongly correlated with new urban developments (as examples, see [31] and [32]). In order to take into account these local dynamics of land use/cover change, a distance to road map has been produced. This has been computed with respect to TeleAtlas 2008 network, including motorways, major roads of high importance, secondary roads, local connecting roads, local roads of high importance and local roads.

Besides driving forces strongly related to transportation systems, other explanatory variables for LUCC dynamics are particularly relevant to the location of semi-natural classes (e.g. semi-natural vegetation, woods, agricultural land). The European Soil DataBase (ESDB) constitutes a reliable point of reference for representing soil properties: this database contains a list of Soil Typological Units (STU), characterizing distinct soil types. Each STU is then described by attributes (variables) specifying the nature and properties of the soils, e.g. texture, moisture regime, stoniness, etc. The richness in information allows, depending on the specific land use/cover taken into account, to include as a respective relevant factor map a (combination of) specific soil characteristics. Detailed factor maps have been prepared also regarding other geomorphological properties, such as the slope (from Global Digital Elevation Model - SRTM).

The so called neighbourhood effect is in charge of modelling proximity interactions (i.e. interactions between neighbouring land use/cover types) between land use/cover types. As many authors have pointed out (see [41], [11], [20] and [19]), neighbourhood plays a major role in the land use/cover change dynamics. The definition (characteristics/parameters) of the neighbourhood function is a demanding task. In fact, there are region and temporal variability issues: the neighbourhood effect may change among different regions and across time [41]. Another issue can be pointed out: depending on the scale of analysis, different neighbourhood relations emerge. As a consequence of the previous issues, particular attention has to be paid to the necessity of including in the modelling framework only independent variables: as the time scale lengthens, many explanatory factors become endogenous [11].

Location Specific Drivers

Suitability maps for each modelled land use/cover class can be altered where a policy or a combination of policies applies. This concept, which was first implemented in the CLUE-s model [49] and [48], was carried over to EUCS100 in order to reflect spatially explicit European policies. It is possible to combine the consequences of policies which award subsidies/incentives for the presence of a specific land use/cover in a unique land use/cover specific map thus increasing the respective local suitability, as well as policies targeted at discouraging the presence of a land use/cover: in the latter case the local suitability will be lowered.

Conversion Settings and Conversion Costs

The matrix responsible for setting the allowed conversions between land-use/cover classes overrules the computation of the overall suitability. It is possible to make land use/cover changes more or less dependent on the local land use/cover history [6]. A certain land-use/cover configuration/pattern can be made stable so to take into account the previous states of the modelled system [49].

Current Land use/cover Input

The starting point of the simulation is given by CLC 2006. The CLC is the only European wide land-use/cover map that is consistent across space and inter-comparable among temporal releases (1990, 2000, and 2006). This is due to the common guidelines that have been followed by the image interpreters since the first version of the CLC. The minimum mapping unit (MMU) of CLC was initially set to 25 ha. This characteristic constitutes a limitation of CLC. In practical terms, it means that any object on the Earth's surface with less than 25 ha in size is not mapped. This parameter has a considerable implication in the identification of urban areas, especially in low density territories with sparse and/or sprawled settlements, where part of the urban land-use/cover is, in fact, not captured. Ideally, all existing urban land should be accounted for modelling purposes. The dissimilarity between the reality and the model's land-use/cover map input is probably the very first contributor for error propagation.

To address this issue, an improvement of the initial CLC map was designed and prepared using finer datasets produced and/or available at JRC. The improvement targets spatial detail by reducing the minimum mapping unit to 1 ha (1 raster cell) whenever possible for the same legend nomenclature. Datasets used in this refinement are: Soil Sealing Layer, Urban Atlas, TeleAtlas and SRTM water bodies.

3.3 Indicators Module

A subset of the indicators described in [14], derived from the EURURALIS framework, is currently implemented in EUCS100. These indicators are embedded inside of the GeoDMS environment and can be generated as maps or as tables and can be aggregated from cell-based indicators to indicators at any regional configuration, allowing the possibility to compare the impacts of the different policy alternatives.

There are two large families of indicators: Land use related and thematic. The land use related indicators put into evidence the most prevalent *changes from* and *changes to* land-use/cover classes and shows the land change hotspots for agricultural land abandonment or expansion and urban expansion. The thematic indicators calculated at the 100m level in EUCS100 are land cover connectivity potential, soil sealing, river flood risk and urban sprawl. Like the land use/cover related indicators, the thematic ones are calculated on a per cell level and can be aggregated at various levels. A full description of these indicators is in [30].

4 Calibration and Validation

As it is to be expected that spatial processes may have different impacts at different scales, the new 100m resolution version of the model needed a revision of the calibration that was initially performed at a 1km scale.

In order to ensure a realistic distribution of land cover classes throughout time, given an evolving demand set and dynamic neighbourhood, EUCS100 is calibrated according to the situation depicted by CLC in 1990. The model calibration is performed using a multinomial logistic regression to establish the most suited parameter sets to describe the probable location of a certain land cover class among all of the endogenous land cover classes. The dependant variable is therefore an endogenous land cover class and the independent variables are the series of factor maps described in section 3.2. The neighbourhood effect is taken into account separately in a second multinomial logistic regression that assesses the land composition of the three rings of cells directly neighbouring the observed (to be explained) cells of each endogenous land cover class. This separate assessment of more general driving forces thought to be operating at a coarser scale and the local neighbourhood relations has the advantage of accounting for spatial autocorrelation.

The analysis is done at national level, with the exception of small countries which are grouped in order to assure an adequate amount of samples. Thus the calibration parameters are country specific in order to reflect the different driving forces behind the land use patterns and changes. The parameters are stored in text files in their respective country folders and are read and implemented when a simulation is run.

In order to validate the calibration parameters set through the process described above, the year 2000 was simulated starting with CLC 1990 and compared with the CLC 2000. For the simulation run, the claims per land use type were obtained from the observed amount of that land use type for 2000, as taken from the CLC 2000. For the validation we followed a similar approach to that carried out in a previous assessment of contemporary land-use change models [33] (see also [34] and [38]). That study distinguished three possible two-map comparisons using observed and simulated land use for two time steps:

- observed 1990-observed 2000 characterising the observed land-use change;
- observed 1990-simulated 2000 referring to the land-use change predicted by the model;
- observed 2000-simulated 2000 determining the accuracy of the prediction.

Based on the available maps it is also possible to perform a three-map comparison that allows distinguishing the correct prediction due to persistence of land use from well-predicted changes in land use. This is an important distinction as the amount of change is normally limited. Comparison methods that compare the complete simulation map with the complete observation map are strongly influenced by the static patterns and tend to overestimate the predictive power of land-use models. Therefore it is more meaningful to focus on the amount of correctly predicted change.

Based on the comparisons of the three available maps we can determine both quantity and location agreement between observed and simulated land use patterns. Quantity agreement is obtained by comparing the aggregate totals of all land-use classes in the simulated and observed land use. To describe location agreement we first create the following maps: (A) well-predicted change; (B) observed change predicted, but as a wrong gaining class; (C) observed change predicted as persistence and; (D) and observed persistence predicted as change. Based on these, location agreement was quantified using the following three statistic measurements:

- figure of merit = A / (A+B+C+D), which measures the accuracy of the model in predicting land-use change;
- producer accuracy = A / (A+B+C), which gives the proportion of pixels that the model predicts accurately as change, given the observed change;
- user accuracy = A / (A+B+D), which calculates the proportion of pixels predicted accurately as change, given the model predicted change.

A first validation was carried out for Belgium and Ireland. These two countries were selected as they show a substantial difference in the amount of actual change between

1990 and 2000 (in Belgium only 1.7% of the total land surface changed its use, while in Ireland 8.2 % changed). For reference purposes two statistics have been added to the three measures proposed by [33]: the proportion of well-predicted persistence and the overall model performance indicating the total amount of correctly predicted pixels (thus including correct persistence).

The preliminary results show that well-predicted persistence ranges from 95.6% to 98% and the overall model performance ranges from 88.5% to 96.5% for Ireland and Belgium, respectively. However, figure of merit, producer and user accuracies are below 10%. The preliminary results indicate the relation between model performance and observed amount of change. With more observed change the model is able to obtain higher user accuracy, but this comes at the cost of predicting persistence less well. However, this work is on-going and results are most likely to change. Further work will focus on the inclusion of explanatory variables at a higher resolution. In addition a longer observation period will be included in upcoming validation efforts, making use of the newly released CLC 2006.

5 Conclusions and Way Forward

The EUCS100 model is at the core of the multi-scale, multi-model framework referred to as the European Land Use Modelling Platform (LUMP) that aims to produce policy-relevant information related to land use/cover dynamics. It can support the creation of *ex-post* and *ex-ante* assessments of policies that impact spatial developments. The EUCS100 is a land allocation model that bridges sector models and indicator models and has the capacity to connect Global and European scale to the local level of environmental impacts.

The model is undergoing continuous development in order to answer to specific questions related to different sectors for assessing policy alternatives. In particular, the current version has been being developed in response to increasing requests for a tool capable of 1) modelling at the same resolution as the best current pan-European input maps (i.e. CLC) while being capable of using other sources of input maps; 2) producing meaningful indicators for the monitoring of Europe's territory, requiring high spatial and flexible thematic precision, especially of urban areas; 3) modelling different regionally-defined areas, independently of national borders.

EUCS100 is on the path towards being fully operational. Throughout the development phase, the model has been modified both conceptually and technically. The conceptual modifications include multiple land use typologies and multiple regional divisions, as well as multiple simulation periods. Technical modifications include the way the model handles images through tiling and re-projection capabilities on the fly to adhere to the Inspire Directive. Current and future developments include 1) a full validation procedure using two test cases with observed land claims at NUTS2 level; 2) the development of new pan-European predicted demand sets at NUTS2 level, driven by sector-specific models in forestry, agriculture and built-up classes; and 3) the increase in the flexibility of the system through the reduction in the current model structure complexity, which is a product of years of evolution.

Acknowledgments. Part of the work herein reported is financed by the Directorate General Environment, European Commission. The authors also express their thanks to Vasco Diogo for his relentless efforts in providing revised calibration results.

References

1. Allen, P.M., Torrens, P.: Knowledge and complexity (introduction). Futures 37, 581–584 (2005)
2. Bakker, M., Verburg, P.: SENSOR Project Deliverable Report D 2.2.2/3; Deliverable title: Scenario based forecasts of land use and land management at 1 km^2 grid and NUTS-X level, based on CLUE-S. In: Helming, K., Wiggering, H. (eds.) SENSOR Report Series 2008/09, ZALF, German (2009)
3. Batty, M., Torrens, P.: Modelling and prediction in a complex world. Futures 37, 745–766 (2005)
4. Britz, W.: Automated model linkages: the example of CAPRI, vol. 57(8), pp. 367–368. Agrarwirtschaft, Jahrgang (2008)
5. Britz, W., Witzke, H.P. (eds.): CAPRI Model Documentation 2008: Version 2. Institute for Food and Resource Economics, University of Bonn (2008)
6. Brown, D.G., et al.: Path Dependence and the validation of agent-based spatial models of land-use. Int. J. Geo. Inf. Sci. 19:2, 153–174 (2005)
7. COM 276 - Communication from the Commission on impact assessment (2002)
8. Dalvi, M.Q., Martin, K.M.: The Measurement of Accessibility: some preliminary results. Transportation 5, 17–42 (1976)
9. De Groot, R.S., et al.: Challenges in integrating the concept of ecosystem services and values in landscape planning, management and decision making. Ecological Complexity 7(3), 260–272 (2010)
10. De Groot, R.S., Wilson, M.A., Boumans, R.M.J.: A typology for the classification, description and valuation of ecosystem functions, goods and services. Ecological Economics 41(3), 393–408 (2002)
11. Dendoncker, N., Rounsevell, M., Bogaert, P.: Spatial Analysis and Modelling of Land Use Distributions in Belgium. Com. Env. Urb. Sys. 31, 188–205 (2007)
12. EEA (European Environmental Agency): Land-use scenarios for Europe: qualitative and quantitative analysis on a European scale, Copenhagen: European Environmental Agency. EEA Technical report no. 9/2007 (2007)
13. EEA (European Environmental Agency): Modelling environmental change in Europe: towards a model inventory (SEIS/Forward), Copenhagen: European Environmental Agency. EEA Technical report no. 11/2008 (2008)
14. Eickhout, B., Prins, A.G.: Eururalis 2.0. Technical background and indicator documentation. Wageningen University and Research and Netherlands Environmental Assessment Agency (MNP), The Netherlands (2008)
15. El-Geneidy, A.M., Levinson, D.M.: Access to Destinations: Development of Accessibility Measures. Final Report for Minnesota Department of Transportation. In: Networks, Economics, and Urban Systems Research Group/Department of Civil Engineering, University of Minnesota, St. Paul (2006)
16. Engelen, G., et al.: The MOLAND modelling framework for urban and regional land-use dynamics. In: Koomen, E., et al. (eds.) Modelling Land-Use Change, Progress and applications, pp. 297–319. Springer, Heidelberg (2007)

17. Geurs, K.T., Ritsema van Eck, J.R.: Accessibility Measures: Review and Applications – Evaluation of Accessibility Impacts of Land-use Transport Scenarios, and Related Social and Economic Impacts. Report for RIVM – National Institute of Public Health and the Environment (2001)
18. Geurs, K.T., van Wee, B.: Accessibility Evaluation of Land-use and Transport Strategies: Review and Research Directions. J. Tra. Geo. 12, 127–140 (2004)
19. Hansen, H.S.: Empirically Derived Neighbourhood Rules for Urban Land-Use Modelling. Environment and Planning B: Planning and Design advance online publication (2010)
20. Hansen, H.S.: Quantifying and Analysing Neighbourhood Characteristics Supporting Urban Land-Use Modelling. In: Bernard, L., Friis-Christensen, A., Pundt, H. (eds.) Lecture Notes in Geoinformation and Cartography, The European Information Society, pp. 283–299. Springer, Heidelburg (2008)
21. Helming, J.F.M., Janssen, S., van Meijl, H.: mpact assessment of post 2013 CAP measures on European agriculture. LEI report, The Hague (2010)
22. Helming, K., et al.: Ex-ante impact assessment of land use change in European regions—the SENSOR approach. In: Helming, K., Perez-Soba, M., Tabbush, P. (eds.) Sustainability Impact Assessment of Land Use Changes, pp. 77–105. Springer, Heidelberg (2008)
23. Hilferink, M., Rietveld, P.: Land Use Scanner: An integrated GIS based model for long term projections of land use in urban and rural areas. Journal of Geographical Systems 1(2), 155–177 (1999)
24. Hoymann, J.: Accelerating urban sprawl in depopulating regions: a scenario analysis for the Elbe River Basin. Reg.Environ.Change (in press)
25. Hoymann, J.: Spatial allocation of future residential land use in the Elbe River Basin. Environment and Planning B: Planning and Design 37(5), 911–928 (2010)
26. Koomen, E., Borsboom-van Beurden, J.: Land-use modeling in planning practice; the Land Use Scanner approach. Springer, Heidelberg (2011)
27. Lavalle, C., et al.: The MOLAND model for urban and regional growth forecast - A tool for the definition of sustainable development paths, European Commission, vol. 21480, p. 22. DG-Joint Research Centre, Ispra, Italy (2004)
28. Lavalle, C., Ehrlich, D., Annoni, A.: Sustainable urban development: the MURBANDY project of the Centre for Earth Observation. In: IEEE International Geoscience and remote sensing symposium proceedings (IGARSS1998), vol. 5, pp. 2571–2573 (1998)
29. Overmars, K.P., Verburg, P.H., Veldkamp, T.A.: Comparison of a Deductive and an Inductive Approach to Specify Land Suitability in a Spatially Explicit Land Use Model. Land Use Policy 24, 584–599 (2007)
30. Pérez-Soba, M., Verburg, P.H., Koomen, E., Hilferink, M., Benito, P., Lesschen, J.P., Banse, M., Woltjer, G., Eickhout, B., Prins, A.G., Staritsky, I.: Land use modelling - implementation; Preserving and enhancing the environmental benefits of land-use services. Final report to the European Commission, DG Environment. Alterra Wageningen UR/ Geodan Next/ Object Vision/ BIOS/ LEI and PBL, Wageningen (2010)
31. Pijanowski, B.C., Brown, D.G., Shellito, B.A., Manik, G.A.: Using neural networks and GIS to forecast land use changes: a Land Transformation Model. Computers, Environment and Urban Systems, 26, 553–575 (2002)
32. Pijanowski, B.C., Shellito, B., Pithadia, S., Alexandridis, K.: Forecasting and assessing the impact of urban sprawl in coastal watersheds along eastern Lake Michigan. Lakes & Reservoirs: Research and Management 7, 271–285 (2002)
33. Pontius, J. R.G., et al.: Comparing the input, output, and validation maps for several models of land change. Annals of Regional Science 42(1), 11–37 (2008)

34. Pontius Jr, R.G., Huffaker, D., Denman, K.: Useful techniques for validation for spatially explicit land-change models. Ecological Modelling 179, 445–461 (2004)
35. Pratt, A.: Social and economic drivers of land use change in the British space economy. Land Use Pol 114, 109–114 (2009)
36. Rich, J., et al.: Report on Scenario, Traffic Forecast and Analysis of Traffic on the TEN-T, taking into Consideration the External Dimension of the Union – TRANS-TOOLS version 2; Model and Data Improvements. DG TREN, Copenhagen, Denmark (2009)
37. SEC1992 Impact Assessment Guidelines (2009)
38. Van Vliet, J.: Assessing the accuracy of changes in spatial explicit land use change models. In: 12th annual AGILE conference on Geographic Information Science (2009)
39. Veldkamp, A., Fresco, L.O.: CLUE-CR: an integrated multi-scale model to simulate land use change scenarios in Costa Rica. Ecol. Model 91 (1996)
40. Veldkamp, A., Lambin, E.F.: Editorial: Predicting land-use change. Agr. Eco. Env. 85(1-3), 1–6 (2001)
41. Verburg, P.H., de Nijs, T.C.M., Ritsema van Eck, J., Visser, H., de Jong, K.: A Method to Analyse Neighbourhood Characteristics of Lan Use Patterns. Computers, Environment and Urban Systems 28, 667–690 (2004)
42. Verburg, P.H., Eickhout, B., van Meijl, H.: A multi-scale, multi-model approach for analyzing the future dynamics of European land use. Annals of Regional Science 42(1), 57–77 (2008)
43. Verburg, P.H., Overmars, K.: Combining top-down and bottom-up dynamics in land use modeling: exploring the future of abandoned farmlands in Europe with the Dyna-CLUE model. Landscape Ecology 24(9), 1167–1181 (2009)
44. Verburg, P.H., Rounsevell, M.D.A., Veldkamp, A.: Scenario-based studies of future land use in Europe. Agriculture, Ecosystems & Environment 114(1), 1–6 (2006)
45. Verburg, P.H., Shot, P.P., Dijst, M.J., Veldkamp, A.: Land Use Change Modelling: Current Practice and Research Priorities. Geo.J. 61, 309–324 (2004)
46. Verburg, P.H., Soepboer, W., Limpiada, R., Espaldon, M.V.O., Sharifa, M.A., Veldkamp, A.: Modelling the spatial dynamics of regional land use: The CLUE-S model. Environmental Management, 30, 391–405 (2002)
47. Verburg, P.H., van de Steeg, J., Veldkamp, A., Willemen, L.: From land cover change to land function dynamics: A major challenge to improve land characterization. J. Environ. Manag. 90, 1327–1335 (2009)
48. Verburg, P.H., van Eck, J.R.R., de Nijs, T.C.M., Dijst, M.J., Schot, P.: Determinants of land-use change patterns in the Netherlands. Environment and Planning B: Planning and Design 31(1), 125–150 (2004)
49. Verburg, P.H., Veldkamp, A.: Projecting land use transitions at forest fringes in the Philippines at two spatial scales. Landscape Ecology 19, 77–98 (2004)

An Adaptive Neural Network-Based Method for Tile Replacement in a Web Map Cache

Ricardo García, Juan Pablo de Castro, María Jesús Verdú, Elena Verdú, Luisa María Regueras, and Pablo López

Higher Technical School of Telecommunications Engineering,
University of Valladolid, Valladolid, Spain
{ricgar,juacas,marver,elever,luireg}@tel.uva.es,
plopesc@ribera.tel.uva.es
http://www.tel.uva.es

Abstract. Most popular web map services, such as Google Maps, serve pre-generated image tiles from a server-side cache. However, storage needs are often prohibitive, forcing administrators to use partial caches containing a subset of the total tiles. When the cache runs out of space for allocating incoming requests, a cache replacement algorithm must determine which tiles should be replaced. Cache replacement algorithms are well founded and characterized for general Web documents but spatial caches comprises a set of specific characteristics that make them suitable to further research. This paper proposes a cache replacement policy based on neural networks to take intelligent replacement decisions. Neural networks are trained using supervised learning with real data-sets from public web map servers. Hight correct classification ratios have been achieved for both training data and a completely independent validation data set, which indicates good generalization of the neural network. A benchmark of the performance of this policy against several classical cache management policies is given for discussion.

Keywords: Web map service, Tile cache, Replacement policy, Neural networks, benchmark.

1 Introduction

The Web Map Service (WMS) standard of the Open Geospatial Consortium (OGC) offers a standardized and flexible way of serving cartographic digital maps of spatially referenced data through HTTP requests [1]. However, spatial parameters in requests are not constrained, which forces images to be generated on the fly each time a request is received, limiting the scalability of these services.

A common approach to improve the scalability of the map server is to improve the cachability of requests by dividing the map into a discrete set of images, called tiles, and restrict user requests to that set. Several specifications have been developed to address how cacheable image tiles are advertised from server-side and how a client requests cached image tiles. The Open Source Geospatial

B. Murgante et al. (Eds.): ICCSA 2011, Part I, LNCS 6782, pp. 76–91, 2011.

Foundation (OSGeo) developed the WMS Tile Caching (usually known as WMS-C) proposal [2], while the OGC has recently released the Web Map Tile Service Standard (WMTS) [3] inspired by the former and other similar initiatives.

The problem that arises when deploying these server-side caches in practical implementations is that the storage requirements are often prohibitive for many organizations, thus forcing to use partial caches containing just part of the total tiles. Even if there are enough available resources to store a complete cache of the whole map, many tiles will never be requested, so it is not worth it to cache those "unpopular" tiles because no gain will be obtained.

When the cache runs out of space it is necessary to determine which tiles should be replaced by the new ones. The cache replacement algorithm proposed in this paper uses a neural network to estimate the probability that a request of a tile occurs before a certain period of time. Those tiles that are not likely to be requested shortly are good candidates for replacement (assuming the cost to fetch a tile from the remote server is the same for all tiles).

The rest of the document is organized as follows. Section 2 provides a brief background about how the map is tiled in order to offer a tiled web map service. Related work in replacement algorithms is presented in Section 3. Section 4 describes the trace files used for training and simulation. In Section 5 the proposed replacement algorithm based on neural networks is discussed. In Section 6 this algorithm is benchmarked against several classical cache management policies in a simulated Proxy Web Cache. Finally, the conclusions of the paper are gathered together in Section 7.

2 Tiling Space

In order to offer a tiled web map service, the web map server renders the map across a fixed set of scales through progressive generalization. Rendered map images are then divided into tiles, describing a tile pyramid as depicted in Fig. 1.

For example, Google Maps uses a tiling scheme where level 0 allows representing the whole world in a single tile of 256x256 pixels (where the first 64 and last 64 lines of the tile are left blank). The next level represents the whole world in 2x2 tiles of 256x256 pixels and so on in powers of 2. Therefore, the number of tiles n grows exponentially with the resolution level l. Using this tiling scheme, to cover the whole world with a pyramid of 20 levels, around 3.7×10^{11} tiles are required and several petabytes of disk are used.

3 Related Work

Most important characteristics of Web objects used in Web cache replacement strategies are: *recency* (time since the last reference to the object), *frequency* (number of requests to the object), *size* of the Web object and *cost* to fetch the object from its origin server. Depending on the characteristics used, replacement strategies can be classified as recency-based, frequency-based, recency/frequency-based, function-based and randomized strategies [4].

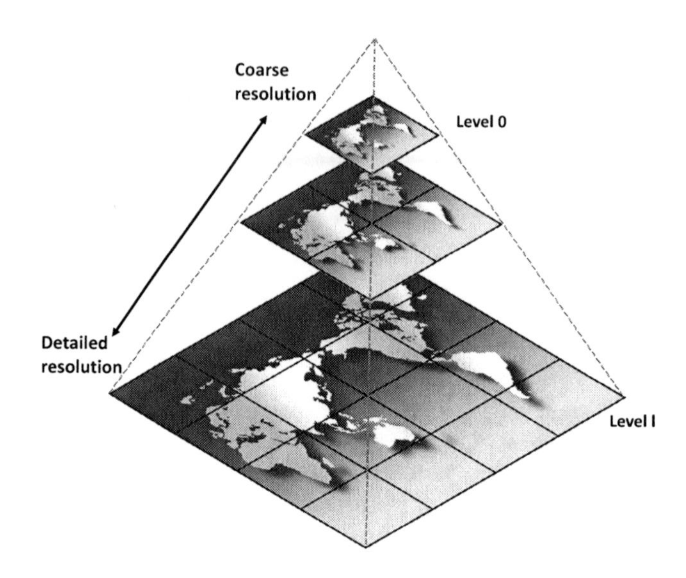

Fig. 1. Tile pyramid representation

Recency-based strategies exploit the temporal locality of reference observed in Web requests. These are usually extensions of the well-known LRU strategy, which removes the least recently referenced object. Pyramidal Selection Scheme (PSS) [5] algorithm uses a pyramidal classification of objects depending upon their size, and objects in each group are maintained as a separate LRU list. Only the least recently used objects in each group are compared, and the object selected for replacement is which is maximizing the product of its size and the number of accesses since the last time it was requested.

Frequency-based strategies rely on the fact that popularity of Web objects is related to their frequency values. These strategies are built around the LFU strategy, which removes the least frequently requested object.

Recency/frequency-based strategies combine recency and frequency information to take replacement decisions.

Function-based strategies employ a general function of several parameters to make decisions of which object to evict from the cache. GD-Size [6] strategy uses a function of size, cost and an aging factor. GDSF [7] is an extension of GD-Size that also takes account for frequency. Least-Unified Value (LUV) [8] rates an object based on its past references to estimate the likelihood of a future request, and normalizes the value by the cost of the object per unit size.

Randomized strategies use a non-deterministic approach to randomly select a candidate object for replacement.

For a further background, a comprehensive survey of web cache replacement strategies is presented in [4]. According to that work, algorithms like GD-Size, GDSF, LUV and PSS were considered "good enough" for caching needs at the time it was published in 2003. However, the explosion of web map traffic did not happen until a few years later.

Despite the vast proliferation of web cache replacement algorithms, there is a reduced number of replacement policies specific to map tile caches that benefit of spatial correlation between requests.

Map tile caching can benefit from spatial locality principle as stated by the Tobler's first law of geography, which states that *"Everything is related to everything else, but near things are more related than distant things"* [9].

The use of neural networks for cache replacement was first introduced by Khalid [10], with the KORA algorithm. KORA uses backpropagation neural network for the purpose of guiding the line/block replacement decisions in cache. The algorithm identifies and subsequently discards the dead lines in cache memories. It is based on previous work by Pomerene et al. [11], who suggested the use of a shadow directory in order to look at a longer history when making decisions with LRU. Later, an improved version of the former, KORA-2, was proposed [12,13]. Other algorithms based on KORA were also proposed [14,15].

A survey on applications of neural networks and evolutionary techniques in web caching can be found in [16].

[17,18,19,20,22] proposes the use of a backpropagation neural network in a Web proxy cache for taking replacement decisions.

A predictor that learns the patterns of Web pages and predicts the future accesses is presented in [21].

[23] discusses the use of neural networks to support the adaptivity of the Class-based Least Recently Used (C-LRU) caching algorithm.

The novelty and significance of this work resides on the target scenario where it is applied. No similar studies about the application of neural networks to take replacement decisions in a Web map cache have been found in the literature.

Although the underlying methodology has already been discussed in related work on conventional web caching, web map requests' distributions and attributes are very different from those of traditional web servers. In this context, the size of the requested object is heavily related to its "popularity" so it can be used to estimate the probability of a future request. Traditional web caching replacement policies commonly use this parameter only to evaluate the trade off between the hits produced on that object and the space required to store it.

4 Training Data

Simulations are driven by trace files from three different tiled web map services, CartoCiudad, IDEE-Base and PNOA, provided by the National Geographic Institute (IGN) of Spain.

4.1 Analyzed Web Map Services

CartoCiudad is the official cartographic database of the Spanish cities and villages with their streets and roads networks topologically structured. PNOA serves imagery from the Aerial Ortophotography National Plan, which updates every two years Spanish covers by aerial photography, high resolution and accuracy digital ortophotography, and high density and accuracy Digital Terrain

Model. IDEE-Base allows viewing the Numeric Cartographic Base 1:25,000 and 1:200,000 of the IGN.

All these services conform to the OSGeo WMS-C recommendation by using the TileCache 2.0 cache implementation from Metacarta [24].

4.2 Data Collection

The access logs of the three web map services described before have been collected for this study. Access information is recorded in the Common Log Format (CLF), the most commonly used log format provided by Web servers. An example web map request recorded in this format looks as follows:

```
193.144.251.29 - - [01/Jun/2008:02:04:11 +0200]
"GET /wms-c/CARTOCIUDAD/CARTOCIUDAD?SERVICE=WMS&VERSION=1.1.1
&REQUEST=GetMap&LAYERS=Todas&FORMAT=image/png
&EXCEPTIONS=application/vnd.ogc.se_inimage
&SRS=EPSG:4326&BBOX=-2.109375,37.265625,-1.40625,37.96875
&STYLES=&WIDTH=256&HEIGHT=256 HTTP/1.1" 200 10495
```

The log entry contains the requestor's IP address, the *userid* of the person requesting the document as determined by HTTP authentication, a timestamp when the server finished processing the request with seconds precision, the request line, the status code and the size of the returned object in bytes.

These access logs contain information on all client requests received along several months, as reflected on Table 1.

Table 1. Summary of Access Log Characteristics (Complete DataSet)

Service	IDEE_BASE	CARTOCIUDAD	PNOA
Total Requests	16978535	3778369	9816747
% Valid	99.49%	89%	98.59%
% Invalid	0.51%	11%	1.41%
GetMap	99.997%	97.479%	99.997%
GetCapabilities	0.003%	2.521%	0.003%
Start Date	2010-03-01	2009-12-09	2010-03-01
End Date	2010-06-03	2010-06-20	2010-05-30
Duration	93 days	193 days	89 days
Frequency	182565 reqs/day	19577 reqs/day	110300 reqs/day
Object Size	6365.574 bytes	29181.239 bytes	12549.867 bytes
Log Size	5.437 GB	1.862 GB	3.810 GB

Trace files were filtered to contain only valid web map requests, so the simulation data only contains requests that would actually be cached. In the first place, all requests with a status code different of $2XX$ were discarded. In the same way, only GetMap requests have been considered, thus discarding GetCapabilities requests. All GetMap requests that are not compliant with the WMS-C recommendation have also been discarded for this study.

Filtered data is then inserted into a PostgreSQL database with the spatial extension PostGIS which adds support for geographic objects. The bounding box of the map request (*BBox* parameter) is stored as a geographic object, so spatial queries can be made taking advantage of the spatial indexes provided by the database.

Integer indexes x, y and z are extracted from the bouding box and inserted in the database. These indexes identify the row, column, and resolution level of the requested tile in the tile pyramid.

Two map requests correspond to the same object if both share the same values for the following parameters: *Layers*, *Format*, *SRS*, *Width*, *Height*, *Styles* and *BBox* (or, equivalently, x, y and z). To speed-up the simulation process, a unique identifier for each object has been created by applying a hash function to the previous parameters.

4.3 Requests Distribution

It must be noted that the performance gain achieved by the use of a tile cache will vary depending on how the tile requests are distributed over the tiling space. If those were uniformly distributed, the cache gain would be proportional to the cache size. However, it has been found that tile requests describe a Pareto distribution, as shown in Figure 2. Tile requests to the CartoCiudad map service follow the 20:80 rule, which means that the 20% of tiles receive the 80% of the total number of requests. In the case of IDEE-Base, this behaviour is even more prominent, where the 10% of tiles receive almost a 90% of total requests. PNOA requests are more scattered. This happens because about the 90% of requests belong to the two higher resolution levels (19 and 20), the ones with larger number of tiles.

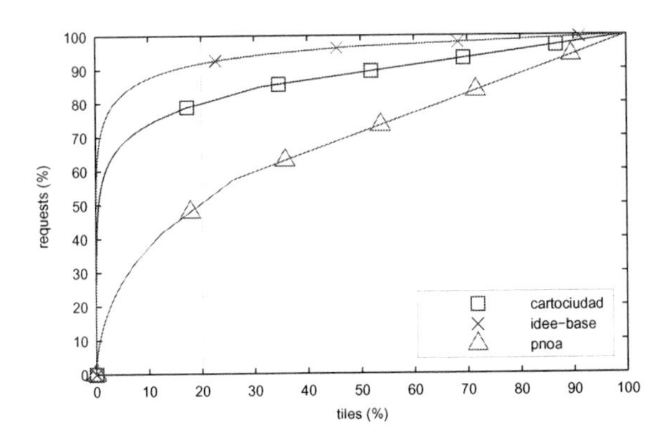

Fig. 2. Percentile of requests for the analyzed services

Services that show Pareto distributions are well-suited for caching, because high cache hit ratios can be found by caching a reduced fraction of the total tiles.

5 Neural Network Cache Replacement

Artificial neural networks (ANNs) are inspired by the observation that biological learning systems are composed of very complex webs of interconnected neurons. In the same way, ANNs are built out of a densely interconnected group of units. Each artificial neuron takes a number of real-valued inputs (representing the one or more dendrites) and calculates a linear combination of these inputs, plus a bias term. The sum is then passed through a non-linear function, known as *activation function* or *transfer function*, which outputs a single real-value, as shown in Fig. 3.

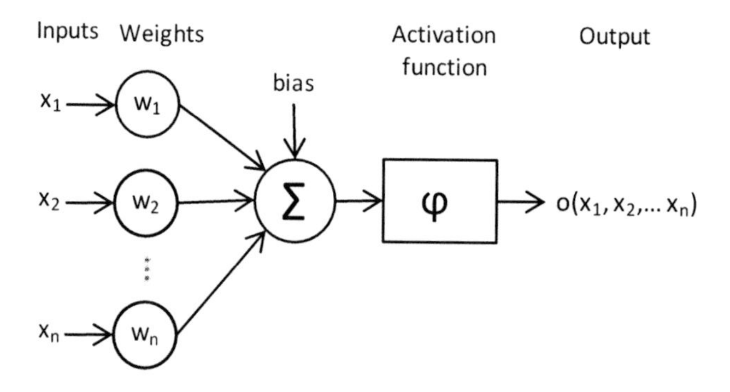

Fig. 3. Artificial neuron

In this work, a special class of layered feed-forward network known as multi-layer perceptron (MLP) has been used, where units at each layer are connected to all units from the preceding layer. It has an input layer with three inputs, two-hidden layers each one comprised of 3 hidden nodes, and a single output.

Learning an artificial neuron involves choosing values for the weights so the desired output is obtained for the given inputs.

Network weights are adjusted through supervised learning using subsets of the trace data sets, where the classification output of each request is known. Backpropagation with momentum is the used algorithm for training. The neural network parameters, such as the learning rate, momentum constant or network size, have been carefully chosen through experimentation. The parameters that provide the best convergence of the neural network are summarized in Table 2.

Table 2. Neural network parameters

Parameter	Value
Architecture	Feed-forward Multilayer Perceptron
Hidden layers	2
Neurons per hidden layer	3
Inputs	3 (recency, frequency, size)
Output	1 (probability of a future request)
Activation functions	Log-sigmoid in hidden layers, Hyperbolic tangent sigmoid in output layer
Error function	Minimum Square Error (mse)
Training algorithm	Backpropagation with momentum
Learning method	Supervised learning
Weights update mode	Batch mode
Learning rate	0.05
Momentum constant	0.2

5.1 Neural Network Inputs

The neural network inputs are three properties of tile requests: recency of reference, frequency of reference, and the size of the referenced tile. These properties are known to be important in web proxy caching to determine object cachability.

Inputs are normalized so that all values fall into the interval $[-1, 1]$, by using a simple linear scaling of data as shown in Equation 1, where x and y are respectively the data values before and after normalization, x_{min} and x_{max} are the minimum and maximum values found in data, and y_{max} and y_{min} define normalized interval so $y_{min} \leq y \leq y_{max}$. This can speed up learning for many networks.

$$y = y_{min} + (y_{max} - y_{min}) \times \frac{x - x_{min}}{x_{max} - x_{min}} \tag{1}$$

Recency values for each processed tile request are computed as the amount of time since the previous request of that tile was made. Recency values calculated this way do not address the case when a tile is requested for the first time. Moreover, measured recency values could be too disparate to be reflected in a linear scale.

To address this problem, a sliding window is considered around the time when each request is made, as done in [17].

With the use of this sliding window, recency values are computed as shown in Equation 2.

$$recency = \begin{cases} max(SWL, \Delta T_i) & if\ object\ i\ was\ requested\ before \\ SWL & otherwise \end{cases} \tag{2}$$

where ΔT_i is the time since that tile was last requested and SWL is the sliding window length.

Recency values calculated that way can already be normalized as stated before in Equation 1.

Frequency values are computed as follows. For a given request, if a previous request of the same tile was received inside the window, its frequency value is incremented by 1. Otherwise, frequency value is divided by the number of windows it is away from. This is reflected in Equation 3.

$$frequency = \begin{cases} frequency + 1 & if\ \Delta T_i \leq SWL \\ MAX\left[\frac{frequency}{\frac{\Delta T_i}{SWL}}, 1\right] & otherwise \end{cases} \quad (3)$$

Size input is directly extracted from server logs. As opposite to conventional Web proxies where requested object sizes can be very heterogeneous, in a web map all objects are image tiles with the same dimensions (typically 256x256 pixels). Those images are usually rendered in efficient formats such as PNG, GIF or JPEG that rarely reach 100 kilobytes in size.

As discussed in [25], due to greater variation in colors and patterns, the popular areas, stored as compressed image files, use a larger proportion of disk space than the relatively empty non-cached tiles. Because of the dependency between the file size and the "popularity" of tiles, tile size can be a very valuable input of the neural network to correctly classify the cachability of requests.

5.2 Neural Network Target

During the training process, a training record corresponding to the request of a particular tile is associated with a boolean target (0 or 1) which indicates whether the same tile is requested again or not in window, as shown in Equation 4.

$$target = \begin{cases} 1\ if\ the\ tile\ is\ requested\ again\ in\ window \\ 0\ otherwise \end{cases} \quad (4)$$

Once trained, the neural network output will be a real value in the range [0,1] that must be interpreted as the probability of receiving a successive request of the same tile within the time window.

A request is classified as *cacheable* if the output of the neural network is above 0.5. Otherwise, it is classified as *non cacheable*.

5.3 Training

The neural network is trained through supervised learning using the data sets from the extracted trace files. Each trace comprises requests received from the 1th to 7th of March in 2010. Between these dates, a total of 25.922, 94.520, and 186.672 valid map requests were received respectively for CartoCiudad, IDEE-Base and PNOA. The trace data is subdivided into training, validation, and test sets, with the 75%, 15% and 15% of the total requests, respectivelly. The first one is used for training the neural network. The second one is used to validate that the network is generalizing correctly and to identify overfitting. The final one is used as a completely independent test of network generalization.

Each training record consists of an input vector of recency, frequency and size values, and the known target. The weights are adjusted using the backpropagation algorithm, which employs the gradient descent to attempt to minimize the squared error between the network output values and the target values for these outputs [26]. The network is trained in batch mode, in which weights and biases are only updated after all the inputs and targets are presented.

The pocket algorithm is used, which saves the best weights found in the validation set.

5.4 Training Results

Neural network performance is measured by the correct classification ratio (CCR), which computes the percentage of correctly classified requests versus the total number of processed requests.

Table 3. Correct classification ratios (%) during training, validation and testing for the different services

	CartoCiudad	PNOA	IDEE-Base
training	76.5952	96.5355	75.6529
validation	70.2000	97.1985	77.5333
test	72.7422	97.4026	82.7867

The neural network is able to correctly classify the cachability of requests, with CCR values over the testing data set ranging between 72% and 97%, as shown in Table 3.

6 Experimental Design

While it has already been proved that the proposed neural network can effectively predict whether a map tile will be requested again before a certain time or not, it is not a reliable metric to evaluate the performance of a replacement policy. To obtain a better judgment, this algorithm has been tested in a simulated cache against other popular replacement policies, using the *hit-rate* and *byte-hit-rate* as performance metrics.

6.1 Simulator Prototype

A custom cache simulator has been developed for this study. This prototype simulates the following replacement process, typical of practical implementations, as stated by Podlipnig and Böszörmenyi [4]: when a cache miss occurs, the requested object is stored in the cache. When the cache size exceeds a given limit, a certain number of objects are evicted from the cache to make room for incoming requests. The cache uses two marks *HM* (high mark) and *LM* (low mark), with $HM > LM$. When the cache size exceeds *HM*, a replacement policy evicts objects until it reaches the low mark *LM*.

The simulator uses a *hashmap* to maintain the metadata associated to each individual tile stored in the cache, indexed by the unique object identifier discussed in Section 4.2. This metadata contains the required information for the replacement algorithms: size of the object, time of the last access, frequency of reference, etc.

6.2 Performance Metrics

In this work, we employ the two most commonly used performance metrics found in the literature [7,4]: *Hit-rate* and *Byte-hit-rate*.

- *Hit-rate*. It refers to the ratio of objects retrieved directly from the cache (*cache hits*) versus the total number of requests made to the cache.
- *Byte-hit-rate*. It is computed as the amount of bytes served directly from the cache versus the total number of bytes served.

Another popular metric, known as *delay-savings-ratio* takes also into account the cost to fetch the object from the original server. It has not been considered here because this information can not be inferred from the analyzed logs.

6.3 Proxy Cache Simulation

In order to evaluate the performance of the proposed algorithm, it has been compared with the following classical cache replacement policies: *Least Recently Used* (LRU); *Least Frequently Used* (LFU); a randomized strategy (Random) which randomly selects objects for replacement; the *Size* algorithm (MaxS) which removes the largest object from the cache [27], and its binary negation *Min Size* (MinS); and the Belady's Optimal algorithm (OPT), which removes the object that will remain unused for the longest time [28], not implementable in practice.

The neural networks used are those with the best weights found during the validation set, as described in Section 5.3.

These algorithms have been benchmarked against a subset of the trace requests collected for the different services, as described in Section 4. A total of 311767, 606220 and 393777 requests from Cartociudad, IDEE-Base and PNOA, respectively, have been used in the simulation.

Figures 4-6 show the hit-rates for the different algorithms. Byte-hit-rates are shown in Figures 7-9. The algorithms have been evaluated for different High Marks. In all cases, the Low Mark value corresponds to the 90% of the High Mark.

The dotted horizontal line corresponds to the performance achieved with an infinite-sized cache. In this scenario, cache misses correspond to the initial request of each object (*cold miss*).

Results show that the proposed neural network outperforms the rest of the algorithms under test in most scenarios, closely followed by LFU.

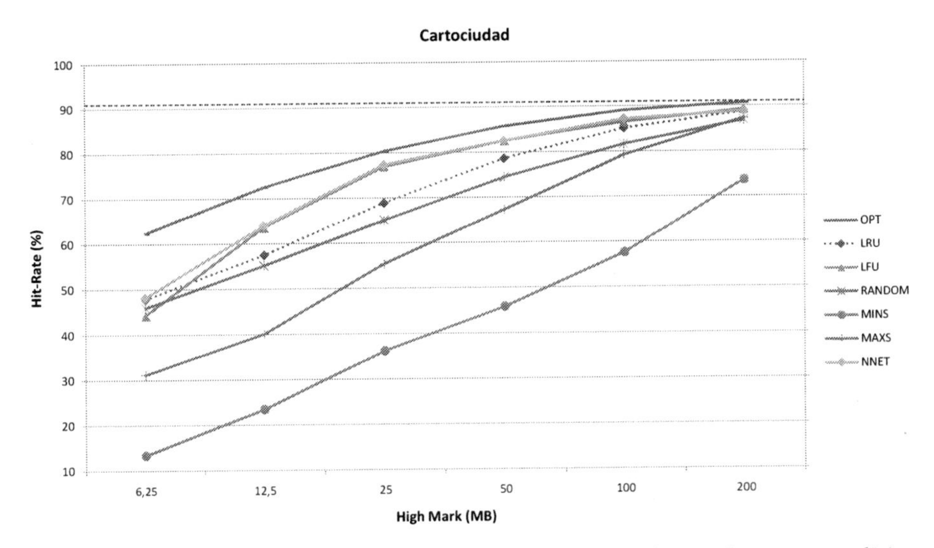

Fig. 4. Hit Ratios for Cartociudad with different cache replacement policies. LM=0.9×HM.

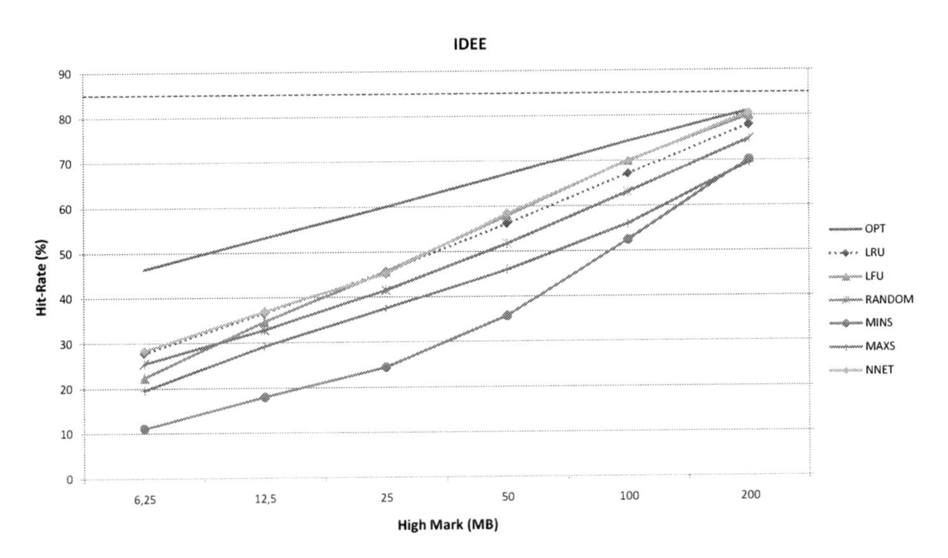

Fig. 5. Hit Ratios for IDEE with different cache replacement policies. LM=0.9×HM.

Despite LRU is the policy most often used in practice, it achieves worse results than LFU. As shown in Figure 2, some "popular" objects are requested frequently, while many others are accessed rarely. This workload distribution encourages the use of the frequency of reference as a relevant characteristic to be considered in a cache replacement decision.

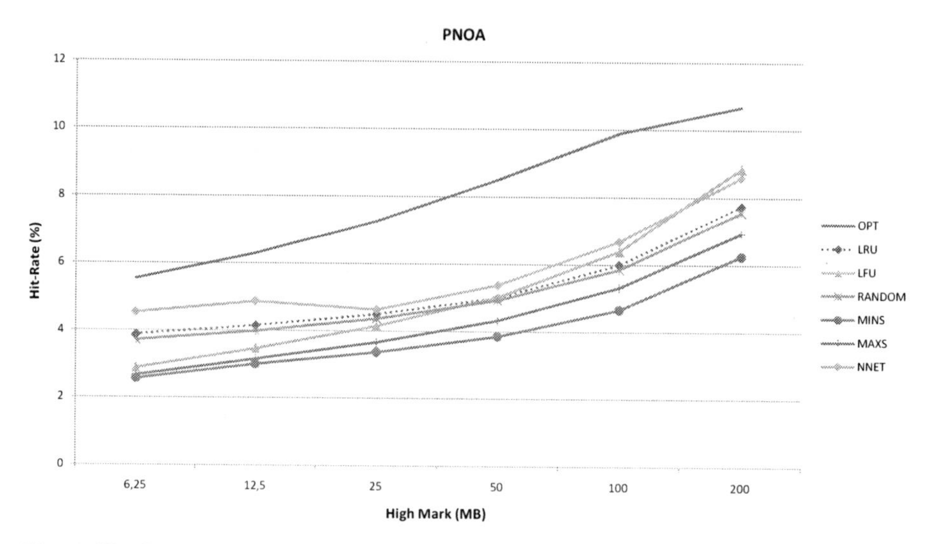

Fig. 6. Hit Ratios for PNOA with different cache replacement policies. LM=0.9×HM.

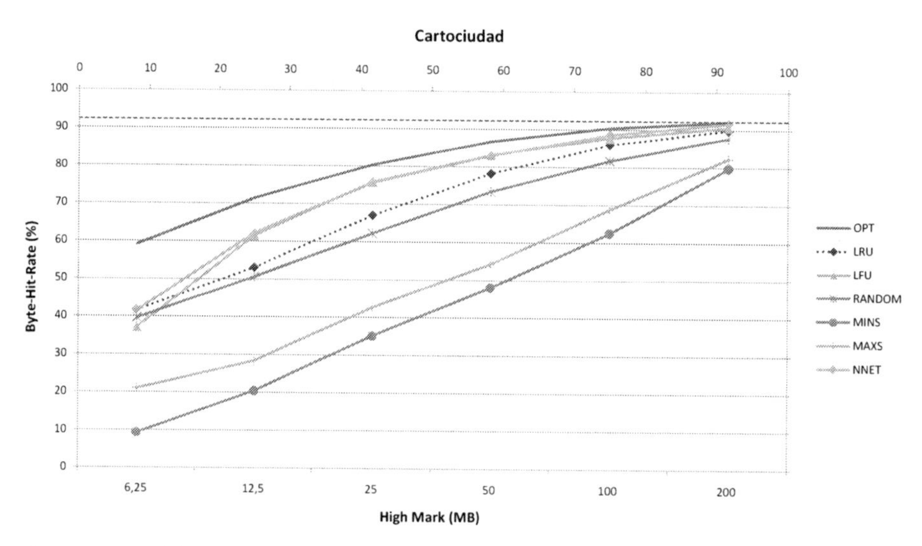

Fig. 7. Byte Hit Ratios for Cartociudad with different cache replacement policies. LM=0.9×HM.

Size-based policies have consistently the worst performance, for both hit-rate and byte-hit-rate, which reflects that to just consider size as unique parameter is not a good approach for taking replacement decisions. In fact, as can be shown, it is even worse than randomly select objects for replacement, the most simple approach.

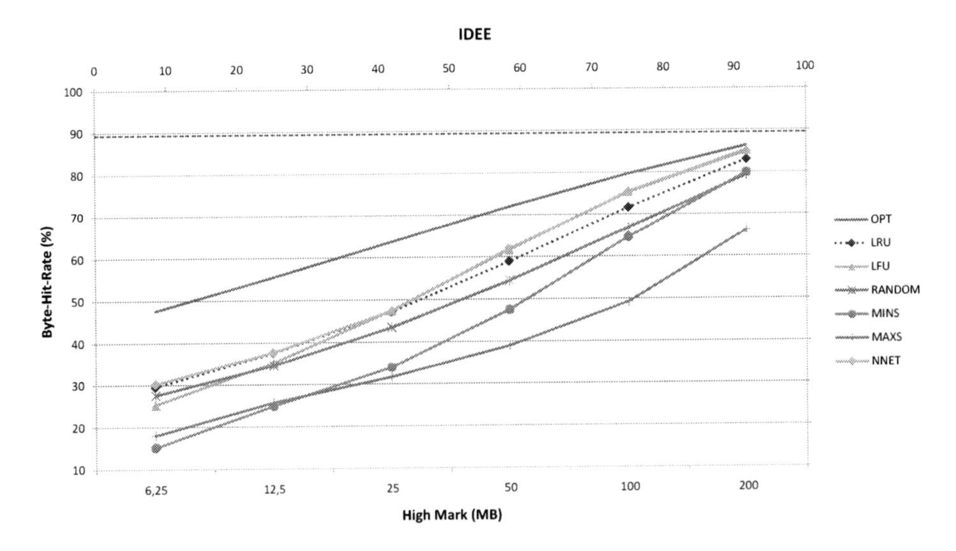

Fig. 8. Byte Hit Ratios for IDEE with different cache replacement policies. LM=0.9×HM.

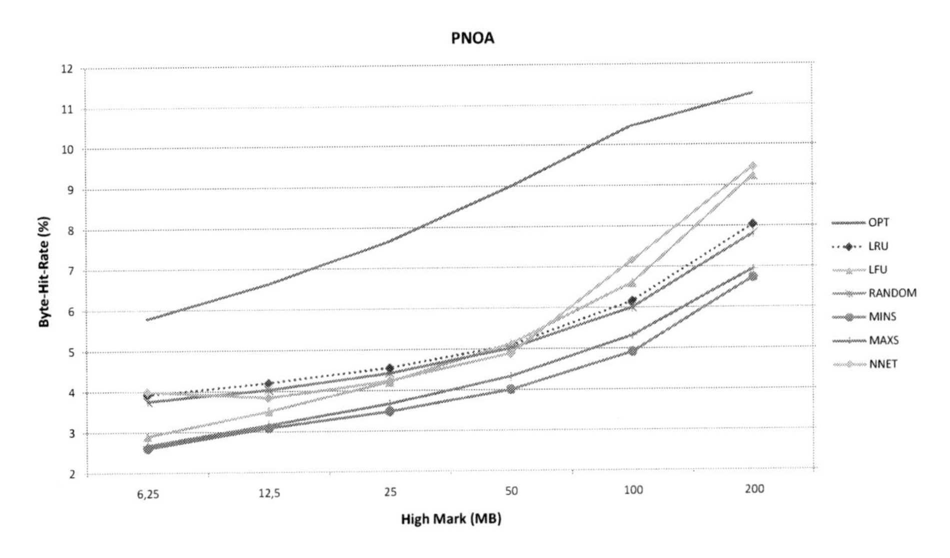

Fig. 9. Byte Hit Ratios for PNOA with different cache replacement policies. LM=0.9×HM.

7 Conclusions

Serving pre-generated map image tiles from a server-side cache has become a popular way of distributing map imagery on the Web. However, storage needs are often prohibitive which forces the use of partial caches. In this work, a cache replacement policy based on neural networks has been proposed. The network is a feed-forward multilayer perceptron with two-hidden layers. Inputs to the neural network are recency, frequency and size of the requested tiles, and it outputs a real value in the range [0,1] which can be interpreted as the probability of receiving a future request of the same object before a certain period of time. It has been trained with backpropagation through supervised learning using real-world trace requests from different map server logs. Trace data sets have been divided into three groups; one for training the neural network, another one for validating network generalization and a completely independent dataset used for testing. The results show that the proposed neural network is able to accurately classify cachability of requests in several scenarios from the real world with different workloads and request distributions. Simulations demonstrate that the proposed solution can achieve good performance, for both hit-rate and byte-hit-rate metrics, when running in a proxy.

Acknowledgments. This work has been partially supported by the Spanish Ministry of Science and Innovation through the project "España Virtual" (ref. CENIT 2008-1030), National Centre for Geographic Information (CNIG) and the National Geographic Institute of Spain (IGN).

References

1. de la Beaujardiere, J.: OpenGIS Web Map Server Implementation Specification, pp. 6–42. Open Geospatial Consortium Inc. (2006)
2. Open Geospatial Foundation. WMS-C wms tile caching - OSGeo wiki(2008) `http://wiki.osgeo.org/wiki/WMS_Tile_Caching`
3. Juliá, N., Masó, J., Pomakis, K.: OpenGIS Web Map Tile Service Implementation Standard, pp. 7–57. Open Geospatial Consortium Inc. (2010)
4. Podlipnig, S., Böszörmenyi, L.: A survey of web cache replacement strategies. ACM Computing Surveys (CSUR) 35(4), 374–398 (2003)
5. Aggarwal, C., Wolf, J.L., Yu, P.S.: Caching on the world wide web. IEEE Transactions on Knowledge and Data Engineering 11(1), 94–107 (1999)
6. Cao, P., Irani, S.: Cost-aware www proxy caching algorithms. In: Proceedings of the USENIX Symposium on Internet Technologies and Systems, pp. 18–18. USENIX Association, Berkeley, CA (1997)
7. Arlitt, M., Cherkasova, L., Dilley, J., Friedrich, R., Jin, T.: Evaluating content management techniques for web proxy caches. SIGMETRICS Perform. Eval. Rev. 27, 3–11 (March 2000)
8. Bahn, H., Koh, K., Noh, H., Lyul, S.M.: Efficient replacement of nonuniform objects in web caches. Computer 35(6), 65–73 (2002)
9. Tobler, W.R.: A computer movie simulating urban growth in the Detroit region. Economic geography 46, 234–240 (1970)
10. Khalid, H.: A new cache replacement scheme based on backpropagation neural networks. SIGARCH Comput. Archit. News 25, 27–33 (1997)
11. J.Pomerene, T.R. Puzak, R.Rechtschaffen, F.Sparacio.: Prefetching mechanism for a high-speed buffer store. *US patent*(1984)

12. Khalid, H., Obaidat, M.S.: Kora-2: a new cache replacement policy and its performance. In: The 6th IEEE International Conference on Electronics, Circuits and Systems,Proceedings of(ICECS1999) (1999)
13. Khalid, H.: Performance of the KORA-2 cache replacement scheme. ACM SIGARCH Computer Architecture News 25(4), 17–21 (1997)
14. Obaidat, M.S., Khalid, H.: Estimating neural networks-based algorithm for adaptive cache replacement. IEEE Transactions on Systems, Man, and Cybernetics, Part B: Cybernetics, 28(4), 602–611 (1998)
15. Khalid, H., Obaidat, M.S.: Application of neural networks to cache replacement. Neural Computing & Applications 8(3), 246–256 (1999)
16. Venketesh, P., Venkatesan, R.: A Survey on Applications of Neural Networks and Evolutionary Techniques in Web Caching. IETE Technical Review 26(3), 171–180 (2009)
17. ElAarag, H., Romano, S.: Training of nnpcr-2: An improved neural network proxy cache replacement strategy. In: International Symposium on Performance Evaluation of Computer Telecommunication Systems (SPECTS 2009), vol. 41, pp. 260–267 (2009)
18. Romano, S., ElAarag, H.: A neural network proxy cache replacement strategy and its implementation in the Squid proxy server. In: Neural Computing & Applications, pp. 1–20.
19. ElAarag, H., Cobb, J.: A Framework for using neural networks for web proxy cache replacement. Simulation Series 38(2), 389 (2006)
20. ElAarag, H., Romano, S.: Improvement of the neural network proxy cache replacement strategy. In: Proceedings of the Spring Simulation Multiconference, pp. 1–8. Society for Computer Simulation International (2009)
21. Tian, W., Choi, B., Phoha, V.: An Adaptive Web Cache Access Predictor Using Neural Network. Developments in Applied Artificial Intelligence, 113–117 (2002)
22. Cobb, J., ElAarag, H.: Web proxy cache replacement scheme based on back-propagation neural network. Journal of Systems and Software 81(9), 1539–1558 (2008)
23. El Khayari, R.A., Obaidat, M.S., Celik, S.: An Adaptive Neural Network-Based Method for WWW Proxy Caches. IAENG International Journal of Computer Science 36(1), 8–16 (2009)
24. MetaCarta Labs. Tilecache(2011), `http://tilecache.org/`
25. Quinn, S., Gahegan, M.: A predictive model for frequently viewed tiles in a web map. T. GIS 14(2), 193–216 (2010)
26. Tom Mitchell, M. (ed.): Machine Learning. McGraw-Hill, New York (1997)
27. Williams, S., Abrams, M., Standridge, C.R., Abdulla, G., Fox, E.A.: Removal policies in network caches for World-Wide Web documents. Computer Communication Review 26(4), 293–305 (1996)
28. Belady, L.A.: A study of replacement algorithms for a virtual-storage computer. IBM Systems Journal (1966)

Mapping the Anthropic Backfill of the Historical Center of Rome (Italy) by Using Intrinsic Random Functions of Order k (IRF-k)

Giancarlo Ciotoli, Francesco Stigliano, Fabrizio Marconi, Massimiliano Moscatelli, Marco Mancini, and Gian Paolo Cavinato

Consiglio Nazionale delle Ricerche - Istituto di Geologia Ambientale e Geoingegneria
Area della Ricerca di Roma 1, Via Salaria 29,300
00015 Monterotondo Stazione, (Italy) Roma
giancarlo.ciotoli@igag.cnr.it

Abstract. The historical centre of Rome is characterized by the presence of high thickness of anthropic cover with scarce geotechnical characteristics. This anthropic backfill could induce damages in urban areas, i.e. mainly differential settlements and seismic amplifications. About 1400 measurements from boreholes stored in the UrbiSIT database have been used to re-construct the anthropic backfill bottom surface by geostatistical techniques. The Intrinsic Random Functions of order k (IRF-k) was employed and compared with other interpolation methods (i.e. ordinary kriging and kriging with external drift) to determine the best spatial predictor. Furthermore, IRF-k allows to estimate by using an external drift as secondary information. The advantage of this method is that the modeling of the optimal generalized covariance is performed by using an automatic procedure avoiding the time-consuming modeling of the variogram. Furthermore, IRF-k allows the modeling of non stationary variables.

Keywords: backfill mapping, geostatistics, IRF-k, Rome (Italy).

1 Introduction

One major concern in land management and conservation planning in urban areas is the reduction of the risk due to differential settlements and seismic amplifications that can cause numerous damages to private and public buildings. At this regard the role played by anthropogenic deposits and artificial ground (i.e., "anthropic backfill units"), is of critical importance due to their poor geotechnical characteristics [1], [2], [3], [4]. On the subsoil modeling of urban areas, the spatial reconstruction of the urban backfill units is a very complex task and must be conducted by means of a detailed analysis of the subsurface data, mostly because of the lack of exposures [1], [5]. Differently from the natural sedimentary deposits, it is difficult to reconstruct a detailed internal stratigraphic architecture of the anthropic backfill, as they are the result of a lot of numerous mixed and discontinuous anthropic processes, such as excavation, backfilling, and remobilization of material. For these reasons such terrains

B. Murgante et al. (Eds.): ICCSA 2011, Part I, LNCS 6782, pp. 92–102, 2011.

are not generally easily mapped in urban areas. The need for reconstructing a spatial continuous backfill bottom surface plays a significant role in urban planning, risk assessment, and decision making in environmental management. Collected borehole data are at the base of this spatial reconstruction, and are typically from point sources. However, environmental managers often require spatial continuous data over the region of interest to make effective and confident decisions, and scientists need accurate spatial continuous data across a region to make justified interpretations.

Discrete spatial information from borehole could be used to estimate the values of an attribute at unsampled points by using different spatial interpolation methods. In particular, the application of a probabilistic model by using geostatistical techniques could help in the definition of the general 3D geometry of the backfill units, and in the reconstruction of its thickness and its bottom surface (with the latter corresponding to the boundary between the anthropic stratum and the geological substratum). This work presents an application of geostatistical techniques to elaborate subsurface data for the reconstruction of the bottom surface of the anthropic backfill unit in the historical centre of Rome. In particular, results obtained from ordinary kriging (OK) and kriging with external drift (KED) have been compared with the Intrinsic Random Functions of order k (IRF-k) to determine the best spatial predictor. This last method results to be the most suitable for mapping the anthropic backfill. The advantage of this method is that the modeling of the optimal generalized covariance is performed by using an automatic procedure avoiding the time-consuming modeling of the variogram. Furthermore, IRF-k allows the modeling of non stationary variables.

2 Material and Methods

Spatial interpolation methods are techniques that predict the value at a given location by using values of the same property sampled at scattered neighboring points [6], [7]. This allows the estimation of the attribute at any location within the data boundary under the assumption that the attribute is spatially dependent, indicating that the values closer together are more likely to be similar than the values farther apart. The most common interpolation techniques calculate the estimates for a property at any given location by a weighted average of nearby data. Weights are assigned either according to deterministic or probabilistic criteria. The goal is to create a surface to best represent empirical reality, but as a number of factors affect map quality (i.e., statistical distribution of the studied variable, sampling density, the applied interpolation method, etc.), the selected method must be assessed for accuracy.

All interpolation methods share the underlying assumption that sample points that are closer to the interpolated location will influence the interpolated value more strongly than sample points which are further away. A key difference among these approaches is the criterion which is used to weight the values of the sample points. Criteria may include simple distance relations (e.g., inverse distance methods), minimization of curvature, and enforcement of smoothness criteria (splining), minimization of variance (e.g., kriging and co-kriging) [8].

Estimations of nearly all spatial interpolation methods can be represented as weighted averages of sampled data, as follows:

$$\hat{z}(x_0) = \sum_{i=1}^{n} \lambda_i z(x_i) \quad (1)$$

where $\hat{z}$ is the estimated value of an attribute at the point of interest x_0, z is the observed value at the sampled point x_i, λ_i is the weight assigned to the sampled point, and n represents the number of sampled points used for the estimation [9].

The interpolation methods can be classified in two major groups depending on the nature of the function that is used to interpolate the values: deterministic and geostatistical methods. The first group uses mathematical formulas to estimate unmeasured values at any point across a given surface. The weight that is assigned to each known value for the interpolation of unknown values depends only on the distance between sample points and the location of the interpolated point. The second group of methods is called geostatistical or stochastic methods. Geostatistics uses kriging probabilistic interpolator, which is based on the theory of regionalized variables [6], [10], [11], [12]. Kriging is a generic name for a family of generalised least-squares regression algorithms. It is increasingly preferred because it allows the study of the spatial correlation between neighboring observations to predict attribute values at unsampled locations [12], [13], [14]. The application of geostatistical methods assumes that the spatial variation of the study variable is too irregular to be modeled by a continuous mathematical function, and it could be better predicted by a probabilistic surface. Variography has been used to describe the way in which similar observation values are clustered in space, in accordance with Tobler's first law of geography [15], giving the measure of the dissimilarity of data pairs as the spatial separation between them increases. Variogram is a graphical representation of spatial autocorrelation created by means of the semivariance (γ) of Z between two data points:

$$\gamma(h) = \frac{1}{2N(h)} \sum_{N(h)} \left[Z(x_i) - Z(x_{i+h})\right]^2 \quad (2)$$

where N is the number of pairs of sample points x(i) and $x(i+h)$ separated by distance h and $\gamma(h)$ is the semivariogram (commonly referred to as variogram) [9], [16]. Variogram modelling and estimation is extremely important for structural analysis and spatial interpolation [16].

Geostatistical interpolation includes different types of kriging-based techniques (i.e., ordinary kriging (OK), universal kriging (UK) and simple cokriging (CoK) often used for spatial analysis [6], [10], [17], [18]). All kriging estimators are variants of the basic equation (3), which is a slight modification of equation (1), as follows:

$$\hat{Z}(x_0) - \mu = \sum_{i=1}^{n} \lambda_i \left[Z(x_i) - \mu(x_0)\right] \quad (3)$$

where μ is a known stationary mean, assumed to be constant over the whole domain and calculated as the average of the data [19]; λ is the kriging weight; n is the number of sampled points used to make the estimation and depends on the size of the search window; and $\mu(x_0)$ is the mean of samples within the search window. The variety of available interpolation methods has led to questions about which is most appropriate in different contexts and has stimulated several comparative studies of relative accuracy. Cross-validation technique (CV) is often used to compare the performance of interpolation algorithms [20], [21]. Furthermore, all kriging interpolators provide a measure of the prediction error (i.e. kriging standard deviation map). However, a major advantage of kriging over simpler deterministic methods is that sparsely sampled observations of a primary attribute can be complemented by secondary attributes that are more densely sampled. Thus, the application of cokriging (CoK) and kriging with external drift (KED) seem to provide the best results. Furthermore, the notion of intrinsic random function of order *k* (abbreviated as IRF-k) constitutes a natural generalization of the intrinsic random functions (i.e., with stationary increments) of traditional geostatistics and was introduced to extend the scope of kriging to *non* stationary cases [22]. An IRF-k is simply a random function with stationary increments of order k; the usual intrinsic model of geostatistics corresponds to k = 0, while a SRF (Stationary Random Function) is intrinsic at all orders, so formally it would correspond to the case k = -1. The basic tool of structural analysis in the stationary case (i.e., variogram), in the IRF-k framework is substituted by a more general one: the generalized covariance function.

In this paper the backfill surface have been calculated by comparing OK and KED with the IRF-k techniques. Sample points (geological borehole data) are stored in the geodatabase organized by the Istituto di Geologia Ambientale e Geoingegneria del Consiglio Nazionale delle Ricerche (i.e. Environmental Geology and Geoengineering Institute of the National Research Council of Italy) for the Dipartimento di Protezione Civile Nazionale (i.e. Italian Civil Protection) in the framework of the UrbiSIT project (www.urbisit.it). As a good correlation exists between the backfill basal surface morphology and the present day topography of the historical center, digital elevation model (DEM) data at the spatial resolution of 20x20m were used as external drift variable (i.e., secondary attribute) for KED interpolation.

Data were described by means of conventional statistics (i.e., mean, maximum, minimum, median, standard deviation, etc.) by using the Statistica software package (StatSoft Inc, 2004); geostatistical analysis was performed using the ISATIS software package from Geovariances for the variogram modeling and the ArcGIS (ERSI Inc.) for the data management and the mapping reconstruction.

3 Results and Discussion

Data of the backfill bottom elevation measured in 1436 boreholes have been used to estimate the continuous basal surface of this anthropic cover. The wellheads are located at an elevation ranging between 12 and 88 m a.s.l. and are uniformly distributed in the study area, whereas the backfill bottom varies between -10 and 87 m a.s.l. Table 1 shows summary statistics of the anthropic backfill bottom measured at the available boreholes of the UrbiSIT database and the elevation data (as the related variable) known in the whole study area according to a 20 x 20 m DEM (Fig. 1).

For this reason the resulting maps have been constructed according to a 20 x 20 m grid of estimated values.

Table 1. Descriptive statistics for anthropic backfill bottom measured in the boreholes, as well as DEM values of the studied area

	N	Mean	Min	Max	Std.Dev.	CV
Backfill (m a.s.l.)	1436	7.6	0.3	38.5	4.88	64.5
Elevation (m a.s.l.)	19188	40.3	11.0	117.0	18.3	45.6

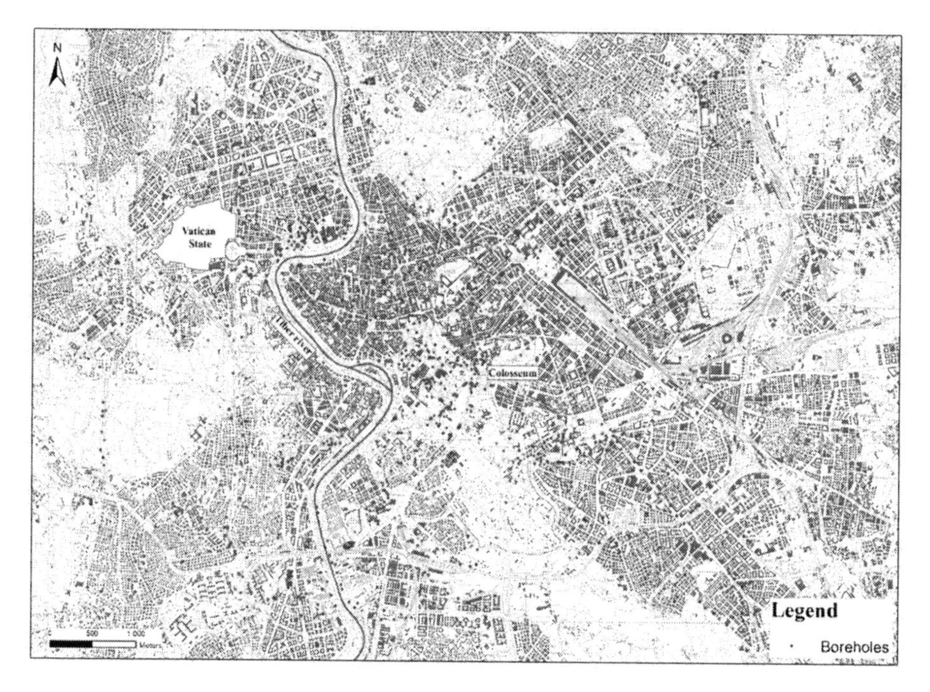

Fig. 1. Map of boreholes location in the study area

Statistical parameters suggest that both variables are not normally distributed according to Kolmogorov-Smirnov test ($p<0.01$). Standard deviations are below the mean value and do not strongly dispersed data.

The histograms of figure 2 show the presence of two populations one with a mean value of about 10 m a.s.l. and the other with a mean value of about 40m a.s.l. for backfill data, and 17 m and 45 m a.s.l. for the elevation data. This statistical behavior suggests that both variables are not stationary in the study area, and that there are trends in the spatial distribution of data values.

As a good correlation ($R^2= 0.928$) between the DEM values and the backfill bottom measured in the boreholes exists (Fig. 3), the modeling of the variogram was applied by using 5% of the total data (i.e. training data, 72 samples), whereas the remaining 95% of the data has been used as control points (test data, 1364 samples).

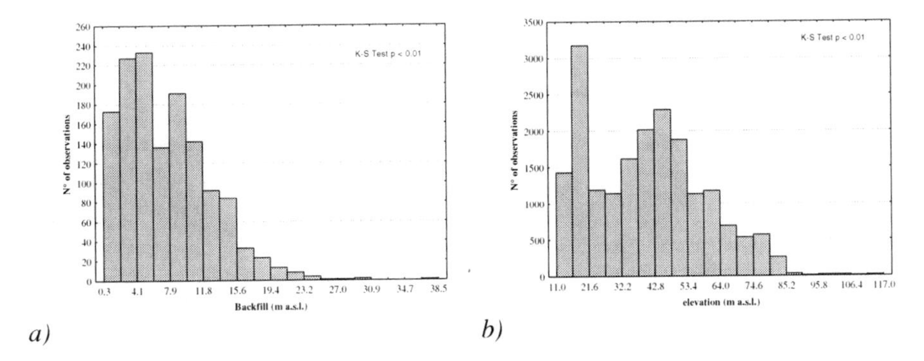

Fig. 2. Frequency histograms of the anthropic backfill values (*a*) observed at the borehole locations and elevation values considered as the covariable for the application of kriging with external drift (*b*)

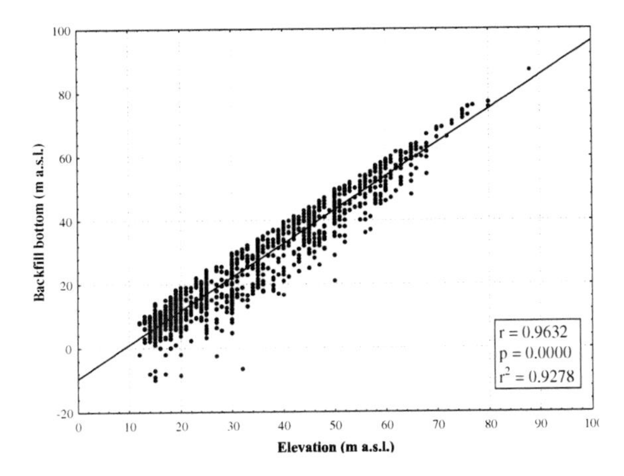

Fig. 3. Scatterplot between the elevation data obtained from the DEM versus the backfill bottom measured in the boreholes

The subset of data (5%) was selected to test the algorithms at conditions typical for urban areas with few scattered borehole data (about 1 borehole per km^2). As the goal of spatial interpolation is to create a surface that is intended to best represent empirical reality, the selected method must be assessed for accuracy. Different measures of fit were used to determine how well an interpolated map represents the observed data. The Root Mean Square Error (RMSE), calculated by the regression between the estimated values obtained by modeling of the training data and the set of the test data, has been used as a measure of the accuracy of the interpolation. The larger the value of the RMSE, the greater the difference between two sets of measurements of the same phenomenon. Its widespread use can be attributed to the relative ease of calculation and reporting and to the ease with which the concept can be understood by most users. The equation of the RMSE is (4):

$$RMSE = \frac{1}{n}\sqrt{\sum_{i=1}^{n}\left[\hat{Z}(s_i) - z(s_i)\right]^2} \quad (4)$$

where $\hat{Z}$ (s_i) is the predicted value at the location i, z(s_i) is the observed value at i location and n is the sample size. Furthermore, the regression coefficient (R^2) between predicted and observed data (test data), as well as the t-test statistics for the significance of the simple linear regression have been calculated (Tab. 2).

The backfill bottom surface was calculated by using different geostatistical interpolators: OK, KED, IRF-k and IRF-k ED. The kriging weights came from a semi-variogram that was developed by looking at the spatial structure of the data.

Table 2. Statistical parameters of the different kriging interpolators

Interpolation Technique	Mean Standard Error	RMSE	Error Variance	R^2	t-test $p(\alpha=0,05)$
OK	-0.35	0.31	1.31	0.792	0.000
KED	0.07	0.14	0.32	0.919	0.000
FAI-k	-0.31	0.48	0.8	0.345	0.000
FAI-k ED	-0.12	0.13	1.06	0.928	0.000

The experimental variograms calculated along four directions do not show the presence of an anisotropy in the spatial distribution of data (Fig. 4a). Therefore, the predictions of the estimated values were based on the semi-variogram and on the spatial arrangement of nearby measured values. The spatial variation depicted by the semivariogram models is shown in figure 4b.

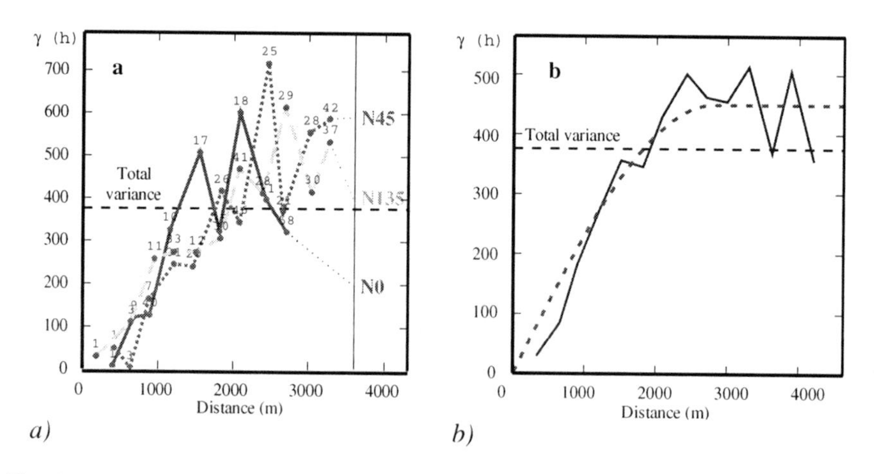

Fig. 4. Experimental variograms (a) calculated along specific directions to investigate the presence of anisotropy. Data do not highlight an anisotropic behavior of the anthropic backfill values in the study area. Figure b shows the experimental isotropic variogram (continuous line), as well as the model (dotted line) used in the kriging algorithm: γ(h) = 449.9 Spherical (2800 m).

The estimated surface obtained by using OK and IRF-k provides cross-validation results with a regression coefficient (R^2) of 0.792 and a RMSE of 0.31 m, as well as a regression coefficient (R^2) of 0.345 and a RMSE of 0.48 m, respectively.

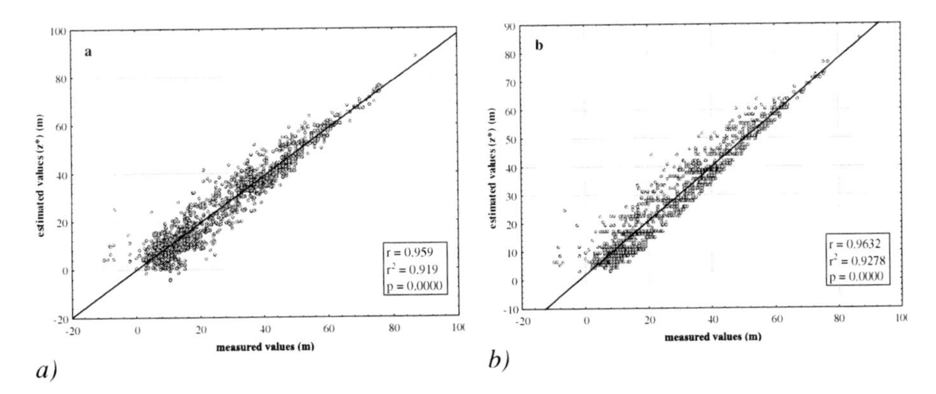

a) *b)*

Fig. 5. Cross-validation results of KED (a) and FAI-k with ED (b)

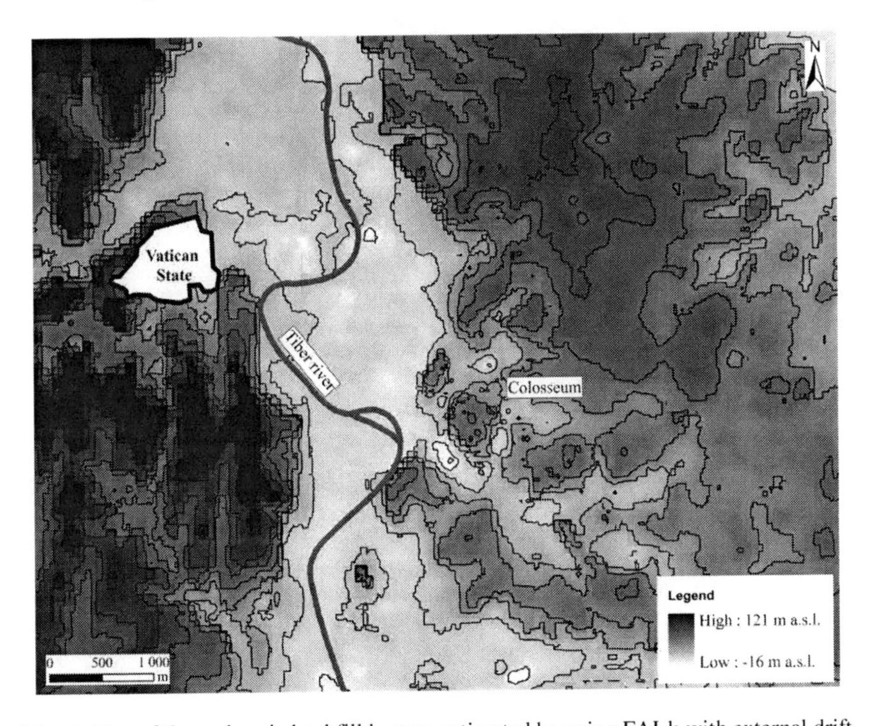

Fig. 6. Map of the anthropic backfill bottom estimated by using FAI-k with external drift

The good correlation between the backfill bottom surface and the present day topography of the historical center of Rome (Fig. 3) allows the use of Digital Elevation Model (DEM) data as the secondary variable to derive the local mean of backfill bottom (primary variable). By adding into the ordinary kriging system the additional conditions provided by the external drift variable (i.e., DEM), the estimation of the backfill-bottom surface has been improved by using KED. Cross-validation results highlights a regression coefficient (R^2) of 0.919 (Fig. 5). A further attempt by using DEM as ED in the IRF-k algorithm provides the best regression coefficient ($R^2 = 0.927$), as well as the best RMSE (0.13 m).

Table 2 shows the comparison among some statistical parameters calculated for the different interpolators. In particular, the best result in terms of RMSE and regression coefficient is provided by the application of the IRF-k ED technique. The method was then used to produce a map of the backfill bottom surface (Fig. 6). A thickness map of the anthropic backfill was finally calculated by subtracting the interpolated surface from DEM (Fig. 7).

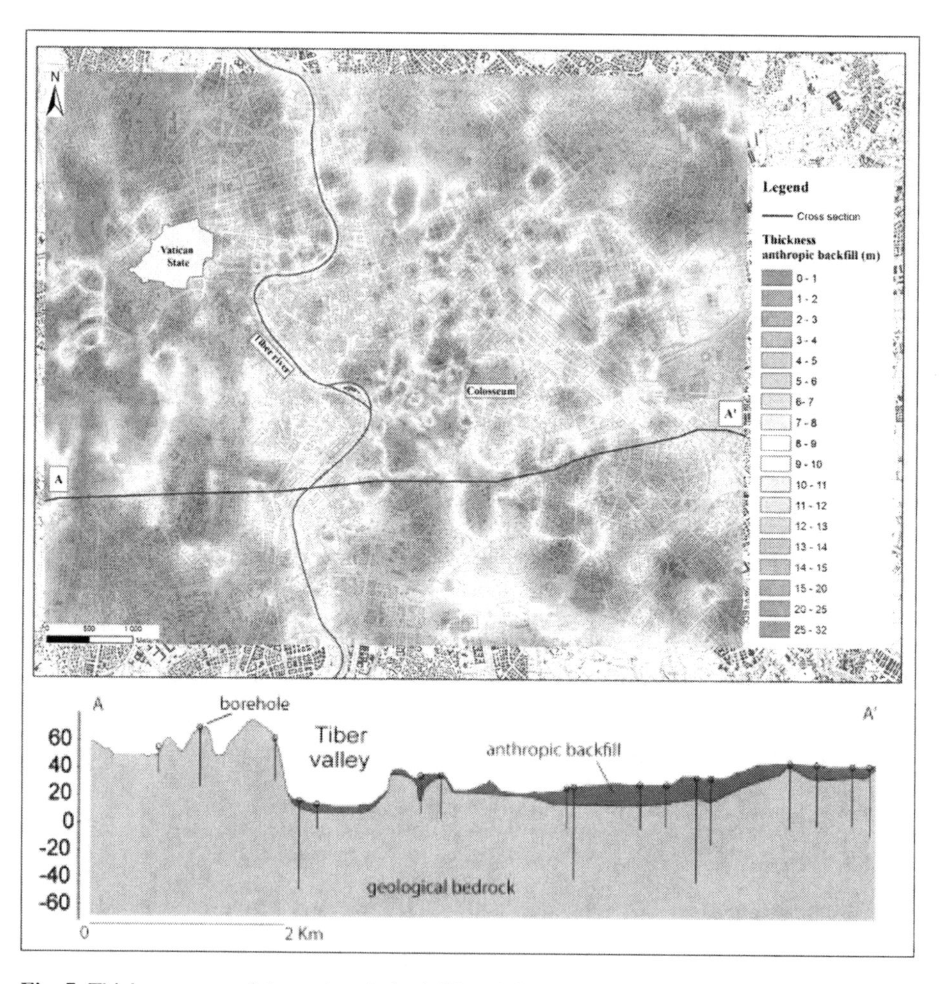

Fig. 7. Thickness map of the anthropic backfill and the cross section shows a comparison with the results estimated by using Kriging with external drift

4 Conclusions

Results from the application of different geostatistical algorithms provided some insights in terms of strength and weakness, and in term of their applicability to the reconstruction of geological surfaces.

This process is very important because there is little evidence that any one method is optimal across the range of conditions. This study suggests that KED seems to give the most coherent results in accordance with cross-validation statistics with respect of OK. Due to the high correlation of the anthropic backfill bottom with topography, it seems logical that the integration of topographic information should improve backfill estimates. KED has also the advantage of requiring a less demanding variogram analysis than other techniques as cokriging for example.

However, when the studied variable is not stationary in its spatial domain, the application of a numerical technique based on the theory of Intrinsic Random Functions of order k (IRF-k) provides further improvements of the estimated surface due to the construction of geo-statistical generalized estimators composed by two functions: a random portion, and a second deterministic portion containing the spatial trend of the non-stationary parameter. The application of IRF-k with ED leads to a smaller RMSE of the estimates and to a better regression coefficient of the cross-validation results.

References

1. Rosenbaum, M.S., Mc Millan, A.A., Powell, J.H., Culshaw, M.G., Northmore, K.J.: Classification of artificial (man-made) ground. Engineering Geology 69, 399–409 (2003)
2. Makedon, T., Chatzigogos, N.P., Spandos, S.: Engineering geological parameters affecting the response of Thessaloniki's urban fill to a major seismic event. Engineering Geology 104, 167–180 (2009)
3. Katz, O., Crouvi, O.: The geotechnical effects of long human habitation (2000<years): Earthquake induced landslide hazard in the city of Zefat, northern Israel. Engineering Geology 95, 57–78 (2007)
4. Danese, M., Lazzari, M., Murgante, B.: Kernel Density Estimation Methods for a Geostatistical Approach in Seismic Risk Analysis: The Case Study of Potenza Hilltop Town (Southern Italy). In: Gervasi, O., et al. (eds.) ICCSA 2008, Part I. LNCS, vol. 5072, pp. 415–429. Springer, Heidelberg (2008)
5. Bouldreault, J.P., Dubé, J.S., Chouteau, M., Winiarski, T., Hardy, E.: Geophysical characterisation of contamineted urban fills. Engineering Geology 116, 196–206 (2010)
6. Journel, A.G., Huijbregts, C.J.: Mining Geostatistics. Academic Press, New York (1978)
7. Jones, T.J., Hamilton, D.E., Johnson, C.R.: Contouring Geologic Surfaces with the Computer, p. 314. Van Nostrand Reinhold, New York (1986)
8. de Ks , Beurs., A, Stein., Hartkamp, A.D., White, J.W.: Interpolation techniques for climate variables. In: NRG-GIS Series, pp. 91–99. CIMMYT, Mexico (1999b)
9. Matheron, G.: Principles of geostatistics. Economic Geology 58, 1246–1266 (1963)
10. Webster, R., Oliver, M.A.: Geostatistics from Environmental Scientist, p. 271. Wiley & Sons, Chichester (2007)
11. Goovaerts, P.: Geostatistics for Natural Resources Evaluation, p. 483. Oxford press, Oxford (1997)

12. Goovaerts, P.: Geostatistical approaches for incorporating elevation into the spatial interpolation of rainfall. Journal of Hydrology 228, 113–129 (2000)
13. Weber, D.D., Englund, E.J.: Evaluation and comparison of spatial interpolators II. Mathematical Geology 26, 589–603 (1994)
14. Zimmerman, D., Pavlik, C., Ruggles, A., Armstrong, P.: An experimental comparison of ordinary and universal kriging and inverse distance weighting. Mathematical Geology 31, 375–390 (1999)
15. Tobler, W.R.: A Computer Movie: Simulation of Population Change in the Detroit Region. Economic Geography 46, 234–240 (1970)
16. Burrough, P.A., McDonnell, R.: A Principals of Geographical Information Systems. Oxford University Press, Oxford (1998)
17. Deutsch, C.V.: Geostatistical Reservoir Modeling. Oxford University Press, Oxford (2002)
18. Cressie, N.: The origins of kriging. Mathematical Geology 22, 239–252 (1990)
19. Wackernagel: Multivariate Geostatistics: An Introduction with Applications. Springer,GmbH & Co., Berlin,Heidelberg (2003)
20. Isaaks, E.J., Srivastava, R.M.: An introduction to Applied Geostatistics, p. 561. Oxford University Press, Oxford (1989)
21. Davis, B.M.: Uses and abuses of cross-validation in geostatistics. Mathematical Geology 19, 241–248 (1987)
22. Myers, D.E.: To be or not to be.stationary? That is the question. Mathematical Geology 21, 347–362 (1989)

Spatio-Temporal Analysis Using Urban-Rural Gradient Modelling and Landscape Metrics

Marco Vizzari

Department of Man and Territory, University of Perugia,
Borgo XX Giugno 74, 06121 Perugia
marco.vizzari@unipg.it

Abstract. Urbanization can be considered as a particular environmental gradient that produces modifications in the structures and functions of ecological systems. In landscape analysis and planning there is a clear need to develop specific and comparable indicators permitting the spatio-temporal quantification of this gradient and the study of its relationships with the composition and configuration of other land uses. This study, integrating urban gradient modelling and landscape pattern analysis, aims to investigate the spatiotemporal changes induced by urbanization and by other anthropogenic factors. Unlike previous studies, based on the transect approach, landscape metrics are calculated diachronically within five contiguous zones defined along the urban to rural gradient and characterized by decreasing intervals of settlement density. The results show that, within the study area, urban sprawl and agricultural land simplification remain the dominant forces responsible for the landscape modifications that have occurred during the period under investigation.

Keywords: urban-rural gradient, urban spatial modelling, urban fringe, agricultural landscapes, landscape metrics, kernel density analysis, GIS.

1 Introduction

The view of landscapes as continua and spatial gradients represents a challenge to the conventional view of how the natural (and human) environment is organized [1]. Urbanization can be considered as a particular environmental gradient that produces modifications in the structures and functions of ecological systems [2, 3] with a magnitude that depends on the steepness of the gradient itself [4]. Along this gradient, urban fringes represent spaces with undefined boundaries [5] inside of which transitions and changes in equilibrium and relationships can be observed [6, 7]. These contexts show a particular vulnerability towards urban sprawl and structural and functional modifications occurring in rural spaces [8-13]. As a consequence, proper landscape analysis, planning and management of peri-urban interfaces become crucial not only for the quality of life of those living in such areas, but also for the entire sustainability of urban and rural development [14]. In this context GIS spatial analysis and modelling can support the definition and calculation of continuous indicators

B. Murgante et al. (Eds.): ICCSA 2011, Part I, LNCS 6782, pp. 103–118, 2011.

allowing better assessment and interpretation of the gradients characterizing landscape [15]. These capabilities improve the understanding of the specific characteristics of sites and the nature of the interactions between human and natural actions in landscape configuration [16].

One effective method for analysing the effects of urbanization on ecosystems is studying the changes of ecosystem patterns and processes along the urban to rural gradient [3]. The quantification of spatial heterogeneity is necessary in order to explore relationships between ecological processes and landscape spatial patterns [17]. To this end, a great variety of metrics for analysing landscape composition and configuration have been developed for the analysis of categorical data [18]. These metrics can be calculated efficiently using specific software packages such as FRAGSTATS [19], and, despite all their conceptual flaws, risks of improper use, and known limitations [20], they are widely used in quantitative landscape ecology studies for characterizing and describing landscape structure.

Within this framework, the transect method has been widely used for the analysis of urban to rural gradient [2, 21-24] also for spatiotemporal pattern detection [23, 25], but has been criticized due to its simplistic spatial approach since most ecological gradients are complex and involve several contrasting variables [4]. Urbanization certainly represents a dominant driving force in land use changes. It results in erosion and progressive modification of the surrounding land uses and, from this point of view, urban fringes can be considered as a moving friction area, to varying extent, in which these processes are potentially stronger. Thus the comparison of metrics referred to the same extents (as conducted, by example, in transect methods), even if it allows a simpler and more intuitive interpretation of changes within sites, results in insufficient understanding of the evolution of landscape configuration and, specifically, of the spatial structure of urban fringes. Indeed, the traditional methods tend to compare multi-temporal data that are spatially coincident but often functionally and ecologically incomparable.

This study intends to develop a methodology, based on urban gradient modelling and landscape pattern analysis, aimed at the investigation of spatiotemporal changes induced by urbanization and other anthropogenic factors. In particular, the method proposed in this study, applied to Umbrian landscapes around Perugia (Italy) within the period 1977 – 2000, attempts to overcome the transect method limitations by analysing metrics referring to different spatial extents but located at the same "functional position" along the urban-rural gradient. The purpose is a more effective analysis of the structural transformations along the urban-rural gradient, and particularly within urban fringes, and a deeper comprehension of the spatial and functional relationships between urban intensity and composition-configuration of the other land uses.

2 Methods

The study area includes the Umbrian municipalities of Perugia, Corciano, Torgiano and Deruta which encompass an urban and productive tissue of high territorial continuity around the city of Perugia [26] (Fig. 1). Land use data, retrieved from the Region of Umbria Land Information System at a scale of 1:10000 for the years 1977 and 2000, was used for the analysis of landscape transformations. Currently, the two datasets, are the only detailed scale resources available for the land use of the area under investigation. The built-up class contained in the land use data was subjected to extensive processing based on morphological analysis methods aimed at segmenting binary patterns of settlements into mutually exclusive categories: core, islet, loop, bridge, perforation, edge, and branch [27] (Fig. 2). The pixels classified as bridge, edge and loop categories were eliminated because of their coincidence with road infrastructures. In the other categories, using the orthophoto as background, a subsequent visual selection of settlements was conducted (urban centres, commercial and production areas, inhabited nuclei and dispersed settlements) within other areas, classified as "built-up" in the 1977 and 2000 land use data.

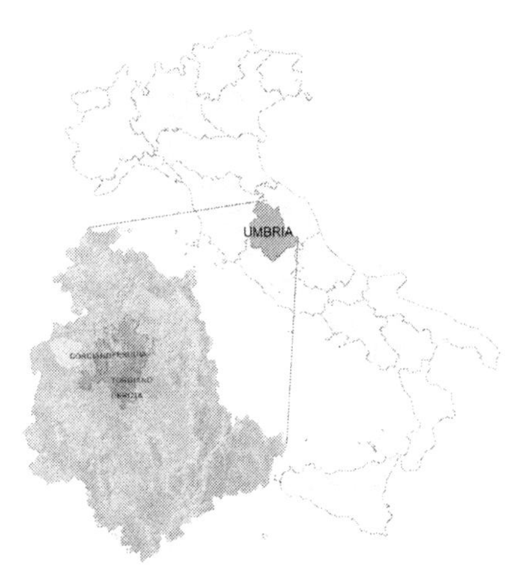

Fig. 1. Location of the area of interest

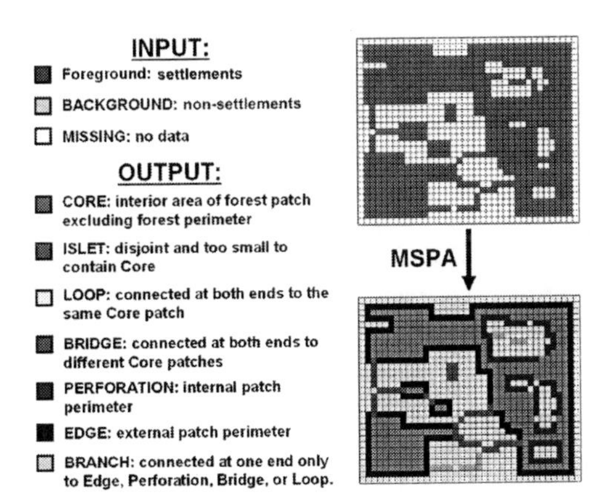

Fig. 2. Output of MSPA procedures (adopted from Soille & Vogt, 2009)

Using GIS density analysis techniques, a spatial index was calculated measuring the density of areas occupied by settlements. GIS density analysis takes measured quantities associated with point data and spreads them across the landscape to produce a continuous surface [28]. Unlike simple density, kernel density estimation (KDE) produces smoother surfaces, more representative of landscape gradients [15]. In KDE a moving window or kernel function is superimposed over a grid of locations and the (distance-weighted) density of point events is estimated at each location, with the degree of smoothing controlled by the kernel bandwidth [29]. The first contribution on KDE is attributed to Fix and Hodges which introduced several basic concepts for the development of nonparametric density estimation [30]. Starting from this pioneer study, many improvements of this smoothing technique, particularly within spatial analysis applications, have been developed by several authors [31 and reference therein]. The application of KDE requires the choice of the type of kernel function (e.g.: Gaussian, triangular, quartic) and the definition of three key parameters: bandwidth, cell size and intensity. Generally bandwidth definition represents the most problematic step, but also the most useful for exploratory purposes, since a wider radius shows a more general trend over the study area, smoothing the spatial variation of the phenomenon, while a narrower radius highlights more localized effects such as 'peaks and troughs' in the distribution [32, 33]. Two main approaches to determinate bandwidth can be found in the literature: the first, more frequently utilized, uses a fixed bandwidth to analyse the entire distribution, while the second implements a local adaptive bandwidth. Jones et al. [33] conducted a comparison of the main methods for the choice of fixed bandwidth, grouping them in two generations. Despite all these approaches the examination of resulted surfaces for different values of bandwidth remains a common method supporting the definition of this parameter [29, 34].

In this study bandwidth selection was not particularly problematic since it was defined with the aim of obtaining a sufficiently generalized surface, for both years, satisfactory for subsequent landscape subdivision and diachronic metrics comparison. After different tests (bandwidths of 500, 750, 1000, 1250, 1500), a bandwidth of 1000 m was adopted in the analysis of settlement density at the working scale. Prior to calculating density, polygonal landscape elements were converted to a mesh of points spaced at intervals equal to the size of the cells used for density analysis. This has allowed a proper resolution to be set, maintaining adequate polygonal shape detail in accordance with the scale of the analysis. KDE was performed using a quartic function, a cell size of 50 m and the unit intensity associated with every point. The calculated spatial index (hereinafter referred to as SDI – Settlement Density Index) expresses settlement concentration as the km^2 of surface occupied by settlements over the km^2 of the territorial surface. It gives a simplified, but reliable, representation of the gradient of urbanization: it assumes maximum values in the central portions of more extensive urban areas, decreasing progressively as we move towards rural areas (Fig. 3).

For the two years under investigation, five urban zones have been defined, each one characterized by a specific SDI interval (Fig. 4). The five zones assume a different spatial extent in the two time periods, but are characterized, as specified, by the same settlement cover intervals (Fig. 5). Despite their different spatial configuration, according to the objectives of this study, these territorial contexts were considered ecologically comparable.

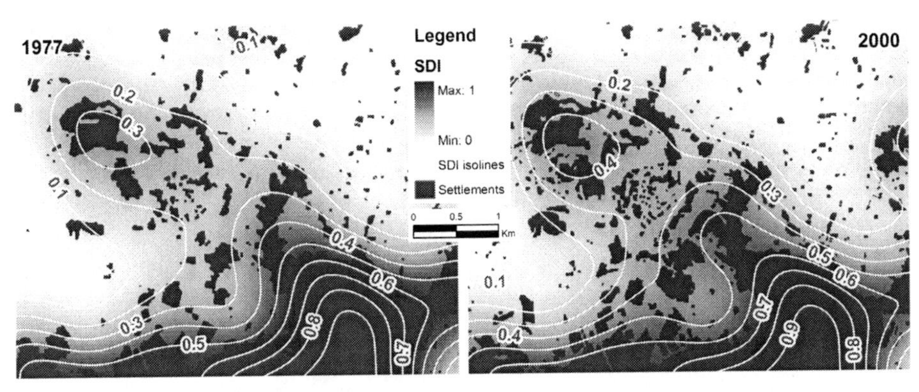

Fig. 3. SDI of years 1977 - 2000 and relative isolines. Detail in the area to the north of Perugia.

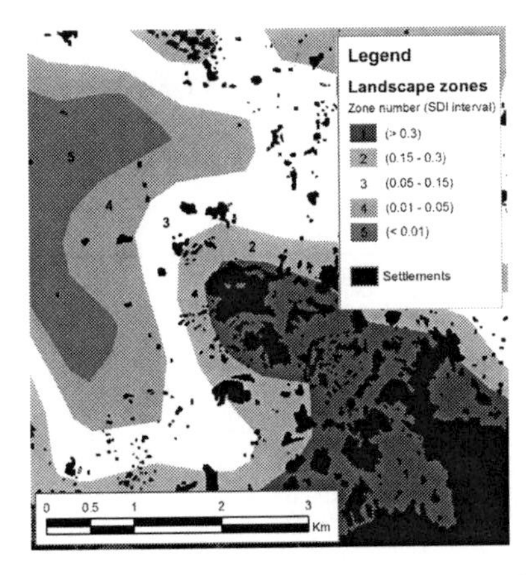

Fig. 4. Landscape subdivision and relative SDI intervals (year 2000)

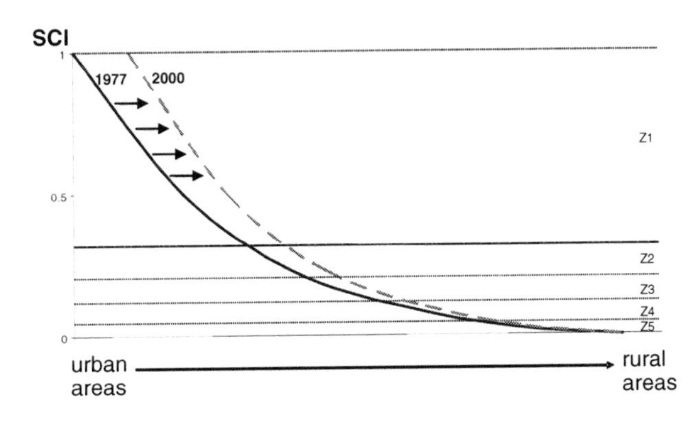

Fig. 5. Spatial shifting of the five landscape zones *(z1 – z5)* along a hypothetical section of the urban-rural gradient

Land-use data for the two time periods, originally available in vector format, was converted to raster format, with a 10 m cell size, to be processed by FRAGSTATS 3.3 [19]. As indicated by many authors [18, 35-37] the results of this type of analysis are conditioned by the input pixel size, especially in an urban environment where the landscape is highly dissected by linear features [38]. Thus, after a preliminary test of the effects of scale on metric analysis, a pixel size of 10 m was chosen since it allowed proper representation of the features and calculation of the landscape metrics, also in consideration of the scale chosen for analysis [39].

Landscape metrics calculation was executed, for the two periods in analysis, on the entire landscape and through all of the five SDI zones. A specific set of landscape metrics was used for this analysis: percentage of landscape (PLAND), patch density (PD), mean patch size (MPS), landscape shape index (LSI) and Shannon's diversity index (SHDI) (Tab. 1). All computable by FRAGSTATS, these metrics appear very reliable in the analysis of urban-rural gradient and give results comparable with other studies concerning the urban to rural gradient. The results of this analysis are two chronological sets of landscape pattern data, at landscape and at class level, along the urban-rural gradient described by the five zones with decreasing SDI intervals.

Table 1. Landscape metrics used for landscape characterization *(*L = Landscape, C = Class)*

Abbreviation	Metric	Description	Range	Level*
PLAND	Percentage of landscape	The proportion of total area occupied by a specific land use class	$0 < PLAND \leq 100$	C
PD	Patch density	Number of patches on a 100 hectares area basis.	$PD > 1$	L, C
MPS	Mean patch size	Average area of patches in a landscape	$MPS > 0$	L, C
LSI	Landscape shape index	Standardized measure of total edge or edge density. Gives a measure of patch shape and indirectly of aggregation or disaggregation.	$LSI \geq 1$	L, C
SHDI	Shannon's diversity index	Measure of diversity in landscape.	$0 \leq SHDI < 1$	L

3 Results

Within the entire study area, during the period under investigation, several significant landscape modifications have occurred (Tab. 2 and 3), such as:

- Consistent increase in built-up areas, mostly to the detriment of peri-urban agricultural areas occupied by sowable land with trees, vineyards, olive groves, ordinary sowable lands.

- Conversion of sowable lands with trees into ordinary sowable lands, mainly due to progressive simplification of the agricultural systems found in Umbria since the late 1950s;
- Significant decrease in vineyards in favour of sowable lands mostly due to changes in European Common Agricultural Policy (CAP);
- Decrease in olive groves due to their progressive transformation into woodland and pasture (especially in steeper area) and their conversion into simple croplands;
- Expansion of woodlands mainly in steeper areas occupied by pastures.

Table 2. Main modifications in land use composition between 1977 and 2000

Land use	Area 1977 *(ha)*	Area 2000 *(ha)*	Variation *(ha)*	Variation *(%)*
Built-up	5616.06	6817.71	1201.65	21.4
Sowable lands	24410.58	28271.23	3860.65	15.8
Woodlands	12975.97	13536.63	560.66	4.3
Pastures	3869.18	3901.54	32.36	0.8
Olive groves	3614.12	3390.66	-223.46	-6.2
Vineyards	2789.45	1336.46	-1452.99	-52.1
Sowable lands with trees	5338.68	1327.03	-4011.65	-75.1

Table 3. Changes in land use, as percentage area, from year 1977 to year 2000

From \ To	Built-up	Olive groves	Vineyards	Sowable lands with trees	Sowable lands	Pastures	Woodlands
Built-up	99.90	0.00	0.01	0.01	0.03	0.02	0.01
Olive groves	3.59	79.23	0.38	1.64	7.59	3.89	3.55
Vineyards	3.64	1.82	37.60	2.83	51.07	0.84	2.04
Sowable lands with trees	5.27	3.97	1.00	16.99	63.44	4.72	3.97
Sowable lands	2.60	0.68	0.84	0.95	91.20	1.28	1.99
Pastures	0.80	1.35	0.11	0.64	12.65	69.76	14.34
Woodlands	0.10	0.35	0.07	0.19	2.38	3.54	92.98

Outputs from FRAGSTATS in relation to the entire study area show a general increase of PD and a reduction of MPS, LSI and SHDI (Tab. 4). These trends indicate a modest increment of patch fragmentation, a little reduction in shape complexity and a consistent reduction in landscape diversity.

The landscape metrics calculated at class level allow a better comprehension of the overall structural transformations occurring for the different land uses (Tab. 5). The higher variations in PD, MPS and LSI are associated with woodlands, built-up areas, vineyards and sowable lands (with and without trees). In some cases, the significant reduction in these metrics is due to the decrease in PLAND of sowable lands with trees and vineyards. The increase for woodlands PD, MPS and LSI (and of the relative value of PLAND) clearly indicates an increase in fragmentation for these land use patches. On the other hand, the increases in the same metrics relating to built-up areas clearly indicate a consistent and diffuse occurrence of urban sprawl, while class metrics calculated for ordinary sowable lands and pastures show a tendential aggregation and simplification of these patches. The modifications of urban gradients, the consequent "urban to rural intrusion" and the "spatial shift" of urban fringes, can be observed by comparing the five equal interval SDI zones for both years (Fig. 6 and 7).

Table 4. Variation of main landscape metrics within the entire study area

Landscape metric	1977	2000	Variation	Variation *(%)*
PD	24.47	26.81	2.34	8.7
MPS	4.09	3.73	-0.36	-9.6
LSI	95.13	93.13	-2.00	-2.1
SHDI	1.69	1.51	-0.18	-12.1

Table 5. Variation of main class metrics within the entire study area

Landscape metric	Class	1977	2000	Variation	Variation *(%)*
PD	Woodlands	1.94	6.67	4.73	70.9
	Built-up	6.04	8.11	2.07	25.5
	Olive groves	2.16	2.10	-0.06	-2.9
	Pastures	2.28	2.37	0.09	3.8
	Sowable lands with trees	3.79	1.37	-2.42	-176.6
	Sowable lands	3.07	2.94	-0.13	-4.4
	Vineyards	4.36	2.54	-1.82	-71.7
MPS	Woodlands	11.28	3.43	-7.85	-228.9
	Built-up	1.56	1.41	-0.15	-10.6
	Olive groves	2.84	2.74	-0.10	-3.6
	Pastures	2.87	2.78	-0.09	-3.2
	Sowable lands with trees	2.37	1.65	-0.72	-43.6
	Sowable lands	13.43	16.29	2.86	17.6
	Vineyards	1.08	0.89	-0.19	-21.3
LSI	Woodlands	74.68	92.32	17.64	19.1
	Built-up	94.82	101.33	6.51	6.4
	Olive groves	57.05	59.67	2.62	4.4
	Pastures	62.30	61.46	-0.84	-1.4
	Sowable lands with trees	78.98	48.92	-30.06	-61.4
	Sowable lands	80.97	82.39	1.42	1.7
	Vineyards	64.71	48.32	-16.39	-33.9
PD	Woodlands	1.94	6.67	4.73	70.9

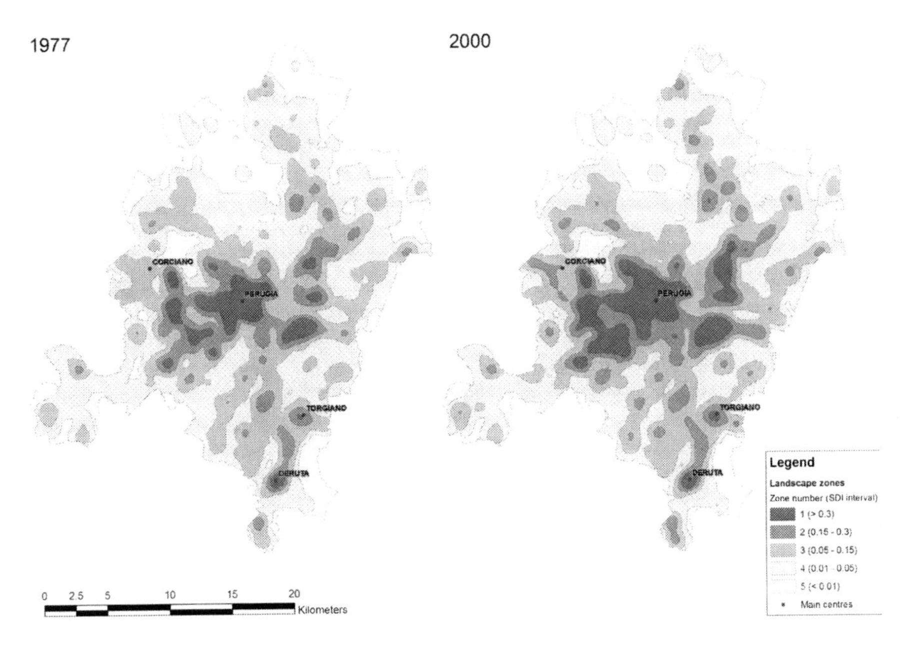

Fig. 6. SDI landscape zones, years 1977 *(left)* and 2000 *(right)*

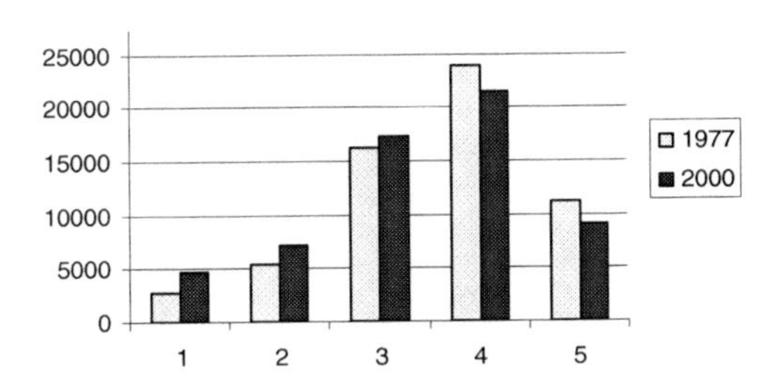

Fig. 7. Area variations *(hectares)* for the five SDI zones between 1977 and 2000

Metrics calculated at landscape level, for both years, within the five zones show the overall pattern of transformations occurring along the urban to rural gradient (Fig. 8). In the interpretation of the results, it is important to consider that the five landscape zones assume a slight different spatial extent in the two years considered, but maintain the same settlement coverage, expressed through the SDI index. PD and MPS curves indicate a growth in fragmentation moving to z2 and a decreasing patch size moving towards rural areas. Both curves, on zone 2, assume a relative maximum and a relative minimum respectively indicating that this zone is the most fragmented. For the year 2000, fragmentation appears higher in urban and near peri-urban areas (z1 and z2), but decreases more rapidly moving towards rural areas. LSI curves also reflect the increase in disaggregation and shape complexity occurring from 1977 to 2000 in urban and fringe areas and the simplification of land uses in outer agricultural lands. Both SDHI curves assume a maximum in z2 indicating a maximum level of landscape diversity in z2, which is, as already mentioned, the most fragmented zone. During the period under investigation, SDHI values decreased considerably from z2 to z5 (with a maximum decrease in z3) indicating a reliable loss of landscape diversity along the whole urban to rural gradient.

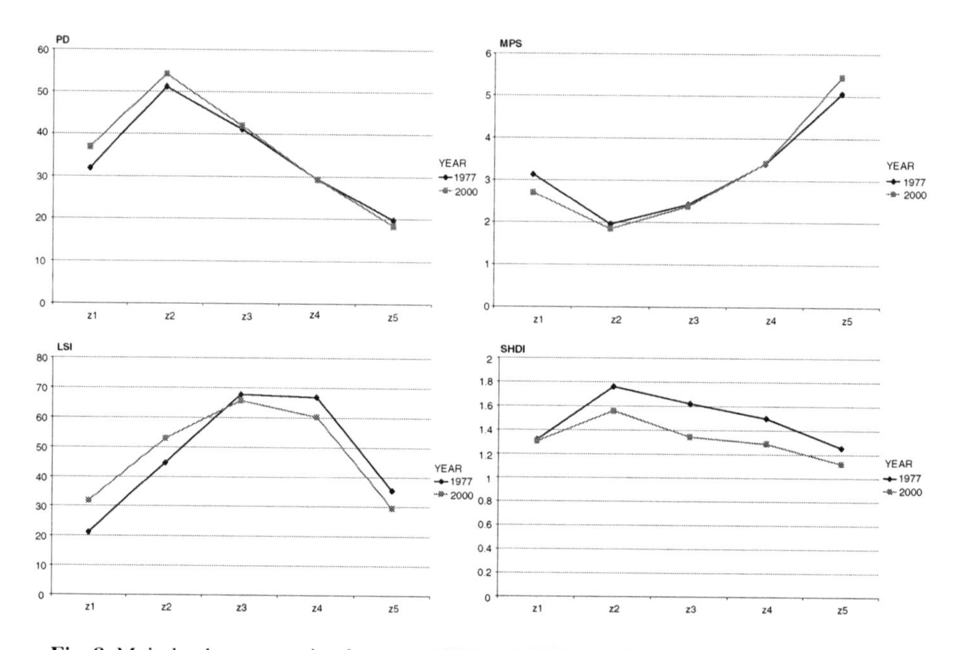

Fig. 8. Main landscape metrics for years 1977 and 2000 calculated along the five SDI zones

The land use transformations shown above can be analysed in detail by exploring the results at the level of the main land use classes. During the period under consideration, the decreasing PLAND of built-up areas remains apparently stable along the urban-rural gradient (Fig. 9). This unusual trend clearly depends on the methodology adopted in this study and in particular on the definition of landscape zones that, as described, were based on diachronically equal SDI intervals. On the other hand, PLAND curves of sowable lands with and without trees diachronically show their relative incidence on land uses along the gradient and reveal the progressive eradication of trees (in most cases vines trained up elm trees) within agricultural fields. In addition, the curves assume a typical agricultural trend along the gradient, indirectly indicating the agriculture intensity, increasing until reaching zone 3 (the most intensive agricultural area) and decreasing as we move towards z5 (the most natural area). The PLAND curve of olive groves shows an equally-sized, small reduction along the gradient, while the curve of vineyards reveals that the decline of this land use type increases until reaching z3 and then decreases as we move towards z5. This tendency confirms how the reduction in vineyards, observed between 1977 and 2000, concerns the most specialized, but very often old cultivations, located mainly in agricultural areas of the plain. These cultivations have progressively been converted into sowable lands as a result of them being economically unsustainable due to their lack of compatibility with mechanization and the introduction of restrictions in the European CAP. As expected, the PLAND diachronic curves for pastures and woodlands show increasing incidence moving from urban to rural areas. The curves also indicate a slight increase in pastures in z3 and z4 and a progressive increase in woodlands moving from z2 to z5.

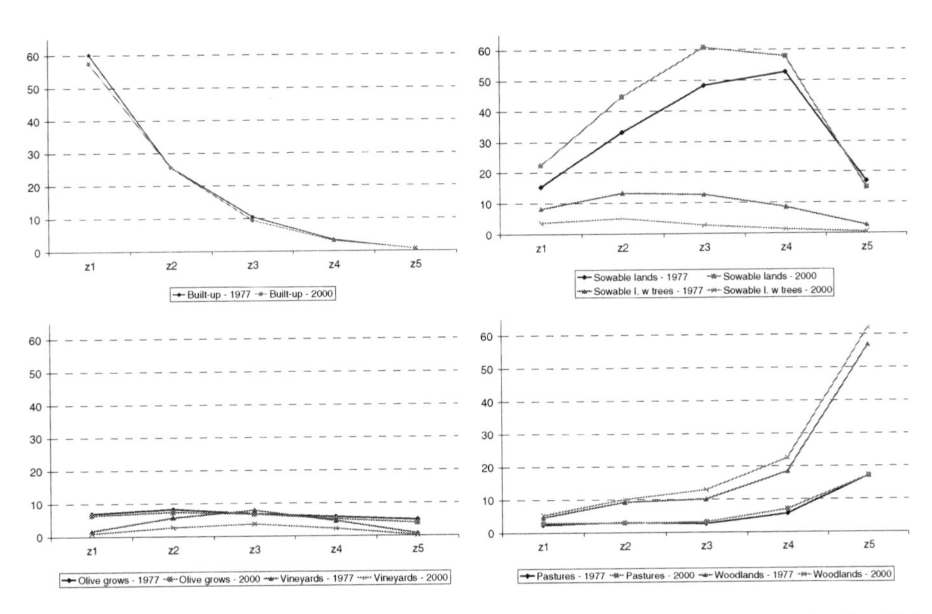

Fig. 9. Percent of Landscape (PLAND) of years 1977 and 2000 calculated along the five SDI zones

As previously specified, PD curves (Fig. 10) can be interpreted together with mean patch size (MPS) (Fig. 11). PD values for built-up areas increased considerably during the period under investigation (with a peak increase in z2), while the related mean patch size decreased progressively along the gradient. This tendency confirms that, from 1977 to 2000, the area of Perugia underwent diffuse urban sprawl, with magnitude increasing from z1 to z2 and decreasing progressively towards z5. On the other hand, PD and MPS curves for sowable lands, vineyards and olive groves generally reflect the landscape transformations described above. Comparing 1977 and 2000 PD curves for sowable lands reveals a particular inversion between z2 and z3. This tendency is due to landscape fragmentation caused by urban sprawl in z1 and z2 and to the aforementioned agricultural land simplification occurring from z3 to z5. The relevant MPS curves also confirm this trend. PD curves for woodlands show a consistent increase along the gradient due mainly to the considerable fragmentation of wooded patches, also highlighted by the relevant MPS curves.

LSI diachronic curves for built-up areas show a peak in z3 in both years with higher shape complexity in year 2000 from z1 to z3 and lower complexity from z4 to z5 (Fig. 12). The shape of both curves confirms, on one hand, the "scattered and branched" urban sprawl occurring around Perugia and, on the other, the tendential patch aggregation in more rural areas (z4 and z5). LSI curves for sowable lands indicate the changes in complexity of patch configuration described above: higher disaggregation in the vicinity of main centres (z1 to z3) and tendential aggregation in the rural areas (z3 to z5). LSI curves for the other land uses generally reflect the landscape modifications occurring in the whole area. In particular, diachronic LSI curves of wooded areas indicate, from z1 to z2, an increase in shape complexity due to the aforementioned fragmentation process, occurring with decreasing magnitude as we move from urban towards rural areas.

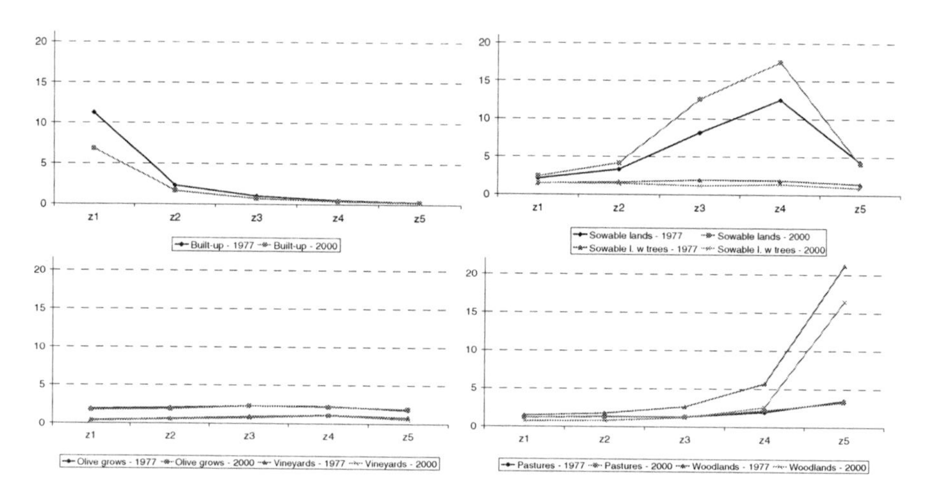

Fig. 11. Mean Patch Size (MPS) for years 1977 and 2000 calculated along the SDI zones

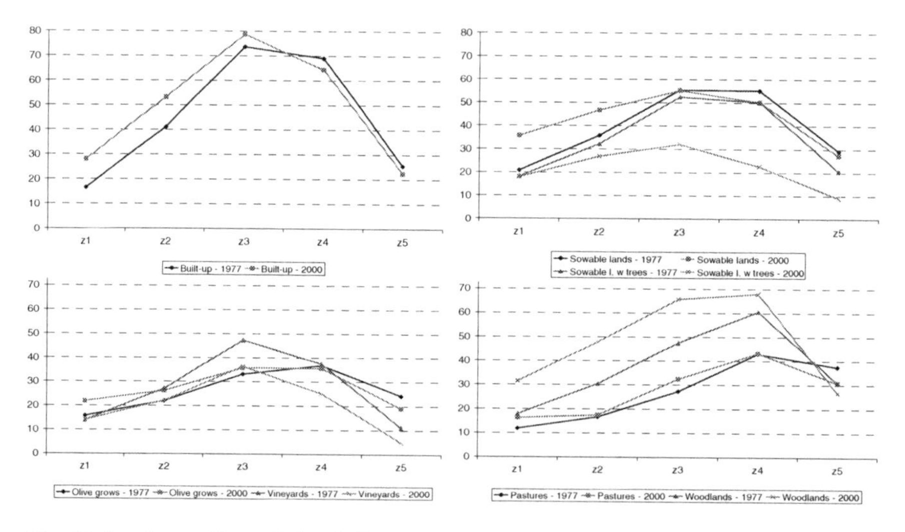

Fig. 12. Landscape Shape Index (LSI) of years 1977 and 2000 calculated along the SDI zones

4 Discussion

The results confirm the modifications occurring extensively throughout central Italy for the period under investigation: a high increase of peri-urban fragmentation due to erosive and diffuse urban sprawl and loss of landscape diversity due to progressive simplification of the agricultural system. For the Perugian landscape in question, the period 1977 to 2000 saw consistent and irregular expansion of built-up areas generating erosion and fragmentation of traditional peri-urban agricultural land uses. The consistent expansion of zones 1, 2 and 3, to the detriment of less urbanized zones 4 and 5, confirms how built-up areas are progressively penetrating into rural areas, resulting in fragmentation and erosion of more traditional land uses. Indeed, as expected, the degree of fragmentation is positively related to the level of urbanization as also demonstrated by the decrease in PD and increase in MPS moving from urban towards rural areas. Fragmentation induced by urbanization, together with simplification of rural land uses, resulted in the consequential loss of landscape diversity along the entire gradient.

By analysing all the results relating to the five SDI zones, it may be observed that, in both years, the high density urban areas (z1) are dominated by aggregated built-up patches and appear to be surrounded principally by mixed land use patches (sowable lands, olive groves and woodlands). Peri urban areas (z2), along the gradient, remain the most fragmented and the most vulnerable to urban sprawl and to rural transformations. Outer areas (z3 and z4) continue to be dominated by agricultural land uses, but progressively become more homogeneous due to a consistent decrease in vineyards, olive groves and to the conversion of traditional sowable lands with trees. Zone 4, in particular, is characterized by many fragmented natural and semi-natural elements that increased between 1977 and 2000. As a consequence this zone represents an ecologically important area of interaction between agricultural uses and wildlife habitats. In zone 5, which includes the main wooded areas of the landscape, there is diffuse expansion of wooded areas due mainly to the abandonment of cultivation in steeper agricultural areas.

Urbanization and agricultural simplification remain the dominant forces responsible for the landscape modifications occurring in Perugian landscapes between 1977 and 2000. The former is produced an eroding and invasive force on peri-urban land uses while the latter, driven by multiple external factors, resulted in progressive simplification of landscape structure. Unfortunately, as already mentioned, the main outcome of these processes was the definitive alteration of the characteristics of the typical peri-urban Umbrian landscapes, characterised by high agricultural diversity. Undoubtedly in Umbria, and in Italy in general, diversity is an important component of landscape quality, thus, the aforementioned loss of diversity inevitably lead to a consistent loss of landscape value.

5 Conclusions

This study confirmed the effectiveness of the combined method of gradient analysis and landscape metrics for interpreting the changes in landscape pattern in response to urbanization and other anthropogenic factors. Unlike previous similar studies, a settlement density index (SDI), calculated using GIS kernel density, has been used to allow analysis of spatial changes in relation to the urbanization gradient, and the related modifications of the other land uses. SDI, in particular, has been used successfully for landscape subdivision for the two periods under investigation with the aim of defining ecologically comparable areas along the urban to rural gradient. Unlike other previous studies conducted on territories surrounding Perugia, the calculation of landscape metrics has allowed deeper comprehension of the modifications occurring along the urban to rural gradient and offers specific information for potential comparison with other similar studies.

In order to study the structural changes in these landscapes more thoroughly, it will be possible to include more specific data (including land registry and census data) relating to the agricultural and urban systems. GIS modelling of urban gradients may be further enhanced by introducing other territorial variables and applying evaluation methods based on multicriteria techniques and Fuzzy logic [40-42]. Nevertheless, as this study demonstrates, many analytical objectives can also be achieved using simplified approaches in modelling of urbanization.

As pointed out in this study, because of their vulnerability, peri-urban landscapes play an increasingly significant role in the planning and management of the Umbrian territory. This is confirmed by the progressive expansion of these contexts and the reliable transformations to which they are subject. Despite all these transformations, agriculture and related activities continue to have a key function in the diversification of peri-urban and rural areas and in preserving the local and regional landscapes' identity.

References

1. Bridges, L., Crompton, A., Schaffer, J.: Landscapes as gradients: The spatial structure of terrestrial ecosystem components in southern Ontario, Canada. Ecological Complexity 4, 34–41 (2007)
2. Luck, M., Wu, J.: A gradient analysis of urban landscape pattern: a case study from the Phoenix metropolitan region, Arizona, USA. Landscape Ecology 17, 327–339 (2002)
3. McDonnell, M.J., Pickett, S.T.A.: Ecosystem structure and function along urban rural gradients - an unexploited opportunity for ecology. Ecology 71, 1232–1237 (1990)
4. McDonnell, M.J., Hahs, A.K.: The use of gradient analysis studies in advancing our understanding of the ecology of urbanizing landscapes: current status and future directions. Landscape Ecology 23, 1143–1155 (2008)
5. Burrough, P.A., Frank, A.U.: Geographic objects with indeterminate boundaries. Taylor & Francis, Abington (1996)
6. Cavailhès, J., Peeters, D., Sékeris, E., Thisse, J.-F.: The periurban city: why to live between the suburbs and the countryside. Regional Science and Urban Economics 34, 681–703 (2004)
7. Valentini, A.: Il senso del confine – Colloquio con Piero Zanini. Ri-Vista Ricerche per la progettazione del paesaggio 4, 70–74 (2006)
8. Baker, W.L.: A review of models of landscape change. Landscape Ecology 2, 111–133 (1989)
9. Pryor, R.J.: Defining the rural-urban fringe. Social Forces 47, 202–215 (1968)

10. Thapa, R., Murayama, Y.: Land evaluation for peri-urban agriculture using analytical hierarchical process and geographic information system techniques: A case study of Hanoi. Land Use Policy 25, 225–239 (2008)
11. Wehrwein, G.S.: The rural-urban fringe. Economic Geography 18, 217 (1942)
12. Brook, R., Davila, J.D.: The peri-urban interface: a tale of two cities. Development Planning Unit, UCL (2000)
13. Tacoli, C.: Rural-urban interactions: a guide to the literature. Environment and Urbanization 10, 147–166 (1998)
14. Allen, A.: Environmental planning and management of the peri-urban interface: perspectives on an emerging field. Environment and Urbanization 15, 135–148 (2003)
15. Vizzari, M.: Spatial modelling of potential landscape quality. Applied Geography 31, 108–118 (2011)
16. Blaschke, T.: The role of the spatial dimension within the framework of sustainable landscapes and natural capital. Landscape and Urban Planning 75, 198–226 (2006)
17. Turner, M.G.: Spatial and temporal analysis of landscape patterns. Landscape Ecology 4, 21–30 (1990)
18. Uuemaa, E., et al.: Landscape Metrics and Indices: An Overview of Their Use in Landscape Research. Living Reviews in Landscape Research 3 (2009)
19. McGarigal, K., Cushman, S., Neel, M.: FRAGSTATS: Spatial pattern analysis program for categorical maps (2002)
20. Li, H., Wu, J.: Use and misuse of landscape indices. Landscape Ecology 19, 389–399 (2004)
21. Hahs, A.K., McDonnell, M.J.: Selecting independent measures to quantify Melbourne's urban–rural gradient. Landscape and Urban Planning 78, 435–448 (2006)
22. Wang, Y., Li, J., Wu, J., Song, Y.: Landscape pattern changes in urbanization of Pudong New District, Shanghai. Chinese Journal of Applied Ecology 17, 36–40 (2006)
23. Weng, Y.: Spatiotemporal changes of landscape pattern in response to urbanization. Landscape and Urban Planning 81, 341–353 (2007)
24. Yang, Y., Zhou, Q., Gong, J., Wang, Y.: Gradient analysis of landscape pattern spatial-temporal changes in Beijing metropolitan area. Science in China. Series E, Technological sciences 53(1), 91–98 (2010)
25. Fichera, C.R., Modica, G., Pollino, M.: Remote sensing and GIS for rural/urban gradient detection. In: Fichera, C.R., Modica, G., Pollino, M. (eds.) XVIIth World Congress of the International Commission of Agricultural and Biosystems Engineering (CIGR), Québec City (2010)
26. Romano, B., Ragni, B., Vizzari, M., Orsomando, E., Pungetti, G.: Rete Ecologica regionale della Regione Umbria. Petruzzi editore, Perugia (2009)
27. McCoy, J., Johnston, K.: Using ArcGIS Spatial Analyst. Environmental Systems Research Institute, Redlands (2002)
28. Soille, P., Vogt, P.: Morphological segmentation of binary patterns. Pattern Recognition Letters 30, 456–459 (2009)
29. Bailey, T.C., Gatrell, A.C.: Interactive spatial data analysis. Longman Higher Education, Harlow (1995)
30. Silverman, B.W., Jones, M.C., Fix, E., Hodges, J.L.: An Important Contribution to Nonparametric Discriminant Analysis and Density Estimation. International Statistical Review 57, 233–247 (1989)

31. Danese, M., Lazzari, M., Murgante, B.: Kernel density estimation methods for a geostatistical approach in seismic risk analysis: The case study of Potenza hilltop town (southern Italy). In: Gervasi, O., et al. (eds.) ICCSA 2008, Part I. LNCS, vol. 5072, pp. 415–429. Springer, Heidelberg (2008)
32. Jones, M.C., Marron, J.S., Sheather, S.J.: A Brief Survey of Bandwidth Selection for Density Estimation. Journal of the American Statistical Association 91, 401–407 (1996)
33. Borruso, G.: Network Density Estimation: A GIS approach for analysing point patterns in a network space. Transactions in GIS 12, 377–402 (2008)
34. Lloyd, C.D.: Local models for spatial analysis. CRC Press, Boca Raton (2007)
35. Wickham, J.D., Riitters, K.H.: Sensitivity of landscape metrics to pixel size. International Journal of Remote Sensing 16, 3585–3594 (1995)
36. Wu, J., David, J.L.: A spatially explicit hierarchical approach to modeling complex ecological systems: theory and applications. Ecological Modelling 153, 7–26 (2002)
37. Imre, A.R., Rocchini, D.: Explicitly accounting for pixel dimension in calculating classical and fractal landscape shape metrics. Acta Biotheoretica 57, 349–360 (2009)
38. Zhu, M., Xu, J., Jiang, N., Li, J., Fan, Y.: Impacts of road corridors on urban landscape pattern: a gradient analysis with changing grain size in Shanghai, China. Landscape Ecology 21, 723–734 (2006)
39. Hengl, T.: Finding the right pixel size. Computers & Geosciences 32, 1283–1298 (2006)
40. Eastman, J.R.: Multicriteria evaluation and GIS. In: Longley, P.A., Goodchild, M.F., Maquire, D.J., Rhind, D.W. (eds.) Geographical Information Systems, pp. 493–502 (1999)
41. Malczewski, J.: GIS and multicriteria decision analysis. John Wiley and Sons, West Sussex (1999)
42. Murgante, B., Casas, G.L., Danese, M.: The use of spatial statistics to analyze the periurban belt. In: Wachowicz, M., Bodum, L. (eds.) The european information society: leading the way with geoinformation. Proceedings of the 10th Agile International Conference on Geographical Information Science, Aalborg, Denmark (2007)

Urban Development Scenarios and Probability Mapping for Greater Dublin Region: The MOLAND Model Applications

Harutyun Shahumyan[1], Roger White[2], Laura Petrov[1], Brendan Williams[1], Sheila Convery[1], and Michael Brennan[1]

[1] University College Dublin Urban Institute Ireland
Belfield, Dublin 4, Ireland
{Harutyun.Shahumyan,Brendan.Williams,Sheila.Convery, Michael.Brennan}@ucd.ie, Laura.Petrov@gmail.com
[2] Memorial University of Newfoundland
St. John's, Newfoundland, A1B 3X9, Canada
Roger@mun.ca

Abstract. The MOLAND land use model was used in several studies to simulate possible scenarios of future settlement patterns in the Greater Dublin Region (GDR). This paper compares the results of three different research outputs with ten possible scenarios for GDR urban development. Brief descriptions of the scenarios and probability maps combining these scenarios are presented. The suggested approach of scenario analysis can be used by planners and decision makers to get an idea of the most likely development areas in the region if several scenarios are under consideration. In addition, probability maps help to find areas where the decisions could have the most influence on development patterns with minimal efforts.

Keywords: scenario, simulation, model, MOLAND, probability maps, urban development, land use, weighted scenario comparison, Greater Dublin Region.

1 Introduction

The management of resources is complex and demanding, particularly where multiple public and private sector institutions are involved. It often involves difficult decisions which have to be made based on limited or incomplete evidence. The management of the environment and in particular land resources can be greatly assisted through the ability to analyse the likely implications of policies, planning and development decisions into the future. The use of scenarios or storylines is a powerful means of exploring possible futures and helping to define common or shared visions of the future [1], [2]. It also assists in the analysis of the likely implications of different decisions. Spatial decision support systems are designed specifically to provide an inherently spatial representation of the future. Thus, they are particularly helpful in the domains of land use and environment which are also inherently spatial.

B. Murgante et al. (Eds.): ICCSA 2011, Part I, LNCS 6782, pp. 119–134, 2011.

Dynamic urban land use models such as MOLAND (Monitoring Land Use / Cover Dynamics) seek to provide the capacity to simulate alternative scenarios of urban development into the future. Recent models explicitly reject the principle of attempting to produce a single absolute projection of the future in favour of numerous alternative visions. The aim of scenario development through land use modelling is therefore to promote structured discussion and awareness among stakeholders of future possibilities and alternatives rather than to provide predictions of future development patterns [3]. The proper use of a dynamic land use simulation model reflects the fact that the future is to some useful degree predictable, but never entirely, or even mostly, so, and that one of the sources of unpredictability is the fact that the planning and policy choices that will affect the future have not all yet been made.

The MOLAND simulation model incorporates, both through its formulation and its proper mode of use, several kinds of uncertainty. At the lowest level, it incorporates a stochastic element that perturbs the otherwise deterministic nature of its representation of urban and regional processes. The stochastic element ensures that every run of the model gives a different output. While most of these will be only trivially different, due to dynamic nonlinearities, bifurcations may appear whereby several different *classes* of outcomes are generated. The differences between these classes may not be unimportant, and hence of interest to model users because they may correspond to actual possible alternative futures. At the next level, the values of the fixed parameters can be varied from one run to another to explore the effects of our ignorance of the "correct" specification of the model; this is the role of sensitivity analysis.

At a still higher level, the model is (or should be) run under various scenarios. Some of these scenarios represent uncertainty about the context within which the modelled phenomena will unfold; for example, how fast will the Irish economy grow in the next 20 years? Such an assumption, or scenario, is not part of the model itself, but determines what is input into the model, and thus affects the output. Other scenarios, more interesting from a user's point of view, represent possible policy or planning options. These scenarios are used to allow the user to explore the consequences of the various options under consideration. The usefulness of the output depends on the fact that the model is *to a degree* predictive; but analysis of the variations in the output due to the stochasticity, uncertainty of parameter values, and varying context scenarios also gives the model user some indication of the degree to which confidence can be placed in the policy implications. In other words, the model, when properly used, in a sense mediates on behalf of the user between the predictability and unpredictability of the future, in order to arrive at a better idea of the future consequences of present actions.

Finally, the model can be used in a participatory context for scenario or storyline development. For example, an individual or group may develop a storyline that seems plausible as one possible future, but which in fact contains inconsistencies. The model can be used to show the possible futures that could follow from the set of assumptions underlying the storyline. If these do not include the future specified in the storyline, then the storyline is probably inconsistent, and should be modified. A model similar to MOLAND has been used to this end in order to develop storylines for the Visions project of the EU [2]; the Visions project supported the development of integrated visions for a sustainable Europe for the years 2020 and 2050.

Within the scope of the Urban Environment Project (UEP) implemented in UCD Urban Institute Ireland (UII) several scenarios have been developed for the Greater Dublin Region (GDR, Figure 1) and simulated by the MOLAND model in recent years. Initially simple scenarios were tested where only one or two elements were changed in each scenario. Then, based on the lessons learnt, more realistic and complex scenarios were simulated with key input from different thematic groups, researchers and officials. Some of these scenarios resulted in practical application, forming an important part of the latest review of the Regional Planning Guidelines (RPG) published in 2010 [4].

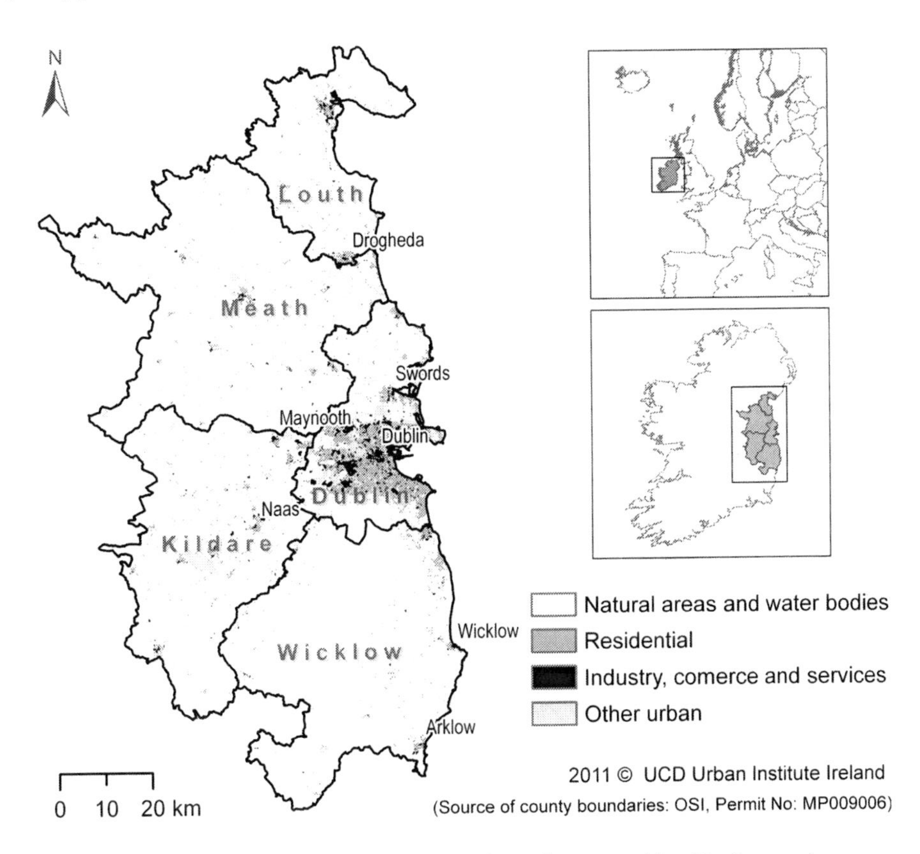

Fig. 1. Greater Dublin Region (2006 land use classes combined in 4 groups)

The MOLAND model and studies based on various scenarios are described in several publications [3], [5]. Specifically, the scenarios discussed in this paper are thoroughly explained in [6], [7] and [8]. Therefore, here only a brief description of the model and the scenarios, indicating their main characteristics and outputs, is given. The focus of this paper is on combining these outputs to analyse the future development patterns in the GDR if the scenarios were to occur.

2 The MOLAND Model

The MOLAND model was developed as part of an initiative of the European Commission's Joint Research Centre as a response to the challenge of providing a means for assessing and analysing urban and regional development trends across European Member States [5]. It comprises two dynamic sub-models with a common temporal increment of one year, but working at different scales:

- At the macro scale, the model takes as input the population and the economic activity (number of jobs) in the GDR, for each year of the simulation period. The model then allocates and re-allocates the activities among the sub-regions of the GDR (counties Louth, Meath, Dublin, Kildare and Wicklow) on the basis of competition among the regions for population and jobs.
- At the micro scale the provision for population and economic activities is translated into demand for the various land uses; for example, the population will be accommodated with residential land use types and the jobs will be provided within commercial, industrial and service land uses.

The micro model is based on a cellular automaton. The land use type assigned to any given cell is determined by the value of that cell for each possible use: each cell is assigned the land use for which it has the most value, until demands for all land uses are met. Values depend on land uses within the neighbourhood of a cell — some land uses in the neighbourhood make the cell more attractive for a particular activity, while others may tend to repel the activity. They also depend on accessibility, suitability, and zoning. The micro-model updates the land use map each year in response to both new demands received from the macro-model and local changes in the values of particular cells caused by previous changes of land use. The model is thus spatially explicit and dynamic; and it provides a means of representing the various social, economic and environmental interactions which together determine land use.

Description of the dataset preparation process as well as the model adaptation and calibration for GDR are provided by Shahumyan et al [9], [10].

3 Urban Development Scenarios for the Greater Dublin Region

All GDR scenarios described hereafter were simulated for the period of 2006-2026. The actual land use map of the GDR for 2006 with 23 land use classes and 200m cell size was the base map used in the model at the starting year of 2006. Different suitability, zoning and transport network maps as well as different population and employment projections were used in the scenarios depending on the assumptions used in each of them.

3.1 Scenarios for Wastewater Treatment Capacity Study (WWT)

Wastewater treatment provision is long term, regional and decided at central government level with inputs by local authorities within the region. As such it represents the context within which long term future planning and settlement patterns are based. The MOLAND model was used to simulate the spatial distribution of new urban development using three population projections for the GDR and to examine how this could impact planned future wastewater treatment capacity and defined catchment areas [6]. The scenarios were based on the Irish Central Statistics Office's (CSO) regional population projections for 2011-2026 [11]. Particularly:

- CSO projection "MRF1 Traditional" was used for population medium growth scenario (WWT2). It combines a continuing decline in international migration with constant fertility and a return to the traditional pattern of internal migration by 2016.
- For population low growth scenario (WWT1), 15% less population than "MRF1 Traditional" was projected.
- For population high growth scenario (WWT3), 15% more population than "MRF1 Traditional" was projected.

3.2 Scenarios for Strategic Environmental Assessment (SEA)

In collaboration with the Dublin & Mid East Regional Authorities (D&MERAs) the MOLAND model was used to generate scenarios illustrating the effects of future policy directions on the GDR [7]. Following extensive consultations with D&MERAs and stakeholder focus groups four scenarios were constructed for evaluation as part of the SEA process:

- Baseline/Continued Trends Scenario (SEA1): exploring the consequences of continuing the current settlement patterns, whereby actual settlement patterns are somewhat at odds with Regional Planning Guideline policy [12].
- Finger Expansion of Metropolitan Footprint Scenario (SEA2): Development is focused within the Metropolitan Footprint (MF) of Dublin city, with minimal growth in other areas and expansion of the MF along key transport corridors.
- Consolidation of Key Towns & the City Scenario (SEA3): Explores a settlement pattern similar to that proposed in the original Strategic Planning Guidelines published in 1999 [13]. This settlement pattern requires development to be consolidated within the existing MF and a small number of development centres along major transport routes. The MF does not expand along these routes.
- Managed Dispersal Scenario (SEA4): Dispersal of development is managed by focusing growth within the existing MF and several development centres across the region. Strictly enforced strategic green belts are used to prevent the merger of towns and ensure corridors remained between urban and rural natural areas.

3.3 Scenarios for Environmental Impact Assessment (EIA)

In addition to the Baseline scenario described above (SEA1), three other scenarios of GDR urban development were produced by early-stage and senior scientists and policy-makers during a summer school and workshop in 2009 held at UII, Dublin. The aim of this exercise was to bridge the scientific and stakeholder communities in order to collaborate around spatial models and produce future land use maps which should have a clear and accepted interpretation, be robust, statistically validated and respond to policy interventions [8].

By using five driving forces (population, economic trends, urbanisation, transport and overall trends), the following qualitative scenarios were realised using the MOLAND model:

– Recession Scenario (EIA1): focusing on urban development during economic recession, including a recovery by 2016.
– Compact Development Scenario (EIA2): demonstrating less pressure on natural land uses, exploring urban growth and urban/regional development in the frame of a strong environmental protection policy.
– Managed Dispersed Scenario (EIA3): The 2025 scenario produced by the European Environment Agency [14] suggests strong urban development along the Dublin-Belfast corridor due to benefits of proximity to the capital or other urban areas. Additionally, personal housing preferences play an important role in rural living in Ireland [15]. Therefore, in the EIA3 scenario the growth and sprawl of rural towns and villages in open countryside, particularly along the Dublin-Belfast motorway, were investigated in more detail. The realisation of this scenario is facilitated by a planning regime which imposes few constraints on the conversion of agricultural areas to low-density housing areas [16].

3.4 Scenario Comparison

All ten scenarios described above were simulated by the MOLAND model running from 2006 to 2026. Though most of the parameters in the model were not changed, there were specific modifications in the input maps of suitability, zoning and the transport network as well as in the projected population and employment numbers. As a result the land use maps of 2026 generated by the model for different scenarios vary substantially. However, the aims of the described three studies are convergent, generating scenarios for future policy directions and urban development (e.g. with impact on future wastewater treatment), and linking the scientific and stakeholders community and therefore, the maps retrieved can be used complementarily. Figure 2 shows a comparison of the maps for each scenario created using the Map Comparison Kit software [17]. Specifically, the residential areas of the simulated maps in 2026 are compared with the actual residential areas in 2006.

The comparison maps confirm that the GDR can have substantially different development patterns depending on the decisions made. Specifically, in some scenarios we have large urban development in the north of Dublin County (WWT2, WWT3); while in other scenarios urban sprawl is directed to the west of Dublin (SEA2, SEA3, EIA2). In some cases a few hinterland towns are developed broadly (SEA3, SEA4, EIA2); while in other cases more dispersed development in the hinterlands is observed (WWT1, WWT2, SEA1, EIA3). It is also noticeable that the most development occurs in the scenario with the highest projected population (WWT3); while the least development and even shrinkage of residential areas occur in the case of the recession scenario (EIA1).

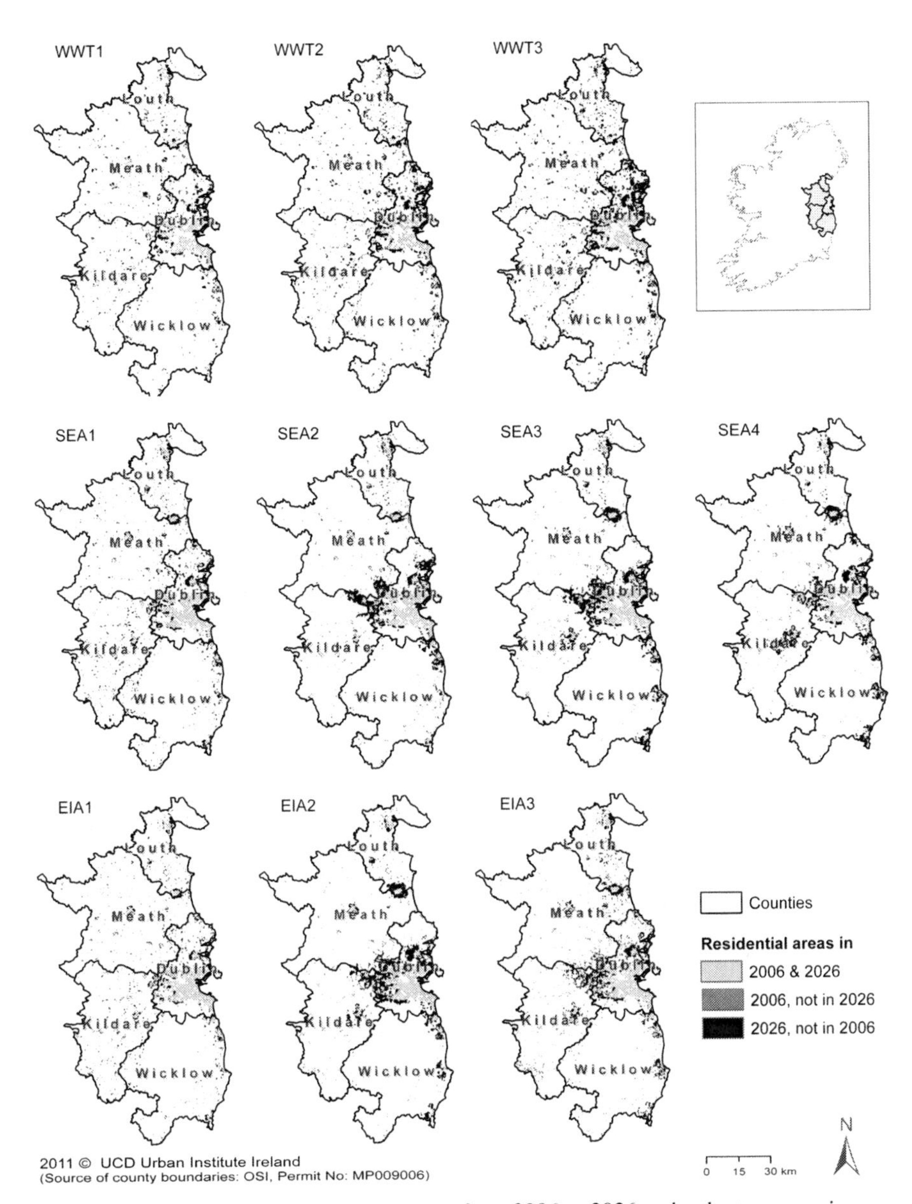

Fig. 2. Residential development patterns from 2006 to 2026 under the ten scenarios

The present simulations concur with [14] showing that the main development axis is to the north from the GDR along the seashore as well as inland (WWT3, SEA3, SEA4, EIA2, EIA3). To the south little new residential or industrial or commercial development will take place because of the physical constraints of upland areas. The scenarios also suggest the development of Dublin City to the northwest along the line of the Dublin-Belfast corridor. SEA scenarios indicate urban sprawl towards west of Dublin and also broadly development.

Land use statistics for each scenario were calculated. Figure 3 presents the residential areas in hectares for each county in 2006 and 2026. The statistics show that most of the development takes place in Dublin County; and the maximum is reached in scenario WWT3.

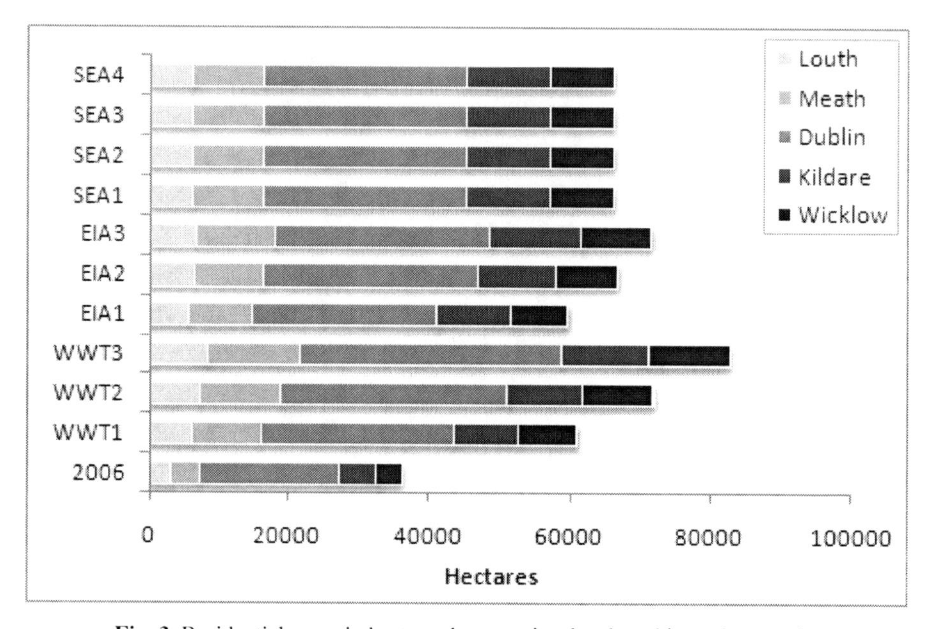

Fig. 3. Residential areas in hectares by counties developed in each scenario

The minimum and maximum possible increase of residential areas if any one of the scenarios occurs was also calculated. Thus, compared with the residential areas as they were in 2006, the maximum increase of total residential areas by 2026 could be about 128% (WWT3) and the minimum increase could be about 65% (EIA1). The average estimated increase from all ten scenarios is about 87%.

Figure 4 shows five statistics (minimum, first quartile, median, third quartile, and maximum) as well as outliners for residential areas in hectares by county. Here again it is noticeable, that for all scenarios the most residential areas are developed in Dublin County. In addition, the residential areas in scenario WWT3 are substantially greater than in any other scenario in almost all counties. Therefore, it appears as an outlier in the boxplot (marked as 'o'). Similarly, residential areas in 2006 are substantially smaller than in 2026 for any scenario, making them outliers too (marked as '*').

Figures 5 and 6 show the industrial areas in 2026 for each scenario in each county compared with 2006 actual areas. Thus, in the case of the EIA2 scenario, Dublin and Kildare get significantly more industrial areas compared with other scenarios and counties (outliers in the boxplot). For industrial areas the maximum increase by 2026 is 101% in the EIA2 scenario while the minimum increase is 11% in the EIA1 scenario. The average estimated increase is 47%.

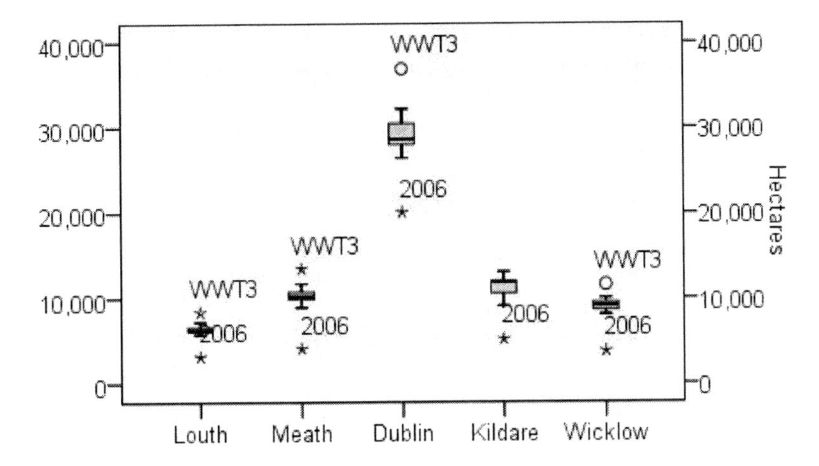

Fig. 4. Boxplot of all ten scenarios of GDR development for residential areas

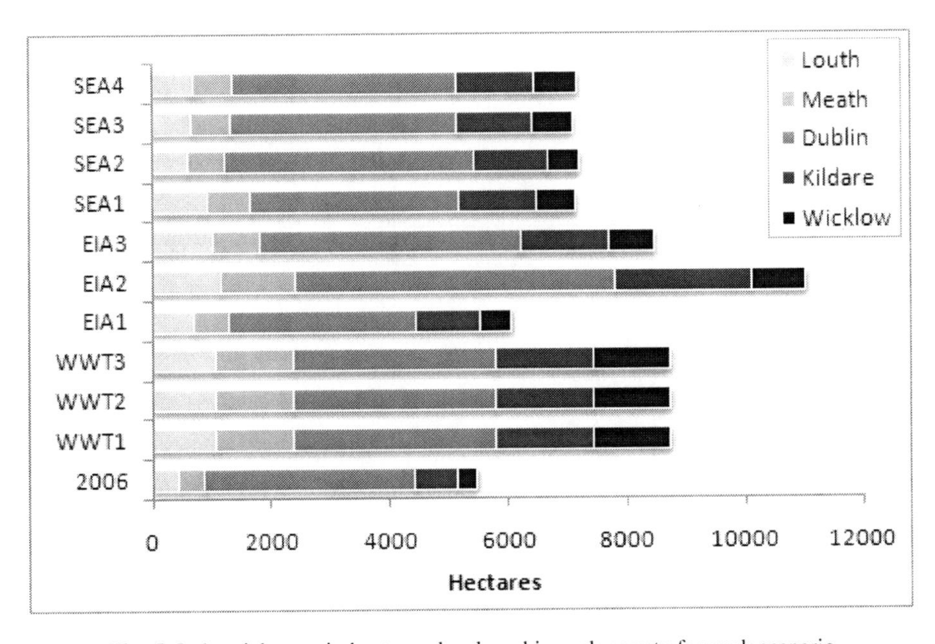

Fig. 5. Industrial areas in hectares developed in each county for each scenario

Similar analyses were done for commercial and service areas in the GDR. For commercial areas, a maximum increase of 150% is estimated in scenarios WWT2 and WWT3; while in the EIA1 scenario about 2% decrease of commercial areas is foreseen. A mean increase of 84% is estimated for commercial areas in GDR by 2026 compared with 2006. For service areas a maximum increase of 110% is estimated for WWT1, WWT2 and WWT3 scenarios while in EIA1 scenario a 6% decrease of service areas in GDR is forecast. Finally over all scenarios, a mean increase of approximately 60% in service areas by 2026 compared with 2006 data was foreseen.

The results show that residential and commercial uses almost double their areas while services and industry areas rise by half. Generally, the maximum increase of residential, commercial and services are foreseen by WWT scenarios while the minimum increase is represented by EIA scenarios. In case of industrial areas, the maximum and the minimum are represented by EIA scenarios where economic activity focuses on industry. Also the Dublin-Belfast corridor is identified as a core axis on the East Coast of Ireland with the potential to attract inward investment flows from the economies of Europe and the US; it also increases the economic importance of towns that exist along the axes.

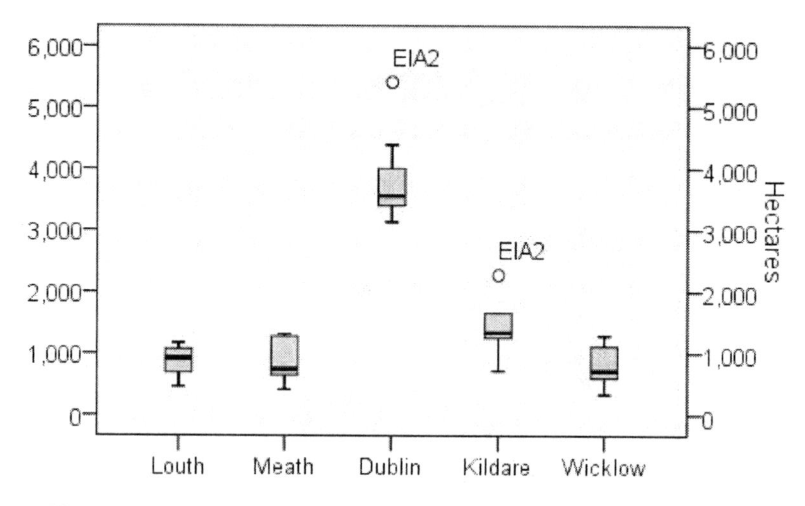

Fig. 6. Boxplot of GDR 10 possible scenarios for industrial areas by counties

4 Probability Mapping

The MOLAND model helps us to understand trends we are interested in and provides predictions of future land use changes. But in reality it offers not predictions of the future, but perspectives or alternative possible futures [18]. Indeed, each time the model runs it gives different predictions, both because of random elements and bifurcations inherent in the dynamics of the model. Therefore, the proper way to view the output of the model is probabilistically. To do this, a simulation should be run a sufficient number of times and a map of all the output possibilities produced [19]. Of course some possible outcomes will be very similar, and some can be quite different. The ability to know the range of future possibilities, and perhaps their relative probabilities of occurring, is very useful, and it is this sort of knowledge that a good model can offer.

4.1 A Single Scenario Probability Map

Returning to the GDR, rather than showing a single land use map for a scenario as the prediction for 2026 (e.g. Figure 2), a series of probability maps can be presented one for each land use class. For example, we have made 10 runs of the model to 2026 for the WWT3 scenario (population high growth)[1]. As a result, 10 distinct land use maps for 2026 were produced. Though the maps look similar, as a result of the random factor and possible bifurcation in the model, there are some differences between them.

Combining the 10 land use maps of scenario WWT3 from different runs using the Raster Calculator in ArcGIS Spatial Analyst, a probability map for residential land use was created. Figure 7 (left image) shows the probability that a specific location will become a residential area by the year 2026 in case of WWT3 scenario.

4.2 Combined Scenario Probability Map

Probability mapping of a single scenario is used often in urban modelling practice. It is an effective approach to assist decision makers to understand the most likely development patterns of a particular scenario. However, if there are several scenarios, it is often difficult to justify the preference of a particular one. A solution can be a combined probability map of several different scenarios. In principle a composite probability map generated from the output of several different appropriate scenarios is not qualitatively different from a probability map representing the effect of the stochastic perturbations within a single scenario. For example, in the case of three growth scenarios — low, medium, and high — the combined probability map of urbanisation is essentially equivalent to a map generated from model runs in which the growth rate parameter varies stochastically. But whether it makes sense to combine scenarios depends also on the point of view — i.e. on the user of the probability map. For example, to combine the output from three different planning scenarios, corresponding to different land use zoning schemes or transportation policies, would make no sense from the point of view of the planner, who would be using the model to examine the consequences of alternative policies with the aim of choosing one of them. But from the standpoint of a developer, who can't know what policy will be adopted in the future, or to what degree a policy, if adopted, will be enforced, combining the scenarios is reasonable because the combined probability map would represent the uncertainty of the future land use environment given the information available to the developer.

In any case some scenarios are more likely than others, and so the composite probability map should be constructed by weighting the various scenarios by their estimated likelihoods. The weighting factors themselves constitute a higher level scenario. For example, instead of separate scenarios for each level of population growth, low (WWT1), medium (WWT2) and high (WWT3), each to be examined separately, we now have a scenario stating how likely we consider each of these growth levels to be, relative to each other. The probability map generated from a weighted combination of scenarios in effect provides planners and the public with a picture of the consequences of ignorance or inaction.

---

[1] For illustration purposes, 10 runs were considered to be a reasonable number; however, for more reliable results the number of runs should be around 100 or more. However, it is time consuming and needs a special tool to simplify the process. For example, Monte-Carlo methodology is often used for such tasks. As the version of MOLAND software used in this research omitted the Monte-Carlo function, we have run this exercise manually based on fewer simulations.

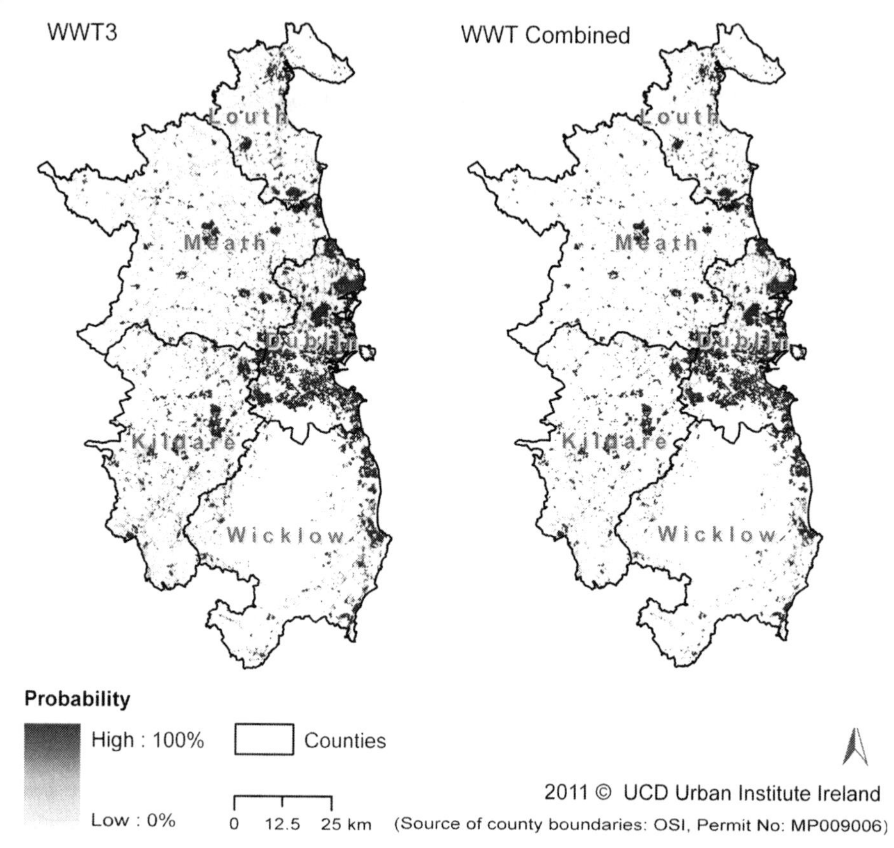

Fig. 7. Probability map of residential areas of the GDR in 2026 based on WWT3 (left) and combined WWT 1-3 (right) scenarios

For illustration purposes the WWT scenarios described above were used for combined probability mapping. These scenarios are especially appropriate because of their similarity and simplicity. In particular, the WWT scenarios vary only by population projections. Using the methodology described in the previous section we have created a composite probability map using the three WWT scenarios. Specifically, each of the WWT scenarios was run 10 times, resulting in 30 land use maps of the GDR for 2026. Based on [11] and discussions with several researchers and officials, the following weights were defined for the WWT scenarios: 0.2 (WWT1), 0.5 (WWT2), 0.3 (WWT3). The residential development probability map was generated from the weighted sum of the 30 land use raster maps in ArcGIS using the specified weights. The result is shown in the right image of Figure 7, which represents the likelihood of residential sprawl in the GDR in 2026 given the assumption that three WWT scenarios have the specified likelihoods (weights).

The maps in Figure 7 show that in the case of combined scenarios, the probability of residential development is decreased in some areas. More spatial statistical analyses of these maps is presented below.

4.3 Analysing Probability Maps

Probability maps show that most areas around Dublin are relatively predictable in terms of future urban land use: either they are likely to be developed or they are unlikely to be developed. However, if one land use class is equally likely as another of being present, there is a high degree of uncertainty related to the modelled class transition. Thus, many areas are not easily predictable (e.g. areas presented by the middle colours from the legend scheme). These areas are approximately equally likely to be developed or not and therefore the model is not capable of predicting accurately what may happen. In spite of this, for planners and decision makers these results still contain useful information (as these areas are capable of change and being influenced to change in various ways). It is useful to know, in a spatially explicit sense, where the probabilities of certain land use transitions are intermediate, because in these areas the future land uses can be influenced by small interventions in the present. In contrast, in the highly predictable area, major efforts would have to be made to alter the future land use patterns.

Figure 8 presents the areas in hectares by counties where there is no likelihood of residential development for scenario WWT3 as well as for all three WWT scenarios combined. Total area in each county with no development is larger in case of WWT combined scenarios compared with a single WWT3 scenario. This is the result of the variation of population projections used in the scenarios. While WWT3 reflects the population high growth scenario, WWT combined scenario includes weighted scenarios with lower population growth. Therefore, in case of combined scenarios we obtain less residential development than in WWT3.

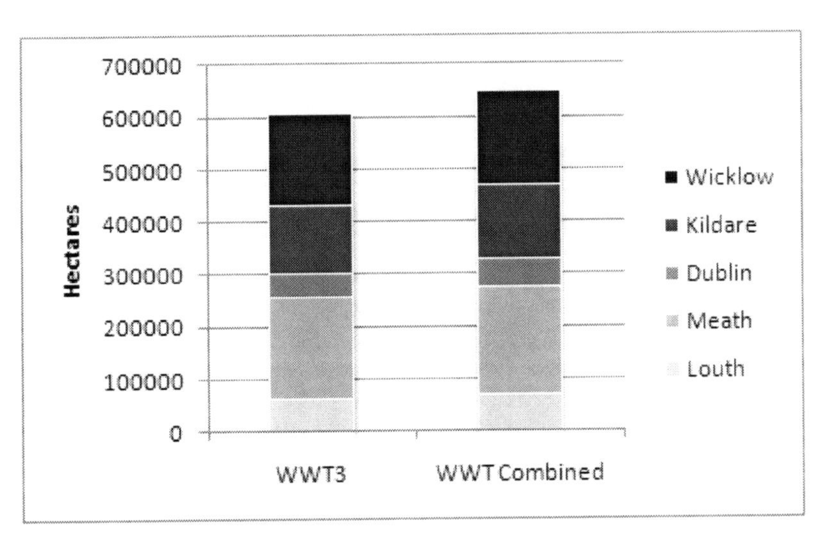

Fig. 8. Areas which will have no residential development in any of considered scenario

More interesting are the areas with some likelihood of becoming residential. Figure 9 presents the areas with 10% to100% probability of residential development in the GDR. It should be noted that the numbers with 100% development include also residential areas already existing in 2006 the start year of the simulations (marked by 'o' in the figure). This shows that the vast majority of areas with 100% predictability are areas which already were residential in 2006 and in case of combined probability mapping new development with 100% probability is essentially smaller than in the case of WTT3 scenario. Indeed, uncertainty is higher in the combined probability mapping. Combining three different scenarios includes some scenario-specific assumptions, making the results more general. Therefore, the combined scenario probability map in general has less area where the model predicts residential development by 2026.

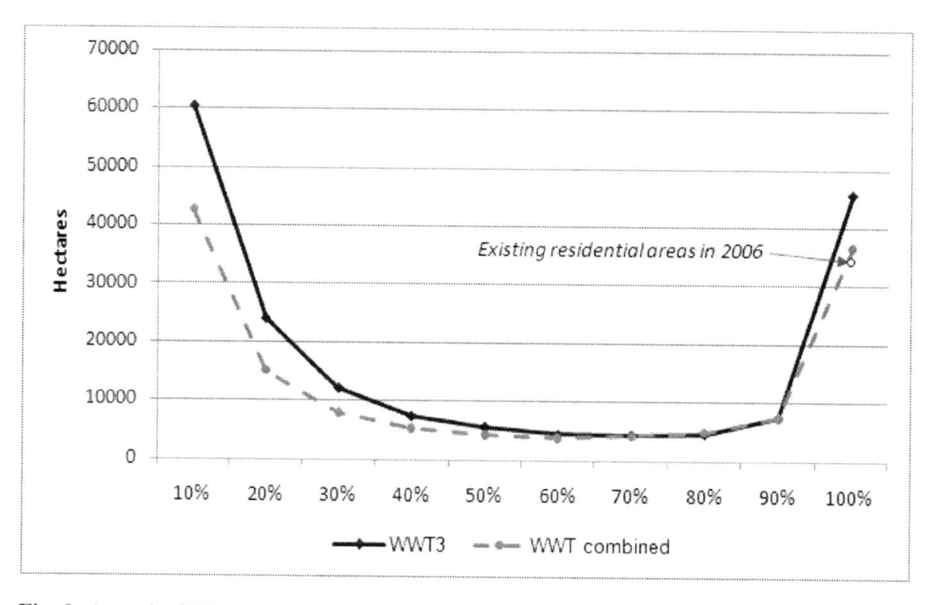

Fig. 9. Areas in GDR by their likelihood of becoming residential by 2026 for WWT3 and for three WWT scenarios combined

5 Conclusions

The purpose of the paper was not to identify or advocate which scenario should be adopted as the growth vision for GDR's 2026 but to generate interesting implications for spatial planning processes. MOLAND assumes that at some level, urban areas are fundamentally similar. They evolve by the same process. However, it is almost impossible to take into consideration all possible elements that can contribute to changes in the region. For example, the landscape value ('scenic beauty') that could be reduced because of urban area expansion or altered hydrological regimes or natural hazards. Also, it is obvious that unpredictable elements such as extreme climate change, global recession, etc. can occur whatever we decide.

The study has shown how the MOLAND model may be usefully applied in exploring the spatial distribution of land uses under a range of scenarios. Scenario comparison and the probability mapping allow estimates to be produced of the likelihood of certain land use transitions. They provide a valuable tool to describe predicted land cover change and its uncertainty. Spatial context can be given to land use change predictions and the associated level of confidence can be assessed.

Difficulties encountered in this research included the absence of Monte-Carlo functionality in the version of the MOLAND available for use in this project; and probability maps were created with limited number of iterations. Notably, the ArcGIS software was used to fill the gap, but it consumes high amounts of time and resources. Furthermore, for more reliable results, sensitivity analysis of different parameters of the model is necessary. Particularly, the effect of changing zoning and suitability maps or accessibility parameters on probability maps should be investigated. This also was not feasible because of the missing Monte-Carlo functionality.

These limitations emphasise the importance of further development of the MOLAND model and the need to introduce the Monte-Carlo functionality for more advanced studies on this topic.

Finally, using the results of the three different research outputs proved that the analysis of complementary scenarios can be used by planners and decision makers for getting a better insight of a region development. In fact, the more scenarios are considered, the more accurate the decisions. Also by using complementarily the experience of several projects, we can see clearer the technical aspects of a model that need to be ameliorated for better performances and/or more precise future estimations.

Acknowledgement

The Urban Environment Project is generously sponsored by the Irish Environmental Protection Agency as part of the ERTDI programme which is funded through the National Development Plan. 2005-CD-U1-M1 "Decision support tools for managing urban environment in Ireland'. All work undertaken on the MOLAND model, for the GDR is subject to the license conditions of the software developers, Research Institute Knowledge Systems b.v. (RIKS b.v.) and the data set owners, DG JRC under license no. JRC.BWL.30715.

References

1. Shearer, A.W.: Approaching scenario-based studies: three perceptions about the future and considerations for landscape planning. Environment and Planning B: Planning and Design 32, 67–87 (2005)
2. White, R., Straatman, G., Engelen, G.: Planning Visualization and Assessment: A Cellular Automata Based Integrated Spatial Decision Support System. In: Goodchild, Janelle, eds (eds.) Spatially Integrated Social Science, pp. 420–442. Oxford University Press, Oxford (2004)
3. Barredo, J., Lavalle, C., Demicheli, L., Kasanko, M., McCormick N.: Sustainable urban and regional planning: The MOLAND activities on urban scenario modelling and forecast. Office for Official Publications of the European Communities, Luxembourg (2003)
4. Dublin & Mid East Regional Authorities: Regional Planning Guidelines for the Greater Dublin Area 2010-2022. The Regional Planning Guidelines Office (2010)

5. Engelen, G., Lavalle, C., Barredo, J., van der Meulen, M., White, R.: The MOLAND Modelling Framework for Urban and Regional Land use Dynamics. In: Koonen, E., Stillwell, J., Bakema, A., Scholten, H. (eds.) Modelling Land-Use Change: Progress and Applications, Netherlands, Springer, Heidelberg (2007)
6. Williams, B., Shahumyan, H., Boyle, I., Convery, S., White, R.: Adapting an Urban-Regional Model (MOLAND) for Supporting the Planning and Provision of Strategic Regional Infrastructure: Providing Wastewater Treatment Capacity in the Dublin Region 2006 – 2026, pp. 2006–2026. UCD Urban Institute Ireland Working Paper Series, Dublin, Ireland (2009)
7. Brennan, M., Shahumyan, H., Walsh, C., Carty, J., Williams, B., Convery, S.: Regional Planning Guideline review: using MOLAND as part of the Strategic Environmental Assessment Process. UCD Urban Institute Working Paper Series, Dublin, Ireland (2009)
8. Petrov, L., Shahumyan, H., Williams, B., Convery, S.: Scenario development and indicators to explore the future of GDR in the context of European impact assessment. In: Submitted to the conference. Spatial Thinking and Geographic Information Sciences, Tokyo (2011)
9. Shahumyan H., Twumasi B., Convery S., Foley R., Vaughan E., Casey E., Carty J., Walsh C., Brennan M.: Data Preparation for the MOLAND Model Application for the Greater Dublin Region. UCD Urban Institute Ireland Working Paper Series, Dublin, Ireland (2009)
10. Shahumyan, H., White, R., Twumasi, B., Convery, S., Williams, B., Critchley, M., Carty, J., Walsh C., Brennan M.: The MOLAND Model Calibration and Validation for Greater Dublin Region. UCD Urban Institute Ireland Working Paper Series, Dublin, Ireland (2009)
11. Central Statistics Office: Regional Population Projections 2011-2026. CSO, Dublin (2008)
12. Convery, F., McInerney, D., Sokol, M., Stafford, P.: Organising Space in a dynamic economy - insights for policy from the Irish experience. Built Environment 32, 172–183 (2006)
13. Dublin & Mid-East Regional Authorities: Strategic Planning Guidelines for the Greater Dublin Area. Dublin: Regional Planning Guidelines Project Office (1999)
14. EEA: Urban sprawl in Europe sprawl in Europe, The ignored challenge, EEA report No 10, ISSN 1725-9177, Copenhagen (2006)
15. Mitchell, C.A.: Making sense of counter urbanization. Journal of Rural Studies 20, 15–34 (2004)
16. Williams, B., Hughes, B., Redmond, D.: Managing an unstable housing market, vol. 10/02. UCD Urban Institute Ireland Working Paper Series, Dublin, Ireland (2010)
17. RIKS: Map Comparison Kit User manual. RIKS, Maastricht, the Netherlands (2009)
18. Barredo, J., Petrov, L., Sagris, V., Lavalle, C., Genovese, E.: Towards an integrated scenario approach for spatial planning and natural hazard mitigation. European Commission, DG- Joint Research Centre, EUR 21900 EN (2005)
19. White, R.: Experimenting with the Future ...Now. Meeting the Challenge of Climate Change Seminar Series. CD Earth Systems Institute, Dublin (2008)

Hierarchical Clustering through Spatial Interaction Data. The Case of Commuting Flows in South-Eastern France

Giovanni Fusco and Matteo Caglioni

UMR ESPACE, University of Nice Sophia-Antipolis, Nice, France
{giovanni.fusco,matteo.caglioni}@unice.fr

Abstract. Regional scientists' methods to partition space in functional areas meet complex system analysts' methods to detect communities in networks. A common concern is the detection of hierarchical sets of clusters representing underlying structures. In this paper modularity optimization in complex networks is compared to polarized functional area definition through dominant flows. Different approaches to the significance of dominant flows are also tested, namely threshold and Multiple Linkage Analysis approaches.

Both methods are applied recursively in order to obtain a hierarchical clustering of municipalities in the PACA region (France) based on commuting flows in 1999. The comparison focuses on the geographical meaning of the results of the analyses. Modularity optimization and dominant flow results agree in many points and highlight the inadequacy of official methods integrating administrative boundaries in functional area definition. When they differ, they offer complementary views on the urban structure of the PACA region.

Keywords: Functional Areas, Complex Networks, Modularity Optimization, Significant Dominant Flows, Multiple Linkage Analysis, PACA Region.

1 Introduction: The Partitioning of Space

The partitioning of space in functional regions is a main issue in regional science research. Functional regions are sub-spaces characterized by strong interaction among human activities taking place within them. The increasing role of urban activities during the twentieth century, lead geographers and regional economists to identify urban phenomena as the main force defining and shaping functional regions (among classical works in this disciplinary tradition see Berry 1973, Dauphiné 1979, Fujita 1989). In particular, flow data (i.e. commuting, migration, etc.) have been used to develop a large variety of empiric methods to partition wider geographic systems in functional regions.

Literature in geography and in the regional science on the partitioning methods based on flow data is really extensive, and encompasses different approaches. We can mention for example the clustering theory based on economic interaction among firms (for an overview, see Karlsson 2007), partitioning around medoids (Kaufman and

B. Murgante et al. (Eds.): ICCSA 2011, Part I, LNCS 6782, pp. 135–151, 2011.

Rousseeuw, 1990, Van der Laan et al. 2002), local labour market area definition (Goodman 1970, Johansonn 1998, Casado-Diaz, 2000), nodal regions definition through dominant flows (Nystuen and Dacey 1961). Van Nuffel (2007) gives a good overview of the different methods and proposes a classification of literature on functional region definition. It is possible to divide these works into three large groups based on their operational approach: deductive methods, inductive methods, and hybrid deductive/inductive methods.

Within deductive methods, centres are defined *a priori*, independently of the actual division method and the goal of the partitioning algorithm is to determine the functional region around these centres. A further distinction within this approach can be made according to whether the centres are defined on the base of internal criteria (i.e. using statistics on commuting flows) or based on external criteria (i.e. major economic or demographic centres, central places of service supply, etc.). Examples of the former can be found in Lagnerö (2003) and Andersen (2000); for the latter the reader can see Willaert et al. (2000) and Van der Haegen and Pattyn (1980).

Inductive methods do not start from a priori given centres as the centres are determined together with the definition of functional region from spatial interaction data. The centres appear in the course of the partitioning process, and their designation is an integral part of the process. Also in this case it is possible to make a further distinction between methods where the centres are explicitly searched for (Nystuen and Dacey 1968) and methods where detecting polarizing centres is not necessary within the partitioning algorithm (Brown and Holmes 1971, Tolbert and Sizer 1996). Inductive methods not based on the determination of polarizing centres can produce non-polarized regions. These methods use the whole data matrix of flows, whereas all the other methods take into account only a subset of flow data (the largest flows, the first two largest flows and/or flows beyond a given threshold value).

A third type of approach is the hybrid one, which combines characteristics of both families of methods (Casado-Diaz, 2000). The procedure starts with a stage in which potential centres are deductively determined, but some of the centres may disappear during the further partitioning process.

Overall, the main feature of geographic approaches to partitioning space in functional regions has traditionally been the use of centres acting as focal points in the structuring of the functional area. Inductive methods not based on the determination of polarizing centres constitute an exception, nearing the geographic problem to approaches developed more recently by research on complex network analysis.

Complex network analysts have developed during the last twenty years methods aimed ad detecting communities in networks (Fortunato 2010, Porter et al. 2009, for methods in social network analysis see Wasserman and Faust 1994). Intuitively, communities are clusters of nodes within a network having stronger ties (modelled as edges or weighted edges) among them than with the rest of the network. The analogy with the geographic problem of functional regions definition is evident, as spatial interaction matrices define complex relational networks among spatial units, the latter representing the nodes of the network. More interestingly, following approaches which are typical of mechanical statistics, communities are seen by complex network analysts as mesoscopic structures averaging microscopic properties of individual nodes within them. These mesoscopic entities interact in order to explain macroscopic structures of the network as a whole.

Not surprisingly, network analysts have developed a variety of operational methods in order to detect communities in complex networks. Fortunato (2010) ranges these methods in three categories: local, global and based on nodes similarities. Local definitions are based on topological properties of nodes within their local neighbourhood, network modifications concerning distant nodes and edges do not have an impact on the belonging of the node to a particular community. Global definitions take into account topological properties of nodes within the whole network, making the belonging of nodes to a particular community dependent on the whole set of edges within the network. Finally, methods based on nodes similarities can take into account a variety of properties making each couple of nodes more or less similar with respect to other couples, regardless of the fact that nodes are connected or not within the network. Classical clustering techniques, like principal component analysis, multivariate hierarchical clustering, k-means clustering, neural network clustering and multidimensional scaling, belong to this category (Porter et al. 2009). These techniques also have a long tradition in regional studies. Centrality-based community detection (Girvan et Newman 2002, where the betweenness centrality measure is calculated over the whole network as proposed by Freeman 1977), modularity optimisation (Blondel et al., 2008) and spectral partitioning are global methods. Algorithms as the k-clique percolation (Palla et al. 2005) are local methods. All these methods would fall in the inductive approach to functional areas detection within the regional science tradition.

A shared interest of regional analysis and network community analysis is the hierarchical structure of the partitioning. The goal is to detect nested partitions of the whole system, where regions / communities of nodes are further grouped in order to form structures of higher hierarchical level. The final result is thus a "hierarchical clustering" of nodes within region/communities (not to be mistaken for the multivariate technique of hierarchical clustering). Porter and al. (2009) remark that the notion of network structure is composed of the two elements of communities (also referred to as modules or clusters) and hierarchy. In geography and regional science, the existence of hierarchical levels of organisation within urban systems is at the base of classical central place theories developed by Christaller and Lösch and is revealed by methods detecting nested levels of polarisation within functional urban areas (as Nistuen and Dacey 1961). Partitioning algorithms are nevertheless seldom applied in recursive ways in order to detect connected clusters of functional areas.

Comparing the contribution to the understanding of urban systems of methods issued of the regional analysis tradition and of the more recent complex network analysis tradition is an interesting, albeit seldom attempted, endeavour. In this paper we will focus our attention on two inductive methods, issued of the two different scientific traditions: community detection in complex networks through modularity optimisation and functional regions definition on the base of significant dominant flows. The two methods can be easily implemented in a recursive way, in order to detect the hierarchical properties of urban systems. First applications of this kind have already been proposed at a regional scale. De Montis et al. (2011) used modularity optimisation to analyse hierarchical urban systems in the island of Sardinia, Italy. Fusco (2009) determined recursively significant dominant flows to detect hierarchical urban systems in the Provence-Alpes-Côte d'Azur (PACA) region in South-Eastern France. In this paper, the comparison of the two methods will be made possible by the implementation over the same empirical data set, describing the 1999 commuting flows within the PACA region.

2 Modularity Optimization and Significant Dominant Flows

The starting point of both modularity optimisation and significant dominant flows analysis is empirical data on the spatial interaction among the units of the study area.

Spatial interaction is a key concept in geographical analysis. Flows between spatial units can be viewed as a kind of spatial interaction. Regional space is concerned with different kinds of spatial interaction flows: commuter flows, tourists' flows, migration flows, information flows, economic flows, etc. They can be taken into account by a spatial interaction matrix, reporting the volumes of the flows taking place between origins (the rows) and destinations (the columns).

2.1 Determining Communities in Networks through Modularity Optimization

Viewing a spatial interaction matrix within a geographic system as a relational network opens the way to applying network analysis techniques in order to regionalise the system. The problem of community detection in complex networks consists precisely in the partition of the network in communities of densely connected nodes, while nodes belonging to other communities are only sparsely connected to them. An analytical solution is computationally intractable. Therefore several algorithmic solutions have been proposed in order to find good partitions in a reasonably fast way. Detection algorithms can be divided in 3 groups: divisive, optimisation and spectral algorithms (De Montis et al. 2011).

Optimization methods for community detection are based on the maximisation of an objective function. The modularity of the partition is one of the most widely used objective functions and is defined as follows (Newman 2004):

$$Q = \Sigma_C \left(\Sigma_{C\ in} / 2m - (\Sigma_{C\ tot} / 2m)^2 \right) \quad (1)$$

where $\Sigma_{C\ in}$ is the sum of the weights of the links falling entirely within the community C, $\Sigma_{C\ tot}$ is the sum of the weights of the links incident to a node within C, and m is the sum of the weights of all the links in the network. Modularity is a scalar value, between -1 and 1, which measures the density of links inside communities compared to links between communities. The fastest approximation algorithm for optimizing modularity on large networks was proposed by Clauset et al. (2004).

Blondel et al. (2008) later developed an algorithm to find high modularity partitions of large networks in a fast way, which unfolds a complete hierarchical community structure for the network. The algorithm has two phases, which are repeated iteratively. It starts with a weighted network with N nodes. Firstly, each node of the network is assigned to a different community. Then, for each node i it is considered the neighbours j of i and evaluated the gain of modularity that would take place by removing i from its community and by placing it in the community of j. The node i is placed then in the community for which this gain is maximum, but only if this gain is positive. If no positive gain is possible, i stays in its original community.

The gain in modularity ΔQ obtained by moving an isolated node i into a community C can easily be computed by:

$$\Delta Q = \left[\frac{\Sigma_{in} + k_{i,in}}{2m} - \left(\frac{\Sigma_{tot} + k_i}{2m}\right)^2\right] - \left[\frac{\Sigma_{in}}{2m} - \left(\frac{\Sigma_{tot}}{2m}\right)^2 - \left(\frac{k_i}{2m}\right)^2\right] \quad (2)$$

where Σ_{in} is the sum of the weights of the links inside C, Σ_{tot} is the sum of the weights of the links incident to nodes in C, k_i is the sum of the weights of the links incident to node i, $k_{i,in}$ is the sum of the weights of the links from i to all nodes in C, and m is the sum of the weights of all the links in the network. A similar expression is used in order to evaluate the change of modularity when i is removed from its community.

The second phase of the algorithm consists in building a new network, whose nodes are now the communities found during the first phase (Figure 1). In order to do so, the weights of the links between the new nodes are given by the sum of the weights of the links between nodes in the corresponding two communities. Flows falling entirely within the communities can either be removed (as it will be the case in the applications of Section 3) or kept for modularity optimization at the following step. Once this second phase is completed, it is possible then to reapply the first phase of the algorithm to the resulting weighted network and to iterate it. Communities of communities defined at step 2, 3, … of the analysis form with the communities of the original nodes a hierarchical set of nested partitions of the spatial units within the study area. At every step of the analysis, the set of communities form an exhaustive and disjunctive partition of the study area, which can be easily map, even if no hierarchical structure is evident among the elements forming a given community. The use of Blondel's algorithm for modularity optimization was motivated both by its recursive nature and by the fact that it was already used in regional studies (De Montis et al. 2011). Modularity can nevertheless be optimized by different algorithms (Fortunato 2010); their comparison exceeds the scope of this paper.

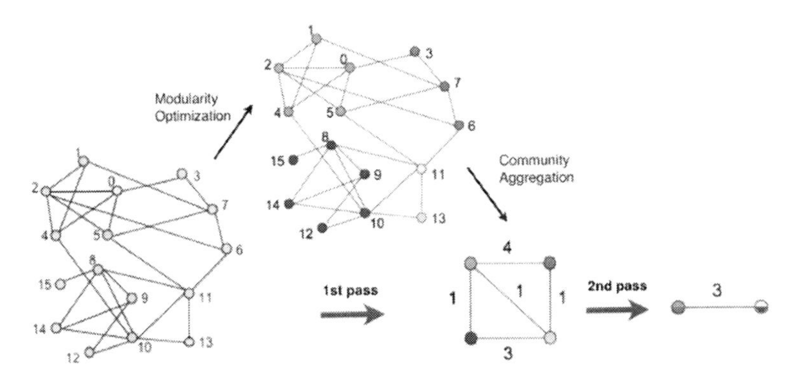

Fig. 1. Recursive use of modularity optimization in networks (Blondel et al., 2008)

2.2 Determining Urban Networks through Significant Dominant Flows

Firstly proposed by Nystuen and Dacey (1961) and later developed by several authors (Kipnis, 1985; Rabino and Occelli, 1997; Berroir et al., 2006), the dominant flows approach aims to extract the skeleton of an urban network (the dominant flows) from a spatial interaction matrix within a study area. As we will see, the regionalisation of the study area will be a by-product of the method.

The dominant flows approach needs an a priori rank among spatial units reflecting their absolute dimension in terms of mass (population, jobs or total exchanged flows are the most common mass criteria). Exchanged flows are then used to produce a hierarchical network among the spatial units, using the notion of "largest flow". This notion can have various definitions (Nystuen and Dacey, 1968), such as the largest out-flow, in-flow or total flow. Out-flows are the most commonly used in defining urban networks through commuter flows (Rabino and Occelli 1997, Berroir et al. 2006). According to this approach, a spatial unit is dominated if it sends its largest flow towards a centre of higher rank. In its classical approach (Rabino and Occelli, 1997) dominant flows detection can also integrate the significance of the flow. The maximal flow towards a higher rank unit is significantly dominant only if it is higher than an absolute threshold and a relative threshold (e.g. a certain percentage of outgoing flows or of the unit's population).

It is thus possible to extract a primary graph from the spatial interaction matrix, which is made of the significant dominant flows within the study area. The graph is rarely totally connected. Several independent treelike sub-networks dominated by different centres form an exhaustive and disjunctive partition of the study area.

The dominant flows algorithm can be applied recursively. All the spatial units linked in a sub-network of significant dominant flows can be aggregated at the next stage of the analysis (internal flows are often eliminated at this point) and new significant dominant flows can be determined among these larger spatial units. We thus obtain networks of networks of spatial units linked by second level dominant flows, and so on. The final result is a hierarchy of nested tree-like network.

This notion of external hierarchy among spatial unit clusters defined at different steps of the analysis (which is common to the modularity optimization method) is doubled by a notion of internal hierarchy that dominant flows define within each cluster at any given step of the analysis. The tree-like networks emerging from the dominant flows have an internal hierarchy as relations of domination link the units. Clusters of units defined through these networks are domination basins depending, directly or indirectly, from a single dominant and non-dominated unit (the inductively determined centre of the functional area). We can attribute hierarchical levels to non-dominant spatial units as well, distinguishing first level dominant centres from second or third level relay centres and so forth down to last level dominated spatial units.

2.3 A New Approach to Significant Dominant Flows: Multiple Linkage Analysis

Applying absolute and relative thresholds is the traditional way to insure that dominant flows are significant for the concerned spatial units (Rabino and Occelli 1998, Fusco 2008, 2009). In their seminal work on urban networks defined through telephone communication flows in the state of Washington, Nystuen and Dacey (1961) justified the use of the first flows as dominant flows inasmuch second flows were in most cases substantially smaller than first ones. No question of significance was than arisen within their work. We can thus propose a formalised method in order to verify Nystuen and Dacey's hypothesis on the distribution of empirical flows justifying the use of first flows as dominant flows.

Empirical flows will first be analysed through the MLA (Multiple Linkage Analysis, Hagget et al. 1977) algorithm. MLA is used to determine how many flows (in-flows, out-flows or total flows) are significant for a given node regardless of the dominance approach. To separate significant from non-significant flows, all flows concerning a spatial unit are ranked from largest to smallest, and compared to theoretical profiles of flows, where the total flow of the unit is concentrated respectively on only one, two, three, … , flows (mono-polarized, bi-polarized, tri-polarized, … model). The goodness-of-fit between the set of observed (empirical) flows and each of the sets of expected (theoretical) flows is measured through the coefficient of determination (R^2). The theoretical profile which presents the highest R^2 with the empirical profile will determine the number of significant flows for that spatial unit. MLA is an extension and improvement of Weaver's combination index (Weaver, 1954; Haggett et al., 1977). The Weaver method uses the sum of the squared differences (instead of the coefficient of determination) between the theoretical and the empirical flows. The number of significant flows corresponds then to the theoretical profile with the smallest sum of the squared differences.

The first flow (which is a dominant flow according to Nystuen and Dacey's approach if it goes towards a spatial unit of bigger size) is always significant according to the MLA algorithm. Combining the dominant flows and the MLA approaches, we will define significant dominant flows as being dominant flows (first flows towards bigger units) which are at the same time the only significant flow according to the MLA. In other words, a spatial unit, in order to be dominated by a bigger centre, must be mono-polarized by this centre. All the spatial units which send MLA significant flows towards different centres will be considered as multi-polarized and hence non-dominated. This approach for the determination of the significant dominant flows restricts considerably the number of dominant flows within a study area, as first flows normally exceed the relative thresholds commonly used (i.e. 5%, 10% or even 20% of outflows) without always being the only MLA significant flows. At the same time, the MLA approach to significance avoids the problems linked to determining the right significance threshold (Kipnis 1985).

3 Hierarchical Functional Areas in the PACA Region

The study area of our research is the Provence-Alps-Côte d'Azur region in South-Eastern France, including the city-state of Monaco. As the third region in France for population and economic activity, PACA has been affected over the past few decades by the emergence of two metropolitan systems in the coastal area: the metropolitan area of Provence, including Marseille, Aix-en-Provence and Toulon, and the metropolitan area of the French Riviera around Nice, Monaco and Cannes (Decoupigny and Fusco 2008). The emergence of metropolitan systems is reshaping the flows exchanged between cities, suburban areas, retail and office concentrations and rural villages. The northern alpine part of the region seems less affected by the emergence of metropolitan systems. The network of commuting flows within the PACA region is constituted by 964 nodes and 34527 relations. It shows an exponential distribution for the node degrees (figure 2), which suggests a random growth. The main urban centres, showing positive deviations from the exponential curve, introduce hierarchical relations going beyond the random model. Dominant flow distribution (see 3.2), with 114 dominant nodes and 816 relations, account for one third of the total flows and is scale free, hinting to self-organized hierarchical structure (Barabasi 2003).

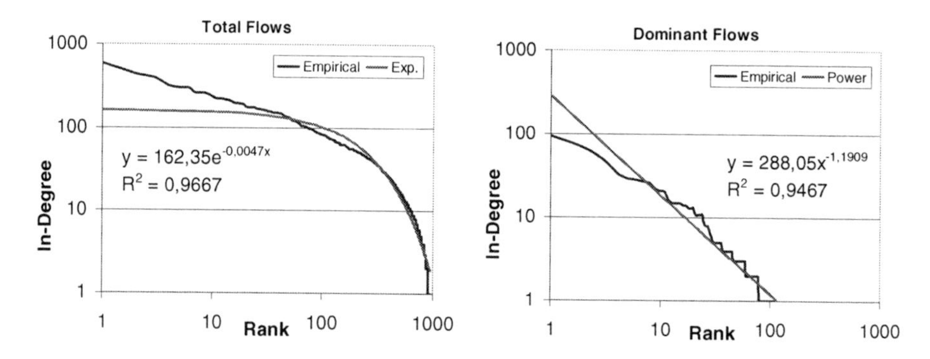

Fig. 2. The relational network of communing flows in the PACA Region

The database we made use of is the 1999 census produced by the French statistical institute (INSEE), completed with data from Monaco 2000 census. More specifically, we used commuting journeys among the 964 municipalities of the PACA region (including Monaco). Data also include flows between the study area and the rest of France (this information was only used in the dominant flows method).

Despite the increasing importance of leisure and commercial trips (not included in the census), daily commuter trips are still the most important flows of personal mobility in defining and structuring urban systems (Berroir et al. 2006). They convey information on the daily functioning of the urban systems and are not surprisingly the basis of official functional area definitions in many countries (namely in France).

Traditionally, INSEE defines employment areas within administrative department borders based on commuter flows from census data. For 1999, the PACA region is thus composed of six departments which are in their turn subdivided in 25 employment areas. Employment areas are an approximation of functional areas defined through a spurious deductive method, taking into account main job centres, commuter flows containment and administrative boundaries. Though practical from an administrative point of view, official employment areas do not necessarily correspond to functional areas defined by commuter flows. Their hierarchical clustering in the six departments can be seen as a mere political necessity.

It is thus interesting to apply methods for partitioning the regional space in hierarchical sets of functional areas which do not take into account internal administrative boundaries. Inductive methods will be preferred, as we want the number and the very nature of the functional areas to be a result of the applied method. We thus decided to define functional areas first as communities of municipalities strongly connected by commuters' flows (modularity optimization) and then as networks of municipalities directly/indirectly dominated by an urban centre (dominant flows). These methods will be applied recursively in order to obtain a hierarchical clustering of functional areas.

In what follows, we will comment the results obtained by the application of the two methods (modularity optimization and dominant flows) in order to determine a hierarchical set of functional areas in the PACA region. As far as dominant flows are concerned, their significance is evaluated both through the threshold and the MLA approaches. Both modularity optimization and dominant flows determination have been applied recursively with the removal of internal flows.

3.1 Modularity optimization: Five Large Functional Basins

It is well known from literature that several clustering methods strongly depend by the order by which the spatial units are considered by the algorithm, especially when they implement a pair-wise comparison. Blondel's algorithm is affected by this problem (whereas the dominant flows method is indifferent to input sorting). Before proceeding to modularity optimization the municipalities within the study area had therefore to be sorted by decreasing total flows, the same treatment was made for the aggregations at the successive levels of the analysis. Sorting municipalities in descending way of spatial interaction favours the aggregation of small municipalities to the main urban centres and subsequently of small clusters to bigger ones.

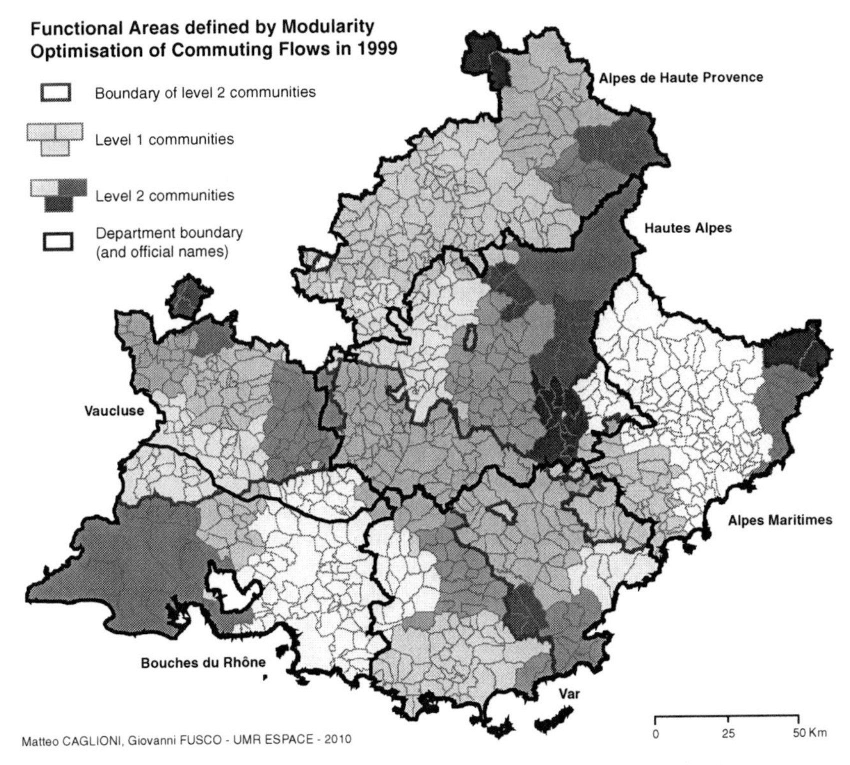

Fig. 3. Hierarchical Clustering through Modularity Optimization

Figure 3 shows the results of the first two steps of modularity optimization of commuter flows in the PACA region. At both levels of the analysis, communities within the relational network are compact areas with very few enclaves. Five communities emerge at the second level, corresponding to large functional basins which have a strong geographical meaning, even if they transgress administrative boundaries. Two communities correspond roughly to the Department boundaries of *Alpes Maritimes* (the metropolitan area around Nice, Cannes and Monaco) and *Vaucluse*. The widest community both in terms of surface and population corresponds to the metropolitan area of Marseille, Aix-en Provence and Toulon; it includes the Department of *Bouche du Rhone* as well as the western part of the *Var* Department and the southernmost part of the *Alpes de Haute Provence* Department. A fourth community encompasses most of the Alpine heartland of the region (*Alpes de Haute Provence* and *Hautes Alpes* Departments). The sixth and last community corresponds to the eastern section of the *Var* Department, among the two main metropolitan areas.

At level 1 we can highlight the existence of sub-basins within the two metropolitan areas (i.e. a central, eastern and western basin within the French Riviera corresponding to the Nice-Antibes, Cannes-Grasse and Menton-Monaco functional areas, a tripartite structure consistent with the official employment areas). The Marseille-Aix-Toulon community is internally much simpler than what the official employment areas suggest: a vast functional area around Marseille and Aix-en-Provence, surrounded by two smaller communities to the west and four on the east, out of the Department boundaries. It is also worth remarking that the vast alpine space is more fragmented at level 1 than the official employment areas.

Modularity optimisation can be further applied to level 2 communities. The result is a bipartite structure of the regional space: Marseille-Aix-Toulon, Vaucluse and the Alpine Space coalesce, as well as Alpes Maritimes and Est Var. At the next step, these two level 3 communities aggregate to cover the whole regional space.

3.2 Dominant Flows with Significance Thresholds

Clustering municipalities through dominant flows does not depend from the order by which the algorithm considers them. Dominant flows definition depends nevertheless from an *a priori* ranking of the spatial units (the first outflow is dominant only if it is directed towards a bigger spatial unit). In our analysis, resident population was used in order to produce the desired *a priori* ranking (total flows could have been used instead, with a few minor differences in the final results). In what follows, dominant commuter flows within the PACA region will be determined though different approaches to dominant flow significance.

Figure 4 shows the results of urban networks defined by significant dominant flows in the first two steps of the analysis, when significance is determined by an absolute threshold of 5 commuters and a relative threshold of 15% of the total outflow for each spatial unit[1]. The absolute threshold is only used to filter noise from the results, as dominant flows defined by extremely small first flows can depend from random

[1] The dominant flows analysis was also carried out without applying significance thresholds. The results were partially consistent with the threshold approach but also contained a few artefacts. Determining dominant flows without any consideration of the significance of the flows has been often criticized (Kipnis 1985, Rabino and Occelli 1998, Fusco 2009).

spatial behaviour of very few commuters (this problem concerns a few rural municipalities in the mountain areas). As a result, a few "holes" can be observed in the partitioning of the regional space through dependency basins. The relative threshold is much more important in shaping the results of the analysis.

The relative threshold of 15% of total outflows results in the emergence of seven domination basins at the second level of the analysis, with a few independent units or basins which do not evolve from the first level of the analysis. These results are in agree with the modularity optimization results, with a few noticeable exceptions.

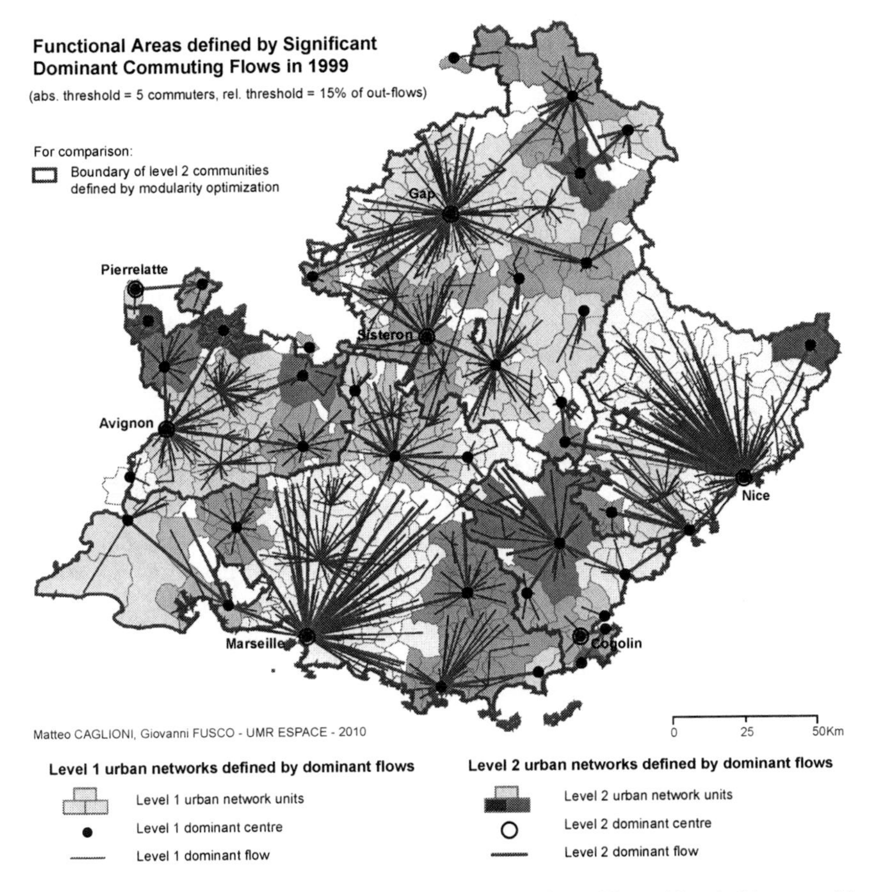

Fig. 4. Hierarchical Clustering through Significant Dominant Flows (threshold approach)

The two methods show a very good agreement for the definition of level 2 functional areas around Marseille and Avignon. The level 2 urban network dominated by Nice is instead larger than the level 2 community of *Alpes Maritimes*, as it includes the eastern section of the *Var* Department. Within eastern *Var*, only a small urban network around Cogolin keeps its independence at level 2. Contrary to the modularity optimization result, the alpine space does not aggregate in one whole network, but is structured in two large independent networks (around the cities of Gap and Sisteron) with a few smaller networks which do not evolve among the two steps of the analysis.

Nevertheless, the main differences among the two methods lie elsewhere. Dominant flows give the possibility to determine dominant centres at the different levels of the analysis. It is thus possible to identify the main urban centres around which the functional area is organized. Even more, it is possible to study internal network configurations of domination basins. Functional areas with direct domination of their urban centre over the other municipalities can thus be distinguished from more complex functional areas, where the main urban centre dominates secondary or tertiary centres which, in their turn extend their influence over more peripheral municipalities. In this respect, the two main metropolitan networks, around Marseille and Nice, show important morphological differences. The Marseille metropolitan area has a complex network structure already at the first step of the analysis, favouring the emergence of secondary and tertiary centres in the hinterland. The network is less articulated along the coast. The metropolitan network around Nice is weakly structured in its hinterland as villages depend directly on the main centre. The network is better relayed along the coast, with the secondary centre of Antibes.

At the second step of the analysis, hierarchical polycentrism is further strengthened in the west around Marseille and Avignon, each of them dominating new urban networks. In the north, polycentrism is less structured as independent networks are not necessarily dominated by Gap. The eastern part of the study area is marked by the weaknesses of its polycentrism in the hinterland at the second level of the analysis, as well. Hierarchical polycentrism is on the contrary better articulated along the coast, where the domination of Nice is relayed by several urban centres which where independent at the first stage of the analysis (Cannes, Frejus, Draguignan).

The dominant flows method can be further applied, until all the regional space (but a few independent units) is covered by a unique urban network. At the third step, the alpine basins, the *Vaucluse* and the *Alpes Maritimes* aggregate to the Marseille dominated network. Only the small network around Cogolin preserves its independence, which is lost at the fourth and last step of the analysis.

3.4 Significant Dominant Flows - MLA Approach

The goal of the MLA approach to significant dominant flows is less to determine a complete partition of the regional space than to define the core of functional areas which are strictly dominated (see 2.3), directly or indirectly, by an urban centre. By no surprise, the resulting regional map (Figure 5) is characterised by the presence of several independent units not belonging to any of these functional areas, as they show significant outflows to several destinations. The regional space is segmented in a bigger number of functional areas at both steps of the analysis (60 level 1 networks and 19 level 2 networks). Only the attractive power of Marseille and Nice is capable of establishing vast functional areas at the first step of the analysis. These networks are nevertheless unable to extend their attraction at the second level of the analysis to reach the perimeter of modularity optimization level 2 communities.

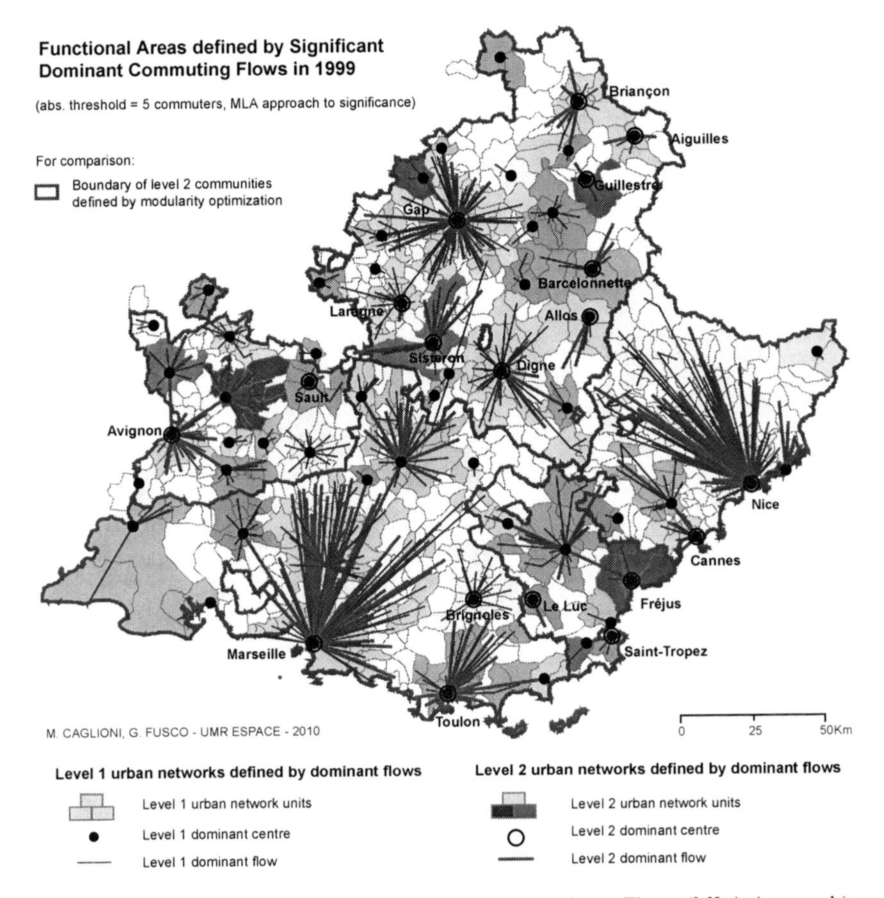

Fig. 5. Hierarchical Clustering through Significant Dominant Flows (MLA Approach)

The slower convergence of spatial units in wider clusters when using the MLA approach to significant dominant flows makes it interesting to further iterate the algorithm. Indeed, the evolution of urban networks stops at level 9 of the analysis (Figure 6). The results are worth commenting. First of all, at the difference of modularity optimisation and of other approaches to dominant flows, the method does not result to a unique functional area covering the whole regional space. At the same time, at the difference of both modularity optimisation and dominant flows with thresholds, a vast functional area covers the south-western half of the regional space, encompassing the Departments *Bouches du Rhone* and *Vaucluse*, as well as part of the Departments *Var* and *Alpes de Haute Provence*. At the other extreme, the Alpine space, resulting in one only level 2 community (modularity optimization) and of two main urban networks (significant dominant flows) remains fragmented in 14 different small to medium urban networks. Finally, the eastern part of the region, partitioned between *Alpes Maritimes* and eastern *Var* by modularity optimization, and between a vast French Riviera network and a much smaller eastern network from significant

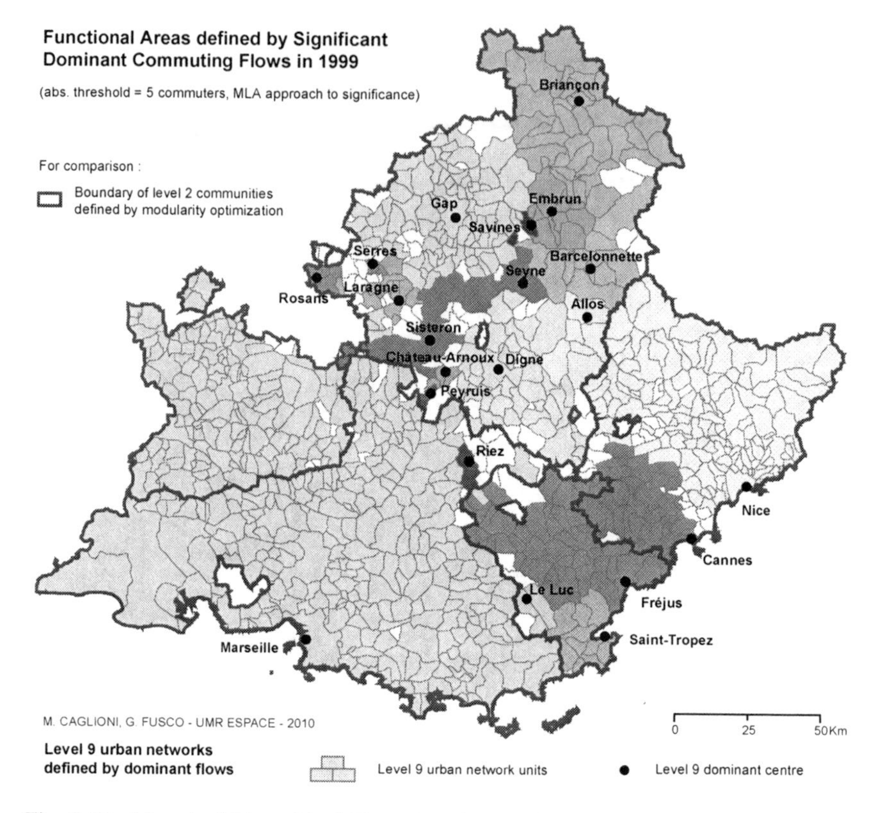

Fig. 6. Final Level of Hierarchical Clustering through MLA Significant Dominant Flows

dominant flows, is now structured by four main networks (dominated by Nice, Cannes, Frejus and Saint-Tropez, respectively). The MLA approach thus highlights the existence of three different spatial structures within the region: a hierarchical set of networks coalescing in a vast metropolitan area dominated by Marseille in the south-west, a highly fragmented alpine space in the north and a limited number of independent networks structuring the French Riviera metropolitan area.

4 Conclusions

Both modularity optimization and significant dominant flows confirm the inadequacy of department administrative boundaries to cluster functional areas in a hierarchical way. Beyond cluster boundaries, the main difference between the two methods is that dominant flows detect polarized regions, where it is possible to attribute different hierarchical levels to the centres and different degrees of complexity to the treelike network structures. Moreover, dominant flows can be used to understand the role of every single spatial unit within the network. This difference arises from the very methodological bases: modularity optimisation is aimed at optimizing the partition of space (under the assumption that modularity is the good function to optimize), dominant flows are aimed at determining the influence areas of main centres (centres which are found inductively within the same process that defines functional areas). The partitioning of space is a by-product of dominant flows approaches.

As a local method, dominant flows are capable of detecting the percolation of domination basins at their borders. The structure of the metropolitan area of the French Riviera is a good example, showing the percolation of dominance relationships from Cannes (the westernmost urban centre in the *Alpes Maritimes*) to the neighbouring urban centres of eastern *Var*. Significant dominant flows through the MLA approach hinder percolation detection through the mono-polarization constraint. Buffer zones or strong relations with several centres preserve the independence of networks at the second level of the analysis. This example can be generalized. With the MLA approach the formation of functional areas is somehow hindered as soon as interface elements, bi- or tri-polarised by different centres are detected. MLA can thus be used to explicitly detect interface structures among major functional areas.

In this respect, the two approaches to significant dominant flows do not necessarily show different realities. They rather cast a different light on the same geographical reality. For the French Riviera area, the MLA approach points to the existence of three main centres defining domination basins around them, whereas the threshold approach highlights the unity of the metropolitan area, defining a hierarchy within a unique polycentric structure. Nevertheless, in comparison to the Marseille metropolitan area (converging to a unique network through iterated MLA), the French Riviera shows a lesser degree of hierarchical domination among the urban centres.

Finally, as a local method, dominant flows (however their significance is defined) result in space partitioning which do not necessarily optimize flow containment, but which are locally robust (i.e. independent from changes concerning distant units). At the opposite, modularity optimization is a global method, where the belonging of a given unit to a community depends from the whole set of flows within the study area.

As already observed, combining modularity optimization and dominant flows allows a deeper understanding of the geographical reality under investigation. The comparison carried out in this paper did not address the computational performance of the different methods as other authors have already proposed (Lancichinetti and Fortunato 2010). Our main interest was comparing the results produced by the different methods for a given study area in order to understand its internal organisation in terms of hierarchical sets of functional areas. Beyond the dismissal of dominant flows used without any concern of significance, the main results produced by modularity optimization and by significant dominant flows (both through thresholds and MLA) could always be supported by expert knowledge of the study area. When conflicts arose among the results of the different methods, we could appreciate their complementarity in moderating the conclusions obtained from a single method. The resulting view offered by all the methods employed can hopefully near the insight of the analysis to the complexity of the geographical reality under investigation. This opinion is shared by authors like Porter et al. (2009), underlining the considerable value of having multiple computational heuristics available, as they are likely to give complementary views over the same dataset.

A major perspective of research is the extension of the comparative analysis of methods for the partitioning of space, derived from both regional science and complex network analysis traditions. Global optimality of traditional regional science methods will have to be assessed, as well as the pertinence of the methods developed from complex network analysts for the study of geographical space. In this respect, local methods of network analysis seem particularly interesting, as k-clique percolation (Palla et al. 2005) which share with the MLA approach to dominant flows the ability to detect interface / overlapping structures between larger clusters.

References

Andersen, A.: Commuting Areas in Denmark. AKF, Copenhagen (2000)

Barabási, A.: Linked: How Everything is Connected to Everything Else. Plume, 304 p (2003)

Berroir, S., Mathian, H., Saint-Julien, T., Sanders, L.: Mobilités et polarisations, In: Bonnet, M., Aubertel, P. (eds.) La ville aux limites de la mobilité, pp. 71–82. PUF, Paris (2006)

Berry, B.: Growth Centers in the American Urban System. Ballinper, Cambridge (1973)

Blondel, V., Guillaume, J., Lambiotte, R., Lefebvre, E.: Fast unfolding of communities in large networks. Journal of Statistical Mechanics: Theory and Experiment 10(10008), 12 (2008)

Brown, L., Holmes, J.: The delimitation of functional regions, nodal regions, and hierarchies by functional distance approaches. Journal of Regional Science 1, 57–72 (1971)

Casado-Diaz, J.: Local labour market areas in Spain. Regional Studies 34, 843–856 (2000)

Clauset, A., Newman, M., Moore, C.: Finding community structure in very large networks. Phys. Rev. 70(066111), 1–6 (2004)

Dauphiné, A.: Espace, Région et Système. Economica, Paris (1979)

De Montis, A., Caschili, S., C.: Commuter networks and community detection: a method for planning sub regional areas. Physics and Society 19 (2011) arXiv:1103.2467

Fortunato, S.: Community detection in graphs. Physics Reports (2010) arXiv:0906.0612v2

Freeman, L.: A set of measures of centrality based betweenness. Sociometry 40, 35–41 (1977)

Fujita, M.: Urban Economic Theory. Cambridge University Press, Cambridge (1989)

Fusco, G.: Modelling Urban Networks from Spatial Interaction Data. In: Scarlatti, F., Rabino, G. (eds.) Advances in Models and Methods for Planning, Pitagora, pp. 63–72 (2009)

Fusco, G., Decoupigny, F.: Logiques réticulaires dans l'organisation métropolitaine en région Provence-Alpes-Côte d'Azur. In: XLVe colloque de l"ASRDLF, Rimouski (2008), http://asrdlf2008.uqar.qc.ca/Papiers%20en%20ligne/FUSCO%20G.%20et%20DECOUPIGNY%20F._texte%20ASRDLF%202008.pdf

Girvan, M., Newman, M.: Community structure in social and biological networks. In: Proceedings of the National Academy of Science, vol. 99, pp. 7821–7826 (2002)

Goodman, J.: The Definition and Analysis of Local Labour Markets: Some Empirical Problems. British Journal of Industrial Relations 8(2), 179–196 (1970)

Haggett, P., Cliff, A., Frey, A.: Locational Analysis in Human Geography. Edward Arnold, London (1977)

Johansson, B. (ed.): Infrastructure, Market Potential and Endogenous Economic Growth. Department of Civil Engineering. Kyoto University, Japan (1998)

Karlsson, C.: Clusters, Functional Regions and Cluster Policies. JIBS and CESIS Electronic Working Paper Series (84) (2007)

Kaufman, L., Rousseeuw, P.: Finding Groups in Data: An Introduction to Cluster Analysis. John Wiley & Sons, England (1990)

Kipnis, B.: Graph Analysis of Metropolitan Residential Mobility: Methodology and Theoretical Implications. Urban Studies 22, 179–187 (1985)

Lagnerö, M.: Local Labour Markets. Working Paper. Statistics, Stockholm (2003)

Lancichinetti, A., Fortunato, S.: Community detection algorithms: a comparative analysis. Physics and Society (2010) arXiv:0908.1062v2

Lombardo, S., Rabino, G.: Urban Structures, Dynamic modelling and Clustering. In: J, Hauer (eds.) Urban Dynamics and Spatial Choice Behaviour, pp. 203–217 (1989)

Newman, M.: Fast algorithm for detecting community structure in networks. Phys. Rev. 69(066133), 1–5 (2004)

Nystuen, J., Dacey, M.: A Graph Theory Interpretation of Nodal Regions. Papers and Proceedings of the Regional Science Association 7, 29–42 (1968)

Palla, G., Derényi, I., Farkas, I., Vicsek, T.: Uncovering the overlapping community structure of complex networks in nature and society. Nature 435, 814–818 (2005)

Porter, M., Onnela, J., Mucha, P.: Communities in Networks. Notices of the AMS 56(9), 1082–1097 (2009)

Rabino, G., Occelli, S.: Understanding spatial structure from network data: theoretical considerations and applications. Cybergeo, n°29 (1997), http://cybergeo.revues.org/2199

Tolbert, C., Sizer, M.: US Commuting Zones and Labour Market Areas. A 1990 Update. US Department of Agriculture, Washington DC (1996)

Van der Haegen, H., Pattyn, M.: An operationalization of the concept of city region in West-European perspective: the Belgian city regions. Tijdschrift voor Economische en Sociale Geografie 61, 70–77 (1980)

Van der Laan, M., Pollardy, K., Bryanz, J.: A New Partitioning Around Medoids Algorithm. University of California, Berkeley, Working Paper Series, Paper 105, 1–12 (2002)

Van Nuffel, N.: Determination of the Number of Significant Flows in Origin-Destination Specific Analysis: the case of Commuting in Flanders. Regional Studies 41.4, 509–524 (2007)

Wasserman, S., Faust, K.: Social Network Analysis: Methods and Applications. Cambridge University Press, Cambridge (1994)

Weaver, J.: Crop combination regions in the Midwest, Geog. Review 44, 175–200 (1954)

Willaert, D., Surkyn, J., Lesthaeghe, R.: Stadsvlucht, verstedelijking en interne migraties in Vlaanderen en Belgie. Vrije Universiteit Brussel, Brussels (2000)

Analyzing and Testing Strategies to Guarantee the Topological Relationships and Quality Requirements between Networks Inside BTA and BTU Data Models

Francisco Ruiz-Lopez, Eloina Coll, and Jose Martinez-Llario

Instituto de Restauración del Patrimonio, Universitat Politècnica de València, Camino de Vera s/n, Valencia 46022, Spain
fraruilo@gmail.com, {ecoll,jomarlla}@cgf.upv.es

Abstract. Along this article we are going to explain some interesting strategies to reach the quality required for the phenomena considered inside the feature catalogue for the new BTA[1] data model proposed by the Spanish Cartographic Norms Committee. We will focus on guaranteeing network connectivity, data accuracy and topological relationships between geometries of different phenomena. To reach our goal we will use ArcGIS software suite and its spatial analyst extension, building our own rules trying to automate the process and avoiding spending time to other interested users reusing our data model topology. The final goal of this work is to promote the migration to the new data model to earn interoperability between every cartographic producer by providing tools for potential users.

Keywords: Topology; data model; BTA; cartography; network; spatial analysis.

1 Introduction

Nowadays a lot of cartography producers are available, but usually there are no restrictions about how the spatial databases should store or manage the information, usually every producer has its own data model. This situation reduces the interoperability and makes more difficult exchanging and mixing datasets from different sources.

As a consequence, the European Parliament has approved the INSPIRE[2] Directive[1] and many governments and Public Administrations have joined this philosophy, which tries to get a unified data model suitable for mostly every standard producer.

Due to this, the Spanish Cartographic Norms Committee has produced the BTA [2] data model and is elaborating the specifications document for the official proposed

---

[1] BTA is the Spanish abbreviation for Harmonized Topographic Database, a data model for intermediate scales (1:5.000 – 1:10.000).

[2] INSPIRE is the acronym of Infrastructure for Spatial Information in Europe.

B. Murgante et al. (Eds.): ICCSA 2011, Part I, LNCS 6782, pp. 152–161, 2011.

data model for urban scales, the BTU[3][3], this new data models pretend to unify rationally the new cartographic datasets according to the standards of organization and quality of spatial information that the society demands, trying to follow as much possible the INSPIRE recommendations.

In this paper we will focus on hydrograph and transport network, two of the eight families which contain all the phenomena considered in the catalogue.

The considered data model requires, the same way as INSPIRE does, that all lineal geometries belonging to the transport or hydrograph configure a continuous network.

Moreover there are more specific requirements related with logical coherence, spatial accuracy and diverse cartographic quality parameters which have to be checked before considering the dataset finished.

Hence we will explain how the dataset should be organized in BTA, check the connectivity and then ensure the particular quality controls for the phenomena in such datasets.

The maintenance of a network topology becomes a hard work task when the network reaches certain level of extension and complexity for the mentioned BTA and BTU data models. Acquiring software capable to perform this task from the standard datasets which come from the official cartographic datasets it would entail big economic resources and only the bigger and stronger Public Administrations are capable of such efforts. Because of that we propose an alternative method within reach of town councils and other users with fewer resources, because it only requires the GIS expert, who will use the spatial analysis tools available in the most part of the software available nowadays, avoiding unnecessary expenses in its exploiting.

Actually, there are some teams researching different faces of topology applied to networks, mainly related with services like the electrical network[4], although our research is focused on the geometry of the hydrograph and transport network because those are the mandatory considered in the data model as networks for facilitate future applications of routing and management.

2 Starting Dataset

The attached figure shows the phenomena present inside those families, those are (hierarchically speaking) "parent" features, inside them there are more "son" phenomena, for instance parent railroad has *Underground*, *Tram*, *Funicular*, etc.

We will use ArcGIS for all the spatial analysis so first of all we are going to create an ESRI Geodatabase to store all the information and topologies. Then we will create two new feature datasets, one for transport features and another one for hydrographic elements.

In this process we have to be cautious because our spatial data is prepared for scale 1:5.000, and the software uses cluster, XY and Z tolerances so we have to use

[3] BTU is the Spanish abbreviation for the Urban Topographic Database, a data model for big scales (1:500 – 1.1.000).

[4] There are a lot of topology applications, but nowadays due to its complexity, topology helps managing large electrical installations and electronic circuits, we have some examples like [4] and [5]. Researchers also work with transport network, trying to represent and predict traffic flows with GIS tools [6].

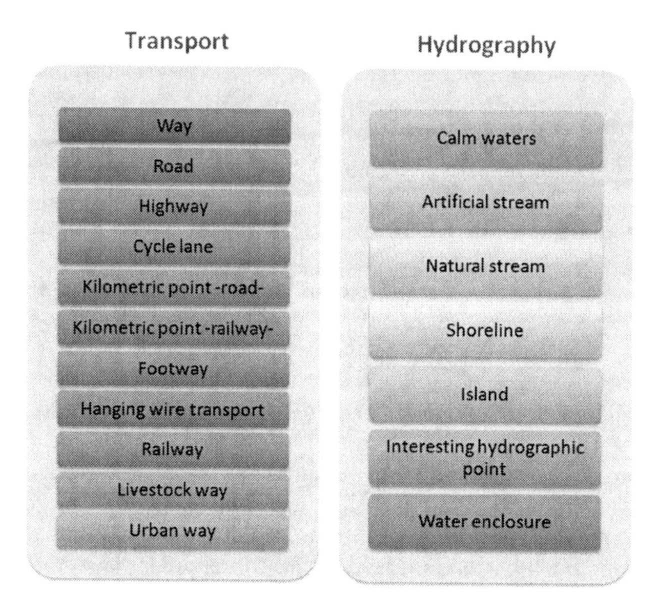

Fig. 1. Parent phenomena for the analyzed features

appropriate values for such parameters [7]. The reference system used is the official by law which is considered in the specifications of the BTA and also in the new BTU [8].

Once the feature datasets are created we should load the geometries to our data model. There are some data models already materialized in an ArcGIS Geodatabase that follows BTA and BTU specifications, so we are going to reuse one which is already developed by our researching workgroup.

Due to reusing the Geodatabase we are saving a considerable amount of time and avoid accidental mistakes creating the database structure, to know more about how to create and reutilize Geodatabases in BTA format there are some papers about it [9].

3 Preparing the Spatial Datasets

3.1 General Splitting

First of all, it is mandatory to split all lineal features inside the dataset because we are going to do spatial analysis checking relative positions between features, and if we do not split the arcs into segments between nodes, those features could share some part of the arc but not the entire one. So it would be much more difficult to do such analysis in a right way.

It is mandatory to do that, even if such features belongs to different layers, because in this data model we consider the topological relations between all the layers due to the representation hierarchy is explicit in one field for all lineal and polygonal phenomena. This way allows us the use of automatic modeling for overlapping entities, and use this field as base for the cartographical representation in map layouts or online servers.

The attached image (Fig.2) contains a graphical example of a small polygon (a building) being "cut" by the cartographic sheet line. It should be split at the points marked with the small black arrows because that segment should be classified as *Cartographic Sheet Contour Line* (case A), however, by default, the full contour is classified as *Contour* (Case B). Owing to the fact that the producer did not consider our final data model as production data model when the dataset was digitalized.

There are two ways to fix this classification mistake, by hand or using spatial analysis tools. Of course the best choice is automating all the process with scripts and taking profit of the tools provided with ArcGIS suite or any other GIS Suite.

But for processing automatically all this steps we have to be able to assign different values to the same polyline, and the only way to do this is, as we said before, is by splitting the contour geometry, and then reclassifying attributes.

Moreover there is no need to differentiate between elements in this step, so we just split and update the feature classes in the database.

Then we are ready to continue with the creation of new derived feature classes.

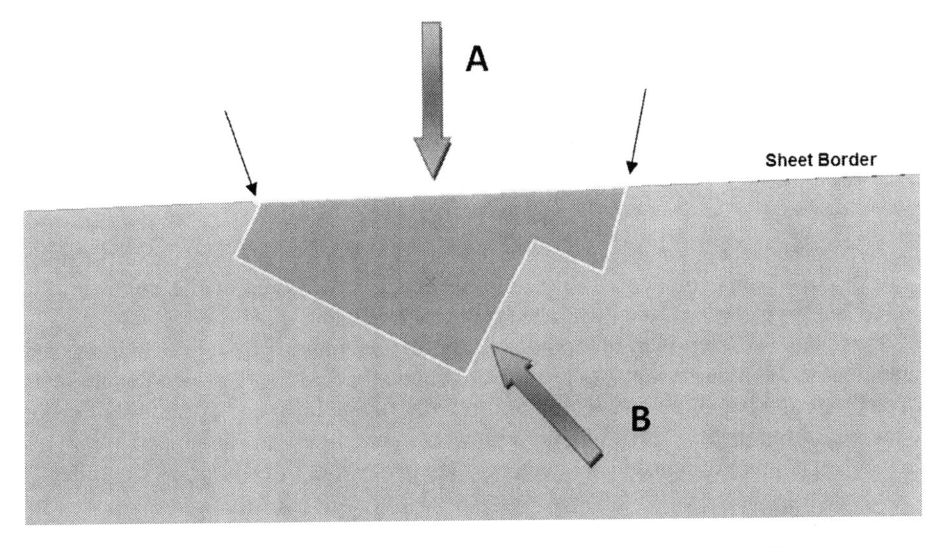

Fig. 2. Example of mistaken classification and the need of splitting and reclassification process

3.2 Feature Class Subdivision

Secondly, it is important to point out the representation attributes of these phenomena which belong to the feature catalogue.

There are four primitive geometries allowed, polygon (2D), polyline (1D), point (0D), and texts.

However only two of them have attributes designed to control the cartographic representation by showing topological relations between elements of different layers, particularly polygon and polyline.

It is mandatory that polygon layers have "Componen2D" field that shows the hierarchical relation between other polygons in the same spatial neighborhood and polyline layers have "Componen1D", which shows the different types of lines which are needed to represent properly the spatial data considering hiding between elements, virtual borders to maintain topology and consistency or linear elements to configure and ensure connectivity in networks.

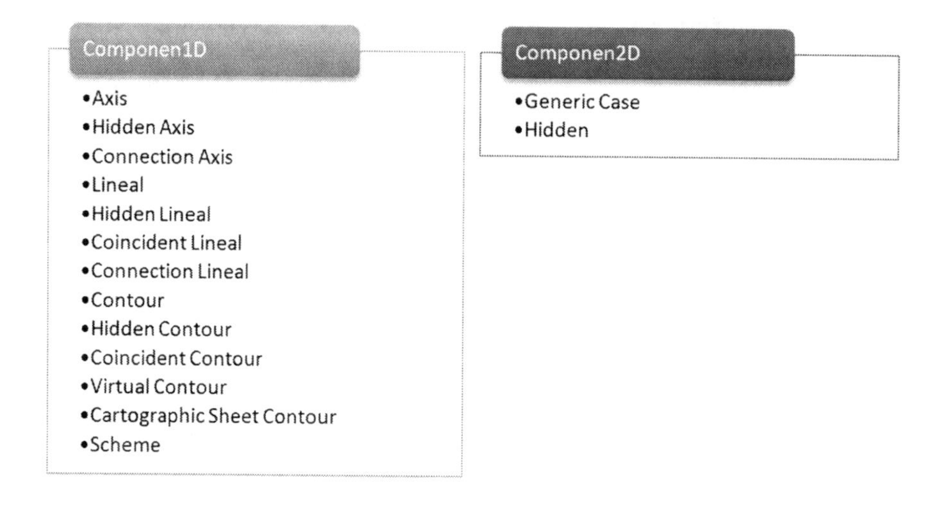

Fig. 3. Allowed values for specific representation fields in the spatial database

As a consequence, we should not consider all the geometries of one layer the same way, because some of them belong the network or to represent the contour of a polygonal geometry, and they will need different treatment.

Then, the next step is to split and separate the geometries by representation, and attending to the requirements for the topology, mainly creating, for each feature, axis geometries and lineal geometries for connectivity in one layer and different layers for contour, cartographic sheet contour, virtual contour, hidden contour and other for coincident contour (scheme is not contemplated in network or hydrography families).

3.3 Applying Topology

We will consider two sets depending on the purpose of such geometries. The first one is composed by Axis and Lineal families with all the secondary options (hidden coincident, etc), and their purpose is to create the network. The other geometries are considered for controlling the cartographic representation of areal phenomena by modeling the hierarchical levels between touching elements.

Between network elements three-dimensional connectivity is required (we have to keep in mind that the cartographic datasets should be in 3D).

Rarely we can find crossroads with disjointed ways, or a river that stops its flow suddenly and disappears. To fix it we have to apply topology rules to detect and fix *Dangle errors* which are segments that are "hanging" and do not intersect with the node which in real world is connected.

Of course there are exceptions, like the end of a road, the cut river with the sheet border or even a water flow arrives to a sinkhole and disappears to continue underground (frequent in Karst terrains), those cases should be marked as exceptions.

There is another interesting topological rule, related with the cartographic sheet border. All the network elements which finish in the border should be cut by it, they could not finish near, just in case this element finish also in the real world. Also this topological rule works to find disconnected crossroads between elements from different phenomena, like connections between roads and streets or ways. To avoid mistakes this correction should be helped with orthophotographs to check the geometry in the real world.

At the next image we can see two examples, the left one is the graphical definition of dangle error, there is a crossroad and the two axis don not touch each other, so there is no network node and it should be, so "Road B" must connect with "Road A" extending the axis and creating a network node.

The right image is related with the other rule, *Endpoints must be covered by*, this rule has to consider all the possible interactions between every phenomenon, so the amount of rules inside this topology will be large, but it is not an inconvenient because we can save the rules in a file and access it always we need. There are a way and a road at a crossroad. We can see how the darker arrow points the connection axis, usually classified as dangle, however it must connect with the road, and finish "covered" by the road axis feature class hence the lighter arrow does not point a dangle, points to a correct topological interaction once the topology is validated and corrected.

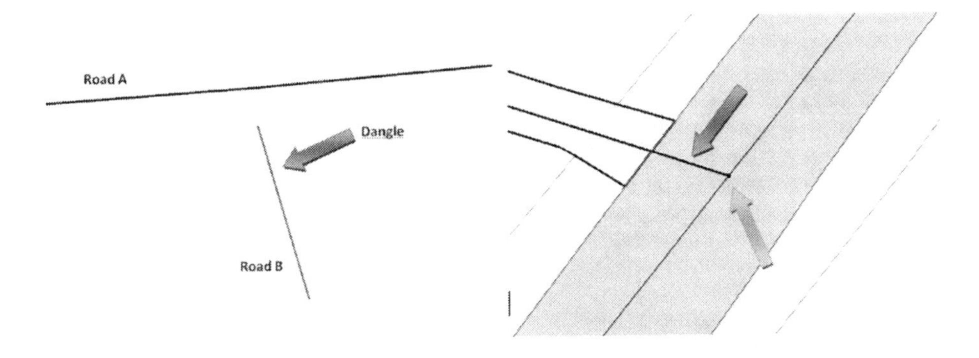

Fig. 4. Graphical examples for frequent topological errors in network datasets

Connection axis also should be covered by polygons, so we have to check it with the appropriate rule, if that rule does not exist we have to program it.

Also hidden axis should be covered by polygons from other layers, and it is important to establish the hierarchical levels for all the phenomena to model in an accurate way the hiding geometries. Considering it inside the program is the best choice, applying ranks and subtypes to model it.

Creating our own topology rules in ArcGIS is possible by using the integrated development environment of Visual Basic for Applications or by programming in another language compatible COM. Although we recommend VBA because it is easy, fast and works properly with applications of standard requirements. We can find an example of topological programming with ArcGIS in the tool developed for managing cartography datasets at the Cartociudad project [10].

In other cases, better than VBA, it would be better using JAVA, because has some advantages (multiplatform, security, extended...), but it depends on the software used by the technician, a good example of a GIS applied JAVA software is JASPA[5] [11].

There is another topology rule that can be used, this rule is not directly related with network topology, but helps with the data model consistency. It is important to associate the kilometric points to the correspondent network axis. So we can check if such kilometric points are inside the polygon of the network communication, taking into account that some elements haven't got superficial representation, like street way, and then we will not be able to generalize this process for all phenomena in the dataset.

Then, in a first approach we will check if these points are inside polygons with *Contains point* rule, fixing all those points which are too far from the road nad helping with orthophotographs.

After that, we will use another rule to check that all points are over the axis, not just inside the road, moving the point to the axis following the perpendicular projection, this way fixes completely the position of these kilometric points.

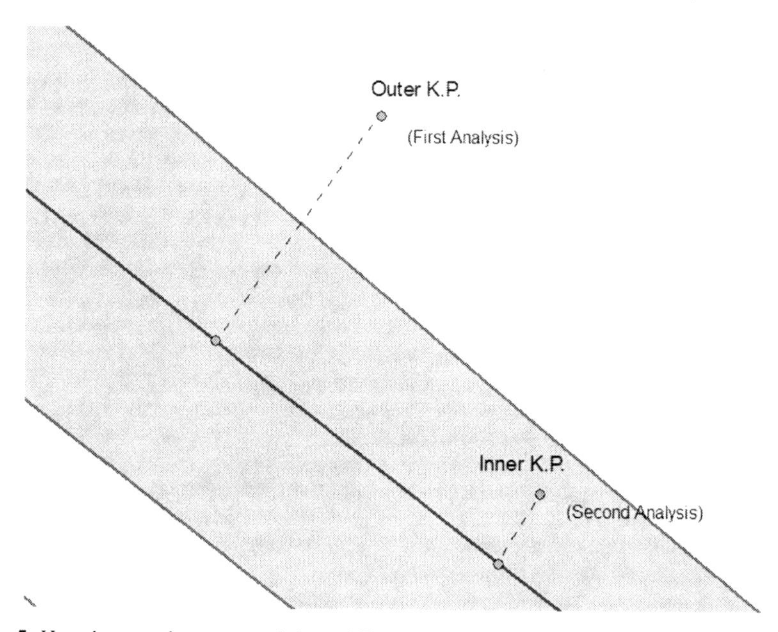

Fig. 5. How have to be corrected those kilometric points which are out of the communication axis. The outer will be corrected in the first process, helping to not confuse the communication way which owns the kilometric point. The second one fixes all points more accurate than first, and faster. The dashed line shows the projection over the axis. Using one or both of them depends on the quality of the dataset.

---

[5] JASPA is the acronym for JAva SPAtial, which is a spatial extension for relational databases. It is under development in the Universitat Politècnica de València.

3.4 Alternative Way for Kilometric Points

Better tan establishing the kilometric points by hand it probably should be a good idea establishing the fourth dimension (M-dimension) to our axis and lineal feature classes.

Using this extra dimension the information about so far is a point of the road is far from the origin of the dataset is implicit in the own geometry. Avoiding mistaken kilometric points or extra work.

Moreover, it gives to us better information along all the way and there is no need to interpolate values. Also analysis makes easier with this dimensional component. The negative point consists in the small amount of software available to manage this dimension. And the difficulty to add such information to all the axis and linear geometries presents in the dataset.

There is another rule that should be programmed, this is much more difficult and depends on the existence of kilometric points properly stored. BTA data model requires checking the direction of axis and lineal phenomena for main communication networks, like roads or highways.

This direction should be coincident with the kilometric points value, if it is positive the road should "advance" and the other way around.

This tool requires that the kilometric points are already over the axis and their existence. Checking this control is much more difficult than the rest.

3.5 Topology about Boundary Elements

The contour geometries should have similar topology rules, based on hiding and modeling supposition of elements. Also considering the cartographic sheet border for reclassify it properly.

The main difference is that in this case it might be other features from other families than interact with such boundaries. So we cannot limit the involving feature class to transport or hydrography, we have to follow the same process than before but all the features present in the data base.

This is a very big amount of job, but it should be done once, because the data model (BTA or BTU) will not change, or at least not too much and then it would be easy to readapt the topology. And we can store the topology and the topology rules inside the Geodatabase. The other software, even free software, also allows this, so we can reuse this topology for every dataset as long as our information is those data models.

We can also do it with information which follows other data models, but then it is mandatory to adapt first to our base data model and the apply topology. There are available some paper about this topic to help the reader to establish correspondence tables and loading data process [12].

The topological relationships between elements from other layers should consider the vertical component relative to contour level lines. These lines have to cross the boundary of transport and hydrography network elements.

4 Alternative Way

We can also use the Network Analyst tools provided by ESRI to create a new network dataset and take profit of those automatic routines already programmed to control the network topology. In fact, this could be considered the second stage for our goal, because with this tools we can also calculate routes, times, optimize the network even considering topographical influence or the real path followed by the car instead the line between network nodes.

This choice is better if our goal is calculating or optimizing routes but is particular for this proprietary software, therefore is not a good idea because the quality controls for those networks should not be restricted by the producer's software.

We have to keep in mind that the data model has been developed for being used independent of the software or the data format.

Although in first option we used topology rules of ArcGIS such rules are available in a similar way in other software, like gvSIG [13], with the advantage that this software is completely free and customizable.

This way does not consider the topological relationships of the boundaries.

5 Conclusions

Although topology is under investigation nowadays by some researchers, our approach focus on the geometric aspects applied to the geographical representation of the network elements, which is much less investigated than the other faces of the topology. Ensuring the cartographic quality components by guaranteeing the consistency between features (i.e.: the axis of the road must be contained in the polygon of such road, interconnecting features from its own layer and other layers of the data model and storing the right relationship in the representation field associated to such entities).

Guarantee the quality is the basement of the new data models which are being developed nowadays. These processes here commented are suggestions to the reader in order to help and optimize the job. There are more tasks which are being investigated actually and developing.

The main inconvenient is the difficulty to manage the large amount of feature classes and datasets that the new data models referred, which are multiplied to classify the representation.

Also the big effort needed to review and fix such amount of topological errors.

The advantage remains in the quality of the actual GIS tools provided by software, which helps considerably with such tasks. The ArcGIS Model Builder tool is especially interesting for earning time. However we are not forced to use the topology tools for ArcGIS, there are other software capable of performing those tasks, and some of them even are free, like the previously mentioned gvSIG, with its topology extension.

Furthermore once the topology is ready inside the database is ready to share and everyone can use it because the data model is the same. Also the topological rules can be saved and shared easily.

The actual society demands quality and tries to reduce costs, so this way to work increases interoperability and reduce extra work, by reusing the same model and sharing it.

This paper has been written from a research which is still in early phases to automate the quality controls inside data models.

Acknowledgements. This project is part of the research project "MOCAIDE", Creation and cartographic feeding of spatial data infrastructures at the Local Administration trough a data model integrating cadastre, urban planning and historic heritage, with reference CSO2008-04808 and financed by the CICYT and European funds.

References

1. INSPIRE Directive website, http://inspire.jrc.ec.europa.eu/
2. Harmonized Topographical Database specifications document(BTA), Special Committee of Cartographic Norms, http://www.csg-cnc.es/web/cnccontent/bta.html
3. Dimas, A., Aguirre, S., Coll, E., Martinez-Llario, J.: Modelo de datos de la Base Topográfica Urbana para la IDE local. In: JIDEE 2010 - Jornadas Ibéricas de la Infraestructura de Datos Espaciales, Lisboa (2010) http://www.idee.es/resources/presentaciones/JIIDE10/ID419_Modelo_de_datos_de_la_Base_Topografica_Urbana_para_la_IDE_local.pdf
4. Cao, Y., Zheng, W., Wang, F.: Study on Power Distribution Network Topology Model Based on GUID. In: 2nd International Conference on Information Science and Engineering (ICISE 2010), vol. (5690764), pp. 4070–4073. IEEE Press, New York (2010)
5. Deng, Q., Zhou, H., Chen, Y.: Topology analysis and modeling for smart distribution network based on complicated equipments. In: International Conference on Electrical and Control Engineering (ICECE 2010), vol. (5629715), pp. 4494–4497. IEEE Press, New York (2010)
6. Jiang, B., Liu, C.: Street-based topological representations and analyses for predicting traffic flow in GIS. International Journal of Geographical Information Science 23(9), 1119–1137 (2009)
7. ESRI ArcGIS 9.2 Desktop help, http://webhelp.esri.com/arcgisdesktop/9.2/index.cfm?
8. Real decreto de cambio de sistema de referencia cartográfico en España, http://www.boe.es/aeboe/consultas/bases_datos/doc.php?id=BOE-A-2007-15822
9. Ruiz, F., Coll, E., Martinez-Llario, J.: Análisis y estructura de una Geodatabase BTA para su utilización en la Administración Local. In: Jornadas Ibéricas de la Infraestructura de Datos Espaciales (JIDEE 2010), Lisboa (2010) http://www.idee.es/resources/presentaciones/JIIDE10/ID414_Analisis_y_estructura_de_una_Geodatabase_BTA.pdf
10. Cartociudad Project website, http://www.cartociudad.es/portal/1024/index.htm
11. JASPA project website, http://forge.osor.eu/projects/jaspa/
12. Ruiz, F., Coll, E.: Producción de Cartografía 1:5.000 ICV en formato BTA. Universidad Politécnica de Valencia, Valencia (2010)
13. gvSIG, http://www.gvsig.org/web/

An Assessment-Based Process for Modifying the Built Fabric of Historic Centres: The Case of Como in Lombardy

Pier Luigi Paolillo[1], Alberto Benedetti, Umberto Baresi, Luca Terlizzi, and Giorgio Graj

[1] Polytechnic of Milan
pierluigi.paolillo@polimi.it
http://webdiap.diap.polimi.it/paolillo/

Abstract. The assignment of admissible degrees of change to the building stock of an historic centre is one of the most important factors to be taken into account when developing an urban plan, because it establishes the possibilities of making changes to an existing urban fabric, or not; and since urban cores formed in ancient times contain the most significant concentrations of high-quality historic/architectural heritage, it is necessary to identify a way of specifically breaking down the changes to each building that are possible. In the case of Como, analysis of the morphological and socioeconomic characteristics of the fabric of the walled town and its extramural historic villages, and then the application of multivariate geostatistics, produced one such classification of this context in the form of six possible scenarios for change, of which the most significant was selected as the most effective for setting in train a profound upgrading of Como's built fabric, which is currently in a state of advanced degradation.

Keywords: Geostatistics and spatial simulation, Spatial statistical models, Spatial data analysis, Urban modelling applied to Italian historic centres.

1 The Built Fabric of Italian Historic Centres: Numerous Unresolved Issues

The question of what to do with urban centres that acquired their basic form in ancient times has always been much debated in Italy and from time to time, convictions have prevailed that often stemmed from defensive ideological positions, indifferent to the complexity of what they implied for urban fabrics so steeped in history and problems.

Whilst the literature is rich in such past experiences, it contains very few stimuli that could suggest a different method for assessment that might cause such anachronistic and misleading positions to be overcome and make it possible to revitalise the present unsatisfactory, and for the most part openly problematic, conditions in which most of Italy's historic centres now find themselves.

B. Murgante et al. (Eds.): ICCSA 2011, Part I, LNCS 6782, pp. 162–176, 2011.

This state of affairs suggests the need for a completely different approach to these matters, above all with regard to: i) awareness of the unsolved problems left behind by past efforts that were initiated and taken through to completion, but failed, ii) the technical possibilities that have now been introduced into urban planning by the use of territorial IT systems that make it possible to experiment with different types of analysis no longer based on functional zoning, but on the identification of urban areas that in themselves are homogeneous from numerous interdependent points of view (and which Geographical Information Systems now make it possible to identify).

With these considerations in mind we began to build up a procedure for analysing and assessing the value of historically valuable built heritage, as a way of defining criteria for establishing the permissible types of change to buildings, distancing ourselves from attitudes adopted since the end of the Second World War, which rightly were intended to safeguard a built heritage that was at risk of disappearing, but were based on merely identifying building types and on preventing changes to their basic structure and volumetry and, in general, any changes that went beyond straightforward maintenance and reinstatement, in the bizarre and wrong-headed belief that all of the built heritage within historic settings should be frozen in its current state, regardless of the actual characteristics of the buildings.

Whilst it is true that immediately after the Second World War the problem in Italy was to protect these historic centres from processes of replacing them with modern building and from the menace of demolition and reconstruction that had already been induced by the ravages of the war, 60 years on this overly protectionist and radical attitude has worryingly now left us with an historic built environment that has been immobilised and, indeed, for the most part, abandoned by those who originally populated it, attracted as they have been to new out-of-town settlements (which amongst other things have generated sprawl, reduced our agricultural space to a moth-eaten patchwork, shifted the residential parts of cities out to the edge, and banalised the language of architecture); these populations have been superseded by an ingress of marginalised social groups, often originating from outside the EU and in all cases indifferent to the intrinsic value or meaning of the history or the original roots of the urban process, motivated as they are to simply find a place, any place, in which to live and certainly not inclined to get involved in discussing its cultural and typological stratification.

2 Methodological Framework: Selecting Variable Factors and Extrapolating Suitable Scenarios for Making Changes to the Existing Built Heritage

Our examination of the past planning of historic centres, which has in fact crystallised them in a state of immobility whilst perpetuating (and in some cases accentuating) their degradation, enables us to avoid the same mistake and to move away from that approach by introducing a different method, which although it is still evolving (and has thus far been in experimental mode), introduces an analytical method into urban planning that is specifically designed to reuse the vast amounts of analytical baggage prepared by the many generations that up to now have investigated the historic building stock (especially in Como) but in our case orientating it towards a completely different type of geostatistically-based outcome concerned with the degrees of suitability for change that can be assigned to individual parts of built fabric.

This methodology consists of a sequence of analytical and assessment steps:

i) identifying descriptors on the basis of their legitimacy in relation to the Single Law on Building, on planning law, on the technical implementation regulations of municipal plans already in force, and on the most significant past experiences in the planning of Italy's historic centres from 1945 until now;
ii) consequently, identifying variables derived from legal regulations or significant past experience. So far as quantitative data for the specific urban situation of Como is concerned, this is obtainable from an already existing database;
iii) developing summary indicators that are able to describe, in both socioeconomic and historic/architectural terms, the spatial phenomena that occur in the historic core;
iv) complex processing of the indices thus produced, using multidimensional analyses and applying algorhythms to summarise these phenomena as we identified them, and to delineate possible scenarios for carrying out alterations to the built fabric;
v) by comparing these various scenarios for change, selecting the one that most closely accords with a previously chosen strategic characterisation for the historic centre.

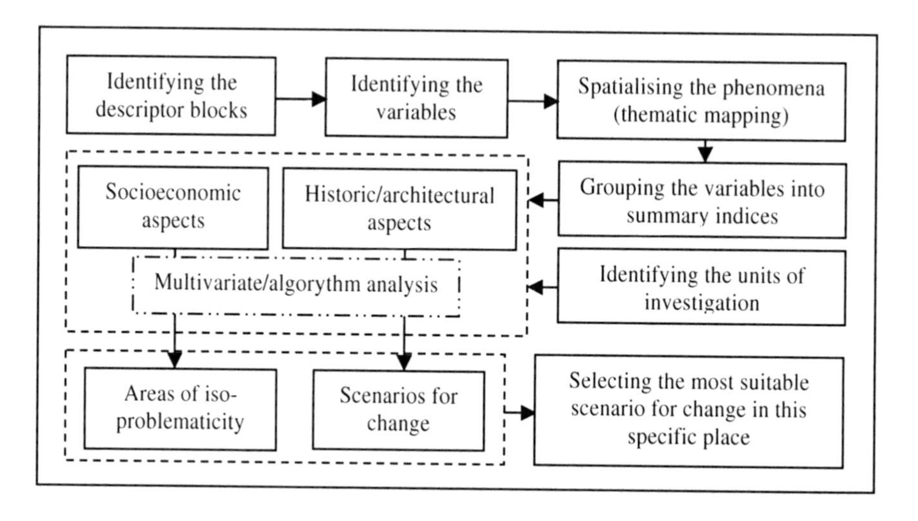

Fig. 1. Procedural diagram of the method, as applied to the case of Como

3 Method for Analysing and Assessing the Characteristics of Centrality and Socioeconomic Vitality

Our analyses of the relational space are concerned with an agglomerate of various factors that on the one hand relate to the intrinsic potentialities of this particular and deeply meaningful place, whose urban form was historically determined by the original "commune" structure of Como's nucleus, and on the other hand its social, demographic and economic aspects. fabric.

By identifying some innovative steps in previous plans for Como and its valley, we were able to take our analyses to a particularly advanced level; these processes, and the most important computer applications used, are described below.

3.1 Indices and Descriptive Variables of Currently Existing Spatial Phenomena

So far as defining the indicators to be adopted for investigating the territorial, configurational, and relational structures is concerned, the following function (1) indicates how various phenomenical aspects flow together as:

$$f(A,B,C,D,E) = f(a_1,a_2,b_1,b_2,b_3,c_1,c_2,d_1,d_2,d_3,e_1,e_2,) \qquad (1)$$

where:

A = tendency to accumulated centrality of the historic centre, as: $a_1, a_2 \in A$, where a_1 = degree of interaction of the street network, a_2 = degree of vitality of the economic fabric;

B = instability of the fabric due to monofunctionality, as: $b_1, b_2, b_3 \in B$, where b_1 = index of ageing, b_2 = residential density, b_3 = index of heterogeneity of non-residential activities;

C = density of localisation of the most strongly attractive activities, as: $c_1, c_2 \in C$, where c_1 = density of commercial activities, c_2 = density of service sector activities;

D = stability of the inter-relating residential fabric, as: $d_1, d_2, d_3 \in D$, where d_1 = incidence of the non-Italian resident population, d_2 = incidence of residential voids, d_3 = distribution of the population within buildings (underoccupation – overcrowding);

E = the urban landscape importance of buildings, as $e_1, e_2 \in E$, where e_1 = perception of a building, e_2 = characterisation of the perception.

3.2 Innovative Use of Methods, Tools, and Applications

The set of indicators we identified was arrived at not only by spatialising the various demographic, social, territorial and economic phenomena, but also by making use of special tools and applying innovative theories.

3.2.1 Use of the Configurational Approach to Extrapolate Urban Centrality

By using applications that identified the character of the urban fabric within the historic centre in relation to how the infrastructure network is configured, and that are designed to reveal its tendency to centrality and accessibility, we found it possible to obtain some extremely important input. The key reference in this regard is the work of Space Syntax, developed by Prof. Bill Hillier[1], whose fundamental ideas were incorporated into our configurational analyses and developed as MCAs (Multiple Centrality Assessments)[2].

Using these MCA applications, implemented as appropriate within a GIS environment, we assessed the intensity of those indicators that play a useful role in determining the tendency to accumulated centrality of the historic centre, specifically by calculating the degree of interaction between the individual units of investigation that we adopted and the street network, in terms of their Betweenness[3] and Local Closeness[4], the outcomes of which (in the case of Como) are as follows:

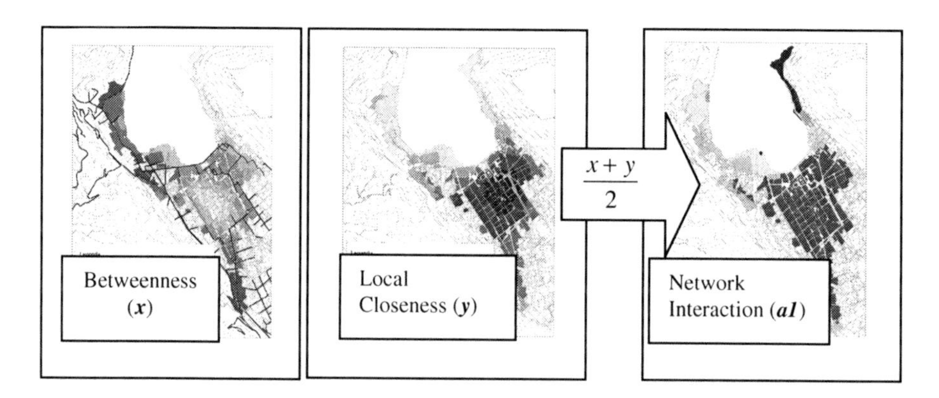

Fig. 2. Use of MCA indices to assess the tendency to centrality

In thus investigating how the functions in this urban armature are localised, our use of configurational analysis applications made it possible to identify those spaces with the greatest central vocation (and which also makes it possible to generate significant impacts on planning decisions) as compared to other areas in which the absence of functions, or a more problematic layout of the street pattern, have over time caused them to become marginalised, so far as their vitality and socioeconomic quality are concerned.

3.2.2 Analysing Urban Landscape by Studying Visibility and Perception of the Fabric

Another tool we used within the GIS environment for analysing the built fabric is the "viewshed", a series of applications that construct scenarios showing how buildings

[1] Space Syntax Laboratory, http://www.spacesyntax.org/.

[2] (Porta *et al.*, 2006a, b, c; Cardillo *et al.*, 2006; Crucitti *et al.*, 2006a, b; Scellato *et al.*, 2006).

[3] Betweenness makes it possible to identify the streets most strongly characterised as "spines" of urban quarters and that act as delivery centres for basic services and as the ordering elements of the urban configuration as a whole.

[4] Local Closeness is the measure that comes closest to the traditional idea of "accessibility" commonly used in transport and geoeconomics; it identifies the most compact and interconnected spatial systems that are relevant for planning the public transport network and the armature of services, at the urban scale.

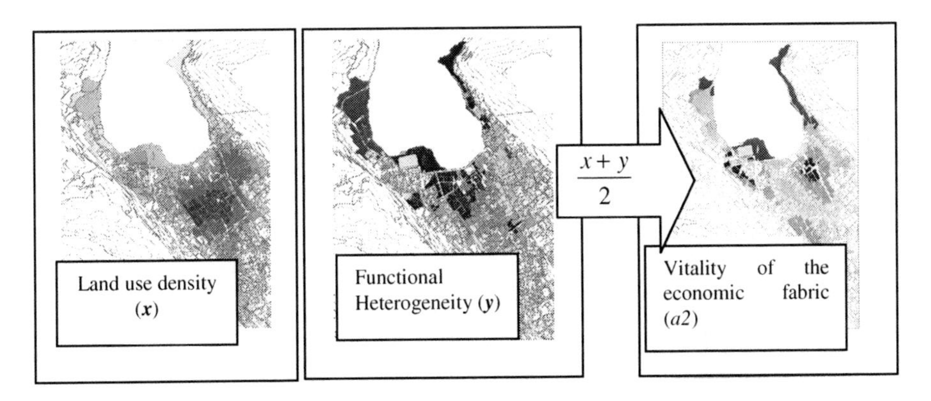

Fig. 3. Analysis of how functions are located in relation to density and heterogeneity of the economic fabric, in order to construct its index of vitality

are visualised (and consequently, perceived); the viewshed approach derives from attempts to subject the visual quality index (Visual Quality Condition) to quantitative and geostatistical analyses. In the case of Como we developed an index of the "landscape importance" for buildings, by taking the degree of perception of a building and its character and relating it to the connotating elements of its external appearance.

The following diagram shows the steps carried out to construct the index of perception e1.

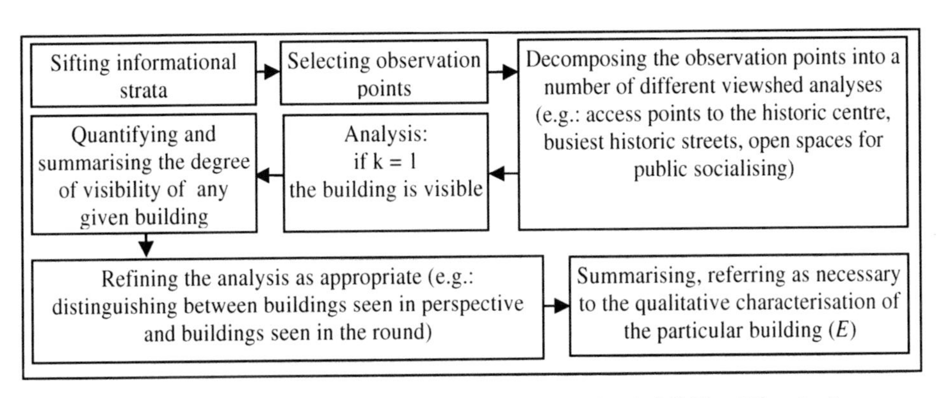

Fig. 4. Procedural diagram for constructing an analysis of visibility (Viewshed)

3.2.3 Applied Multidimensional Geostatistic Analysis

The applications discussed so far, derived from estimating $f\ (A, B, C, D, E)$ using aggregated indicators, were then summarised by means of applied geostatistics procedures, in particular multivariate analysis, using dedicated AddaWin software in the GIS environment. Inputting these variables into the units of investigation required the application: ***i***) of correlational analysis, aimed at reducing the number of variables to be tested and used in the subsequent stages; ***ii***) of non-hierarchical analysis, which makes it possible to identify the spaces that are homogeneous in terms of the variables adopted for describing the model.

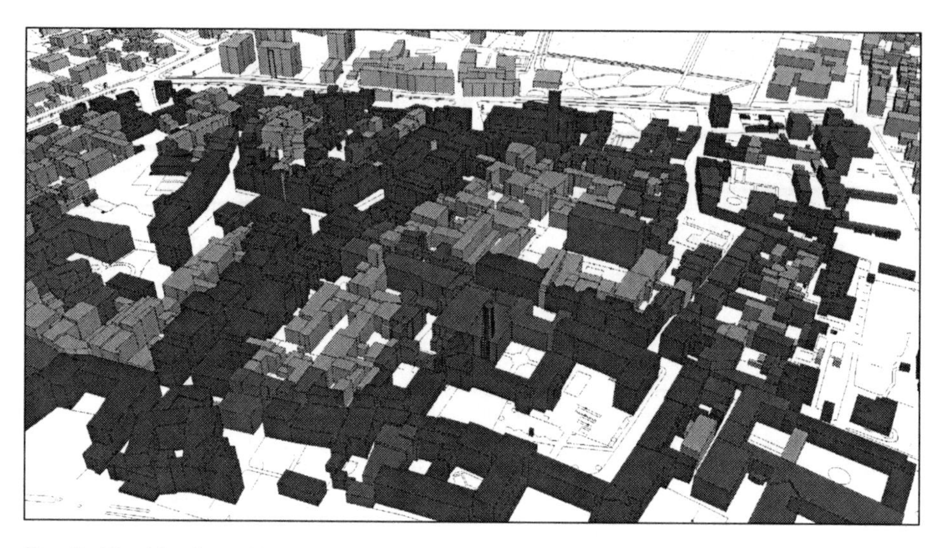

Fig. 5. Classification of buildings by degree of cumulative perception, from light to dark as intensity increases

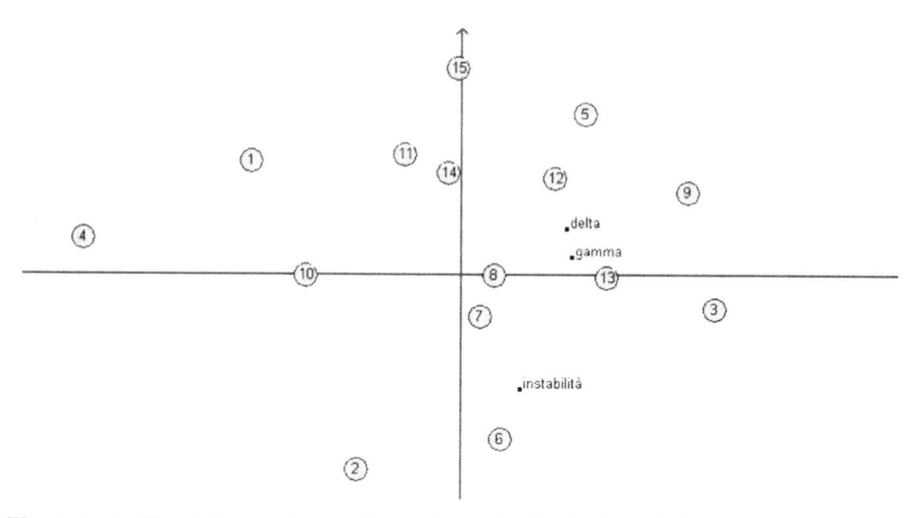

Fig. 6. A significant Facplan image that analyses the distribution of the centres of the stable profiles with respect to the two axes that most effectively explain the multivariate situation under investigation

Thereafter the pathway becomes a matter of constructing homogeneous clusters and then interpreting the relationships between classes and variables. By examining where the variables are located along pairs of factor axes (in the example shown here, attractiveness for sensitive population groups, centrality and vitality in terms of the commercial and service sectors, and conditions of instability due to monofunctionality) we obtain a two-dimensional distribution of the individual classes of homogeneous profiles generated by the non-hierarchical analysis. By developing these axes and the non-hierarchical analyses, the characteristics of each of the classes that emerged for each of the variables adopted for analysis can now be identified.

As necessary, several classes can then be bundled together into a single cluster that behaves homogeneously with respect to the three components of multidimensional analysis that were taken into consideration.

4 Assessing the Values of the Built Fabric

Our analyses of this urban place are based on the character of the buildings in relation to a number of their conformative factors and intrinsic within them, the possibility of extrapolating the types of alteration to individual buildings that could be permitted[5].

4.1 Setting Up Indices and Variables to Explain Physical/Structural Characteristics

For establishing indicators, of types that lend themselves to investigation, about the forms of buildings, the following formula (2) shows how a range of different phenomenical aspects flow together, as:

$$f(F,G,H,I) \tag{2}$$

where:
F = suitability for retention, intrinsic in the building itself, as:

$$f_1, f_2, f_3, f_4, f_5 \in F ,$$

where f_1 = the presence of environmental and historic/architectural constraints, f_2 = the persistence of buildings over time, f_3 = the state of conservation of buildings, f_4 = the presence of facades to be retained, f_5 = the presence of excrescences;
G = suitability for retention, in the relationship between building and urban block, as:

$$g_1, g_2, g_3 \in G ,$$

where g_1 = the presence of transformed areas (causing discontinuities in the urban fabric), g_2 = the degree of uniformity with respect to the first historic threshold adopted, g_3 = the degree of uniformity with respect to the number of storeys;
H = the urban landscape importance of the context in which the building stands, as:

$$h_1, h_2, h_3, h_4 \in H ,$$

[5] We based this analytical framework on inputting the information collected into a database, and reprocessing it to classify many different aspects of the built heritage in the Como valley, including its threshold date of construction, its degree of persistence, its degree of uniformity in relation to the number of floors, as well as the presence of any high-quality parts in a particular building, and their state of conservation.

where h_1 = the presence of elements of quality, h_2 = the presence of elements of contrast, h_3 = the index of perception of the building, h_4 = the incidence of the building on the public space;
I = suitability for increases in volume, as:

$$i_1, i_2 \in I$$

where i_1 = the character of the existing built frontage, i_2 = the possibility of adding extra storeys, after further analytical investigations have been carried out within the urban plan.

Table 1. The dataset adopted for assessing built heritage in relation to historic/architectural values

<table>
<tr><th>Unit</th><th>Descriptor</th><th>Variable</th></tr>
<tr><td rowspan="12">Single building</td><td>A – Finishes of the building</td><td>A1 – Facades to be retained</td></tr>
<tr><td>F – Condition of the building</td><td>F1 – State of conservation of the building</td></tr>
<tr><td>G – External profile of the building</td><td>G1 – Number of storeys</td></tr>
<tr><td rowspan="3">O –Historic or architectural constraints</td><td>O1 –Subject to architectural constraint</td></tr>
<tr><td>O2 –Architectural constraint zone</td></tr>
<tr><td>O3 –Environmental constraint zone</td></tr>
<tr><td rowspan="2">N – Elements not related to the original characteristics (excrescences)</td><td>N1 – Additional storeys added</td></tr>
<tr><td>N2 – Additional volumes added</td></tr>
<tr><td rowspan="2">P – Buildings of architectural quality</td><td>P1 – Presence of high-quality built fabric</td></tr>
<tr><td>P2 – Presence of buildings constructed before 1860</td></tr>
<tr><td rowspan="3">R – Built space (not the same as architectural space)</td><td>R5 – Presence of alterations (discontinuities with the rest of the fabric)</td></tr>
<tr><td>R6 – Persistence of the buildings (in relation to historic thresholds)</td></tr>
<tr><td rowspan="2">Urban block</td><td>R7 – Degree of uniformity in relation to the earliest historic threshold adopted</td></tr>
<tr><td>T – Formal elements (arrangement of volumes, architectural elements, particular finishes)</td><td>T3 – Degree of uniformity in relation to the number of storeys</td></tr>
</table>

4.2 Algorhythm for Assessing Suitability for Change

Applying multivariate analysis made it possible to avoid placing unreasonable constraints and limits on the alteration of buildings within the historic centre, objectifying (and thus making possible) a comparison of the constituent factors making up the map of suitability for change, which was the final outcome of these analytical and evaluational operations. Thus in the case of Como, our investigation of the built fabric, which was not necessarily limited to currently existing building types, brought out the actual conformation of the buildings and identified their specific characteristics by means of the following algorhythm:

$$f = \frac{1}{U_{aa} * \left[2 + U_{bb} + U_{cc} + U_{dd}\right]^{-1} + U_{ee}} \tag{3}$$

where:

U_{aa} = the indicator of reference for buildings classified as of historic/architectural value;

U_{bb} = the summary indicator of the variables of the physical characteristics of the building;

U_{cc} = the behaviour indicator for the individual building in relation to the block as a whole;

U_{dd} = corrective indicators in relation to given phenomena;

U_{ee} = a specific indicator for buildings of historic/architectural value.

Using ArcGIS Natural Breaks, the results were then bundled into homogeneous classes of suitability for change. ArcGIS Natural Breaks uses Jenks' algorhythm to identify break points in the distribution of the units of investigation and then breaks them down into differentiated classes for each portion of built fabric.

5 Interactions between Socioeconomic and Historic/Architectural Configurations: Mapping Suitability for Change

Having thus described the complex factors that populate the many-sided historic cores of Como, we now describe the steps taken from running the initial algorhythm to constructing the final map setting out the permissible degrees of urban transformation.

Identifying a base-scenario was the starting point for activating the summary indicators, which estimate the contribution made to the final assessment by the variable factors; the suitability of buildings for change derives from the various combinations of persistence of the parts of urban fabric over time, the presence of high-quality elements, the insistence of excrescences that compromise the original framework, and the other indicators.

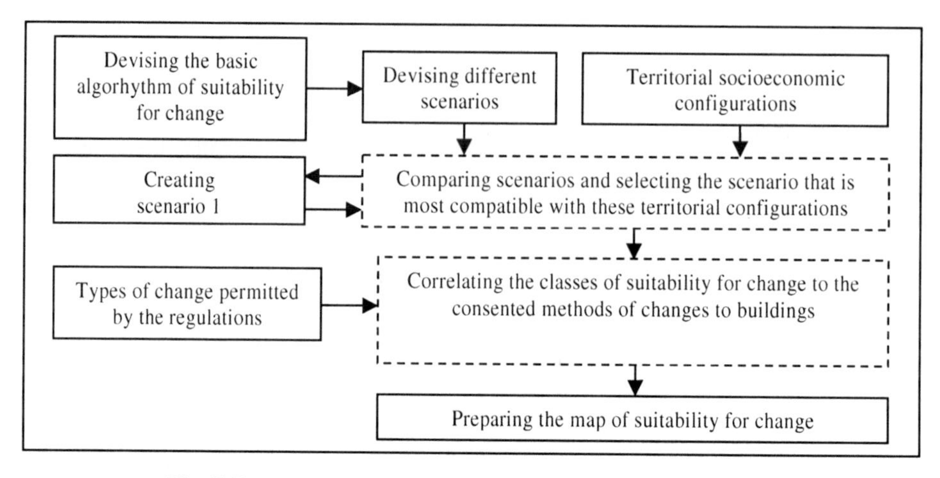

Fig. 7. Procedural diagram summarising the analyses carried out

This makes it possible to begin configuring heterogeneous scenarios for change which in turn can be evaluated using processes of internal and external consistency, where:

i) internal consistency is the level of correlation between the classes of suitability for change and the dataset of variables adopted, identifying any excessively high interdependencies that would indicate a flattening of the outcomes of the multivariate analysis in relation to any single factor of those examined;

Table 2. Outcome of the correlations that emerged from the 5 scenarios initially formulated for Como, of which the summary indicates the preferred scenario as no. 3

Correlated variables	*Sc. 1*	*Sc. 2*	*Sc. 3*	*Sc. 4*	*Sc. 5*
State of conservation Suitability for change	*	+	–	+	*
High persistence of the parts of built fabric Low and Medium/Low suitability for change	+	–	+	–	+

ii) external consistency is the correspondence between the general character of the scenario, in terms of suitability for change, and the real social, economic, and demographic characteristics of the territorial configurations of the historic core.

Whilst in *Scenario 1* the assessment of the high-quality buildings along the lake front is penalised, in *Scenario 2* the greater incidence given to "high-quality construction" as a variable makes it possible to identify buildings of quality, even if they are not subject to any architectural or environmental constraint, but the class of suitability for change is less here than that given by Scenario 1. *Scenario 3* permits the greatest degree of change, since the modifications with respect to the basic algorhythm classify as "particularly suitable for change" most of the buildings not subject to any constraints or not recognised as "high-quality construction". The

Table 3. Selecting the preferable scenario with respect to the emerging socioeconomic and demographic problems in Como

Suitability for change	*Scenario*	*Units with highsuitability for change*		*Units with medium/high suitability for change*		*Units suitable for significant change*	
		Tot.	*%*	*Tot.*	*%*	*Tot.*	*%*
(+ +)	**3**	**210**	**18**	**538**	**46**	**748**	**64**
(+)	5	153	13	482	42	635	55
Median	1	150	13	484	41	634	54
(–)	2	173	15	440	38	613	53
(– –)	4	132	11	401	34	533	45

Table 4. Selection of the scenario definitively adopted as preferable (no. 6)

Correlated variables	**Sc. 1**	**Sc. 2**	**Sc. 3**	**Sc. 4**	**Sc. 5**	**Sc. 6**
Buildings constructed before 1860 High and Medium/High suitability for change	– –	–	– –	–	– –	– – –
Good state of conservation Low and Medium/Low suitability for change	*	–	+	–	*	+
Buildings of historic importance Low and Medium/Low suitability for change	+ +	*	+	*	+	++
Summary	*	–	*/+	–	*	+

reduction in the incidence of the excrescences in buildings of "high-quality construction" has a positive influence on the classification given by *Scenario 4*, which makes it possible to take account of the presence of high-quality buildings on the lake shore, whilst *Scenario 5* recognises that most of the buildings on the lake shore fall within the Medium/Low or Low classes of suitability for change; however, the general tendency is in favour of change.

But when this process had been completed, for the case of Como it seemed opportune to develop one more scenario for which the algorhythm was modified and which produced a result that, as shown in the following table, led to the identification of this sixth scenario as "preferable".

As a final step we classified the parts of agglomerated urban fabric that derive from the classes of transformation, using methods of change that harmonised with the degrees of suitability for change previously identified, taking as our most important reference the Single Law on Building (Presidential Decree 380/2001, updated), in respect of which it was possible to draw up the matrix of changes set out in the following table, and which then made it possible to classify the parts of fabric.

Table 5. Matrix of interconnection between suitability for change, degree of suitability for change, and types of change as per Presidential Decree 380/2001

Class of suitability for change		*Map of suitability for change in the built fabric of the Como valley*	*Types of change*
High		***High suitability for change*** Changes permissible to buildings up to and including total replacement	T5 – Urban upgrading
Medium/high		***Medium/high suitability for change*** Changes permissible to buildings up to and including partial replacement	T4 – Architectural upgrading
Medium/low		***Medium/Low suitability for change*** Changes permissible to buildings up to and including renovation, but without demolitions or reconstruction	T3 – Renovation
Low		***Limited suitability for change*** Changes permissible to buildings up to partial or total reinstatement back to their original condition	T2 – Reinstatement
Historic buildings of architectural merit to be upgraded and conserved		***Extremely limited suitability for change*** Changes to listed buildings of historic and architectural quality limited to conservation and restoration	T1 – Consolidation

Thus carried out, this analytical investigation has made it possible to greatly expand the possibility of making changes in the Como valley, which were previously limited by the strict rules laid down in the previous plan and were based mainly on restrictive morphological and typological considerations. By making an assessment of the suitability for change of the building stock, this present analysis loosens the grip of those constraints by introducing numerous variables that we elaborated beginning from a set of strategic values that now make it possible to rediscover Como and revitalise it, as compared to its current state of accentuated stasis. To change that condition (by bringing into play the strategic role of urban landscape factors, the centrality of sites, their economic potentialities, and their degree of social stability) and in view of the generalised needs for change that the rebirth of Como would in fact require, the variety of the phenomena considered here argues for the reinstatement of those wider operating margins that the valley of Como really requires.

Fig. 8. A three-dimensional view of the built fabric within the historic core of Como and its valley, classified according to the degrees of suitability for change that were adopted

References

1. Astengo, G.: Assisi: salvaguardia and rinascita. Urbanistica 24/25 (1958)
2. Coppa, M.: Vicenza nella storia della struttura urbana: piano del centro storico. Cluva, Venezia (1969)
3. Cervellati, P.L., Scannavini, R.: Interventi nei centri storici: Bologna, politica e metodologia del restauro. Il Mulino, Bologna (1973)
4. Cox, J., Thurstain-Goodwin, M., Tomalin, C.: Town Centre Vitality and Viability: A Review of the Health Check Methodology. National Retail Planning Forum, London (2000)
5. Crucitti, P., Latora, V., Porta, S.: The Network Analysis of Urban Streets: A Dual Approach. Physica A, Statistical mechanics and its applications 369(2) (2006)
6. Fabbris, L.: Statistica multivariata. Analisi esplorativa dei dati. McGraw Hill, New York (1997)
7. Fiale, A., Fiale, E.: Diritto urbanistico. Edizioni giuridiche Simone, Napoli (2008)
8. Fraire, M., Rizzi, A.: Statistica. Metodi esplorativi e inferenziali. Carocci, Roma (2005)
9. Hillier, B., Hanson, J.: The social logic of space. Cambridge University Press, Cambridge (1984)

10. Hillier, B.: Space is the machine: a configurational theory of architecture. Cambridge University Press, Cambridge (1999)
11. Gambino, R., Massarella, G.: Centro storico città regione. Franco Angeli Editore, Milano (1978)
12. Griguolo, S.: Addati. Un pacchetto per l'analisi esplorativa dei dati – Guida all'uso. Istituto Universitario di Architettura di Venezia, Venezia (2008)
13. Mioni, A., Pedrazzini, L.: Valorizzazione dei centri storici. Franco Angeli Editore, Milano (2005)
14. Nijkamp, P.: Soft multicriteria analysis as a tool in urban land-use planning. Environment and Planning B 9(2), 197–208 (1982)
15. Nijkamp, P.: Multicriteria analysis; a decision support system for environmental management. In: Archibugi, F., Nijkamp, P. (eds.) Economy and Ecology; Towards Sustainable Development. Kluwer, Dordrecht (1989)
16. Paolillo, P.L.: La misura della sostenibilità dei vincoli insediativi: un'applicazione di supporto alla Vas. Territorio (25), 65–76 (2003)
17. Paolillo, P.L.: Acque suolo territorio. Esercizi di pianificazione sostenibile. Franco Angeli Editore, Milano, pp. 11–147 (2003)
18. Paolillo, P.L.: New survey instruments: studies for the environmental assessment report of the general plan in a case in Lombardy. In: INPUT 2008, Conferenza nazionale in Informatica and Pianificazione Urbana and Territoriale, pp. 1–10. Politecnico di Milano, Facoltà di Ingegneria, Polo regionale di Lecco (2009)
19. Paolillo, P.L.: New survey instruments: studies for the environmental assessment report of the general plan in a case in Lombardy. In: Rabino, G., Caglioni, M. (eds.) Planning, Complexity and New Ict, pp. 215–224. Alinea, Firenze (2009)
20. Paolillo, P.L.: Sistemi informativi e costruzione del piano. Metodi e tecniche per il trattamento dei dati ambientali. Maggioli, Rimini (2010)
21. Rozzi, R.: Il nuovo Prg di Novara. Urbanistica Informazioni (76) (1984)
22. Terranova, A.: Il piano particolareggiato per il centro storico di Melzo. Urbanistica Informazioni (74) (1982)
23. Thurstain-Goodwin, M.: Data surfaces for a new policy geography in Longley, P.A. and Batty, M. Advanced Spatial Analysis. The CASA book of GIS, ESRI Press, Redland (2003)
24. Voogd, H.: Prescriptive analysis in planning. Environment and Planning B: Planning and Design 12(3), 303–312 (1985)

XPath for Querying GML-Based Representation of Urban Maps*

Jesús M. Almendros-Jiménez, Antonio Becerra-Terón, and Francisco García-García

Information Systems Group
University of Almería
04120-Spain
{jalmen,abecerra,paco.garcia}@ual.es
http://indalog.ual.es

Abstract. *Geography Markup Language* (GML) has been established as the standard language for the transport, storage and modelling of geographic information. In this paper we study how to adapt the XPath query language to GML documents. With this aim, we have defined a *XPath based query language* which handles the *"semantic structure"* of GML. Our approach focuses on querying urban maps whose representation is based on GML. We have developed a system called *UALGIS*, in order to implement the approach. Such system stores GML documents by means of the PostGIS RDBMS. In order to execute semantic-based XPath queries we have defined a translation of the queries into SQL. Such translation takes into account the GML schema. Finally, the system allows to visualize the result. With this aim, the result of a query is exported to the *Keyhole Markup Language (KML)* format.

Keywords: Query languages, GML, XPath, KML, GIS, SQL.

1 Introduction

The *Geography Markup Language* (GML) [7,19,20,8] has been established as the standard language for the transport, storage and modeling of geographic information. GML is a dialect of the *eXtensible Markup Language (XML)* [26] for Geo-spatial data. XML allows to describe the structure of Web data by means of a tree. The tree structure is used to describe relations between data: for instance, a tree node with tag paper contains author, title and publisher as subtrees tags and the subtree publisher can be described by means of name of the journal, country, editors, etc. The need for querying XML documents has motivated the design of the *XPath* query language [3]. The XPath language allows to specify the path of the XML tree to be retrieved. In addition, XPath allows to constraint the query by means of boolean conditions about the attributes and tags of the

* This work has been partially supported by the Spanish MICINN under grant TIN2008-06622-C03-03.

B. Murgante et al. (Eds.): ICCSA 2011, Part I, LNCS 6782, pp. 177–191, 2011.

selected nodes. For instance, we can query the editors of the journals in which "Becerra" has published a paper. It can specified as follows:

$$/\texttt{papers}/\texttt{paper}[\texttt{author}/\texttt{last} = \text{“}\texttt{Becerra}\text{”}]/\texttt{journal}/\texttt{editor}$$

GML allows to describe spatial objects, including their geometry, the coordinate reference system, etc. However, in practice, GML does not use the tree-based structure of XML documents. In most of cases, GML documents store spatial data as a sequence of XML elements, and they are stored as children of the tree root [17,28]. It makes XPath not useful for GML querying. In other words, the tree structure is not used for representing *spatial relations*. One could think that the subtrees of a node represent, for instance, the spatial objects enveloped by the node. However, it is not true in general.

GML allows the specification of relations between spatial data. In particular, the *GML schema* allows to define a vocabulary of relations. For instance, the European INSPIRE Directive [12] has defined a certain vocabulary for GML whose aim is to create a spatial data infrastructure in the European Union in order to share information through public organizations. However, the syntatic structure of GML does not necessarily take into account the "semantic structure" of such vocabulary. In practice, GML documents describe spatial relations by means of the linking mechanism (i.e., *xlink:href*) of XML documents. Using the standard XPath, we could follow the links of the GML document in order to retrieve relationships between spatial objects, however, it makes XPath queries very sophisticated.

In this paper we study how to adapt the XPath query language to GML documents. With this aim, we have defined a *XPath based query language* which handles the *"semantic structure"* of GML documents. Our approach focuses on querying urban maps whose representation is based on GML. In such a case, urban maps describe city elements: streets, buildings, buildings of interest, shops, etc. Such city elements can be related by several spatial relations: located at, next to, belongs to, etc. Such relationships can be modelled by means of a suitable GML based data model. We have developed a system called *UALGIS*, available via Web in `http://indalog.ual.es/ualgis/index.jsp` in order to implement the approach. Such system stores GML by means of the PostGIS [23]. In order to execute semantic-based XPath queries we have defined a translation of the queries into SQL. Such translation takes into account the GML schema. Finally, the system allows to visualize the result. With this aim, the result of a query is exported to the *Keyhole Markup Language (KML)* format [6].

1.1 Related Work

Spatial data can be handled by well-known relational database management systems (RDBMS) like: *SpatialSQL* [11], *GeoSQL* [13], *Oracle Spatial* [24] and *PostGIS* [23]. Basically, they are based on extensions of the relational model for storing spatial objects, and provide an enriched SQL query language for the retrieval of spatial queries.

In the case of GML data, the *Web Feature Service (WFS)* is a standard of the *OpenGis Consortium (OGC)* [21] for data manipulation of geographic features stored on a Web site (i.e., a *Web Feature Server*) using *http* requests. The expressiveness of WFS is very poor compared with query languages like SQL. *GQuery* [4] is a proposal for adding spatial operators to *XQuery* [5], the standard XML query language, which includes XPath as sublanguage. Manipulation of trees and sub-trees are carried out by means of *XQuery*, while spatial query processing is performed using geometric functions of the *JTS Topology Suite* [25]. JTS is an open source API that provides a spatial model and a set of spatial operators. The *GeoXQuery* approach [14] extends the *Saxon XQuery* processor [16] with function libraries that provide geo-spatial operations. It is also based on *JTS* and provides a GML to *Scalable Vector Graphics (SVG)* [27] transformation library in order to show query results.

With respect to *GQuery* and *GeoXQuery*, our proposal can be seen as a specific query language for GML instead of considering ad-hoc mechanisms for querying GML in *XPath* and *XQuery*. Our approach is focused on the *XPath* query language which is a sublanguage of the *XQuery* language. Our work can be seen as the first step to use *XQuery* as query language for GML. However, our proposal is based on the semantic structure of GML documents, instead of the syntactic one used in *GQuery* and *GeoXQuery*. With respect to geometric operations, we are not still interested in such queries. We are mainly interested to query semantic spatial relations expressed in GML, which do not depend on the geometry of the objects, such as located at, next to, etc. In any case, geometric operations are considered as future work, which will be included thanks to the *PostGIS* libraries. Finally, our system is able to visualize query results, but instead of using SVG like in GeoXQuery, we export to KML.

GML Query [18] is also a contribution in this research line, storing GML data in a spatial RDBMS. This approach, firstly, performs a simplification of the GML schema, and secondly, the GML schema is mapped to the corresponding relational schema. The basic values of spatial data are stored as values of the tables. Once the document is stored, spatial queries can be expressed using the *XQuery* language with spatial functions. The queries are translated to SQL which are executed by means of the spatial RDBMS. This approach has some similarities with our. Firstly, the storage of GML data by the spatial RDBMS taking into account the GML schema. We store GML documents in the *PostGIS* RDBMS. Secondly, in our approach the queries are expressed in XPath and translated into SQL. In contrast, our XPath-based query language is properly based on the GML schema, which makes easier the specification of queries and the translation to SQL.

Another problem related to GML is how to visualize documents. There are several technologies (i.e., SVG, VRML, HTML, among others) to specify how to show the content of GML documents. KML [6] is an XML-based language focused on geographic data visualization, including annotation of maps and images, as well as controlling the display in the sense of where to go and where to look. From this perspective, KML is complementary to GML and most of the major standards of the OGC including WFS and the *Web Map Service* (WMS) [22]. We have decided to export query results to KML due to the advantages that offer this technology (i.e., APIs, WFS, WMS). In addition, KML has been approved by the OGC as standard for the exchange and representation of geographic data in three dimensions. In this way, the results can be interpreted by different GIS or Earth browsers. Some of the quoted approaches return the query results in SVG format which is just a graphical format. KML can display the results without losing GML semantics, allowing to include meta-data/schema.

The structure of the paper is as follows. Section 2 will present the GML data model and schema. Section 3 will define the semantic-based XPath language. Section 4 will describe the system and, finally, Section 5 will conclude and present future work.

2 GML Data Model

Next, we show an example of GML document representing an urban map:

```
<CityCenter gml:id="C1">
  <gml:name>London </gml:name>
  <geometry>
    <gml:Point>
      <gml:pos>45.256 -71.92 </gml:pos >
    </gml:Point >
  </geometry >
  <cityCenterMember>
    <Building gml:id="B1">
      <gml:name>Great Building </gml:name>
      <belongsTo>
        <Block xlink:href="BL1"/ >
      </belongsTo>
    </Building>
  </cityCenterMember>
  <cityCenterMember>
    <Building gml:id="B2">
      <gml:name>Small Building </gml:name>
      <belongsTo>
        <Block xlink:href="BL2"/ >
      </belongsTo>
    </Building>
  </cityCenterMember>
  <cityCenterMember>
    <Block gml:id="BL1">
      <gml:name>Grey Block </gml:name>
      <nextTo>
        <Way xlink:href="W1"/ >
```

```
      <Way xlink:href="W2"/ >
    </nextTo>
  </Block>
</cityCenterMember>
<cityCenterMember>
  <Block gml:id="BL2">
    <gml:name>Green Block </gml:name>
  </Block>
</cityCenterMember>
<cityCenterMember>
  <Way gml:id="W1">
    <gml:name>6th Street </gml:name>
  </Way>
</cityCenterMember>
<cityCenterMember>
  <Way gml:id="W2">
    <gml:name>5th Street </gml:name>
  </Way>
</cityCenterMember>
</CityCenter>
```

This document describes the center of a city. This description includes tags for *CityCenter* and *cityCenterMember*'s of a city, such as *buildings*, *blocks* and *ways*. However, the same semantic content can be expressed in several syntactic ways. For instance, the previous GML document can be also represented as follows:

```
<CityCenter gml:id="C1">
  <gml:name>London </gml:name>
  <geometry>
    <gml:Point>
      <gml:pos>45.256 -71.92 </gml:pos>
    </gml:Point>
  </geometry>
  <cityCenterMember>
    <Building gml:id="B1">
      <gml:name>Great Building</gml:name>
      <belongsTo>
        <Block gml:id="BL1">
          <gml:name>Grey Block</gml:name>
          <nextTo>
            <Way gml:id="W1">
              <gml:name>6th Street</gml:name>
            </Way>
            <Way gml:id="W2">
              <gml:name>5th Street</gml:name>
            </Way>
```

```
        </nextTo>
      </Block>
    </belongsTo>
  </Building>
</cityCenterMember>
<cityCenterMember>
  <Building gml:id="B2">
    <gml:name>Small Building</gml:name>
    <belongsTo>
      <Block gml:id=BL2>
        <gml:name>Green Block</gml:name>
      </Block>
    </belongsTo>
  </Building>
</cityCenterMember>
</CityCenter>
```

where instead of using linking mechanisms of XML (i.e., *xlink:href*), the elements of the document are nested. However, it is normally recommended to avoid nesting in order to do not increase the complexity of GML documents. From the point of view of using XPath for querying GML documents, it makes not easy to express complex queries. Our approach aims to propose a semantic version of XPath in which the result of a query does not depend on the syntactic structure. In other words, nesting of elements is not relevant in our approach because we will follow the GML schema to define queries. Next, will present a standard of GML schemas used in our approach.

2.1 INSPIRE Directive

The *European INSPIRE Directive* [12] aims to create a spatial data infrastructure in the European Union to share information through public organizations and facilitate public access across Europe. Furthermore, the spatial information considered under this policy is extensive and covers a wide range of areas and topics. INSPIRE is based on several common principles:

- Data should be collected once and should be stored where they can be maintained more efficiently.
- It must be able to easily combine the spatial information from different sources across Europe and share it with other users and applications.
- It should be possible to collect information at some detail level and share it with all levels, e.g. detailed for local analysis, general for global strategic purposes,. . .
- Geographic information should be transparent and readily available.
- It must be easy to find which geographic information is available, how it can be used to meet a specific need, and under which conditions can be acquired and used.

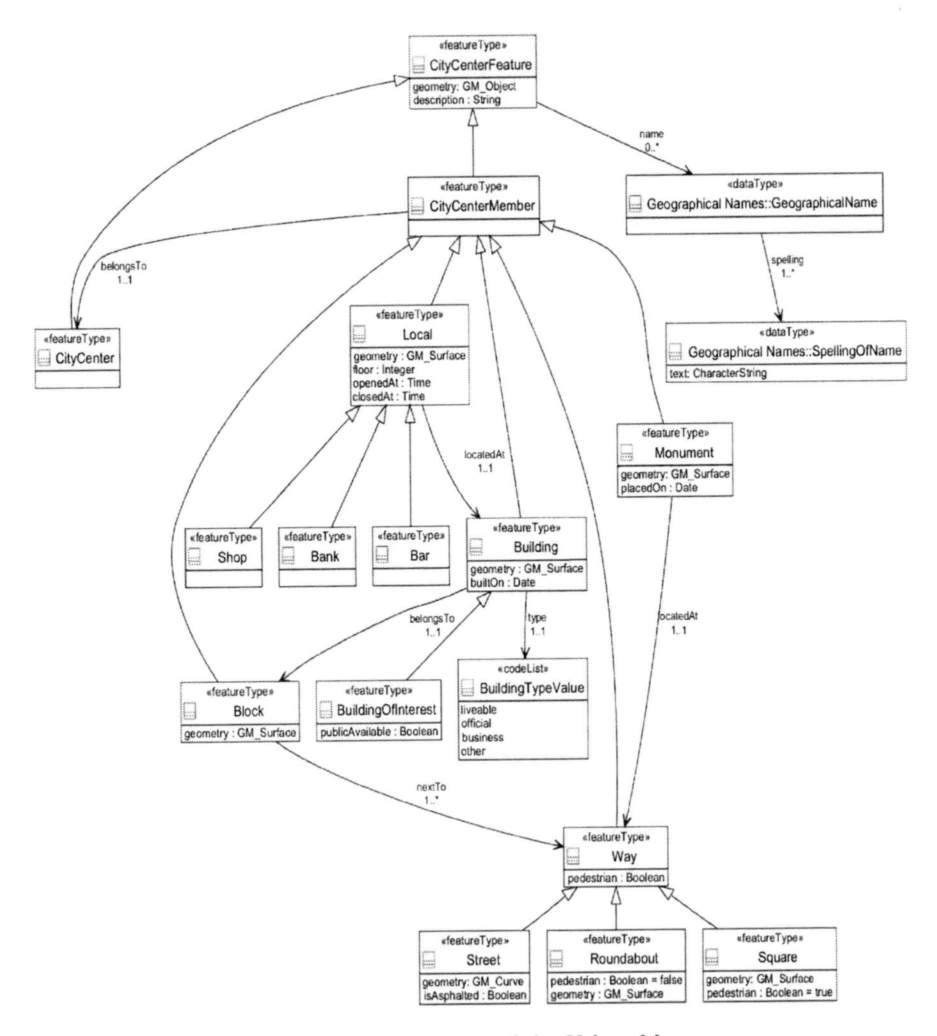

Fig. 1. GML Schema of the Urban Map

The INSPIRE directive defines 34 topics on spatial data needed for application development. The INSPIRE directive defines GML schemas for each one of these topics. We have followed and extended this directive in our approach in order to make our proposal more interesting in the real world.

For instance, Figure 1 shows an example of GML schema for urban maps. This schema uses an UML Profile for its definition called HollowWorld [9]. It is based on Table E.1 of ISO 19136:2007 (GML 3.2.1) [1] and is used by the INSPIRE Directive. Figure 1 describes the elements (i.e., members) of a *city center*. A city center contains different entities like *locals*, *buildings*, *blocks*, *monuments* and *ways*. Locals can be *bars*, *shops* and *banks*. Buildings can be *buildings of interest*, and ways can be *roundabouts*, *streets* and *squares*.The GML schema also includes *spatial relationships* between the entities: locals are *"locatedAt"* buildings that *"belongsTo"* blocks which can be *"nextTo"* Ways. Monuments can be also *"locatedAt"* a way. GML schema allows to describe entities by means of the *Feature type*. In addition, the INSPIRE directive provides *data types* for geographical entity naming.

3 XPath for Querying GML-Based Representation of Urban Maps

Now, we present the proposed XPath based query language. Basically, the path of the query has to follow the GML schema of Figure 1, and the query can include boolean conditions on the elements of the GML schema. For instance, we can express the following queries w.r.t. the running example:

Query 1. *Buildings of the Block named "Grey Block"*
/Building[belongsTo/Block/gml:name="Grey Block"]
Query 2. *(Pieces of) Ways called "5th street" next to some Block*
/Block/nextTo/Way[gml:name="5th Street"]
Query 3. *Ways next to a Building called "Great Building"*
/Building[gml:name="Great Building"] /belongsTo/Block/nextTo/Way

The semantic version of the *XPath* query language is syntactically similar to the tree-based version. However, the semantic version can specify paths starting from any point of the GML schema (i.e., the root of the XPath expression can be any of the features of the GML document). The XPath expression alternates features with spatial relations. For instance, starting from *Building* we can build the following (semantic) XPath expression:

/Building[gml:name="Great Building"]/belongsTo/Block/nextTo/Way

by following the sequence *Building*, *belongsTo*, *Block*, *nextTo* and *Way*, where *Building*, *Block* are Features and *belongsTo* and *nextTo* are relationships among Features.

Finally, let us see the syntactic version of the above *Query 1*, which is considerably more complex than the proposed semantic one:

/CityCenter/cityCenterMember/Building[belongsTo/Block
[@xlink:href=/CityCenter/cityCenterMember/Block[gml:name="Grey
Block"]/@gml:id]]

In summary, the proposed semantic version of XPath makes easier to the user the formulation of queries: (s)he does not need to know the physical structure of the GML document. Rather than it is enough to known the GML schema. The translation of the semantic XPath expression to a syntactical one is hidden to the user.

3.1 Translation XPath to SQL

In order to implement the proposed XPath-based query language, we have to proceed as follows:

1. The GML schema is transformed into a Relational schema.
2. The GML document is stored in the spatial RDBMS.
3. The XPath query is translated into an equivalent SQL query.
4. The result of the query is exported to GML or KML format.

Transforming the GML Schema into a Relational Schema. For data storage the GML schema has to be transformed into a relational schema of PostGIS. For this transformation we proceed as follows:

- A table is created for each element of Feature type of the GML schema.
- Attributes of elements of Feature type are mapped to columns of the corresponding tables.
- Geometric attributes of elements of Feature type are mapped to columns of PostGIS geometry type.
- Spatial relations between elements of Feature type are represented as follows:
  - A one to one relationship is mapped to a column that references the primary key of the elements in the spatial relation.
  - A one to many relationship is mapped to a table, named as the name of the feature + name of spatial relation, with columns containing the primary keys (i.e., foreign keys) of the elements in the spatial relation.
  - A many to many relationship is mapped as two one-to-many relationships.
- Feature inheritance is represented by table inheritance in PostGIS.

Figure 2 shows the result of the transformation of some elements of the GML schema represented in Figure 1.

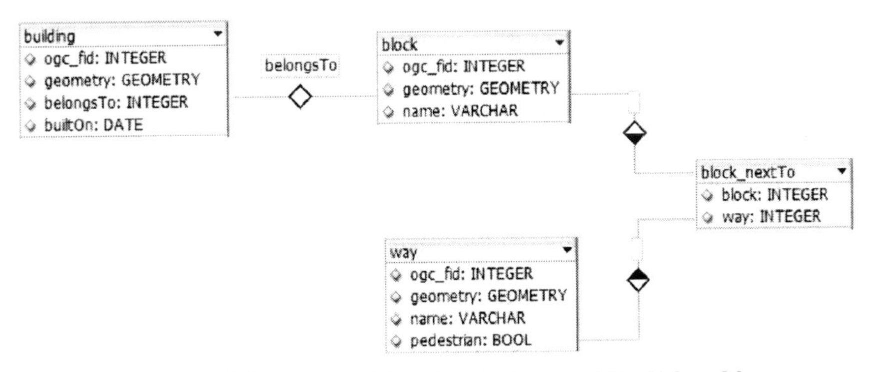

Fig. 2. A Fragment of Relational schema of the Urban Map

Storage of GML Documents in the Spatial RDBMS. The GML documents are stored in the spatial RDBMS as follows. Firstly, features instances are added to tables. Secondly, spatial relations are added to columns (in the case of one to one relationships) and to tables (in the case of one to many and many to many relationships). Next, we show the table instances of the running example:

Table: Building

ogc_fid	name	belongsTo
B1	Great Building	BL1
B2	Small Building	BL2
...	...	...

Table: Block

ogc_fid	name
BL1	Grey Block
BL2	Green Block
...	...

Table: Way

ogc_fid	name
W1	6th Street
W2	5th Street
...	...

Table: Block_nextTo

Block	Way
BL1	W1
BL1	W2
...	...

Translation of XPath into SQL. We can now define the translation of XPath-based queries into SQL queries. It is based on the transformation of the GML Schema into the Relational Schema. The translation is as follows:

1. Case */A/p/B* where p is a one to one relationship: *Select B.* From A,B Where A.p = B.ogc_fid*
2. Case */A/p/B* where p is a one to many relationship or a many to many relationship: *Select B.* From A,A_p,B Where A_p.A = A.ogc_fid and A_p.B = B.ogc_fid*
3. Case */A[cond1]/p/B[cond2]* where p is a one to one relationship: *Select B.* From A,B Where A.p = B.ogc_fid and cond3 and cond4*, where *cond3* and *cond4* are the translation of *cond1* and *cond2*, respectively.
4. Case */A[cond1]/p/B[cond2]* where p is a one to many relationship: *Select B.* From A,A_p,B Where A_p.A = A.ogc_fid and A_p.B = B.ogc_fid and*

cond3 and cond4, where *cond3* and *cond4* are the translation of *cond1* and *cond2*, respectively.

5. Similarly, the rest of the cases

Now, we show the translation into SQL expressions of the XPath queries of Section 3 w.r.t. the GML schema of Figure 1.

Query 1. *Buildings of the Block named "Grey Block"*

XPath Query	**SQL Expression**
/Building[belongsTo/Block /gml:name="Grey Block"]	Select Building.* from Building,Block where Building.belongsTo = Block.ogc_fid and Block.name = "Grey Block"

The result of the query in GML format is as follows:

GML Output

```
<Building gml:id="B1">
   <gml:name>Great Building <gml:/name>
   <belongsTo>
      <Block xlink:href="BL1"/ >
   </belongsTo>
</Building>
```

Query 2. *(Pieces of) Ways called "5th Street" next to some Block*

XPath Query	**SQL Expression**
/Block/nextTo /Way[gml:name="5th Street"]	Select Way.* from Way, Block_nextTo, Block where Block_nextTo.Block = Block.ogc_fid and Block_nextTo.Way = Way.ogc_fid and Way.name="5th Street"

Query 3. *Ways next to a Building called "Great Building"*

XPath Query	**SQL Expression**
/Building[gml:name="Great Building"] /belongsTo/Block/nextTo/Way	Select Way.* from Way, Block_nextTo, (Select Block.* from Building, Block where Building.belongsTo = Block.ogc_fid and Building.name = "Great Building") Block_0 where Block_nextTo.Block = Block_0.ogc_fid and Block_nextTo.Way = Way.ogc_fid

Exporting to GML and KML. *PostGIS* natively provides several functions for the conversion of stored geometries to GML and KML formats:

- AsGML: Returns the geometries as GML elements, allowing the selection of the spatial reference system.
- AsKML: It works similarly to AsGML but returning the geometries as KML geometries. We cannot select the spatial reference system because it is fixed to *WGS84*.

These functions are only responsible for generating the geometries in GML / KML format but it is still required the exporting of tables as GML and KML elements. With this aim, *PostGIS* provides two tables as follows:

- *Geometry_columns* table: It stores a catalog with the schema and table names, and column names of geometric data and their spatial reference system.
- *Spatial_ref_sys* table: It contains a collection of spatial reference systems and stores information about projections for transforming from one system to another.

4 *UALGIS* System

For validating the proposed query language a *Web Geographic Information System*, called *UALGIS (University of ALmería Geographic Information System)*, has been implemented. It is available from `http://indalog.ual.es/ualgis/index.jsp`. In summary, the main features of the UALGIS system are the following:

- XPath-based querying.
- *Google Maps*-based client for displaying the resut of queries.
- Querying of elements of *feature* type of the database.

For testing our system, we have used data available from *IDEAndalucía* [15] which is part of the geo-services of the INSPIRE directive. This is an *Andalusian Cartographic System Geo-portal* available to search, locate, view, download or request geographic information referring to the territory of Andalusia in Spain. IDEAndalucía provides several services, such as WMS (Web Map Server) and WFS (Web Feature Service), with data available in GML format. By a *GetFeature* request to the WFS server a GML document is obtained, with all the features enclosed in a rectangle (i.e., *GML Bounding Box*) of a selected type.

The system has been built by using the *Java Eclipse Galileo* [10] as *IDE*. For project management, we have used *Maven 2* [2] for handling library dependences. Figure 3 shows the system architecture.

- The system architecture includes the *PostGIS* server, version *1.4*, and the *Tomcat* server, version *6.0*, for handling logic and presentation layers. For the presentation layer, pages are programmed using *JSP*, *HTML* and *AJAX*.

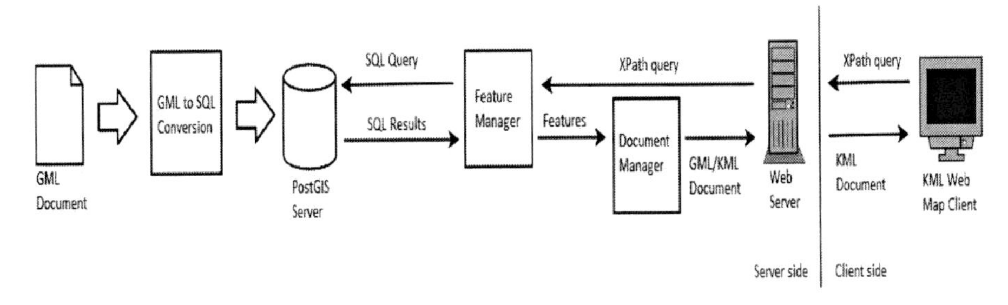

Fig. 3. *UALGIS* System Architecture

Fig. 4. Snapshot of *UALGIS*

- The *FeatureManager* component is responsible of the management of the features stored in the database. It is also responsible of the translation of *XPath* queries into *SQL* queries, and the transformation of the database rows to an object model that can be processed by the *DocumentManager* component.
- The *DocumentManager* component is responsible for handling documents in *GML/KML* format. It is also responsible for transforming the objects obtained from the *FeatureManager* into KML / GML documents. It also supports the *reverse* process, i.e., the extraction of features from GML/KML documents.

- For creating and reading documents the *dom4j* library is used. This is an *open source XML framework* for *Java* that allows reading, writing and navigation of XML documents.
- On the client side, the *Web GIS* is a map browser with support for KML. It is based on the *Google Maps API* including a fast and efficient *2D* browser that provides features like zooming, panning, searching and displaying geographical names and information about geographic entities. Figure 4 shows a snapshot of the *UALGIS* system.

5 Conclusion and Future Work

In this paper we have studied how to adapt the XPath query language to GML documents representing urban maps. With this aim, we have defined a XPath-based query language which handles the semantic structure of GML documents. We have developed a system called *UALGIS*, in order to implement the approach. Such system stores GML documents by means of the PostGIS RDBMS. In order to execute semantic-based XPath queries, we have defined a translation of the queries into SQL. Such translation takes into account the GML schema. Finally, the system allows to visualize the result in KML format. As future work, firstly, we would like to integrate GML filtering and indexing, among others, in the UALGIS system in order to improve the performance. Currently, UALGIS has been tested with small examples. We would like to have a good benchmark study to compare UALGIS with similar systems. Secondly, we would like to extend our work to the *XQuery* language. Finally, we would like to combine our GML query language with spatial ontologies. Its use would improve the kind of queries and answers obtained from GML documents.

References

1. Iso 19136, Encyclopedia of Database Systems (2009)
2. Apache Software Foundation. Apache Maven, `http://maven.apache.org/`
3. Berglund, A., Boag, S., Chamberlin, D., Fernandez, M.F., Kay, M., Robie, J., Siméon, J.: XML path language (XPath) 2.0. W3C (2007)
4. Boucelma, O., Colonna, F.M.: GQuery: a Query Language for GML. In: Proc. of the 24th Urban Data Management Symposium, pp. 27–29 (2004)
5. Chamberlin, D., Draper, D., Fernández, M., Kay, M., Robie, J., Rys, M., Simeon, J., Tivy, J., Wadler, P.: XQuery from the Experts. Addison Wesley, Boston, USA (2004)
6. OpenGIS Consortium. KML 2.2 Reference - An OGC Best Practice (2008), `http://www.opengeospatial.org/standards/kml/`
7. OpenGIS Consortium. GML Specifications (2010), `http://www.opengeospatial.org/standards/gml/`
8. Córcoles, J.E., González, P.: GML as Database. Handbook of Research on Geoinformatics (2009)
9. Simon Cox. HollowWorld (2009), `https://www.seegrid.csiro.au/twiki/bin/view/AppSchemas/HollowWorld`

10. Eclipse Foundation. Eclipse, http://www.eclipse.org/
11. Egenhofer, M.J.: Spatial SQL: A Query and Presentation Language. IEEE Transactions on Knowledge and Data Engeneering 6(1), 86–95 (1994)
12. European Union. InspireE, http://inspire.jrc.ec.europa.eu/
13. Wang, F., Sha, J., Chen, H., Yang, S.: GeoSQL:a Spatial Query Language for Object-Oriented GIS. In: Proc. of the 2nd International Workshop on Computer Science and Information Technologies (2000)
14. Huang, C.H., Chuang, T.R., Deng, D.P., Lee, H.M.: Building GML-native web-based geographic information systems. Computers & Geosciences (2009)
15. Junta de Andalucia. IDEAndalucia, http://www.andaluciajunta.es/IDEAndalucia/IDEA.shtml
16. Kay, M., Limited, S.: Ten reasons why Saxon XQuery is fast. IEEE Data Engineering Bulletin (1990)
17. Lake, R.: Geography mark-up language (GML). Wiley, Chichester (2004)
18. Li, Y., Li, J., Zhou, S.: GML Storage: A Spatial Database Approach. In: ER (Workshops) On Spatial Database Approach. LNCS, vol. 3289, pp. 55–66. Springer, Heidelberg (2001)
19. Lu, C.T., Dos Santos, R.F., Sripada, L.N., Kou, Y.: Advances in GML for geospatial applications. Geoinformatica 11(1), 131–157 (2007)
20. Need, A.P.: Querying GML. Handbook of Research on Geoinformatics, 11 (2009)
21. OpenGis Consortium (OGC). OpenGis Specifications (2003), http://www.opengeospatial.org
22. OpenGis Consortium (OGC). WMS Specifications (2008), http://www.opengeospatial.org/standards/wms
23. PostGis. PostGis, Geographic Objects for Postgres (2003), http://postgis.refractions.net
24. Ravada, S., Sharma, J.: Oracle8i Spatial: Experiences with Extensible Databases. In: Güting, R.H., Papadias, D., Frederick Lochovsky, H. (eds.)SSD, LNCS, vol. 1651, pp. 355–359. Springer, Heidelberg (2001)
25. Shekhar, S., Xiong, H.: Java Topology Suite (JTS). In: Encyclopedia of GIS, p. 601. Springer, Heidelburg (2008)
26. W3C. Extensible Markup Language (XML). Technical report, W3C (2007)
27. W3C Recommendation. Scalable Vector Graphics (SVG) 1.0 Specification (2001)
28. Wei, S., Joos, G., Reinhardt, W.: Management of Spatial Features with GML. In: Proceedings of the 4th AGILE Conference on Geographic Information Science, pp. 370–375. Brno, Czech Republic (2001)

Accounting for Spatial Heterogeneity in Educational Outcomes and International Migration in Mexico

Edith Gutiérrez, Landy Sánchez, and Silvia Giorguli

El Colegio de México. Camino Al Ajusco 20. 10740,
Mexico, DF. Mexico
{egutierrez,lsanchez,sgiorguli}@colmex.mx

Abstract. This paper analyzes the link between international migration and educational attainment of the Mexican youth at the municipality level in 2000. This approach examines spatial heterogeneity in such relationship by testing two regionalization proposals, through spatial regime models. On one hand, we test well-known hypothesis that geographical differences obey to the historical-migratory trajectory of each region. On the other hand, we propose a model that accounts for the spatial differences based on the interface of migration and labor markets performance. Results suggest a large, negative effect of international migration on educational achievement, and strong spatial heterogeneity in that association. Results from groupwise heteroskedastic spatial regimes support the second hypothesis, since it captures better the spatial variability, as well as the behavior of the international migration variable across these regimes. These outcomes highlight the need to use proper geostatistical methods to examine territorial disparities.

Keywords: regional variations, spatial regimes, heterogeneity, migration, educational attainment, Mexico.

1 Introduction

Studies of international migration effects on educational achievement are still inconclusive. On one hand, remittances could affect positively youth's educational trajectories delaying their school to work transition, especially in poor families [4][8][18][22]. On the other hand, exposure to international migration decreases the probabilities of attending school because of an expanding "culture of migration": joining the stream is a better mechanism of social mobility than education [18][19][25][29]. Therefore, migration disincentivizes adolescents to stay in school [26][29].

This paper contributes to the discussion by analyzing the link between international migration and educational attainment of the Mexican youth at the municipality level. This aggregated approach allows us to consider dimensions that have been underexplored. First, we want to examine *regional variations* in educational achievement and international migration rates, accounting for local job opportunities and educational services. Second, we test the well-known hypothesis that negative

B. Murgante et al. (Eds.): ICCSA 2011, Part I, LNCS 6782, pp. 192–206, 2011.

effects depend on the historical migration trajectory of a region -a spatial heterogeneity hypothesis- through proper geostatistical models. Finally, we propose a model that accounts for the spatial differences in the effects of international migrations on educational achievement based on the interface of migration and labor markets performance.

2 Theoretical Background and Hypothesis

Mexican migration to USA is the largest cross-border migration flow in the world [24]. Despite the wide spread of migration along the Mexican territory, it is known that the stream has a strong regional component due to historical trajectories, the proximity to the border and the variations in the prevalence at the local level [13][15]. Such regional pattern, however, is becoming more diversified since 1990s, when migration rates grew to unprecedented levels and regions without any previous migration experience to US incorporated into the flows: in the 1987-1992 period only 8.8% of migrant population was from the South, while ten years later this percentage grew 3.1 points [28]. Within this emerging scenario, it is likely that the relationship between educational attainment and international migration varies significantly at the regional level.

Previous studies suggest that international migration could positively impact educational achievement, increasing households' resources through remittances of both money and transferred knowledge. Those resources would allow children and adolescent to stay in school longer, and this may also have a demonstration effect in nearby localities [4][8][22]. Some studies suggest that even in places of strong migration tradition, educational attainment would not decrease because the expected returns to education are greater in US. Still, other studies suggest that, in order to be competitive in the American labor markets, the Mexican youth would have to complete at least upper secondary education, a level unattainable for most Mexicans; which suggest that migration expectations are unlikely to incentive youth's investments in education [18].

In contrast, other studies indicate that international migration may have a negative effect on educational achievement since it disincentives adolescents to stay in school, given that migrating is a better alternative for social mobility [18][19][25][29]. These studies concur with the culture of migration theory, where "going north" is part of an economic strategy as well as a cultural experience for the youth [11][20][26][29]. Migration negative effects on education, therefore, will be largely dependent on the regional history of the stream and its prevalence rate, since how much migration is framed as an alternative to education depends on those characteristics. Durand and Massey [15] define four regions based on the historical intensity of Mexicans flows to the United States and on migration prevalence ratio (see Figure 2, top map): traditional, border, central and the southeast region. The authors suggest that each region reflects unique cultural characteristics, migrant social networks and a migration system built upon the historical resettlement flow. Although Massey and Durand [15] do not analyze educational achievement, their overarching argument about how spatial differentiation is shaped by the historicity of the migration stream holds for this variable too. Under this argument, international migration is expected to have a negative effect on educational achievement but such effect will differ significantly across historical regions: it will be stronger at the traditional and the border ones, while it will be smaller in the regions where outflow started recently, such as the southeast. This is a spatial heterogeneity hypothesis which must be tested through spatial regime models, as we explain in the methods section of this paper.

Disparities in educational achievement are likely to be impacted also by labor markets opportunities as well as the availability and quality of educational services. Existing educational services, in particular the possibility of continuing into high school, as well as the type and quality of services, influence educational expectations and achievement of the youth [21][31]. Moreover, such infrastructure varies deeply across Mexico, for example, while 87% of the municipalities at the center have a high school, only 41% at the south have it. Labor markets dynamics also differ importantly across space, as well as their impact on adolescents and parents decisions about expected schooling returns [21]. Specifically, studies show that wage levels, types of jobs, and how dynamic a labor market is will impact educational achievement since local and regional employment options will shape educational expectations and requirements to compete in those markets [6][21].

Although most authors recognize that international migration effects on education are mediated by labor markets opportunities or educational services, the spatial variability of these variables has not been taken into account neither to define regional frontiers nor to explore their heterogeneous effects across the territory. Therefore, we propose to define regions based on Geographically Weighted Regression (GWR) results, which will account for the observed spatial heterogeneity in the relationship between migration, education, and labor markets. Then, we test that these regions behave as spatial regimes, therefore, we hypothesize that international migration will be associated to lower expected years of schooling but its effect will differ across regimes, reflecting not only historical migratory trajectories but also employment and educational characteristics.

3 Data and Methods

To explore the link between education and international migration and to understand how this link varies according to regional context, as suggested by previous works [7][15], we implement spatial analysis methods that allow coefficients to diverge across the space, testing if regional definitions previously proposed accurately capture the differences documented by literature. We use three data sources: the 2000 Census Sample [23], indicators produced by the National Population Council [12], and the Administrative Records of the Educational Ministry. We also use the 2000 census cartography at the municipal level[1].

For measuring educational attainment, we apply survival analysis to estimate expected years of schooling after elementary school. Based on the concepts of events and population exposed to risk [27], we use the 2000 Census sample data of the child's last year approved and current enrollment to calculate the probability of

[1] The cartography was projected with the Nad 1927 UTM Zone 14N coordinate system.

attending the next grade, according to age's standards. With those probabilities, we build schooling life tables at the municipality level under the assumption of closed populations[2] to obtain the expected years of schooling. This measure is a schooling life expectancy and refers to the mean years that a child expects to attend after approving a grade. One of the advantages of using this indicator is that it eliminates the effect of age population distribution and it is especially sensitive to the initial and ending conditions of the attendance structure. This condition is not accounted for in traditional measures of educational outcomes, such as attendance rates or raw proportions. Mainly, the schooling life expectancy allows us to capture more precisely absorption and retention capacities of the educational system, because it is strictly calculated with the population exposed to the risk.

To incorporate international migration into the analysis, we use the 2000 intensity index of international migration produced by the National Population Council at municipal level [12]. This indicator is a good measure of migration for the phenomenon studied in this paper, because it considers the main migration effects suggested by the literature –the community migratory experience, the internalization process of migration as a transition to adulthood or as possible labor alternative, and the cultural and economic repercussions of remittances in the sending communities. The index measures the prevalence of migration in the households either because a family member ever migrated or because they receive remittances.

On the discussion about migration patterns of the youth, the labor market has a fundamental role. Literature suggests that even in the 21st century, employment search is the main reason for Mexican migration. Thus, in order to capture the youth's employment opportunities and the jobs quality in their communities, we use three indicators: the female participation rate (as a proxy of market dynamics), the proportion of employed population who earns less than two minimum wages and the proportion of population working in the manufacturing sector. All are obtained from the 2000 Census sample data [23].

The availability and quality of educational services are factors that strongly impact adolescents' possibilities for continuing their educational careers and by consequence will influence the community's expected years of schooling. To measure these, we use two variables: the educational profile of the teachers as a proxy of educational quality [31] and the availability of general track high schools in the municipality. In terms of indicators, the first variable is the proportion of teachers with a bachelor's degree or more, and the last one is a categorical variable divided in three: without schools (reference category), only technical schools, and academic and technical track schools available in the municipality. Both are obtained from administrative records from the Education Ministry.

Finally, using the 2000 Census sample data, we control for internal migration with the inter-municipality migration rate and the levels of urbanization of the

---

[2] This assumption could be a limitation because of the extended mobilization of young people inside and outside the country. Therefore, in addition to the international migration variable, which is useful for controlling the effects on migration on the probabilities structure, in the regression analysis we also control for internal migration. In addition, mortality is not taken into account in our estimate; nonetheless, it does not significantly interfere with the educational outcomes, since mortality in those ages is low.

municipalities. With this information, we estimated all the variables mentioned for 2443 municipalities in Mexico[3].

The spatial distribution of our main variables is shown in Figure 1. These and the following maps are shaded in quintile ranges, where pink colors represent the lowest quintile and green colors the highest one[4]. The first map shows that, on average, educational attainment is above compulsory education[5] –nearly five years of schooling after elementary education. But, it also shows that in almost 10% of the municipalities children will not finish middle school and young people will get tertiary education only in 25% of the municipalities. The highest concentration of municipalities with the lowest educational achievement is located in the South-Pacific, in the area with extreme poverty conditions. In contrast, it is possible to identify that the Center of the country, the Northwest coast and the Southwest, are regions with higher levels. An explanation for the central region could be –as Mexico has been a centralized country –the concentration of resources in this area that is also accompanied by a concentration of commodities, services and opportunities. So, it is clear that the educational attainment is heterogeneous across the country; these spatial differentiations are expressions of the educational inequality in Mexico.

The spatial distribution of the migration intensity index (Figure 1, bottom map) is characterized by marked regional patterns, showing a historical path of migration in Mexico. Municipalities with a high prevalence are located in the "traditional region", defined by Durand and Massey [15] (see Figure 2). They are concentrated mainly in the West of the country, crossing the territory from the Pacific coast to the Central-North. In contrast, municipalities with low rates of migration intensity are mostly located in the South. This region includes the Central and the Southeast and, although patterns are not exactly homogeneous within the region, the range in which the index varies is small (from 0.11 to 0.20). The Northern region is probably the most unequal in terms of behavior and exposure to migration: generally is characterized by municipalities with median levels (colored in white and light green), but also, there are small patches shaded in pink indicating the presence of low exposure municipalities.

Joining both distributions, the results show that international migration is associated with lower educational attainment. We clearly see that the regions where educational achievements are high (Northwest Coast, Southeast and the Central part), migration intensity rates are fairly low, while rates in the West and the Central-North

[3] A municipality is a single political-administrative unit, but in our cartography 33 of 2443 are constructed as two objects, and another one as three objects, since their geographies are not continuous. After several sensitivity tests of the spatial exploratory statistics, and because we did not have information at a lower scale, we decided to consider these objects as separated units. Under the assumption of uniformity inside the municipality, we impute them their municipality indicators. So, our size sample is 2478 instead 2443.

[4] Colors from www.ColorBrewer.org by Cynthia A. Brewer, Geography, Pennsylvania State University.

[5] By law, children in Mexico are expected to complete up to middle school, which is equivalent to three years after completing elementary school (six years).

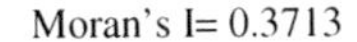

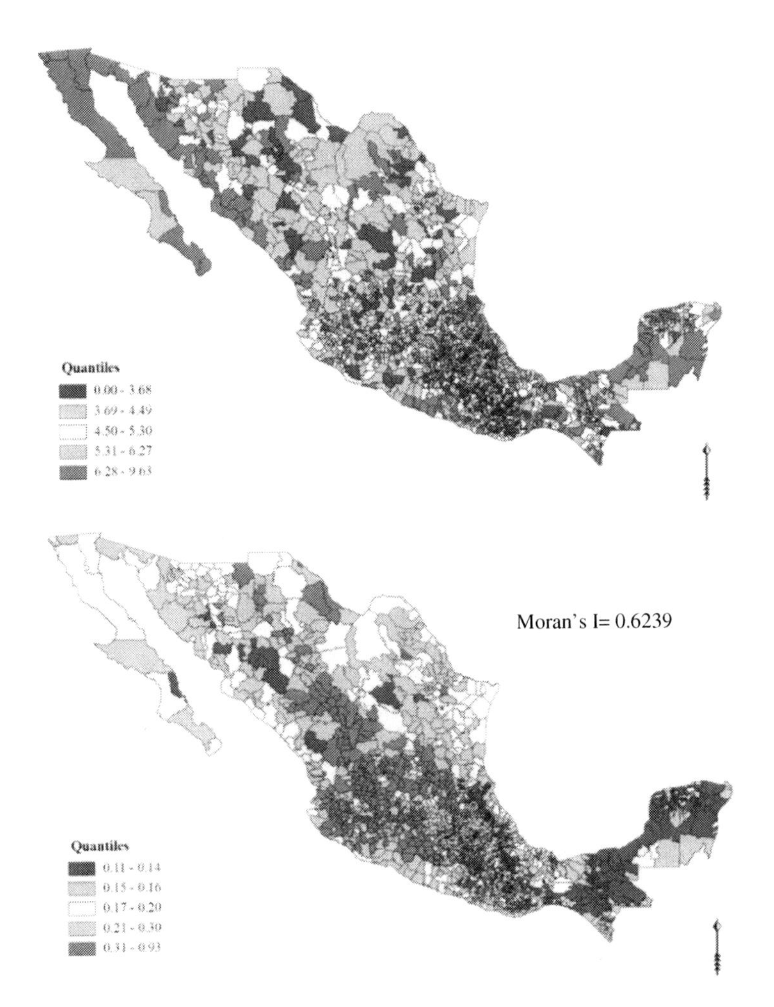

Fig. 1. Mexico, 2000. Spatial distribution by quintile ranges for: expected years of schooling after elementary education (top), and intensity index of international migration (bottom)

are the highest in the map and the attainment is in the opposite level. Finally, "emerging migration areas" located in the South show low achievement that may be due to the mediating effects, such as labor markets and educational opportunities.

From these results, we can expect a regional effect in the relationship between educational attainment and international migration. Thus, we run a spatial regime model using the most common regionalization proposal in migration literature to test whether this partitioning helps to understand the regional behaviors of education and migration.

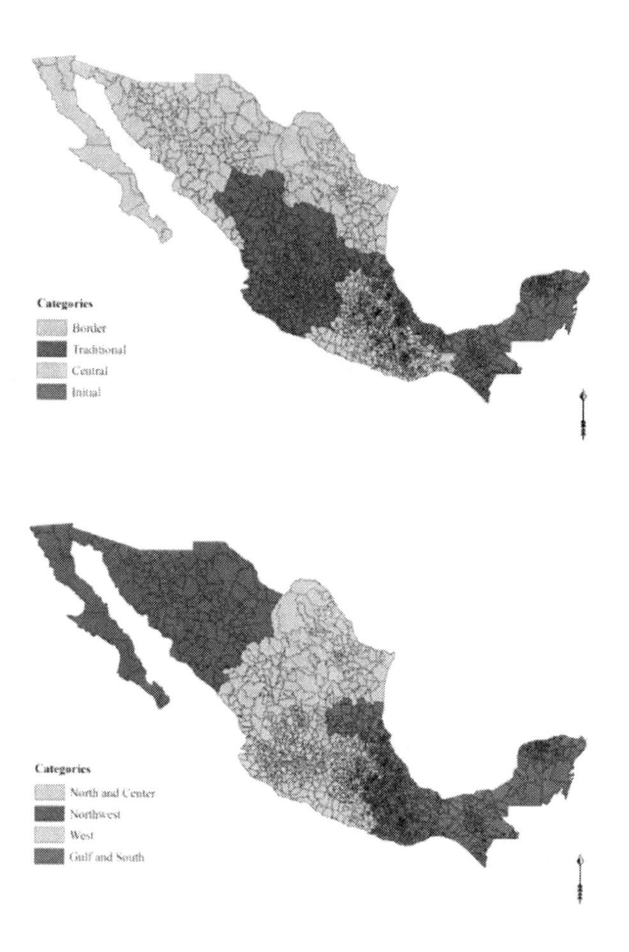

Fig. 2. Mexico, 2000. Regions proposed by Durand & Massey [15] (top) and new regions based on GWR results (bottom)

To corroborate spatial effects on the phenomenon studied, we first run an OLS regression, Moran's I, and Local Indicators of Spatial Association[6]. Results show evidence of spatial dependence and presence of heteroskedasticity, and some significant clusters across the country, which suggest: 1) significant variations in the relation modeled that were not accounted for, and 2) the need to use a technique that helps to capture heterogeneity in a more accurate form. Based on this, the error variances were assumed to be dissimilar; we introduce in all structural instability models a groupwise correction, which considers distinct error variances across regimes, but constant within each of them [3]. In addition, we expect significant spatial dependence. The two hypotheses tested in this paper suggest a substantive spatial autocorrelation process (autoregressive lag model) having place within each

[6] These estimations are available per request. Please contact the authors.

Table 1. Mexico, 2000. Spatial Error Model with Structural Change and Groupwise Heteroskedasticity for the Regionalization of Durand and Massey [15].

For first-order standardized weights matrix				
Dependent Variable	Expected years of schooling			
No. of observations	2478		DF	2430
No. of variables	48		Sq. Corr.	0.50
R^2	0.49		LIK	-3535.94
AIC	7167.88		SC	7447.01
λ	0.3487***			
Parameters and Significance				
Variable	*Regime 1*	*Regime 2*	*Regime 3*	*Regime 4*
CONSTANT	5.9523 ***	5.3396 ***	8.4059 ***	9.9628 ***
International migration	-3.3901 ***	-2.4448 ***	-1.6794 ***	-2.3856 *
Female participation rate	2.1380 *	4.5737 ***	1.6383 ***	-0.2163
Low-income workers	-2.4715 ***	-1.4827 **	-5.6792 ***	-6.2572 ***
Industrialization level	-2.7595 ***	-1.9265 ***	-0.5960 *	0.8803
Educational profile of teachers	0.8001 *	0.1028	0.4648 ***	0.7976 *
Without high schools vs technical schools	0.2423	0.1868	0.4771 ***	-0.1561
Without high schools vs technical-general track schools	0.4912 **	0.3459 **	0.5742 ***	0.3040 *
Inter-municipal migration	-0.4465	-0.1273	-0.0893	-0.1796
Rural vs Rural-Urban	0.2363	-0.0085	0.2672 **	0.0301
Rural vs Urban	0.4379 *	0.2120	0.2141 *	0.2683 *
Rural vs Metropolitan area	0.8950 **	0.4538	0.2108	0.0130
n	***300***	***478***	***1222***	***478***
Group Variances				
Variable	*Coeff.*	*S. D.*	*z -Value*	*Prob*
Regime 1. Border	0.9112	0.0745	12.2309	0.0000
Regime 2. Traditional	0.6694	0.0434	15.4335	0.0000
Regime 3. Central	1.2416	0.0504	24.6158	0.0000
Regime 4. Initial	0.9896	0.0641	15.4299	0.0000
Test on Structural Instability for 4 Regimes				
Test		*DF*	*Value*	*Prob*
Chow - Wald		36	276.91	0.00
Stability of Individual Coefficients				
Test		*DF*	*Value*	*Prob*
CONSTANT		3	45.97	0.00
International migration		3	3.94	0.27
Female participation rate		3	27.91	0.00
Low-income workers		3	73.08	0.00
Industrialization level		3	19.36	0.00
Educational profile of teachers		3	3.35	0.34
Without high schools vs technical schools		3	4.84	0.18
Without high schools vs technical-general track schools		3	4.01	0.26
Inter-municipal migration		3	0.39	0.94
Rural vs Rural-Urban		3	4.34	0.23
Rural vs Urban		3	0.89	0.83
Rural vs Metropolitan area		3	4.58	0.21
Test on Groupwise Heteroskedasticity				
Test		*DF*	*Value*	*Prob*
Likelihood Ratio Test		3	112.99	0.00

Significance Level: * p<0.1 , ** p<0.01 , *** p<0.001

region between municipalities educational levels, after accounting for explanatory variables. Such process could suggest a demonstration effect of the trade-offs between education, migration or employment.

To examine the historical hypothesis, we specify a spatial regimes model using the regions defined by Durand and Massey [15]. The spatial diagnostics in the OLS regime models favored the autoregressive error model, contrary to the proposed dependence process[7]. This model would suggest that spatial clustering of educational achievement would be accounted for the geographical pattern of explanatory variables and the error term (unobserved variables) [5]. Based on these results, a spatial error model with structural instability and groupwise heteroskedasticity was estimated (Table 1).

While the Chow-Wald test suggests regions vary in the explanatory factors accounting for educational attainment, the individual test of the coefficients show that such regionalization does not account for migratory dynamics: across the regimes the migration index is highly not significant, suggesting that its effects remain invariant across the regions; even though, migration itself is a significant predictor of educational achievement within each region. In contrast, the instability of the labor markets coefficients implies that their effects vary spatially. Moreover, the size of the migration coefficient across regions does not behave as expected under the traditional region hypothesis. International migration has the largest effect in the border region, but its effect is smaller in the traditional region –where it was expected to be the largest. Moreover, in the Southeast, an emerging migratory region, its impact is almost as large as in the traditional region. These results suggest that such regionalization does not capture the spatial variability between migration and educational achievement (see Table 1).

4 Geographical Weighted Regression

As it was mentioned before, one of the goals of this paper is to propose a regionalization based on the links between educational attainment, international migration and labor markets dynamics that illustrates actual structural changes between defined regions. We do so by running a Geographical Weighted Regression (GWR) which let us observe the spatial heterogeneity of the explored relations. As it is a changing model over space, GWR results would assess whether and where the effects of the variables differ, to what extent and at what significance levels. These variations would suggest regions delimitations, allowing us to examine the arguments that regions are characterized by distinct migration effects[8].

GWR is a technique that estimates local parameters to better describe a non-stationary spatial process; it is useful to identify spatial heterogeneity in each of the explanatory variables [10]. In this model the calibration of the parameters assumes that data observed near point i have more influence in the estimation of the parameters for location i than those located farther away from it. Hence, each observation is weighted in accordance with its proximity to point i [16]. In our

---

[7] The robust LM error statistic for spatial dependence was significant at a 0.001 level versus a robust LM lag statistic that only achieved a 0.01 level.

[8] Because we are interested in the *effect* of the variables and not the clustering of observation with similar characteristics, we implemented a GWR test instead of following clustering techniques such as those suggested by Duque, Anselin and Rey [14]. A reviewer suggested we perform a cluster analysis of the GWR coefficients and we will explore this suggestion in a later paper.

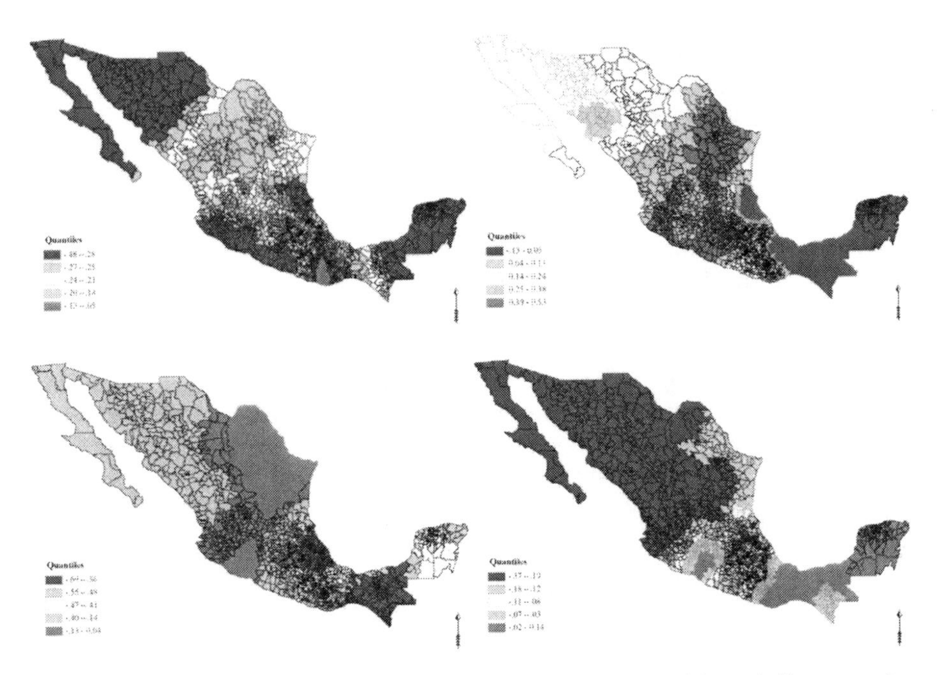

Fig. 3. Mexico, 2000. Parameters of Geographical Regression Model by quintile ranges for: intensity index of international migration (top left), female labor participation (top right), low-income workers (bottom left), and industrialization level (bottom right)

analysis, we use a weighting matrix function with an adaptive bandwidth because our units vary greatly in density across the country [9]. Results were mapped following the same criterion as the exploratory maps[9] (Figure 3).

Figure 3 illustrates the variation and significance of the regression parameters by location for the international migration in one map and for the labor market variables in the rest. As we expected, GWR results for international migration index (see top left map) show spatially differentiated behaviors identified by changes in the parameters, which have grouping effects across the country. This map points out two things: 1) heterogeneity is an important component in the relationship between educational attainment and international migration and 2) the observed behavior does not respond to the traditional regions, since its coefficient (size and significance) does not follow the frontiers of such regions. The model suggests that while regional variations do reflect historical migration municipality experiences, labor markets conditions and educational services shape the boundaries of educational attainment

[9] The model has an acceptable goodness of fit with the local values of R^2 that varies across the country from moderate levels (0.38) to good (0.71) and the residuals did not show a particular spatial pattern. On the other hand, the parameters' spatial variability Monte Carlo test shows that almost all of the variables present strong spatial variation, except for four control variables. Since only control variables are stationary, we followed Fortheringham, Brunsdon and Charlton's advice and estimate a standard GWR model [17].

inequality. These GWR results reinforce the need for creating a new regionalization that captures the joint spatial effects of migration and labor markets.

According to the strength and significance of the relationship between those dimensions, the GWR maps (figure 3) allow identifying four distinct areas. In the Northwest of the country, educational achievement is strongly and negatively related to exposure of international migration and to industrialization and, although weakly, negative and significantly related to the proportion of low-income workers. It is also linked to positive-medium levels of female labor force participation.

Looking at the North, the Center and the West of the country we see that the female participation rate has a strong and positive relationship with educational achievement, but the low-income indicator has a weak or even null relationship with the dependent variable across the three regions. However, when considering international migration and industrialization effects such territorial homogeneity loses strength. The migration variable has contiguous median values in the Northern and the Central regions, where the industrialization parameter fades from strong to medium levels; both effects are always negative. On the other hand, in the West, migration highly reduces the educational attainment, but industrialization has a negative uneven effect.

Finally, it is possible to identify another region integrated by the Gulf and the South. Here the labor market variables have more homogenous effects than migration. The female participation rate and the industrialization level have the smallest and less significant effect, while the proportion of low-income workers strongly reduce the educational attainment. Although migration parameter varies within the region, it shows a clear geographical relationship with the labor markets variables, particularly with industrialization. It is notorious that they run in opposite directions: when migration has a strong effect, industrialization reduces its weight, and vice versa.

Based on these results and on tests for spatial dependency, we define and test a spatial regimes model that accounts for regional variations in the relationship between expected schooling, exposure to international migration and labor market conditions. We do so with a spatial autoregressive model, testing for the presence of spatial regimes.

5 Spatial Regime Model for the New Regionalization

The spatial regions identified by GWR results (see Figure 2, bottom map) are tested in this section as spatial regimes models, in order to examine if there are structural differences in the factors that account for educational achievement in each region. As in the traditional region hypothesis, it is expected that the international migration variable will have a negative and dissimilar effect across regions; but here we expect such regional dynamic will be shaped by labor markets characteristics: international migration prevalence matters according to the jobs availability. For example, we anticipate that in regions where the labor opportunities are poor, although the outflow maybe small or recent, international migration effect would be stronger.

We first run an OLS with structural instability for the four regimes defined. As expected, the results show there is evidence of spatial autocorrelation in the dependent variable and heteroskedasticity issues[10]. We estimate a spatial lag model with structural change and groupwise heteroskedasticity, considering the error variance to be different among regimes.

The Chow-Wald test proves that regimes are significant, and the coefficient stability test show that international migration and labor markets variables vary significantly across regimes, accounting for educational attainment heterogeneity (see table 2). In all the regimes, international migration strongly and significantly depresses educational achievement, supporting the hypothesis of negative effects[11]. However, its relevance does not depend only on the migrant historical trajectory, but also on the accounted labor dynamics. Thus, the strongest effect of migration occurs in the Northwest region (regime 2), with medium migration levels but a polarized labor market that reinforces the negative effects of migration on educational attainment. In contrast, in the North-Center (regime 1), although it also has medium migration levels, the effect of this variable is the smallest since the labor markets conditions are good and more homogenous within the region [1][30]. Thus, it diminishes the relevance of migration as a determinant of educational achievement. In regime 4, the Gulf and South, the low prevalence of migration is enhanced by the poor labor markets conditions, which is evident in the effects of earnings and female labor force participation, as well as the null impact of industrialization –given its minimal and concentrated presence in the region. In the West (regime 3), the highest and longest migration stream does not correspond with the strongest effect, since labor markets indicators across municipalities attenuates its impact on education. Thus, although the migration coefficient is significant and large, it is smaller than in other regions with a shorter migration tradition.

Results suggest that this second model accounts better for spatial heterogeneity between educational attainment and migration. This claim is supported by the fact that the migration parameter shows significant variability (see the stability of individual coefficients test, Table 2). In addition, the regimes defined based on GWR results require to run a spatial lag model, which accounted for the spatial autocorrelation in educational achievement between neighboring municipalities. In contrast, when using the historical regimes it is necessary to specify a spatial error model, which often evidences an ill adjustment between the actual frontiers of the social process and the boundaries fixed by the specified model [2].

---

[10] The robust LM lag statistic for spatial dependence was significant at 99% of confidentiality versus a robust LM error statistic that only achieved a 90%. The Koenker-Bassett test for heteroskedasticity has a 1% significance level.

[11] If we run "by steps" this model we observe that the migration variable remained unstable as we added the labor market and control variables with the current specified regions. However, in the spatial error regime models for the historical hypothesis, migration loses its spatial variability as we added other variables. This suggests that the later specification does not capture heterogeneous effect of migration.

Table 2. Mexico, 2000. Spatial Lag Model with Structural Change and Groupwise Heteroskedasticity for the New Regionalization

For first-order standardized weights matrix			
Dependent Variable	Expected years of schooling		
No. of observations	2478	DF	2429
No. of variables	49	Sq. Corr.	0.53
R^2	0.51	LIK	-3580.62
AIC	7259.24	SC	7544.18
ρ	0.3246***		

Parameters and Significance

Variable	*Regime 1*	*Regime 2*	*Regime 3*	*Regime 4*
CONSTANT	2.8903 ***	4.6963 ***	3.9882 ***	6.8904 ***
International migration	-1.8532 ***	-3.7145 ***	-3.3764 ***	-2.3594 ***
Female participation rate	5.3197 ***	0.0373	3.5359 ***	0.8843 ***
Low-income workers	-1.3066 ***	-2.3537 **	-1.3685 **	-5.1795 ***
Industrialization level	-2.4191 ***	-1.8162 *	-1.6712 **	-0.1766
Educational profile of teachers	0.5682 *	0.7801 *	0.2493	0.5163 ***
Without high schools vs technical schools	0.1958	0.3404	0.1036	0.5950 ***
Without high schools vs technical-general track schools	0.4498 ***	0.3420 *	0.2926 *	0.6266 ***
Inter-municipal migration	-0.2217	-1.4272 *	-0.4503	0.1342
Rural vs Rural-Urban	0.1107	0.3993 *	0.2865 *	0.1333
Rural vs Urban	0.3964 **	0.6365 *	0.3079	0.2257 *
Rural vs Metropolitan area	0.4435 **	1.3557 **	0.6156 *	0.0200
n	***585***	***169***	***383***	***1341***

Group Variances

Variable	*Coeff.*	*S. D.*	*z -Value*	*Prob*
Regime 1. North and Center	0.7220	0.0423	17.0820	0.0000
Regime 2. Northwest	0.9794	0.1066	9.1879	0.0000
Regime 3. West	0.8794	0.0636	13.8270	0.0000
Regime 4. Gulf and South	1.2736	0.0493	25.8191	0.0000

Test on Structural Instability for 4 Regimes

Test	*DF*	*Value*	*Prob*
Chow - Wald	36	240.65	0.00

Stability of Individual Coefficients

Test	*DF*	*Value*	*Prob*
CONSTANT	3	66.78	0.00
International migration	3	7.71	0.05
Female participation rate	3	58.82	0.00
Low-income workers	3	87.93	0.00
Industrialization level	3	22.38	0.00
Educational profile of teachers	3	0.94	0.82
Without high schools vs technical schools	3	5.20	0.16
Without high schools vs technical-general track schools	3	6.08	0.11
Inter-municipal migration	3	3.16	0.37
Rural vs Rural-Urban	3	2.14	0.54
Rural vs Urban	3	2.16	0.54
Rural vs Metropolitan area	3	6.68	0.08

Test on Groupwise Heteroskedasticity

Test	*DF*	*Value*	*Prob*
Likelihood Ratio Test	3	68.63	0.00

Significance Level: * p<0.1 , ** p<0.01 , *** p<0.001

6 Conclusions

This paper addresses the spatial relationship between educational achievement and international migration in Mexico at the municipality level. The descriptive results confirm strong geographical differences in the expected years of schooling for the youth, as well as important differences in migration prevalence despite the increase and geographical spread of the outflows during the 1990s. The exploratory analysis finds a close negative association between migration and educational attainment, but it also confirms the spatial heterogeneity in that relationship. To explain such pattern, we explore two hypotheses. First, that spatial heterogeneity responds to the historical migratory trajectory of regions, which created a unique internal dynamic that impacts its developmental process, as well as youth schooling expectations –after controlling for other local characteristics. A second hypothesis is that geographical heterogeneity emerges from the link between migration prevalence and regional labor markets conditions, which jointly generate a setting that shapes people's educational investment.

We examine both hypotheses with groupwise heteroskedastic spatial regime models. Results supports the second hypothesis, since our regionalization based on spatial-varying links between education, migration and labor is more appropriate than those regions defined previously by migration historicity. This claim is supported by the significant spatial variability reflected by the spatial-regimes estimations, as well as by the behavior of the international migration variable across these regimes. These outcomes highlight the need and usefulness of proper geostatistical methods to test and develop hypotheses that imply spatial effects. Moreover, in the definition of regions within countries, it is essential to consider how the relationships between sociodemographic variables shape geographical disparities.

References

1. Alba, F., Banegas, I., Giorguli, S., De Oliveira, O.: El bono demográfico en los programas de las políticas públicas de México (2000-2006): un análisis introductorio. In: Poblacion, C.N.d. (ed.) (ed.) La situación demográfica de, pp. 107–129. Consejo Nacional de Poblacion, Mexico (2006)
2. Anselin, L.: Spatial econometrics: Methods and Models. Kluwer Academic Publishers, Massachusetts (1988)
3. Anselin, L.: SpaceStat tutorial: A workbook for using SpaceStat in the analysis of spatial data. University of Illinois, Urbana (1992)
4. Antman, F.M.: Parental Migration and Children's Education in Mexico: How Important is Child Age at the Time of Parent's Migration?, `http://paa2008.princeton.edu/download.aspx?submissionId=80796`
5. Baller, R., Anselin, L., Messner, S.F., Deane, G., Hawkins, D.F.: Structural Covariates of U.S. County Homicide Rates: Incorporating Spatial Effects. Criminology 3, 561–590 (2001)
6. Beck, E., Colclough, G.: Schooling and Capitalism. The effect of urban economic structure on the value of education. In: Frankas, G., England, P. (eds.) Industries, firms and jobs. Sociological and economic approaches, pp. 113–139. Plenum Press, Nueva York (1988)
7. Bracho, T.: Perfíl educativo regional en México. Estudios Sociologicos 51, 703–742 (1999)
8. Borraz, F.: Assessing the Impact of Remittances on Schooling: the Mexican Experience. Global Economy Journal 5.1, 1–30 (2005)
9. Brunsdon, C., Fotheringham, S., Charlton, M.: Geographically weighted regression: a method for exploring spatial nonstationarity. Geographical Analysis 28.4, 281–298 (1996)
10. Brunsdon, C., Fotheringham, S., Charlton, M.: Geographically weighted regression – modeling spatial non-stationarity. The Statician 47, 431–443 (1998)
11. Bustamante, J.A.: NAFTA and labour migration to the United States. In: Bulmer-Thomas, V., Craske, N., Serrano, M. (eds.) Mexico and the North American Free Trade Agreement Who will Benefit?, pp. 79–94. Saint Maritn's Press, New York (1994)

12. Consejo Nacional de Poblacion: Índice de Intensidad Migratoria México-Estados Unidos, 2000. Consejo Nacional de Población, Mexico (2002)
13. Consejo Nacional de Poblacion: La nueva era de las migraciones. Consejo Nacional de Poblacion, Mexico (2005)
14. Duque, J., Anselin, L., Rey, S.: The Max-p-region Problem. Working Paper, GeoDa Center on Geospatial Analysis and Computation, http://geography.sdsu.edu/.../Duque_the_max-p-value_problem_draft.pdf
15. Durand, J., Massey, D.S.: Migración México-Estados Unidos en los albores del siglo XXI. Miguel Angel Porrua, Mexico (2003)
16. Fotheringham, A.S., Brunsdon, C.: Local forms of Spatial Analysis. Geographical Analysis 31, 340–358 (1999)
17. Fotheringham, A.S., Brunsdon, C., Charlton, M.: Geographically Weighted Regression: the analysis of spatially varying relationships. John Wiley and Sons, England (2002)
18. Giorguli, S., Serratos, I.: El impacto de la migración internacional sobre la asistencia escolar en México:'paradojas de la migración? In: Leite, P., Giorguli, S. (eds.) Las políticas públicas ante los retos de la migración mexicana a Estados Unidos, pp. 313–344. Consejo Nacional de Poblacion, Mexico (2009)
19. Giorguli, S., Vargas, E., Salinas, V., Hubert, C., Potter, J.: La dinámica demográfica y la desigualdad educativa en Mexico. Estudios demograficos y urbanos 73, 8–44 (2010)
20. Goldscheider, C.: Migration and Social Structure: Analytical Issues and Comparative Perspectives in Developing Nations. Sociological Forum 2.4, 674–696 (1987)
21. Gutierrez, E., Sanchez, L., Giorguli, S.: The Spatial Dimension in Educational Inequality in Mexico at the Beginning of the Twenty-First Century, http://paa2010.princeton.edu/abstractViewer.aspx?SubmissionId=101821
22. Hanson, G., Woodruff, C.: Emigration and Educational Attainment in Mexico, http://cpe.ucsd.edu/assets/022/8772.pdf
23. Instituto Nacional de Estadistica y Geografia, http://inegi.org.mx/est/contenidos/Proyectos/ccpv/cpv2000/default.aspx
24. International Organization for Migration: World Migration Report 2010 - The Future of Migration: Building Capacities for Change, http://publications.iom.int/bookstore/free/WMR_2010_ENGLISH.pdf
25. Kandel, W.: Temporary U.S. Migration and Children's Educational Outcomes in Three Mexican Communities. PH.D. Dissertation. The University of Chicago (1998)
26. Kandel, W., Massey, D.S.: The Culture of Mexican Migration: A Theoretical and Empirical Analysis. Social Forces 80.3, 981–1004 (2002)
27. Leguina, J.: Fundamentos de demografía. Siglo XXI, Madrid (1981)
28. Leite, P., Angoa, M.A., Rodríguez, M.: Emigración mexicana a Estados Unidos: balance de las ultimas decadas. In: de Poblacion, C.N. (ed.) La Situacion Demografica de Mexico 2009, pp. 103–123. Consejo Nacional de Poblacion, Mexico (2009)
29. Meza, L., Pederzini, C.: Migración internacional y escolaridad como medios alternativos de movilidad social: el caso de México. Estudios Económicos, special issue, 163-206 (2009)
30. Sanchez, L.: Activo demográfico y calidad del empleo en México: situación en las entidades federativas del país, 2000. M.D. Dissertation. El Colegio de Mexico (2006)
31. Schmelkes, S.: La desigualdad en la calidad de la educación primaria. Revista Latinoamericana de estudios educativos 24.1, 13–18 (1994)

Accessibility Analysis and Modelling in Public Transport Networks – A Raster Based Approach

Morten Fuglsang[1,2], Henning Sten Hansen[1], and Bernd Münier[2]

[1] Aalborg University Copenhagen, Lautrupvang 2B
2750 Ballerup, Denmark
{mofu,hsh}@plan.aau.dk
[2] National Environmental Research Institute, Aarhus University,
Frederiksborgvej 3994000 Roskilde, Denmark
{mofu,bem}@dmu.dk

Abstract. Accessibility is an important factor in the development of land-use patterns and urban settlements, and these two components are considered linked close together. With the increasing attention payed to reduction of CO2 emissions, focus has been turned towards increased accessibility to public transportation. The work conducted in this study proposes a simplistic raster based model of accessibility to workplaces based on the public transportation services. The raster GIS based approach was chosen due to the need to combine this model with a raster based land-use simulation framework, where simulations of future urban development could be modelled. It has been shown that the proposed GIS model developed is suitable for representing the spatial difference in terms of accessibility for commuters by public transport within the study area. Clear differences between the Copenhagen Metropolitan area and the outskirts of the region can be seen, and are very much in line with statistics on commuting activities for the municipalities involved.

Keywords: Accessibility, GIS, raster-based modelling, public transportation.

1 Introduction

The concept of accessibility is a key element in urban, regional and transportation planning, and accessibility is generally considered as one of the most important determinants of urban land-use patterns. On the other hand land-use has a strong impact on accessibility through the spatial distribution of human activities. Thus accessibility and land-use are strongly interlinked. In most recent research, accessibility analysis and modelling has predominantly been related to transport on road networks by private cars, but public transport by busses and trains has gained enhanced attention due to the efforts of reducing greenhouse gas emissions by concentrating and replacing use of private cars with public transport – last not least for daily commuting between home and workplace [1].

Accessibility can be defined as the ease with which activities at one place can be reached from another place through a transport network. Accordingly, accessibility is

B. Murgante et al. (Eds.): ICCSA 2011, Part I, LNCS 6782, pp. 207–224, 2011.

dependent on the spatial distributions of the destinations (centres of opportunities) relative to a given location, the magnitudes, quality and character of the activities found at the destinations, the efficiency of the transportation system, and the characteristics of the traveller.

Several studies have indicated that people's choice of residential sites and companies' location depends on the level of accessibility to work, shopping and public service for people, and accessibility to raw materials, goods, labour force, market and public service for companies. Therefore, accessibility to some degree determines the development of land-use patterns. In his pioneering work, Hansen analysed the relationship between residential development and accessibility to employment and shopping in Washington DC [2], and he found strong support for the linkage between accessibility and urban development.

Traditionally, accessibility analysis has been performed using network analysis software well suited for estimating accessibility within an existing transportation network. Some efforts have been made to utilise a raster-based approach to accessibility modelling for non-public transport [3] [4], but until now, to our knowledge there have been no attempts to utilise a raster-based approach aiming at analysing accessibility related to public transport networks.

In this research, a raster-based approach was chosen because the output of this model was to be applied within a wall-to-wall covering raster based land-use simulation framework, where impacts of transportation strategies should be analysed. The objective of the current research project has been to develop a raster based accessibility model for public transport in order to serve the assessment of the environmental impact of various urban development strategies and hereby contribute to the efforts on developing models and scenarios for sustainable urban development. Furthermore, the construction of the analysis should be based on relatively simple geographic data, since it should be reproducible for other regions of Europe or other pan-European implementation.

The paper is divided into 5 parts. After the introduction we describe and discuss various concepts of accessibility and describe the developed approach to accessibility modelling and its implementation. In part 3 and 4 we describe the results and discuss their implications respectively. The paper ends with some conclusions on the current model, and an outline for subsequent work.

2 Theory and Methods

Accessibility is a relative quality assigned to a piece of land through its relationships with the transport network and the system of opportunities represented mainly by the facilities in urban centres. Thus we can state that all locations are endowed by a degree of accessibility, but some locations are more accessible than others [5]. Obviously, accessibility is related to the principle of movement minimisation – particularly when accessibility is represented by some kind of distance measure. Traditionally, locational decisions have been made on the principle of minimising the frictional effects of distance. This principle was originally expressed by [6] as the *principle of least effort* and by [7] as the *law of minimum effort*. The underlying reason for these principles is the general tendency for human activities to agglomerate in order to save operation costs by taking advantage of economies of scales [5]. Thus settlements can be considered as a manifestation of the economies of scale, and the intra-urban concentration of industrial districts and shopping centres are even stronger manifestations of this principle. At larger spatial scale we can observe that more accessible locations appears to be sites for larger agglomerations, which further on leads to a hierarchical urban structure as described by Christaller in his Central Place Theory [8].

Accessibility analysis involves selecting among existing measures or developing new measures according to the purpose of the analysis. Below we give short descriptions of various measures of accessibility and discuss their appropriateness in the current research.

2.1 Distance Based Accessibility

Relative accessibility is the simplest measure of accessibility. Basically, relative accessibility is represented by the distance between one location and different opportunities – or centres [9]. The closer a given location is to the destination, the higher the accessibility. The calculation of these distances can be performed in several different ways – from simple Euclidian straight line distances to more complicated infra-structure based accessibility measure using average travel speed or average travel time between two locations or, most advanced, including variation in travel speed along a transport route.

Considering distance as the only relevant factor is a rather simple approximation to the complicated task of measuring accessibility in the real world and pure distance based accessibility measures have mainly been used in various studies analysing problems like maximum distance or travel time to locations as hospitals [10] or infrastructure points like railway stations [11].

2.2 Gravity Based Accessibility

The gravity measure of accessibility was originally developed by Hansen [2], which used the concept to explain why land values are high in the central areas of cities and at other easily accessible points. The pure distance based measure of accessibility disregards the attractiveness of the centres. A gravity based measure appears to be able to capture and interrelate both elements: a) distance – the farther places, people or activities are apart, the less they interact; b) attractiveness – where large cities tend to generate more activities and attract more people and companies than smaller cities. This model can be put into a generalised from, aggregating interaction between single pairs of centres into a set of interactions among all centres in a region.

The destination centres have many attributes that make them attractive to visitors, but very often population has been used as a surrogate variable for the attractive mass due to data availability. Employment is another popular measure of attractiveness, because employment can be regarded as an indicator of economic development [12]. Besides, the impact of distance may vary depending on the transport mode in question. Besides Hansen's original work [2], the potential measure of accessibility has been rather popular and used widely to analyse accessibility to jobs [12], retail services [13], and infrastructure improvements [14]. However, several critics have been raised. The measure does not take into account possible limitations of the available opportunities (e.g. jobs) in the centres [15]. Furthermore, it is clear that the distance-decay function has significant impact on the accessibility values estimated, and generally it is difficult to give precise estimates of the shape of this function. Although the potential accessibility measure obviously seems most advanced, the many uncertainty components may obscure our goal to analyse the relationship between contemporary urban development and accessibility. Therefore, we decided in the current research to use the less sophisticated distance based measure.

In order to describe the measure of accessibility applied to the study, accessibility can be broken down into four components:

1. The land-use component:
   Describing the land-use system, consisting of the spatial distribution opportunities supplied at each destination (jobs, shops, health, social and recreational facilities etc.), or by the demand for these opportunities at origin locations (e.g. where inhabitants live),
2. The transportation component:
   Describes the transport system, by measuring the time consumption required for reaching services specific transportation modes. This component can include various numeric measures such as the amount of time, the costs of the travel or the effort such as reliability and ease of use.
3. The temporal component:
   Reflects the availability of opportunities at different times of the day, and the time constraints available for participation in activities.
4. The individual component:
   Reflects the needs, abilities and opportunities of individuals. These characteristics influence a person's level of access to transport modes and spatially distributed opportunities [16] [17].

Based on this description of components, the measure of accessibility that will be used in this study describes three of the four components. Land-use is described through distribution of centres and location of jobs, the transportation component is described through the modelling of the transportation network and the individual component is described through the job opportunities that the individuals can reach through the transportation network.

In the following paragraphs, the implementation of methodology applied to create the transportation models will be presented. All aspects of data acquisition and preparation as well as model construction will be described.

2.3 GIS Based Modelling of Accessibility

The usage of Geographical Information Systems (GIS) in transport planning has been linked closer and closer together with the development of better tools, data and IT-infrastructure. This is because the nature of the problems concerning accessibility is well handled in the GIS systems, which are capable of handling large amounts of georeferenced spatial data, coupling them with socio economic data about preferences and choices of people [18]. In the last two decades, much focus has been given to the field of geographical based analysis of accessibility. The dominant GIS software providers have included powerful tools for accessibility analysis, as well as many open source based GIS suits are developing network based analysis tools. Examples of powerful network analysis tools are ESRI's Network Analysis extension, and the GRASS GIS Network analysis tools, which both can conduct advanced spatial analysis of accessibility patterns using vector networks.

Many different frameworks for accessibility measures have been developed in recent years. A general framework for accessibility analysis, focusing on the practical development of the concept was proposed by [19]. The framework focuses on the concept of accessibility in terms of developing measures of accessibility into a complete framework [19]. Yang et al. conducted a comparison of spatial based measures of accessibility, where the technological development of GIS systems is used as mean to evaluate new powerful measures such as floating catchment and Kernel density methods [20].

Methodologies for determining the access to neighbourhood goods and services through GIS are also widely conducted using accessibility measures. Examples of these analysis are [21], where accessibility to green parks are assessed using GIS analysis, [22] has developed algorithms for non-motorised accessibility, and [23] has studied the accessibility of neighbourhood facilities in Teheran, suggesting a methodology applying GIS based analysis with fuzzy logic approach to demographic and socio economic data [23] Finally, [24] have analysed the accessibility to airport transportation systems in the United States, as an example of high gravity potential services, where accessibility potential is a high determinant for travel.

2.4 Modelling of Transportation Access

The accessibility concept is furthermore directly applicable to the measures of connectivity to transportation networks, which is the fundamental aspect of this analysis. As mentioned in the introduction, [4] has conducted raster-based analysis of accessibility via non-public transportation options in Portugal. Other examples of coupling transportation and accessibility are recently published work by [25] utilising network based accessibility measures of transportation networks and [26] who was modelling time accessibility in car transportation networks. Access to healthcare has been an area where modelling of network accessibility has gained increasing focus. Examples on these analyses are [27], who compared accessibility by car and by public transportation to healthcare services, and [28] created accessibility maps on access to public healthcare in United Kingdom.

Based on the different implementations of accessibility calculations, the methodology for this study will be constructed using a location based accessibility measure, and creating a raster based model for access to public transportation. The choice of distance-based accessibility is based on the scope of the analysis. Since the construction of the model is focused on general accessibility to all public transportation possibilities, it is the distance function that is the main parameter. Should the model utilise gravitational accessibility, the competition between transportation services would have to be accessed, which is not intended in this analysis, focusing on the technical GIS modelling-tasks. The result of the accessibility analysis will then be combined with spatial data about workplace location and travel time, to create a combined map showing the accessibility to workplaces.

3 Methodology

Modelling the availability of workplaces by transportation mode, requires input data from a wide array of data providers, since parts of the transportation network are operated by government bodies whereas others by private companies. Data included in the analysis was provided by the Statistics Denmark, who provided population and workplace statistics, the Danish National Survey and Cadastre who provided road networks and road classification from the national topographic database KORT10, and finally Movia – the company responsible for the bus services on Zealand - who provided vector based bus networks.

3.1 Creation of Input Data

From the KORT10 database a dataset containing centre points was acquired as basic input for the centre dataset. The dataset contains the coordinates of the town centre point, a (town) name and an ID for each feature. This town centre dataset was afterwards joined to tables with population data acquired from the Statistics Denmark. Information regarding workplaces for the selected towns was not directly available, since these data where recorded on municipality level. The data on workplaces was distributed among town centres applying the population of the town centres as distribution key. The centres included in the analysis, and a classification based on number of workplaces can be seen in figure 1.

In order to create a functional and logical cost surface for public transportation, several adjustments between data were necessary. First, no data describing the Copenhagen Metro-lines where available for the analysis, meaning that these had to be digitized manually. Using the maps from the Metro Company, and the location of stations on Google Earth, this dataset was created in vector format. The available railway datasets consist of maps with railway lines and stations from the KORT10 database, and these data had to be adjusted regarding topological relationships before being used in the analysis. The dataset contains information of old decommissioned train lines, as well as service areas along the train lines. Furthermore the KORT10 database is based on visual interpretations of aerial photos, resulting in inappropriate behaviour. Thus a railway line is discontinued when the railroad line enters a tunnel. In order to create perfect connectivity and precision, a manual interpretation of the railway lines were carried out, adding or removing features where it was necessary.

The final part of the public transportation dataset was information on bus lines and stops. Movia, the bus company of Zealand, provided us with vector-based datasets about lines and station locations. The datasets however contained no information regarding road types or drive time, so this information needed to be created for the bus lines. Information regarding road types is available in the KORT10 network dataset, where each road is classified according to their width and type. The data regarding the railway lines did not share line topology with the vector data from KORT10, so a spatial join with distance tolerance was applied to adapt the KORT10 information to the railway line dataset.

In order to determine the average drive speed of the different road classes, drive speeds where obtained from the national emission inventories [29]. The values used for the drive speed calculations for busses can be seen in table 1.

Table 1. Applied drive speeds for the road network

Object Code (from Kort10)	Road type	Transport Speed (km / h)	Cell crossing time (CCT) minutes / 100m
2111	Highway	100	0,06
2112	Main roads	70	0,086
2115	Roads + 6m	50	0,12
2122	Roads 3 - 6 m	40	0,15
2123	Other roads	25	0,24

Drive speed for trains and metro lines are generated based on information on average trip speed, derived from the service providers, and can be seen in table 2 together with walking pace for pedestrians assumed for the rest of the area.

Table 2. Applied drive speeds for the non-road based transportation forms

Travel mode	Transport Speed (km / h)	Cell crossing time (CCT) minutes / 100m
S-train	120	0,05
Intercity trains	160	0,0375
Local trains	100	0,06
Metro	80	0,075
Walking	6	1

Once all three public transportation datasets where calculated, with drive speeds and classifications, each theme was converted to raster format, applying the official Danish reference grid with 100-meter spatial resolution, using drive speeds as main parameter.

To make the model as precise as possible, it was important to model the drive time of lines in relation to the stations. Whereas you can enter a car-based network at all nodes and lines, a public transportation network is limited to only permit entering the network at railway stations and bus stops. In order to model these elements, constraints where added along the lines of the transportation networks. The workflow of the cost surface creation is illustrated at figure 2.

Using a function that expands specified zones of a raster by a specified number of cells a one-cell buffer was applied to the raster datasets representing the public transportation network, and these buffers where given the value 999. The resulting

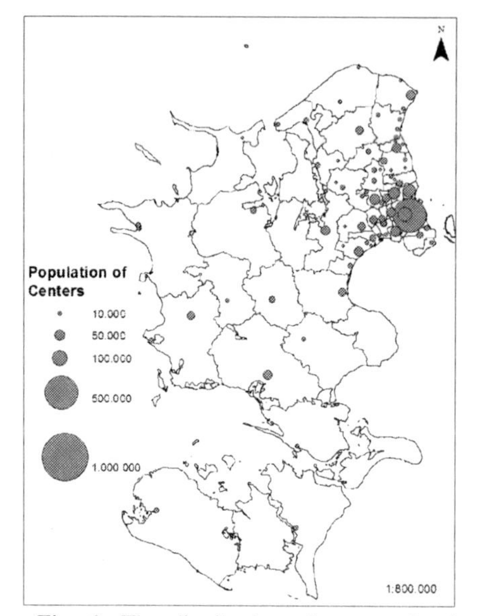

Fig. 1. The distribution of centres in the Zealand region

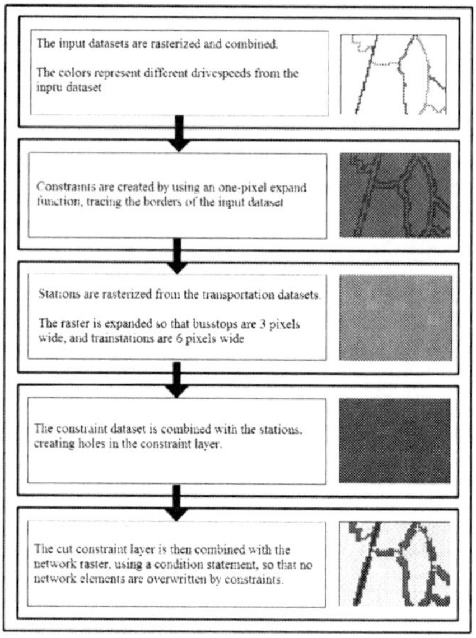

The input datasets are rasterized and combined.

The colors represent different drivespeeds from the inpru dataset

Constraints are created by using an one-pixel expand function, tracing the borders of the input dataset

Stations are rasterized from the transportation datasets.

The raster is expanded so that busstops are 3 pixels wide, and trainstations are 6 pixels wide

The constraint dataset is combined with the stations, creating holes in the constraint layer.

The cut constraint layer is then combined with the network raster, using a condition statement, so that no network elements are overwritten by constraints.

Fig. 2. The process used to create the Cost surface layer for the analysis

raster where then processed using conditional statements ensuring that the buffers where combined properly, maintaining the correct information from the original raster dataset. Next, the stops and stations had to be added to the layer, enabling entry to the transportation network. The stations from all three networks where converted into raster, and a two cell expand was applied to cut the holes in the constraint layer.

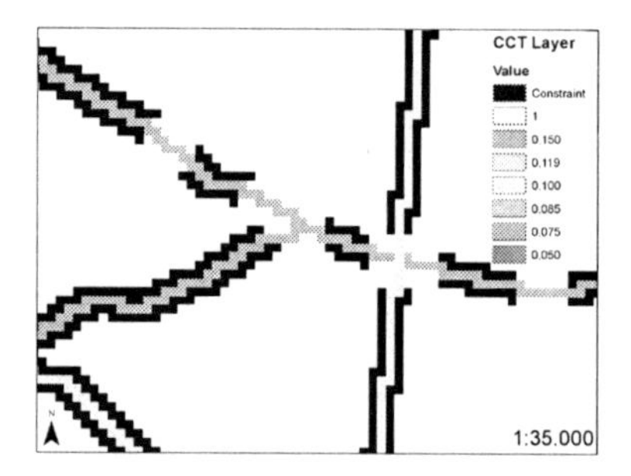

Fig. 3. The final CCT cost surface. The cut-out gaps at stations are visible as breaks in the constraints.

Figure 3 illustrates how the cost surface raster was constructed. The black 999 values in the image are the constraint zones, which the model will avoid to pass. At locations with stations or bus stops, the constraint layer is pierced, making it possible for the model to transits between travel modes.

Cells without transportation lines are assigned a travel speed of 1 CCT, which at a 100-metre grid resolution translates to a speed of 6 km/h, and this is on average equivalent to walking pace of pedestrians.

With all the information regarding lines and constraints combined into one raster layer, the cost surface could be generated. The calculation utilises a Cell Crossing Time (CCT) algorithm developed by [4] which determines the travel time value for each cell in the raster dataset. The calculation of the CCT is based on the following algorithm:

$$CCT = \frac{C_s * 60}{T_s * 1000} \tag{1}$$

Where CCT is the cell crossing time expressed by the cell size of the raster dataset (C_s) and the travel speed (T_s) derived from the network.

3.2 Model Construction

With the two cost surfaces constructed, the centres dataset and the two surfaces where combined in a Python based model using ArcGIS 9.3.1, conduction a ratio distance analysis on the dataset. The algorithm for the analysis is based on the work of [30], where the use of the employment / population ratio is discussed for identifying urban employment centres. The ratio is for our purpose then combined with the cost distance calculation into the following combined equation:

$$\sum \frac{(\frac{W_i}{P_i})}{T_i} \tag{2}$$

Where the travel time for a given cell specified by the cost surface is T for the centre i, W is the number of workplaces for the centre i, and P is the population of the centre i.

The modelling of the result surface required a clean-up function to be applied after the execution of the model, where a replace-function removed the influence of the constraint pixels on the output map. The function replaces cells of a raster map, highlighted in a mask map, with the values of the nearest neighbours. The cells with the constraint showed a low accessibility due to the calculation procedure, but by creating a mask with the constraint zones, and replacing the values of the constraint cells with their lowest neighbouring value, the visual noise from the constraints where removed.

A graphical presentation of the processing steps included in the calculation of the public transportation accessibility can be seen in figure 4.

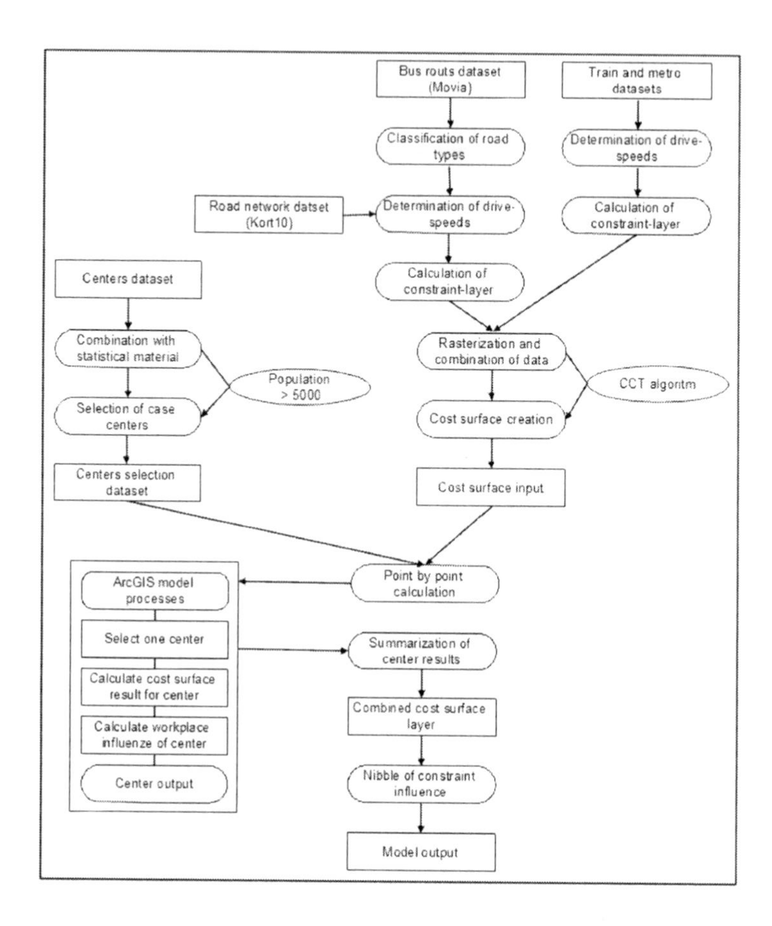

Fig. 4. The structure of the model applied, from input data, over cost surface creation to the final raster based analysis

4 Results

The model produces one raster image covering the entire Zealand region, where each pixel summarises the relative accessibility of all centres from the specific pixel in terms of distance and weight. An example of the result of one centre can be seen on figure 5. The result of the model can be seen on the right side of figure 6.

The red colour indicates high accessibility representing areas where the overall accessibility to the jobs in centres is best. The result describes where the travel time to jobs is best in terms of connectivity to the public transportation network. This means that in terms of potential, the result of the analysis describes where the best connectivity to jobs can be found in, based on where the accessibility to transportation measures are best. It is evident that since there is a much higher density of public transportation options in the Copenhagen Metropolitan area, the accessibility is the highest in this region. The medium sized cities in Zealand outside the Copenhagen Metropolitan area are all scoring relatively high accessibility, since many busses and

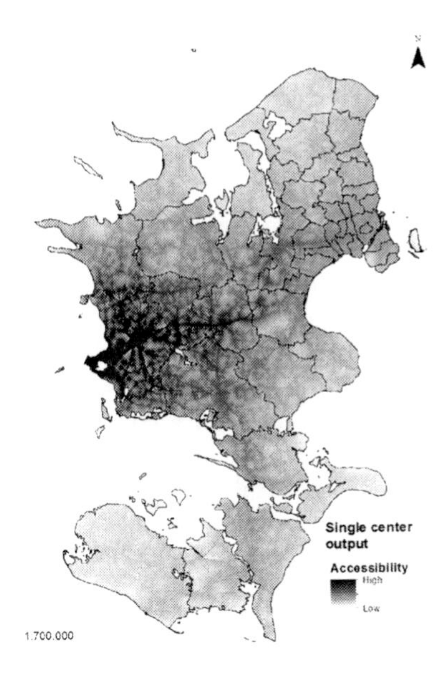

Fig. 5. An intermediate result from the model, showing the result for one centre as high to low accessibility values

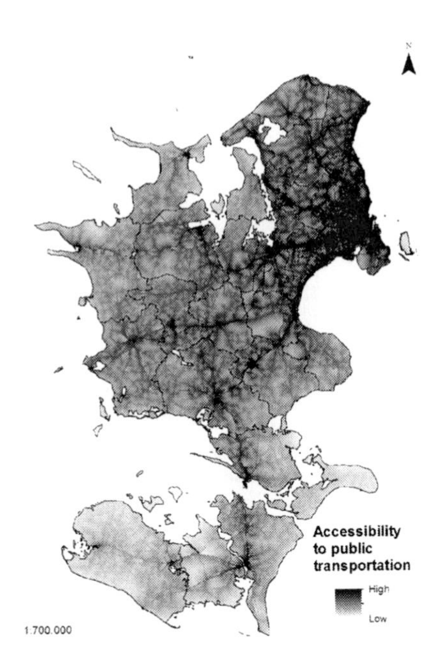

Fig. 6. The complete model output

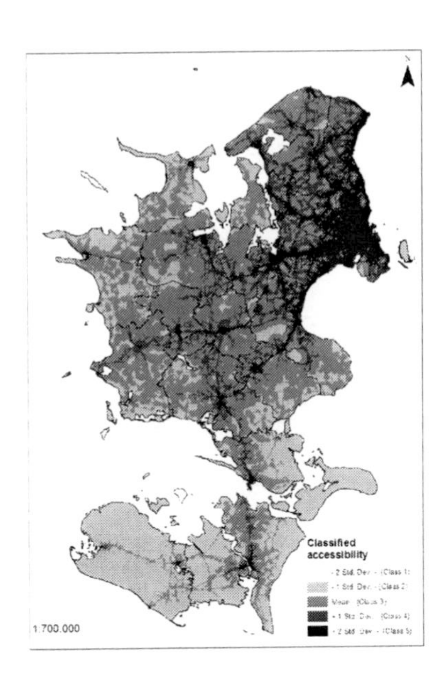

Fig. 7. The model output visualized by +/- 2 standard deviations

railway lines pass through these cities. Besides the positive influence on the S-train on the accessibility is clearly visible by the 'fingers' outgoing from Copenhagen.

The resulting map was afterwards classified using the standard deviation method in ArcMap, and 2 standard deviations around the mean were plotted. The result of the statistical measurement of distribution can be seen on the left side of figure 6. The yellow class describes the average accessibility for the entire region, based on the mean value for the entire dataset. The red areas have better than average accessibility, and the blue areas have lover than average accessibility, expressed by one and two standard deviations above and beyond the mean value.

The statistical description of accessibility was then summarised for each of the 45 municipalities in the region, and the average accessibility score was calculated for each municipality, based on the class value of figure 6. For the total number of pixels in the municipalities, an average value by summarising the value of each pixel, and dividing it by the total number. The result of the average calculation can be seen in figure 7. The figure summarises the standard deviation scores of the entire municipality, describing the general accessibility to public transportation for the entire municipality.

Based on data from the Statistics Denmark, the commuting between the municipalities could be calculated, and for each municipality, the ratio between outgoing and incoming commuting could be determined. The map in figure 7 shows the percentage of jobs in the municipalities occupied by commuters crossing a municipality border in their travel to and from work.

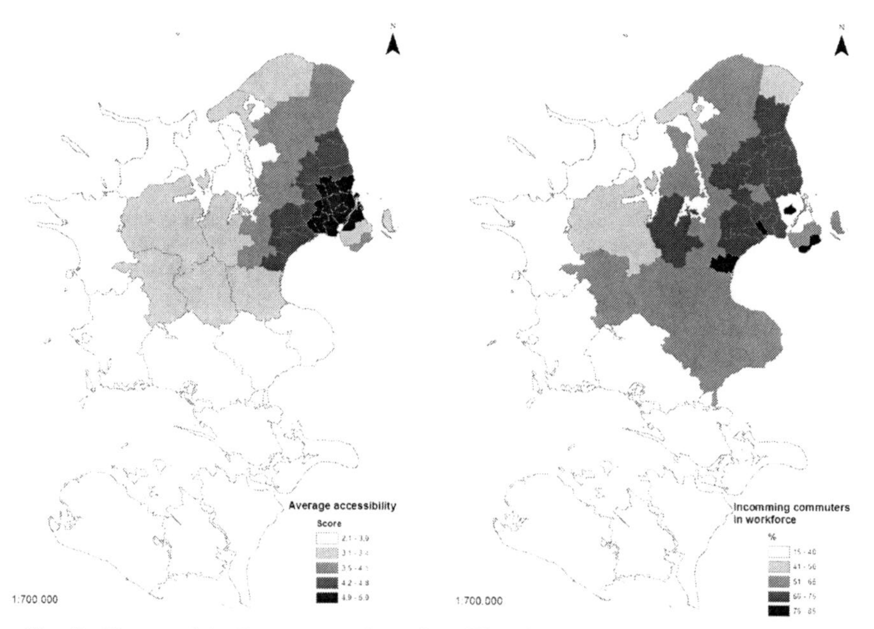

Fig. 8. The municipality mean value of accessibility

Fig. 9. The statistical number of jobs occupied by commuters in the municipalities

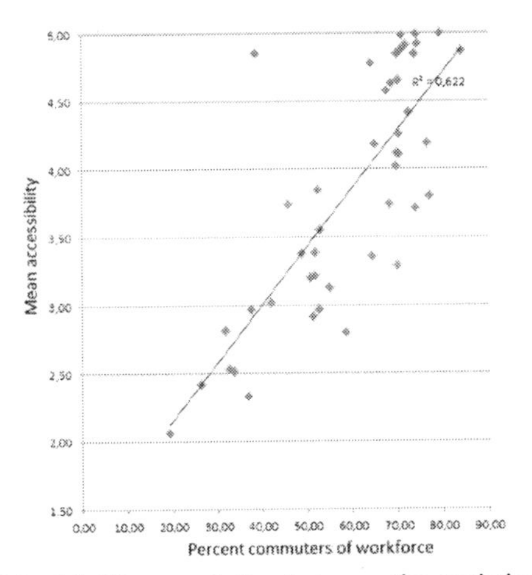

Fig. 10. The correlation between the statistical evidence and the mean accessibility

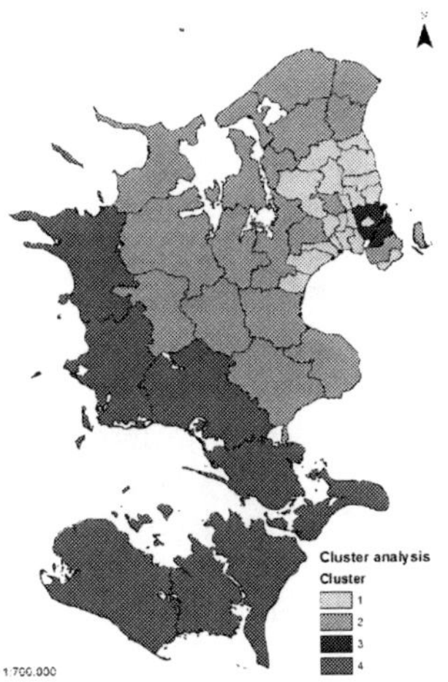

Fig. 11. the clustering analysis of the accessibility

In figure 8, the correlation between the data from figure 7 is shown, combining the mean accessibility with the per cent commuters. The graph shows a strong correlation between the two variables. To elaborate further on the data, the two variables were then combined in a clustering analysis using the ArcGIS 10 "Geostatistical Analyst" to analyse the patters. The results of the clustering analysis can be seen on figure 8, where the regional differences observed in the 6a and 6b datasets are clearly distinguishable. The result creates four clusters if we allow the city of Copenhagen to become a cluster of its own (cluster #3). The values for Copenhagen are different from the other towns regarding both variables, because it has a high accessibility and a relatively low percentage of commuters. The other three clusters are relatively easy to identify based on figure 8.

5 Discussion

5.1 Modelling of Accessibility

The aim of the current research project has been to develop a modelling concept for accessibility by public transportation for a large region, giving a general overview of public transportation access, based on relatively simple data requirements. Thus, the aim of the analysis was to illustrate the combined connectivity of the transportation network in terms of transporting commuters to job locations. We decided to apply a raster based modelling approach, even though a model build upon a vector network could produce more detailed results, but has much higher requirements on input data and computing time. Each pixel is assigned a nominal value, reflecting the influence on the accessibility from all centres in the analysis. A cell with high accessibility will have good connectivity to public transportation services and to jobs. By using the raster domain for the analysis, the output of the model becomes a simple to interpret continuous surface, combining the results from each centre, where high and low values indicate good and bad accessibility respectively.

The result of the accessibility analysis is to some extent influenced by the landscape construction of the case area. In the highly populated eastern Zealand region, the number of centres is much higher than in the remaining Zealand surface. The high density of centres is also part of the explanation as to why the public transportation line density is much higher in this region than in the rest of the region. This will of course have implications on the results, meaning that the result is expected to be much higher in this region. The cluster analysis emphasises and illustrates the regional differences of the Zealand region, and illustrates how periphery regions are less well serviced in terms of public transportation options. The need for, and availability of transportation offers are often linked together. The share of jobs occupied by commuters and the accessibility modelled is in this analysis turned out of being capable to illustrate this point. The correlation between the statistical commuting patterns and the accessibility to public transportation shows that in areas where the public transportation services are spars, people are less inclined to commute.

In political and public discussions, the term 'Outskirt Denmark' has been given much focus during recent years, and is used to describe the migration of both individuals and opportunities away from the outermost municipalities in Denmark. Together with figures from the official statistics for Denmark, our model illustrates some of the issues involved in this problem. The outer regions have public transportation infrastructure that seem to meet the current demand, since a majority of the jobs in the municipality are occupied by local workforce, living within the municipality.

The inter municipality connectivity with public transportation is much better in the Greater Copenhagen area, making it possible for a large number of individuals to work outside their home municipality. There are commuters coming to Copenhagen area from the western and southern regions of Zealand, travelling more than one hour in each direction every day. These either live close to a railway station with good connections or commute by car for at least part of the journey, since transportation opportunities by busses often are sparse and time consuming.

The key element of better accessibility to public transportation in the Copenhagen region is a network of integrated regional trains, S-trains and metro lines, connecting large areas and many municipalities with fast and efficient train transportation. The bus systems of the region is furthermore build to enhance the connectivity of train lines and cities, creating a system with good interconnectivity between municipalities, and thereby connecting the workforce and workplaces in a large area. In our model, the travel time is determined by the speed of the transportation services. The good interconnectivity between the busses and the trains makes the general accessibility better. The travel times are improved by the possibility to access faster transportation modes, and with the good connectivity to the train stations by busses, the accessibility is greatly improved by the good train system.

The raster model has several temporal issues that are to be addressed through further research, since the depiction of time in the analysis is somewhat inadequate. However, this has not been addressed in this work, since the construction of the travel time is homogenous throughout the entire area, meaning that the influence is expected to be even over the entire cost surface in relation to line-travel times. First, the model utilises functions from the ArcMap toolbox for calculations and output adjustments. Centrally the model uses several functions from the Spatial Analyst toolbox. A clear weakness of using these tools is, that the documentation of calculation procedures is often quite vague. However we suggest that the uncertainty introduced to the modelling is of limited consequence to the output. The main calculation function is the cost surface tool that works closely together with the CCT surface, summarising the pixel inputs. The nibble function used to clean the output is of relatively little importance to the output, as it basically only replaces the single pixel-width constraint zones with the nearest neighbour average. If complete transparency in terms of calculations where to be obtained, a complete python based model should be developed. Second, if the time modelling of the transportation networks should be improved, several issues regarding connectivity and stops in terms of time consumption need to be modelled. As stated in the theory chapter, our model does not describe the temporal aspects of accessibility. Connectivity in our model is reduced to stops through the constraints in the cost surface and no time is added for change of transport lines, and the passing of bus stops and train stations is also 'costless' in terms of time consumption. The model should be fitted for further use with information about stop types, and a time constraint should be modelled for connectivity. However the model as it is constructed in raster format has no information about line numbers, meaning that the model cannot distinguish between a line passing through, and connectivity between lines. In that case, a network based vector model would be much more capable of handling the precision modelling of travel time in relation to stops and connectivity, however also much more complex to set up, and the demand for detailed input data regarding lines and connectivity would be much higher. Detailed information regarding line specific travel time information would need to be gathered, describing the actual time consumption of all line elements for all transportation modes, as well as a complex rule set would need to be created, describing transactions between lines at all stop. These data might not be directly obtainable for the entire region, making it difficult to conduct the more precise vector based analysis.

The strength of the model presented here is that it can be easily transferred and set up for other regions, since the demand for input data is quite low. The information required is limited to vector based information about lines and stops, and a travel speed attribute for each line segment. As shown in our analysis the travel speed can be calculated using simple statistical data for roads and rail, meaning that they should be easy obtainable for any region.

The final issue to be discussed here is that the model indicates 'perfect' availability of all lines. No waiting for trains or busses are modelled, which would increase the travel time much further. The availability is however hard to model since it requires detailed information regarding every transportation line. The wait time at stations could be modelled, as average waiting time, for each line – but this would need to be incorporated into the model for each line.

5.2 Potential Applications

In the western and southern parts of the case region where the distribution of lines and stations are much sparser than in the Copenhagen region, this model will be a tool to evaluate the impacts of new public transportation investments, since the enhanced accessibility provided by new lines and stops would be clearly visible here.

Furthermore the model could be used as an input to land use modelling in terms of residential location. Based on the model output, it would be possible to add the information regarding public transportation accessibility to an evaluation of suitability for land use change potential, dictating that new urban land must be in zones with good public transportation accessibility.

6 Conclusion

For very detailed accessibility modelling the network model is superior to the raster based approach applied in the current project. However, the aim of the current project has been to develop a raster based accessibility model supporting raster based urban land-use modelling at the regional scale. Focus of our modelling efforts has been to develop a model, which can support our research on sustainable urban development. Therefore, it has been important to analyse the accessibility for a larger region in terms of accessibility to goods, services, and not at least work places, which is of great importance for most people, and which accounts for a major share of the transport demand in Denmark. Addressing the general accessibility through a public transportation network is thus a major issue for developing a sustainable urban structure. Furthermore, it was important to build a model that was based on simple geographic data, making the requirements of the model as low as possible. In order for the model to be used on other European regions, it was important that data input should be data that was considered general enough to be available in most countries.

In its current form, the model utilizes vector based lines and stops, and statistical data regarding jobs in the main centres, which is the most simple data requirements that can comprise the modelling task.

A more advanced model that utilises information of availability and quality of the transportation services in a raster based framework should also be one of the next steps of the model, however this intensifies the data requirements of the model to an extent, where it might not be reproducible for other regions, due to missing information regarding route travel-times, line-directions etc. Being out of scope of the current projects, the task to create an accurate space-time representation of public transportation services in the raster domain is a challenge that would be beneficiary for many raster based land use change models.

The developed approach to assess the accessibility for the Danish island of Zealand, which includes the Copenhagen Metropolitan area, is a first attempt to use raster techniques to analyse and model accessibility through a public transportation network. The model outcome illustrates the regional differences in the case region, where both the distribution of population, jobs and public transportation services vary greatly. The analysis has shown that the modelled access to public transportation can be coupled closely with the statistical data on commuting, meaning that there is evidence that the access to public transportation is coupled to the commuting choice of individuals.

Further research, will also aim at modelling individual road transport by cars based on the same raster principles, as well as integrating these two modes for commuting according to the park and ride principles.

Acknowledgement

Funding for this study was provided by the European Union, Seventh Framework Program, Theme 8, Small or medium-scale focused research project, Grant agreement no.: 244766 - PASHMINA (PAradigm SHifts Modeling and INnovative Approaches).

The authors would like to thank the many contributors of materials that have been used in the analysis. These include the Danish statistical office for commuting data, Gitte Schwartz at Movia for bus line data and Morten Winter at the National environmental research institute, Aarhus University for the average drive speeds on roads. Finally the National Survey and Cadastre (KMS) provided road data and administrative boundaries.

References

1. EEA: Climate for a transport change. Report No. 1/2008. European Environment Agency (2008)
2. Hansen, W.G.: How Accessibility shapes land-use. Journal of the American Institute of Planners 25, 73–76 (1959)
3. Hansen, H.S.: An Accessibility Analysis of the Impact of major Changes in the Danish Infrastructure. In: Harts, J., Ottens, H., And Scholten, H. (eds.) – Proceedings of the 4th European Conference and Exhibition on Geographical Information Systems (EGIS 1993)Genova, pp. 852–860 (1993)
4. Juliao, R.P.: Accessibility and GIS, pp. 1–11. European Regional Science Association (1999)
5. Garner, B.J.: Settlement Locations. In: Chorley, R.J., Haggett, P. (eds.) Socio-economic Models in Geography, Taylor & Francis, Abington (1967)
6. Zipf, G.: Human Behaviour and the Principle of Least Effort. Addison-Wesley, Reading (1949)
7. Lösch, A.: The Economics of Location. Yale University Press, London (1954)
8. Christaller, W.: Die Zentralen Orte in Südeutchland, Jena (1933)
9. Ingram, D.R.: The Concept of Accessibility: A Search for an operational Form. Regional Studies 5, 101–107 (1971)
10. Love, D., Lindquist, P.: The geographical Accessibility of Hospitals to the Aged: A Geographic Information Systems Analysis within Illinois. Health Service Research 29, 629–651
11. Rietveld, P.: The Accessibility of railway Stations: The Role of the Bicycle in the Netherlands. Transportation Research Part D: Transport and Environment 5, 71–75
12. Linneker, B.J., Spence, N.A.: Accessibility measures compared in an analysis of the impact of the M25 London Orbital Motorway in Britain. Environment and Planning A 24, 1137–1154 (1999)

13. Guy, C.M.: The Assessment of Access to Local Shopping Opportunities: A Comparison of Accessibility Measures. Environment and Planning B 10, 219–238 (1983)
14. Sten Hansen, H.: Analysing the Role of Accessibility in Contemporary Urban Development. In: Gervasi, O., Taniar, D., Murgante, B., Laganà, A., Mun, Y., Gavrilova, M.L. (eds.) ICCSA 2009. LNCS, vol. 5592, pp. 385–396. Springer, Heidelberg (2009)
15. Shen, Q.: Location characteristics of inner city neighbourhoods and employment accessibility of low-wage workers. Environment and Planning B 25, 345–365 (1998)
16. Geurs, K.T., Ritsema van Eck, J.R.: Accessibility measures: review and applications. Urban Research Centre, Utrecht University (2001)
17. Geurs, K.T., Van Wee, B.: Accessibility evaluation of land-use and transport strategies: Review and research directions. Journal of Transport Geography 12, 127–140 (2004)
18. Liu, S., Zhu, X.: Accessibility Analyst: an integrated GIS tool for accessibility analysis in urban transportation planning. Environment and Planning B: Planning and Design 31, 105–124 (2004)
19. Liu, S., Zhu, X.: An Integrated GIS Approach to Accessibility Analysis. Transactions in GIS 8, 45–62 (2004)
20. Yang, D.-H., Goerge, R., Mullner, R.: Comparing GIS-Based Methods of Measuring Spatial Accessibility to Health Services. Journal of Medical Systems 30(1), 23–32 (2006)
21. Comber, A., Brunson, C., Green, E.: Using a GIS-based network analysis to determine urban greenspace accessibility for different ethnic and religious groups. Landscape and Urban Planning 86(1), 103–114 (2008)
22. Iacono, M., Krizek, K., El-Geneidy, A.: Measuring non-motorized accessibility: issues, alternatives, and execution. Journal of Transport Geography 18(1), 133–140 (2008)
23. Lotfi, S., Koohsari, M.J.: Measuring objective accessibility to neighborhood facilities in the city (A case study: Zone 6 in Tehran, Iran). Cities 26(3), 133–140 (2009)
24. Matisziw, T.C., Grubesic, T.H.: Evaluating locational accessibility to the US air transportation system. Transportation Research Part A: Policy and Practice 44(9), 710–722 (2010)
25. Chen, A., Yang, C., Kongsomsaksakul, S., Lee, M.: Network-based Accessibility Measures for Vulnerability Analysis of Degradable Transportation Networks. Networks and Spatial Economics 7(3), 241–256 (2007)
26. Hudecek, T.: Model of time accessibility by individual car transportation. Geografie 113, 140–153 (2008)
27. Lovett, A., Haynes, R., Sünnenberg, G., Gale, S.: Car travel time and accessibility by bus to general practitioner services: a study using patient registers and GIS. Social Science & Medicine 55, 97–111 (2002)
28. Nettleton, M., Pass, D.J., Walters, G.W., White, R.C.: Public Transport Accessibility Map of access to General Practitioners Surgeries in Longbridge, Birmingham, UK, pp. 64–75. Journal of Maps (2006)
29. Winther, M.: Analyse af emissioner fra vejtrafikken: Sammenligning af emissionsfaktorer og beregningsmetoder i forskellige modeller, Faglig rapport fra DMU, no. Danmarks Miljøundersøgelser, Aarhus Universitet (1999)
30. Coffey, W.J., Shearmur, R.G.: The identification of employment centers in Canadian metropolitan areas: the example of Montreal. Canadian Geographer 45, 371–386 (2001)

MyTravel: A Geo-referenced Social-Oriented Web 2.0 Application

Gabriele Cestra, Gianluca Liguori, and Eliseo Clementini

University of L'Aquila, Department of Electrical and Information Engineering
Via Campo di Pile, 67100 L'Aquila, Italy
gabriele.cestra@bluedeep.it, gianluca.liguori@bluedeep.it,
eliseo.clementini@univaq.it

Abstract. The paper describes the architecture and main concepts of a geo-referenced web 2.0 application with a strong social-network component and possibly integrated with handheld devices. The application, called MyTravel, allows the users to share their travel experiences, in terms of geographic information and publishing contents like photos, notes and comments. In the paper we present the domain model of the application, associated to the functional prerequisites; then, we describe the architecture, technologies and the integration techniques adopted in the project, that have been validated through the development of the application prototype.

Keywords: GIS, Web 2.0, Geo Social Network, geo-tag, Android, Google Maps, Facebook.

1 Introduction

During the last few years the growth of spatial data availability has provided a strong stimulus to the development of geo-referenced applications, not only in traditional fields such as Geographic Information Systems (GISs) or Car Navigation Systems, but also in completely different fields, related to leisure or common daily activities [1].

At the same time, the rise of "social-oriented" applications introduced a strong innovation in the Information Technology (IT) field. While the volunteered activity of information feeding is already a consolidated reality among some specialized communities of users, the majority of them consumes the information in a passive way. The spread on a global scale of Web 2.0 social applications and the amazing growth of social network communities, such as Facebook[1] or Twitter[2], have convinced the mass of users to become active feeders of information. The development and wide diffusion of "geo-reference enabled" electronic devices is contributing to add spatial data to the flow of information daily shared by users.

---

[1] http://www.flickr.com/
[2] http://twitter.com/

B. Murgante et al. (Eds.): ICCSA 2011, Part I, LNCS 6782, pp. 225–236, 2011.

One of the most important challenges in current application development is the integration of heterogeneous technologies into a flexible and functional architecture in order to provide appealing social applications with a strong spatiotemporal support.

The paper is organized as follows. In Section 2, we discuss the context of the research, presenting the motivations behind the development of MyTravel. We also explain the main prerequisites that guided the design and development phases and must be satisfied by the application. In the last part of the section we give an overall view of the technologies and tools used during the project realization. Section 3 focuses on the domain model of MyTravel, showing the core use cases and classes defined during the research, along with the solutions adopted to manage the spatio-temporal aspects of the application. In section 4, we show the application prototype developed to validate the architectural and technological choices made during the project. The paper ends with some considerations and proposes possible further developments.

2 The Context

The aim of the work described in this paper is the definition of a prototypal architecture to develop web 2.0 application with a strong spatiotemporal component. The architecture should be flexible and should allow the integration with heterogeneous technologies, such as handheld geo-referenced devices and consolidated geographic information services such as Google Maps[3].

2.1 Background and Related Work

The convergence between the web 2.0 paradigm, along with the social networking phenomenon, and GISs has set a new trend in the IT field, leading to so-called Volunteered Geographic Information (VGI). VGI is the harnessing of tools to create, assemble, and disseminate geographic data provided voluntarily by individuals [2].

There are many applications that use geo-referenced data in various ways in order to achieve different goals. Flickr [3], for example, uses geo-tagging in order to link photos uploaded by its users to specific places. A similar approach has been followed by Wikipedia[4], which now contains over 1 million of geo-referenced articles [4].

While applications like Flickr and Wikipedia focus their core business in photo-sharing and document-sharing respectively, other systems, such as Google Latitude[5], Foursquare[6] and Gowalla[7] are entirely built around geo-referenced information. For example Google Latitude allows the user to share his position in real-time with his friends, while Foursquare and Gowalla are focused on the review of places like shops, pubs and other public places.

The application presented in this paper has a strong relation with the second group of applications just cited, in that MyTravel makes a heavy use of geo-referenced data.

---

[3] http://maps.google.it/
[4] http://it.wikipedia.org/
[5] https://www.google.it/latitude/
[6] https://foursquare.com/
[7] http://gowalla.com/

Nevertheless, there is an important difference between MyTravel and other applications: MyTravel allows us to acquire not only geo-referenced points, but also trajectories, making possible to store routes together with points of interests (POIs).

Another difference, between MyTravel and the applications previously described, is the presence of the time or temporal component. Many geo-referenced applications do not provide an integrated visualization and interaction between events and POIs [5], while MyTravel allows the user to register a path or POI described by time and space. Other applications, such as PhotoBrowser [6] and the one presented by Hertzog and Torrens [7], relate concepts as photos, meetings and business travels to the temporal component. In MyTravel the temporal dimension is linked both to punctual elements like POIs and journeys/trajectories. For each travel, MyTravel reports the temporal interval in which it occurred and associates to every POI the acquisition date.

There are many affinities between MyTravel and another application known as the STEVIE system [5]. The STEVIE system is a mobile application that allows users to manage POIs and events. In STEVIE the time is linked to an event, while in MyTravel is associated to a journey, but in both applications the temporal dimension is crucial. Another system that proposes an architectural approach similar to MyTravel is SeMiTri [8], which provides a framework to query, analyze, and visualize trajectories.

Like other social-oriented applications, our project aims to encourage a growing community with a common interest in this case the passion for travels-- to share their information by building a dataset of user-generated content [9].

2.2 MyTravel

The vision of the internet as a platform, according to the web 2.0 trend [10], is one of the most important innovation occurred in the IT field. The definition of the web as a platform implies that the applications evolve to become complex systems with a high degree of user interaction, instead of being simple web sites showing static information.

This vision of the internet can be summarized with the definition "participatory web" [11]: the web and its applications provide a participatory instruments to the internauts. In this way, a relation is created between users and the application: the application grows in importance only if the users are willing to share their information; nevertheless the application has a strong appeal on the users only if it is able to provide relevant information to them. User participation is a *condicio sine qua non* for the application success. These considerations have represented the *incipit* for MyTravel project.

In the beginning the attention has been focused on the research of the application context, trying to identify what kind of information to manage in order to attract a certain amount of users. After this phase, we decided to propose the users to share their journeys, both those they have already done and those they would like to do. MyTravel allows us to share a journey experience through photos, comments about visited places and mainly through a geographic map that shows places and point of interest reached during the journey.

The application can be defined as a geo-social network, where the geo-referenced component plays an important role. Another important feature of MyTravel is the opportunity to describe and share the journey experience in real time through a smartphone. We are convinced that the spatio-temporal dimension, seen as real-time geo-localization, is one of the most innovating elements of web 2.0.

2.3 MyTravel Requirements

The project MyTravel's final intent is to lead to the implementation of a geo-social network. We define geo-social network an application that allows its users to show to other users their geographical position. Starting from the concept of position, MyTravel develops the concept of journey seen as a group of positions ordered through the temporal dimension. Users can build a journey simply registering a set of positions.

Obviously, since we are talking about a geo-social network, the user should be able to share his/her information quickly and easily. MyTravel provides two ways to insert and manage the journeys: we have called these modalities atHome and onRoad, respectively. In the first case, users can insert their journeys through the web, allowing them to setup an already done journey or one that they would like to carry on in the future. In the second case, users aim to share their journey experience in real time, while it is happening. It is evident that in the second case the application has to support the use of mobile devices such as smartphones.

Another important application prerequisite is the registration of a timestamp for every geo-referenced information, in order to allow the manipulation of both the spatial and temporal dimensions.

The concept of time is necessary to define the interaction between different users during their journeys: the simplest question that may arise is "did user1 meet user2, since they visited the same place during their journeys?".

Taking in account the temporal dimension allows the users to know the current or past position of their friends, and to know their future position in the context of planned journeys, in order to organize their journeys in accordance to this information.

2.4 Technologies and Tools

For the development of the project we needed to choose and integrate different technologies. An important preliminary step was the creation of a Rich Internet Application (RIA) and we decided to develop it using the Java programming language and Oracle JEE environment.

To speed up the design and development process we decided to adopt a methodology [12] [13] [14] and a framework [15] [16] that allow us to realize Ajax-enabled, full featured applications writing only Java code. The framework, based on an extended Model-View-Controller architecture [17] [12], provides elements and services to manage the main parts of an application, such as graphical interface management, use case lifecycle management and domain objects persistence (Fig. 1).

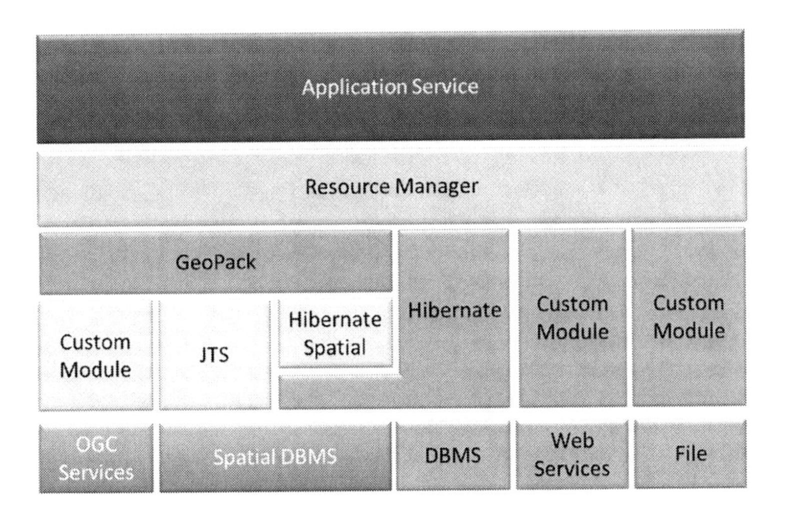

Fig. 1. Overall view of the framework architecture

Moreover, the framework offers a set of tools and services to manipulate spatial data. This set of components is called GeoPack [18] [19] and allows the communication between the application domain model and the underlying spatial database.

As the reference DBMS for the application, we chose PostgreSQL 8.4, along with the PostGIS spatial extensions [20]. The communication between the framework and the database has been implemented using the ORM open-source library Hibernate [21] and its optional model devoted to the spatial data manipulation: Hibernate Spatial [22].

We decided to rely on the Google Maps platform for geographic data visualization and tracking features. This choice has been made thanks to the free use granted by Google and to the richness and flexibility of the Maps API, which allowed a seamless integration of the requested features into the application architecture.

Another important element needed to complete the application structure is the interaction with handheld devices with a GPS device, in order to allow users to acquire their position both in spatial and temporal dimensions. Once acquired, the spatio-temporal data need to be sent to the server deploying the web application. In order to accomplish these tasks we decided to use a PocketPC device, working with GpsGate and Netfront browser. We also used the Google Android SDK to develop an application that runs on Android devices.

3 The Development

In this section, we will show the main elements designed and developed to implement MyTravel. The chosen methodology [13] [12] adopts a strong Model-Driven approach [16], along with a Use-Case centric flow of analysis [23], that evolves seamlessly through design and development. For these reasons, we propose the application Use Case Model, followed by the Domain Model, and explain the integration of this model with the components provided by the GeoPack module.

3.1 The Use Case Model

The main use cases (UC) that have been defined and implemented are those related to the journey management. The most important among others are the spatial-temporal data acquisition UC and the UC related to create, edit and manage the journeys and attached information, such as points of interest.

3.2 The Domain Model

Given the strong MDA [24] orientation of the methodology and the related framework, the domain model can be considered the foundation of the entire application. In Fig. 2, we show the core elements of MyTravel domain model.

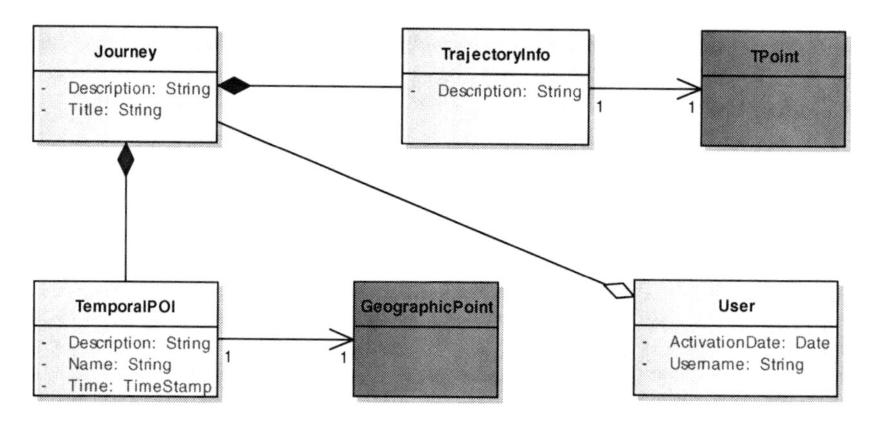

Fig. 2. MyTravel core domain model

The central element of the model is the journey. Every journey is associated to a user. In fact we must remember the strong social orientation of the application and take into account that almost every information inserted into the system comes from MyTravel community. We decided to model the journey as a composition of trajectories: in fact we can assume that a journey can be divided into one or more trips between two geo-referenced points, which will be respectively the trip starting point and the trip end point. The trip itself is then composed by a certain amount of intermediate geo-referenced points: we can say, under a geometrical point of view, that a trip could be considered as a polyline, and a journey can be modeled as an ordered set of polylines, in which the starting point of the first polyline and the ending point of the last polyline represent respectively the journey starting point and ending point. Another important element that can be noticed in the class diagram is the TemporalPOI. Since the points of interest are inserted by users during the tracking of their journeys, the spatial dimension alone could be inadequate, so we needed to introduce the temporal dimension. We must notice that in the last diagram we cannot find details about the spatial-temporal elements, but can identify two elements, TPoint and GeographicPoint, that are colored in a darker tone respect to the others domain objects. These elements come from the GeoPack module and provide structures and facilities to manage the spatial-temporal features described above.

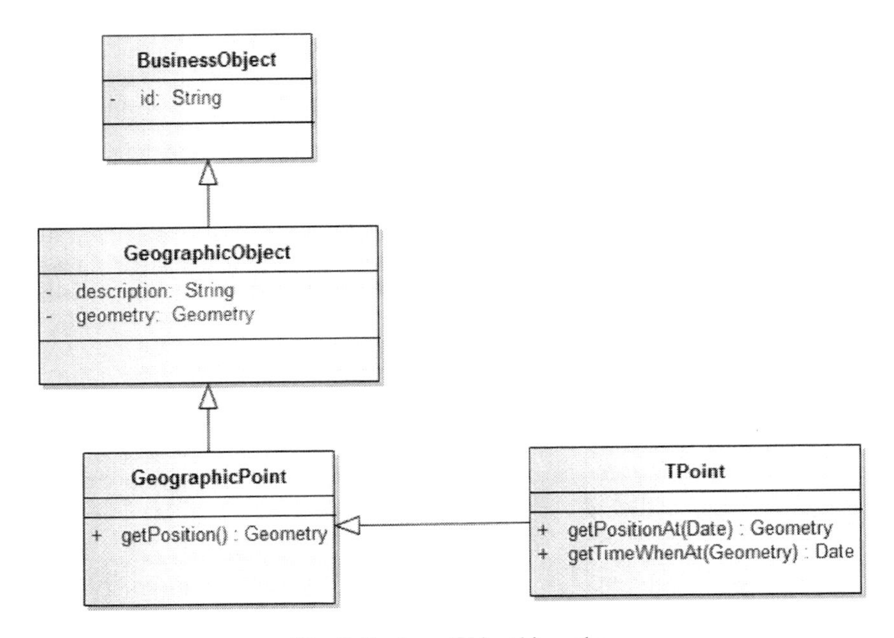

Fig. 3. Business Object hierarchy

Since the framework adopted during the development of the projects provides facilities to realize not only Geographic-aware applications, the meta-model class hierarchy starts from BusinessObject which represents the most generic kind of domain object and the specializes it enriching the meta-model and adding the features need to manage spatial-temporal data (Fig. 3).

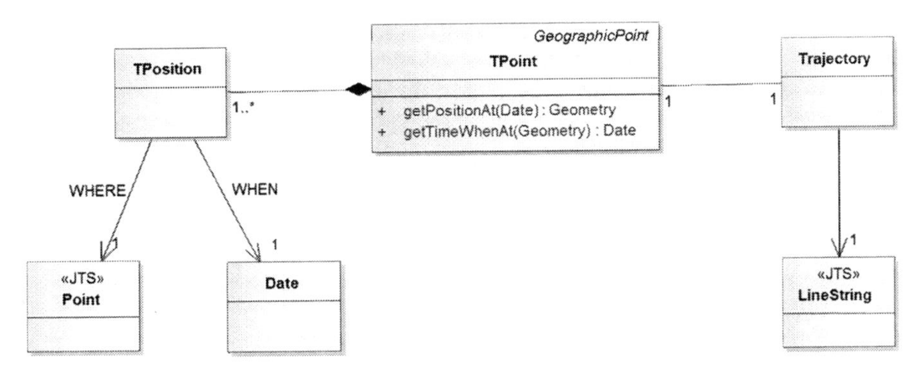

Fig. 4. The TPoint infrastructure in detail

To allow the management of moving points that change their position while time passes, the Geopack provides the above infrastructure, which allowed us to model the trips without information loss ensuring, at the same time, the use of well-known operators between spatial objects, such as Linestrings intersections and length calculation (Fig. 4). It is interesting to notice that the GeoPack relies upon elements from the Java Topology Suite [25], such as Points and Linestrings. This ensures robustness and efficiency during spatial data manipulation, granting at the same time compatibility with third-part libraries, such as Hibernate Spatial [22] and RDBMS.

4 The Prototype

In this section, we will show the prototype realized during our experience. We can see in Fig. 5.5 MyTravel HomePage. The application, as any web application, allows the user to register and login. The home page presents a list of Travels sorted by date added. Each Travel is described by the traveller photo, the date and the start city and the end city.

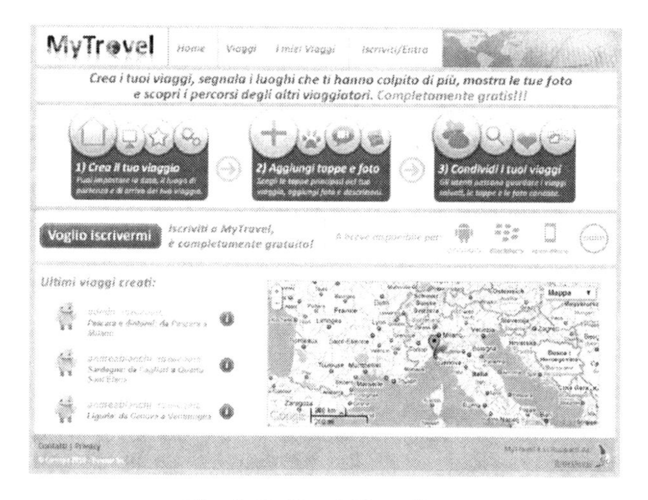

Fig. 5. MyTravel Home Page

The registered user can insert a new travel or can see a friend's travel. In Fig. 6.6, we show how a user visualizes a journey. In the upper of the page there are the name of traveller and travel. The little button on the upper-corner allows the user to choose another couple of travel/traveller.

The principal part of the page is the map, through it the user can see the trajectory of a travel and the POI that the user has reported. With a simple click on a Google Marker the user can access to an informative section about POI. Here he can find a Photo, a description, and in the future version of MyTravel a Rank. The possibility for the user to insert a rating for a POI represents an important new feature.

Fig. 6. The user can see or insert a travel

On right of the page, the user can find a descriptive list of a travel, in term of POIs. The user can insert POIs using Google Maps in the middle of the page. Each travel can be made up of several stages, and each stage aggregates POIs and Trajectories. Generally, we might confront the concept of stages with a day trip.

The web application is not the only way to insert or view a travel, but it can be done also through a mobile device application. In the following, we will show the Android version of the application.

Fig. 7. Some screenshots of the Android application

The user, during the travel, can share his/her experience through his/her smartphone. In Fig. 7.7, we propose a few screenshots of the mobile application. The Android application allows the user to take a photo, add descriptions and share it with other users. In this way, users can share the points of interest of their journey in real time. Another feature of the MyTravel Mobile app is the tracking option. If the user turns it on, during a trip, the app records all the movements of the user and shows the route effectively taken: the user can share his path with his friends. If the user agreed to the sharing of his information, the app automatically publishes the journey (the path and POIs), on the MyTravel web application.

Another important feature of the MyTravel Mobile is the integration with social networks. In this first version, the application can communicate with Facebook only. As shown in Fig. 8.8, when a user registers a POI, automatically the photo and the description are published on Facebook and on the MyTravel Web App.

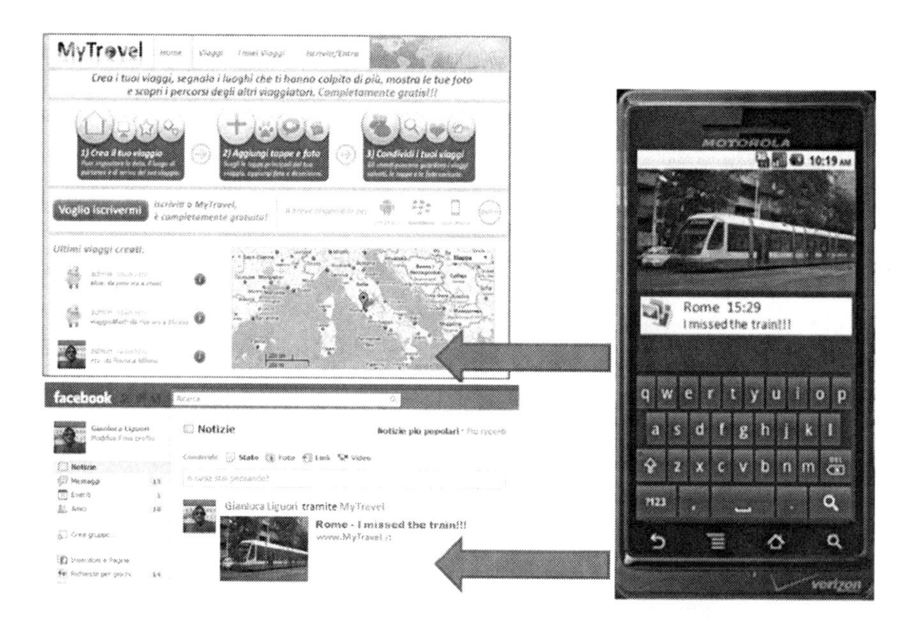

Fig. 8. Integration among the web app, Android app and Facebook

5 Conclusions and Further Work

The design and development activities carried out during the project described in this paper have led to the implementation of a working prototype that has been tested by a restricted community of users. The prototype allowed us to validate the technological and architectural choices, along with the integration with the tools. Assuming that the core element of a geo social network like MyTravel is its own community, the next step in the research will be the publication of the application on the internet, in order to verify its appeal to the internauts and its robustness and scalability under heavy load.

We are planning to extend the research, and consequently the application, toward the integration with other social networks, such as Twitter, in order to enrich the user experience and provide, through an information mesh approach, the shared user information base, which represents the key of success of such applications.

Another important task that we will carry on is the integration of other mobile device technologies, like iOS and Blackberry, in order to enlarge the candidate user base. All these activities will be executed in parallel with the development of new features, that will involve the enrichment of the domain model and the implementation of new operators that can manipulate such a model.

References

1. Fischer, F.: Learning in Geocommunities: An explorative view on geosocial network communities. In: Jekel, T., Koller, A., Donert, K. (eds.) Learning with Geoinformation IV, Heidelberg (2009)
2. Goodchild, M.F.: Citizens as sensors: the world of volunteered geography. GeoJournal 69(4), 211–221 (2007)
3. Kennedy, L., Naaman, M., Ahern, S., Nair, R., Rattenbury, T.: How flickr helps us make sense of the world: context and content in community-contributed media collections. In: Proceedings of 15th Annual ACM International Conference on Multimedia (MM 2007), pp. 631–640. ACM, New York (2007)
4. Kennedy, L., Naaman, M.: Generating diverse and representative image search results for landmarks. In: Proceeding of the 17th international conference on World Wide Web (WWW 2008), pp. 297–306. ACM, New York (2008)
5. Schmeiß, D., Scherp, A., Staab, S.: Integrated Mobile Visualization and Interaction of Events and POIs. In: Proceedings of the international conference on Multimedia (MM 2010), pp. 1567–1570. ACM, New York (2010)
6. Harada, S., Naaman, M., Song, Y.J., Wang, Q., Paepcke, A.: Lost in memories: interacting with photo collections on PDAs. In: ACM/IEEE-CS joint conference on Digital (JCDL2004), pp. 325–333. ACM, New York (2004)
7. Hertzog, P., Torrens, M.: Context-aware mobile assistants for optimal interaction: a prototype for supporting the business traveler. In: Proc. 9th International conference on Intelligent user interfaces (IUI 2004), pp. 256–258. ACM Press, New York (2004)
8. Yan, Z., Chakraborty, D., Parent, C., Spaccapietra, S., Aberer, K., S.: A Framework for Semantic Annotation of Heterogeneous Trajectories (EDBT 2011), Uppsala, Sweden, pp. 259–270 (2011)
9. Graham, M.: Neogeography and the Palimpsests of Place. Tijdschrift voor Economische en Sociale Geografie 4, 101 (2010)
10. O'Reilly, T.: What Is Web 2.0. In: O'Reilly Network, O'Reilly, Sebastopol (2005), `http://oreilly.com/web2/archive/what-is-web-20.html`
11. Decrem, B.: Introducing Flock Beta 1. Flock official blog (Reported 13.06.2006)
12. Liguori, G.: PhD thesis. Una Metodologia per lo sviluppo di sitemi Web-Gis. University of L'Aquila, L'Aquila, Electrical and Information Engineering Department (2010)
13. Paolone, G., Liguori, G., Clementini, E.: A methodology for building enterprise Web 2.0 Applications. In: MITIP 2008 The Modern Information Technology in the Innovation Processes of the Industrial Enterprises (2008)
14. Paolone, G., Di Felice, P., Liguori, G., Cestra, G., Clementini, E.: A Business Use Case Driven Methodology: a step forward. In: 5th International Conference on Evaluation of Novel Approaches to Software Engineering, Athens Greece (July 2010)
15. Paolone, G., Liguori, G., Clementini, E.: Design and Development of web 2.0 Applications. In: ITAIS 2008, France, December 13-14 (2008)
16. Paolone, G., Liguori, G., Cestra, G., Clementini, E.: Web 2.0 Applications: model-driven tools and design. In: Costa Smeralda, A., D'Atri, M., De Marco, A.M., Braccini, F. (eds.) ITAIS 2009 Management of the Interconnected World, October 2-3, 2009, Springer, Heidelberg (2009)

17. Fowler, M.: Patterns of Enterprise Application Architecture, November 15, 2002. Addison-Wesley, Reading (2002)
18. Liguori, G., Clementini, E.: GeoPack: a Java wrapper for geographic web services. In: Towards INSPIRE International Workshop, Hungary, March 18-19 (2009)
19. Liguori, G., Cestra, G., Di Felice, P.: Un'architettura software per lo sviluppo di applicazioni riguardanti dati spazio-tempo dipendenti. In: Congresso nazionale AICA (October 2010)
20. Refractions Research, PostGIS 1.5.1 Manual (Reported 02.05.2010)(2010), http://postgis.refractions.net/documentation/
21. King, G., Bauer, C., Andersen, M.R., Bernard, E., Ebersole, S.: Hibernate Reference Documentation 3.5.1-Final - HIBERNATE Relational Persistence for Idiomatic Java. In: King, G., Bauer, C., Andersen, M.R., Bernard, E., Ebersole, S. (eds.) Hibernate Relational Persistence for Java &.NET (2010), http://docs.jboss.org/hibernate/stable/core/reference/en/pdf/hibernate_reference.pdf
22. Geovise Hibernatespatial (Reported 14.06.2010) http://www.hibernatespatial.org/
23. Paolone, G., DiFelice, P., Liguori, G., Cestra, G., Clementini, E.: Use Case double tracing: Linking business modelling to software development. In: 7th Conference of the Italian Chapter of AIS, Napoli (October 2010)
24. Object Management Group, MDA, Model Driven Architecture, MDA Guide Version 1.0 (2003), http://www.omg.org/mda/mda_files/MDA_Guide_Version1-0.pdf.
25. Vivid solutions, JTS Topology Suite Technical Specifications Version 1.4. JTS Topology Suite (Reported 14.06.2010), http://www.vividsolutions.com/jts/jtshome.htm

GIS and Remote Sensing to Study Urban-Rural Transformation During a Fifty-Year Period

Carmelo Riccardo Fichera[1], Giuseppe Modica[1], Maurizio Pollino[1,2]

[1] 'Mediterranea' University of Reggio Calabria,
Department of Agroforestry and Environmental Sciences and Technologies (DiSTAfA), Loc. Feo di Vito – 89122 Reggio Calabria, Italy
{cr.fichera, giuseppe.modica}@unirc.it
[2] ENEA - National Agency for New Technologies, Energy and Sustainable Economic Development - "Earth Observations and Analyses" Lab (UTMEA-TER) Casaccia Research Centre - Via Anguillarese 301, 00123 Rome, Italy
maurizio.pollino@enea.it

Abstract. A relevant issue in Remote Sensing (RS) and GIS is related to the analysis and the characterization of Land Use Land Cover (LULC) changes, very useful for a wide range of environmental applications and to efficiently undertake landscape planning and management policies. The methodology described has been applied to a case-study conducted in the area of the Province of Avellino (Southern Italy). Firstly, aerial photos and Landsat imagery have been classified to produce LULC maps for a fifty-year period (1954÷2004). Then, through a GIS approach, change detection and spatiotemporal analysis has been integrated to characterize LULC dynamics, focusing on the urban-rural gradient. This study has shown that LULC patterns and their changes are linked to both natural and social processes whose driving role has been clearly demonstrated: after the disastrous Irpinia earthquake (1980), local specific zoning laws and urban plans have significantly addressed landscape changes.

Keywords: GIS, Remote Sensing, Satellite imagery classification, Land Use Land Cover (LULC) changes, Urban sprawl, Urban/Rural fringe areas.

1 Introduction

Monitoring Land Use Land Cover (LULC) changes is fundamental for a wide range of environmental applications, especially in the case of natural resource management, urban planning or local/regional policy programmes. Therefore, Central and Local Administrations require timely and accurate information on existing LULC, that conventional survey and mapping methods cannot deliver in the same timely and cost-effective way. The design of new techniques for spatial data processing and the integration of remotely sensed (RS) imagery (aerial and satellite) with other data sources, during the last years, have strongly improved quality, efficiency, timeliness and cost-effectiveness of the LULC monitoring process. In this framework, Geographic Information Systems (GIS) represent the most significant technological

B. Murgante et al. (Eds.): ICCSA 2011, Part I, LNCS 6782, pp. 237–252, 2011.

and conceptual approach to spatial data analysis, in order to provide reliable information for both planning and decision-making tasks [1]. As a matter of fact, GIS techniques are efficiently exploited to analyse the effects of various factors on LULC changes: those factors include population density, terrain slope, proximity to roads, and surrounding land use. The availability of time-series dataset is fundamental to understand and monitor the urban expansion process, in order to characterize and locate the evolution trends at a detailed level. Hence, satellite RS technologies give a valuable contribution, thanks to their capability to provide both historical data archive and up-to-date imagery [2]. In fact, during the last three decades, satellite time series as Landsat images have been exploited during several studies [3, 4, 5] to evaluate built-up expansion and to assess urban morphology changes. The main goal of those studies was the spatiotemporal analysis of LULC dynamics, focusing on urban growth/sprawl phenomenon and loss of rural land. The term "sprawl" is often used to describe the awareness of an unsuitable development, as a disordered growth of urban areas [6]. Sprawl is the consequence of many individual decisions and among the possible causes of this phenomenon we can find population growth, economy and proximity to resources and basic facilities [7].

The case-study described in this paper has concerned the use and the integration of RS with GIS techniques for LULC classification and change detection analysis, focusing on the urban sprawl/growth phenomenon, especially along urban-rural fringe areas. By means of that integration, it is possible to analyse and to classify the changing pattern of LULC during a long time period and, as a result, to understand the changes within the area of interest. In fact, is well known that the development of the urban areas is able to transform landscapes formed by rural into urban life styles and to make functional changes, from a morphological and structural point of view [8, 9]. Historically, urban development (driven by the population increase) and agriculture are competing for the same land: cities expansion has typically take place on former agricultural use. Just to mention some data, the amount of land consumed by urban areas and associated infrastructure throughout Europe was about 800 $km^2 \cdot year^{-1}$ between 1990 and 2000 [10]. Changes in landscape take continuously place, with significant repercussions for quality of life and natural habitat ecosystems, especially through their impacts on soil, water quality and climatic systems [11]. For all those factors, it is very important to recognize the factors that driving the dynamic processes of rural-urban areas transformation and to explore their relevance, in order to give an efficient effort to sustainable landscape planning and monitoring activities.

2 Materials and Methods

2.1 The Case-Study

The study area is located within the Province of Avellino, in the Campania region (Italy). It is characterized by many small towns and settlements scattered across the Province; its capital city, Avellino (40°5'55"N 14°47'23"E, 348 m a.s.l., 42 km NE of Naples, Total population: 52,700), is situated in a plain called "Conca di Avellino" (Fig. 1) and surrounded by mountains: Massiccio del Partenio (Monti di Avella, Montevergine e Pizzo d'Alvano) on NW and Monti Picentini on SE. Due to the

Motorway A16 and to other main roads (S.S. 7 and S.S. 7bis), Avellino also represents an important hub on the road from Salerno to Benevento and from Naples.

Avellino was struck hard by the disastrous Irpinia earthquake of 23 November 1980. Measuring 6.89 on the Richter Scale, the quake killed 2,914 people, injured more than 80,000 and left 280,000 homeless. Towns in the province of Avellino were hardest hit and the Italian Government spent during the last thirty years around 30 billion of Euros on reconstruction. Consequently to the earthquake and to regulate the reconstruction activities, several specific acts, decrees, zoning laws and ordinance have been issued: the first one was the Law n. 219/1981 that entrusted the urban planning to the damaged municipalities, under the coordination of the Campania Region. From 2006 the urban planning issues of Avellino and neighbour areas are regulated by two instruments: P.I.C.A. (Italian acronym that stands for Integrated Project for Avellino City) and P.U.C. (Master Plan for Avellino Municipality).

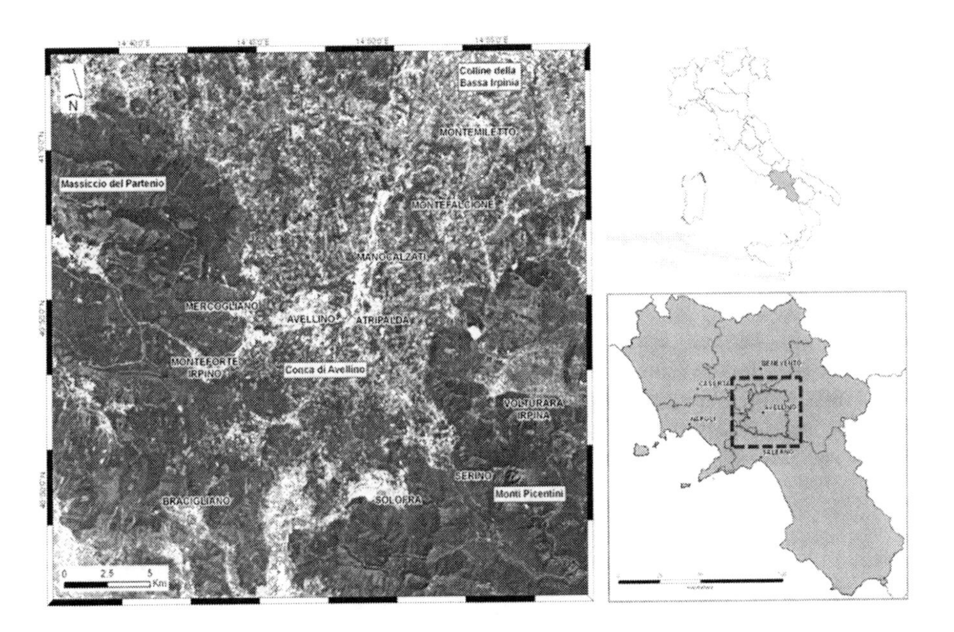

Fig. 1. Geographic location of the study area

Considering this general framework, the analysis here presented pertains to the Conca di Avellino area (extended 57,355 ha), in consequence of its particular location: a built-up area between the two natural protected zones of Regional Park of Partenio (14,870 ha) and Regional Park of Picentini Mountains (62,200 ha).

2.2 Land Cover Classification

A multi-temporal set of RS data of the area of interest has been used to study and classify LULC [12]. This dataset included:

- Aerial photos: 1954, 1974 and 1990 surveys carried by Italian Military Geographical Institute (IGMI);
- Landsat images: MSS 1975, TM 1985, TM 1993 and ETM+ 2004 (source: Global Land Cover Facility, GLCF, http://glcf.umiacs.umd.edu);
- Digital aerial orthophotos 1994 and 2006 (available for consultation and visualization at www.pcn.minambiente.it, the National Cartographic Portal).

Digital image-processing software ERDAS Imagine (v.2010) has been used to process, analyse and integrate the spatial data and geographic information so as to achieve the above mentioned goals. Further, all the aerial photos, satellite images and maps produced have been georeferenced in UTM-33N projection, Datum WGS84, using the 2006 orthophotos as thematic and geometric reference. First of all, the multi-temporal dataset has required a geometric registration, in order to decrease the distortions effects and to reduce pixel errors that could be interpreted as LULC changes. Then, Landsat images have been atmospherically corrected by means the tool ATCOR for ERDAS Imagine, in order to take into account the variations in solar illumination conditions, the atmospheric scattering and absorption: those factors, in fact, could cause differences in radiance values unrelated to the reflectance of land cover [13]. To mitigate the seasonal effects and uncertainness of inter-annual variability, which often lead to errors in change detection, only the imagery acquired during the summer period have been used.

Using as thematic reference the 2006 orthophotos, the earliest information about LULC has been extracted from black-white (BW) monoscopic aerial frames taken in 1954 (1:35,000 flight scale). Then, to obtain information about LULC for the 1975÷2004 time interval, the Landsat images have been processed and classified [14]. Using the supervised approach (Maximum Likelihood Classification algorithm, MLC), four classes have been defined: Urban, Woodland, Cropland and Grassland/Pasture. In addition to the 2006 orthophotos, 1974 and 1990 aerial photos and 1994 orthophotos have been used as a reference material for the classification procedures. To evaluate the user's and the producer's accuracy, a confusion matrix was applied to the classified images [15, 16]: for each Landsat image, the LULC class assigned to 256 pixels (selected using a stratified random sample) was visual compared with the equivalent area in the aerial frames closer to the same period. The overall accuracy values of each classified image are reported in Table 1.

Table 1. Accuracy assessment for the classified images

Reference Year	Classified image	Overall Classification Accuracy	Overall Kappa Statistics
1975	Landsat MSS	86.72%	0.7478
1985	Landsat TM	82.42%	0.6863
1993	Landsat TM	83.20%	0.7060
2004	Landsat ETM+	95.70%	0.9285

After all classification procedures, five different LULC maps have been produced: 1954 (from aerial photos), 1975, 1985, 1993 and 2004 maps (from Landsat data) [17]. The 1975÷2004 maps, originally in raster format (30 m pixel resolution), have been converted into the shapefile (*.shp) vector format, whereas the 1954 one has been directly produced into this format. Vector format is more suitable for the change-detection procedures subsequently carried out, according to the GIS-based approach pursued [18]. The advantage of using vector format it's not only linked to exploitation of GIS database capabilities, but also to the ability to manage different LC maps by "intersect" and "union" vectorial operation, in order to easily evaluate the amount of change [19].

2.3 Change Detection

Starting from the above described dataset of multi-temporal classified images, the process of digital change detection developed has allowed to determine and describe changes in LULC between four fundamental time interval periods: 1954÷1975, 1975÷1985, 1985÷1993 and 1993÷2004. In fact, RS data opportunely processed and elaborated, jointly with GIS techniques, are fundamental in change detection tasks to monitor the differences of LULC at different times [20]. A lot of change detection methods have been proposed [21] and each one presents variations depending on the imagery type, final purpose for the change image and the type of change to be detected. In the present case-study the approach followed has been the so called "Post-classification comparison" [22, 23], that allows to determine the difference between independently LULC map in vector format from each of the dates in question. This is the only method in which "from" and "to" classes can be calculated for each changed feature. The main advantage of this method is to easily create and update GIS databases, as class/categories are given, and manage quantitative values of each class. In order to efficiently integrate LULC maps and to quantitatively outline the dynamic of change in each category, the post-classification method has been combined with a GIS approach [24]. According to that approach, by comparing the LULC data, it has been possible to directly manage the tables containing the spatial information of each class (area, perimeter, etc.) and the information about amount, location, and typology of change [19, 25]. The analysis of the dynamics has been carried-out comparing each classified map with the successive, in order to determine the changes in LULC at different years from 1954 to 2004. To validate this step, the accuracy assessment of the above described change detection procedures has been performed following the approach proposed by Congalton & Macleod [26]: the error matrix normally used for the single-date classification is modified, so that it has the same characteristics as the single-date classification error matrix, except that it also assesses errors in changes between two time periods and not simply in a single classification. Considering the classes defined in this case-study, the single classification matrix is 4x4, whereas the change detection error matrix is 16x16 (the size of the number of categories squared): this matrix concerns changes between two different maps generated at different times (t_1 and t_2) in assessing change detection [16]. Then, the change detection error matrices have been simplified by collapsing into no-change/change error matrices: the upper left box reports the areas that did not change in either the classification or reference data; the upper right box indicates the areas that the classification detected no change and the reference data considered

Table 2. No-change/change error matrices for the change-detection technique

1975÷1985				**1985÷1993**				**1993÷2004**			
	NC	C	Total		NC	C	Total		NC	C	Total
NC	**181**	56	237	NC	**179**	54	233	NC	**188**	34	222
C	6	**12**	18	C	9	**14**	23	C	14	**20**	34
Total	187	68	**255**	Total	188	68	**256**	Total	202	54	**256**
Overall accuracy 75.7%				*Overall accuracy 74.5%*				*Overall accuracy 81.3%*			

changed. Those matrices have been produced for the intervals 1975÷1985, 1985÷1993 and 1993÷2004 (LULC maps extracted from Landsat data) and the relative values are reported in Tab. 2.

2.4 Spatial Analysis of Land Cover Transformations

LULC changes are clearly connected to the spatiotemporal growth of urban areas [27]. The dynamics are influenced by several social, economic, and political causes. The driving factors are various and include the effects of natural environment, demography, economy, transportation network, preference (by people) for proximity, neighbourhoods, and central/local policies [28]. Those factors are able to produce a clustering and localized pattern of urbanization, where new development has tended to infill around existing development, as well as a dispersed trend, wherein urban land uses increasingly spread out across a metropolitan area. Verburg et al. [29] related the physical growth of a city directly to its population growth, as an increase in population size that encouraged the agglomeration of businesses and new urban development. Moreover, the presence and accessibility of transportation routes force patterns of urban growth. To understand the dynamics of the urban sprawl/growth phenomenon [30] connected to LULC changes in the land surrounding Avellino, different GIS spatial analysis [31] have been performed. To reach this goal, various geographic layers have been taken into account: main road network, railway, administrative boundaries, Digital Terrain Model (DTM), Census data from the ISTAT [32]. Those layers, then, have been combined with the LULC maps produced through image classification, by means of spatial overlay functions performed using ESRI ArcGIS Desktop (v. 9.2). Analysing and integrating LULC maps with other spatial data, it has been possible to produce the GIS layers describing transformation and dynamics in consequence of LULC changes. Further elaborations has allowed to locate the land areas belonging to the urban-rural gradient and representing the spatial location of the urban-rural fringe [33]. In this way, it has been possible to extract from all the LULC maps the information about the dynamics of change and, then, to produce four transition matrices (1954÷1975, 1975÷1985, 1985÷1993 and 1993÷2004) of the LULC changes in the study area. The amount of the change for each category analysed is given in Tab. 3, in which are reported the relative statistics, aggregated for LULC class. The values (in hectares) reported along the diagonal express the area of the unchanged LULC types; the other cells contain the measurement of the areas that have transformed from a LULC class to another.

Table 3. Total LULC changes for the types defined: dynamics from 1954 to 1975 (A), from 1975 to 1985 (B), from 1985 to 1993 (C) and from 1993 to 2004 (D). Area values in [ha]

A - Dynamics 1954÷1975	Urban	Grassland pasture	Cropland	Woodland	**1954**
Urban	893.45	0.00	0.00	0.00	**893.45**
Grassland/pasture	28.77	561.88	837.25	159.61	**1587.51**
Cropland	342.21	42.06	29347.26	1238.67	**30970.19**
Woodland	29.11	23.09	1002.13	21399.38	**22453.71**
1975	**1293.53**	**627.03**	**31186.64**	**22797.67**	**55904.86**
B - Dynamics 1975÷1985	Urban	Grassland pasture	Cropland	Woodland	**1975**
Urban	1256.84	0.00	0.00	0.00	**1256.84**
Grassland/pasture	30.06	425.13	137.54	34.48	**627.22**
Cropland	1482.36	649.65	28516.42	573.14	**31221.57**
Woodland	45.05	162.97	435.72	22155.50	**22799.23**
1985	**2814.31**	**1237.76**	**29089.67**	**22763.12**	**55904.86**
C - Dynamics 1985÷1993	Urban	Grassland pasture	Cropland	Woodland	**1985**
Urban	2754.19	0.00	0.00	0.00	**2754.19**
Grassland/pasture	30.35	433.72	622.02	181.78	**1267.87**
Cropland	584.30	3.71	27327.98	1193.36	**29109.35**
Woodland	21.56	9.61	569.83	22172.45	**22773.45**
1993	**3390.42**	**447.04**	**28519.82**	**23547.58**	**55904.86**
D - Dynamics 1993÷2004	Urban	Grassland pasture	Cropland	Woodland	**1993**
Urban	3321.87	0.00	0.00	0.00	**3321.87**
Grassland/pasture	3.50	412.87	12.67	19.65	**448.68**
Cropland	1774.19	1101.12	20994.56	4711.31	**28581.19**
Woodland	35.19	489.22	907.97	22120.74	**23553.12**
2004	**5134.75**	**2003.21**	**21915.20**	**26851.70**	**55904.86**

The column on the right sum up the LULC areas at the beginning of all the intervals examined, while the last row sums up the LULC areas at the end. Moreover, by the selection of the "Urban" LULC features from each map produced through image classification, it has been obtained the diachronic expansion of urban areas during the period examined. In particular, this information has been compared, via GIS analysis, with the other spatial data available. The spatial overlay with the Census data, has allowed evaluating the relationship between population growth trends and urban expansion during the examined period. Further, following the same approach, it has been investigated the relationship between the urban expansion and the terrain slope (extracted from DTM) and the interaction with the transportation infrastructure getting across the area of interest (buffer analysis).

3 Results and Discussion

Analysing the results coming from the procedures above described, the urbanization has considerably modified the LULC of the study area, with significant land conversions. Urbanization is a complex diffusion process that is spreading dramatically and which affecting rural landscape differently in space and at different scales [5]. Figure 2 depicts changes and dynamics of LULC happened during the overall period (1954÷2004), while the unchanged areas are blank filled. In particular, during the five decades analysed, the Urban LULC type has almost quintupled passing from 901 hectares to 5,140 (from the 1.6% to the 8.6% of the total of study area), mostly at the expense of the cropland areas, which have most suffered the effects of the expansion of the built-up areas. Woodland and Grassland/Pasture LULC types have, instead, shown a relatively lower change rate, although the first one category has recorded a valuable 16% increment between the overall period 1954÷2004 (Tab. 2).

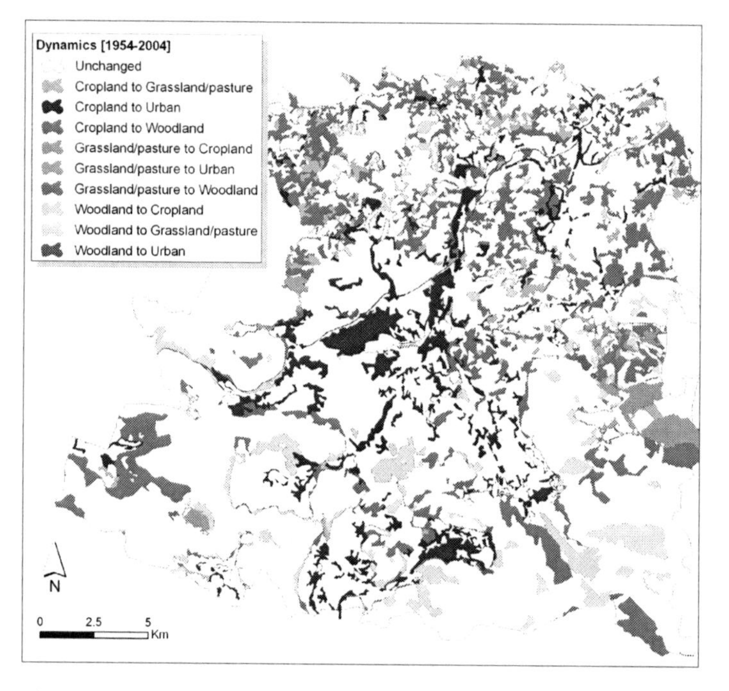

Fig. 2. LULC changes and dynamics within the study area for the overall time span 1954÷2004

Urbanized areas represent the LULC type with the largest growth rate (Fig. 3): the maximum increase has occurred in the period after the year 1985 when reconstruction process started in consequence of the 1980 earthquake. In fact, this phase was characterized by significantly suburban spreading of some residential and industrial areas, leading to attrition process (gradual loss of remaining fragments). Among the leading factors able to drive the LULC change, the first to be considered is the terrain slope, a physical characteristic of the landscape: flat areas are generally the first to be annexed to urban expansion. So occurred in the study area, where the 78% of new development is located in land with a terrain slope between 0 and 10 degrees (Fig. 4).

Cropland, the largest class at the beginning of the study period, was mainly distributed in the lowland area in the centre of the Conca di Avellino and has changed the most because of human activities (Fig. 5). It is also possible to partially retrieve such reduction of cropland LULC type into the census data achieved from ISTAT in relation to agriculture activities. The last two surveys available (ISTAT General Agriculture Census 1990 and 2000) recording a decrease of permanent farming and arable areas, with rates of 13% and 55% respectively, and are very useful to statistically explain and understand the LULC change detected. Thus, considering the data stored into the Table 3 and depicted in Fig. 2, the majority of rural land transformed was converted into urban areas, with a valuable loss of cropland. Such conversion is mainly located on urban-rural fringe, whose spatio-temporal evolution (Fig. 6) has been forced by the sprawl process that interests the study area [7, 34]. Urban fringes can be considered as transition landscapes characterized by fuzzy boundaries and are very sensitive to the spatiotemporal variations of built-up (due to expansion and sprawl) and - in general - to land use changes [35]. The analysis of the evolution of those areas represents one of the future research directions utilising, among others, VHR satellite images and very detailed digital cartography (scale 1:5,000) as reference data.

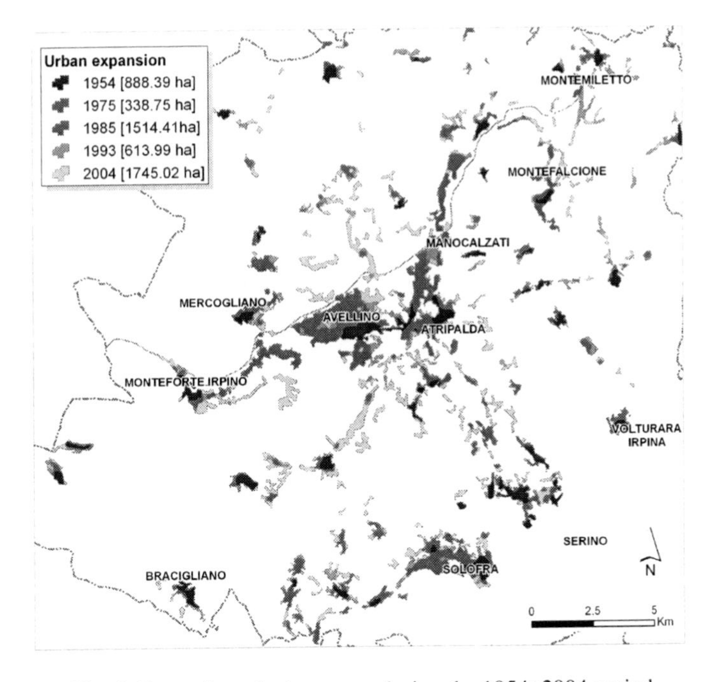

Fig. 3. Expansion of urban areas during the 1954÷2004 period

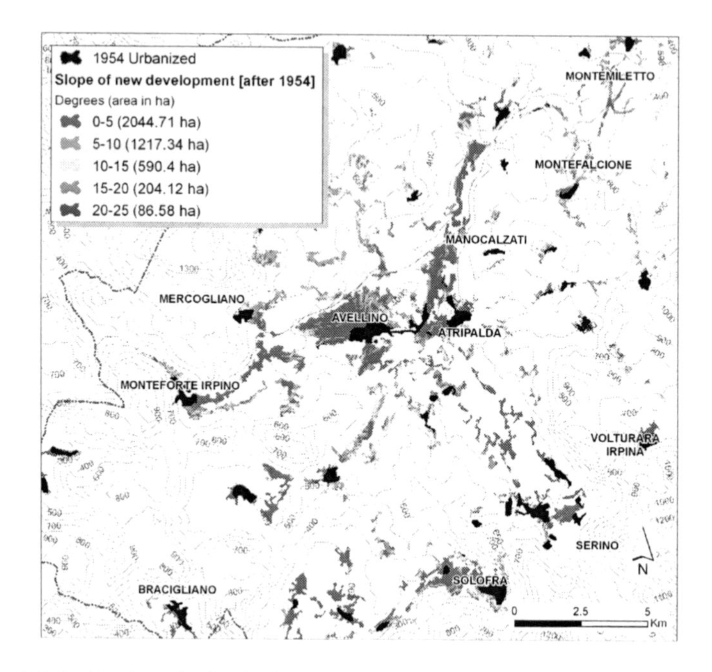

Fig. 4. Spatial distribution of urbanized areas classified according to the terrain slope values (in degrees) overlapped to the main contour lines (at 100-meter intervals)

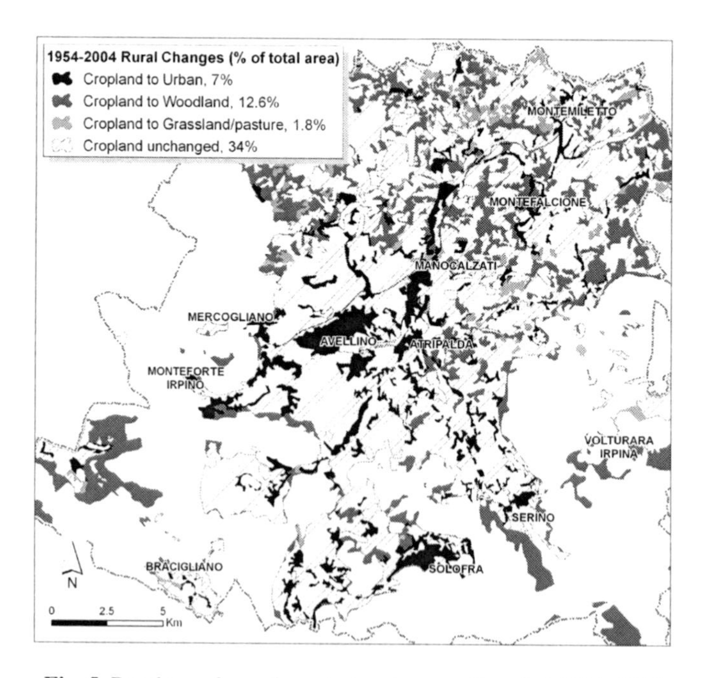

Fig. 5. Rural transformations during the overall period 1954÷2004

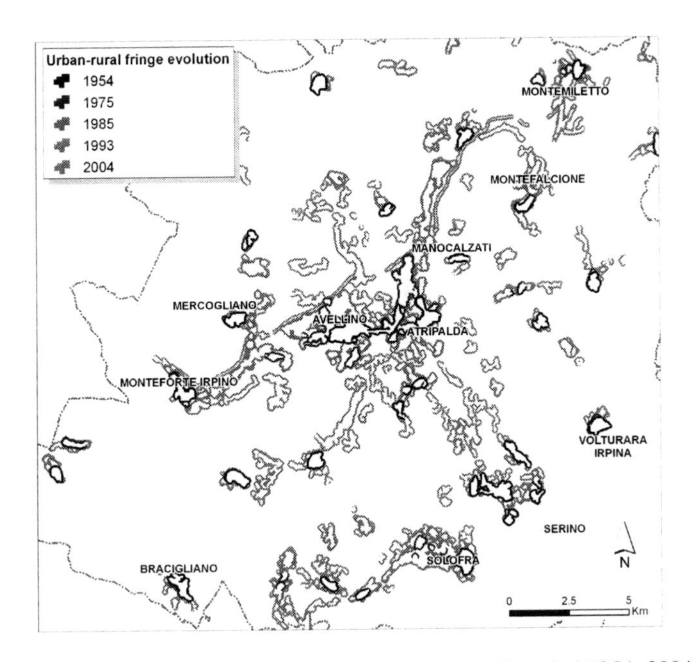

Fig. 6. Urban-rural fringe evolution during the overall period 1954÷2004

One of the causes of the variations among LULC patterns is due to the proximity of Avellino to other urban centres (Monteforte Irpino, Mercogliano, Atripalda, Manocalzati and Montefredane). Those towns, situated in the central zone of the study area and placed along the SW-NE direction, are currently in a territorial continuity with the main settlement of Avellino: this interaction has given rise, during the last years, to the urban sprawl phenomenon highlighted by the above mentioned spatial analysis. The growth of urbanized areas is generally related to population growth that drives the built-up area to expand [36]. It is possible to identify the sprawl by a careful comparison of urbanized area expansion rate and population growth rate. Therefore, changes in LULC have been compared whit the demographic data achieved from the ISTAT (Fig. 7): while the total population within the study area increased by 14.4% between 1971 and 2001, urbanized areas increased by 75.5% between 1975 and 2004 (time interval of the RS data comparable with the Census data available). In fact, the comparison between 1975÷2004 LULC maps and 1971÷2001 Census data (ISTAT) clearly indicates that growth rate of urbanized areas is always higher than the growth rate of population, during all the time periods considered (Fig. 7). In this case, the population displacement was the contributory cause of the urban expansion in the area surrounding Avellino: the inhabitants of the capital city were 36,965 in 1951, 41,825 in 1961, 52,382 in 1971, 56,862 in 1981 (ISTAT Census data).

At a certain moment, between 1981 and 2001 it is possible to observe for Avellino Municipality a decreasing trend (55,662 in 1991 and 52,703 in 2001), due to the transfer of many people from the main urban settlement to the above-mentioned neighbouring towns: the consequence is an "extended" urban area, with around

90,000 inhabitants. This remark is corroborate considering the census data from 1971 to 2001: adjacent Municipalities have demonstrate high growth rates of population (i.e., Mercogliano and Monteforte Irpino: +62% and +57%, respectively), instead of the above mentioned situation of Avellino. To analyse the relationship between population and urban growth, the attention has been just focused to the six Municipalities pointed out. Although their extension (1063.15 ha) represents the 18.5% of the entire study area, around the 40% of the total urbanized LULC type is here concentrated ("Central zone"). An useful measure to quantify the sprawl is the proportion of the total population in a zone to the total built-up area of the zone itself [8]. So, the proportion of population (Census: 1971, 1981, 1991 and 2001) and proportion of urbanized area (LULC Maps: 1975, 1985, 1993, 2004) has been calculated by dividing respectively population and urbanized area of the two defined zone (the "Central zone" and the remaining area) by the total population and total extension of the study area. By subtracting the proportion of population from the proportion of urbanized area, it has been obtained a result in a range from –1 to 1, as shown in Table 4, where negative values indicate population crowding (which may cause environmental problems), positive values indicate higher per capita consumption of built-up area (that are a sign of a better environmental situation) [30].

Table 4. Proportion of urbanized area minus proportion of population

Reference Year	**[A] % urbanized**		**[B] % population**		**[C] Proportion** *([A]-[B])*	
(Census/LCMap)	*Central zone*	*Other zones*	*Central zone*	*Other zones*	*Central zone*	*Other zones*
1971/1975	0.40	0.60	0.43	0.57	-0.02	0.02
1981/1985	0.42	0.58	0.45	0.55	-0.03	0.03
1991/1993	0.40	0.60	0.46	0.54	-0.05	0.05
2001/2004	0.38	0.62	0.45	0.55	-0.07	0.07

This urban sprawl is also a direct consequence of the course of the state-road S.S. 7bis ("Terra di Lavoro") that, along the SW-NE direction, connects Avellino to Monteforte Irpino, Mercogliano, on the West side and Atripalda, Manocalzati and Montefredane on the East side, underlining the relation between place of residence and place of work [37]. Transportation routes are responsible for the so called "linear branch" development [34, 7] and represent a key catalyst of sprawl.

Urban expansion is also connected to economic growth [38], assuming that there is a relationship between people's economic status, available areas to be built up and expansion of urbanized area. To establish this relationship, numerous economic parameters can be considered, like per capita income, which however presents the disadvantage to average the data. To overtake this problem, it is possible to relate the built-up area with the number of working persons only, for the specific reason that they are mainly responsible for new construction. Referring to the study area, the total amount of workers have been extracted from the last four census data available (ISTAT, Industry and services Census: 1971, 1981, 1991 and 2001): these data clearly show that the urbanized areas have grown at a similar rate as working persons and this factor is probably connected to the special laws [39] promulgate after the 1980 earthquake (Law n. 219/1981 and later). These laws were the general framework of P.I.C.A. and P.U.C. plans, from whose has come the indication to place the areas devoted to the industrial use in the northern zone of Avellino. This factor has represented another important push to the urban expansion along the SW-NE direction, that includes the new industrial estate of Avellino.

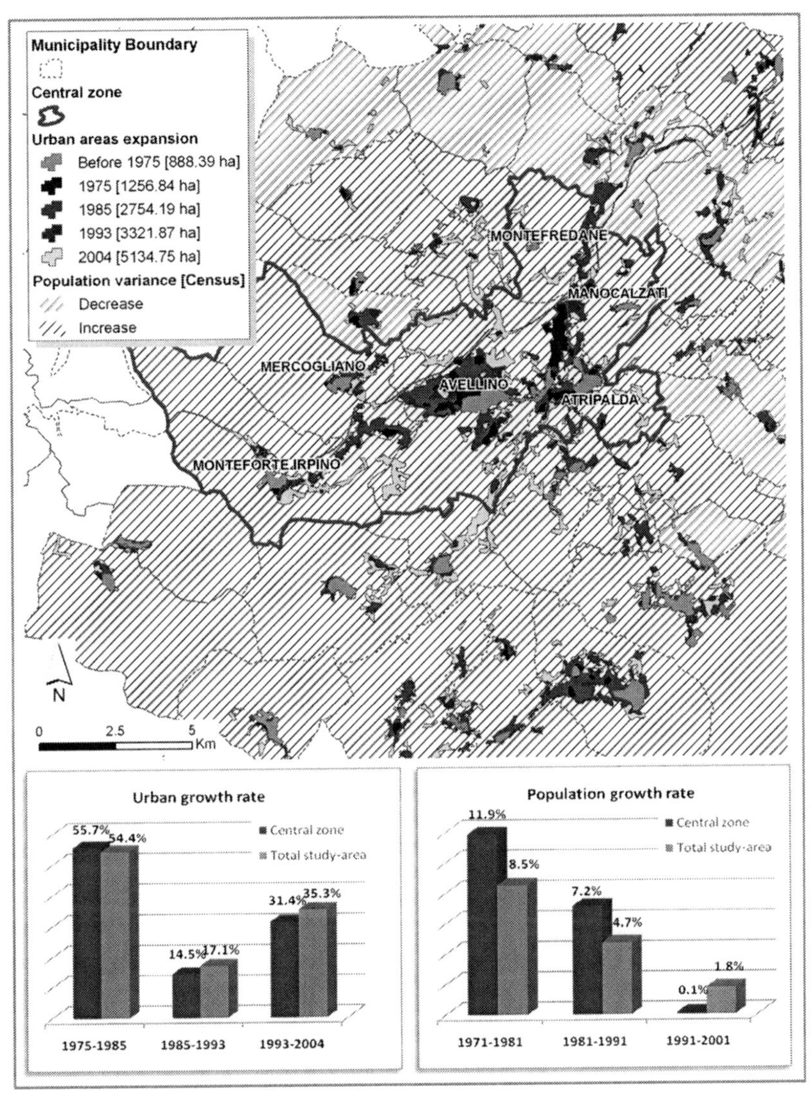

Fig. 7. Comparison among urban expansion and population variance subdivided for Municipality areas (outlined in red the "Central zone"). The graphs below show the growth rate comparisons respectively for population and urban areas.

4 Conclusions

In summary, the results achieved have demonstrated that aerial and satellite imagery classifications can be successfully used to produce accurate LULC change maps and statistics. The approach pursued was articulated in:

- Land cover classification in the Conca di Avellino area (urban, woodland, cropland, grassland/pasture) during the period from 1954 to 2004;
- Classification and change detection accuracy assessment of maps produced;
- Statistical estimation of the amount of change through ''from–to'' information derived from the classifications maps;
- Spatial analysis and LULC dynamics characterization, focusing on urban areas change patterns in relation to morphology, transportation network, population growth, Population Growth Rate (PGR), etc.

The present paper is part of a wider research concerning the analysis and interpretation of urban-rural gradient at regional scale. After examined the usefulness of landscape metric analysis [17], we have analysed the relationship between demographic and physical feature with the urban sprawl phenomenon. In accordance with other experiences in different areas [36, 40] the present research found that the demographic and physical measures of urbanisation can outline some aspects of urbanisation that was not captured by the landscape metrics and vice versa. The results confirm the capability of multi-temporal RS data to provide accurate and cost-effective tools to understand LULC changes, through detailed spatiotemporal analysis. This approach, applicable to studies at various locations, can be used to improve land management policies and decisions [41]. Moreover, it represents a valid contribution to land-use planning, especially considering the necessity to cope with matters related to the sustainable urban development [42]. Finally, mapping periodically the structure of urban growth and the LULC changes via GIS and spatial analysis is fundamental to forecast future development and in order to monitor and to assess the effectiveness of planning policies [43]. The analysis of the evolution urban-fringe areas represents one of the future research directions utilising, among others, VHR satellite images and very detailed digital cartography as reference data.

References

1. Michalak, W.Z.: GIS in land use change analysis: integration of remotely sensed data into GIS. Appl. Geogr. 13, 28–44 (1993)
2. Telesca, L., Coluzzi, R., Lasaponara, R.: Urban Pattern Morphology Time Variation in Southern Italy by Using Landsat Imagery. In: Murgante, B., Borruso, G., Lapucci, A. (eds.) Geocomputation & Urban Planning. SCI, vol. 176, pp. 209–222. Springer, Heidelberg (2009)
3. Masek, J.G., Lindsay, F.E., Goward, S.N.: Dynamics of urban growth in the Washington DC metropolitan area, 1973-1996, from Landsat observations. Int. J. Remote Sens. 21, 3473–3486 (2000)
4. Yang, X., Lo, C.P.: Using a time series of satellite imagery to detect land use and land cover changes in the Atlanta, Georgia metropolitan area. Int. J. Remote Sens. 23, 1775–1798 (2002)

5. Yuan, F., Sawaya, K.E., Loeffelholz, B.C., Bauer, M.E.: Land cover classification and change analysis of the Twin Cities (Minnesota) Metropolitan Area by multitemporal Landsat remote sensing. Remote Sens. Envir. 98(2&3), 317–328 (2005)
6. Sudhira, H.S., Ramachandra, T.V., Jagdish, K.S.: Urban sprawl: metrics, dynamics and modelling using GIS. Int. J. Appl. Earth Obs. 5, 29–39 (2004)
7. Wilson, E.H., Hurd, J.D., Civco, D.L., Prisloe, S., Arnold, C.: Development of a geospatial model to quantify, describe and map urban growth. Remote Sensing of Environment 86(3), 275–285 (2003)
8. Antrop, M.: Changing patterns in the urbanized countryside of Western Europe. Landscape Ecolo 15(3), 257–270 (2000)
9. Antrop, M.: Landscape change and the urbanization process in Europe. Landscape Urban Plan 67, 9–26 (2004)
10. EEA (European Environment Agency): Urban sprawl in Europe: The ignored challenge. EEA Report No. 10/2006, Copenhagen, Denmark (2006)
11. Antrop, M.: Why landscapes of the past are important for the future. Landscape Urban Plan 70, 21–34 (2005)
12. Lucas, R., Rowlands, A., Brown, A., Keyworth, S., Bunting, P.: Rule-based classification of multi-temporal satellite imagery for habitat and agricultural land cover mapping. ISPRS J. Photogramm. 62, 165–185 (2007)
13. Song, C., Woodcock, C.E., Seto, K., Lenney, M.P., Macomber, S.: Classification and change detection using Landsat TM data: when and how to correct atmospheric effects. Remote Sens. Environ. 75, 230–244 (2001)
14. Lillesand, T., Kiefer, R., Chipman, J.: Remote sensing and image interpretation. Wiley, New York (2003)
15. Congalton, R.G.: A review of assessing the accuracy of classifications of remotely sensed data. Remote Sens. Environ. 37(1), 35–46 (1991)
16. Congalton, R.G., Green, K.: Assessing the Accuracy of Remotely Sensed Data: Principles and Practices. CRCPress Taylor & Francis Group, Boca Raton (2009)
17. Fichera, C.R., Modica, G., Pollino, M.: Characterizing land cover change using multi-temporal remote sensed imagery and landscape metrics. In: Las Casas, G., Pontrandolfi, P., Murgante, B. (eds.) Informatica e Pianificazione Urbana e Territoriale - Atti della Sesta Conferenza Nazionale INPUT 2010, Potenza (2010)
18. Lo, C.P., Shipman, R.L.: A GIS approach to land-use change dynamics detection. Photogramm. Eng. Rem. S. 56, 1483–1491 (1990)
19. Petit, C.C., Lambin, E.F.: Integration of multi-source remote sensing data for land cover change detection. Int. J. Geogr. Inf. Sci. 15(8), 785–803 (2001)
20. Singh, A.: Digital Change Detection Techniques using Remotely Sensed Data. Int. J. Remote Sens. 10, 989–1003 (1989)
21. Lu, D., Mausel, P., Brondízio, E., Moran, E.: Change detection techniques. Int. J. Remote Sens. 25(12), 2365–2401 (2004)
22. Jensen, J.R., Ramsat, E.W., Mackey, H.E., Christensen, E.J., Sharitz, R.P.: Inland wetland change detection using aircraft MSS data. Photogramm. Eng. Rem. S. 53, 521–529 (1987)
23. Dimyati, M., Mizuno, K., Kobayashi, S., Kitamura, T.: An analysis of land use/cover change using the combination of MSS Landsat and land use map-a case study in Yogyakarta, Indonesia. Int. J. Remote Sens. 17, 931–944 (1996)
24. Taylor, J.C., Brewer, T.R., Bird, A.C.: Monitoring landscape change in the national parks of England and Wales using aerial photo interpretation and GIS. Int. J. Remote Sens. 21, 2737–2752 (2000)

25. Fichera, C.R., Modica, G., Pollino, M.: Remote sensing and GIS for rural/urban gradient detection. In: XVIIth World Congress of the International Commission of Agricultural Engineering (CIGR), Québec City, Canada (2010)
26. Congalton, R.G., Macleod, R.D.: Change detection accuracy assessment on the NOAA Chesapeake Bay Pilot Study. In: International Symposium on the Spatial Accuracy of Natural Resources Data Bases, pp. 78–87. ASPRS, Williamsburg (1994)
27. Huang, B., Zhang, L., Wu, B.: Spatiotemporal analysis of rural-urban land conversion. Int. J. Geogr. Inf. Sci. 23(3), 379–398 (2009)
28. Mayer, C.J., Somerville, C.T.: Land use regulation and new construction. Reg. Sci. Urban Econ. 30, 639–662 (2000)
29. Verburg, P.H., Koning, G.H.J., Kok, K., Veldkamp, A., Priess, J.: The CLUE modelling framework: an integrated model for the analysis of land use change. In: Singh, R.B., Jefferson, F., Himiyama, Y. (eds.) Land Use and Cover Change, Science Publishers, Enfield (2001)
30. Bhatta, B.: Analysis of Urban Growth and Sprawl from Remote Sensing Data. Springer, Heidelberg (2010)
31. Fotheringham, S., Rogerson, P.: Spatial analysis and GIS. Taylor & Francis, London (1994)
32. Italian National Institute of Statistics, `http://www.istat.it`
33. Murgante, B., Las Casas, G., Sansone, A.: A spatial rough set for locating the periurban fringe. In: Batton-Hubert, M., Joliveau, T., Lardon, S. (eds.) SAGEO 2007, Rencontres internationales Géomatique et territoire (2007)
34. Forman, R.T.T.: Land Mosaics: The Ecology of Landscapes and Regions. Cambridge University Press, Cambridge (1995)
35. Vizzari, M.: Spatial modelling of potential landscape quality. Appl. Geogr. 31(1), 108–118 (2011)
36. Bhatta, B.: Analysis of urban growth pattern using remote sensing and GIS: a case study of Kolkata, India. Int. J. Remote Sens. 30(18), 4733–4746 (2009)
37. Roca, J., Burnsa, M.C., Carreras, J.M.: Monitoring urban sprawl around Barcelona's metropolitan area with the aid of satellite imagery. In: Proceedings of Geo-Imagery Bridging Continents, XXth ISPRS Congress, Istanbul, Turkey, July 12–23 (2004)
38. Almeida, C.M., Monteiro, A.M.V., Mara, G., Soares-Filho, B.S., Cerquera, G.C., Pennachin, C.S.L., Batty, M.: GIS and remote sensing as tools for the simulation of urban land-use change. Int. J. Remote Sens. 26, 759–774 (2005)
39. Harvey, R.O., Clark, W.A.V.: The nature and economics of urban sprawl. Land Economics 41(1), 1–9 (1965)
40. Hahs, A., Mcdonnell, M.: Selecting independent measures to quantify Melbourne's urban–rural gradient. Landscape Urban Plan 78(4), 435–448 (2006)
41. Saizen, I., Mizuno, K., Kobayashi, S.: Effects of land-use master plans in the metropolitan fringe of Japan. Landscape Urban Plan 78, 411–421 (2006)
42. Zhang, Y., Guindon, B.: Using satellite remote sensing to survey transport-related urban sustainability: Part 1: Methodologies for indicator quantification. Int. J. Appl. Earth Obs. 8(3), 149–164 (2006)
43. Hardin, P.J., Jackson, M.W., Otterstrom, S.M.: Mapping, Measuring, and Modeling Urban Growth. In: Jensen, R.R., Gatrell, J.D., McLean, D. (eds.) Geo-Spatial Technologies in Urban Environments, Springer, Heidelberg (2007)

Spatial Clustering to Uncluttering Map Visualization in SOLAP

Ricardo Silva[1], João Moura-Pires[1], and Maribel Yasmina Santos[2]

[1] Departamento de Informática, Universidade Nova de Lisboa
Quinta da Torre. 2829–516 Caparica, Portugal
ricardofcsasilva@gmail.com, jmp@di.fct.unl.pt

[2] Departamento de Sistemas de Informação, Universidade do Minho
Campus de Azurém, Guimarães, Portugal
maribel@dsi.uminho.pt

Abstract. The main purpose of SOLAP concept was to take advantage of the map visualization improving the analysis of data and enhancing the associated decision making process. However, in this environment, the map can easily become cluttered losing the benefits that triggered the appearance of this concept. In order to overcome this problem we propose a post-processing stage, which relies on a spatial clustering approach, to reduce the number of values to be visualized when this number is inadequate to a properly map analysis. The results obtained so far show that the usage of the post–processing stage is very useful to maintain a map suitable to the user's cognitive process. In addition, a novel heuristic to identify the threshold value from which the clusters must be generated was developed.

Keywords: SOLAP, spatial clustering, DBSCAN.

1 Context and Motivation

Most OLAP applications are focused on textual data and numerical measures even though studies have concluded that 80% of data refer to spatial information [4]. Consequently, the integration of spatial data within the multidimensional model was envisaged. Rivest et. al. [23] defined the Spatial OLAP (SOLAP) concept as "*a visual platform built especially to support rapid and easy spatio–temporal analysis and exploration of data following a multidimensional approach comprised of aggregation levels available in cartographic displays as well as in tabular and diagram displays*". The concept of SOLAP emerged from the integration of the spatial data included either in dimensions (spatial dimensions) or in fact tables (spatial measures), enabling cartographic displays in the OLAP applications [23]. This way, thematic maps are produced by using the members of spatial dimensions and the numerical measures, combining them with the visual variables [21].

In this new environment for the analysis of spatial data, several benefits have been mentioned from thematic maps visualization enabled by SOLAP systems:

B. Murgante et al. (Eds.): ICCSA 2011, Part I, LNCS 6782, pp. 253–268, 2011.

better and faster global perception of query results, and the possibility to discover correlations between phenomena, as detailed in [4].

It is worth to mention that there is a difference in the results that can be analyzed by an OLAP user compared to a SOLAP user. A typical result of the former involves one to two dozen lines with the aggregated data. However, the latter may involve hundreds of lines if the user is interested in the analysis of data at a lower level of granularity (ex: customer level). If this is the case, and depending on the geographical distribution and the spatial objects' representation, a thematic map can easily become cluttered and hard to analyze, as illustrated in Fig. 1 (where the points are airports locations and the brightness is given by the value of the numerical measure).

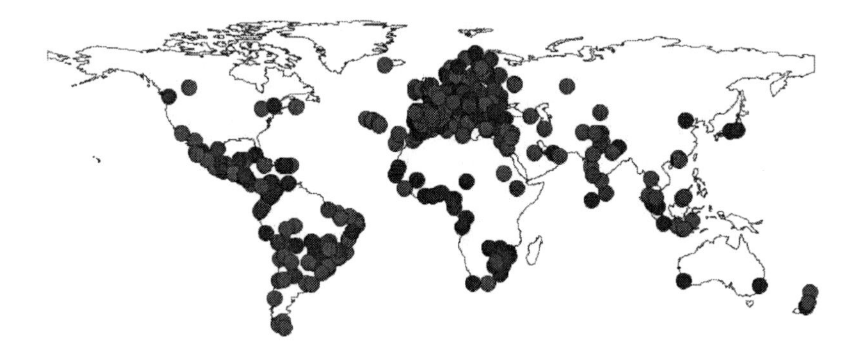

Fig. 1. Example of a cluttered thematic map

A SOLAP user shouldn't lose the power of maps, so it is necessary to control the number of results returned to the user. In order to maintain the benefits that come from map visualization, we propose a post-processing stage, which relies on a spatial clustering approach, that is applied before the results (maps, data tables and graphics) are presented to the user.

Our solution takes into account the amount of spatial information to be displayed on the map and the possible overlapping between the different representations. Moreover, it gives to the user the ability to control the existence, or not, of the post-processing stage. When the post–processing stage is used, the user can also control the clustering level based on the proposed heuristic to determine the threshold value used to generate the clusters and perform the clustering process constrained by a spatial hierarchy defined in the multidimensional model. Additionally, this process is query-aware, i.e. takes in account: (*i*) the type of spatial objects selected from dimensions (points or polygons); (*ii*) the numerical measures and the used aggregation operator; (*iii*) the semantic attributes and their relation (of granularity) with the spatial attributes.

The work developed so far is aligned with the research agenda for *Geovisual analytics for spatial decision support* area presented by Andrienko et. al. [2], since we are dealing with a visualization problem in a spatial decision support tool.

This paper is organized as follows. Section 2 presents some related work associated with the area of SOLAP and Spatial Clustering Algorithms. Section 3 addresses the post-processing model proposed in this work to deal with the huge complexity that can emerge in the analysis of spatial data in a SOLAP system. Section 4 presents the prototype that was developed to implement the post-processing model. Section 5 concludes with some remarks about the undertaken work and some guidelines for future work.

2 Related Work

In this section, we present some of the research literature related to SOLAP and spatial clustering algorithms.

2.1 SOLAP Concepts and Systems

The integration of spatial data into a multidimensional model adds it two new main concepts: spatial dimensions and spatial measures. The use of spatial measures is a widely discussed subject but it is far from having an agreement on it [17][22][6]. A spatial dimension is a dimension where one or more attributes are spatial objects. Each dimension can index data at several detail/aggregation levels. Hierarchies can then be defined using the levels of a dimension. Malinowsky and Zimányi [16] have defined different kinds of spatial hierarchies in spatial dimensions. In general, the most common hierarchies are those whose relationships between their members can be represented as a tree, designated *simple spatial hierarchies*. For example, the date dimension can have day, week, month, year as a *simple spatial hierarchy*.

A SOLAP system should incorporate, according to [23], three main areas: visualization, exploration and structure of data. For data visualization, the authors argue that cartographic displays should allow adequate exploration of the geometric component of the spatial data being analyzed (from spatial dimensions members), as well as the use of contextual information. Also, they refer the need to include statistical diagrams into cartographic displays, such as bar charts, pie charts, among others, in order to obtain summarized information of the data being analyzed. This combination of elements, which can be present in a map, call our attention to the need of maintaining a legible and organized map. There are more guidelines related to the other two areas, but they are out of the scope on this paper.

Based on these concepts and guidelines, the SOLAP+ system was developed [11] [26]. It will be used to implement the post-processing model proposed in this paper (and detailed in section 3 and section 4). Other examples of SOLAP tools are described in [5].

The SOLAP+ framework includes three main components: the map, the support table and the detail table, as shown in Fig. 2. In the map, thematic maps, spatial objects and other spatial information are presented to the user. The support table is used to show the data in an alphanumeric format from which the map representation depends on. The detail table is a tool that provides to the user a more in–depth analysis.

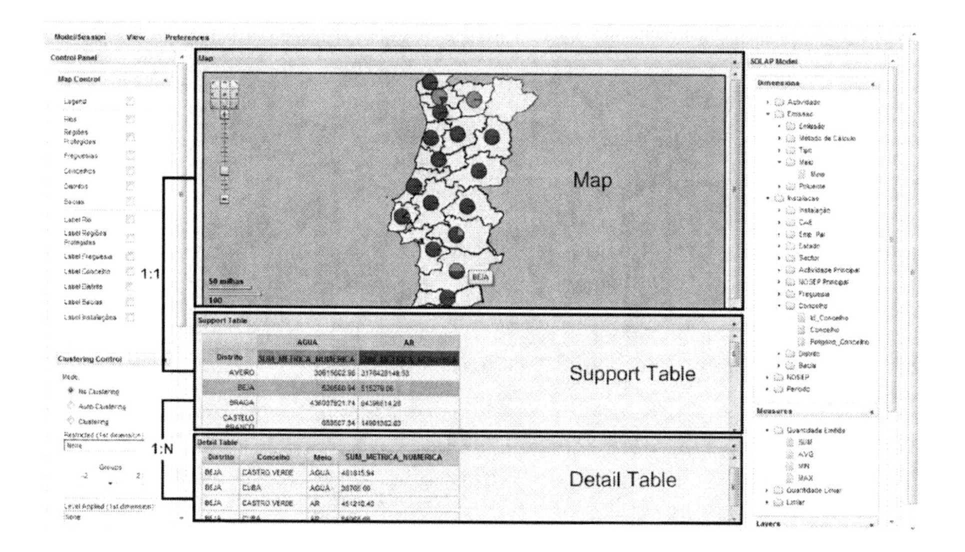

Fig. 2. The SOLAP+ interface

There is one line in the support table to each spatial object displayed on the map (from the query result). This 1:1 relationship is maintained in any situation and should always be kept. Its purpose is to facilitate the user's cognitive process relating the alphanumeric data and their spatial location/representation. The detail table is used to do drill-down operations on a line of the support table. As a result, we have n lines in the detail table for each line in the support table. Further details about the SOLAP+ system can be found in [26].

2.2 Spatial Clustering Algorithms

Clustering is the process of grouping a set of objects into clusters in such a way that objects within a cluster have high similarity with each other, but are as dissimilar as possible to objects located in other clusters [18]. Spatial clustering techniques have emerged to deal with the growing amount of spatial data that have been stored in spatial databases. Those techniques revealed great potentialities in the generalization of the spatial component present in spatial databases, reducing the number of elements to be observed and represented.

To uncluttering map visualization in a SOLAP context, a spatial clustering algorithm should verify a set of requirements. From the mentioned in [15], the most meaninful, from our point of view, are: (*i*) no *a-priori* knowledge, like the number of clusters, should be required because the request of several input parameters will turn the application very demanding, from a user point of view; (*ii*) the algorithm must be able to quickly process large amounts of data, avoiding the introduction of high delays in the data visualization process that results from the proposed post-processing model; (*iii*) the algorithm should identify groups with arbitrary shapes since it is expected that real datasets contain clusters with irregulars shapes.

Despite the high number of existing spatial clustering algorithms, they can be categorized in four methods [18]: (*i*) partitioning; (*ii*) hierarchical; (*iii*) density-based; (*iv*) grid-based.

In the partitioning method, k partitions are created forming k groups of data. The partitioning is created by attempting to optimize an objective partitioning criterion, such as the distance between the objects. The number of groups must be given in advance. The partitioning methods include algorithms like k-Means [10], k-Medoid (PAM) [14] and CLARANS [20].

In the hierarchical method, a hierarchical decomposition is performed from an initial dataset. From a hierarchical decomposition results a dendrogram that shows the hierarchical structure of clusters. As a result, typically, there is no need to specify the number groups, yet it is required to set the end condition for the decomposition process. In this process, a key decision is related to whether the objects must be, or not, unified. A wrong decision cannot be corrected later on. To overcome this issue other clustering techniques were integrated with hierarchical algorithms, emerging the multi-phase hierarchical algorithms, including Chameleon [13]. Other examples of hierarchical algorithms are CURE [9] and BIRCH [27].

Concerning the density-based algorithms, they adopt the straightforward idea of identify cluster as dense regions of objects that are separated from clusters with lower density of objects. The main advantage of this approach is the ability to discover clusters with irregular shapes. Some relevant examples of density-based clustering algorithms are DBSCAN [8], P-DBSCAN [12] and SNN [7].

Regarding the grid-based method, its algorithms quantize the original space into a finite number of cells, creating a multi-level grid structure. The clustering operations are performed on the quantized space. The main advantage of this approach is its fast processing time, which is typically independent of the number of data objects as it only depends on the number of cells in each dimension in the quantized space. Examples of algorithms of this approach are CLIQUE [1] and WaveCluster [25].

In the context of this work, the partitioning algorithms are not considered since they require as input parameter the number of clusters. Looking at the grid-based algorithms, we may have efficient algorithms but they also need several input parameters, for which no heuristic approach is available to make them user independent. This is also true for the hierarchical algorithms. Excluding these three types, remain the density-based algorithms. These algorithms also require input parameters. However, some studies have been carried out to define heuristics that estimate the values of the input parameters [8] [24]. Although user interaction continues to be needed, this paper proposes a novel user independent heuristic (presented later in section 4).

From the several density-based algorithms, this work adopted DBSCAN. No benchmarking was carried out to select this algorithm. This selection was only guided by the fact that there is a variant of DBSCAN to cluster regions (P-DBSCAN). This is relevant in the context of this work, as we can have spatial objects represented as points and others represented as regions.

3 Post-processing Model

Given a query in a SOLAP context, the goal of the post-processing model is to display the data in an organized and understandable way for the user, maintaining the benefits that emerge from map visualization and its advantages to the user's analysis. An overview of the defined post-processing model is presented in Fig. 3.

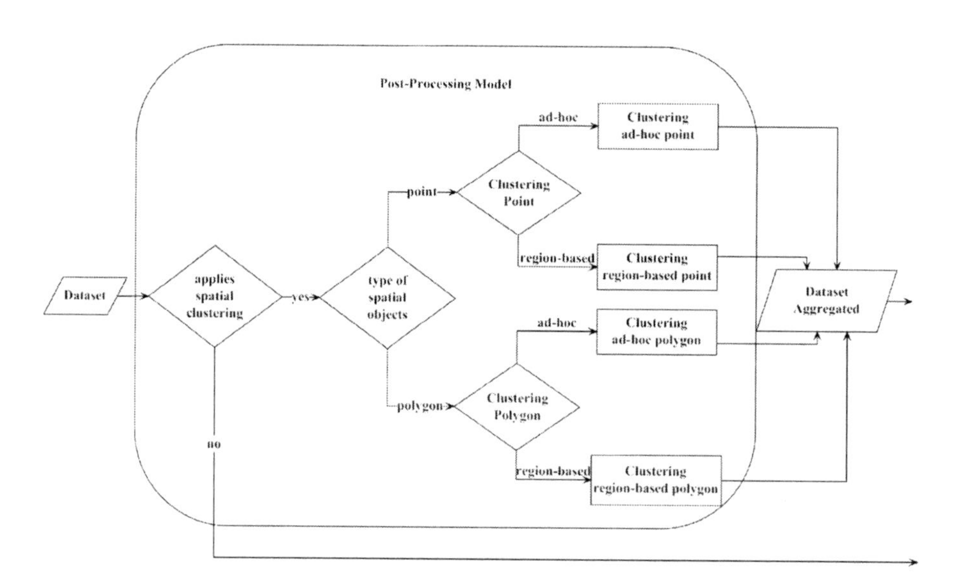

Fig. 3. The Post-processing model

The first step consists on evaluating if the spatial clustering process should be applied or not. There are three options: (*i*) the user does not apply; (*ii*) the user does apply; (*iii*) the post-processing component decides automatically.

For the latter option, a heuristic is needed to make that evaluation. Our main objective is to ensure proper objects visibility, but the performance should also be taken into account since the clustering process will introduce a new processing time component. This component may introduce a significant time, or not, in the overall time to display the results. Therefore, a heuristic should compute the decision, based on indicators about the legibility and the performance. If one of them is verified, then the post-processing should be performed. The heuristic used to obtain these indicators should introduce a very small extra processing time, when compared with the time of the clustering process.

Regardless the spatial clustering algorithm used, the algorithm will be applied to the real coordinate space (geographical coordinates). The input parameters are influenced by the zoom level determining a clustering with more or less clusters. Even if the results are not all displayed, due to the zoom level, the clustering process is applied to all the input data. In the visualizations, a panning doesn't change the clusters as a zooming (in or out) does.

The most common spatial objects associated to spatial attributes are points (ex: customers locations) or polygons (ex: country administrative divisions). At this stage, it is necessary to identify the type of spatial object we are dealing with, as the post-processing model needs to choose the proper clustering algorithm and the respective distance function. Despite the type of spatial object, the spatial clustering algorithm should include the requirements presented and discussed in section 2.

In order to accomplish our goal, we propose two approaches in the post-processing model: an ad-hoc or a region-based clustering. The first approach creates clusters without any semantic meaning apart from the geographical proximity among objects. Alternatively, the region-based clustering creates groups that also consider the geographical proximity, but the groups that can be created are constrained by a spatial hierarchy. Therefore, the clustering approach to be applied depends not only on type of spatial object but also in the selected approach. Nevertheless, a spatial clustering algorithm is always applied to the spatial objects.

After the clusters identification, a new representation for each cluster is generated, decreasing the number of spatial objects that need to be placed on the map. If a new spatial representation is generated for each cluster, the non-spatial data should also be aggregated allowing the synchronization between the map and the tabular display, as it is recommended in [23]. As result, this process has an important impact in the tabular display, as the data can be presented at different levels of granularity.

As it was mentioned previously, the post–processing model is query-aware and it is defined based on the next notation:

- **Semantic Dimension (sD)** is a dimension where all levels contain semantic attributes.
- **Spatial Dimension (spD)** is a dimension with one or more levels that contain spatial attributes.
- **Spatial Attribute (spA)** is a spatial attribute of a *spD* level.
- **Semantic Attribute (sA)** is a textual or numerical attribute from a *sD* or *spD*.

- **Numerical Measure (nM)** is a numerical value associated to a fact and stored in the fact table. The numerical measure associated to an aggregation operator is represented as $nM(aggO)$.
- **Query (Q)** defines a representative query. It could have spatial attributes, semantic attributes and numerical measures. We assume that Q may have : (i) one or two spA and the corresponding $sA(spA)$; (ii) zero or more sA; (iii) one or more nM, each one with an associated aggregation operator.
- **Spatial Hierarchy (spH)** is a simple spatial hierarchy composed by n levels.

In the following sections, it will be detailed the post–processing model based on representatives queries.

3.1 One Spatial Attribute: Point

Whenever the query result involves a spatial point attribute, the clustering point flow of the post-processing model will be performed. For now, we only consider queries that fit into the following representative query:

$$Q\{spA_1, sA_1(spA_1), nM_1(aggO_1), \ldots, nM_n(aggO_n)\}$$

This representative query contains one spatial attribute and the associated semantic attribute, and one or more numerical measures. An example of such query could be: *What is the total amount of carbon dioxide emissions by facilities that are within 5 Km radius from a city?* (where spA_1 is the facilities locations, $sA_1(spA_1)$ is the facilities names and $nM_1(SUM)$ is the corresponding total amount of the carbon dioxide emissions).

Initially, all spA_1 values resulting from Q are extracted and a spatial clustering algorithm is applied to them. The algorithm result will be the spA_1 values associated to one or no cluster. After that, two actions are performed for each group: the definition of a new representation and the aggregation of data with the appropriate aggregation operator, i.e the data must be aggregated using the aggregation operator associated to each nM.

There are several possibilities to create a new representation, such as: centroid, center of gravity, convex or concave hull. Each one of them has its advantages in some specific applications. If we use a polygon solution as a new representation then it would restrict the number of possible thematic maps that can be created, since it will be impossible to use the visual variable size to express some attribute or numerical measure. Therefore, our proposal for a new cluster representation is to combine a polygon with a point representation. This way, the possibility to apply the visual variable size is maintained and at the same time the user has information about the area covered by the cluster. Once it is expected clusters with arbitrary and irregular shapes, the use of a concave hull algorithm [19], when compared with a convex hull approach, fits better.

The other approach for clustering point data is the region-based clustering. This method is very similar to the previous. However, the clusters that result

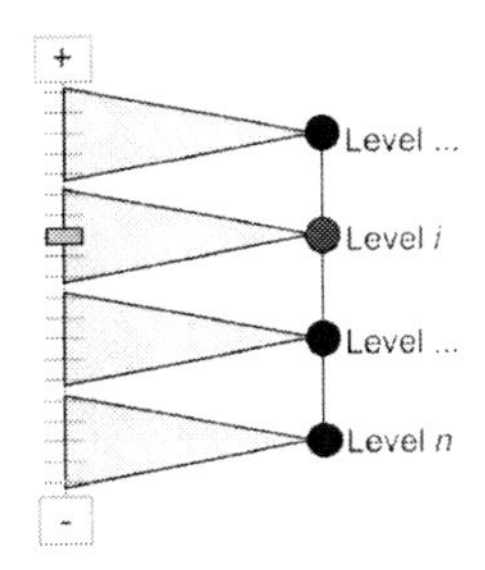

Fig. 4. Illustrates how the level to constraint the clusters is computed (level i)

from the spatial clustering algorithm share the spatial attribute value that comes from the computed level.

Consider that spH is the hierarchy chosen by the user, which must include spA_1. The considered spH levels have to be at a higher level of granularity compared to the spA_1 level and have to be represented by polygons. The level that constraint the clusters is computed in the following way: the map zoom levels are divided equally by the hierarchy levels and the resulting level is given by the mean zoom level as illustrated in Fig. 4. Despite the proposed approach, the user has the possibility to select it manually.

3.2 One Spatial Attribute: Region

In this section, we assume the same representative query but in this case spA_1 values are polygons. A possible query could be: *What is the total amount of carbon dioxide emissions by counties that have inside at least one protected region?*

When the spatial objects are represented by polygons, which cannot typically overlap, a negative impact on map visualization can arise not only by the polygons visualization, but also by the visualization of charts or other elements associated with them.

The flow for clustering polygons is similar to the flow for clustering points. Initially, all the spA_1 values are obtained. Then, it is applied a spatial clustering algorithm suitable for polygons. Finally, for each group of polygons, a cluster, a new representation is computed and the data is aggregated in a proper way as we mentioned in the previous section. In this case, the new representation is the union of polygons.

New issues arise in the process of clustering polygons. In the clustering of points, the Euclidean distance or other similar metric is used. Whatever the distance function is, it is expected the time complexity to be constant. The same is not true for polygons. A good way to measure the distance between two polygons is through the Hausdorff distance [3]. Unfortunately, the computation of the Hausdorff distance is expensive from the computational point of view, even though there are works that attempt to minimize it, such as [3]. Although those efforts, there is no work, to the best of our knowledge, that achieved a constant time complexity.

To overcome this issue we propose the pre-computing of the distances among the polygons in the spatial dimensions. When a spatial clustering algorithm is applied to the query result, only the distances already computed are needed. Through this solution the time complexity for the distance function goes constant.

3.3 More Complex Queries

So far, we have excluded the semantic attributes from the query. Including the semantic attributes into the query gives us the following representative query:

$$Q\{spA_1, sA_1(spA_1), sA_1, \ldots, sA_j, nM_1(aggO_1), \ldots, nM_n(aggO_n)\}$$

An example of a query in this context could be: *What is the total amount of air and water pollutant emissions by counties that have inside at least one protected region?* (where spA_1 is the counties polygons, $sA_1(spA_1)$ is the counties names, sA_1 describe the mean value of the pollutant emissions and $nM_1(SUM)$ is the corresponding total amount of the pollutant emissions).

When we are dealing with semantic attributes we need to consider two distinct cases: (i) sA_i comes from the same spD that spA_1; (ii) sA_i comes from another dimension (where $1 \leq i \leq j$).

In the first case, it is important to look at the level at which both spA_1 and sA_i are. If sA_i is at the same or at a higher level than the spA_1 then there is only one value of sA_i for each spA_1. For these cases we introduce a new constraint to the clustering process: each cluster must share the sA_i value (regardless the selected clustering approach). Consider the following representative

$sA_1(spA_1)$	sA_i	$nM_1(SUM)$
name1	x	10
name2	y	10
name3	x	30
name4	w	30
name5	w	30
name6	w	40

a) →

$sA_1(spA_1)$	sA_i	$nM_1(SUM)$
cluster1	x,y	50
cluster2	w	100

b) →

$sA_1(spA_1)$	sA_i	$nM_1(SUM)$
cluster1	x	40
name2	y	10
cluster2	w	100

Fig. 5. The tabular form after the post-processing stage is applied in first case: a) without constraint; b) with constraint

query: $Q\{spA_1, sA_1(spA_1), sA_i, nM_1(SUM)\}$. In Fig. 5 the sA_i verifies the first case and suppose that the *name1*, *name2* and *name3* forms one cluster and the remaining values form another. Without the proposed constraint, the correspondence between the values of spA_1 and sA_i could change after the post-processing stage as illustrated in Fig. 5a. This restriction was introduced to maintain the analysis with the same tabular representation. This way we maintain the 1:1 relationship between the tabular form and the map (Fig. 5b).

In the second case, the post-processing stage applies a straightforward approach that maintains the relations between the attribute values as depicted in Fig. 6, illustrating the previous Q.

$sA_1(spA_1)$	sA_i Value 1	sA_i Value 2
	nM_1(SUM)	nM_1(SUM)
name1	10	20
name2	5	8
name3	30	14

$sA_1(spA_1)$	sA_i Value 1	sA_i Value 2
	nM_1(SUM)	nM_1(SUM)
cluster1	45	42

Fig. 6. The tabular form after the post-processing stage is applied in the second case

In the previous sections we have assumed queries with only one spatial attribute. However, we may have two spatial attributes. In this case, the representative query is:

$$Q\{spA_1, sA_1(spA_1), spA_2, sA_2(spA_2), \ldots, sA_i, \ldots, nM_n(aggO_n)\}$$

In [11] it is presented a generic case where there are two spA in the query. These attributes belong to different $spHs$ but are from the same spD. For these cases the authors proposed, under some conditions (see [11]), the intersection between spA_1 and spA_2. A possible query in this context could be: *What is the total amount of air and water pollutant emissions by counties and watersheds?* (where spA_1 is the counties polygons, spA_2 is the watersheds polygons, $sA_1(spA_1)$ and $sA_2(spA_2)$ are the respective counties and watersheds names, sA_1 describe the mean value of the pollutant emissions and $nM_1(SUM)$ is the corresponding total amount of the pollutant emissions). These cases also work well in the proposed approach since the original Q becames similar to a context presented in section 3.2: $Q\{spA_1 \cap spA_2, sA_1(spA_1), sA_2(spA_2), nM_1(aggO_1), \ldots, nM_n(aggO_n)\}$.

4 Prototype

The SOLAP+ is a generic system that relies on a three tier architecture composed by Oracle as the Database Management System (gives support to spatial data), a SOLAP server, and a client coupling the OLAP features with maps. The server was implemented from scratch, in Java, and it is responsible for listening to client requests, processing them and retrieving the appropriate results. The client handles all user interaction, data presentation and request generation. It was implemented in Java Server Faces (JSF) and the communication with the server is performed based on the XML protocol. Also, it uses Oracle Maps JavaScript API (to enhance the functionality of the Oracle MapViewer) in order to support map visualization and interaction.

Our post-processing model was included in the SOLAP server tier. Also, the DBSCAN and the P-DBSCAN were implemented and included in this tier. To the chosen algorithms, the object neighborhood is based on some radius (*Eps*) and an object in the cluster has to contain at least $MinElements$ of elements. Thus, appropriate values for *Eps* and $MinElements$ need to be identified in order to implement the post-processing model in a user-friendly approach. For the latter parameter, we follow the formula: $MinElements = 2 * Dimensionality - 1$, proposed in [24]. For the former we propose a novel user independent heuristic presented in the next section.

4.1 User Independent Heuristic to Determine Eps

In [24] it is proposed a $k.distance$ function mapping each object to the distance from its k-th nearest neighbors. Based on these $k.distance$ values it is created a plot with those values sorted in descending order, called a *sorted* $k.distance$ *plot* where the k value corresponds to the $MinElements$ value (in 2D space $k = 3$). That plot gives some hints concerning the objects density distribution.

When choosing an object obj, it is assigned to the *Eps* parameter the value $k.dist(obj)$, and all objects at right will be considered as core objects, while the objects at left are labeled as noise. In this heuristic, the authors proposed that the user would choose the object at the first "valley", as illustrated in Fig. 7 obtained from [24].

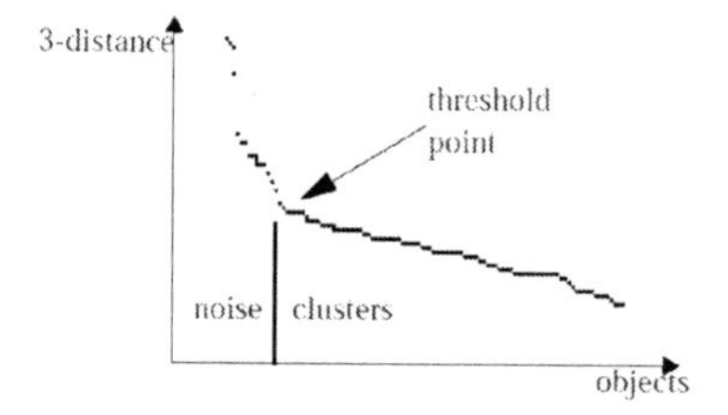

Fig. 7. Sorted 3-distance plot

However, this heuristic is not user independent. To make the process user independent we developed a new heuristic. The proposed one aims to find not only an appropriate value, but more than one value allowing the user to choose between a result with more or less clusters (for the same map zoom level).

Therefore, we propose a heuristic that searches for "breaks" in the $3.distance$ function. It is on the several breaks that the turning points are found, those that produce clusters with different densities.

The breaks are obtained as follows: initially, the objects are sorted in an ascending order by the value of $3.distance$ function. Then, in each iteration, it is calculated the following $\delta_i = 3.distance(obj_{i+1}) - 3.distance(obj_i)$ and the average of this value is updated $\Delta_i = \sum_{0<j<i} \Delta_j / i$. When $\delta_{i+1}/\Delta_i \geq \alpha$ it is considered a break where α is an arbitrary value. In such cases, the

$3.distance(obj_i)$ value is stored and the average is initialized to zero. Furthermore, the $3.distance(obj_i)$ value will be used as the *Eps* parameter. At the end of this process, a set of breaks are obtained, given the $break_i$ value more clusters than the $break_{i+1}$. This set must have at least three values to allow the user to choose between less or more clusters. Otherwise, the previous process is repeated with a decremented α until the resulting set of breaks verifies the previous condition. In our implementation, the α value is initialized to three and it is decremented from one value if the set of breaks has less than three values.

4.2 Demonstration Case

The demonstration case is based in a real data set about pollutant emissions in Portugal. The relevant information about the multidimensional model can be summarized as:

1. Contains the spatial dimension: Facility. This dimension includes five spatial attributes: location, drainage basin, *Freguesia*, *Concelho*, *Distrito* and the semantic attribute: facility name.
2. The Facility dimension contains the Portuguese administrative divisions as a spatial hierarchy.

In this demonstration case the data is sliced with respect to three districts. From $Q\{facility\ location, facility\ name, nM_1(SUM)\}$ result, without performing the post-processing stage, the obtained results are displayed in Fig. 8a.

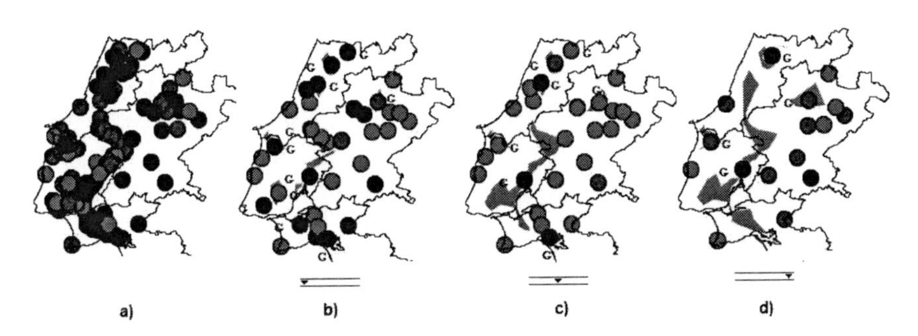

Fig. 8. The result using clustering ad–hoc approach. The parameters applied are: b) *Eps* = 0.050; c) *Eps* = 0.059; d) *Eps* = 0.071. The value of *MinElements* is 3.

Fig. 8 also shows the result obtained after the application of the ad–hoc clustering approach. In Fig. 8b, Fig. 8c and Fig. 8d are used the *Eps* values returned by the proposed heuristic allowing the user to choose between a result with more or less clusters. These results can be changed if the user is not satisfied with them. As the user is unaware from the *Eps* values, he/she only has to drag the slider to the left (more groups) or to the right (less groups) to change the detail associated to the results.

In Fig. 9 is displayed the result of the post-processing stage using the region-based approach (based on Fig. 8d). The *spH* chosen is the hierarchy of the Portuguese administrative divisions. The level computed to restrict the cluster from the map zoom level is the District level. Also, the partial results of the support table and the detail table (detailing the *Group0*) are displayed.

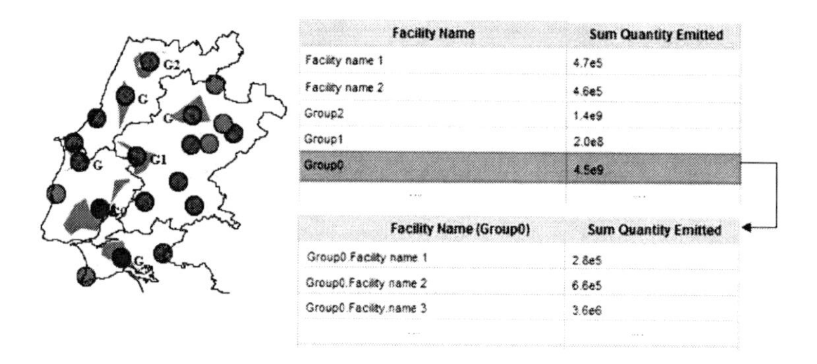

Facility Name	Sum Quantity Emitted
Facility name 1	4.7e5
Facility name 2	4.6e5
Group2	1.4e9
Group1	2.0e8
Group0	4.5e9
...	...

Facility Name (Group0)	Sum Quantity Emitted
Group0.Facility name 1	2.8e5
Group0.Facility name 2	6.6e5
Group0.Facility name 3	3.6e6
...	...

Fig. 9. The result using clustering region–based approach with the Fig. 8d context. The support table is above and the detail table bellow.

For each cluster it is displayed a delimited area that is covered by it. These areas are computed through the concave hull algorithm that uses as input the coordinates of the objects that are inside each cluster. The numerical measure value is displayed through a marker positioned at the cluster centroid (and over the covered area) that has the same representation of non-clusters markers. The facilities' names are not displayed as this information is confidential.

5 Conclusions and Future Work

This paper presented a post-processing model to maintain the benefits that emerge from map visualization in a SOLAP environment. The post-processing model relies on a spatial clustering technique to generalize the spatial component, attached to the SOLAP queries, in order to reduce the number of elements observable in the map.

The proposed post-processing model attends to balance the amount of spatial information to be displayed and the possible overlapping between representations. Additionally, it gives to the user the capability to control the existence, or not, of the post-processing stage, the level of the clustering process (based on our novel heuristic to estimate the *Eps* DBSCAN/P-DBSCAN parameter), and if the clusters are restricted by a spatial hierarchy. All this data analysis process in driven by a user specified query. To the best of our knowledge, there is no other SOLAP system with a mechanism to control the map visualization.

Future work can be directed to the proper evaluation of the post-processing model. We envisage two levels of evaluation. First, we should evaluate the spatial clustering approach to reduce the clutter in the map. Second, we should perform a comparative evaluation about the novel user independent heuristic face to the classic heuristic proposed by the DBSCAN authors. Regarding the first level, we should compare several spatial clustering algorithms in terms of the requirements discussed in section 2.2 and the quality of the results. Concerning the second indicator, it can be subdivided into a set of sub indicators that characterize the quality of the results, such as: number of clusters, average spacing between clusters, etc. This evaluation should be carried out both with synthetic and real data sets. Also, the level of users' satisfaction in the presence and absence of the post-processing model and the effectiveness of the novel heuristic regarding the number of generated clusters should be evaluated.

Additionally, future work also includes: the automatic identification of the legibility and performance indicators to be used by the post–processing component to automatically decide if the spatial clustering process should be applied or not; the application of the spatial clustering process based on map representations (instead real coordinate space); and, finally, the extrapolation of this approach to other contexts beyond SOLAP, or SOLAP contexts but with spatial measures.

References

1. Agrawal, R., Gehrke, J., Gunopulos, D., Raghavan, P.: Automatic subspace clustering of high dimensional data for data mining applications. In: Proceedings of the ACM SIGMOD Int'l Conference on Management of Data, Seattle, Washington, June 1998, pp. 94–105. ACM Press, New York (1998)
2. Andrienko, G.L., Andrienko, N.V., Jankowski, P., Keim, D.A., Kraak, M.J., MacEachren, A.M., Wrobel, S.: Geovisual analytics for spatial decision support: Setting the research agenda. International Journal of Geographical Information Science 21(8), 839–857 (2007)
3. Atallah, M.J.: A linear time algorithm for the hausdorff distance between convex polygons. Inf. Process. Lett. 17(4), 207–209 (1983)
4. Bédard, Y., Rivest, S., Proulx, M.J.: Spatial on-line analytical processing (solap): Concepts, architectures, and solutions from a geomatics engineering perspective. In: Data Warehouses and OLAP: Concepts, Architecture, pp. 298–319 (2006)
5. Bimonte, S.: On Modeling and Analysis of Multidimensional Geographic Databases. In: Data Warehousing Design and Advanced Engineering Applications: Methods for Complex Construction, chap. 6 (2010)
6. Bimonte, S., Wehrle, P., Tchounikine, A., Miquel, M.: GeWOlap: A web based spatial OLAP proposal. In: Meersman, R., Tari, Z., Herrero, P. (eds.) OTM 2006 Workshops. LNCS, vol. 4278, pp. 1596–1605. Springer, Heidelberg (2006)
7. Ertöz, L., Steinbach, M., Kumar, V.: Finding clusters of different sizes, shapes, and densities in noisy, high dimensional data. In: Proceedings of Second SIAM International Conference on Data Mining (2003)
8. Ester, M., Kriegel, H.P., Sander, J., Xu, X.: A density-based algorithm for discovering clusters in large spatial databases with noise. In: Proc. of 2nd International Conference on Knowledge Discovery, pp. 226–231 (1996)

9. Guha, S., Rastogi, R., Shim, K.: CURE: an efficient clustering algorithm for large databases. In: Haas, L., Drew, P., Tiwary, A., Franklin, M. (eds.) SIGMOD 1998: Proceedings of the 1998 ACM SIGMOD International Conference on Management of Data, pp. 73–84. ACM Press, New York (1998)
10. Hartigan, J.A., Wong, M.A.: A K-means clustering algorithm. Applied Statistics 28, 100–108 (1979)
11. Jorge, R.: SOLAP+: Extending the Interaction Model. Master's thesis, FCT / UNL, João Moura-Pires (superv.) (July 2009)
12. Joshi, D., Samal, A., Soh, L.K.: Density-based clustering of polygons. In: CIDM, pp. 171–178. IEEE, Los Alamitos (2009)
13. Karypis, G., Han, E.H., Kumar, V.: Chameleon: hierarchical clustering using dynamic modeling. Computer 32(8), 68–75 (1999)
14. Kaufman, L., Rousseeuw, P.: Finding Groups in Data: An Introduction to Cluster Analysis. Wiley Interscience, New York (1990)
15. Kolatch, E.: Clustering algorithms for spatial databases: A survey. Tech. rep. (2001)
16. Malinowski, E., Zimányi, E.: Spatial hierarchies and topological relationships in the spatial MultiDimER model. In: Jackson, M., Nelson, D., Stirk, S. (eds.) BNCOD 2005. LNCS, vol. 3567, pp. 17–28. Springer, Heidelberg (2005)
17. Malinowski, E., Zimányi, E.: Logical representation of a conceptual model for spatial data warehouses. Geoinformatica 11, 431–457 (2007)
18. Miller, H.J., Han, J.: Geographic Data Mining and Knowledge Discovery, 2nd edn. (2009)
19. Moreira, A., Santos, M.Y.: Concave hull: A k-nearest neighbours approach for the computation of the region occupied by a set of points. In: GRAPP (GM/R), pp. 61–68. INSTICC - Institute for Systems and Technologies of Information, Control and Communication (2007)
20. Ng, R.T., Han, J.: Clarans: A method for clustering objects for spatial data mining. IEEE Transactions on Knowledge and Data Engineering 14(5), 1003–1016 (2002)
21. Rivest, S., Bédard, Y., Proulx, M., Nadeau, M., Hubert, F., Pastor, J.: SOLAP technology: Merging business intelligence with geospatial technology for interactive spatio-temporal exploration and analysis of data. ISPRS Journal of Photogrammetry and Remote Sensing 60(1), 17–33 (2005)
22. Rivest, S., Bédard, Y., Proulx, M.J., Nadeau, M.: Solap: a new type of user interface to support spatio-temporal multidimensional data exploration and analysis. In: Proceedings of the ISPRS Joint Workshop on Spatial, Temporal and Multi-Dimensional Data Modelling and Analysis (2003)
23. Rivest, S., Bédard, Y., March, P.: Towards better support for spatial decision-making: defining the characteristics. Geomatica, the Journal of the Canadian Institute of Geomatics 55(4), 539–555 (2001)
24. Sander, J., Ester, M., Kriegel, H.P., Xu, X.: Density-based clustering in spatial databases: The algorithm gdbscan and its applications. Data Mining and Knowledge Discovery 2(2), 169–194 (1998)
25. Sheikholeslami, G., Chatterjee, S., Zhang, A.: WaveCluster: a wavelet-based clustering approach for spatial datain very large databases. The VLDB Journal 8(3-4), 289–304 (2000)
26. Silva, R.: SOLAP+. Master's thesis, FCT / UNL, João Moura-Pires (superv.) (October 2010)
27. Zhang, T., Ramakrishnan, R., Livny, M.: BIRCH: an efficient data clustering method for very large databases. In: SIGMOD, pp. 103–114 (1996)

Mapping the Quality of Life Experience in Alfama: A Case Study in Lisbon, Portugal

Pearl May dela Cruz, Pedro Cabral, and Jorge Mateu

Instituto Superior de Estatística e Gestão de Informação, ISEGI,
Universidade Nova de Lisboa, 1070-312 LISBOA, Portugal
Departament de Matemàtiques, Universitat Jaume I,
Campus Riu Sec 12071 Castelló, Spain
pearlmayd@yahoo.com, pcabral@isegi.unl.pt,
mateu@mat.uji.es

Abstract. This research maps the urban quality of life (QoL) in Alfama, Lisbon (Portugal) through objective and subjective measures. A survey of 69 respondents and locations of social services were gathered suggesting the subjective and objective QoL respectively in the physical, economic, and social domains. The relationship between the two measures is examined using correlation analysis. It was determined that the association between them is weak and not significant, which could have been caused by the geographic scale and the sample size. These two factors also affected the spatial autocorrelation check implemented to the 15 subjective indicators using the Moran's I test. Out of 15, only 3 indicators were spatially autocorrelated. These 3 indicators were interpolated using Ordinary Kriging (OK). The rest is interpolated using the Voronoi polygon. All 15 prediction maps were used to create the overall subjective QoL using Weighted Sum procedure.

Keywords: Spatial Prediction Methods, Ordinary Kriging, Weighted Sum, Voronoi Polygon, Moran's I Test, Quality of Life.

1 Introduction

Quality of Life (QoL) is a multifaceted concept that has been a significant feature in research. Numerous authors focus their attention in establishing its definition and how it can be measured ([1]; [2]; [3]; [4]; [5]; [6]; [7]). In fact, [4] noted that it has been an explicit or implicit goal from an individual to the world.

QoL in urban areas is increasingly recognized by planners in assessing the urban environment. This is one of the major objectives of urban policies; create a better quality of life for the residents. As the level of urbanization has been increasing relentlessly at least in one part of a country, certain to transpire are its positive (i.e. employment creation) and negative aspects (i.e. growing issues of disorder, environmental degradation) that tend to influence the quality of life of residents. A rising need of QoL evaluation is on the mark.

Based on the worldwide Mercer 2010 Quality of Living Survey, Lisbon, the capital and the wealthiest part of Portugal, ranked the 45th out of 420 cities. Mercer (2010)

B. Murgante et al. (Eds.): ICCSA 2011, Part I, LNCS 6782, pp. 269–283, 2011.

defines Quality of Living based on indicators that are 'objective, neutral and unbiased'. QoL on the other hand, may also include subjective indicators based on people's perceptions and opinions. In this study, QoL is defined as a measure of objective and subjective features of life. The former involves concrete objects such as employment, level of education, family income and other physical, social and economic aspects that are quantifiable in nature; while the latter involves background perception of QoL based on personal and life history, attitudes, goals in life and emotional and physical well-being [8].

With Lisbon covering an area of 84.8 km^2, variability across the city exists. This in turn led to an interest to study Alfama which is described as a poor neighborhood. The main purpose is to assess the urban QoL in Alfama using objective and subjective QoL measures. Two null hypotheses are formulated prior to the study; that there is no linear relationship between objective and subjective QoL measures, and that subjective QoL measures are not spatially dependent. Consequently, the relationship between objective and subjective QoL measures are determined, and if the latter measure is spatially autocorrelated. The subjective indicators which have high priorities for each domain (physical, social, and economic) given by the respondents are also identified.

2 Study Area

Alfama is one of the districts of Lisbon that has no clearly defined boundaries, identifiable based on living standards, local characteristics, or structural morphology. In this case, since it has no clearly defined boundaries, only the two civil parishes São Miguel and Santo Estevão were included in the study, covering most parts of Alfama, see Fig. 1. The two parishes in this case are assumed to be equivalent to Alfama.

Alfama is known for its exceptional structural morphology due to its labyrinth-like narrow streets. It is recognized as the oldest district in Lisbon. But other than these distinctive characteristics, it is described to cater poor neighborhood. It has poor housing conditions where numerous are abandoned. The 2001 statistics shows that out of 655 building, 363 were already built before 1919. Although Alfama is a priority area for rehabilitation and urban renewal, degradation of houses is insurmountable and further rehabilitation is of great need.

With a total area of 254,593.4 m^2, it has a population of 3,726 as of 2009. It has a relatively old population to which 1,089 of the inhabitants have ages of 65 years and up. Concerning Alfama's educational attainment as of 2001, inhabitants of 3,188 were in the following categories: "don't know how to read and write", completed 1°/2 °/3 ° cycles of basic education or secondary education, signifying poor level of education.

Although with these dismal features, Alfama is one of the tourist attractions in Lisbon with several places to visit such as viewpoints (i.e. Miradouro de Santa Luzia), churches (i.e. Cathedral Sé de Lisboa), and various shops and Fado restaurants. It is also close to Lisbon commercial center that makes it in close proximity to trams, buses, metro and train stations. On the contrary, inside Alfama is the opposite. The central area is in fact not accessible by vehicles due to its narrow streets. These ambiguities and contradicting conditions of Alfama make it interesting as a study area for QoL assessment.

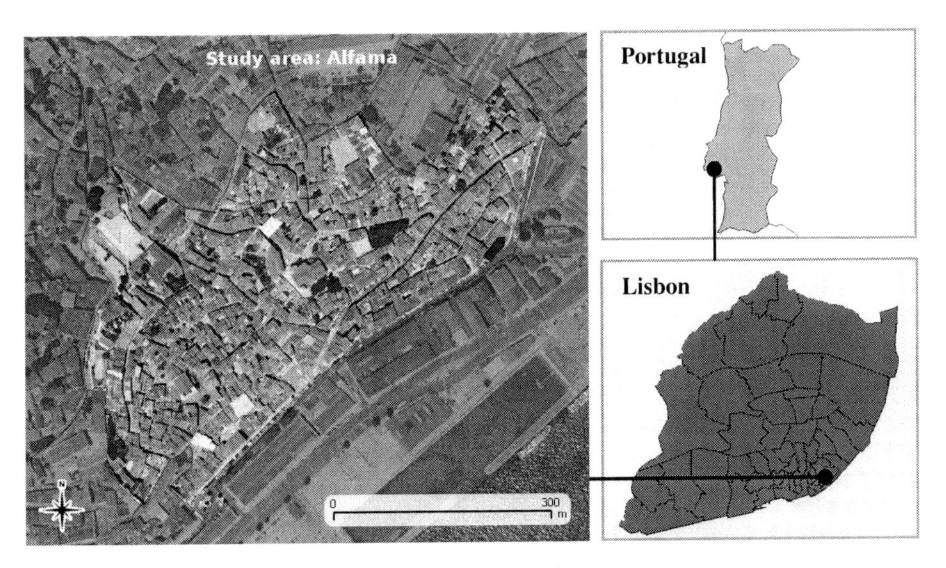

Fig. 1. Study area- Alfama

3 Data and Methods

3.1 Data

Two types of data were collected during the fieldwork, namely subjective and objective data. The former was gathered using a residential survey, which represented the perceptions of respondents towards the pre-selected 15 indicators of QoL (Table 2), abstracted using the Likert scale [7]. These 15 indicators/variables were grouped into three domains, the physical, social, and economic domains that are used for the creation of overall QoL map. The latter was provided through the GPS collection of service locations within and close to Alfama. These service locations were grouped together, which formed a total of 11 objective variables, see Table 1. The distances of these 11 groups of services from the respondents' residences were then established. The two types of data were tested for correlation, which determines the inclusion or exclusion of objective data in the creation of QoL map. The open-source software QuantumGIS (http://www.qgis.org/) and R (http://www.r-project.org/) are used simultaneously for the correlation, spatial autocorrelation, and spatial prediction analyses.

3.2 Correlation

A test of correlation is performed in the pursuit to interpolate QoL. It is a degree of relationship between two or more variables and represented using the correlation coefficient. The Spearman's rho is used to detect inter-correlations between subjective variables since these variables are ordinal and Spearman's rho is good for ordinal data. The formula (1) illustrates this correlation where r_s is the Spearman's rho, D^2 is the squared differences between ranks, and N is the number of cases [9].

$$r_s = 1 - \frac{6\sum D^2}{N^3 - N}. \tag{1}$$

The Polyserial correlation on the other hand is used to correlate objective and subjective variables since it is good for correlating a continuous variable with an ordinal one, and a certain degree of precision is needed. The formula is taken from the work of [10].

Significance testing is done to determine the statistical significance of the Polyserial correlation; that it occurs not by chance. With the hypotheses established previously and a significance level (α-level) set to 0.05, a p-value lower than 0.05 means the rejection of the null hypothesis (no linear relationship between objective and subjective data).

3.3 Spatial Autocorrelation

The variables are also tested for spatial autocorrelation, which determines the patterns of the values collected at locations [11]. It is calculated using the formula of Moran's I test (2).

$$I = \frac{n}{\sum_{i=1}^{n}\sum_{j=1}^{n} wij} \frac{\sum_{i=1}^{n}\sum_{j=1}^{n} w_{ij}(y_i - \overline{y})(y_j - \overline{y})}{\sum_{i=1}^{n}(y_i - \overline{y})^2}. \tag{2}$$

The formula consists of y_i which is the ith observation, $\overline{y}$ which is the mean of the variable of interest, and w_{ij}, which *is* the spatial weight of the link between i and j.

3.4 Spatial Prediction

Selecting the spatial prediction method used for each indicator is based on the modified decision tree (Fig. 2) of [6]. The Voronoi polygon is integrated in the model for non-spatially autocorrelated variables since it does not take into account neighborhood values (Costa, verbal communication, December 14, 2010). Since no correlations were found and only 3 variables were spatially autocorrelated, the Voronoi polygons and Ordinary Kriging (OK) are used for interpolation.

Voronoi polygon is used for the 12 non-spatially autocorrelated variables. It is the resulting area from a Voronoi diagram. In mathematical terms, say s_i is the nearest point from s or vice versa, the relation between is defined by formula (3) below [14]:

$$V(s_i) = \{x|\ \|x - x_i\| \leq \|x - x_i\| for\ j \neq I, j \in I_n\}. \tag{3}$$

where n is the finite number of points in the Euclidean plane, assumed to be $2 \leq n < \infty$; x is the Cartesian coordinates or (location vectors); and p is the n points to which the Euclidean distance from s to s_i is given by

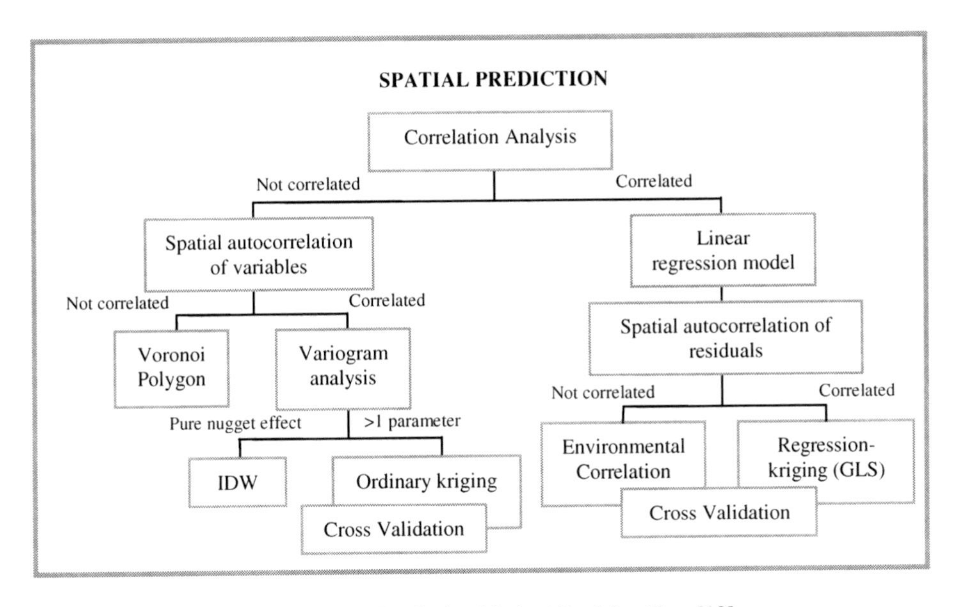

Fig. 2. Spatial Prediction Method Decision Tree [13]

$$d(s, s_i) = \|x - x_i\| = \sqrt{\left[(x_1 - x_{i1})^2 + (x_i - x_{i2})^2\right]}. \quad (4)$$

The Voronoi diagram then created by s is illustrated below (5):

$$v = \{V(s_i), \ldots, V(s_n)\}. \quad (5)$$

Ordinary Kriging (OK) is used for the 3 variables proven to be spatially autocorrelated. Since a variogram model is needed for this, a sample variogram is created, which is an estimate of the variogram using the N_h sample data pairs $Z(s_i)$, $Z(s_i + h)$ for a number of distances h_j by formula (6) [12].

$$\hat{\gamma}(\tilde{h}_j) = \frac{1}{2N_h} \sum_{i=1}^{N_h} (Z(s_i) - Z(s_i + h))^2, \quad \forall h \in \tilde{h}_j. \quad (6)$$

The variogram composed of the values of range, sill, and nugget is created by fitting a parametric model to it and fitted in the sample variogram. Since no variograms are created with a pure nugget effect, OK is used for the interpolation.

OK assumes an unknown constant trend and a known variogram, assuming mean and covariance stationarity and a normal distribution of values. It is based on the formula (7):

$$Z(s) = \mu + \varepsilon'(s). \quad (7)$$

where μ is the constant stationary function (global mean) and $\varepsilon'(s)$ is the spatially correlated stochastic part of variation [6]. With formula (7) as the base model, the prediction is done using formula (8):

$$\hat{Z}_{OK}(s_0) = \sum_{i=1}^{n} w_i(s_0) \cdot Z(s_i). \tag{8}$$

where w_i are the kriging weights [6].

The OK results are cross-validated to check the quality of maps, which works by producing two sets from dividing the data set consisting of the modeling set and validation set. The former is used for variogram modeling and kriging on the locations of the validation set, and the validation measurements are compared to their predictions [12]. The cross-validation (CV) method used is the leave-one-out, which uses each sampling point to determine the accuracy of prediction. Each sampling point is evaluated against the whole data set [13]. If the CV residuals are small, and the mean is close to zero and no evident structure [12], then it means the prediction performed is relatively accurate. The z-score of validation that takes into account the kriging variance (standardized residual) is also obtained to determine the performance of the variogram model in predicting the variables. If the z-score shows a mean and variance close to 0 and 1 respectively, it means that the variogram is correct. Otherwise, other factors such as anisotropy and nonstationarity might have not been considered. The z-score is calculated using formula (9):

$$Z_i = \frac{z(s_i) - \hat{z}_{[i]}(s_i)}{\sigma_{[i]}(s_i)}. \tag{9}$$

where $\hat{z}_{[i]}(s_i)$ is the cross validation prediction for s_i, and $\sigma_{[i]}(s_i)$ is the kriging standard error.

3.5 Weighted Sum

Using all prediction maps for the 15 QoL indicators using Voronoi polygons and OK, four maps are produced. These are the Physical, Social, Economic, and Overall QoL maps of Alfama. The Multi-Criteria Decision Method (MCDA) method called Weighted Sum is used; a way to transform the variables relative to its importance, by using 'weights' that are scaled, see formula (10) [15]:

$$J_{weighted\ sum} = w_1 J_1 + w_2 J_2 + \cdots + w_m J_m. \tag{10}$$

where w_i $(i = 1, \ldots, m)$ is the weighting factor for the ith objective function, and J are the variables. Note that the weights are acquired from the survey in which a column in the questionnaire is allocated for the scale of priority of respondents for each indicator and each domain and hence, the weights are subjective and relative to each indicator or each domain. These weights are normalized to get a sum of 1 for each domain and for the overall QoL.

4 Results

4.1 Survey Statistics

A total of 69 surveys for the subjective data were considered for analysis to which the respondents' residences are GPS located. There are 25 females and 44 males; most of them are 'Clerks and related workers' and 'Retired', with total individuals of 23 and 16 respectively. Respondents of 50 are also in the categories of 'did not study' or completed basic or secondary education, implying low educational attainment despite the age range of mostly 20-55.

The service locations on the other hand which represents the objective data were collected within and close Alfama. A total of 310 points were collected showing that only within the study area, 85 restaurants, 15 mini-markets, 4 bus and tram stops, 24 high and low-order shops, and other services were found, illustrating the touristic Alfama. On the contrary, based on the 2001 statistics, Alfama is said to be suffering from poverty with 58.6% of the residents having no economic activities.

4.2 Correlation

No significant correlations were found between objective and subjective QoL measures, see Table 1. It illustrates that the Polyserial correlations betweenn the 4 subjective physical indicators against the 11 objective variables show that p-values above 0.05 are obtained. Note that due to numerous indicators considered, not all results are shown in the paper including results from the Social and Economic indicators.

Table 1. Polyserial Correlation of Physical Indicators

Objective	**Subjective**							
Distance from a nearest service	Street cleanliness		Car circulation		Parking space		Green Space	
	Rho	P-value	Rho	P-value	Rho	P-value	Rho	P-value
Recycling bins	0.06	0.128	0.09	0.126	0.04	0.128	-0.05	0.130
Parking lots	-0.04	0.128	0.08	0.129	<-0.0	0.130	<0.0	0.131
Police stations	0.10	0.126	0.02	0.131	0.01	0.129	0.06	0.128
Recreational centers	0.05	0.127	0.26	0.119	0.07	0.127	0.05	0.127
Markets	0.18	0.123	0.16	0.125	-0.13	0.125	<0.0	0.129
Urban open spaces	0.10	0.126	-0.02	0.128	0.02	0.127	0.09	0.128
Main streets	-0.02	0.128	0.23	0.119	-0.02	0.128	-0.10	0.130
Transport facilities	-0.03	0.127	-0.16	0.124	-0.10	0.126	-0.10	0.128
Restaurants	0.13	0.125	0.20	0.12	-0.26	0.118	0.09	0.128
Institutions	-0.01	0.128	-0.16	0.124	0.04	0.128	0.33	0.112
High and low order shops	0.07	0.127	-0.07	0.126	-0.12	0.124	-0.02	0.129

The Spearman's rho results on the other hand, which are used to detect inter-correlations among the 15 subjective variables shows relatively moderate to high correlations with 9.4% of them having correlation coefficients higher than 0.40.

4.3 Spatial Autocorrelation and Spatial Prediction Maps

Out of the 15 subjective indicators, only 3 were spatially autocorrelated, highlighted in Table 2. These are Safety at Home, Public Transport Facilities Accessibility, and Recycling Bin Accessibility, which were interpolated using OK. The rest was done using Voronoi polygon.

The 12 non-spatially autocorrelated variables were predicted using the Voronoi polygon. Fig. 3 shows four of these maps under the Physical domain. Note that all maps in Fig. 3 were on a Likert scale with red areas meaning "extremely poor" and green areas meaning "excellent". Fig. 3 illustrates that the variable Street Cleanliness gained better assessment out of the other 3 variables. The variable Availability of Green Space on the other hand is largely poorly assessed.

Table 2. Moran's I Test for the 15 subjective indicators

Subjective Indicator	Moran's Index	P-value
Physical Domain		
Street cleanliness	-0.075618	0.453490
Car Circulation	0.084046	0.226065
Parking Space	0.040147	0.500866
Green Space	0.029617	0.584764
Social Domain		
Safety at Home	0.209053	0.005651
Safety at Streets	0.107389	0.134920
Health Care center Accessibility	-0.082397	0.405184
Supermarket Accessibility	-0.056243	0.608571
Public Transport Facilities Accessibility	0.215310	0.004544
Recreational Center Accessibility	0.027370	0.606232
Recycling Bin Accessibility	0.234164	0.002270
Neighborhood Interaction	-0.084477	0.388314
Economic Domain		
Level of Education	0.020318	0.664140
Affordability of Housing Cost	-0.011197	0.965526
Housing Quality	0.102984	0.147231

The experimental variograms necessary for OK are given in Fig. 4, showing that Safety at Home fits better with an Exponential model, a partial sill of 0.099, and a range of 12. The Public Transport Facilities Accessibility uses also an Exponential model with a partial sill of 0.0437, and a range of 26. The Recycling Bin Accessibility on the other hand fits better with a Spherical Model and a partial sill of 0.258, and a range of 38. All three models use 0 nuggets.

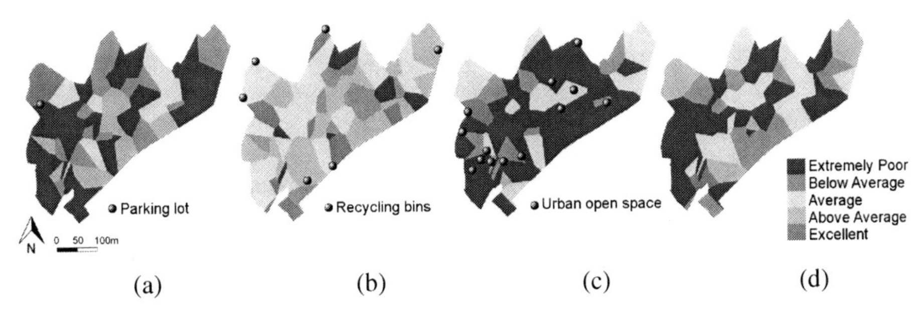

Fig. 3. Voronoi Polygons for Physical Indicators (a) Availability of Car Parking Space with the Parking Lot Location (b) Street Cleanliness with Recycling Bin Locations (c) Availability of Green Space with Urban Open Space Locations and (d) Car Circulation.

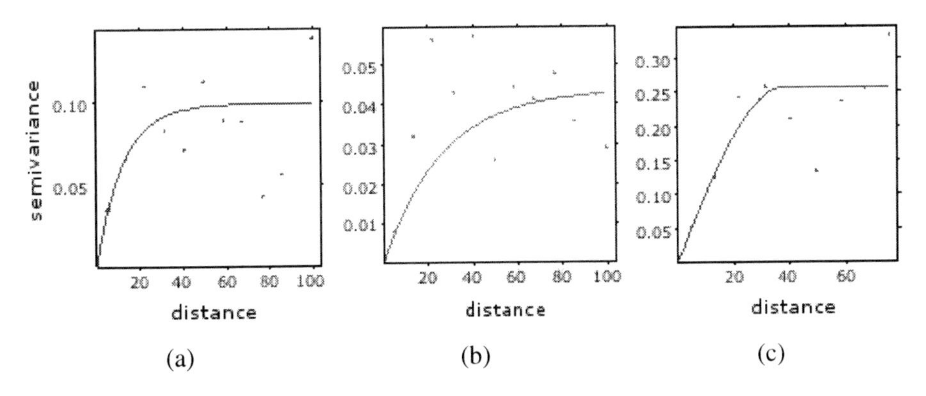

Fig. 4. Experimental Variogram Models for (a) Safety at Home (b) Public Transport Facilities Accessibility and (c) Recycling Bin Accessibility

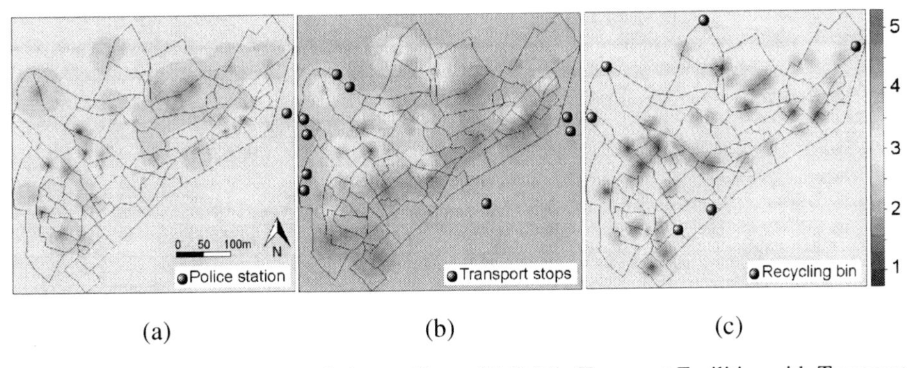

Fig. 5. Ordinary Kriging for (a) Safety at Home (b) Public Transport Facilities with Transport Stop Locations (c) and Recycling Bin Accessibility with Recycling Bin Locations

The three experimental variograms were used to create the OK for the variables in Fig. 5. Note that all have similar definition of scale with 1 indicating "extremely poor" and 5 indicating "excellent" assessments. Fig. 5a shows a relatively average

Safety at Home with above average assessments on the north and northeastern parts of the study area. Some random extremely poor areas also exist in the northwest and northeast. This means that the residents feel unsafe in these areas. Fig. 5b which illustrates the Public Transport Facilities Accessibility shows an above average assessment. Some extremely poor areas are in the northwest, and average and excellent assessments located in the northwest and in the south respectively. Fig. 5b shows that transport accessibility in Alfama is comparatively good. Fig. 5c on the other hand shows contradicting area assessments. With mostly average assessments, extremely poor and excellent areas exist together.

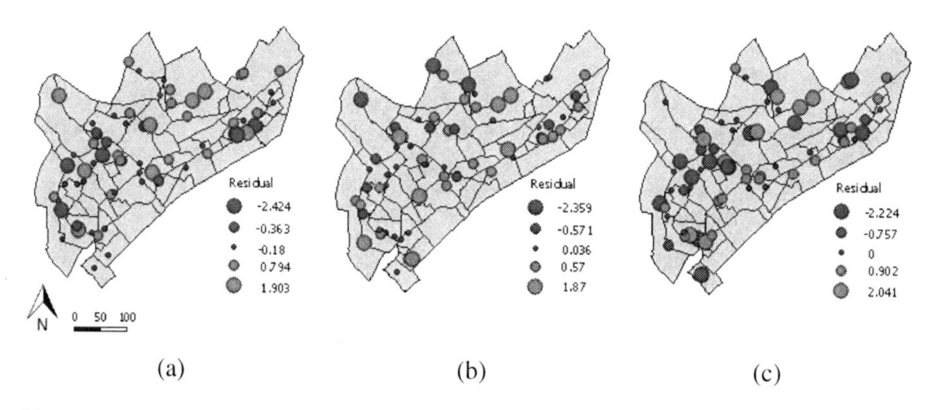

Fig. 6. Cross-validation for (a) Safety at Home (b) Public Transport Facilities (c) and Recycling Bin Accessibility

Table 3. Cross-validation summary results for (a) Safety at Home (b) Public Transport Facilities (c) and Recycling Bin Accessibility

	Prediction	Prediction Variance	Observed Values	Residual	Z-score
		Safety at Home			
Min.	1.705	0.02796	1.000	-2.42358	-10.056
Mean	3.222	0.08013	3.232	0.01018	0.01915
Max.	4.088	0.09329	5.000	1.90268	6.38808
		Public Transport Facilities			
Min.	2.440	0.009422	1.000	-2.35904	-11.301
Mean	3.681	0.038150	3.681	0.000343	0.009
Max.	4.682	0.051803	5.000	1.870253	8.942
		Recycling Bin Accessibility			
Min.	2.356	0.02796	1.000	-2.18341	-9.1994
Mean	3.160	0.08013	3.116	-0.04447	-0.0822
Max.	4.705	0.09329	5.000	2.06028	8.07613

The CV is done for the three OK prediction maps. Fig. 6 shows the CV residual maps to which the size of the points denotes the size of the residuals (i.e. the difference between original and predicted value. The green and red points indicate under and over-prediction respectively. Table 3 on the other hand shows the CV summary results for Fig. 6 which includes the prediction, prediction variance, observed values, residuals and z-score. Fig. 6a shows that the Safety at Home has the highest over-prediction residual of -2.42358 (Table 3a). Fig. 6b on the other hand gives better mean and z-score of 0.0003426 and 0.009 (Table 3b) respectively since it is closer to zero. Fig. 6c provides the highest under-prediction and z-score of 2.06028 and -0.0822 (Table 3c) correspondingly.

4.4 Weighted Sum

Fig. 7a which shows the Economic QoL formed from the weighted sum of 3 indicator maps shows below average assessment. It means that the majority of respondents perceive that the Economic QoL of Alfama is poor. Fig. 7b on the other hand illustrates good assessments with mostly average to above average Social QoL, formed from the weighted sum of 8 indicator maps. Some below average to extremely poor areas are also found in the north and northwest. Fig. 7c which shows the Physical QoL created from the weighted sum of 4 indicator maps, illustrates the worst evaluation compared to Economic and Social QoL. It shows large areas of below average and extremely poor Physical QoL. The Overall QoL in Fig. 7d, which is formed from the weighted sum of the 3 previous domain maps, show roughly below average to average assessment. It means that the Overall QoL of Alfama is poor based on the indicators considered and assessed by the respondents and needs further growth.

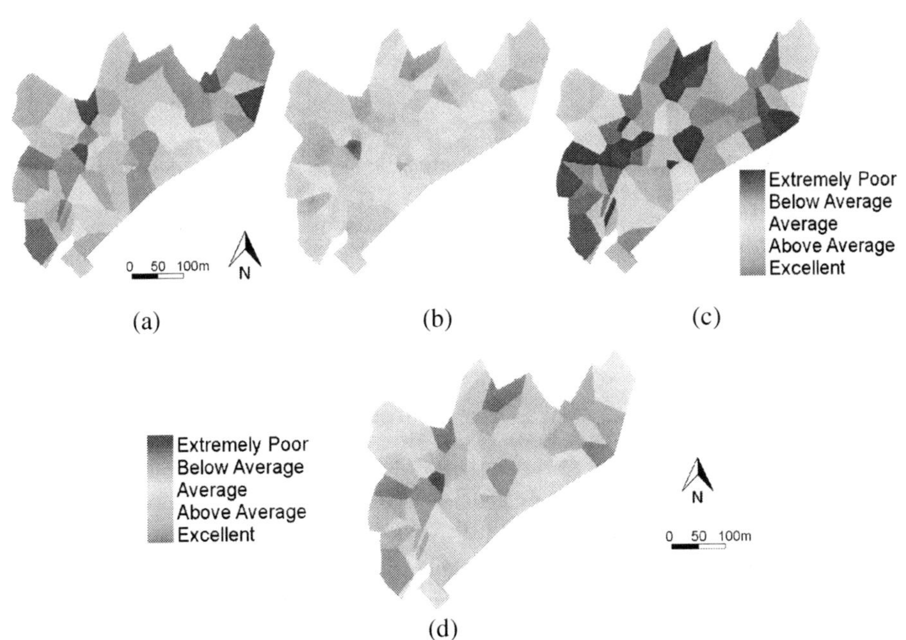

Fig. 7. Weighted Sum results for (a) Economic QoL (b) Social QoL (c) Physical QoL and (d) Overall QoL

5 Discussion and Conclusion

Similar to existing studies, the objective and subjective QoL measures have weak correlations. The significance testing resulted also to non-significant polyserial correlations. This however does not mean that correlations are equal to zero. It is just that based on the data, the study cannot tell whether they are different from zero. Remember the dictum "Absence of evidence is not evidence of absence [16]. Assuming that a Type II error (when the null hypothesis is accepted when in fact it should have been rejected) occurred in the hypothesis testing, several problems may have taken place. First is the sample size. With time limitation, only 69 respondents are collected, which is not even half of the mean sample size of 337 computed with a confidence level of 95%, a confidence interval of 5.1, and a total population of 3,726. Note that even more samples are necessary for spatial prediction. Enough and well-distributed samples are needed since larger sample size would mean smaller confidence interval, higher statistical power, and more reliable estimates.

Another problem is the issue of geographical scale. Costanza et al. [4] stated that choosing the scale is a matter of constructing the objectives in relation to identifying the composites and structures. They suggested that the focus should be on larger spatial regions or longer temporal scales to find statistical ensembles for which observations become more regular. They asserted that "moving between scales trade off the loss of heterogeneity for the gain of predictability". It means that the average within a large area is more precise than prediction at many locations. Although QoL assessment in small area such as Alfama is not a problem, precision of prediction is still affected since heterogeneity increased especially with the small sample size available. Notice that the Voronoi polygon maps in Fig. 3 and OK predictions in Fig. 5 show as well the objective data locations. These maps particularly the subjective Street Cleanliness (Fig. 3b), Safety at Home (Fig. 5a), and Recycling Bin Accessibility (Fig. 5c) somehow show visual correlation with the objective data, the closer the objective data (service locations), the better the subjective assessment. This on the other hand is affected by the heterogeneity of the answers and/or noises included in the analysis.

Particular problem on subjective QoL measurement also existed in the study. Aside from the fact that people have different criteria, scale of measurement and abrupt altering of answers, validity doubts also transpired in the experiment; a problem where questions in the survey may have suggested responses that are different to what the author has in mind [17]. Adding the fact that the survey is conducted in Portuguese, which is not the local language of the interviewer, communication was restricted. This however, does not mean that subjective QoL measure is unreliable and invalid since experimental studies such as of [18] suggested that population levels of life satisfaction were consistent.

Assuming that the results are true; that there is no significant correlation between objective and subjective QoL, it actually follows results of existing studies. This means that subjective measure has dissimilar criteria as compared to the objective measure. They are not comparable in this case. The Spearman's rho inter-correlation between subjective indicators also suggested similar results with existing studies since 9.4% of it gets at least 0.40 coefficient; indicating stronger correlation than between objective and subjective indicators. For the spatial autocorrelation on the other hand, similar problems were experienced akin to correlation.

The results for Voronoi polygon showed effectively the outliers and homogeneity or heterogeneity. It also visualizes abrupt change in the values giving a discontinuous surface, which is not realistic. The OK on the other hand shows continuous surface since it considers neighboring values in the prediction. The CV resulted to mean residuals and variances roughly close to 0 and 1 but elevated values of under-and over-prediction. This however does not mean that the variogram model is wrong. It can also be due to reasons such as anisotropy or nonstationarity that are not considered in the study, which are highly possible to exist in the data. Variogram model might also be improved by experimenting on the values of cutoff and lag distance for experimental variogram, and partial sill and range for semivariogram.

In the Physical Domain, the Car Circulation, Availability of Car Parking Space, and Green Space indicators follows the objective condition of Alfama since it is characterized by narrow streets impassable by vehicles, and roughly no space for parking or green areas. Street Cleanliness on the other hand shows better assessment since objectively, street cleaning schedule provided by the Municipal Office exists. In general, the Physical QoL of Alfama gets the worst assessment out of all the domains since the comprising indicators are also poorly assessed.

Given that the Social domain consists of 8 indicators, 3 out of 8 prediction maps were shown in the paper. OK predictions on Safety at Home, Public Transport Facilities Accessibility, and Recycling Bin Accessibility somehow show consistencies with the objective data; that the closer the objective data, the higher the assessments are.

The prediction results for the indicators of Economic Domain are not shown in the paper. Nevertheless, the map in Fig. 7a illustrates poor assessment, which follows the 2001 statistics of Alfama that the level of education, residents' economic activities, and housing quality, which are the indicators comprising this domain, are in poor conditions.

The overall QoL of Alfama, which is the combination of Physical, Economic, and Social domains shows below average to average assessments. The Physical and Economic domains mainly contributed to this poor assessment result, following the 2001 statistical description of Alfama that physical and economic conditions in particular need total improvements for the QoL development of the residents.

In general, QoL changes over time and thus, is only a depiction at a particular time. QoL is an intricate concept to which an actual definition and measurement are yet to be disputed. Continuous research particularly experimental studies are needed to explore and confirm existing theories and establish concrete methodology for QoL measurement.

6 Future Work and Recommendation

For future work, multiple correlations might improve the relationships between objective and subjective QoL. After using the entire samples, reduction of noises seen might also be done to determine if weak correlations were caused by these noises, which were included in the study due to lack of samples. Experimental and semivariogram models might also be improved by playing with cutoff and lag width, and sill and range respectively. Note that increasing or decreasing these parameters would also have consequences (i.e. decreasing lag width increase the detailed estimates but increasing as well the noise). Considering anisotropy and stationarity might also be done. Lastly, cluster analysis can be done for supplementary research, which will determine the group of people based on their profile which have similar perceptions on QoL.

Since the geographical scale and sample size have been the greatest issues in this research, it is recommended to have larger spatial region and enough and well-distributed samples. This would give higher predictability for the results. It is also suggested to involve the local government in determining which indicators are must-puts in the survey, which likely depends on the type of study area.

Acknowledgements

The work has been supported by the European Commission, Erasmus Mundus Programme, M.Sc. in Geospatial Technologies, project NO. 2007-0064. It could not have been written as well without the generous help of Kristina Helle (Institute for Geoinformatics, University of Muenster, Germany) and Oscar Vidal (Instituto Superior de Estatística e Gestão de Informação, ISEGI, Universidade Nova de Lisboa, Portugal.

References

1. Campbell, A., Converse, P., Rodgers, W.: The Quality of American Life: Perceptions, Evaluations and Satisfaction. Russel Sage Foundation, New York (1976)
2. Rogerson, R.: Quality of life and city competitiveness. Urban Studies 36(5), 969–985 (1999)
3. Tuan Seik, F.: Subjective assessment of urban quality of life in Singapore (1997-1998). Habitat International 24(1), 31–49 (2000)
4. Costanza, R., Fisher, B., Ali, S., Beer, C., Bond, L., et al.: Quality of lif: An approach integrating opportunities, human needs, and subjective well-being. Ecological Economics 61(2-3), 267–276 (2007)
5. Bonnes, M., Uzzell, D., Carrus, G., Kelay, T.: Inhabitants' and experts' assessments of environmental quality for urban sustainability. Journal of Social Issues 63(1), 59–78 (2007)
6. Das, D.: Urban quality of life: A case study of Guwahati. Social Indicators Research 88(2), 297–310 (2007)
7. Tesfazghi, E., Martinez, J., Verplanke, J.: Variability of quality of life at small scales: Addis Ababa, Kirkos sub-city. Social Indicators Research 98(1), 73–88 (2010)
8. Bowling, A.: Measuring Health: A Review of Quality of Life Measurement Scales, American Journal of Physics, Berkshire, UK (2005)
9. Paler-Calmorin, L., Calmorin, M.: Statistics in Education and the Sciences: With Application to Research, Rex Bookstore Inc., Quezon City (1997)
10. Olsson, U., Drasgow, F., Dorans, N.: The polyserial correlation coefficient. Psychometrika 47(3), 337–347 (1982)

11. Upton, G., Fingleton, B.: Spatial Data Analysis by Example: Point Pattern and Quantitative Data. Wiley, University of Michigan (1985)
12. Bivand, R., Pebesma, E., Gomez-Rubio, V.: Applied Spatial Data Analysis with R. Springer, Heidelberg (2008)
13. Hengl, T.: A practical Guide to Geostatistical Mapping, Office for Official Publications of the European Communities, Luxembourg (2009)
14. Okabe, A., Boots, B., Sugihara, K.: Spatial Tessellations: Concepts and Applications of Voronoi Diagrams. John Wiley & Sons, Chichester (1992)
15. Kim, I.Y., de Weck, L.: Adaptive weighted sum method for multiobjective optimization: a new method for Pareto front generation. Structural and Multidisciplinary Optimization 31(2), 105–116 (2005)
16. Altman, D., Bland, J.: Statistics notes: Absence of evidence is not evidence of absence. BMJ 311, 485 (1995)
17. Veenhoven, R.: Why social policy needs subjective indicators. Social Indicators Research, 33–45 (2004)
18. Cummins, R.: Objective and subjective quality of life: An interactive model. Social Indicators Research 52(1), 55–72 (2000)

Evolution Trends of Land Use/Land Cover in a Mediterranean Forest Landscape in Italy

Salvatore Di Fazio, Giuseppe Modica, and Paolo Zoccali

'Mediterranea' University of Reggio Calabria, Department of Agroforestry and Environmental Sciences and Technologies (DiSTAfA) www.distafa.unirc.it, loc. Feo di Vito c/o Facoltà di Agraria, 89122 Reggio Calabria, Italy
giuseppe.modica@unirc.it

Abstract. To understand the evolution trends of landscape, in particular those linked to urban/rural relations, is crucial for a sustainable landscape planning. The main goal of this paper is to interpret the forest landscape dynamics occurred over the period 1955÷2006 in the municipality of Serra San Bruno (Calabria, Italy), an area of high environmental interest. The peculiarity of the analysis is the high level of detail of the research (minimum mapping unit 0.2 ha). Data were obtained through the digitisation of historical aerial photographs and digital orthophotos by homogenising Land Use/Land Cover (LULC) maps according to the Corine Land Cover legend. The investigated period was divided into four significant time intervals, which were specifically analysed to detect LULC changes. The use of thematic overlays and transition matrices enabled a precise identification of the LULC changes that had taken place over the examined period. As a result, detailed description and mapping of the landscape dynamics occurred over the same period were obtained.

Keywords: Mediterranean forest landscape, Aerial photographs, Land Use/Land Cover (LULC) changes, Transition matrices, GIS, Calabria (Italy).

1 Introduction

Most of the actions leading to Land Use/Land Cover (LULC) changes have a local origin; yet, because of their speed, extension and intensity, they have numerous and important global implications, in particular on the carbon [15] and water cycles [41]. Today, these changes are considered as real environmental indicators [23] and are, therefore, the object of many researches. Though influenced by biophysical factors (soil conditions, topography, etc.), the result of LULC changes is mostly due to the human activity [15]. Over the last three centuries, Ramankutty and Foley [32] have observed that 1.2 million km² of forests and woods and 5.6 million km² of grasslands and pastures were converted into other land use types. Consequently, in order to understand LULC changes, the integration and interaction of human and biophysical factors must be taken into account [37]. Over the last decades, the progressive world population growth has implied an increasingly intensive exploitation of resources, with a growing demand for residential space and a corresponding increase in

B. Murgante et al. (Eds.): ICCSA 2011, Part I, LNCS 6782, pp. 284–299, 2011.

industrial activities. In this regard, in 1994, the International Geosphere-Biosphere Programme (IGBP) and the International Human Dimensions Program (IHDP) started the LUCC (Land-Use and Land-Cover Change) project to pursue two objectives: to identify the dynamics of LULC changes through regional comparative studies and to evaluate their impacts [20, 21, 41]. As far as landscape structure and configuration are concerned, certain LULC changes cause landscape fragmentation while others tend to increase homogeneity [14, 16]. LULC changes affect dramatically ecosystem structure and functioning [12, 45]; therefore, their analysis is one of the main research topics of ecology [7] and landscape ecology [47]. Moreover, the estimation of LULC status and change can provide crucial ecological information for science-oriented resource management and policy-making, according to a wide range of human activities [9]. The Mediterranean area is one of the most significantly altered hotspots on Earth [30]; the agricultural lands, evergreen woodlands and maquis habitats, so widespread in the whole Mediterranean basin, are the result of anthropogenic disturbances occurred over centuries or even millennia [6]. However, over the last decades, the most significant LULC changes have occurred as a consequence of a series of widespread and often connected phenomena: urban sprawl; agricultural intensification in the most suitable areas and agricultural abandonment in marginal areas; more frequent and more intense summer forest fires; and the rapid expansion of tourist activities and infrastructures, above all along the coasts [4, 8, 19]. Owing to this continuous anthropic pressure and to the different physical and climate conditions, further researches are needed to monitor and analyse, in quantitative and qualitative terms, the ongoing LULC changes [3, 34]. If, until some time ago, researches on the dynamics of LULC changes focussed on the urban landscape or on the consequences of deforestation [20, 43], nowadays researchers from various disciplines are conducting a great number of studies to understand the evolution of the rural landscape in relation to LULC changes [1, 11, 29, 36, 37, 46, 49, 50, 51] and are paying particular attention to the forest [17, 40] and agricultural aspects [48].

2 Material and Methods

2.1 Study Area

The research was carried out in the territory of the municipality of Serra San Bruno (Fig. 1), located in *Serre Vibonesi*. It is a mountainous area of Calabria (a region in the South of Italy), presenting many heritage resources of great natural, historic and architectural interest. The municipality has an area of 4,037.60 ha, a residential density of 178.5 inhab.$\cdot$km^{-2} and is situated at an average altitude of 980 m a.s.l. (min. 733 m, max. 1418 m). Century-old woods, characterized by the prevalence of chestnut (*Castanea sativa* Miller), beech (*Fagus sylvatica* L.) and silver fir (*Abies alba* Miller subsp. *apennina* Brullo, Scelsi and Spampinato) population, highlight the intrinsic suitability of this land for forestry. Starting from the 1950s, reforestation has been practised by planting conifers, such as the European black pine (*Pinus laricio* Poiret) and fir-douglas (*Pseudotsuga menziesii* (Mirb.) Franco), and then broad-leaved trees such as chestnut. For centuries, the forests were owned by the

Charterhouse of Serra San Bruno, which was founded by St Bruno of Cologne in the 11th century. They were managed efficiently and sensitively according to the methods of the Carthusian monks. Count Roger II gave St Bruno of Cologne the territories of *Serre Vibonesi* plateau to build his hermitage, the Charterhouse of Santo Stefano del Bosco, first Carthusian monastery in Italy and second in Europe after the one in Grenoble (France). The high environmental value of this area motivated the institution of two National Natural Reserves (now SCIs - Sites of Community Importance - of the Natura 2000 network, the centrepiece of EU nature and biodiversity policy) and of the Regional Natural Park of Serre in 2004.

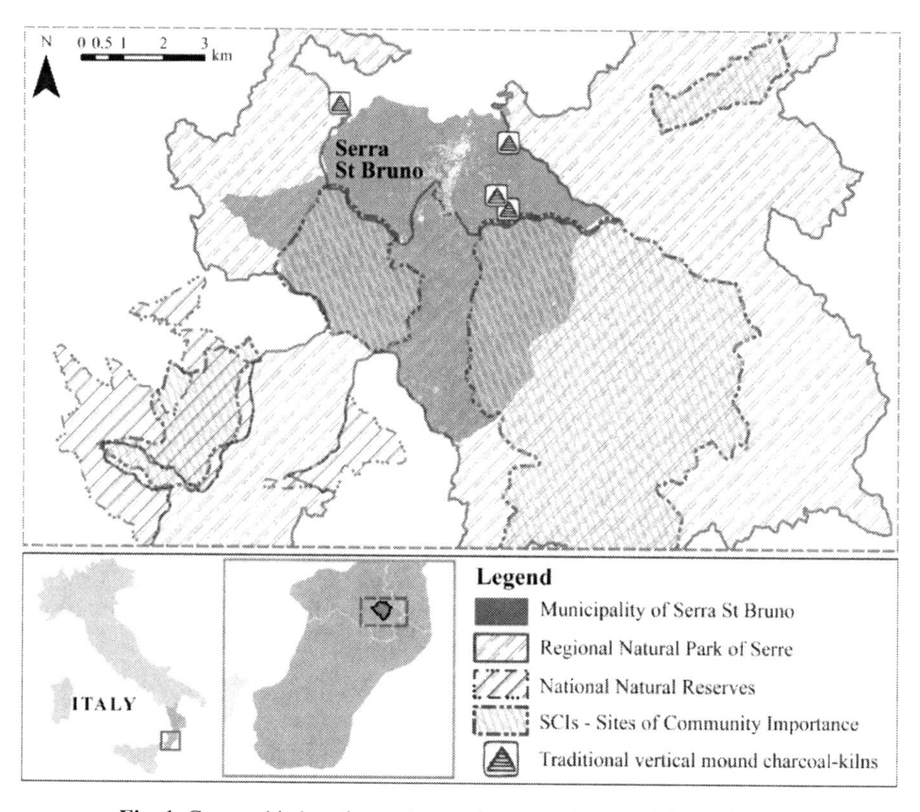

Fig. 1. Geographic location and natural protected areas of the study area

The local industrial archaeology heritage is also very significant. It is related to the utilization of water and wood as energy sources and dates back to the time of Bourbon rule in Calabria (XVIII-XIX centuries). Worth mentioning, in the same area, are the many historic monastic complexes, some of which are still inhabited by the original religious orders, either Catholic or Greek-Orthodox. They transferred in the area not only their spirituality, but also technological knowledge, thus becoming centres of cultural irradiation and civilization.

2.2 Aerial Image Processing and GIS Database Preparation

After identifying the study area, the first step towards the definition of LULC changes was the acquisition of the LULC maps referred to the different periods under investigation. All the cartographic data were implemented in an ArcGIS™ environment and anchored on the Gauss-Boaga conformal representation of the Italian National Geodetic System with UTM orientation at Monte Mario (Datum Rome 1940). LULC Changes were analysed using aerial photographs from the Italian Military Geography Institute (IGMI), for the years 1955 and 1983, and digital georeferenced orthophotos from the National Cartographic Portal (NCP) of the Italian Ministry of the Environment, for the years 1994 and 2006 (Tab. 1).

Table 1. Characteristics of the aerial photographs and orthophotos used in the research

Year	Frame data	Flight data	Source
1955	Sheet n° 246 Strip/Frames: 241/ 10369 10370 10371 Strip/Frames: 242/ 10418 10419 10420 Format: Digital – 600dpi	Height: 6000 m Scale: 1:36000 Date: 02 July 1955	Italian Military Geography Institute (IGMI) [www.igmi.org]
1983	Sheet n° 246 Strip/Frames: 16/ 534 535 536 537 538 Strip/Frames: 17/ 573 573 575 576 577 Format: Digital – 600dpi	Height: 4800 m Scale: 1:30000 Date: 06 Sept. 1983	
1994	B/W Aerial Orthophotos via GIS Server catalogue	Date: May 1994	National Cartographic Portal (NCP) of the Italian Ministry of the Environment, Land and Sea [www.pcn.minambiente.it]
2006	Colour Aerial Orthophotos via GIS Server catalogue	Date: 18 May 2006	

The 1955 and 1983 aerial photographs consisted of 9 contact copies (23 cm x 23 cm) scanned at 600x600 dpi resolution by IGM and provided in non-georeferenced and non-compressed TIFF graphic format. All the aerial photographs were georeferenced by using 20÷30 ground control points (GCP) each; the spatial resolution was 1.37 m with RMSE (Root Mean Square Errors) less than 6.5 m. For the purpose of georeferencing, a set of 1998 orthophotos and a 1:10,000 numerical topographic map were used as reference material. Geographical information was obtained from 1:25,000 and 1:10,000 topographic maps, in raster and vector format respectively, and from Digital Elevation Model (DEM) with 10 m x10 m spatial resolution. The thematic cartography was produced with a 1:10,000 nominal scale.

2.3 LULC Mapping

Commonly, LULC maps are produced either by aerial photo-interpretation or by classification of multi-spectral remotely sensed images. Although visual interpretation is a time-consuming technique requiring skilled analysts, it is still widely used [5] also because of the scarcity, if not lack, of historic satellite imagery. This method is characterised by a qualitative approach and is based on the visual recognition of the LULC classes, which have been defined in a specific study in relation to the chosen objectives and scale. The visual photo-interpretation of aerial photographs continues to be a standard tool for landscape mapping and monitoring at a detailed scale [24]. In this research, LULC mapping was conducted by visual photointerpretation, thanks to the availability of aerial photographs, referred to the period investigated, with good frequency and high spatial resolution. As in the case of other qualitative methods, there is no single ''absolute'' way to approach the interpretation process [22]. The basic characteristics of the aerial photo used when carrying out a visual interpretation are: shape, size, pattern, tone or hue, texture, shadows, geographic or topographic site and associations between features [22, 18].

Table 2. Correspondence between CORINE legend nomenclature and LULC classes used for detecting LULC changes. In brackets, the symbols utilized in following figures and tables.

CORINE Land Cover Classes		**LULC Classes used for data analysis**
Artificial surfaces	Historical buildings Continuous urban fabric	Continuous urban fabric (Urb-Cont)
	Discontinuous urban fabric Green urban areas	Discontinuous urban fabric (Urb-Disc)
	Industrial or commercial units Mining areas	Industrial, commercial and transport units (Ind)
Agricultural areas	Land principally occupied by agriculture, with significant areas of natural vegetation Pastures Non-irrigated arable land	Agricultural areas (Agric)
Forests and semi-natural areas	Coniferous forest Coniferous Reforestation	Coniferous forest (Con-For)
	Broad-leaved forest Broad-leaved mix forest Broad-leaved Reforestation	Broad-leaved forest (Blv-For)
	Mixed forest Rows of trees	Mixed forest (Mix-For)
	Transitional woodland shrub Open spaces with little or no vegetation	Shrub and/or herbaceous vegetation associations (Shrub)
Water bodies	Water bodies	Inland waters (Waters)

Photointerpretation was conducted by the same operator for all periods investigated (1955, 1983, 1994 and 2006). Digitization of the maps was made at a fixed scale of 1:1,000÷1:1,500 so as to minimize mapping errors. In order to reduce geometrical discordances, which could be interpreted as LULC changes, the other LULC maps were produced by updating 2006 LULC map, this representing a reference map. The minimum mapping unit was 0.2 ha. In addition, with a view to produce settlement-strata, the buildings in the study area were digitized according to all the time intervals defined. Accuracy assessment is very important to understand and use the results developed as a support of the decision-making process [25]. Most common features of the accuracy assessment include overall accuracy, producer's accuracy, user's accuracy and Kappa analysis [10]. In order to assess the thematic accuracy of the 2006 LULC map, a stratified random sampling was performed. In the summer of 2009, direct surveys were carried out in 100 sampling points distributed among the various LULC classes according to their surface share in landscape mosaic. The overall classification accuracy was 95.85% with a Kappa coefficient (K_{hat}) of 0.94. The reference legend is the one of the EU programme

CORINE (Coordination of Information on the Environment) at III level of detail [13]. In order to improve the interpretation of the results, the original 44 LULC classes of the CORINE classification system were aggregated in 9 classes so as to interpret changes as *change types* (Tab. 2). Urb-Disc and Urb-Cont classes were assigned when the urban structures and the transport network occupy 30÷80% or > 80% of the surface area respectively (Fig. 2).

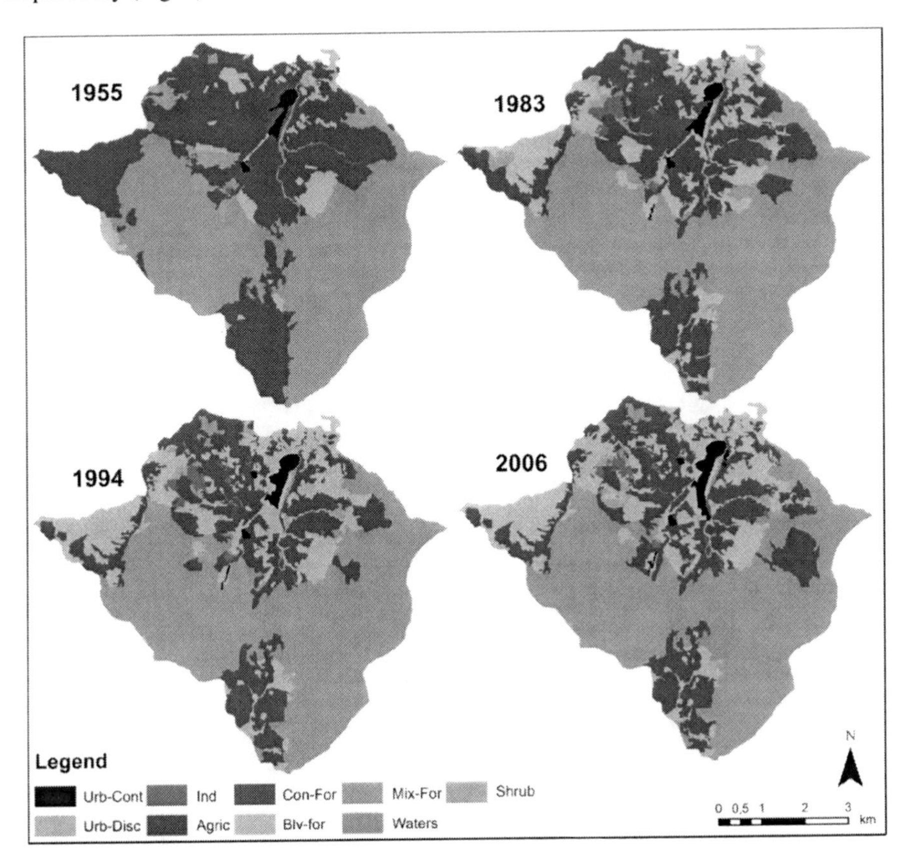

Fig. 2. LULC maps the study area for the limit-years of the time intervals investigated

2.4 Analysis of LULC Changes

After georeferencing and mapping LULC, the spatial comparison of LULC maps was performed for the years 1955, 1983, 1994 and 2006 by means of geoprocessing operations carried out on LULC vector maps. This is a post-classification comparison technique that can provide a complete matrix of change directions [25]. It is a thematic overlay mapping which enables to highlight the changes occurred in time both in qualitative terms, by showing them directly on the map, and in quantitative terms, by allowing to calculate total areal extent of land use change between two successive periods. Furthermore, it is possible to calculate and map which areas and, therefore, which LULC types have changed in the time interval examined. Change detection is the process of identifying differences in the state of an object or phenomenon by observing it at different times [37] and involves the application of multi-temporal datasets to quantitatively analyse the temporal effects of the phenomenon [22]. When implementing a change detection research project the following information should be provided [22]: area change and change rate; spatial distribution of change types; change trajectories of LULC types; and accuracy assessment of change detection results. Several authors [26, 28, 37] stress the importance of some aspects of change detection to be considered when monitoring natural resources: detecting if a change has occurred; identifying the nature of the change; measuring the areal extent of the change; and assessing the spatial pattern of the change. In this research, LULC changes were detected according to the 4 time intervals defined (1955÷1983, 1983÷1994, 1994÷2006 and 1955÷2006) by means of cross-tabulation analysis. All the land use maps were in vector format.

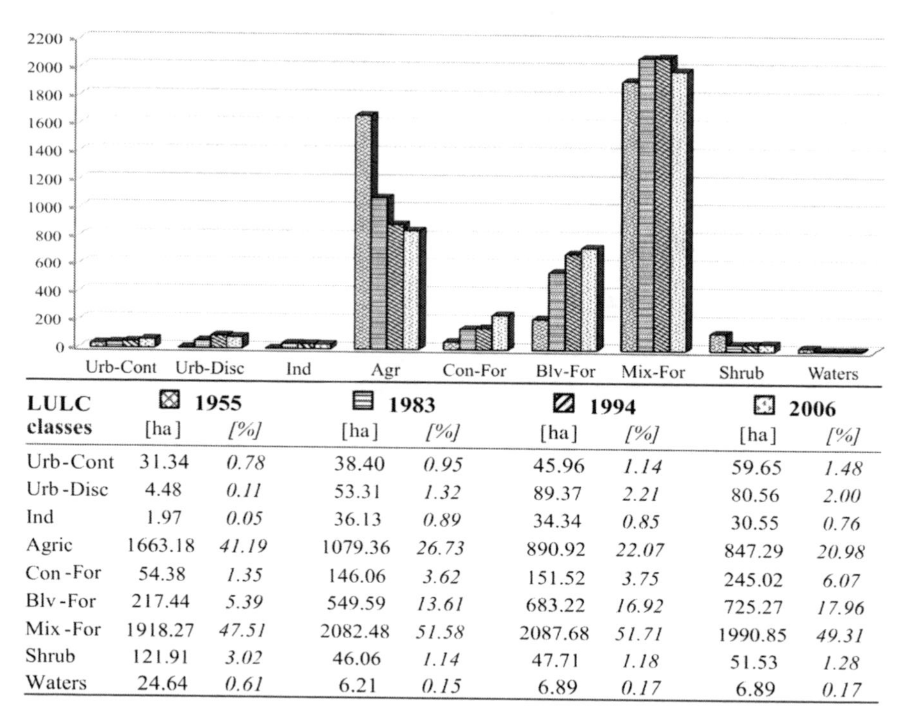

LULC classes	1955 [ha]	1955 [%]	1983 [ha]	1983 [%]	1994 [ha]	1994 [%]	2006 [ha]	2006 [%]
Urb-Cont	31.34	*0.78*	38.40	*0.95*	45.96	*1.14*	59.65	*1.48*
Urb-Disc	4.48	*0.11*	53.31	*1.32*	89.37	*2.21*	80.56	*2.00*
Ind	1.97	*0.05*	36.13	*0.89*	34.34	*0.85*	30.55	*0.76*
Agric	1663.18	*41.19*	1079.36	*26.73*	890.92	*22.07*	847.29	*20.98*
Con-For	54.38	*1.35*	146.06	*3.62*	151.52	*3.75*	245.02	*6.07*
Blv-For	217.44	*5.39*	549.59	*13.61*	683.22	*16.92*	725.27	*17.96*
Mix-For	1918.27	*47.51*	2082.48	*51.58*	2087.68	*51.71*	1990.85	*49.31*
Shrub	121.91	*3.02*	46.06	*1.14*	47.71	*1.18*	51.53	*1.28*
Waters	24.64	*0.61*	6.21	*0.15*	6.89	*0.17*	6.89	*0.17*

Fig. 3. Landscape composition in Serra St. Bruno in the four periods under investigation. Data in [ha] and in [%].

Relative and absolute changes for each of the 9 land cover types were calculated from 9x9 transition matrices. The transition matrices (or cross-tabulation matrices) are built for the time interval t_1 to t_2, wherein the rows display the categories of time 1 and the columns display the categories of time 2. Row and column vectors have a precise meaning: a row vector shows the evolution of a land use type in the period $t_1 \div t_2$. A column vector shows the land use type at time t_1, from which another land use type was generated at time t_2. The data of the main diagonal, shown in bold in the transition matrices, indicate the area that was not subject to LULC changes (persistence). In addition to the absolute and percentage change rates defined for each period under investigation (Tab. 6), the average annual percentage change was calculated for each LULC class. Among the several indices proposed by various authors, due to its explicit ecological meaning this research utilized the *single land-use dynamic degree* (r) proposed by Puyravaud [31]. In formula:

$$r = \frac{1}{t_2 - t_1} \cdot \ln\left(\frac{A_{t_2}}{A_{t1}}\right) \cdot 100 \quad (1)$$

A_{t_1} : Area of the land use type in time t_1 ; A_{t_2} : Area of the land use type in time t_2 .

3 Results and Discussions

By 1955, the study area had a quite homogeneous structure, with two definitely prevailing LULC classes: Agric (41.19%) and Mix-For (47.51%). The urban fabric was almost completely continuous (Urb-Cont) and coincided with the historical nucleus (0.89%). In 2006, the situation was considerably different. Agricultural and forest areas as a whole occupied 94.32% of the study area, even though they were differently distributed in comparison with the first period investigated: agricultural areas passed from 41.19% in 1955 to 20.98% in 2006, while forest areas passed from 54.25% in 1955 to 73.34% of the total area in 2006 (Fig. 3 and 4). The urban fabric extended, thus creating discontinuous belt areas (Urb-Disc) between the rural and the urban space. Overall, continuous and discontinuous urban areas occupied 3.48%.

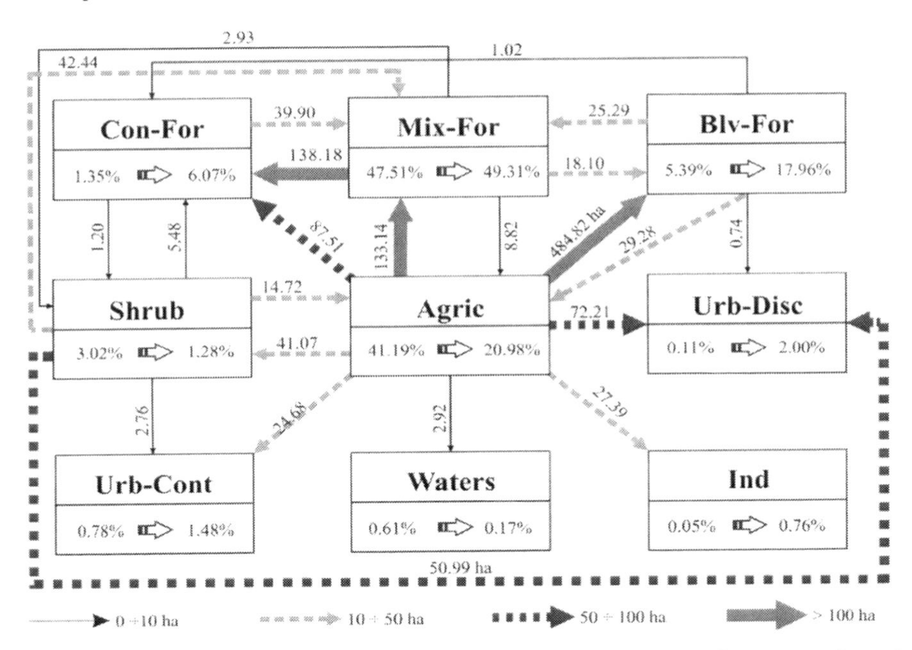

Fig. 4. LULC dynamic networks of 1955÷2006 changes. The sequence of percentage in each LULC class type indicates its share on the entire landscape composition analysed in 1955 and in 2006, respectively. Numbers of [ha] marked above the arrows represent the surface changed from one LULC class type to another.

Table 3. LULC changes for each LULC class defined in the time intervals under investigation. Data in [ha] and in [%].

	1955÷1983		1983÷1994		1994÷2006		1955÷2006	
LULC classes	[ha]	%	[ha]	%	[ha]	%	[ha]	%
Urb-Cont	7.07	*22.55*	7.56	*19.67*	5.59	*12.16*	20.21	*64.49*
Urb-Disc	48.83	*1090.30*	36.06	*67.64*	-8.75	*-9.79*	76.14	*1699.98*
Ind	34.16	*1734.16*	-1.79	*-4.95*	-4.10	*-11.94*	28.27	*1435.21*
Agric	-583.82	*-35.10*	-188.44	*-17.46*	-43.58	*-4.89*	-815.85	*-49.05*
Con-For	91.71	*168.65*	5.43	*3.72*	93.53	*61.73*	190.67	*350.63*
Blv-For	332.15	*152.75*	133.63	*24.32*	42.23	*6.18*	508.02	*233.63*
Mix-For	164.20	*8.56*	5.20	*0.25*	-96.74	*-4.63*	72.67	*3.79*
Shrub	21.40	*86.87*	-39.14	*-85.03*	0.00	*0.00*	-17.74	*-72.03*
Waters	-115.70	*-94.90*	41.50	*668.04*	11.82	*24.78*	-62.38	*-51.17*

3.1 Area Change and Change Rates of LULC Classes

Artificial areas. The urban sprawl of the study area extends along the Ancinale river and two roads, which start from the historical nucleus (Fig. 2), go through Serra San Bruno and connect it with the surrounding urban centres. Urbanization has concerned and is still concerning almost flat areas (slope ≤10%). The urban fabric reveals a complex dynamic throughout the period investigated and only the analysis of intermediate periods allows to fully understand it. The Urb-Disc class is particularly interesting. It is characterised by a dramatic growth and mostly occupies (Tab. 3÷4) former agricultural lands, with an area passing from 4.48 ha in 1955 to 80.56 ha in 2006 (r +5.67) (Tab. 5). In particular, the areas of greatest expansion of the Urb-Disc class are those next to the historic centre, where they form a continuously evolving outer urban belt. Moreover, the discontinuous urban fabric is still changing: since the '90s, it has been tending to be incorporated in the Urb-Cont class and, in the period 1994÷2006, 10.59 ha of this class, i.e. 11.85%, passed to the Urb-Cont class. The trend of the total number of buildings further confirms what has been stated above: 331 in 1955, 1011 in 1983, 1829 in 1994 and, finally, 2057 in 2006. However, this does not correspond to the demographic trend, considering that the resident population passed from 8,517 inhabitants in 1961 to 6,955 in 2010. The highest rate of housing construction was recorded between 1983 and 1994, when, owing to the temporary lack of effective planning regulations, many buildings were constructed in unsuitable areas. Housing construction did not meet a residential need, but rather it answered the purpose to invest in the property market in a period of high monetary inflation. Between 1994 and 2006, a significant downturn in the construction of new buildings was recorded, which did not affect agricultural areas, as it had happened in the past, but rather the open spaces in discontinuous urban areas (10.59 ha of the Urb-Disc class turned into Urb-cont). In 1997, the General Master Plan was adopted as a tool, which has been recently updated, that should help direct new urbanization processes towards areas that are not intended for agricultural and forest use.

Agricultural areas. Agricultural areas were the most concerned by the changes occurred in the period under investigation. They passed from 1663.18 ha in 1955 to 847.29 ha in 2006, with an overall variation of -49.05% (r –1.26%) (Tab. 5).

Considering the composition of the examined landscape, the area occupied by the Agric class almost halved, passing from 41.19% in 1955 to 20.98% in 2006. In particular, the agricultural area was incorporated in two classes (Fig. 3): 484.82 ha passed to the Blv-For class and 133.14 ha passed to the Mix-For class (Table 5). This dynamic is due to the intensification of silvicultural activities (mostly funded at first by national and regional public funds and then by EU funds) and to a gradual agricultural abandonment. Therefore, the LULC changes described above combine with those related to the abandonment of the Agric class, consequently occupied by shrub and herbaceous vegetation (Shrub). Such dynamics are confirmed by the Agricultural Censuses conducted by ISTAT for the years 1970, 1982, 1990 and 2000. The analysis of data shows that the number of farms has halved, passing from 503 in 1970 to 228 in 2000[1] (-54.7%). Correspondingly, UAA (Utilized Agricultural Area) has dropped, though less dramatically, from 1027.63 ha in 1970 to 716.74 ha in 2000 (- 30.25%), with a consequent increase (+53.9%) in the average area per farm (from 2.04 ha to 3.14 ha). This was also due to public measures of financial assistance to specific interventions of land consolidation.

Forestry areas. Forestry areas have always been of crucial importance in the structural dynamics concerning the territory of Serra San Bruno. They are exactly the formation that replaced, in time, a considerable part of the area of the Agric class. Today, 705.48 ha (86.46%) of the 815.94 ha of the Agric class, which have changed over the fifty years investigated, are forestry areas, 52.72 of which are the result of reforestation. In particular, the most significant increase was recorded for the Blv-For class, which passed from 217.44 ha in 1955 to 725.27 ha in 2006 (r +2.36%) (Table 5). The Mix-For class, which is made up of mixed formations of beech and silver fir combined with small nuclei of European black pine, is characterized by the largest area among all those defined in this research. Over the fifty years examined, its area has not been subject to evolution trends as significant as those of the other forestry areas and, therefore, it has kept a relatively constant share in the composition of the *Serre Vibonesi* landscape. It passed from 1918.27 ha in 1955 (47.51% of the whole municipal territory) to 1990.85 in 2006 (49.31% of the total) (Fig. 3) with r equal to +0.07%. The Con-For class is less important than the two classes mentioned above: over the period investigated, it has passed from 54.38 ha in 1955, 1.35% of the total, to 245.05 ha in 2006, 6.07% of the total (r +2.95).

Shrub and Waters. Shrub formations are characterized by the dominance of shrub species typical of the Mediterranean maquis of sub-mountain and mountain areas. Distinctive species are the Common Broom (*Cytisus scoparius* L.), the Common Bracken (*Pteridium aquilinum* (L.) Kuhn), the Hairy Broom (*Cytisus villosus* Pourr.) and the Tree Heath (*Erica arborea* L.). The Shrub class is a LULC type halfway between the dynamics of agricultural areas and those of forestry areas. As a matter of fact, from the point of view of the ecological succession, it is the first phase of the process of regaining abandoned agricultural lands by forest vegetation. In Calabria, this succession has often been accelerated by reforestation, which, after the Second World War, was carried out in large areas. The analysis of data in Fig. 3 shows that this class has passed from 121.91 ha in 1955 to 46.06 ha in 1983, when reforestation was very intensive.

---

[1] At present, these are the latest available data, since those of the 6th census (2010) are not published yet (http://en.istat.it/censimenti).

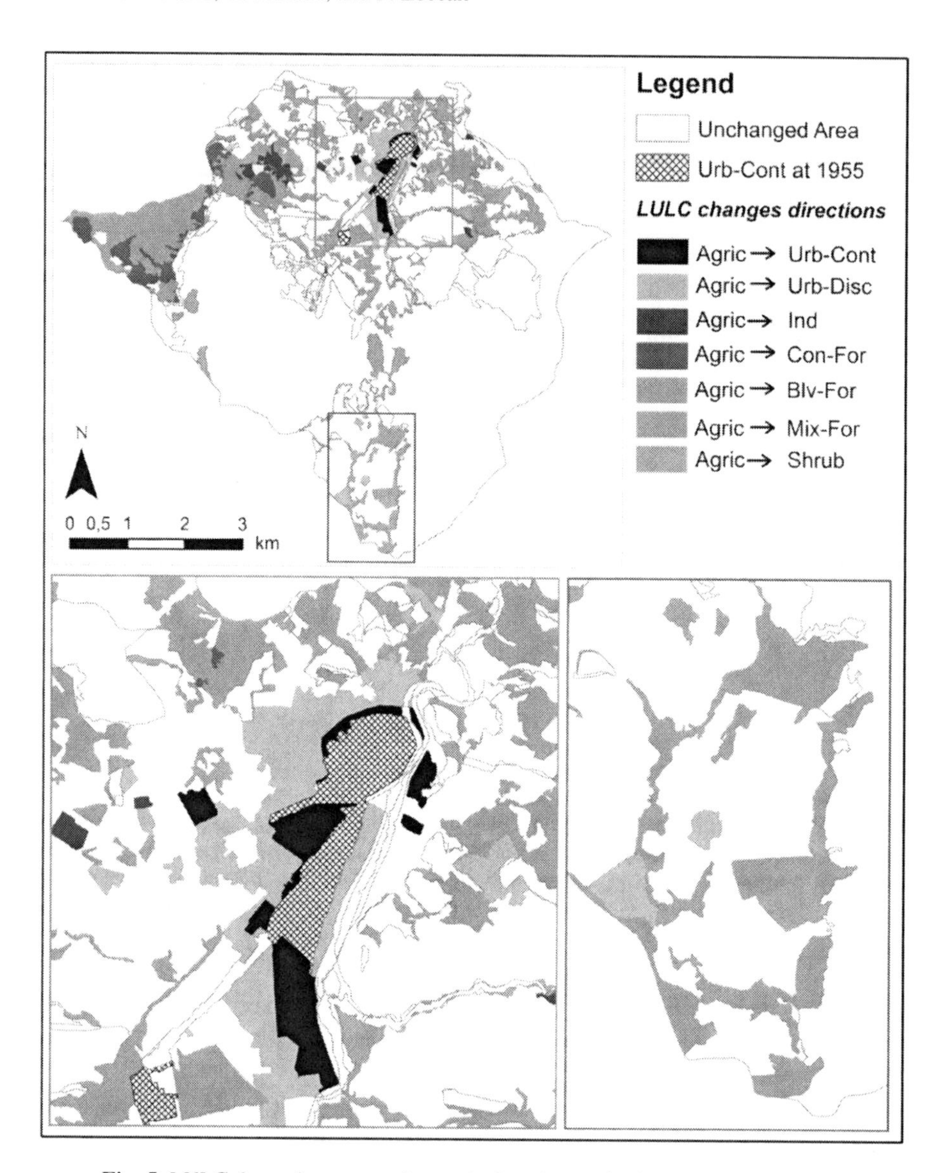

Fig. 5. LULC dynamics concerning agricultural areas in the period 1955÷2006

This reduction was the determinant of the two opposite trends highlighted above. In fact, on the one hand, 39.34 ha have passed from the Agric class to the Shrub class; on the other hand, 94.18 ha of the Shrub class have become forestry areas (Tabb. 4÷5). In 1994, a further drop was recorded (47.71 ha), while, in 2006, there was a slight increase (51.53 ha) (Fig. 3). Finally, as far as the Waters class is concerned, worth mentioning is a phenomenon, which is often disregarded in literature but deserves to be considered in the studies about LULC changes characterized by a high level of detail. Actually, compared to 1955, the area of this class diminished in the following periods. Nevertheless, this is due to the development of thick hygrophilous vegetation that covers most streambeds and is classified as Blv-For in the photointerpretation process.

Table 4. Transition matrices for the intermediate time intervals defined

1955 \ 1983	Urb Cont	Urb Disc	Ind	Agric	Con For	Blv For	Mix For	Shrub	Waters	Sum 1955
Urb-Cont	**27.27**	1.13	-	1.66	-	-	-	-	1.28	*31.34*
Urb-Disc	-	**4.47**	-	-	-	-	-	-	-	*4.47*
Ind	-	-	**1.97**	-	-	-	-	-	-	*1.97*
Agric	8.95	45.85	33.15	**991.46**	78.83	335.52	128.63	39.34	1.45	*1663.18*
Con-For	-	-	-	1.91	**12.17**	1.20	39.10	-	-	*54.38*
Blv-For	-	0.16	-	36.48	1.05	**135.07**	43.18	1.44	-	*217.44*
Mix-For	1.70	-	1.02	15.06	48.56	21.16	**1829.93**	0.84	-	*1918.27*
Shrub	-	-	-	23.53	5.59	47.95	40.64	**4.21**	-	*121.91*
Waters	0.39	1.69	-	9.25	-	8.73	0.84	0.28	**3.44**	*24.63*
Sum 1983	*38.32*	*53.31*	*36.19*	*1079.35*	*146.20*	*549.63*	*2082.31*	*46.11*	*6.16*	***4037.60***

1983 \ 1994	Urb Cont	Urb Disc	Ind	Agric	Con For	Blv For	Mix For	Shrub	Waters	Sum 1983
Urb-Cont	**33.90**	2.65	-	-	-	0.91	0.95	-	-	*38.40*
Urb-Disc	2.60	**48.87**	-	1.39	-	0.45	-	-	-	*53.31*
Ind	-	-	**22.92**	7.40	0.94	1.42	0.26	3.19	-	*36.13*
Agric	6.17	34.21	8.09	**823.58**	22.35	134.77	38.20	8.16	3.81	*1079.36*
Con-For	-	-	-	2.40	**96.29**	20.05	26.37	0.96	-	*146.06*
Blv-For	1.99	2.00	2.47	34.91	5.57	**466.08**	18.95	17.60	-	*549.59*
Mix-For	-	-	0.48	19.53	26.00	31.14	**2002.68**	2.63	-	*2082.48*
Shrub	-	-	0.25	1.52	0.51	28.36	0.26	**15.16**	-	*46.06*
Waters	1.48	1.63	-	-	-	0.11	-	-	**2.99**	*6.21*
Sum 1994	*46.14*	*89.37*	*34.22*	*890.72*	*151.65*	*683.31*	*2087.68*	*47.71*	*6.80*	***4037.60***

1994 \ 2006	Urb Cont	Urb Disc	Ind	Agric	Con For	Blv For	Mix For	Shrub	Waters	Sum 1994
Urb-Cont	**45.51**	-	-	0.45	-	-	-	-	-	*45.96*
Urb-Disc	10.59	**77.52**	-	1.26	-	-	-	-	-	*89.37*
Ind	-	-	**24.01**	0.73	-	5.25	0.61	3.73	-	*34.34*
Agric	2.41	1.91	4.73	**814.54**	0.10	47.35	7.40	12.47	-	*890.92*
Con-For	-	-	-	-	**135.26**	-	16.26	-	-	*151.52*
Blv-For	1.14	1.14	1.25	20.21	0.97	**648.96**	1.60	7.95	-	*683.22*
Mix-For	-	-	0.13	5.55	108.68	8.03	**1964.69**	0.60	-	*2087.68*
Shrub	-	-	0.43	4.54	-	15.67	0.29	**26.77**	-	*47.71*
Waters	-	-	-	-	-	-	-	-	**6.89**	*6.89*
Sum 2006	*59.65*	*80.56*	*30.55*	*847.29*	*245.02*	*725.27*	*1990.85*	*51.53*	*6.89*	***4037.60***

Table 5. Transition matrix for the main time interval (1955÷2006)

1955 \ 2006	Urb Cont	Urb Disc	Ind	Agric	Con For	Blv For	Mix For	Shrub	Waters	Sum 1955
Urb-Cont	**30.98**	-	-	-	-	-	-	-	0.36	*31.34*
Urb-Disc	-	**4.48**	-	-	-	-	-	-	-	*4.48*
Ind	-	-	**1.97**	-	-	-	-	-	-	*1.97*
Agric	24.68	72.21	27.39	**789.43**	87.51	484.82	133.14	41.07	2.92	*1663.18*
Con-For	-	-	-	-	**12.86**	0.36	39.90	1.20	-	*54.38*
Blv-For	-	0.74	-	29.28	1.02	**160.78**	25.29	0.33	-	*217.44*
Mix-For	0.27	-	0.75	8.82	138.18	18.10	**1749.21**	2.93	-	*1918.27*
Shrub	2.76	0.28	-	14.72	5.48	50.99	42.44	**5.25**	-	*121.91*
Waters	1.07	2.76	0.25	4.98	-	10.41	0.78	0.76	**3.61**	*24.64*
Sum 2006	*59.76*	*80.47*	*30.41*	*847.24*	*245.05*	*725.46*	*1990.77*	*51.55*	*6.89*	***4037.60***

Table 6. Mean annual rate of LULC changes calculated by *r* index as suggested by Puyravaud (2003). Data in [%]

LULC classes	1955÷1983	1983÷1994	1994÷2006	1955÷2006
Urb-Cont	0.73	1,63	2,17	1,26
Urb-Disc	8.84	4,70	-0,86	5,67
Ind	10.39	-0,46	-0,97	5,38
Agric	-1.54	-1,74	-0,42	-1,32
Con-For	3.53	0,33	4,01	2,95
Blv-For	3.31	1,98	0,50	2,36
Mix-For	0.29	0,02	-0,40	0,07
Shrub	-3.48	0,32	0,64	-1,69
Waters	-4.92	0,94	0,00	-2,50

4 Conclusions

This study highlights how understanding the evolution trends of landscape, particularly those related to urban/rural relations, is crucial for a sustainable landscape planning. The research is part of a wider study and is aimed at quantifying and interpreting LULC changes in a significant area and time interval in the context of the Mediterranean forest landscape. To that end, in a first phase, the integration of spatial, historical (aerial photographs) and recent (orthophotos) data with socio-economic data in a geodatabase at a detailed scale proved to be of great importance to understand the relations between land-use, -cover and -functions [44]. The observation and temporal analysis of LULC changes represent a procedure that allows to identify on-going trends and to better understand landscape evolution. Therefore, it is fundamental to implement tools that may anticipate future scenarios. Considering that land cover changes did not occur at equal rates during all time intervals, it is important to analyse intermediate time intervals in order to understand the evolution trends of the landscape under investigation. Furthermore, the analysis of LULC changes allows the temporal verification of the positive (e.g. reforestation of agricultural areas, land consolidation, etc.) or negative (lack of landscape planning tools, effects of agricultural abandonment, etc.) effects of past territorial policies. Finally, these analyses can allow to implement landscape management tools and sustainable medium- and long-term landscape planning so as to direct future changes towards those which had positive effects and to contrast the actions which had negative effects on the landscape in the past. This would have significant impacts not only at a local scale, which was the scope of this research, but also on regional-scale plans and programmes. Future developments of the research will include the use of VHR satellite remote sensing data (such as Quickbird, Ikonos, WorldView, etc.) and landscape metrics. Moreover, the comprehension of the influence of climate on LULC changes should be implemented so as to acquire data and information allowing to evaluate the effects of climate change in the long-term [1].

References

1. Acampora, A., Sole, A., Carone, M.T., Simoniello, T., Manfreda, S.: Le Metriche del Paesaggio come Strumento di Analisi del Territorio. In: Casas, L., et al. (eds.) Informatica e Pianificazione Urbana e Territoriale (Informatics and Landscape and Urban Planning), Proceedings of 6th INPUT Conference, Potenza, Italy, vol. 1, pp. 221–231 (2010)
2. Alados, C., Pueyo, Y., Barrantes, O., Escós, J., Giner, L., Robles, A., et al.: Variations in landscape patterns and vegetation cover between 1957 and 1994 in a semiarid Mediterranean ecosystem. Landscape Ecol 19(5), 545–561 (2004)
3. Antrop, M.: Conservation of biological and cultural diversity in threatened Mediterranean landscapes. Landscape and Urban Plan 24, 3–13 (1993)
4. Antrop, M.: Landscape change and the urbanization process in Europe. Landscape and Urban Plan 67(1-4) (2004)
5. Asner, G.P., Keller, M., Pereira, R., Zweede, J.C.: Remote sensing of selective logging in Amazonia—Assessing limitations based on detailed field observations, Landsat ETM+, and textural analysis. Remote Sens. Environ. 80, 483–496 (2002)
6. Blondel, J.: The 'design' of Mediterranean landscapes: a millennial story of humans and ecological systems during the historic period. Hum. Ecol. 34, 713–729 (2006)
7. Braimoh, A.K.: Random and systematic land-cover transitions in northern Ghana. Agric. Ecosyst. Environ. 113, 254–263 (2006)
8. Burke, S.M., Thornes, J.B.: A thematic review of EU Mediterranean desertification research in Frameworks III and IV: Preface. Adv. Envir. Monit. Model, pp. 11–14 (2004)
9. Cihlar, J.: Land cover mapping of large areas from satellites: status and research priorities. Int. J. Rem. Sens. 21, 1093–1114 (2000)
10. Congalton, R.G., Green, K.: Assessing the Accuracy of Remotely Sensed Data: Principles and Practices, vol. 2. CRC Press, Boca Raton (2008)
11. Cousins, S.: Analysis of land-cover transitions based on 17th and 18th century cadastral maps and aerial photographs. Landscape Ecol. 16(1), 41–54 (2001)
12. DeFries, R.S., Foley, J.A., Asner, G.P.: Land-use choices: balancing human needs and ecosystem function. Frontiers in Ecology and the Environment 2(5), 249–257 (2004)
13. European Environment Agency 2000.: CORINE land cover technical guide – Addendum. Technical report n. 40 (2000)
14. Farina, A.: Principles and methods in landscape ecology. Chapman & Hall, London (1998)

15. Foley, J.A., Defries, R., Asner, G.P., Barford, C., Bonan, G., Carpenter, S.R., et al.: Global consequences of land use. Science 309(5734), 570–574 (2005)
16. Forman, R.T.: Land mosaics. Cambridge University Press, Cambridge (1995)
17. Geri, F., Rocchini, D., Chiarucci, A.: Landscape metrics and topographical determinants of large-scale forest dynamics in a Mediterranean landscape. Landscape and Urban Plan 95(1-2), 46–53 (2010)
18. Gomarasca, M.: Basics of Geomatics, p. 656. Springer, Heidelberg (2009)
19. Kosmas, C., Danalatos, N.G., Gerontidis, S.: The effect of land parameters on vegetation performance and degree of erosion under Mediterranean conditions. Catena 40, 3–17 (2000)
20. Lambin, E., Rounsevell, M., Geist, H.: Are agricultural land-use models able to predict changes in land-use intensity? Agr. Ecosyst. Environ. 82(1-3), 321–331 (2000)
21. Lambin, E.F., Geist, H.J. (eds.): Land-Use and Land-Cover Change. Local Processes and Global Impacts. Global Change. The IGBP Series. Springer, Heidelberg (2006)
22. Lillesand, T., Kiefer, R., Chipman, J.: Remote sensing and image interpretation, 5th edn., p. 784. Wiley, New York (2003)
23. Lindquist, E.J., Hansen, M.C., Roy, D.P., Justice, C.O.: The suitability of decadal image data sets for mapping tropical forest cover change in the Democratic Republic of Congo: implications for the global land survey. Int. J. Remote Sens. 29(23–24), 7269–7275 (2008)
24. Loveland, T.R., Sohl, T.L., Stehman, S.V., Gallant, A.L., Sayler, K.L., Napton, D.E.: A strategy for estimating the rates of recent United States land-cover changes. Photogramm Eng. Rem. S. 68(10), 1091–1099 (2002)
25. Lu, D., Mausel, P., Brondízio, E., Moran, E.: Change detection techniques. Int. J. Remote Sens. 25(12), 2365–2401 (2004)
26. Macleod, R.D., Congalton, R.G.: A quantitative comparison of change- detection algorithms for monitoring eelgrass from remotely sensed data. Photogrammetric Photogramm. Eng. Rem. S. 64, 207–216 (1998)
27. Munsi, M., Malaviya, S., Oinam, G., Joshi, P.K.: A landscape approach for quantifying land-use and land-cover change (1976–2006) in middle Himalaya. Reg. Environ. Change 10(2), 145–155 (2009)
28. Malila, W.: Comparison of the information contents of Landsat TM and MSS data. Photogramm Eng. Rem. S. 51(9), 1449–1457 (1985)
29. Matsushita, B., Xu, M., Fukushima, T.: Characterizing the changes in landscape structure in the Lake Kasumigaura Basin, Japan using a high-quality GIS dataset. Landscape and Urban Plan 78(3), 241–250 (2006)
30. Myers, N., Mittermeier, R.A., Mittermeier, C.G.: Biodiversity hotspots for conservation priorities. Nature 403, 853–858 (2000)
31. Puyravaud, J.P.: Standardizing the calculation of the annual rate of deforestation. Forest. Ecol. Manag. 177, 593–596 (2003)
32. Ramankutty, N., Foley, J.A.: Estimating historical changes in global land cover: Croplands from 1700 to 1992. Global Biogeochemical Cycles 13(4), 997 (1999)
33. Read, J.M., Lam, N.S.N.: Spatial methods for characterizing land cover and detecting land cover changes for the tropics. Int. J. Remote Sens. 23, 2457–2474 (2002)
34. Scarascia-Mugnozza, G., Oswald, H., Piussi, P., Radoglou, K.: Forest of the Mediterranean region: gaps in knowledge and research needs. Forest Ecol. Manag. 132, 97–109 (2000)
35. Schneider, L.C., Pontius, R.G.: Modeling land-use change in the Ipswich watershed, Massachusetts, USA. Agr. Ecosyst. Environ. 85, 83–94 (2001)

36. Serra, P., Pons, X., Sauri, D.: Land-cover and land-use change in a Mediterranean landscape: A spatial analysis of driving forces integrating biophysical and human factors. Appl. Geophys. 28(3), 189–209 (2008)
37. Singh, A.: Digital change detection techniques using remotely sensed data. International Int. J. Remote Sens. 10(6), 989–1003 (1989)
38. Sluiter, R., Jong, S.M.: Spatial patterns of Mediterranean land abandonment and related land cover transitions. Landscape Ecol. 22(4), 559–576 (2006)
39. Sommer, S., Hill, J., Megier, J.: The potential of remote sensing for monitoring rural land use changes and their effects on soil conditions. Agr. Ecosyst. Environ. 67, 197–209 (1998)
40. Teixido, A.L., Quintanilla, L.G., Carreño, F., Gutiérrez, D.: Impacts of changes in land use and fragmentation patterns on Atlantic coastal forests in northern Spain. J. Environ. Manage. 91(4), 879–886 (2010)
41. Turner, M.G., Ruscher, C.L.: Changes in landscape patterns in Georgia, USA. Landscape Ecol. (1988)
42. Turner II, B.L., Skole, D., Sanderson, S., Fischer, G., Fresco, L., Leemans, R.: Land-use and Land-cover Change Science/Research Plan, IGBP Report No. 35 (1995)
43. Verburg, P.H., Veldkamp, A., Willemen, L., Overmars, K.P., Castella, J.: Landscape Level Analysis of the Spatial and Temporal Complexity of Land-Use Change. In: DeFries, R., Asner, G., Houghton, R. (eds.) Ecosystems and Land Use Change. Geoph. Monog. Series, vol. 153, pp. 217–230 (2004)
44. Verburg, P.H., van de Steeg, J., Veldkamp, A., Willemen, L.: From land cover change to land function dynamics: A major challenge to improve land characterization. J. Environ. Manage. 90, 1327–1335 (2009)
45. Vitousek, P., Mooney, H., Lubchenco, J., Melillo, J.: Human Domination of Earth's Ecosystems. Science 277(5325), 494–499 (1997), doi:10.1126/science.277.5325.494
46. Weng, Q.: Land use change analysis in the Zhujiang Delta of China using satellite remote sensing, GIS and stochastic modelling. J. Environ. Manage. 64(3), 273–284 (2002)
47. Wu, J., Hobbs, R.: Key issues and research priorities in landscape ecology: an idiosyncratic synthesis. Landscape Ecol. 17, 355–365 (2002)
48. Wu, J., Cheng, X., Xiao, H., Wang, H., Yang, L., Ellis, E.: Agricultural landscape change in China's Yangtze Delta, 1942–2002: A case study. Agr. Ecosyst. Environ. 129(4), 523–533 (2009)
49. Zhang, J., Zhengjun, L., Xiaoxia, S.: Changing landscape in the Three Gorges Reservoir Area of Yangtze River from 1977 to 2005: Land use/land cover, vegetation cover changes estimated using multi-source satellite data. Int. J. Appl. Earth. Obs. 11(6), 403–412 (2009)
50. Zomeni, M., Tzanopoulos, J., Pantis, J.: Historical analysis of landscape change using remote sensing techniques: An explanatory tool for agricultural transformation in Greek rural areas. Landscape and Urban Plan 86(1), 38–46 (2008)
51. Zomer, R.J., Ustin, S.L., Carpenter, C.C.: Land Cover Change Along Tropical and Subtropical Riparian Corridors Within the Makalu Barun National Park and Conservation Area, Nepal. Mountain Research and Development 21(2), 175–183 (2001)

Application of System Dynamics, GIS and 3D Visualization in a Study of Residential Sustainability

Zhao Xu[1] and Volker Coors[2]

[1] Politecnico di Torino, DITAG- Land Environment and Geo-Engineering Department
Via Pier Carlo Boggio, 61, 10138 Torino, Italy
zhao.xu@polito.it
[2] University of Applied Sciences Stuttgart Germany
Faculty Geomatics, Computer Science and Mathematics
Schellingstr. 24, 70174 Stuttgart, Germany

Abstract. Constructing and improving urban residential areas is an eternal critical subject in the process of the whole urban development which is connected with a series of challenges and problems. In this paper, firstly DPSIR (Driving Forces-Pressure-State-Impact-Response) framework has been employed for better systematizing the indicators on residential sustainability. Due to the urban activities cause impacts not only on local level but also a broader scale, a simulation model, using System Dynamics (SD) methodology, is structured to quantitatively investigate the developmental tendency of the indicators. And then the estimated results were shown in 2D density maps in ArcGIS and 3D visualization in CityEngine. The integration of GIS, SD model and 3D, called GISSD system here, can better explain the interaction and the variation in time of the sustainability indicators for residential development. Hence it's able to support the Decision Maker to view the sustainable level of urban residential areas more comprehensively.

Keywords: sustainability, residential areas, System Dynamics model, GIS, density map, 3D visualization, CityEngine.

1 Introduction

Sustainability is a multi-dimensional concept that takes into account different elements of territorial development, such as economic growth, well-being of population, environmental quality, etc [1]. Since the early '90s many countries and international organizations have been working on sustainable development assessment by means of specific indicators [2-4]. With specific reference to urban areas, the indicator approach is useful to give information about the sustainability condition of the system under examination and it can be used in order to make previsions about future trends on sustainability [5-7]. Urban residential areas, facing restriction by social-economic level, environmental pressure, population pressure and traffic pressure etc, also attract growing attentions nowadays as an important component of sustainability study. This is of particular importance in the context of

B. Murgante et al. (Eds.): ICCSA 2011, Part I, LNCS 6782, pp. 300–314, 2011.

emerging countries, where large urban development are going on very quickly and the necessity of tools being able to predict the future sustainability levels is real.

This paper proposes a GISSD system, by integrating Geographic Information System (GIS), System Dynamics (SD) model and 3D visualization, designed for sustainability assessment of urban residential development, linking residential housing prediction and sustainability indicators in four main sectors: housing, society, economics and environment. System Dynamics is a realistic tool for sustainability assessment, utilized to better understand the sustainable development in a considered period and forecast the future trends, while GIS provides a consistent visualization environment for displaying the input data and results of a model which is a very useful ability in a decision-making process. Forecasts for future evolution must be made available for assisting policy makers. Although this is a hard task due to the hypercomplexity of the systems, it is necessary to provide the decision makers with assessments regarding the future [8]. In the context of housing supply as the connection between the two different methods GIS and SD, the paper attempts to provide a tool that is able to inform the Decision Maker on whether residential urban areas develop sustainably or not, and provide more information about housing equilibrium.

2 Background

2.1 A GISSD System

The GISSD system specifically presented in this study, as shown in Figure 1 is mainly composed of two aspect: the forecasting function of the SD model, and then, the visualization of spatial pattern for residential development which is supported in GIS technology. Firstly a sustainability indicator system was constructed in a Driver forces–Pressures–State-Impact-Response (DPSIR) framework with expert opinions. Related statistic data and GIS data were collected from many sources, for example yearly reports on local economic-social development published by the State Statistic Bureau of Baden-Württemberg. Subsequently based on the simulation results obtained in the SD model, on the other hand, the estimated stock of housing supply as one of many simulation results was input into GIS analysis to generate new 2D density maps of residential distribution using ArcGIS 10. Finally 3D visualization of the residential buildings and houses was simulated in a software- CityEngine.

2.2 DPSIR Framework

The DPSIR approach, formerly developed by OECD(1993) in the PSR form, was used to highlight relationships between human activity and environment degradation [9]. The DPSIR framework is able to illustrate the complexities of the system interactions in sustainable development. It is based on a concept of causality: human activities exert pressures on the environment and change its quality and the quantity of natural resources. Society responds to these changes through environmental, general economic and sector policies. The latter form a feedback loop to pressures through human activities [9-11]. According to the DPSIR framework there is a chain of causal links, namely, starting with " Driving forces" through "Pressures" to

"States" and "Impact", eventually leading to "Response" [12]."State" is the result of specific "Driver forces" and "Pressure" which impacts "Environment". "Response" represent the solutions (e.g. policies, investments) for what should then be done to improve or maintain that state.

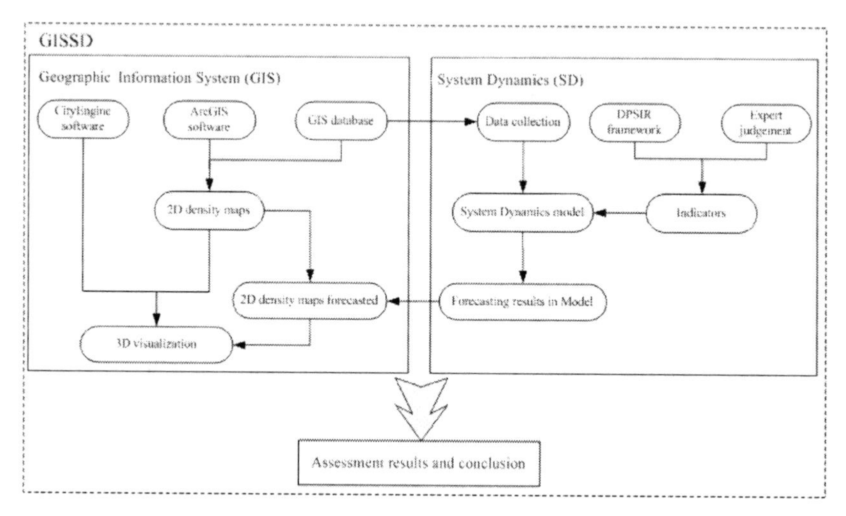

Fig. 1. Conceptual model of GISSD system

3 Study Area and Data

Stuttgart Region as one of the four administrative regions (Regierungsbezirke) of the state, is located in the north-east of Baden-Württemberg in the southern Germany. It is sub-divided into the three regions Heilbronn-Franken, Ostwürttemberg and Stuttgart, among which Stuttgart is the capital city of the state of Baden-Württemberg and the sixth-largest city in Germany with a total of 23 city districts, 5 inner districts and 18 outer districts. As shown in Figure 2, Plieningen is a southern district of Stuttgart city and 10 kilometers from the city center with a district area of around 13.07 square kilometers and 13.035 thousand residents. The population of Stuttgart Region rose from 3,751 million in 1991 to 4.001 million in 2009 according to the population report of the State Statistic Bureau of Baden-Württemberg.

The evaluation of sustainability index and housing equilibrium mainly investigated in this study is strongly related to the economic-environmental-social variables such as disposable income per capita, population density, family size, etc in the whole Stuttgart Region circumstance. All the data used in the modeling was taken from the "Structural and Regional Database" (the State Statistic Bureau of Baden-Württemberg) [13], the "State Database" (the State Statistic Bureau of Baden-Württemberg) [13], the "Statistic report of Baden-Württemberg for housing sample" (the State Statistic Bureau of Baden-Württemberg) [13], the "Economic and social development in Baden-Württemberg" (the State Statistic Bureau of

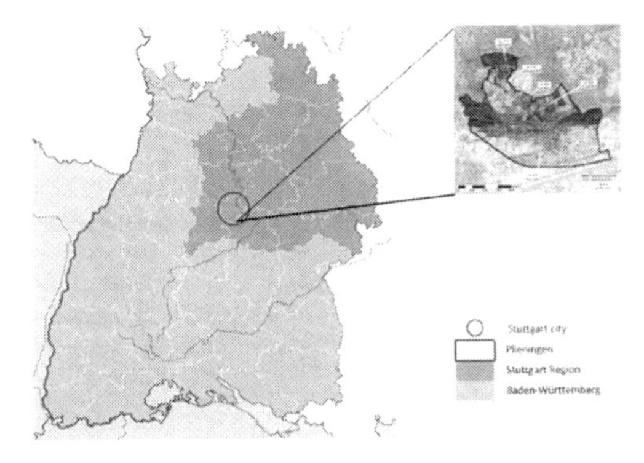

Fig. 2. Location of the study area

Baden-Württemberg) [13], the "Economic facts and figures Baden-Württemberg 2010" (Ministry of Economy Baden-Württemberg) [14], the "Energy report 2010" (Ministry of Economy Baden-Württemberg) [14], and the "State development report Baden-Württemberg (LEB) 2005" (Ministry of Economy Baden-Württemberg) [14], etc.

4 GISSD System

4.1 Sustainability Indicators Selection in DPSIR Framework

An indicator is a parameter which is associated with an environmental phenomenon, which can provide information on the characteristics of the event in its global form [3]. Many indicators are available for sustainable development assessment. Among the several indicator systems, mention can be made to the following three sets related to international and European organizations that work in the field of sustainable development: the Organization for Economic Cooperation and Development (OECD) environmental indicator system [3], the United Nations Commission on Sustainable Development (UNCSD) indicator system [15], and the European System of Social Indicators (EUSI) [16].

The availability of data and the sensitivity of indicators to reflect the underlying social and economical processes were viewed as the criteria to establish the indicator system proposed in this work. Economics, Environment and Society are 3 main aspects considered and analyzed individually in the current field of sustainable development. In this study, Housing aspect is added as the fourth designed to especially emphasize the living quality and housing equilibrium. From indicators identified in the three indicator-systems, we selected a synthetical and concise indicators system which is suitable for dealing with sustainability assessment of urban residential areas. Tab 1 lists all the 24 indicators selected for the application of the sustainability assessment of urban residential areas considering the DPSIR framework. In this way, the overall system is described as different layers: categories of the DPSIR framework, thematic areas and indicators.

Table 1. Indicators selected for the application

DPSIR categories	Indicator	Unit	Thematic area
Driving Forces	• I1 Gross Domestic Product (GDP);	billion €	Economics
	• I2 Disposable income per capita;	€	Economics
	• I3 Investment on the residential projects completed;	€	Economics
	• I4 Urban population;	million inhabits	Society
Pressures	• I5 Housing demand;	million m^2	Housing
	• I6 Family size;	inhabits/household	Environment
	• I7 Population density;	inhabits/km^2	Society
	• I8 Unemployment rate;	%	Society
	• I9 Road traffic accidents.	inhabits	Society
	• I10 Total motor vehicles;	million	Society
State	• I11 Living space per capita;	m^2	Housing
	• I12 Land price per floor area	€ per one m^2	Economics
	• I13 Urban housing price index;	None	Economics
	• I14 rent- price index;	None	Economics
	• I15 Population below the poverty line	inhabits	Society
Impact	• I16 Urban air pollution (NO2;SO2;PM10);	ton	Environment
	• I17 Domestic water consumption;	m^3	Environment
	• I18 Delivering quantity of domestic waste;	ton	Environment
	• I19 Green coverage ratio;	%	Environment
	• I20 Urban area;	km^2	Environment
Response	• I21 Completed area of residential projects;	€	Housing
	• I22 Housing supply;	million m^2	Housing
	• I23 New residential projects investment;	€	Economics
	• I24 Environmental regulation investment	€	Economics

4.2 System Dynamics Model

In this study, a SD model designed for Stuttgart Region was employed to provide a means of simulating urban development and its internal interactions among 24 sustainability indicators. The model has a 30-year time horizon, from 1991 to 2020, with reference to the related statistic reports published by the State Statistical Bureau of Baden-Württemberg. And the indicators are subdivided into four primary model sectors, called sub-systems, which are the Housing sector, the Economics sector, the Environment sector and the Society sector in accordance with the thematic areas in Table 1.

In the developing stage, the structure of the SD model is represented as a set of stocks and information flows, and the according stock-flow diagrams of four sectors shown in Figure 3 were developed using a SD software Vensim from Ventana Systems, Inc. The model is composed of, in the aggregate, 58 variables including 4 stock variables, 6 rate variable and 48 convertor variables in which the 24 indicators selected to develop sustainability assessment are also contained. Each variable in the model was defined with a unique formula expression.

Housing sector

Housing sector is of great significance in this SD model. The demand of house purchasers mainly come from 4 sources: 1, the growth of nonagricultural population (natural demand); 2, the growth of Stuttgart's per capita living area (initiative demand); 3, demolishment and relocation of old houses (passive demand); and 4, housing purchase of immigrated population. According to the national land use planning 2020 from the

Germany Federal Office for Building and Regional Planning (BBR), the number of households in Baden-Württemberg will have a strong rise in medium-term, while the average household size correspondingly decrease further [17]. In 1994 the per capita living area in Baden-Württemberg was only 37.6 m^2, and then the figure rose to 40.9 m^2 at the end of 2003 with an average annual increase of almost 0.4 m^2 between 1994 and 2003, compared to the 70s and early 80s, when the average annual growth was almost 0.7 m^2. In 2002 Stuttgart region had around 151 million m^2 residential buildings and 3.985 million inhabitants. By the end of 2009 the housing supply had already reached a total of 168 million m^2 for its population of 4.001 million. That is to say, the per capita average living area increased from 37.89 m^2 in 2002 to 41.99 m^2 in 2009. However it is currently hard to foresee the end of the growth trend of per capita living area in Stuttgart region considering that it has already reached more than 50 m^2 in some other countries, such as Switzerland or the United States. In the next few years, Stuttgart region will still be in a developing stage of housing construction and real estate to meet increasing housing demand. In 2009 housing supply-demand ratio stood at a 0.989:1 and then is to increase to 0.991:1 in 2020 as the estimated value simulated in the SD model. There is one "Pressure" indicators (I5 "The housing demand"), one "State" indicator (I11 "Living space per capita") and two "Response" indicators (I21 "Completed areas of residential projects", I22 "The housing supply") (Table 1) in this sector.

Economics sector

Baden-Württemberg's strength lies in its high economic performance. This strong export-oriented economy invests enormous amounts in research and development, as well as in innovations. Flagship branches are the technology sectors, such as automobile production, and mechanical and electrical engineering. After the historical economic decline in 2009, the economic indicators at the end of 2010 showed that it recovered much sooner than was expected one year earlier in Baden-Württemberg. In 2010 the GDP of Baden-Württemberg increased by 4.75% which is lower than the growth of 7.4% in 2009. According to the economic indicators of the State Statistic Bureau, a slow GDP growth of 2.5% was predicted in 2011 [18]. Under these conditions the GDP will reach the level of 2008 at the end of 2011.

The GDP of Stuttgart Region rose from 96.776 billion Euros in 1991 to 145.865 billion Euros in 2008 at an annual average growth rate of 2.443%, and fell around 7.4%in 2009, then followed by a slight increase from 2010. By 2020, the GDP of Stuttgart Region is expected to exceed 160 billion Euros as calculated in the model. There are three "Driving Forces" indicators (I1 "GDP", I2 "Disposable income per capita", I3 "Investment on residential projects completed"), three "state" indicators (I12 "Land price per floor area", I13 "Urban housing price index", I14 "Rent price index"), and two "Response" indicators (I23 "Investment on new residential projects", I24 "Environmental regulation investment") in this sector. "Urban housing price index" is subject to three variables: total construction cost, rent price index, and housing supply-demand ratio. As the stock variable, GDP depend on economic growth rate in this model.

Society sector

Society sector mainly refers to the quantity and structure of population, urban traffic, unemployment rate and population below the poverty line. Population comprised of

Usual Residents and Mobile Residents (immigration population and emigration population), in this paper, can cause changes of the indicators in other three sectors. From 1991 to 2007, the population of Stuttgart Region stably rose from 3.751 million to 4.007 million with a positive net migration. The number of population will firstly fall below 4 million in 2011 and then continue decreasing to 3.951 million at the end of 2020, according to the regional population forecast of the State Statistic Bureau. Two variables involved in urban traffic were discussed here, "Total motor vehicles" and "Road traffic accidents". There are one "Driving forces" indicator (I4 "Urban population"), four "Pressure" indicators (I7 "Urban family size", I8 "Unemployment rate", I9 "Road traffic accidents", I10 "Total motor vehicles") and one "State" indicator (I15 "Population below the poverty line") in this sector.

Environment sector

As an important component of this SD model, environment sector mainly refers to one "Pressure" indicator (I6 "Population Density") and five "Impact" indicators (I16 "Urban air pollution (NO2,SO2,PM10)", I17 "Domestic water consumption", I18 "Delivery quantity of domestic waste", I19 "Green coverage ratio", I20 "Urban areas"). Rapid economic recovery in Stuttgart Region and the state of Baden-Württemberg leads to an increase in demand for environmental resources [18]. In a residential context, the statistic data of the energy consumption and air emissions can be differentiated by two sources: households, and small residential consumers (ancillary facilities, public facilities, services, etc) based on the energy balance of Baden-Württemberg [19]. In Stuttgart Region, NOx and SO2 emissions generated from households and small residential consumers separately decreased from 6099 tons and 6645 tons in 1995 to 4017 tons and 2337 tons in 2007. Domestic water supply also dropped from 176.96 million m^3 in 1998 to 167.57 million m^3 in 2007. And domestic waste disposal increased to 1426 tons in 2009 compared to 1282 tons in 1991.

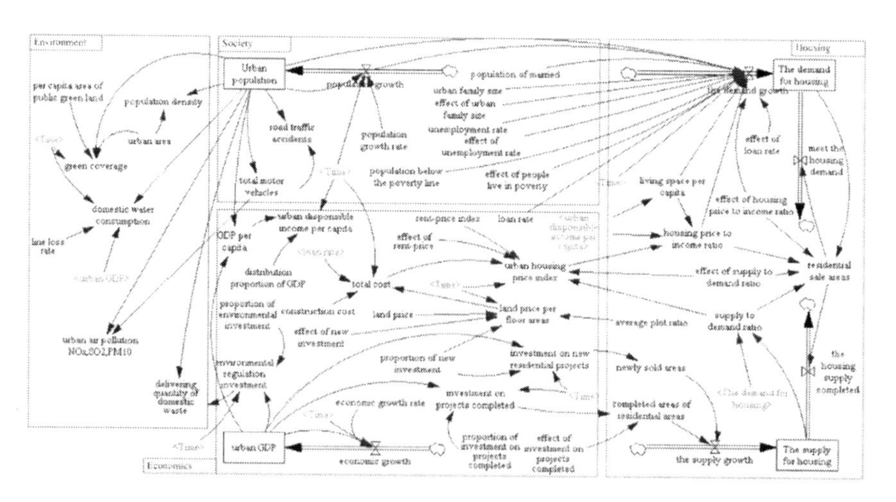

Fig. 3. Stock-flow diagrams of 4 sectors in Vensim

Based on the structure of the SD model and the calculation formulas identified, the simulation corresponds to the historic data quite closely (Figure 4a, Figure 4b). Figure 4a illustrate that the values of housing supply for the initial part of the model are slightly inaccurate (142.94 million m2 in simulation is 5.33 percent lower than the historic data 151 million m2 in 2002) but appear to converge towards the same final value in 2009 and are basically in keeping with the developing trend of the total housing supply in the state of Baden-Wurttemberg. Figure 4b shows the comparative results of housing supply and housing demand in Stuttgart Region calculated in the model; and the variation of supply-to-demand ratio remains stable between around 0.99 and 1.1.

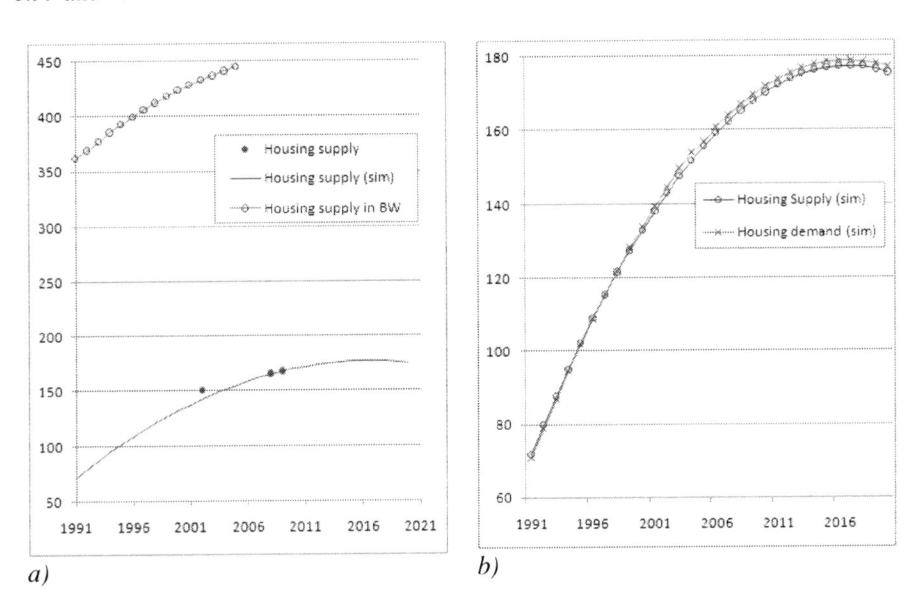

Fig. 4. (*a*) Simulation results of Housing Supply in Baden-Wurttemberg state and Stuttgart Region, and (*b*) Simulation results comparison of housing supply and housing demand in Stuttgart Region

4.3 Geographic Information System (GIS)

GIS technology was used to manage the digital database and to provide a connection to visualization and SD simulation models discussed above. The GIS operations were performed with ArcEditor 10 the product of the Environmental Systems Research Institute (ESRI, Redlands, CA) and CityEngine. In this part, Plieningen was selected to be the sample district in Stuttgart with 2395 buildings in 2009 including 1125 residential buildings and houses. The data was used both as input to the SD model as well as to visualize the results of the SD simulation. Figure 5a shows the ground polygon shapefile representing all buildings in Plieningen in 2009 displayed in ArcMap. And then the ground polygon shapefile was converted to point shapefile (Figure 5b) using "ArcToolBox –>Data Management Tools –>Features –>Feature to Point" in ArcMap. During this process, the attributes of the input features of the buildings are maintained in the output points feature class.

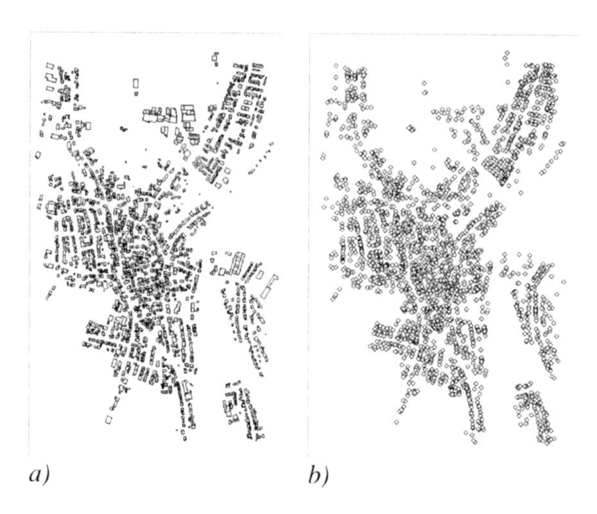

Fig. 5. (a) The ground polygon and (b) point shapefile of the Plieningen data set

4.3.1 2D Density Map in ArcMap

Density analysis is an important step in the GISSD frame which "takes known quantities of some phenomena and spreads it across the landscape based on the quantity that is measured at each location and the spatial relationship of the locations of the measured quantities" [20]. The density function in the spatial extension of ArcEditor distributes a measured quantity of the generated point shapefile (/layer) (in Figure 5b) throughout the sample district to produce a continuous surface. Then density surfaces show where point or line features are concentrated. In this study, different point values for each point (building) representing separately the total number of buildings, location areas, floor areas and volumes will be given and input into the shapefiles in ArcMap to observe more about the spread of buildings over the region. Sine all the buildings can't be located at just one or two building points and distinct building types have respective demand for size and plot ratio, a density surface needs to be created using kernel density calculation function with a fixed search radius in the Spatial Analyst extension to show where points are concentrated. In kernel density, the values of point attributes spread out from the point location with the highest value at the center of the surface (the point location) and tapering to zero at the search radius distance [20]. For a point in the selected region (Plieningen district in this paper), the measure of the density at point "s" can be defined as:

Density = mean point values of events per unit area at point "s" defined as the limit. And the measure of the density at point "s" can also be marked mathematically as:

$$D(s) = \lim_{ds \to 0} Y\left(ds\right) / \, ds \tag{1}$$

D(S): the density values at point "s".
ds: the area of small region around "s".
Y(ds): the given attribute values of events at point "s".

Two assumptions

1, Table 2 shows the estimated results of housing supply, supply-demand ratio and supply variation predicted by SD model for next 10 years in the entire Stuttgart Region. The total residential supply will increase by around 4.46% from 168 million m^2 in 2009 to 175.502 million m^2 in 2020. As a small district in big Stuttgart Region, Plieningen has a residential stock of 335.132 thousand m^2 in 2009 which can be expected to rise to 350.097 thousand m^2 in 2020 by assuming the same average growth rate as in Stuttgart Region.

2, Plieningen district had 1125 residential buildings with a total floor areas of 335.132 thousand m^2 in 2009. The average floor areas per one building was 297.89 m^2 (=335132/1125). In all 1125 buildings, then a standard building identified in "ID_351600539503" with just the average floor areas was found. Meanwhile its location areas were 74.82 m2 and the building volume was 933.92 m3. According to assumption 1, the total residential supply in 2020 will exceed the figure for 2009 by 14965 m2. Then the number of new residential buildings required up to 2020 is obtained using 14965/297.89=50 assuming all new buildings are built in the form of the chosen standard one "ID_351600539503".

Two urban development patterns

Combining housing supply and mobility practices in cities with high or low densities constitutes two main branches which are compact development and outward development in architecture and urban planning. The Compact City is an urban planning and urban design concept, characterized relatively high residential density with mixed land uses. While outward development, also known as Urban sprawl, is a multifaceted concept, which includes the spreading outwards of a city and its suburbs to its outskirts to low-density and auto-dependent development on rural land, high segregation of uses (e.g. stores and residential), and various design features that encourage car dependency.

Bill Randolph (2006) [21] indicated that Successful compact city policies will require a viable and acceptable strata governance framework to minimize conflicts between neighbours and between owners, as well as maximize long term standards in higher density stock. Thinh et al. (2002) [22] discussed two dimensions of the compact city: physical and functional and created a ArcInfo- database of land use patterns and to model the physical compactness of the German Regional Cities. Arriba-Bel et al. (2010) [23] identified the most sprawled areas in Europe characterizing them in terms of population size and used the self-organizing map (SOM) algorithm as a visualization tool to better understand urban sprawl in Europe. Jat et al. (2007) [24] investigated the usefulness of the spatial techniques, remote sensing and GIS for urban sprawl detection and handling of spatial and temporal variability of the same. Sudhira et al. (2004) [25] analyzed and understood the urban sprawl pattern and dynamics to predict the future sprawls and address effective resource utilization for infrastructure allocation.

In this study, 50 points (in green colour shown in the first column of Figure 6) having the same attribute values with the standard building "ID_351600539503" were positioned in the point shapefile in two ways: gathering around the district center representing compact development or dispersing along the district boundary representing outward development.

Table 2. Indicators selected for the application

	Stuttgart Region			Plieningen
	Housing supply (million m^2)	Supply-demand ratio	Supply variation	Housing supply (1000 m^2)
2009	168	0.989373		335.132
2010	170.292	0.989476	+1.36%	
2011	172.262	0.990182	+1.16%	
2012	173.91	0.990224	+0.96%	
2013	175.236	0.990261	+0.76%	
2014	176.24	0.990293	+0.57%	
2015	176.922	0.990321	+0.39%	
2016	177.282	0.990857	+0.2%	
2017	177.32	0.990875	+0.022%	
2018	177.036	0.992702	+0.023%	
2019	176.43	0.996514	+0.039%	
2020	175.502	1.002195	+0.045%	350.097

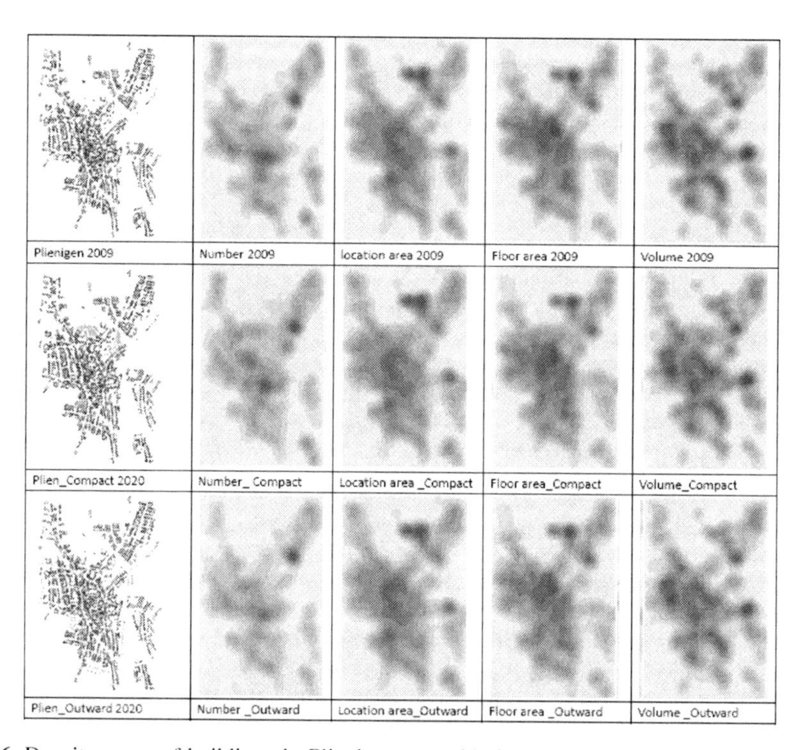

Fig. 6. Density maps of buildings in Plieningen considering four different attributes: the total number of buildings, location areas, floor areas and building volumes in 2009 and 2020

It can be observed in Figure 6 that deep color dots with the attributes of heavy densities are distributed to different parts of the density maps in the four attributes: the total number of buildings, location areas, floor areas and building volumes. Generally speaking, also verified according the shapefile displayed in Figure 6 and the GIS database, in density map of the building quantity, deep colour areas stand for big residential settlements and intensive commercial zones mainly found in the center district; in density map of floor areas, deep colour areas are mainly occupied by the multistory buildings used for residence and office, in contrast with the high location areas of some big commercial architectures and factories in the map of location areas; lastly in density map of the building volumes, the deepest colour areas represent some big-scale buildings usually featured in both of the height and location areas.

Regarding the prediction of two development patterns for Plieningen in 2020, the density maps of outward development do not show significant changes compared to the ones in 2009, especially for the three attributes: the total number of buildings, location areas and floor areas, firstly due to the small number 50 of new buildings relative to 1125 the total quantity of buildings in Plieningen and secondly their scattering distribution along the boundary which also weakens the density growth of attribute values at the points location. However the density map of building volumes in 2020 has an obvious expansion in the outward developing areas which means that the newly increased residential buildings of about 15 thousand m^2 floor areas in 2020 will have greater effects on the entire building volume distribution than the other three attributes in Plieningen in the trend of outward development. And for the compact development pattern, the density maps of the four attributes explicitly indicate the density variation in the specific area where new residential buildings of about 15 thousand m^2 floor areas are located together.

4.3.2 3D Visualization in CityEngine

3D visualization provides a more comprehensive approach to observe the prediction of urban development. Cityengine is a software product which is able to give users in architecture, urban planning, GIS and general 3D content production a design and modeling solution for the efficient creation of 3D cities and buildings. And Cityengine also supports 2D-to-3D conversion of GIS data. The three shapefiles (Plieningen in 2009, Pliningen_Compact development in 2020 and Pliningen_Outward development in 2020) containing the shape data and attributes for each shape were imported into Cityengine using the shapefile importer.

A shape in Cityengine consists of a name, parameters, attributes containing the numeric and spatial description, geometry, scope and pivot. The CGA (Computer Generated Architecture) shape grammar of the CityEngine is a unique programming language specified to generate architectural 3D content based on the shapes which uses a different syntax but provides the same functionality with the widely used GML shape grammar. The idea of grammar-based modeling is to define rules that iteratively refine a design by creating more and more detail [26] and the shapes are replaced by a number of new shapes according to the rules. The rule's script in this paper mainly has 3 parts: extrude the footprint shapes to their specified height, create the roofs and texture the facades which illustrate the implementation of the CGA grammar-based model.

In figure 7 we have two group 3D simulation phenomena on a small scale. On the left we show the 50 new buildings are located centrally in Plieningen. On the right we show a scatter distribution of the 50 buildings in Plieningen.

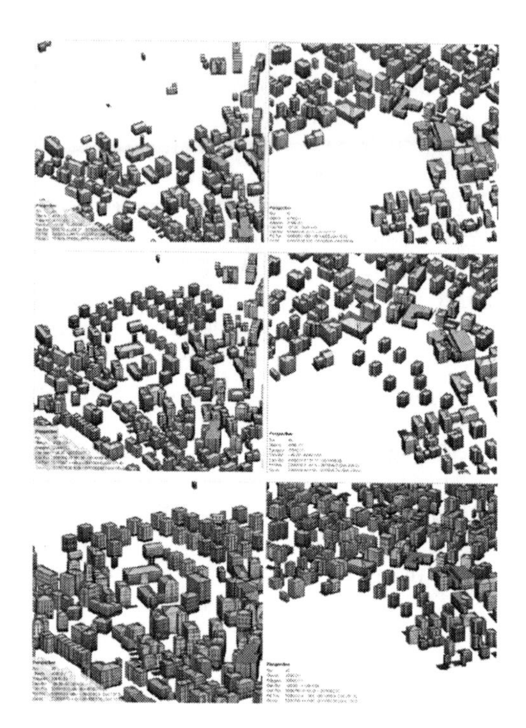

Fig. 7. 3D visualization simulation generated in CityEngine showing different development patterns in Plieningen in 2020

5 Conclusion

In this paper, a GISSD system integrated of system dynamics model, GIS analysis and 3D visualization is developed to assist evaluation of the future trend of the residential development in Plieningen district of Stuttgart Region, Germany. The model examines interactions among five subsystems ("Driving forces", "Pressures", "State", "Impact" and "Responses" categories) within a time frame of 30 years, and then the result of this model is discussed with two possible development patterns – compact development and outward development in Plieningen. It can be also concluded that the integration of System Dynamics theory and GIS is a useful strategy to study the development of urban residential areas in terms of sustainability. In this study, we can find the development process of our residential areas, and put a further concern on the development of the urban residential development from another point of view.

However, there are still a number of opportunities for expanding the study and for validating the results obtained herein. Firstly, only core-indicators were considered in this work. It would be of scientific interest to add other indicators resulting from policies and strategies. Secondly, further research would be required to collect more historic data and optimize the structure of system dynamics model. Finally, in this paper we only considered the a very small district of Stuttgart as the study object and also didn't input terrain data and street data in 3D GIS analysis.

Acknowledgements. This study was partly supported by SCUOLA INTERPOLITECNICA DI DOTTORATO (SIPD) a special project whereby the three Italian Technical Universities, the Polytechnic of Torino (coordinator of the project), the Polytechnic of Bari and the Polytechnic of Milano (2010-2011).

References

1. Bruntland, G.: Our common future. Oxford University Press, Oxford (1987)
2. World Bank: Expanding the measures of wealth: Indicators of environmentally sustainable development. Environmentally sustainable development studies and monographs series, No. 17, Washington DC (1997)
3. OECD Organization of Economic Co-operation and Development. Environmental indicators development, measurement and use. Paris: OECD Environment Directorate Environmental Performance and Information Division (retrieved February 19, 2010)
4. Lisa, S.: Indicators of Environment and Sustainable Development Theories and Practical Experience. The International Bank for Reconstruction and Development, Washington, D.C (2002)
5. Bottero, M., Mondini, G.: The construction of the territorial performance index for testing the environmental compatibility of projects and plans. In: Proceeding of the International Conference on Smart and Sustainable Built Environment, Brisbane, Australia (2003)
6. Brandon, P.S., Lombardi, P.: Evaluating sustainable development. Blackwell Publishing, Oxford (2005)
7. Nessa, W., Montserrat, P.E.: Sustainable Housing in the Urban Context: International Sustainable Development Indicator Sets and Housing. Social Indicators Research 87, 211–221 (2008)
8. Brans, J.P., Macharis, C., Kunsch, P.L., Chevalier, A., Schwaninger, M.: Combining multicriteria decision aid and system dynamics for the control of socio-economic process, An iterative real-time. European Journal of Operational Research 109, 428–441 (1998)
9. Pirrone, N., et al.: The Driver-Pressure-State-Impact-Response (DPSIR) approach for integrated catchment-coastal zone management: preliminary application to the Po catchment-Adriatic Sea coastal zone system. Regional Environmental Change 5, 111–137 (2005)
10. OECD Organisation for Economic Cooperation and Development (OECD) (eds) OECD Core set of Indicators for Environmental Performance Reviews. Environment Monographs 83, Paris (1993)
11. EEA Environmental indicators: typology and overview. In: Smeets E., Weterings R. (eds) Technical report no 25, p. 19 (1999)
12. Kristensen, P.: The DPSIR Framework. Workshop on a comprehensive/detailed assessment of the vulnerability of water resources to environmental change in Africa using river basin approach. UNEP Headquarters, Nairobi, Kenya (2004)
13. State Statistic Bureau of Baden-Württemberg, `http://statistik-bw.de/`
14. Ministry of Economy Baden-Württemberg, `http://www.wm.baden-wuerttemberg.de/sixcms/detail.php/62315`
15. United Nations: Indicators of Sustainable Development: Guidelines and Methodologies. United Nations, New York (2007)
16. Berger-Schmitt, R., Noll, H.: Conceptual framework and structure of a European system of social indicators. In: EuReporting Working Paper 2000, No. 9. Mannheim: Centre for survey research and methodology (2000)

17. Landesentwicklungsbericht Baden-Württemberg 2005. Wirtschaftsministerium Baden-Württemberg (2005)
18. Bauer-Hailer, U., et al.: Wirtschafts- und Sozialentwicklung in Baden-Württemberg 2010/2011. Statistisches Landesamt Baden-Württemberg (2010)
19. John, B.: Die Energiebilanz 2007 für Baden-Württemberg. Statistisches Monatsheft Baden-Württemberg 1/2010, pp. 30–33 (2010)
20. ESRI, ArcGIS desktop 10 Help (2010), http://help.arcgis.com/en/arcgisdesktop/10.0/help/
21. Bill, R.: Delivering the compact city in Australia: Current trends and future implications. City Futures Research Centre, University of New South Wales (2006)
22. Thinh, N.X., et al.: Evaluation of urban land-use structures with a view to sustainable development. Environmental Impact Assessment Review 22, 475–492 (2002)
23. Arribas-Bel, D., Nijkamp, P., Scholten, H.: Multidimentsional urban sprawl in Europ: A self-organizing map approach. Computers Environment and Urban systems (2010) (in press)
24. Jat, M.K., Garg, P.K., Deepak, K.: Monitoring and modeling of urban sprawl using remote sensing and GIS techniques. International Journal of Applied Earth Observation and Geoinformation 10, 6–43 (2008)
25. Sudhira, H.S., Ramachandra, T.V., Jagadish, K.S.: Urban sprawl: metrics, dynamics and modeling using GIS. International Journal of Applied Earth Observation and Geoinformation 5, 29–39 (2004)
26. Müller, P., Wonka, P., Haegler, S., Ulmer, A., Van Gool, L.: Procedural Modeling of Buildings. In: Proceedings of ACM SIGGRAPH 2006 / ACM Transactions on Graphics (TOG), vol. 25(3), pp. 614–623. ACM Press, New York (2006)

The State's Geopolitics versus the Local's Concern, a Case Study on the Land Border between Indonesia and Timor-Leste in West Sector

Sri Handoyo

Researcher, National Coordinating Agency for Surveys and Mapping of Indonesia (BAKOSURTANAL),
yshandoyo@yahoo.com

Abstract. One of geopolitics realizations is the establishment of international borders between countries. Indonesia has international land borders with 3 countries, including the land border with Timor-Leste.

Establishment of the international border between Indonesia and Timor-Leste was a joint mandate of the two Governments, these are the Republic of Indonesia (RI) and the Democratic Republic of Timor-Leste (RDTL) based on the Treaty 1904 and the Arbitrary Awards 1914. Joint border surveys have been in progress achieving 96% of the total length of the border lines. During the surveys there were problematic situations occurred.

A government's geopolitics is not always fortunate to immediately match with the local's concern on the land border line establishment. This is particularly happening at a certain land border line segment between Indonesia and Timor-Leste in West Sector. This paper does not describe which side is right or wrong, instead, a soft approach of solution is underlined.

Keywords: geopolitics, international land border, border line, Indonesia, Timor-Leste, Treaty 1904, Arbitrary Awards 1914, delineation survey, demarcation survey.

1 Introduction

Timor-Leste was the 27th province of Indonesia namely the Province of East Timor. The years of 1997 and 1998 were the time of crisis in Indonesia (in economics, politics, and in socio-cultural) as it also happened in other countries such as in Singapore, Malaysia, Thailand, and the Phillipines. During these years, there was a political pressure from the people in the Province of East Timor to the Indonesian Government demanding a referendum. President B.J. Habibie of RI decided to agree with the referendum with two options, (1) obtain the independent and separate from Indonesia, or (2) stay as part of Indonesia. Further in 1999, the referendum in the Province of East Timor was conducted under the United Nations and around 80% of the Timorese chose to have the independent and separated from Indonesia. Finally, based on the referendum on the 30th August 1999, Timor-Leste is an independent country since the 20th May 2002 [7].

B. Murgante et al. (Eds.): ICCSA 2011, Part I, LNCS 6782, pp. 315–328, 2011.

The workplan to establish the land border between Indonesia and Timor-Leste had been inisiated by Indonesia and the United Nations Transitional Administration in East Timor (UNTAET) with reference to the 1904 Treaty between the Dutch and Portuguese, and to the Permanent Commission Award 1914 (PCA 1914). It was started with the forming of Joint Border Committee (JBC, in Indonesia it is chaired by the Director General for General Governmental Affairs) and Technical Sub-Committee on Border Demarcation and Regulation (TSC-BDR, in Indonesia it is chaired by the Chairman of BAKOSURTANAL) in 2000. This was followed by the JBC and the TSC-BDR meetings in 2001 and 2002. After the independent of Timor-Leste, there was the 1st meeting of the Joint Ministerial Commission (JMC, of both Indonesia and Timor-Leste Ministers of Foreign Affairs) in 2002 that instructed TSC-BDR to carry out joint delineation and demarcation surveys (see Figure 1).

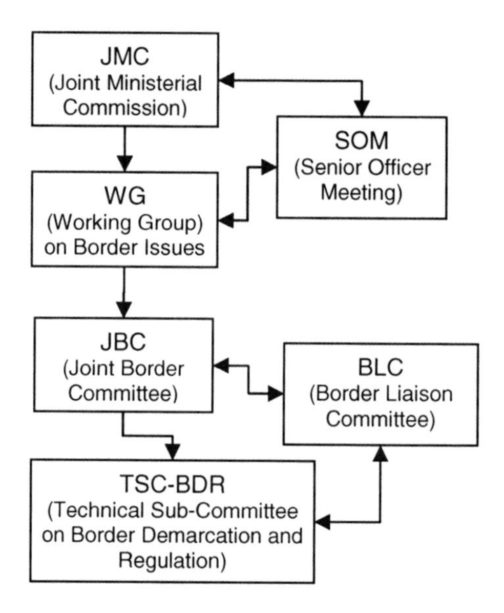

Fig. 1. The organizational chart of RI-RDTL land border establishment

1.1 Geopolitics and the International Boundary

Despite the increasing effects of globalization, the basic unit in the contemporary world political system is the state. It is an internationally recognized political and juridical entity which claims sovereignity over a specific area of land and possibly adjacent sea, the inhabitants of the area and the resources located therein. This area is delimited in the minds of individuals and groups owing allegiance to the state, in most cases on cartographic representations of the state and in some cases on the ground itself, by boundaries. Boundaries indicate the accepted territorial integrity of the state and the extent of government control. In the majority of cases boundaries are legally recognized by the states which share them and also by the international community [1].

In Flint (2006), Strausz-Hupé (1942), and other sources, there are geopolitical theories and histories, such as of: Friedrich Ratzel (1844-1904), Karl Ernst Hausofher (1896-1946), Sir Halford Mackinders (1861-1947), James Burnham (1941), Machiavelli (XVII century), Fuerback and Hegel (XVII century), Napoleon Bonaparte (XVIII century), General Clausewitz (XVIII century), Lenin (XIX century), Lucian W. Pye and Sidney (1972), also Sir Walter Raleigh and Alferd Thyer Mahan and W. Mitchel, A. Seversky, Giulio Douhet, J.F.C. Fuller (view concepts of sea and air), and Nicholas J. Spykman. These are theories and history about power and strength of nation, political and economic power, war and politics, economic strength of nations, cultural politics of nations, political power and geography, state geopolitics and boundaries, trade and power, and arm forces and power [4][8].

However, learning all those mentioned above, and having past history and experiences has led Indonesia to be united in one nation, one territory, and one language of Indonesia. Indonesia has the geopolitical views based on the archipelago concept as the land and sea territory as one nation. The views also based on that Indonesia loves peace with neighbors and with the principles of no conflict and expansionism. Indonesia also has the "Panca Sila" as the five principle of life as nations. The Indonesian is socio-culturally peaceful and religious.

Therefore, the joint establishment of the land border between Indonesia and Timor-Leste is geopolitically an important element in the development of peace and friendship in the area concerned [2].

1.2 It Is Not an Anthropomorphic Boundary

Boundaries drawn according to cultural elements such as language, religion or ethnology, are known as anthropomorphic. Among the best known of such boundaries are those between India and Pakistan and between India and Bangladesh [1]. According to what is written in the Treaty 1904 the boundaries between Indonesia and Timor-Leste are classified as morphological. These border lines consist of watersheds, rivers, and thalwegs of large rivers. In the contrary, both local people of Indonesia and Timor-Leste along the border lines have the same culture. They have the same language i.e. Tetun in the East Sector border and Dawan in the West Sector, and even they are in the same religion as Christians. The followings are sources of information that they are from one ancestor.

1.3 The Social Capital of the Timorese Community

Referring to oral and traditional sources, the people of West Timor who speak Dawan (in the regencies of Timor Tengah Utara, Timor Tengah Selatan, and Kupang) mentioned that the ancestors of the Timorese Kings were originally from Malaka and consisted of four brothers who came and governed the Island of Timor. They are Liurai Sila, Liurai Sonbai, Liurai Benu, and Liurai Afoan. The agreement of the kingdoms division were taken at the peak of Mutis Mountain, i.e. the territory of Liurai Sila was in the East part of Timor Island covering (now) the regency of Belu (Nusa Tenggara Timur) and the entire territory of Timor-Leste; Liurai Sonbai had the territory of the North and the South of Timor Island including (now) the regencies of

Timor Tengah Utara and Timor Tengah Selatan; Liurai Benu had the territory of the district Oecussi (Timor-Leste); and Liurai Afoan had the territory of the West and the South of Timor Island as the regency of Kupang (Abi, 2009 in Handoyo, 2011). While Wala, in Handoyo (2011), mentioned that from many aspects such as language and traditional behaviors, the community of East Timor who live in the area of East Timor and the community of West Timor who live in the area of West Timor are having unbreakable similarity in manners.

Every ethnic has related history with the other ethnic through the relations in marriage and agreement (being united in politics and traditional economy). There are philosophy and cosmology and basic values of every group of ethnic which become a control means of behaviors of the community member so that it will prevent the conflict, strengthen the cooperation, and provide peace. Those basic values together with the social network and organization have build "something" that become a social capital. It seems that the people in Timor Island have enough social capital which tied them one and another both individually and in group. There are three basic values being conceived by the community. In Tetun these are: Taek no Kneter (ethics and moral), Ukon no Badu (forbidden and punishment), and Makerek no Badaen (knowledge and art). Those three basic values were understood by other ethnics (13 ethnics) who lived in Timor Island although in different terms among them. The three basic values have been derived to many other values and norms to maintain peace in life from the family unit to inter-ethnics. The maintenance of peaceful life and the conflict solution were heritage to generations through lyrics, poems, paraphrase. There is a Tetun terms that encourage people to live in unity, such as; "ho ema at malu, ita rua keta; ita rua at malu, rai at ona" (Scribd.com in Handoyo, 2011).

The communities of West and East Timor have the social capitals as the means of solution for their problems as long as there is no strong intervention from the State or Government. The values of living in peace being heritage from their ancestors in the past history in harmony, family relations, feeling of trust and tolerance, togetherness, are part of the social capitals that prevented tension, conflict, and crisis that have happened among them. Besides, the social capital are also explored from the relation between strong religious groups, especially within the diocese of Atambua, covering the area of Belu and Timor Tengah Utara, there is a Religion Leader Communication Forum that has active roles to eliminate potential of conflict between the religious groups and even increase the relation and communication among them.

2 The Border Line Implementation

With reference to the 1904 Treaty between the Dutch and Portuguese, and to the Permanent Commission Award 1914 (PCA 1914), the land border line implementation was realized by delineating on agreed border maps and in the field. This field survey is within the frame work of the CBDRF which was jointly established before hand in 2002-2003. The delineation of the border is entrusted to the TSC-BDR, which was established at the 2nd Meeting of RI and UNTAET held in Jakarta (Indonesia) on 19-20 July 2001. The TSC continued to exercise this mandate after the independent government of the Democratic Republic of Timor-Leste succeeded to the UNTAET. The TSC-BDR held a number of related meetings with the objective to carry out its responsibilities, within which the Standard Operating Procedures (SOP) applied in the field delineation survey was produced.

Up to the year of 2005 the implementation of the land border between Indonesia and Timor-Leste has reached the progress of 96% out of the total length 268,8 kilometers. This was formalized in the Provisional Agreement signed by the Ministers of Foreign Affairs of both countries representing both Governments of Indonesia and Timor-Leste in Dili, April 8th, 2005.

The next Table 1 shows the related field surveys as being carried out jointly in sequence. During the Joint Delineation Survey (June-July 2003) and the Extra Joint Delineation Survey (November 2003) the problem of un-surveyed segment of Subina-Oben firmly arouse.

Table 1. The chronology of related joint field surveys

Survey	Period	Objective
Joint Reconnaissance Survey	April-May 2002	To assess on the conditions of work and visit some of the more complex border segments.
CBDRF survey	May 2003	To establish the CBDRF, intended to support future surveys and to serve as the coordinate reference frame for the border line.
Joint Delineation Survey	June-July 2003	To survey the border line and to survey Ground Control Points.
Extra Joint Delineation Survey	November 2003	To resurvey some segments that were left as unresolved in the previous survey.

The following is the complete stages of the land border activities being conducted jointly by both delegations and the field survey teams of Indonesia and Timor-Leste.

2.1 Stages of the Joint Land Border Activities

The activities to establish the land border between Indonesia and Timor-Leste consist of stages as follows:

a. Studying the documents of Treaty 1904 and other relevant documents. This is to make interpretations of the Treaty verbal description of the border lines.
b. Joint reconnaissance survey. This is to jointly make traces in the field of those border line descriptions.
c. Joint survey and construction of the common border datum reference frame (CBDRF) [3]. This jointly establishes a set of common border datum in both sides of the border line as the reference frame for the border point coordinates measurements.
d. Joint delineation surveys. This activities are to jointly decide the border point positions and measure its coordinates.
e. Joint demarcation surveys. This is the following activities of the joint delineation surveys to establish markers on or between the border points.
f. Joint mapping and reporting. This is to produce joint border maps, depicting the border points and lines, at scale of 1:25.000 covering both the East (main) and the West (Oecussi) sectors.

All those joint activities were carried out and based on related joint technical specifications and standard operational procedures. Unfortunately, during those stages, particularly the field surveys, there were problems arouse.

2.2 Status of the Results

As the results of the bilateral meetings and the joint field surveys between Indonesia and Timor-Leste, from 2001 to present, there are now in existence as follows:

a. Length of the land border lines: the East sector of 149.1 kilometers, and the West sector of 119.7 kilometers, giving the total length of 268.8 kilometers.
b. "Interim Report on the land Border Delineation between Republic of Indonesia and Democratic Republic of Timor-Leste", 2004, consists of three volumes:
   i. Volume 1: Results of the Land Border Delineation.
   ii. Volume 2: Description of Process of Land Border Delineation.
   iii. Volume 3: Joint Compilation of Reference and Auxiliary Documents.
c. "Provisional Agreement between the Government of the Republic of Indonesia and the Government of the Democratic Republic of Timor-Leste on the Land Boundary", 2005, covering nine Articles and Annexes:
   i. Annex A-List of 907 border point coordinates.
   ii. Annex B-1 sheet of General Map at Scale 1:125.000 and 17 sheets of Border Maps at Scale 1:25.000.
   iii. Annex C-Unresolved Segments.
d. 103 demarcated border markers.
e. Documents and Record of Discussion (RoD) of: JMC meetings (2), JBC meetings (2), Special JBC meeting (1), TSC-BDR meetings (23).
   Notes: there were no bilateral activities in 2006 and 2007 due to internal problems in Timor-Leste.

2.3 Problems with the Land Border

As mentioned before, there are problems with the land border in the form of un-resolved segments as the results of the joint field reconnaissance and delineation surveys that annexed in the 2005-Provisional Agreement. There were originally 8 (eight) segments that have problems, mostly due to disagreement, or different interpretations of the Treaty both verbally and in the field, between the two survey teams. At the 15th TSC-BDR meeting in Yogyakarta, on the 29th-30th October 2004, there were 5 (five) segments solved [10]. Up to present there are still 3 (three) un-resolved segments left. This is as stated by Indonesian delegation during the 19th TSC-BDR meeting December 2005 in Surabaya [11] that the positions concerning 3 (three) un-resolved segments namely Manusasi/Oben, Noel Besi/Citrana, and Memo/Dilumil, remained the same as stated at the TSC-BDR Meeting held in Yogyakarta, October 2004. These un-resolved segments could not be solved due to different interpretations of the legal documents, generally related with lack of detail in the Treaty or inconsistencies between present toponimy, Treaty map toponimy, and Treaty text toponimy.

Indeed those 3 (three) un-resolved segments are not easy problems to solve. So far up to the bilateral discussion in the 23rd Meeting of the TSC-BDR RI-RDTL in Bogor, Indonesia, August 2010, there was still an impasse on the final resolution [12]. Both government, Indonesia and Timor-Leste, have now been working hard trying to look for the solutions. However, those are not the problems this paper is discussing about. There is still another critical problem namely un-surveyed segment which is at the land border segment of Subina-Oben. This problem has nothing to do with the differences between the two survey teams since the Treaty interpretation of this particular segment has actually been acceptably depicted on maps. Problem of this Subina-Oben segment is due to the claim of the Indonesian locals as they did not allow the segment to be surveyed by the joint survey team because, according to them, they have farming lands that will be inside Timor-Leste area if the international land border applied based on the Treaty 1904.

To get the comparison view of the segment problems see the next Table 2. Also see Figure 2 and 3 to learn about Timor-Leste and the land border locations.

Table 2. The problems at the land border

Segments	Problem Status	In This Paper
Manusasi/Oben (in West sector)		
Noel Besi/Citrana (in West sector)	Un-resolved	Not discussed
Memo/Dilumil (in East sector)		
Subina-Oben (in West sector)	Un-surveyed	Discussed

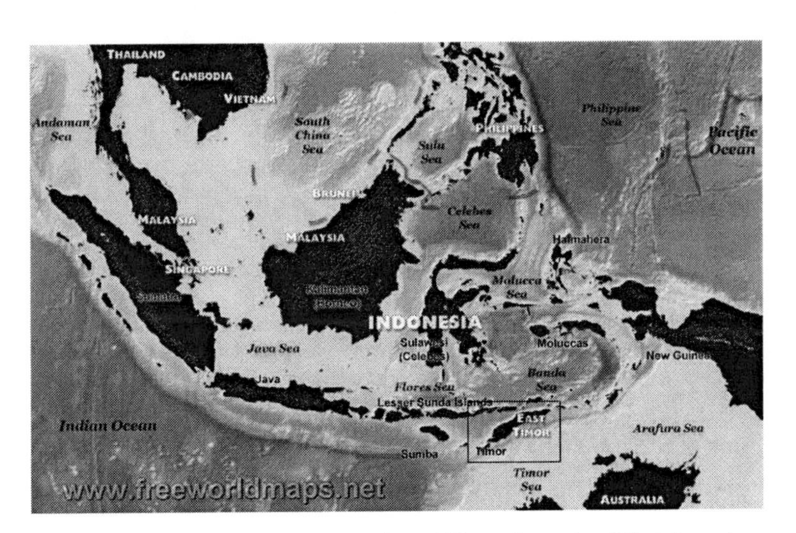

Fig. 2. The illustrative map showing the location of Timor-Leste (modified from the source of www.freeworldmaps.net).

3 The Un-surveyed Segment of Subina-Oben

Subina-Oben is the South-East part of the land border segment in the West sector around 18 kilometers length (see Figure 3). At the segment of Subina-Oben the resistance from the local people of the Timor Tengah Utara (TTU, Indonesia) District population precluded the joint delineation survey (2003) from taking place [9]. This un-surveyed segment have been put aside for submission to the higher levels of which so far no solution yet.

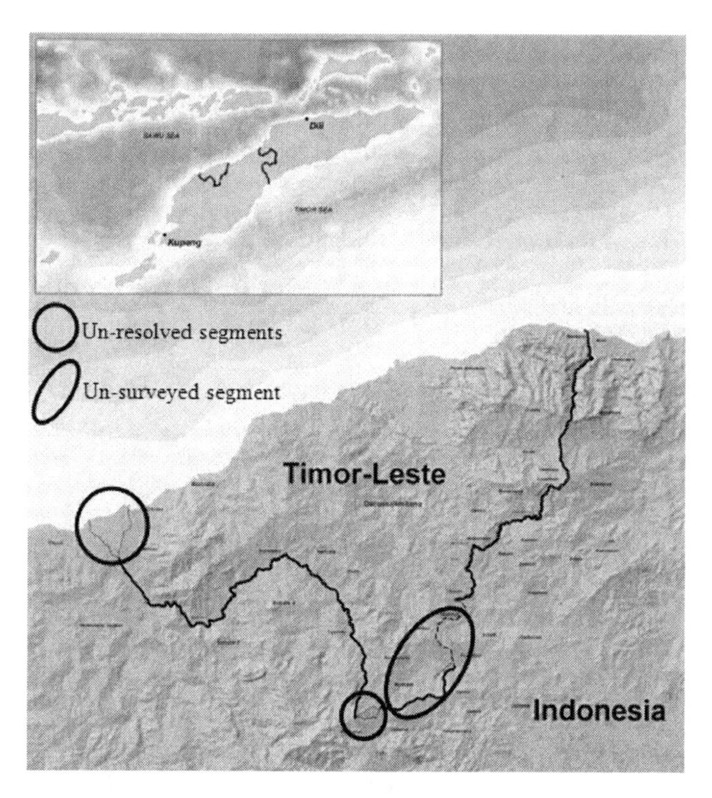

Fig. 3. The illustrative map showing the land border line in the West sector (Oecussi). The areas with circles are the un-resolved segments, while the area with ellipse is the discussed un-surveyed segment of Subina-Oben.

3.1 Version of the Indonesian Locals

The chronological events regarding the land problems according to the TTU local people including the local authorities: (Data and information collected during the unilateral field survey of October-November 2008) [13].

a. Year 1966: Since the Dutch era there was traditional uses of the land, as *ulayat right*, along the border by the people of Lake tribe. Then happened the taking over by the people of Ambeno, Oecussi. The people of Lake tribe did not accept this and it had become a war.

   As a solution, there was a 1966 agreement between the District Governments of TTU and Oecussi, with the land border as claimed by the TTU people now, of which also determined in the "Agreement Statement" letter between J.T. Sonbay as the Committee Chairman of the Changes of the Nilulat Fectorate Government and Tasi Lopo as the *Cheve* of Bobo Meto, in Nanao, dated July the 10th, 1964.

b. Year 1988: The border demarcation of the 27th Province of Timor Timur followed the 1904 Treaty. Those areas as limited by the 1966 case became part of the Timor Timur Province. In this case the people of TTU along the border of the Subina-Oben segment did not have any objection because they could still use the land.

c. Year 2002: Timor-Leste has been an independent state and followed by the delineation of the land border by RI and RDTL. This had an effect that the TTU people along the border of the Subina-Oben segment lost their use of the land both for farm and cattle, therefore they claimed them.

Conclusion according to Indonesian side: The land belongs to the tribe of LAKE, and it has a list of land users (333) within the area around 683 hectares, see Table 3.

Table 3. The claimed land belongs to Lake tribe with its locations and farmers

No	Area	Approx. Area	Villages	Users	Year	Crops and Others
1	Tubu-Nilulat	230 Ha	Tubu	71 men	1916 to 2005	Cattle farming, Crops: Jati, mangga, kepok, ampupu, kapuk. 2006-2008 disputes.
			Nilulat	72 men		
2	Haumeniana-Nifonunpo	107 Ha	Haumeniana	77 men	1916 to 2005	Cattle farming, Crops: Jati, mangga, cemara, ampupu, jambu mete. 2006-2008 disputes.
3	Pistana	142 Ha	Nainaban	33 men	1916 to 2005	Cattle farming, And there were traditional exchanges of land due to marriages, and there were cemetery, Crops: Jati, lamtoro, jambu mete, cemara, kentang. 2006-2008 disputes.
			Sunkaen	27 men		
4	Subina	206 Ha	Inbate	53 men	1916 to 2005	Cattle farming, Crops: Jati, mahoni, ampupu, kemiri, cemara, mangga, 2006-2008 disputes.
	Total:	683 Ha		333 men		

3.2 Version of the Timor-Leste Locals

Timor-Leste delegation has reported in the 21st TSC-BDR meeting in Bandung (2008) and described the activities and the results of the Unilateral Field Survey in the area of Oecussi held in 1-15 June 2008 [12]. The objective of the field survey was to identify land parcels in the border area, their usage and ownership, to assess the clarification of the "social issues" referred by the Indonesian side. The survey team was led by Roberto Soares (Ministério dos Negócios Estrangeiros) with 5 team members and local authorities and local representatives of the population from Naktuka, passabe, and Subina. The result was listed in 68 land parcels with the owners.

The information about the ownership of each parcel was collected from the local people, including the most respected members of the community. For each parcel was possible to collect the information of inheritance for three generations. It was possible to trace the ownership of the land up the grandfathers of the current owners. For some of the parcels it was identified a former "liurai" of Bobometo, named Taque Taiboko, that long time ago was responsible for the distribution of land among the people.

The land parcels are depicted in as a rough sketches in Figure 4 and Figure 5. These maps do not result from an accurate survey, only the relative position of each parcel to the neighboring parcels is intended to be accurate, at it is correct the neighborhood with the border line.

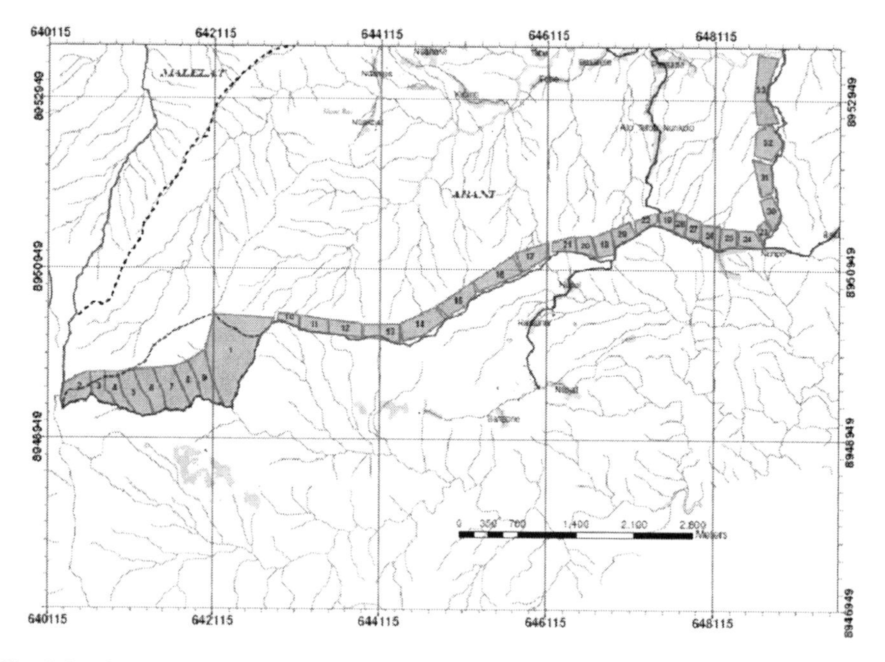

Fig. 4. Land parcels from Bijael Sunan, Oben, Banat, Kita and Nunpo up to Noel Passabe [12]

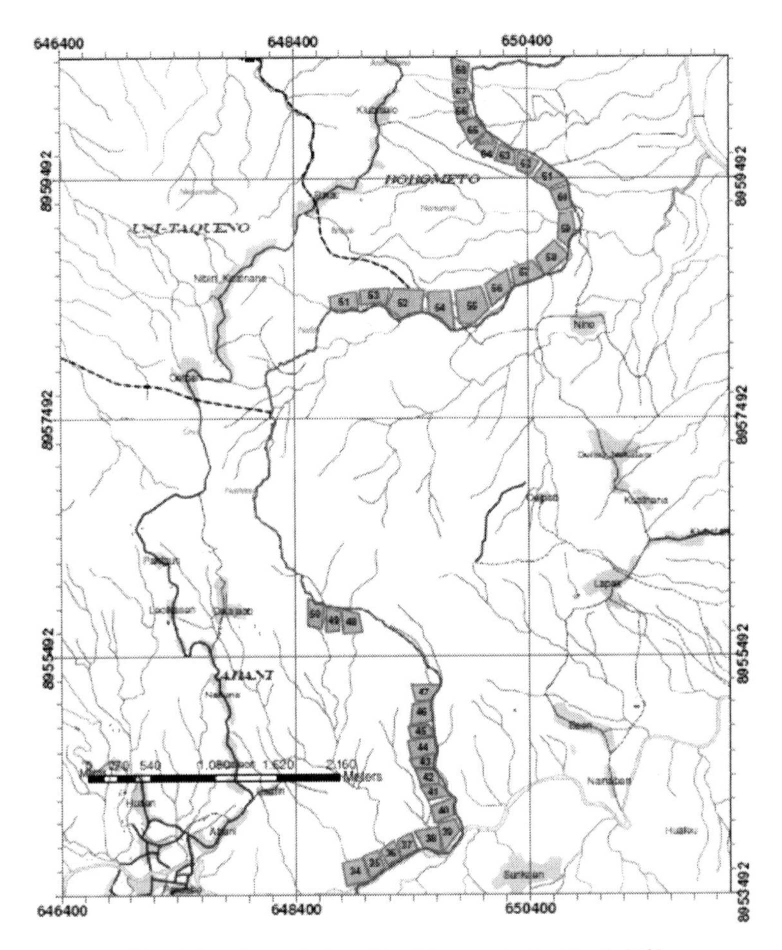

Fig. 5. Land parcels from Noel Passabe to the North [12]

Conclusion according to Timor-Leste side: The land belongs to Timor-Leste local people along the border line.

4 Flow of thoughts for Solution

There is the border line segment (±18 kilometers) in existence from a place named Subina to a place named Oben which has been agreed by both governments Indonesia and Timor-Leste through the Joint Field Survey Teams based on the Treaty 1904. There was no problem with the border line for both Field Survey Teams. However, the Indonesian locals along the border line do not agree with the existence. This is the problem since they claim for the farming and cultivated lands practically along the border line "inside" the Timor-Leste territory.

The recommendation is that both Governments, who have the political will to stay as neighbors in peace and friendship, should persuade and facilitate both locals to meet and talk in traditional manner for they are normally well-known with. It is also strongly advised to utilize their social capitals to support the objective of the meeting. The objective of the meeting are firstly the border line segment should stay as it is as agreed by both Governments. Secondly, try to arrange the cooperation between both locals to use (or manage) the land for farming and cultivation in peace and friendship. The following is the diagram of the flow of thought.

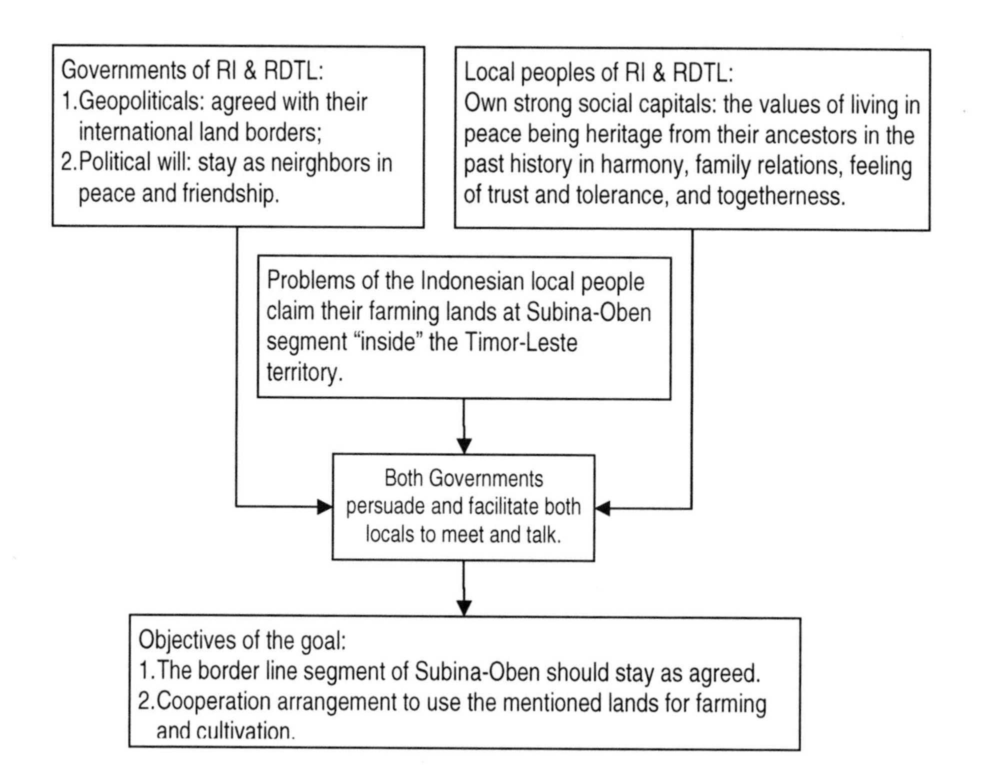

Fig. 6. Diagram of the flow of thought for solution

5 Concluding Remarks

Experiencing both countries of, so far, having the impasse on the final resolution of the un-resolved segments [14], the following positive point of views are underlining the remarks in relation to the problem of the un-surveyed segment. First, it is indeed that both people in either side of the country, Indonesia and Timor-Leste, are actually belonging in one family, who were back in time originated from the same ancestors, and are traditionally tied in the same culture. The values of living in peace being heritage from their ancestors in the past history in harmony, family relations, feeling of trust and tolerance, togetherness, are part of the social capitals that prevented tension, conflict, and crisis that have happened among them. Secondly, Indonesia has the geopolitical views that also based on that Indonesia loves peace with neighbors and with the principles of no conflict and expansionism. Thirdly, both Governments are presently working together within the Commission of Truth and Friendship (CTF) that having the sacred goals of recommendations, namely peace, stability, and well-being of the people of both countries [6]. These facts have to be linked and summarized as supporting factors for the management of the borders, including a soft border management implementation, in such away that lead to the benefit of the people in the border areas.

Finally, it is important to provide the closeness situation of both peoples along the concerned un-surveyed segment, and to enhance people-to-people contact between communities of the two countries, because this will not only create significant multiplier effects, but further enhance closeness and mutual understanding between peoples of the two countries.

Acknowledgement. The Author would like to thank Dr. Sobar Sutisna (Center for Defense Boundary Research, University of Defense, Jakarta, Indonesia), Mrs. Tri Patmasari and Dr. Dewayani (Center for Boundary Mapping and the Geomatics Research Division, BAKOSURTANAL, Cibinong, Indonesia), for the supports. The Author also would like to thank Prof. João Matos from the Instituto Superior Técnico, Lisboa, Portugal, as his field counterpart and colleague of discussion.

References

1. Anderson, E.W.: Geopolitics: International Boundaries as Fighting Places. In: Gray, C.S., Sloan, G. (eds.) Geopolitics, Geography, and Strategy, Frank Cass, London (1999)
2. Hakim, L.: Border Diplomacy Related to Defense Aspect. Boundary Seminar: The Role of Research in Searching for Solution of the Boundary Problems, 23 February. University of Defense, Jakarta (2011)
3. Fernandes, R.M.S., Fahrurrazi, D., Matos, J., Handoyo, S.: The Common Border Datum Reference Frame (CBDRF) between Indonesia and Timor-Leste: Implementation and Processing. In: Cartografia E Geodesia. Actas da IV Conferência Nacional de Cartografia e Geodesia, Coordenação: João Casaca & João Matos, Lisboa (2005)
4. Flint, Colin: Introduction to Geopolitics. Routledge, London (2006)
5. Handoyo, S.: Menelaah Kompleksitas Permasalahan Batas Negara Darat RI-RDTL. In: Seminar Nasional Geomatika,Bakosurtanal. Cibinong, Indonesia (2011)
6. Joint Ministerial Commission: Joint Statement of the Fourth Meeting of the Joint Ministerial Commission for Bilateral Cooperation between The Republic of Indonesia and The Democratic Republic of Timor-Leste. Dili, Timor-Leste (July 2010)
7. Sutisna, S., Handoyo, S.: Delineation and Demarcation Surveys of the Land Border in Timor: Indonesian Perspective. In: Paper presented at The International Symposium on Land and River Boundaries Demarcation and Maintenance in Support of Borderland Development, Bangkok, Thailand (2006)
8. Strausz-Hupé, R.: Geopolitics: The Struggle for Space and Power. G.P. Putnam's Sons, New York (1942)
9. TSC-BDR RI-RDTL: Interim Report on the Land Border Delineation between Republic of Indonesia and Democratic Republic of Timor-Leste, Jakarta, Indonesia, vol. 1,2,3 (June 2004)

10. TSC-BDR RI-RDTL: The 15th Meeting of Technical Sub-Committee on Border Demarcation and Regulation between the Republic of Indonesia and the Democratic Republic of Timor-Leste, Yogyakarta, Indonesia (October 2004)
11. TSC-BDR RI-RDTL: The 19th Meeting of Technical Sub-Committee on Border Demarcation and Regulation between the Republic of Indonesia and the Democratic Republic of Timor-Leste, Surabaya, Indonesia (December 2005)
12. TSC-BDR RI-RDTL: The 21st Meeting of Technical Sub-Committee on Border Demarcation and Regulation between the Republic of Indonesia and the Democratic Republic of Timor-Leste, Bandung, Indonesia (July 2008)
13. TSC-BDR RI-RDTL: The 22nd Meeting of Technical Sub-Committee on Border Demarcation and Regulation between the Democratic Republic of Timor-Leste and the Republic of Indonesia, Dili, Indonesia (2009)
14. TSC-BDR RI-RDTL: The 23rd Meeting of Technical Sub-Committee on Border Demarcation and Regulation between the Republic of Indonesia and the Democratic Republic of Timor-Leste, Timor-Leste (August 2010)

CartoService: A Web Service Framework for Quality On-Demand Geovisualisation

Rita Engemaier and Hartmut Asche

University of Potsdam, Department of Geography,
Karl-Liebknecht-Strasse 24/25, 14476 Potsdam, Germany
{rita.engemaier,hartmut.asche}@uni-potsdam.de
http://www.geographie.uni-potsdam.de

Abstract. The last decades have seen a steady increase of digital spatial data and their effective availability. Embedded in the rapid developments in information and communication technology (ICT) such geospatial data or geodata are globally accessible, mainly via the internet, in a magnitude unseen before. In a parallel development of geographical information systems, computer-assisted cartography and the internet, a vast variety of web-based services have emerged to capture, store, analyse and present geodata. The map output from these systems is frequently suboptimal, lacking graphic expressiveness and effectiveness. This paper discusses a web-based service framework, the CartoService, to improve the geovisualisation quality of mapped geodata and provide laypersons and professionals with quality map graphics.

Keywords: geovisualisation, on demand mapping, web mapping, web cartography, web services, high quality mapping.

1 Introduction

The last decades have seen a steady increase of digital spatial data and their effective availability. Embedded in the rapid developments in information and communication technology (ICT) such geospatial data or geodata are globally accessible, mainly via the internet, in a magnitude unseen before. Today, about 80 percent of all digital data have a spatial dimension [1] and thus are geodata. Spatial data have penetrated our business as well as private life to an extend that Google and its geographical viewing and mapping services (Google Earth, Google Maps, Google Streetview) have become almost iconic representations of the present geoinformation era. At the same time the almost unrestricted availability of both geoinformation (GI) technology and geographic data have massively promoted the use of maps by professional and private users [2].

Today, a broad range of internet-based services for geodata are available on the internet, whether topographic or thematic, map data and maps on various global and regional scales. These services facilitate the acquisition, processing, presentation and dissemination of geographical information. Interactive and dynamic user participation in web-based geoinformation is one key characteristic

B. Murgante et al. (Eds.): ICCSA 2011, Part I, LNCS 6782, pp. 329–341, 2011.

of the so-called Web 2.0 [3]. In fact, web services and the underlying principle of Service Oriented Architectures (SOA) can be identified as the most important driving force in GI development of today.

Motivated by the rich geovisualisation capabilities of thematic (web) mapping techniques, this contribution presents a concept to provide professional cartographic modelling functionality and expertise to the digital production environment of recent maps. The overall objective of the CartoService approach presented here is the improvement of web mapping quality by an informed utilisation of web services for cartographic modelling. Based on a brief overview of relevant concepts of web cartography the respective web services (chapter 2) and the cartographic modelling process is outlined in its components and workflow (chapter 3). Against this background requirements and benefits of a web service for professional cartographic modelling are discussed (chapter 4). From this discussion a concept of the so called CartoService is extracted and presented including its major components and architecture. The present development stage of CartoService is subsequently assessed by relevant use case scenarios (chapter 5). A brief conclusion sums up major finding ot this ongoing research (chapter 6).

2 Cartography 2.0

The ongoing development of GI technology has forever changed and (r)evolutionised the traditional geodata processing disciplines of geodesy and cartography. Cartography has a long and rich tradition in storing and communicating spatial information by analogue map graphics. Originally developed for and applied to paper maps and atlases, cartographic expertise of processing and visualising geodata has accumulated and matured during the last five centuries. Prior to the advent of digital technologies cartographic methods have constantly been adapted to the changing production techniques [4]. Computer-assisted cartography and geoinformation systems have, however, both opened up and required the expansion and re-adjustment of this wealth of geovisualisation expertise. In fact, the adaption of traditional methods and knowledge to present-day geoinformation technology and environment remains the principal challenge of modern cartography, cf. [2], [5], [6].

The global spread of web-based ICT, in particular, has expanded the range of cartographic presentation methods and media, collectively referred to as Web Mapping 2.0 or Web Cartography 2.0 [2]. These and similar terms relate to the underlying ICT development termed Web 2.0 which denotes a "variety of innovative resources, and ways of interacting with, or combining web content" [2]. Accordingly Web Cartography 2.0 includes "Web 2.0 applications that have a spatial frame of reference" [3]. Web Mapping 2.0, in particular, summarises such diverse new geospatial web based applications as GeoTagging, GeoBlogging, Web Map Mashups or interactive geospatial Application Programming Interfaces (APIs) the number of which is still rapidly increasing [3]. The Web 2.0 environment has also been a productive breeding ground for the concept of

Service Oriented Architecture in software development in general as well as in GIS in particular.

To make the innovative potential of the internet platform and architecture fully available to modern cartography, geospatial standardisation organisations like the Open Geospatial Consortium (OGC) are of major importance to the definition and implementation of interoperability standards in web mapping and web cartography. The OGC defines Web services as "self-contained, self-describing, modular applications that can be published, located, and invoked across the Web" [7]. Key functionalities of web services include flexibility (from simple to complex processes) and interoperability (deployed once, discovered, invoked and used frequently). OGC standards, such as the Web Map Service (WMS), Web Feature Service (WFS) or Web Processing Service (WPS), have improved the interoperability of GIS and the dissemination of geoinformation tremendously. Although WMS and WFS are able to respond to the range of clients' requests in different geographical forms and formats, the principle feature of these services is the (technical) interchange of geodata. The informed construction of meaningful maps is neither supported nor supposed. As a consequence, the map output frequently lacks the cartographic modelling quality characteristic of maps generated in accordance with the principles of (thematic) cartography. To specify such deficits and identify potential weak points in recent web map compilation and production, the cartographic modelling process is analysed subsequently.

3 The Cartographic Modelling Process

The main objective for the construction and dissemination of cartographic presentations, e.g. maps, is to communicate geospatial information in an adequate, efficient and intuitive manner. That is why maps are required to have clear-cut content, explicit map symbols, easy-to-comprehend map graphics and an attractive overall map design [8], [9]. It is now generally accepted that maps are models, in particular analogue graphic models, of the environment. Prior to the digital age the analogue map has been the only medium to store and communicate the position of spatial objects (topography) as well as their neighbourhood (topology) at the same time. The spatial structure of any given region can thus be explored at one glance. While the storage of spatial data has become the domain of geo databases, the map graphic has, for the time being, not been substituted by any other medium for the visual analysis of spatial distributions, structures and positions of geographical objects. This is accomplished by generating a graphic spatial model of the environment from a graphic-free data model. Accordingly maps are secondary (graphic) models derived from existing primary (numeric) data models explored and analysed by the individual user. From the map graphic each user creates her individual tertiary (mental) model of the environment (Fig. 1).

Efficient communication of spatial objects and structures via (carto-)graphic models requires a substantial overlap of the reality models involved, particularly of the cartographic and the mental model. The best way to achieve this is to provide the user with a professional graphic representation of the geographical relationships which we term a quality map. Quality maps are composed of abstracted spatial symbols and generalised line and area structures originating from a method-driven cartographic modelling process including technical and mental constraints. Whether conventional or digital, professionally constructed map graphics have a proven record of efficient communication of geographic data, effective visual exploration and analysis facilitating intuitive comprehension and interpretation of complex spatial structures by the user. In the last two decades, this unrivalled capacity of quality maps has successfully been adapted for the "mapping" and representation of non-spatial data by the so-called map metaphor.

3.1 Cartographic Communication Model

It has to be stressed that the cartographic communication model briefly discussed above has been developed in the pre-digital era. In the geoinformation era of almost ubiquitous geodata and geo processing tools it is no longer the trained cartographer who is only able to construct maps and distribute them. It is also the user who can manipulate a map's design, even its data and save as well as disseminate the result, preferably on the internet. As a consequence, the one well-defined roles of map producer and map user have become blurred, which, in turn, necessitates a revision of the classic one-directional cartographic communication model (Fig. 1, a) [10].

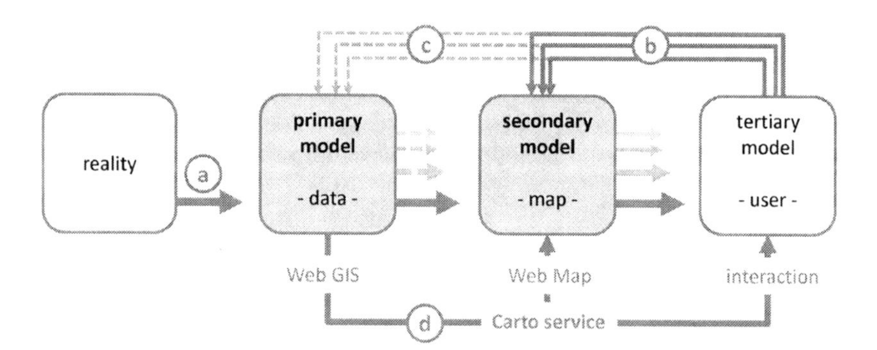

Fig. 1. Cartographic communication model; a: the classic one-directional cartographic information flow; b: interaction; c: information access through the map interface; d: representing the CartoService bus; [10]

Modern web-based maps offer dynamic, interactive (two-way) communication and information exchange [11]. They facilitate extensive interactive map use by providing user access to the map graphic (the secondary model) as well as the interconnected map data (the primary model)(Fig. 1, b, c). In fact, the map graphic can be considered the graphic of the map dataset which allows the user to select and filter objects, properties and relations from the graphic-free primary data. In the wider context, digital networks, such as the internet or intranet, offer numerous alternatives to access and use a wide range of dynamically scalable storage and processing components to process complex tasks interactively in acceptable answer time. Both, the IT environment and the capabilities of the web-based map generation are a prerequisite for the development and provision of a high quality cartographic visualisation service on the internet.

In the conceptual phase of such cartographic visualisation service the process of proper map construction needs to be broken down in to a sequence of components each of which requires further consideration. The resulting component structure is modular and generic and can thus be adapted to changing input data or mapping task requirements. The definition of connections between the map construction components is of particular importance in the conceptual development of such service framework to ensure generation of a meaningful map construction workflow.

3.2 Map Construction Components and Workflow

Map construction, whether conventional or web-based, can be anatomised into five major components, cf. [9], [12], [13]. In the web-based CartoService under development these components are coupled by a one-directional workflow (Fig. 2).

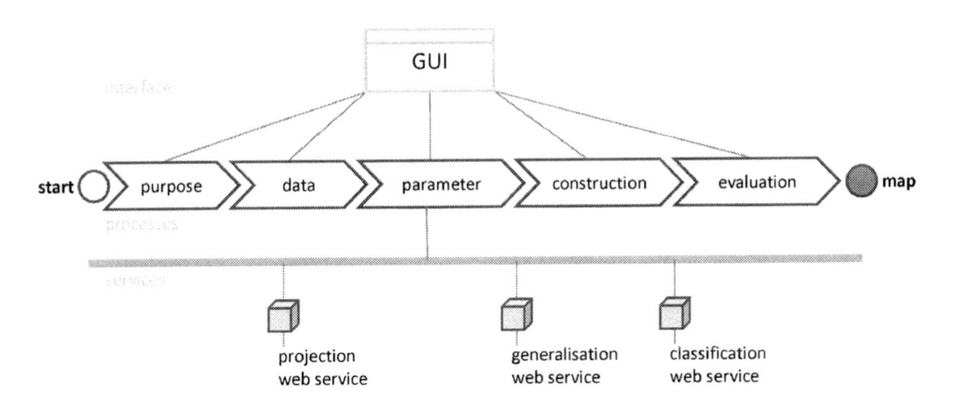

Fig. 2. Modular map construction workflow with coupled web services

To compile a complete map the whole process chain has to be processed sequentially. The map compilation can only be continued with the second component when the first component has been processed etc. Once the map construction process is initiated (Fig. 2: start) the first component is the *map purpose* component necessitating a decision on the map purpose. Parameters of choice include general public map, scientific map; screen map, print map; view-only map, interactive map etc. Depending on the selection of map purpose the *data* component governs the identification and acquisition of the data required using appropriate catalog services. Data sources can range from hard copy files or geospatial databases to certain feature sets provided by web services (e.g. WFS). The data selected are subsequently processed according to the decision taken in the *modelling* component. Defined by the choices made in the map purpose and map data components, respectively, data modelling is governed by parameters such as map scale, classification of the source data or level of semantic and graphic generalisation. The non-graphic modelling of map data is a prerequisite for the actual generation of the map graphic in the *construction* component. This component facilitates the selection of map style, symbol, colour, text and label placement etc. The construction component concludes the map compilation, composition and rendering process. This component provides both a preview and the final map graphic. In the final *evaluation* component the map graphic generated is validated, modified and re-processed, if necessary. The selection parameters of each component can be re-adjusted via a graphical user interface (GUI) and an API as well (Fig. 2: GUI). While the user interface facilitates the interactive visual control of parameters and workflow, the programming interface provides a suitable connection to other external software components or services. The CartoService approach can therefore be understood as a workflow management service which is able to select, orchestrate and chain catalog, data and processing services due to the user and map purpose requirements [14].

4 The CartoService Approach

At present, the concept of a web-based CartoService, as outlined above, is in its development stage. Its sequential ordering provides the layperson with an easy-to-understand guide to quality map production. Such quality web-map service is not available on the internet to date. A brief look at the map production functionalities of current web-based atlas information systems (AIS), such as the national atlases of Canada [15] and the USA [16] or the series of Australian regional atlases [17], will readily confirm this assessment. Basically, web maps created from those AIS are graphic presentations of the non-graphic geodata modelling results (primary model) in the AIS-GIS. To the layperson the resulting map graphic may look like a map. However, a professional appreciation will find that the map-like graphic lacks almost all quality features that define a proper map. The most notable of these is the separate graphic modelling of the map graphic (secondary model) on the basis of the data model. Well known from visual outputs of such AIS and geographical information systems these map-like presentations are referred to as displays [18]. For the time been this secondary (graphic) modelling process can not be automated. Automated transformation of the primary data model into the secondary data model has not been achieved yet and remains an unsolved research problem, cf. [6]. As a consequence, the majority of web maps (like the majority of digital maps) generated by a simple graphic presentation of the data model fall short of cartographic quality as put forward by [18], [19] and [20]. Such suboptimal map graphics, in essence, lack the indispensable potential to communicate, explore and analyse spatial data visually. CartoService aims to address this fundamental cartographic problem by providing a rule-based framework for quality map-production including the proven techniques of thematic cartography and map types. For that purpose, CartoService offers a component-based environment for quality map production. As has been presented above, each component provides the resources or services required to go through the modelling and visualisation process. The architecture, components and processes of CartoService are based on the principles of SOA, in particular the publish-find-bind paradigm and the loose coupling [21].

4.1 CartoService Requirements

CartoService first and foremost attempts to improve the graphic modelling quality of web-based maps by limiting the amount of visual clutter in the map graphic. For that purpose, three different aspects have to be addressed: first, improvement of the map graphic itself (resolution, colour space), second, implementation of principle methods of (thematic and topographic) cartography into the automated construction of internet maps, and, third, integration of interaction and map dynamics into the map construction workflow.

Aspects one and two roughly correspond to the criteria of expressiveness and effectiveness[1] [22], when applied to cartographic symbol language. The first aspect (map graphics quality) is related to improvements in the rendering quality, resolution, data formats and colour space of digital web map graphic. Specific cartography-related map graphic (quality) criteria are object-symbol-relation, graphic density relating to topic and scale, level of graphic generalisation and colour schemes.

The second aspect relates to a particular phenomenon primarily found in web maps which, as yet, has rarely been discussed. This phenomenon can be characterised as the lack of cartographic principles in web map visualisations. The bulk of web maps use rather simple visualisation techniques, as a glance at the AIS mentioned above will confirm. Even well-established visualisation principles like the visual variables [23] or dynamic visual variables described by [24] and [25] are hardly made use of. A cursory analysis of the map graphics of internet maps shows that the vast majority of these are suboptimal to inappropriate. Equivalent to GIS presentations most of those maps are missing generalisation, adequate symbolisation and text and label placement [18]. In fact, the majority of thematic web maps are either of the choropleth type or simple composition of point, line or area symbols. From the ten professional methods to visualise geospatial data ([8],[9]) and a total of eleven graphic variables [26] only a few are applied in web map construction. More complex or sophisticated methods of geodata visualisation are rarely found in recent internet maps. The implementation

[1] Expressiveness determines "whether a graphical language can express the desired information", whereas effectiveness determines "whether a graphical language exploits the capabilities of the output medium and the human visual system."

of thematic mapping methods and techniques in standard web mapping services, in particular, is an ongoing research topic in geovisualisation (e.g. [27], [28], [29], [30]). To make available the entire spectrum of cartographic visualisation methods for automated web map production including a data- and purpose-specific summary, is a major task addressed by CartoService. Through its modular composition CartoService will filter, present and apply relevant cartographic visualisation methods to user specific datasets and presentation purposes, respectively. The third major objective is the integration of interaction and dynamics into the automated map construction process. In the current development and implementation phase of CartoService this last aspect is subordinated.

4.2 CartoService Components and Architecture

The CartoService web map environment is conceptualised as a management web service which itself is able to search, select, orchestrate and chain different web services. These services are integrated as loosely coupled components embedded in the generic architecture of the CartoService framework. The management functionality of CartoService offers user- and purpose-centred support for high quality map compilation. Basic components of CartoService are: the graphical user interface, data filter and analyst, method selector, compiler, the style library and the rendering machine. Each component is linked and managed by CartoService via a service programming interface. Because of its generic architecture the integrated components (services) can be flexible exchanged in the framework if requirement. General service processing and parameter setting is controlled through the GUI and stored in a user and task specific profile. The CartoService processing chain will be initiated by the input of spatial dataset(s) or data service delivered feature set(s). After reading the input dataset(s) CartoService will process a standard workflow which uses default values for each component and component parameters. In the first implementation phase default values are derived from expert knowledge and cartographic expertise. At a later implementation stage feedback from the evaluation of CartoService maps will add information on about cartographic quality to the rule base.

The provision of CartoService functionality is based on the well-known publish-find-bind paradigm in SOAs (Fig. 3). To allow for requests to CartoService a registry is published in machine readable form, which is available to clients via internet. Using the registry, clients are able to locate the service host, link to it and receive a description of the service functionality provided. Subsequently the information required is exchanged and the service can be used to send, process geodata and finally to request appropriate geovisualisation in web map form. Users (human clients) are able to access CartoService through a GUI provided on the CartoService website. An alternative option to use CartoService functionality is through plug-ins or extensions of standard proprietary or open source GI and map construction systems.

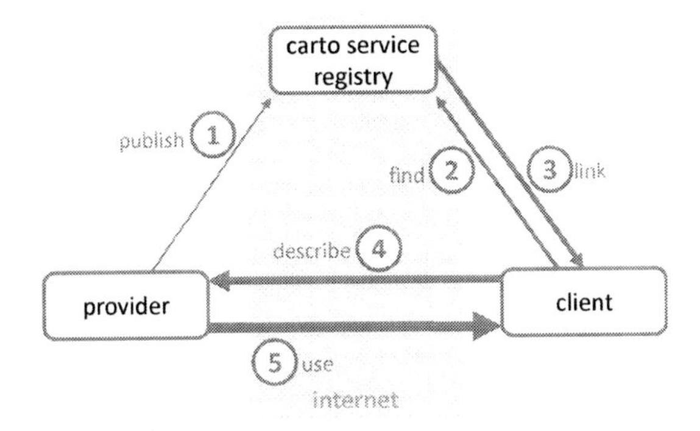

Fig. 3. CartoService architecture

5 CartoService Use Cases and Evaluation

Based on CartoService and its components, architecture and requirements described above a number of use case scenarios for CartoService have been designed and analysed prior to full-scale implementation. In a first stage an operational prototype has been developed and selected features have been evaluated.

5.1 Purpose versus Data-Driven Demand

The two most important use cases for CartoService are: first, the purpose-driven and, second, the data-driven use case. The purpose-driven use case focuses on the task or communication related objectives of the map maker (user). Her task can be described by: I want to communicate geospatial information. What is the most appropriate visualisation method? Where can I find relevant data resources? Which media-specific visualisation modes can be applied? The (second) data-driven use case can be described as: I have particular geospatial data or data services about which I don't know details. Which geospatial information is accessible by the application of geovisualisation methodology?

While the first use case addresses a typical, traditional question in map construction, the second reflects present-day objectives (exploration) and user centricity (layperson) in the usage of geospatial data and map production. It is almost paradoxical that the ubiquitous availability of geodata has not increased the knowledge about the origin and features of such data in the same way, let alone basic knowledge of proper visualisation of this data. The controversial discussion about Google's streetview data, at least in central Europe, provides a clear proof of this knowledge deficit. CartoService advises the user to select and apply the best-fitting visualisation method by analysing the data (with data mining techniques) and providing the adequate geovisualisation (Fig. 4). The resulting quality map can then be further refined by the user in an iterative, interactive manipulation process. However, the underlying basic methodical decisions cannot be revised.

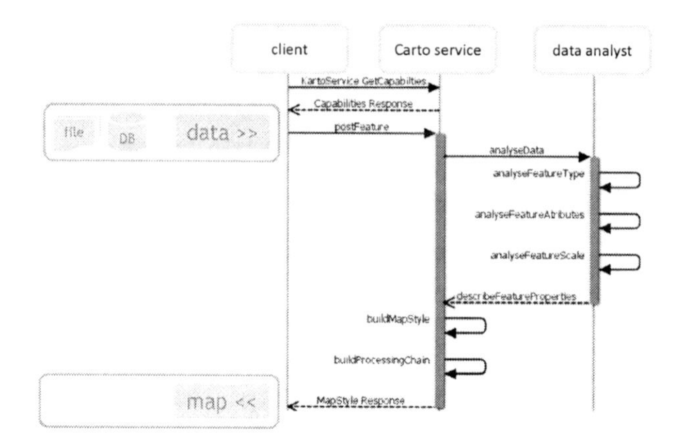

Fig. 4. Part of the simplified CartoService sequence diagram

The map produced can then again be disseminated via standard web map services (WMS). In addition CartoService stores and, on demand, provides the processing and styling information for each component service used in the workflow. Stored as user and workflow profiles such information can be used to establish user, purpose or data specific visualisation patterns. These patterns can be utilised to produce maps with similar (corporate) cartographic design and quality.

5.2 CartoService Process Chain Evaluation

In the ongoing research the development of CartoService has left its conceptual stage. Because of the complexity presented above it has not been implemented yet. However, a first assessment can be made by evaluating the process chain of the service components and the relevant parameter settings. Existing cartography-related web services have been investigated and utilised to simulate the CartoService processing chain. The frames of reference of this research have been the principle components and processes of cartographic modelling. As dynamic binding and linkage of resources is not available to date, data had to be loaded and exchanged manually. Apart from services providing geographical data in vector (WFS) or raster format (WMS) e.g. by Geoserver or Mapserver, web services have been used to generalise data. Other services have been used to detect and apply colour schemes to visualise datasets already classified. The MapShaper [31] and ColorBrewer [32] services were used here. Not all components conceptualised in the CartoService framework are available as web services to date. That is why missing components, e.g. classification, were be substituted by GI software for the time being. As [33] and [34] have shown such GIS functionality can be implemented into a SOA-based environment.

The CartoService process chain has also been evaluated for purpose-driven use case scenarios. For this assessment, the map construction purpose was the visualisation of population density in the German federal state of Brandenburg in

2003 at county level. First results of the application of the CartoService workflow to the test data are shown in figure 5. The results of this first application of CartoService are promising when it comes to improve cartographic quality. A full and detailed evaluation of the CartoService mapping quality has not been done at this stage, further research and implementation efforts are required. The evaluation of this component remains a major issue in the ongoing research.

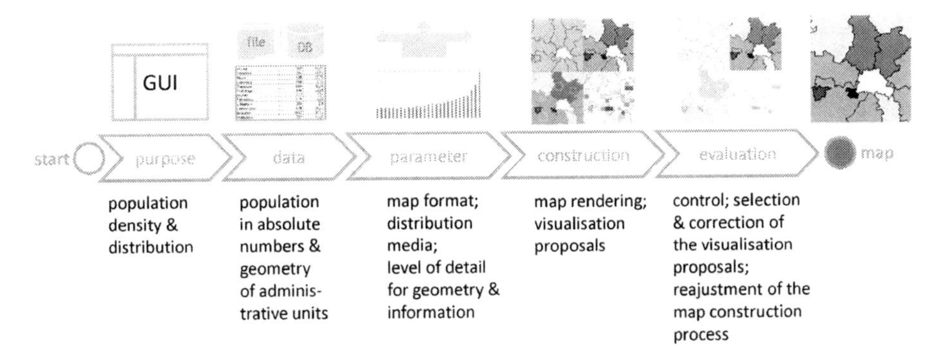

Fig. 5. CartoService process applied to population density visualisation

6 Conclusion

For the time been CartoService has not been implemented completely and evaluated. At the present stage of development it has, however, been demonstrated that the approach presented here offers significant potential to improve web mapping quality in an extensive automated map construction workflow. It has also been shown clear intelligent chaining and orchestration of web mapping related services, completed by functionalities of user interaction and manipulation, can generate a consistent SOA-based web mapping workflow and related service quality. CartoService provides the basic management and control components as well as the architecture to improve geovisualisation quality throughout the complete map construction workflow. Thanks to its modular and generic structure, dynamic functionality and interoperability CartoService empowers laypersons and professional map makers alike to add cartographic visualisation expertise to their mapping purpose. Eventually CartoService will thus help to safeguard minimum map quality standards in any geodata visualisation.

References

1. Franklin, C.: An Introduction to Geographic Information Systems: Linking Maps to Databases. Database, 13–21 (April 1992)
2. Peterson, M.P.: The brave new world of online mapping. In: Proceedings of ICC 2009, Santiago de Chile (2009)

3. Gartner, G.: Applying WebMapping 2.0 to Cartographic Heritage. Online Journal: e-Perimetron 4 (2009)
4. Monmonier, M.: Technological transition in cartography. Madison (1955)
5. Kraak, M.J., Brown, A. (eds.): Web cartography: developments and prospects, p. 213. Taylor and Francis, London (2001)
6. Virrantaus, K., Fairbairn, D., Kraak, M.-J.: ICA research agenda on cartography and GIScience. Cartography and Geographic Information Science 2(36), 209–222 (2009)
7. Open Geospatial Consortium (OGC): Glossary: Web Service (2007), http://www.opengeospatial.org/ogc/glossary/ (accessed 2011-02-11)
8. Dent, B.D.: Cartography: thematic map design, vol. 5. WCB/McGraw-Hill, Boston, Mass (1999)
9. Slocum, T.A.: Thematic Cartography and Visualization, 2nd edn. Pearson Prentice Hall, Upper Saddle River (2004)
10. Dickmann, F.: Einsatzmöglichkeiten neuer Informationstechnologien für die Aufbereitung und Vermittlung geographischer Informationen – das Beispiel kartengestützter Online-Systeme, p. 182 S. Verlag Erich Goltze GmbH & Co. KG, Göttingen (2004)
11. Visvalingam, M.: Cartography, GIS and Maps in Perspective. Cartographic Journal 26(1), 26–32 (1989)
12. Witt, W.: Thematische Kartographie, vol. 2. Jänecke, Hannover (1970)
13. Engemaier, R., Schernthanner, H.: Kartenkonstruktion mit Freier und Open-Source-Software. In: Strobl, J., Blaschke, T., Griesebner, G. (eds.) Angewandte Geoinformatik 2010. Beiträge zum 22. AGIT-Symposium Salzburg, July 7-9, 2010, Universität Salzburg, Heidelberg (2010)
14. Alameh, N.: Chaining Geographic Information Web Services. IEEE Internet Computing 7(5), 22–29 (2003)
15. The Atlas of Canada (2011), http://atlas.nrcan.gc.ca/ (accessed 2011-02-11)
16. National Atlas of the United States (2011), http://www.nationalatlas.gov/ (accessed 2011-02-11)
17. Atlas of South Australia (2011), http://www.atlas.sa.gov.au/ (accessed 2011-02-11)
18. Van der Merwe, F.: GIS, Maps and Visualization. In: Proceedings of ICC, Durban, South Africa (2003)
19. Buckley, A., Frye, C., Buttenfield, B.: An information model for maps: towards cartographic production from GIS databases. Mapping approaches into a changing world. In: Proc. 22nd Intern. Cartogr. Conference A Coruna [CD-ROM], July 9-16, 2005, International Cartographic Association, A Coruna (2005)
20. Hardy, P.: High-Quality Cartography in a Commodity GIS: Experiences in Development and Deployment. In: ICA symposium on cartography for Central and Eastern Europe [CDROM], February 16-17, Technische Universität Wien/ International Cartographic Association, Wien (2009)
21. Melzer, I.: Service-orientierte Architekturen mit Web Services: Konzepte - Standards - Praxis. - 4. Aufl. Spektrum Akademischer Verlag, Heidelberg (2010)
22. Mackinlay, J.D.: Automating the Design of Graphical Presentations of Relational Information. ACM Transactions on Graphics 5(2), 110–141 (April 1986)
23. Bertin, J.: Graphische Semiologie - Diagramme, Netze, Karten. In: Übersetzt und bearbeitet nach der 2. Französischen Auflage, Walter de Gruyter, New York (1974)
24. MacEachren, A.M.: How Maps Work - Representation, Visualization and Design. The Guilford Press, London (1995)

25. Robinson, A.H., Morrison, J.L., Muehrcke, P.C., Kimerling, A.J.: Elements of Chartography. John Wiley & Sons, Chichester (1995)
26. Ellsiepen, I.: Methoden der effizienten Informationsübermittlung durch Bildschirmkarten. Dissertation, Bonn (2004)
27. Dietze, L., Zipf, A.: Extending OGC styled layer descriptor (SLD) for thematic cartography - Towards the ubiquitous use of advanced mapping functions through standardized visualization rules. In: 4th Int. Symp. on LBS and Telecartography (2007)
28. Hugentobler, M., Iosifescu-Enescu, I., Hurni, L.: A Design Concept for Implementing Interoperable Cartographic Services based on reusable GIS Components. In: Proceedings of ICC, Moscow (2007)
29. Rita, E., Borbinha, J., Martins, B.: Extending SLD and SE for Cartograms. In: Proceedings of the GSDI 12 world conference, Signapore, 19-22 October (2010)
30. Sae-Tang, A., Ertz, O.: Towards Web Services Dedicated to Thematic Mapping. OSGeo Journal, Vol. 3 (accessed 2011-02-20) (2007), `http://www.osgeo.org/files/journal/v3/en-us/final_pdfs/ertz.pdf`
31. Harrower, M., Bloch, M.: Mapshaper.org: A map generalisation web service. IEEE Computer Graphics and Applications 26(4), 22–27 (2006)
32. Brewer, C.: A Transition in Improving Maps: The ColorBrewer Example. Cartography and Geographic Information Science 30(2), 159–162 (2003)
33. Brauner, J.: Providing GRASS with a Web Processing Service Interface. In: GI-Days 2008, Münster (2008)
34. Bergenheim, W., Sarjakoski, L.T.: A Web Processing Service for GRASS GIS to Provide on-line Generalisation. In: Proceedings of the 12th AGILE International Conference on Geographic Information Science, pp. 1–10. Leibniz Universität, Hannover (2009)

An Analysis of Poverty in Italy through a Fuzzy Regression Model*

Silvestro Montrone, Francesco Campobasso, Paola Perchinunno, and Annarita Fanizzi

Department of Statistical Science, University of Bari, Via C. Rosalba 53, 70100 Bari, Italy
{s.montrone,fracampo,p.perchinunno,a.fanizzi}@dss.uniba.it

Abstract. Over recent years, and related in particular to the significant recent international economic crisis, an increasingly worrying rise in poverty levels has been observed both in Italy, as well as in other countries. Such a phenomenon may be analysed from an objective perspective (i.e. in relation to the macro and micro-economic causes by which it is determined) or, rather, from a subjective perspective (i.e. taking into consideration the point of view of individuals or families who locate themselves as being in a condition of hardship). Indeed, the individual "perception" of a state of being allows for the identification of measures of poverty levels to a much greater degree than would the assessment of an external observer. For this reason, experts in the field have, in recent years, attempted to overcome the limitations of traditional approaches, focusing instead on a multidimensional approach towards social and economic hardship, equipping themselves with a wide range of indicators on living conditions, whilst simultaneously adopting mathematical tools which allow for a satisfactory investigation of the complexity of the phenomenon under examination. The present work elaborates on data revealed by the EU-SILC survey of 2006 regarding the perception of poverty by Italian families, through a fuzzy regression model, with the aim of identifying the most relevant factors over others in influencing such perceptions.

Keywords: Fuzzy logic, poverty, regression model, Eu-Silc.

1 Introduction

The profound economic and social transformations witnessed in recent decades have underlined the necessity of analyzing the phenomenon of poverty in terms of its multiple facets. The identification of the poor as a subject living on the edge of society (as, for example, the homeless) appears to have already been superseded in favour of a growing academic focus placed on general context, including both economic hardship and social exclusion.

* The contribution is the result of joint reflections by the authors, with the following contributions attributed to S. Montrone (chapter 4), to F. Campobasso (chapter 1 and 2), to P. Perchinunno (chapter 3.1 and 3.2), and to A. Fanizzi (chapter 3.3).

B. Murgante et al. (Eds.): ICCSA 2011, Part I, LNCS 6782, pp. 342–355, 2011.

The numerous definitions found in the literature are almost all retraceable to the traditional distinction between absolute and relative poverty: the first understood as the incapacity to reach an objective level of wellbeing, independent of relevant social and temporal contexts; the second definition is, instead, based on the assumption that the social condition of an individual cannot be adequately defined without taking the environment in which they live as a starting point. In one case an individual is thus considered as poor if he is not be able to satisfy his primordial needs; whilst in the other case, if the individual live is in a worse state than the standard of the particular community in which he is located.

A transversal approach is therefore proposed as an alternative to the above, considered as subjective, through which the poor are defined as those who identify themselves as such, even if this identification is revealed as a result of the comparison that they operate with the rest of society in terms of perceived wellbeing. It is, therefore, the "perception" an individual may have of their own condition which allows for the identification of the measure of poverty to a far greater degree than the assessment of an external observer would allow.

It should therefore be noted that such a line of enquiry is part of the wider trend which attempts to focus, in particular, on the multidimensional nature of poverty i.e. on the necessity to take into consideration not only one single indicator but a group of indicators (considered useful in the definition of greater or lesser degrees of hardship in the individual observed). This approach recalls, in particular, the work of Renè Lenoir [8] on social exclusion, the human poverty index in the United Nations Report on Human Development as well as the work of A. Sen [10]on functioning and capability.

In this context the adoption of a fuzzy numbers theory, introduced by L. A. Zadeh [7,14], is considered as valuable and allows the intrinsic complexity of the phenomenon under investigation to be adequately taken into account. Indeed, Zadeh underlined that nature frequently does not present us with a set composed of clearly separate objects, to which it is possible to apply classical principals of set theory such as that of the principle of non-contradiction or the excluded middle; he thus introduced the concept of the degree of membership which, in opposition to classical theory (according to which a specific property may be proved as either true or false) also allows for possible intermediate values of veracity.

The principal advantage of fuzzy logic lies precisely in its ability to align itself with human interpretation and, in the case in point, allows for the rejection of the "rich/poor" dichotomy in order to take into account the variety of levels which exist between the two extreme conditions (marked material hardship and high-level wellbeing).

In the present work, data are elaborated arising from the EU-SILC survey of 2006 [5], regarding the perception of poverty by Italian families. The interviewees express, in everyday language, their personal opinions regarding the relative burdens of various expenses with which they are faced (for example, in the form of categories, ordered on a scale ranging from "negligible burden" to "heavy burden"). This creates an initial problem associated with the objective difficulty and the inevitable subjectivity that characterizes every attempt at quantification of qualitative variables, given that the arising personal opinions are distributed along an ideal continuum, while the actual number assigned to each mode of a categorical variable is only a value of densification on the considered scale. The categories which constitute the resulting ordinal scale are, in any case, verbal labels that correspond to vague and uncertain sets and find their ideal representation through the use of tools which are fuzzy in nature.

Precisely a fuzzy regression model, considered as more efficient than a traditional logit model, is applied in order to identify the factors that most influence the perception of poverty over others [2,3].

2 A Fuzzy Regression Model

2.1 Introduction

Fuzzy regression techniques can be used to fit fuzzy data into a regression model [6,11,12]. Diamond [4] treated the case of a simple model introducing a metrics into the space of triangular fuzzy numbers; recently we provided some theoretical results about the estimates of the coefficients in a multiple regression model and about the decomposition of the sum of squares of the dependent variable. In this work, as the estimation procedure of a fuzzy regression model changes depending on whether the intercept is fuzzy or not, we explicit the expression of the parameters in the multiple case, starting from the simple model handled by Diamond.

2.2 A Fuzzy Approach to the Least Squares Regression Model

A *triangular fuzzy number* $\tilde{X} = (x, x_L, x_R)_T$ for the variable X is characterized by a function $\mu_{\tilde{X}} : X \to [0,1]$, like the one represented in Fig. 1, that expresses the *membership degree* of any possible value of X to $\tilde{X}$.

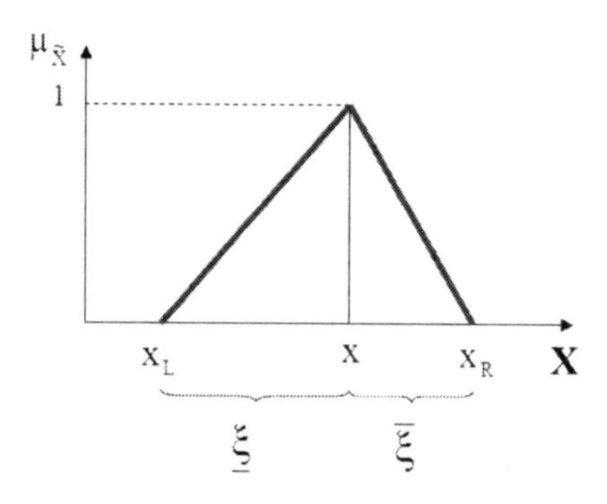

Fig. 1. Representation of a triangular fuzzy number

The x value is considered the *core* of the fuzzy number, while $\overline{\xi} = x - x_L$ and $\underline{\xi} = x_R - x$ are considered the *left spread* and the *right spread* respectively. Note that x belongs to $\tilde{X}$ with the highest degree (equal to 1), while the other values included between the *left extreme* x_L and the *right extreme* x_R belong to $\tilde{X}$ with a gradually lower degree.

The set of triangular fuzzy numbers is closed under addition: given two triangular fuzzy numbers $\tilde{X}=(x,x_L,x_R)_T$ and $\tilde{Y}=(y,y_L,y_R)_T$, their sum $\tilde{Z}$ is still a triangular fuzzy number $\tilde{Z}=\tilde{X}+\tilde{Y}=(x+y,x_L+y_L,x_R+y_R)_T$. Moreover the opposite of a triangular fuzzy number $\tilde{X}=(x,x_L,x_R)_T$ is $-\tilde{X}=(-x,-x_R,-x_L)_T$.

It follows that, given n fuzzy numbers $\tilde{X}_i=(x_i,x_{Li},x_{Ri})_T$, $i=1, 2, .., n$, their average is $\overline{X}=\frac{\sum \tilde{X}_i}{n}=\left(\frac{\sum x_i}{n},\frac{\sum x_{Li}}{n},\frac{\sum x_{Ri}}{n}\right)_T$.

Diamond introduced a metrics onto the space of triangular fuzzy numbers; according to this metrics, the squared distance between $\tilde{X}$ and $\tilde{Y}$ is

$$d(\tilde{X},\tilde{Y})^2=d((x,x_L,x_R)_T,(y,y_L,y_R)_T)^2=(x-y)^2+(x_L-y_L)^2+(x_R-y_R)^2.$$

The same Author derived the expression of the estimated coefficient in a fuzzy regression model of a dependent variable $\tilde{Y}$ on a single independent variable $\tilde{X}$.

In this work we generalize this estimation procedure to the case of several independent variables with a fuzzy asymmetric intercept, which seems more appropriate than the non-fuzzy one as it expresses the average value of the dependent variable (which is also fuzzy) when the independent variables equal zero.

Assuming to regress a dependent variable $\tilde{Y}_i=(y_i,y_{Li},y_{Ri})_T$ on k independent variables $\tilde{X}_{ij}=(x_{ij},x_{Lij},x_{Rij})_T$ in a set of n units, the linear regression model with a fuzzy intercept is given by

$$\tilde{Y}_i^*=\tilde{A}+\sum b_j\tilde{X}_{ij}, \qquad i=1,2,...,n; \quad j=1,2,\dots,k; \quad a, b_j \in \mathbb{R}$$

with * denoting the theoretical value, where $\tilde{A}=(a,a_L,a_R)_T$ and $a_L=a-\underline{\gamma}$, $a_R=a+\overline{\gamma}$ and $\underline{\gamma}$, $\overline{\gamma}>0$. The estimates of the fuzzy regression coefficients are determined by minimizing the sum of Diamond's squared distances between theoretical and empirical values of the dependent variable

$$\sum d(\tilde{Y}_i,\tilde{A}+\sum b_j\tilde{X}_{ij})^2$$

respect to a, b_1, .., b_k, $\underline{\gamma}$, and $\overline{\gamma}$. The function to minimize assumes different expressions according to the signs of the regression coefficients b_1, .., b_k.

In matricial terms, the estimates of the fuzzy regression coefficients are given by

$$\boldsymbol{\beta}=(\mathbf{X}'\mathbf{X}+\mathbf{X_L}'\mathbf{X_L}+\mathbf{X_R}'\mathbf{X_R})^{-1}(\mathbf{X}'\mathbf{y}+\mathbf{X_L}'\mathbf{y_L}+\mathbf{X_R}'\mathbf{y_R}),$$

where:

$\mathbf{y} = [\, y_i \,]$ is the n-dimensional vector of centers of the dependent variable;

$\mathbf{y_L} = [\, y_{Li} \,]$ and $\mathbf{y_R} = [\, y_{Ri} \,]$ are the n-dimensional vectors of the lower extremes and the upper extremes respectively of the dependent variable;
$\mathbf{X}$ is the n×(k+3) matrix formed by vectors **1**, k vectors of cores of the independent variables and two vectors **0**;
$\mathbf{X_L}$ is the n×(k+3) matrix formed by vector **1**, k vectors of lower bounds of the independent variables and vectors **-1** and **0**;
$\mathbf{X_R}$ is the n×(k+3) matrix formed by vector **1**, k vectors of upper bounds of the independent variables and vectors **0** and **1**;
$\boldsymbol{\beta}$ is the vector $(a, b_1, .., b_k, \underline{\gamma}, \overline{\gamma})'$.

The total sum of squares of the dependent variable can be decomposed according to Diamond's metrics. In particular we obtain the expressions of two components, the regression sum of squares and the residual one, like in the OLS estimation procedure for classical variables. This happens only because the intercept has exactly the same form as the dependent variable and, therefore, the theoretical and empirical values of the average fuzzy dependent variable coincide. Synthetically, denoting the fuzzy average of the dependent variable with:

$$\overline{Y} = \left(\frac{\sum y_i}{n}, \frac{\sum y_{Li}}{n}, \frac{\sum y_{Ri}}{n} \right)_T = (\overline{y}, \overline{y}_L, \overline{y}_R)_T ,$$

the expression of the total sum of squares:

$$\text{Tot SS} = \sum d(\tilde{Y}_i, \overline{Y})^2 = \sum [(y_i - \overline{y})^2 + (y_{Li} - \overline{y}_L)^2 + (y_{Ri} - \overline{y}_R)^2]$$

can be written as:

$$\text{Tot SS} = \text{Reg SS} + \text{Res SS}$$

where:

$\text{RegSS} = \sum d(\tilde{Y}_i^*, \overline{Y})^2 = \sum [(y_i^* - \overline{y})^2 + (y_{Li}^* - \overline{y}_L)^2 + (y_{Ri}^* - \overline{y}_R)^2]$ represents the regression sum of squares,

$\text{ResSS} = \sum d(\tilde{Y}_i, \tilde{Y}_i^*)^2 = \sum [(y_i - y_i^*)^2 + (y_{Li} - y_{Li}^*)^2 + (y_{Ri} - y_{Ri}^*)^2]$ represents the residual sum of squares.

A fuzzy version of the index R^2, which may be called Fuzzy Fit Index (FFI), will be used in order to evaluate how the model fits data. Its expression is:

$$\text{FFI} = \frac{\sum d(\tilde{Y}_i^*, \overline{Y})^2}{\sum d(\tilde{Y}_i, \overline{Y})^2} .$$

The more this index is next to one, the better the model fits the observed data.

Finally we propose a stepwise procedure in order to simplify, in terms of computational, the identification of the most significant independent variables. At each iteration, this procedure inserts a variable in the regression equation, according to two fundamental criteria: the significance of the contribution made by single variable, measured by the relative increase of the total deviance in the dependent variable, and its originality, that is the ability to introduce information into the equation other variables have not already introduced (assessed in terms of correlation with the latter).

3 The Application of the Fuzzy Approach

3.1 The Construction of Eu-Silc Indicators

Poverty is an extremely complex and multifaceted phenomenon which cannot be reduced to its purely economic or monetary dimensions. Since 2004 Italy has participated within the *EU-SILC, the European Standard on Income and Living Conditions*, which harmonizes surveys on living conditions and incomes of Italian families at EU level. This sample survey replaces the *European Community Household Panel* (ECHP) and has as its core objective the provision of comparable data, both transversal and longitudinal, for the analysis of income distribution, welfare and standards of living for families as well as economic and social policy [9].

In the present study a range of information resulting from EU-SILC survey of 2006 is considered, of relevance in terms of the difficulties faced by Italian families in "getting through to the end of the month", which were revealed through the completion of a questionnaire presented to 21,499 families.

Table 1. Distribution of households surveyed by level of hardship in terms of "getting through to the end of the month"

Getting through to the end of the month...	Absolute values	%
1 with great difficulty	2,884	13.4%
2 with difficulty	4,181	19.4%
3 with some difficulty	8,643	40.2%
4 fairly easily	4,506	21.0%
5 easily	1,115	5.2%
6 very easily	170	0.8%
Total	**21,499**	**100.0%**

Source: EU-Silc, 2006.

It emerges, in particular, that the majority of households surveyed declared themselves to be in a state of hardship (either in great hardship 13.4%, in hardship 19.4% or in some degree of hardship, 40.2%). There are, however, few (0.8%) families declaring that they get through to the end of the month with absolute confidence.

Table 2. Percentage distribution of households surveyed according to level of hardship in terms of "getting through to the end of the month", by region of residence

Region of Residence	**With great difficulty**	**With difficulty**	**Other answers**	**Total**
Sicilia	24.7%	28.1%	47.2%	100%
Campania	24.1%	24.6%	51.3%	100%
Puglia	22.6%	24.0%	53.4%	100%
Sardegna	22.6%	23.1%	54.3%	100%
Calabria	21.1%	32.1%	46.8%	100%
Basilicata	19.8%	24.9%	55.3%	100%
Lazio	13.3%	20.9%	65.8%	100%
Abruzzo	13.0%	20.0%	67.0%	100%
Piemonte	11.8%	17.6%	70.6%	100%
Toscana	11.7%	17.8%	70.5%	100%
Molise	11.5%	24.3%	64.2%	100%
Marche	11.1%	20.0%	68.9%	100%
Umbria	10.9%	19.7%	69.4%	100%
Veneto	10.8%	15.6%	73.6%	100%
Liguria	10.5%	20.3%	69.2%	100%
Emilia Romagna	10.4%	15.9%	63.7%	100%
Friuli Venezia Giulia	9.6%	17.3%	73.1%	100%
Lombardia	8.8%	15.0%	76.2%	100%
Valle d'Aosta	6.7%	15.4%	77.9%	100%
Bolzano	4.6%	11.6%	83.8%	100%
Trentino Alto Adige	3.9%	10.0%	86.1%	100%
Italy	**13.4%**	**19.4%**	**67.2%**	**100%**

Source: EU-Silc, 2006.

In terms of geographical distribution, such a table confirms that poverty is concentrated within central and southern Italy. In particular, the regions with the highest percentage of families who face getting through to the end of the month with "great difficulty" are Sicilia (24.7%), Campania (24.1%), Puglia and Sardegna (both with 22.6%) and Calabria (21.1%). The same regions have an equally high percentage level of families who responded in terms of "difficulty", particularly Calabria (32.1%) and Sicilia (28.1%).

The EU-SILC survey of 2006 also asks families, divided between those living in rented accommodation and homeowners with a mortgage, the incidence of a range of expenses with which they are faced. It results, in particular, that at least half of families in rented accommodation assessed the cost of rent as a "heavy" burden (49.9%), along with housing costs in terms of electricity, gas and telephone bills and maintenance (59.8%), and other expenses such as the hire-purchase of furniture or other assets (56.4%).

Table 3. Distribution of households surveyed living in rented accommodation according to the perceived burden in relation to each type of expenditure

	Rental costs	**Housing costs**	**Other expenditure**
Heavy	49.9%	59.8%	56.4%
Bearable	47.4%	38.8%	42.0%
Negligible	2.7%	1.4%	1.6%
Total	**100.0%**	**100.0%**	**100.0%**

Source: EU-Silc, 2006.

Families tied to a mortgage on their property assessed mortgage repayment as a "heavy" burden, with other associated housing costs showing a fall in the percentage response rate to 45.4% and 47.8% respectively.

Table 4. Distribution of mortgage-holding homeowner households surveyed according to the perceived burden in relation to each type of expenditure

	Rental costs	**Housing costs**	**Other expenditure**
Heavy	59.1%	45.4%	47.8%
Bearable	39.9%	53.2%	47.6%
Negligible	1.0%	1.4%	4.6%
Total	**100.0%**	**100.0%**	**100.0%**

Source: EU-Silc, 2006.

3.2 Definition of Fuzzy Numbers to Be Applied in Regression Model

The aim of the present work is that of identifying which expenses, and their relative burdens as revealed by the survey (expenses of rent/mortgage payments, for the running of the household and other debts), determine, in comparison to others, and to a greater or lesser degree, the difficulty of "getting through to the end of month".

Such verification could be carried out through a logit model, from the moment in which such variables are taken into consideration on a categorical scale. Nevertheless, the estimates obtained would thus result as highly complex, when both the number of explanatory variables and, above all, the relative modes involved are taken into consideration.

In order to respond to such a dilemma a *fuzzy* regression model is proposed, made possible by the preliminary transformation of the information revealed through the EU-SILC survey in triangular fuzzy numbers [1,13].

Nevertheless, the nuclei of the six response categories to the question of hardship in terms of "getting through to the end of the month" ("very easily", "easily", "fairly easily", "with some difficulty", "with difficulty" and "with great difficulty") are identified in correspondence with natural numbers between 0 and 5. Furthermore, in order to normalize the data collected, a quantification criteria has been maintained of the explanatory variables which are analogous for all: in particular, the three response categories to the question of burden in terms of mortgage payments, rent and household costs (Negligible, Bearable, Heavy) are centred on 1, 3 and 5, whilst the four response categories relating to the burden of expenses for other debts (No expenses incurred, Negligible, Bearable, Heavy) are centred on 0, 1, 3 and 5.

It is noted that all responses received present, having been quantified, a range of variation of between 0 and 6 and, furthermore, that the width of the intervals in which the membership functions vary according to the intensity of the response expressed. The adoption of triangular functions allows for the attribution of a degree of membership within the associated category to each interval point, inversely proportional to the distance from the fuzzy number nucleus.

In particular, with reference to the explanatory variable, the two extreme categories express highly decisive responses and, as such, have corresponding functions of membership which are defined at more contained intervals than the intermediate response categories. This is true to an even greater extent in the case of the "No expenses incurred" category.

The membership functions of all variables included in the model are reported below, defined on the basis of the observations formulated thus far, considerations which are purely heuristic in nature.

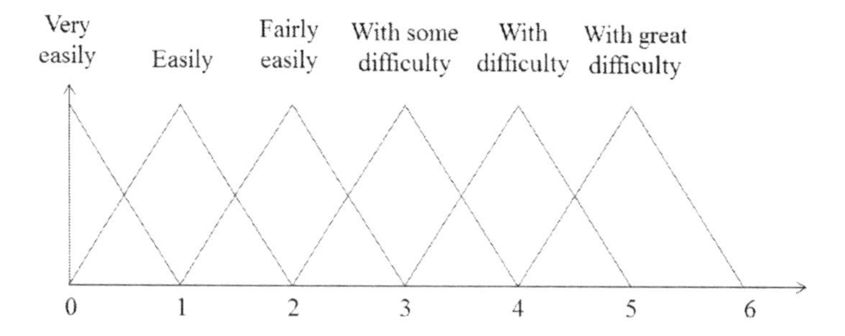

Fig. 2. Representation of the membership function relative to the "difficulty in getting through to the end of the month"

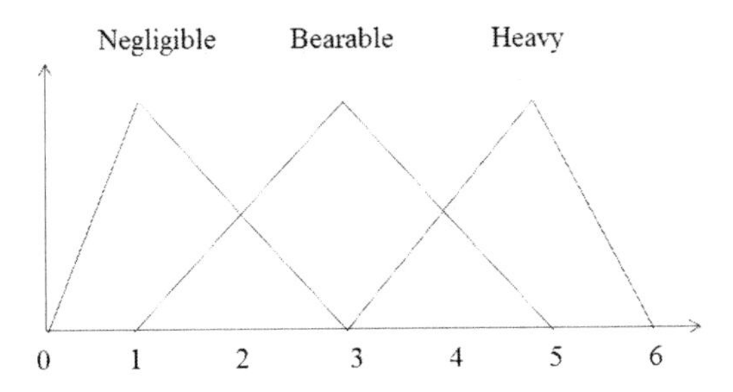

Fig. 3. Representation of the membership function relative to the following expense: "Mortgage", "Rent" and "Household costs"

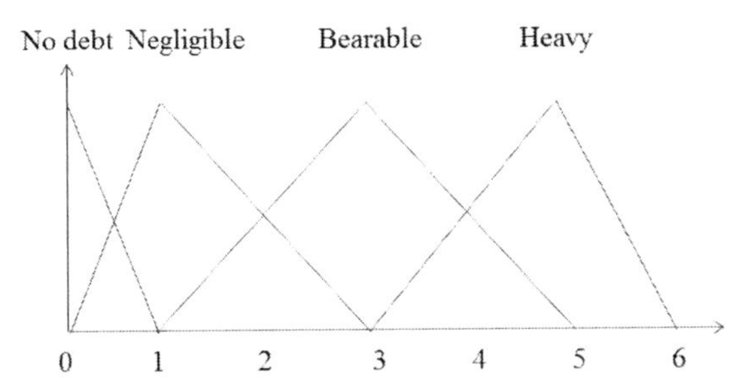

Fig. 4. Representation of the membership function relative to expenses "for other debts"

3.3 Results of the Regression Model

The verification of those expenses which, to a greater or lesser degree than others, determine the degree of difficulty in terms of getting through to the end of the month is conducted by differentiating, firstly, those families residing in rental accommodation as against homeowners living in a property purchased with a mortgage and, secondly, according to geographical area (north of Italy, centre and south and islands). The estimated regression coefficients within the six models considered, each of which has a discrete index of adaptability to the observed data (in all cases greater than 0.5), are reported below in table 4.

Taking the families residing in rented accommodation into consideration, the most relevant expenses in terms of difficulty in getting through to the end of month result as being those relative to monthly rental payments. This is true across all geographic areas under examination and above all in the north, possibly due to the specific characteristics of the housing market.

Expenses for the running of a household (electricity, gas, telephone and general costs) are revealed as being relevant, in this case particularly in the south, due to its generally depressed economic situation.

Expenses for other debts (relative to credit purchases or other goods) are revealed as being of greater relevance in terms of the difficulty of getting through to the end of

Table 5. Estimated regression coefficients in the model relative to families in rented accommodation, by geographical area

Families in rented	North	Centre	South and Islands
Intercept	2.72	2.25	2.48
Spread left	0.32	0.36	0.47
Spread right	0.28	0.30	0.27
Rental expenses	0.22	0.19	0.20
Household expenses	0.14	0.14	0.16
Other debts	0.12		
FFI	*0.52*	*0.52*	*0.53*

Table 6. Estimated regression coefficients in the model relative to mortgage-holding homeowner households, by geographical area

Familes with mortgage	North	Centre	South and Islands
Intercept	2.19	2.26	2.94
Spread left	0.33	0.49	0.30
Spread right	0.67	0.22	0.70
Mortgage expenses	0.20	0.22	0.23
Household expenses	0.13	0.13	0.14
Other debts	0.15		
FFI	*0.56*	*0.52*	*0.54*

the month specifically in the north, presumably due to the reluctance of families in other geographic areas to take on such expenses.

When taking into account the second set of families considered, the expenses relative to the payment of mortgage rates is revealed as relevant in all geographic areas and, differently from the previous model, above all in the south (possibly due to the decreasing ability to guarantee the income-levels necessary to meet a substantial monthly economic payment of this kind); other types of expenses are shown to have an identical degree of relevance in terms of getting through to the end of month as that seen in the previous model.

The validity of the proposed method in terms of analysing poverty levels in accordance with the various types of expenses incurred is verified, however, by proceeding with estimation, as an alternative, of a logit model, taking into account the same variables as considered thus far, without the need to transform them into triangular fuzzy numbers.

Test values of significance associated with the explanatory variables of the logit model, differentiated by geographical area, are reported below, summarized through the omission of the relative coefficients.

Table 7. Significance of the explanatory variables of the logit model relative to families in rented accommodation, by geographical area

Families in rented	North			Centre			South and Islands		
	Test G^2	d.o.f.	Signific.	Test G^2	d.o.f.	Signific.	Test G^2	d.o.f.	Signific.
Intercept	0.000	0	.	0.000	0	.	0.000	0	.
Rental expenses	68.91	10	0.00	72.93	10	0.00	68.03	10	0.00
Household expenses	44.17	10	0.00	52.38	10	0.00	63.61	10	0.00
Other debts	26.67	15	0.03	40.28	15	0.01	27.19	15	0.02

Table 8. Significance of the explanatory variables of the logit model relative to families with mortgage, by geographical area

Families with mortgage	North			Centre			South and Islands		
	Test G^2	d.o.f.	Signific.	Test G^2	d.o.f.	Signific.	Test G^2	d.o.f.	Signific.
Intercept	0.000	0	.	0.000	0	.	0.000	0	.
Mortgage expenses	97.37	10	0.00	67.97	10	0.00	80.03	10	0.00
Household expenses	88.19	10	0.01	48.71	10	0.01	63.61	10	0.00
Other debts	60.63	15	0.00	.	15	.	.	15	.

Similarly, in this case, and regardless of geographical area, the expenses which demonstrate the greatest degree of difficulty in getting through to the end of the month are represented by the payment of a mortgage or by payment of monthly rent, according to the whether families reside in rented accommodation or whether they are homeowners with a mortgage, followed by general household costs and, finally, by other debts.

It may be noted that in the particular case of families residing in rented accommodation, and differently from that emerging from a fuzzy regression model, expenses for other debts are significant not only in the north of Italy but, also, in the centre and south. On closer inspection, however, the estimated coefficients in relation to the four response categories to the question of the burden of expenses for other debts (No expenses incurred, Negligible, Bearable, Heavy) are all close to zero. In other words, whilst seen through the logit model this burden results as significant, its incidence in terms of difficulty in getting through to the end of the month is almost nothing.

The verifications conducted thus far demonstrate the fuzzy regression model to be more efficient with respect to the classic logit model in estimating in what ways the single factors taken into consideration may influence the perception of interviewees regarding their ability to get through to the end of the month.

4 Concluding Remarks

In this work we propose a *Fuzzy Regression Model* in order to identify the factors that most influence the perception of poverty by Italian families.

In literature poverty, with regards to its economic nature, is usually defined as an insufficiency of the resources necessary to guarantee a high level of well-being with respect to certain predefined standards. There is a general agreement that evaluating poverty means measuring the economic resources of individual families with respect to the economic resources of other families. The use of monetary variability (in terms of *consumption* and *income*) is based on the implicit assumption of equivalence between available economic resources and well-being. Such minimum levels of well-being may be expressed in terms of being *absolute* or *relative*. A transversal approach is therefore proposed as an alternative to the above, considered as *subjective*, through which the poor are defined as those who identify themselves as such, even if this identification is revealed as a result of the comparison that they operate with the rest of society in terms of perceived wellbeing. It is, therefore, the "perception" an individual may have of their own condition which allows for the identification of the measure of poverty to a far greater degree than the assessment of an external observer would allow.

The presence of a varied range of definitions on the theme of poverty underlines the necessity of no longer relying on a single indicator, but on a group of indicators which are useful in the definition of living conditions of various subjects (*multidimensional approach)*.

Income and consumption therefore represents only part of the overall dimension of poverty, in as much as this approach focuses on quality of life rather than on wealth, thus allowing for a more accurate description of the phenomenon, a more appropriate explanation of causes.

A subjective approach to poverty suggests the adoption of a fuzzy regression model, made possible by an initial transformation of data into triangular fuzzy numbers, from information gathered from the EU-SILC survey on Income and Living Conditions, carried out by ISTAT in 2006 on a sample of Italian families.

The results obtained allow for the identification of the type of expenses that constitute the most significant determinants of the actual capacity of Italian families to get through to the end of month. This analysis was carried out, differentiated by geographical areas of reference (north Italy, central, south and islands) as well as by types of families both in rented accommodation or homeowners with a mortgage. In particular, it is noted that rental payments significantly affect difficulty in terms of getting through to the end of the month in the north, whilst mortgage payments show similar levels of difficulty in the south.

The validity of this method, proposed in order to analyze poverty levels according to the different types of expenses incurred, is confirmed through the comparison of results obtained from the estimation of a classical logit model applied using the same explanatory variables.

References

1. Cammarata, S.: Sistemi a logica fuzzy. Come rendere intelligenti le macchine, ETAS (1997)
2. Campobasso, F., Fanizzi, A., Tarantini, M.: Una generalizzazione multivariata della Fuzzy Least Square Regression. Annali del Dipartimento di Scienze Statistiche dell'Università degli Studi di Bari 7, 229–243 (2008)
3. Campobasso, F., Fanizzi, A., Tarantini, M.: Some results on a multivariate generalization of the Fuzzy Least Square Regression. In: Proceedings of the International Conference on Fuzzy Computation, Madeira, pp. 75–78 (2009)
4. Diamond, p.M.: Fuzzy Least Square. Information Sciences 46, 141–157 (1988)
5. ISTAT (2006), EU-SILC, the European Standard on Income and Living Conditions, Anno (2006)
6. Kao, C., Chyu, C.L.: Least-squares estimates in fuzzy regression analysis. European Journal of Operational Research 148, 426–435 (2003)
7. Kosko, Bart.: Fuzzy Thinking: The New Science of Fuzzy Logic. Hyperion (1993) ISBN 0-7868-8021-X
8. Lenoir, R.: Les Exclus. Un francais surd ix. Seuil, Paris (1974)
9. Montrone, S., et al.: A Fuzzy Approach to the Small Area Estimation of Poverty in Italy. In: Phillips-Wren, G., et al. (eds.) Advances in Intelligent Decision Technologies, Smart Innovation, Systems and Technologies, vol. 4, pp. 309–318. Springer, Heidelberg (2010), ISSN 2190-3018, ISBN 978-3-642-14615-2, DOI 10.1007/978-3-642-14616-9
10. Sen, A.: Well-Being, Capability and Public Policy, Giornale degli Economisti e Analisi di Economia, vol. 3 (1994)
11. Tanaka, H., Uejima, S., Asai, K.: Regression analysis with fuzzy model. In: IEEE Transactions on Systems, Man, and Cybernetics SMC, vol. 12, pp. 903–907 (1982)
12. Takemura, K.: Fuzzy least squares regression analysis for social judgment study. Journal of Advanced Intelligent Computing and Intelligent Informatics 9(5), 461–466 (2005)
13. Veronesi, M., Visioli, A.: Logica fuzzy. Fondamenti teorici e applicazioni pratiche. Franco Angeli, Milano (2003)
14. Zadeh, L.A.: Fuzzy sets as a basis for a theory of possibility. Fuzzy Sets and Systems 1(1), 3–28 (1978)

Fuzzy Logic Approach for Spatially Variable Nitrogen Fertilization of Corn Based on Soil, Crop and Precipitation Information

Yacine Bouroubi[1,*], Nicolas Tremblay[1], Philippe Vigneault[1], Carl Bélec[1], Bernard Panneton[1], and Serge Guillaume[2]

[1] Horticulture Research and Development Centre, Agriculture and Agri-Food Canada, 430 Gouin Blvd, St-Jean-sur-Richelieu, Qc., Canada, J3B 3E6.
[2] Cemagref, UMR ITAP, F-34196 Montpellier, France
yacine.bouroubi@agr.gc.ca

Abstract. A fuzzy Inference System (FIS) was developed to generate recommendations for spatially variable applications of nitrogen (N) fertilizer using soil, plant and precipitation information. Experiments were conducted over three seasons (2005-2007) to assess the effects of soil electrical conductivity (ECa), nitrogen sufficiency index (NSI), and precipitations received in the vicinity of N fertilizers application, on response to N measured at mid-season growth. Another experiment was conducted in 2010 to understand the effect of water supply (WS) on response to N, using a spatially variable irrigation set-up. Better responses to N were observed in the case of high ECa, low NSI and high WS. In the opposite cases (low ECa, high NSI or low WS), nitrogen fertilizer rates can be reduced. Using fuzzy logic, expert knowledge was formalized as a set of rules involving ECa, NSI and cumulative precipitations to estimate economically optimal N rates (EONR).

Keywords: Variable nitrogen fertilization, fuzzy inference systems, soil electrical conductivity, nitrogen sufficiency index, precipitations, water supply.

1 Introduction

Crop production is a complex system that integrates physical, chemical and biological processes and that is managed under increasing economic and ecologic constraints. Nitrogen management is one of the primary factors affecting crop yield and pollution from agroecosystems. Nitrogen availability to crops is affected by a set of interacting edaphic, climatic and management factors and N fertilizer needs can vary within and among fields. The application of fertilizer N uniformly across fields leads to over- or under-application since requirements are known to vary in space and in time. The variable application of N fertilizers is the only strategy capable of optimizing the overall use of N to reach both economic and environmental objectives [1] [2] [3].

The *economically optimum nitrogen rate* (EONR) can likely be estimated by integrating crop, soil and weather information. Soil properties, soil moisture availability and yearly climatic factors should be considered when making fertilizer

B. Murgante et al. (Eds.): ICCSA 2011, Part I, LNCS 6782, pp. 356–368, 2011.

recommendations [4]. In situ measurements of chlorophyll, leaf area index (LAI), fluorescence properties, biomass status or vegetation indices obtained through a variety of sensors and platforms are also indicators of actual local N availability [5].

Soil texture plays an important role in N availability. Soil *apparent electrical conductivity* (EC_a) has been used as a surrogate parameter to textural properties providing great spatial details. Tremblay et al. [1] reported that best mid-season growth was found in areas of low EC_a, corresponding to soil with coarser textures. The best responses to in-season N fertilization were found in areas with high EC_a corresponding to unfavourable soils for corn growth. Kitchen et al. [6] also showed that crop growth is inversely correlated with EC_a.

The influence of water on N uptake has been reported in several scientific papers. Van Es et al. [7] showed that variations in soil water and texture within fields cause localized differences in soil N availability and therefore are a potential basis for variable N fertilizer application in corn. Hong et al. [8] found that irrigation significantly affected N uptake. Xie et al. [9] showed that rainfall have profound effects on the efficiency of nitrogen fertilization. The best response to N was obtained for high but not excessive values of *cumulative precipitations* (PPT) collected between sowing and in-season nitrogen application dates.

The NDVI acquired prior to N fertilizer application can be used as a valuable indicator of crop performance. It can be translated into a *Nitrogen Sufficiency Index* (NSI; a relative indicator, based on non-limiting N reference [N-rich] areas) to judge the fertilization requirements of the rest of the field [1] [10] [11]. The use of NSI is claimed to remove many of the soil and seasonal variation factors that interfere with the diagnostic assessment [5].

There have been many studies relating plant, soil or weather conditions to EONR, but there are few examples where models have been developed to integrate these effects for estimating EONR. Brentsen et al. [12] proposed a polynomial model for EONR based on properties such as crop biomass and EC_a. However, such a modeling approach is hardly compatible with datasets characterized by strong variability and weak correlations. Thus, it is proposed that other strategies based on artificial intelligence are better adapted than deterministic approaches for predicting EONR or for similar purposes. Among them, Assimakopoulos et al. [13] proposed fuzzy logic as the most flexible and comprehensive approach to use in such a context, particularly where expert knowledge can be included. Jones and Barnes [14] developed a fuzzy logic based decision-support system that integrates remote sensing data and plant growth models to manage within-field spatial variation. Molin and Castro [15] as well as Yan et al. [16] used a fuzzy logic classifier of crop and soil features to define uniform site-specific management zones. They showed that this approach provides valuable information for variable application of inputs. Papageorgiou et al. [17] used fuzzy cognitive maps to model expert knowledge and predict cotton yield from complex interacting soil, crop and weather factors. This technique was capable of dealing with uncertain descriptions like human reasoning of complex data. Krysanova and Haberlandt [18] developed a fuzzy-rule based model simulating nitrogen fluxes for the analysis of nitrogen leaching and the assessment of water quality. This model allowed the establishment of fertilisation schemes for climate zones and soil classes.

The aim of this work was to develop a fuzzy inference system (FIS) to predict EONR for corn plots from soil texture (EC_a maps), crop information (NSI maps) and water availability estimated through cumulative precipitations (PPT) recorded before N fertilizer application. The rules of this FIS are established from the results about soil and weather effects on response to N.

2 Materials and Methods

2.1 Experiments to Study Soil, Weather and Plant Status Effects on Response to N

The experiments were conducted in summer 2005, 2006 and 2007 on corn (*Zea mays* L.) in commercial fields located in the Montérégie region of Quebec, Canada (Table 1). The fields were characterized by highly variable textures (heavy clay, clay-loam, loam, sandy loam and loamy sand). The field was under conventional tillage with a row spacing of 0.75 m. Four in-season N-rate treatments (30 to 158 kg N ha^{-1}) plots were randomized within four blocks. Strips with no N fertilizer for the whole season (Nil-N) and others with non-limiting N fertilizer applications (rich-N strips with 250 kg N ha^{-1}) were also laid out. Soil EC_a (0-30 cm) was measured on bare soil before sowing (2005 and 2006) or after harvest (2007) using a Veris model 3100 sensor (Veris Technologies, Inc., Salina, KS) with a sampling density of one point per 6 to 12 m, and krigged using GS+ software (Gamma Design Software, LLC, Plainwell, MI). EC_a values ranged approximately from 2 mS m^{-1} (in sandy soils) to 30 mS m^{-1} (in clayey soils).

Table 1. Characteristics of the trials in commercial fields located in St-Valentin, QC (2005, 2006 and 2007) and in the AAFC experimental farm of L'Acadie, QC (2010)

	2005	2006	2007	2010
Coordinates (Lat/Long)	45° 5' 28" N 73° 21' 11" W	45° 5' 21" N 73° 21' 3" W	45° 5' 24" N 73° 21' 11" W	45° 17' 52" N 73° 20' 16" W
Field size (ha)	8.66	6.52	8.74	2.49
Sowing date	12 May	9 May	3 May	13 May
Irrigations before side-dress	-	-	-	8, 12 and 15 June (V4*-5)
Early-season plant status observation	$NDVI_1$ 28 June (V7)	$NDVI_1$ 5 July (V7)	$NDVI_1$ 22 June (V5-6)	-
Irrigations after side-dress	-	-	-	22 and 29 June (V6-7)
In-season N application	23 June (V6)	5 July (V7)	22 June (V5-6)	19 June (V5-6)
Mid-season plant status observation	$NDVI_2$, 12 July (V9)	$NDVI_2$, 20 July (V10)	$NDVI_2$, 10 July (V8)	SPAD, 6 July (V9)
Variety	DKC4627 Bt (2950 CHU)	DKC4627 Bt (2950 CHU)	DKC4627 Bt (2950 CHU)	Pioneer 38M58 (2800 CHU)

* Corn growth stages according to [19]

The NDVI measurements were obtained before the application of N (defined as $NDVI_1$) from five GreenSeeker sensors (NTech Industries Inc., Ukiah, CA). The corresponding NSI (NSI = $NDVI_1$ of the area to diagnose/ $NDVI_1$ of nearest neighbourgh in N-rich strips) was calculated for every point. NDVI was recorded again at mid-season (defined as $NDVI_2$) with a Compact Airborne Spectrographic Imager (CASI; ITRES Research Ltd., Calgary, AB) to assess plant response to N fertilizer application and terrain features. Fig. 1 shows 2005 maps. More details about these experiments are given in [1].

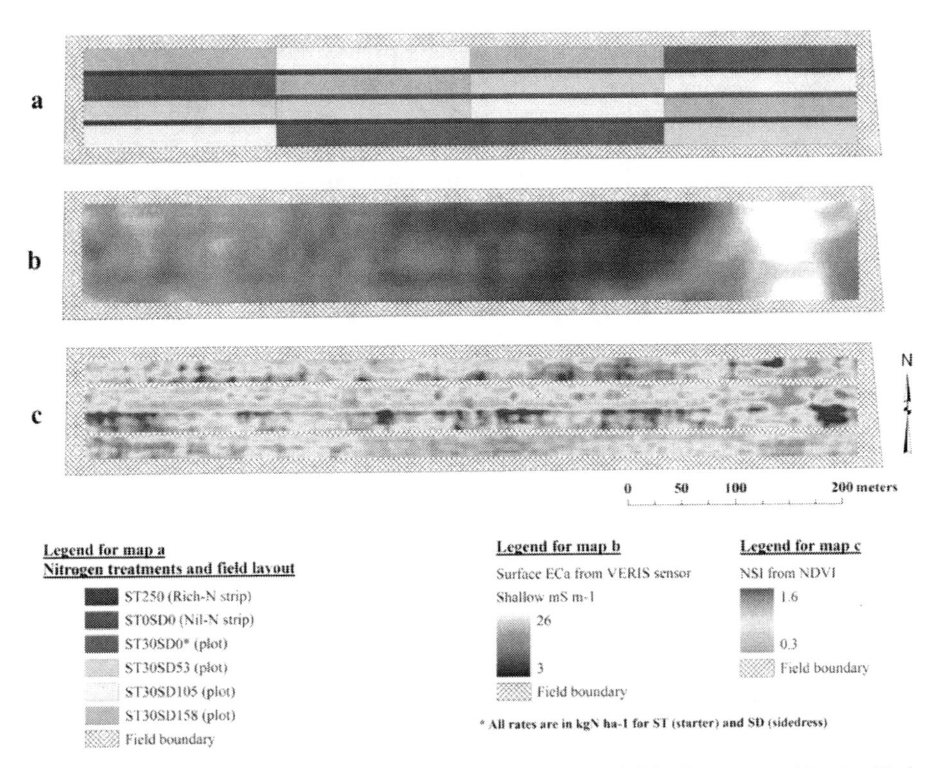

Fig. 1. Maps showing N treatments (a), the spatial variation of EC_a (b) measured by the Veris and the NSI calculated from NDVI measured by GreenSeeker at side-dress stage (c) in 2005

2.2 Experiment to Study Water Supply Effect on Response to N

An experiment was conducted in summer 2010 on corn (*Zea mays* L.) in a field located at the L'Acadie Experimental Farm, Quebec, Canada (Table 1). The field was characterized mainly by clay loams. The EC_a (0-30 cm) was measured using a VERIS model 3100 sensor cart system (Veris technologies, Salina, Kansas, USA) and showed moderate variability from 3 to 10 mS m^{-1}. The EC_a were collected on transects approximately 5 m apart on bare soil after harvest, in 2009 and krigged using GS+ (Gamma Design Software, LLC, Plainwell, MI, USA). The field was under conventional tillage with a row spacing of 0.75 m. Four in-season N-rate (50 to 180 kg ha^{-1}) treatments plots were randomized within four blocks. Plots with no N fertilizer (Nil-N) and with non-limiting N fertilizer applications (Rich-N with 250 kg ha^{-1}) were also included in the experimental design. One irrigation sprinkler line was established transversally in the middle of each block (Fig. 2) and was activated at several dates before and after side-dress (Table 1). This irrigation set-up was expected to produce a spatially variable water supply (WS) as the amount of water normally decreases further away from the sprinkler heads. The actual spatial distribution of the irrigation water applied was measured by a network of pluviometers corresponding to the sampling points. In each plot, 3 sampling points were randomly selected but scattered to encompass spatial variation (Fig. 2). Water supply before side-dress (WS_{BS}) was cumulated over the three irrigation dates closest to side-dressing and water supply after side-dress (WS_{AS}) was cumulated over the two irrigation dates following side-dressing (Table 1). Corresponding LAI measurements were made with a LAI-2200 instrument (Li-Cor, Lincoln, NE, USA) and chlorophyll levels were estimated using a SPAD-502 chlorophyllmeter (Soil Plant Analysis Development, Minolta Camera Co., Ltd., Japan). Shoot biomass was sampled, weighed and dried.

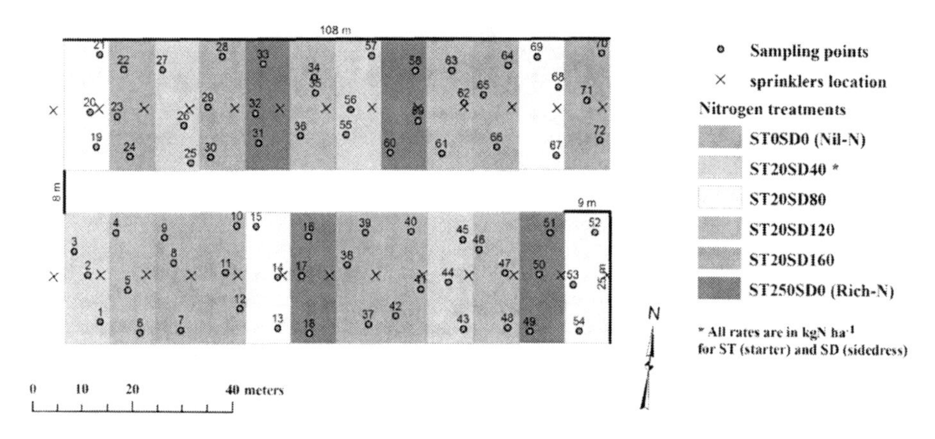

Fig. 2. Maps of the field experiment showing N treatments, field layout, sprinklers and sampling locations in 2010

2.3 Fuzzy Inference System Modeling

Fuzzy logic is widely used as an interface between symbolic and numerical spaces, allowing the implementation of human reasoning in computers. Fuzzy inference systems deal with linguistic variables, this concept is implemented using fuzzy sets, which usually overlap. For example, the EC_a variable can be handled by means of three linguistic terms: *Low*, *Medium* and *High*. A fuzzy set is defined by its membership function (MF). A point, **x**, in the universe, such as a given EC_a (i.e. a real value), belongs to a fuzzy set, A, with the membership degree $0 \leq \mu_A(\mathbf{x}) \leq 1$. On the left side of Fig. 3, the value belongs to the *Medium* set with a degree of $\mu_M(\mathbf{x})$ and to the *High* set with a degree of $\mu_H(\mathbf{x})$. These degrees can be interpreted as the level to which the **x** EC_a should be considered *Medium* or *High*. In this case, it cannot be considered *Low* at all.

The core of the fuzzy system is a set of fuzzy *if-then* rules. When several variables are involved in the rule description, the membership degrees can be combined using an *AND* operator (the most common are the minimum and the product) or an *OR* operator (maximum or bounded sum; the sum is limited to 1) to give the weight of a rule when conclusions from a set of rules are aggregated to infer an output. As a consequence of overlap in the fuzzy partition, several rules are likely to be called by the same input data. The inferred output is the result of the aggregation of all these weighted conclusions.

The design of a fuzzy system involves two main steps: definition of the input and output variables, and rule description. In the present study, both steps were carried out on the basis of expert knowledge and experimental results, as discussed below. Generally, simulations of output values corresponding to combinations of inputs are used to observe the behaviour of the designed model. An example showing details of calculating EONR' using an FIS similar to the one presented here is given in [20].

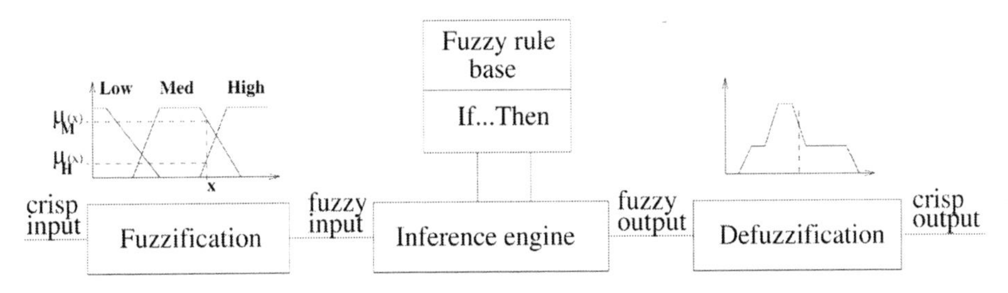

Fig. 1. Principles of the fuzzy inference system FIS

3 Results and Discussion

3.1 Effects of EC_a and Precipitations on Response to N in 2005, 2006 and 2007

Soil EC_a values were arranged into three classes (using histogram segmentation) and corresponding $NDVI_2$ (measured at mid-season) averages were calculated for each N treatment (Fig. 4). In 2005, responses to N were very high under high EC_a. This observation was of less obvious in 2006 and 2007. For low EC_a, NDVI is mostly independent of N rate. For medium EC_a, moderate responses to N rate are observed. Still, responses to N were high in soils unfavorable to mid-season growth (high EC_a) and low in soils more favourable to growth (low EC_a). The difference between responses to N among years may be due to weather conditions.

Seasonal differences in responses to N may be related to the quantity of precipitation received in the vicinity of N side-dress application (Fig. 5, Table 2). In 2005, the soil was moist both at N side-dress (65 mm of rain was received in the previous 10 days) and after side-dress (106 mm between N side-dress and mid-season). In 2006 and 2007, the lack of rain experienced before or after N side-dress

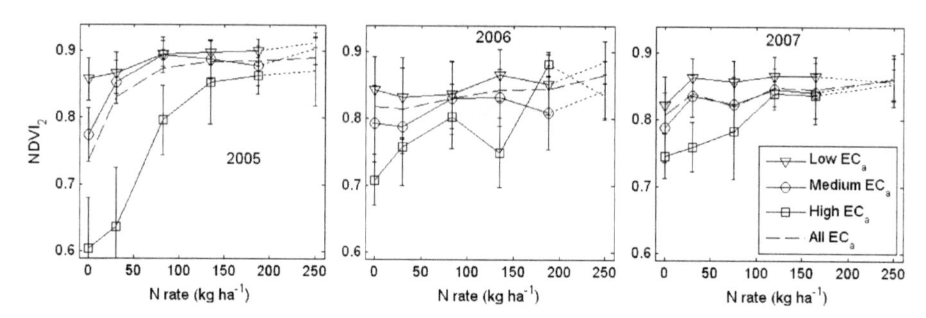

Fig. 2. Corn growth (represented by $NDVI_2$) response to nitrogen treatments for three classes of EC_a (low EC_a: 0 to 9 mS m^{-1}; medium EC_a: 9 to 18 mS m^{-1}; high EC_a: greater than 18 mS m^{-1}). Error bars represent standard deviations

may have interfered with both the release of N provided as fertilizer or the ability of the crop to use it. In 2006, the delay between N application and $NDVI_2$ measurement was also particularly short (15 days as compared to 18 days or more in the other years; Table 2), reducing the time for the N applied to generate measurable changes in corn growth at mid-season. For these reasons, the responses to N in 2006 and 2007 were at smaller magnitudes than in 2005.

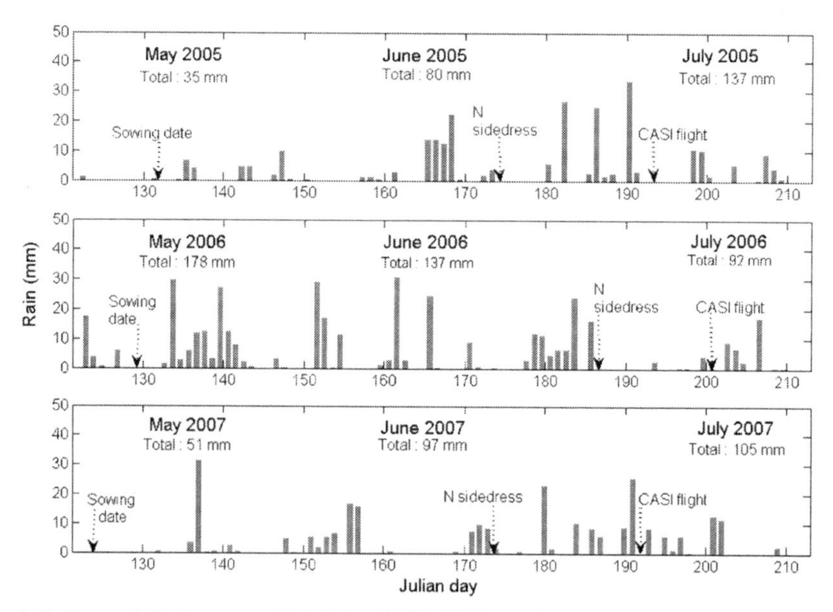

Fig. 3. Daily precipitations recorded during 2005, 2006 and 2007 with dates of interest (sowing, N side-dress application and remote sensing imagery acquisition)

Table 2. Cumulative precipitations (PPT) recorded before and after N side-dress application

Year	PPT_{SD-10} recorded 10 days before N side-dress (mm)	PPT between N side-dress and mid-season observation (mm)	Number of days between N side-dress and mid-season observation	Response to N
2005	65	106	19	High
2006	83	7	15	Low
2007	17	93	18	Low

3.2 Effects of NSI on Response to N in 2005, 2006 and 2007

Plant-status at side-dress measured by NSI affects N requirements (Fig. 6). When NSI is high, dNDVI (dNDVI = $NDVI_2$-$NDVI_1$; measures the growth between N fertilizer application and mid-season observation) is low. The reason is that crop growth was high in early season (at side-dress, before in-season application of N) for all N rates comparatively to non-limiting N supply condition. In this case (high NSI), NDVI increased from side-dress stage to mid-season stage with only 0.3 to 0.4 and almost independently of applied N rate and of soil conditions. When NSI is low, indicating a reduced growth at side-dress stage, dNDVI is high (0.6 to 0.7) in good soil conditions (low ECa) for almost all applied N rates. But, in situation of low NSI combined with bad soil conditions (high ECa), the response to N is very high and dNDVI increases from 0.3 for low N rate (0 or 30 kg ha^{-1}) to 0.7 for high N rate (187 kg ha^{-1}). It follows that crop growth can be enhanced by in-season N applications in areas of high ECa (unfavourable soil conditions) when a reduced growth is observed at side-dress stage.

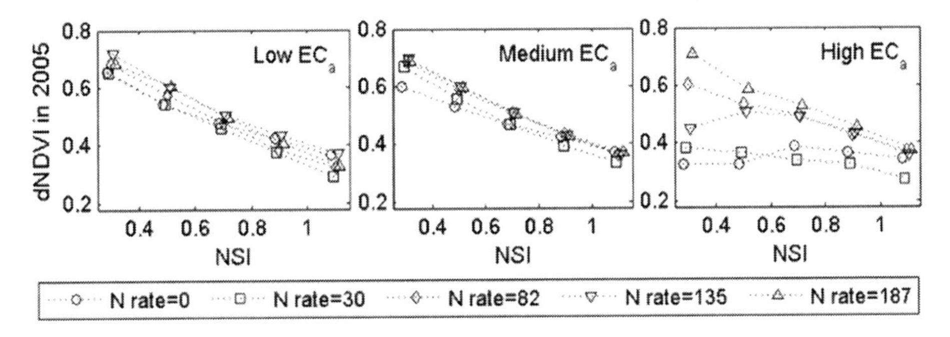

Fig. 4. The effect of NSI on growth (represented by dNDVI = $NDVI_2$-$NDVI_1$) for different rates of N applied (in kg N ha^{-1}) and for three classes of EC_a (year 2005)

3.3 Effect of Water Supply on Response to N in 2010

Water supply from spatially variable irrigation allows for simulating variable precipitation levels. Water supply recorded before (WS_{BS}) and after (WS_{AS}) side-dress were classified into low, medium and high levels, but with different limits since WS_{BS} was cumulated over three days and WS_{AS} was cumulated over two dates.

Areas with low and medium WS_{BS} presented higher SPAD values than those with higher WS_{BS}. Areas with medium WS_{AS} presented higher SPAD values than those with lower or higher WS_{AS} (Fig. 7). Areas with medium and low WS_{BS} and WS_{AS} had a moderate response to N while areas with high WS_{BS} and WS_{AS} responded more to N fertilizer application. These results illustrate that the impact of N rate on chlorophyll levels is dependent on water availability in the vicinity of N side-dress. A low WS, resulted in response to N levelling-off at a relatively low N level, as N was likely not the growth-limiting factor in this context. In contrast, areas with high WS_{BS} and WS_{AS} were not limited by water and responded to relatively high N levels. These results confirm the conclusions of the 2005 to 2007 experiments on effects of water availability on corn response to N fertilizer application.

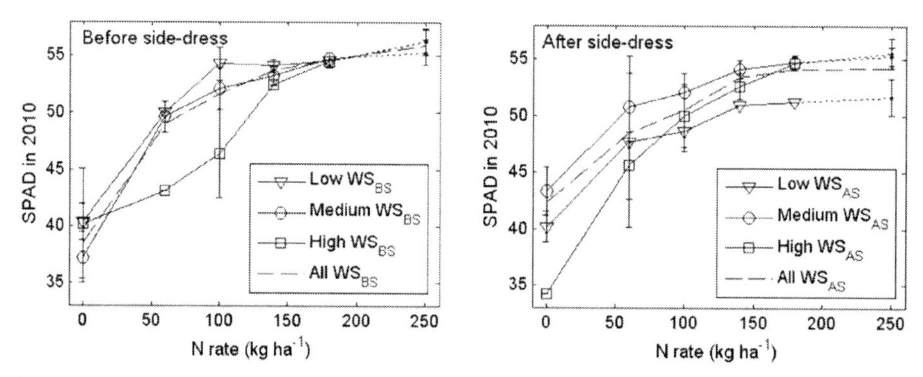

Fig. 5. Corn growth response (measured by SPAD) to N rates for three classes of WS measured before side-dress (WS_{BS}) and after side-dress (WS_{AS}). Low WS_{BS}: 0 to 10 mm; medium WS_{BS}: 10 to 20 mm; high WS_{BS}: greater than 20 mm. Low WS_{AS}: 0 to 15 mm; medium WS_{AS}: 15 to 30 mm; high WS_{AS}: greater than 30 mm. Error bars represent standard deviations

3.4 Fuzzy Sets and Inference Rules from Expert Knowledge

The proposed FIS uses the variables PPT_{SD-10} (PPT recorded 10 days before N side-dress), EC_a and NSI to determine EONR' (a surrogate of the real EONR, based on response of the crop to N rate in the period following side-dressing). Trapezoidal shapes and three fuzzy sets (low, medium and high levels) for EC_a and PPT were used. For NSI, two fuzzy sets (low and high) were used. The EONR' was also fuzzified with trapezoidal MFs into three sets, low, medium or high, to allow more possibilities for rule definition and more precise EONR' recommendations (Fig. 8). The limits of the EC_a and NSI fuzzy sets were determined in accordance with the corresponding classes determined by histogram segmentation. The fuzzy sets for PPT_{SD-10} was determined from values of Table 2 but could be improved from meteorological archives. The MFs of EONR' were also of trapezoidal shape and were defined from expert knowledge.

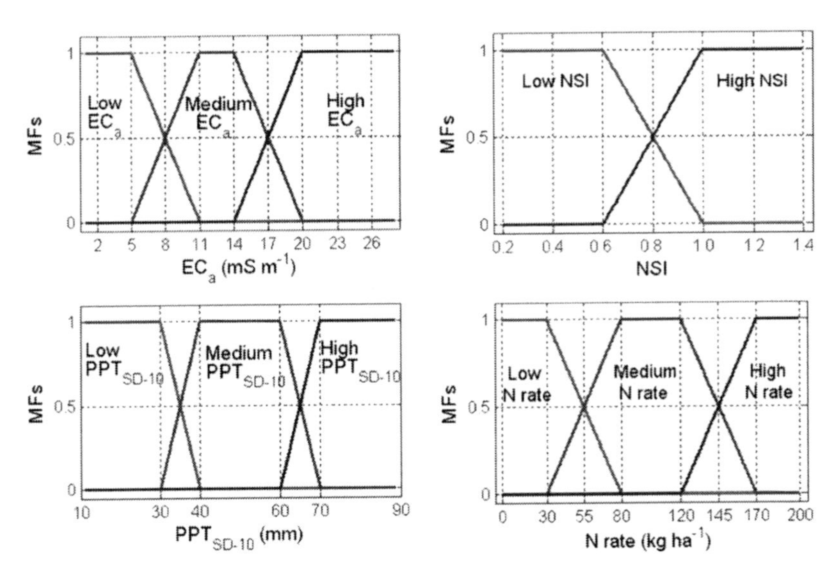

Fig. 6. Membership functions (MFs) for FIS input (EC_a, NSI and PPT_{SD-10}) and output (EONR') variables

On the basis of the expert knowledge gathered from the four experiments (seasons 2005, 2006, 2007 and 2010) through the analysis of the relationship between growth and soil conditions (ECa), plant status (measured by NSI) and water availability (represented by PPT_{SD-10}), the following rules are proposed:

IF (EC_a is med **OR** EC_a is low) **AND** (NSI is low) **THEN** (EONR' is med)
IF (EC_a is med **OR** EC_a is low) **AND** (NSI is high) **THEN** (EONR' is low)
IF (EC_a is high) **AND** (NSI is low) **THEN** (EONR' is high)
IF (EC_a is high) **AND** (NSI is high) **THEN** (EONR' is med)
IF (PPT_{SD-10} is low **OR** PPT_{SD-10} is med) **THEN** (EONR' is med)
IF (PPT_{SD-10} is high) **THEN** (EONR' is high)

These rules can be updated to include local knowledge or new experimental results.

3.5 Simulation of EONR' by the FIS

To observe the behaviour of the FIS output, EONR' was generated for combinations of EC_a (ranging from 2 to 25 mS m^{-1}), NSI (ranging from 0.35 to 1.2) and for two weather conditions: dry with PPT_{SD-10} = 30 mm and wet season with PPT_{SD-10} = 70 mm. The results of EONR' simulations as functions of input parameters are illustrated in Figs. 9a and 9b. As expected, the general surface response of EONR' to EC_a and NSI are similar for dry and wet seasons. However, EONR' is higher in wet season by about 60 kg N ha^{-1} in the case of high EC_a and low NSI.

For wet seasons and unfavourable soil conditions (high EC_a), a maximum of 180 kg N ha^{-1} is recommended if NSI is low (high deficiency), but only 100 kg N ha^{-1} if it is high (low deficiency). Under favourable soil conditions (low EC_a), the model recommends 30 kg N ha^{-1} if NSI is high and 90 kg N ha^{-1} if NSI is low. The model accounts for the influence of EC_a with a 90 kg N ha^{-1} difference between EC_a extremes.

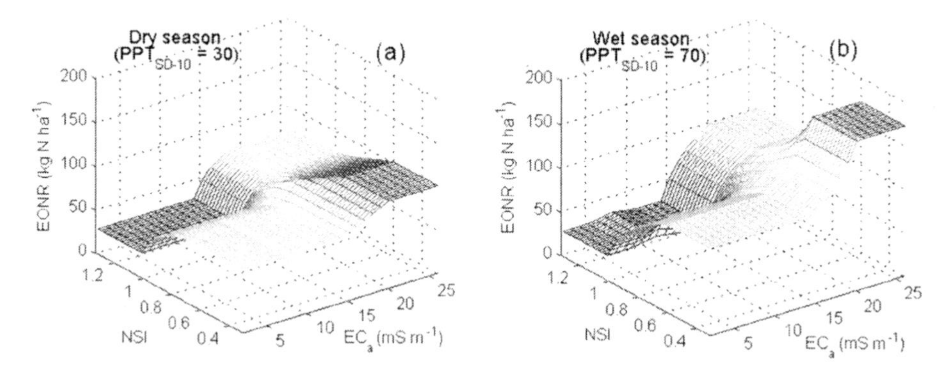

Fig. 7. Simulation of EONR' using the FIS developed for different situations of EC_a and NSI input values under situations of: a) dry season and; b) wet season

3.6 FIS Implementation in a GIS

The developed FIS can be implemented in a GIS following these steps: (1) membership function generation, (2) rules application, (3) aggregation of partial outputs and (4) defuzzyfication of the final output. The implementation of these operations in ArcGIS™ (ESRI, Redlands, CA, USA) is described in [20].

As shown in Fig. 10, the calculated N requirements for 2005 field vary between 35 and 160 kg N ha^{-1}, according to the field attributes and crop status, as measured by $NDVI_1$ and translated into NSI and precipitation records. This map shows the high spatial variability of N requirements. The patterns in EONR' map are similar to those observed in EC_a and NSI maps showed in Fig. 1. Such variable N rate proposed by the output map should help to enhance nitrogen use efficiency (NUE) and reduce the environmental impact of over-fertilization.

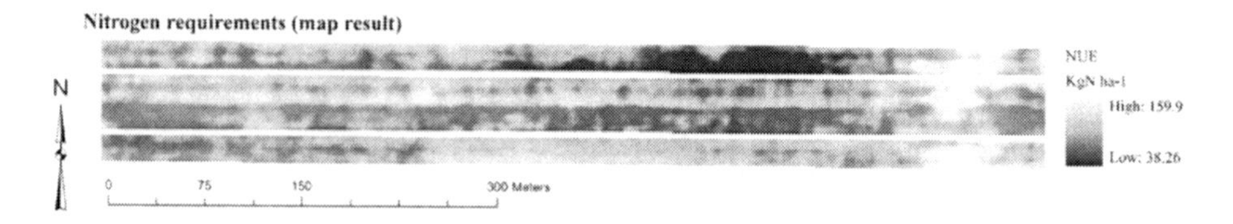

Fig. 8. EONR' calculated using the developed FIS for 2005 field

4 Conclusion

The spatial variability of soil conditions within and among fields and the temporal variability of water availability from season to season complicate the determination of the optimal N fertilizer requirements. Fuzzy inference systems (FIS) can combine plant and soil based spatial information and weather seasonal information to provide nitrogen fertilizer recommendations adapted to conditions prevailing at different locations, from year to year.

This study demonstrates the importance of considering both soil texture and seasonal precipitations for the determination of adequate in-season N supply. Results shows that nitrogen fertilizer rates can be reduced where soil and/or weather conditions are naturally favourable to growth, and increased where soil conditions are unfavourable to growth or when limited water availability does not allow for a high response to N. Since water availability is important both before and after N fertilizer application, precipitation forecasts for few days after side-dress could be added to precipitations recorded before side-dress (the PPT_{SD-10} variable used in the FIS) for a better use of weather information available at the time of side-dress, including forecasts for the period immediately following this time.

The FIS presented in this paper could be adapted to other conditions by adjusting the membership functions (MFs) to the ranges of prevailing input and output variables, and possibly by adapting fuzzy rules from complementary expert knowledge.

References

1. Tremblay, N., et al.: Development and Validation of a Fuzzy Logic Estimation of Optimum N Rate for Corn Based on Soil and Crop Features. Precision Agriculture 11, 621–635 (2010)
2. Welsh, J.P., et al.: Developing Strategies for Spatially Variable Nitrogen Application in Cereals. part II: Wheat. Biosystems Engineering: Special issue on Precision Agriculture - Managing Soil and Crop Variability for Cereals 84, 495–511 (2003)
3. Scharf, P.C., et al.: Spatially Variable Corn Yield is a Weak Predictor of Optimal Nitrogen Rate. Soil Science of American Journal 70, 2154–2160 (2006)
4. Derby, N.E., et al.: Interactions of nitrogen, weather, soil, and irrigation on corn yield. Agronomy Journal 97, 1342–1351 (2005)
5. Tremblay, N.: Determining Nitrogen Requirements from Crops Characteristics. Benefits and Challenges. Recent Res. Devel. Agron. Hortic. 1, 157–182 (2004)
6. Kitchen, N.R., et al.: Soil Electrical Conductivity and Topography Related to Yield for Three Contrasting Soil-Crop systems. Agronomy Journal 95, 483–495 (2003)
7. Van Es, H.M., Yang, C.L., Geohring, L.D.: Maize Nitrogen Response as Affected by Soil Type and Drainage Variability. Precision Agriculture 6, 281–295 (2005)
8. Hong, L., et al.: Multispectral Reflectance of Cotton Related to Plant Growth, Soil Water and Texture, and Site Elevation. Agronomy Journal 93, 1327–1337 (2001)
9. Xie, M., et al.: Weather Effects On Corn Response to in-Season Nitrogen Rates. In: ASA, CSSA, and SSSA International Annual Meetings, long Beach, CA, October 31-November 3 (2010)
10. Samborski, S.M., Tremblay, N., Fallon, E.: Strategies to Make Use of Plant Sensors-based Diagnostic Information for Nitrogen Recommendations. Agronomy Journal 101, 800–816 (2009)

11. Berntsen, J., Thomsen, A., Schelde, K., Hansen, O., Knudsen, L., Broge, N., Hougaard, H., Hørfarter, R.: Algorithms for Sensor-based Redistribution of Nitrogen Fertilizer in Winter Wheat. Precision Agriculture, 7, 65-83 (2006).
12. Doerge, T.: Nitrogen Measurement for Variable-rate N Management in Maize. Communications in Soil Science and Plant Analysis 36, 23–32 (2005)
13. Assimakopoulos, J.H., Kalivas, D.P., Kollias, V.J.: A GIS-based Fuzzy Classification for Mapping the Agricultural Soils for N-fertilizers use. Science of the Total Environment 309, 19–33 (2003)
14. Jones, D., Barnes, E.M.: Fuzzy Composite Programming to Combine Remote Sensing and crop Models for Decision Support in Precision Crop Management. Agricultural Systems 65, 137–158 (2000)
15. Molin, J.P., De Castro, C.N.: Establishing management zones using soil electrical conductivity and other soil properties by the fuzzy clustering technique. Scientia Agricola 65, 567–573 (2008)
16. Yan, L., Zhou, S., Feng, L., Hong-Yi, L.: Delineation of site-specific management zones using fuzzy clustering analysis in a coastal saline land. Computers and Electronics in Agriculture 56, 174–186 (2007)
17. Papageorgiou, E.I., Markinos, A.T., Gemtos, T.A.: Soft computing technique of fuzzy cognitive maps to connect yield defining parameters with yield in cotton crop production in central Greece as a basis for a decision support system for precision agriculture application. Studies in Fuzziness and Soft Computing 247, 325–362 (2010)
18. Krysanova, V., Haberlandt, U.: Assessment of nitrogen leaching from arable land in large river basins. Part I. Simulation experiments using a process-based model. Ecological Modelling 150, 255–275 (2002)
19. Ritchie, S.W., Hanway, J.J., Benson, G.O.: How a Corn Plant Grows. In: Report No. 48. Iowa State University of Science and Technology, Cooperative Extension Service, Ames (1992)
20. Tremblay, N., Bouroubi, M.Y., Panneton, B., Vigneault, P., Guillaume, S.: Space, Time, Remote Sensing and Optimal Nitrogen Fertilization Rates – A Fuzzy Logic Approach. In: GIS Applications in Agriculture Nutrient Management for Improved Energy Efficiency.CRC Press, Boca Raton (2010)

The Characterisation of "Living" Landscapes: The Role of Mixed Descriptors and Volunteering Geographic Information

Ernesto Marcheggiani[1,2], Andrea Galli[2], and Hubert Gulinck[1]

[1] Division of Forest, Nature and Landscape, Department of Earth and Environmental Sciences, Katholieke Universiteit Leuven, Celestijnenlaan 200E, B-3001 Leuven, Belgium
[2] Department SAIFET – Section of Agricultural Engineering and Landscape, Technical University of Marche, Via Brecce Bianche – 60131 – Ancona, Italy
{ernesto.marcheggiani,hubert.gulinck}@ees.kuleuven.be,
{andrea.galli}@univpm.it

Abstract. Over the last decade the need for public bodies to characterise the vitality and degree of sustainability of their territories is well acknowledged. Still it remains unclear how to integrate the different categories of values of our daily life places in a comprehensive way in order to develop appropriate and well balanced policies. An experimental case has been designed to provide novel sets of indicators by integrating information extracted from custom maps, spatial descriptors of land use and land cover and socio-economic indicators. In order to fully grasp the character of a living place, the nuances of less tangible aspects should be also understood. To do so, the results developed during first steps have been subsequently refined by incorporating relevant volunteering geographical information available on Google Earth® platform.

Keywords: Volunteered Geographic Information; Complexity assessment and mapping; Mixed open spaces characterization.

1 Introduction

The role of geographic information in planning is well proven, and it is commonly accepted that GISs are base tools for analysis [4]. Matthews [11], pointed out how "rural land managers in Europe are faced with an increasingly complex decision-making environment where production has to be achieved within narrowing environmental and social limits". Furthermore, he argued that a varied mix of financial, social and environmental goals has to be achieved to allow the managers to evaluate the options and impacts of alternative land use strategies. Expanding the information base is one issue, the question whether and how such information can properly serve planning process is another matter. According to Rizov [18] diversity and sustainability are key issues for rural development policies. As well as local and regional autonomy are currently playing a pivotal role overarching modernist discourses on the alternative concepts of rurality [3]. A debate, with a special regard

B. Murgante et al. (Eds.): ICCSA 2011, Part I, LNCS 6782, pp. 369–380, 2011.

to open spaces of rural-urban mixed settings, should be therefore addressed on "which kind" of information pieces are relevant to spatial management and planning. In particular, which values should have to be considered and which can be neglected? And how should certain values be addressed in our analysis? For instance, how can we record the smells and the sounds, essential landscape qualifiers, and insert this information in our database? Reversely, are the current vocabularies for spatial information and the on-going efforts for standardisation (Inspire Directive for instance), effective to describe attractiveness of places, sense of being home, attachments, ..., all important qualities that people feel about their everyday life places? Do we have to blindly trust on the official spatial information, made, managed and released by governmental or private bodies (remote sensing and mapping companies, rural development agencies, cadastre and census agencies,...) or are there other potential sources? Furthermore, are these last ones unusable in planning processes or some potential uses can be envisaged to address the aforementioned questions?

The emerging issue is "how" to analyse, and asses each landscapes characteristic and value, from both the spatial (structural organization of their spatial patterns) and functional (goods and services provided, ecological, social, cultural, etc.) points of view. As Linehan [8] states, it is important to reassert the value of relationships among physical, cultural, economic and political dimensions of space, to better incorporate the knowledge, perceptions, and practice that exist between the place we study and the people and communities who call this place home.

Moving beyond the traditional targets of landscape policies (cultural heritage, scenic landscapes, the rural sphere, and biodiversity relics), and embracing the landscape relevance of any space, altered landscapes, hybrid rural-urban landscapes, as well as adopting a much broader vision on multiple landscape functions and services such as formulated by the Millennium Assessment, demands constant refuelling of the landscape vocabularies and analytical instruments. Therefore, this present paper aims to further investigate the possibility to improve the methodological approaches for the characterization of territories by enriching analytical tools with interweaved layers of both spatial and socio-economical information, and able to take advantage form the available volunteered and free online information. The case has been designed to try break ground in the methodological principles commonly addressed for landscapes and territorial analysis and is discussed as a way to integrate a mix of indicators with relevant geographical information. The experiment focuses first on land use and land cover indicators (through landscape metrics) in combination with socio-economic indicators. Furthermore the potentialities inherent the reuse of volunteered geographical information available thanks to 3D virtual globe browsers, such as Google Earth ®, have been investigated. Abd-Elrahmana et al., [1] have pointed out as citizens have become more aware of the relationships between human society, geography and natural resources [17]. This desire of involvement in the planning and decision making processes is clear. Now raises the question whether or not participants with minimal background on landscape and land use, spatial information, and GIS and with amateur equipment, such as a handy camera and Google Earth, are useful sources for terrain information. Amichai-Hamburger [2] has recently defined these volunteers of information as "field workers" using their computers to help, communicate and share gathered information with others many thousands of miles away.

2 Case Study Description

The case study area covers 550 sq km including nine municipalities in central Italy. Its boundaries embrace an area suitable for high-quality agriculture markets such as grapes and prized wines (Verdicchio di Matelica) that qualify for the DOC mark. From the morphological point of view the study area lies in a well-defined physiographic region belonging to the wider anticline–syncline system characterizing the innermost part of central Marche region (Fig. 1). The landscape patterns of this area are represented by agro-ecosystems, whereas semi-natural woodland is found only in the limited anticline areas. At lower altitudes and in valleys, semi-natural patches are small, while green corridors made up of riparian vegetation are quite abundant. The area occupied by urban-industrial systems is not large and is fairly evenly distributed, while anthropic corridors (roads and railways) tend to occupy the valley floors. The cultural context is that of characteristic mix of peri-urban and rural landscapes in Italy. The area has undergone considerable changes over years. The sharecropping system abandonment and mechanisation of agriculture and the industrial boom have led to massive migration to coastal areas. Despite the decline of traditional farming activities and a relentless sprawl of artisans as well as a large industrial area, landscape retains much of its authentic rural character. In this ecological, economic and socio-cultural context the landscape could be the central element in a broader strategy of local sustainable development. The storyline of this case fits well with a scenario in which emphasis is on capitalizing on multifunctionality and the landscape diversity triggers to sustainable development in rural areas.

The Italian law on agricultural orientation (DL 228/2001) has recently challenged each regional authority to identify its most suitable rural districts, where to implement actions to sustain local agricultural markets and to improve the quality of landscape according to local stewardship councils, farmers and dwellers. Such areas called Rural Quality Districts have to be compulsorily identified and their boundaries have to be made explicit and drawn in maps. However there are no clear guidelines that explicit criteria to define such RQD, leaving this burden on the shoulders of local regional authorities.

Whilst the ability to characterize rural areas has important implications in spatial planning policies (Mountrakis and Ruskin 2005) the identification of rural districts is usually underrated in Italy. Classifications are usually performed by Agriculture Services at regional level throughout mere county-based GIS applications. As is the classification currently in force in Marche region. Such a layer is not at all a sufficient base to identify RQD in so far as none aspect able to capture the inner characteristics of the places at local scale has been accounted. In 2006 the Marche region has made explicit reference to the OCSE's methodology for the implementation of its Rural Development policy act [15]. Similar approaches can however only grasp certain aspects of rurality. The general purpose is therefore to define a methodological approach to help public land managers to select the best suitable areas for RQD at regional scale.

According to Gilg [6] the definition of rural has to be referred at last to three different aspects involving the ecological, cultural and occupational dimensions. Following this appeal a methodology is proposed inspiring on one side to the holistic principles considering landscape as a complex of self-organizing spatial features (spatial patterns) and, on the other, to try to capture the most relevant socio-economic aspects of a territory as a possible synthesis of the values expressed by people.

Fig. 1. Study area location in central Italy (upper frame), note the Synclinal–Anticlinal structure characterizing the overall morphology of the area (bottom frames)

3 Methodology to Integrate a Mix of Indicators

The purpose consists of a sequence of steps as shown in Fig. 2. Structural landscape patterns were measured through a software module, Maptool [9]. Such a module operates in GIS environment and is inspired to the approach proposed by Fragstats® [12] [13] [14] [7]. This offers the advantage of giving a spatial representation of values calculated per each metric, hence its spatial distribution. The performed metrics (Shannon diversity, patch density, splitting index and contagion) have been chosen as those most appropriate to describe the complexity of the arrangement, size of patches and their spatial distribution. The calculation of metrics (Fig. 3 up) has been performed by windows of 500 m per side with a ground resolution of 25 hectares, being the source land use maps represented at the scale of 1:10’000 with 10 m of original resolution.

To integrate the analysis with structural and social and economic descriptors, a set of indicators able to capture the social structure in terms of ageing and density (generational change ratio, population aging rate, population density) and the sharing of active workers among the three main economic sectors (number of employees in industrial, agriculture and tertiary sectors).

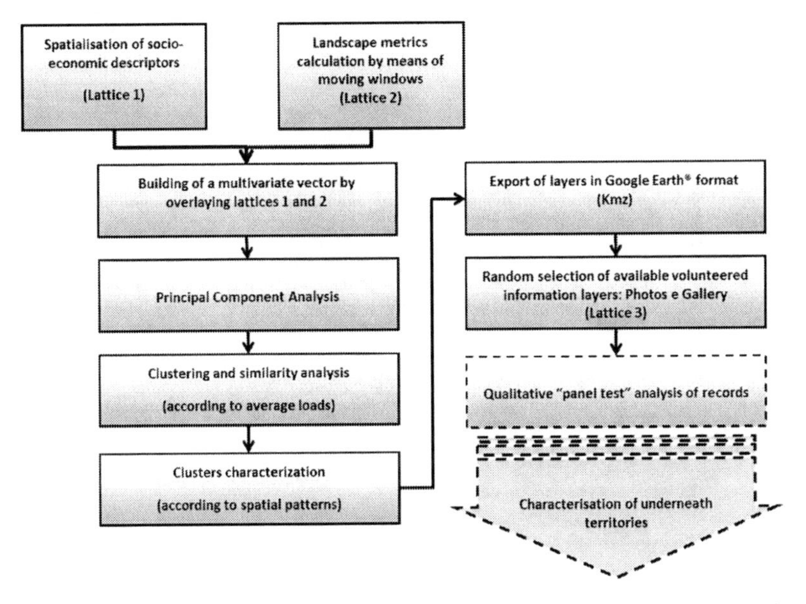

Fig. 2. The flowchart for the proposed methodology, with dashed line the on-going steps

Census information have been spatialised according to a finer level than that of municipality. To do so a collection of census tracts has been acquired in GIS vector format (ArcGis Desktop® shape file format). In the collection every vector polygon represents a sub-parcel of the municipality's territory characterized by specific values for each of the descriptors of the society and of the economic condition. The coordinates of each polygon's centroid have been extracted building an irregular plot of points scattered all over the area (Fig. 3 down). By interpolation (Inverse distance weighted algorithm) socio-economic information have been spatialised according to a lattice (500 m x 500 m) congruent with that of spatial metrics.

To evaluate the degree of redundancies among chosen indicators and in order to define groups of similarly behaving indicators Principal Component analysis (Varimax rotation) has been performed [16] [5]). According to the results, the performed PCA explains more than the 62% of variance and two PCs alone account the half of the existent variability. The first PC, named Economy and Society, is clearly characterized by five latent variables (Generational change ratio, Population aging rate, Population density, Agriculture employees, Employees in other sectors). The second PC refers to the structural aspect of landscape patterns being dominated by only three latent variables all referred to landscape metrics (Shannon diversity, Patch density and Splitting index). Latent components have been clumped according to three classes: high, medium and low standardized values. This has produced an arrangement of six classes (Fig. 4), where e.g., a class embodying a cell, belonging to the first PC, showing high average values would co-occur with one of the second PC with low values, and so on.

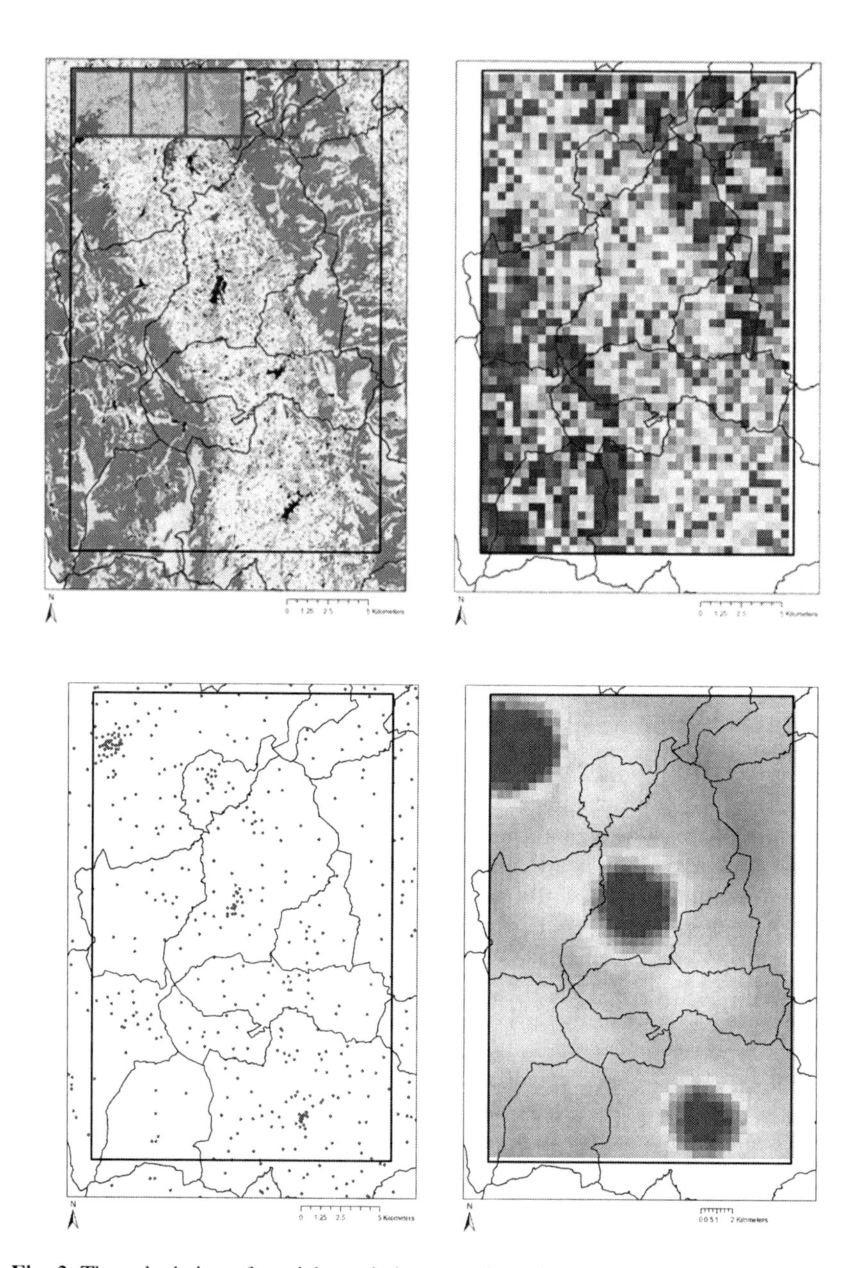

Fig. 3. The calculation of spatial metric by a moving windows technique (up) and (down) the scattering of census tracts nodes (red dots) as a prelude to the fuzzy spatialisation of human society descriptors. As an example, on the right-hand bottom, the resulting patterns for population density is shown (high density in red as well as low one in green)

This way every cell of the multivariate lattice has been labelled as to belong to a specific class of combinations among clusters. The lattice has been classified as follows:

ageing people living in highly fragmented complex settings (red); ageing people living in mild or simplified settings (orange); commuters or farmers living in highly fragmented complex settings (blue); commuters or farmers living in mild or simplified settings (dark blue); intermediate condition characterised by significant sprawl; and Intermediate condition in simplified settings (light green). By displaying the localization of each pair of co-occurrences (Fig. 5), it is possible to account for a rapid appraisal of the sites where specific spatial processes (such as, fragmentation instead of clumping for instance) are occurring together with aggregations of social human behaviours (e.g., areas within which people live in dense or sparse settlements, etc.).

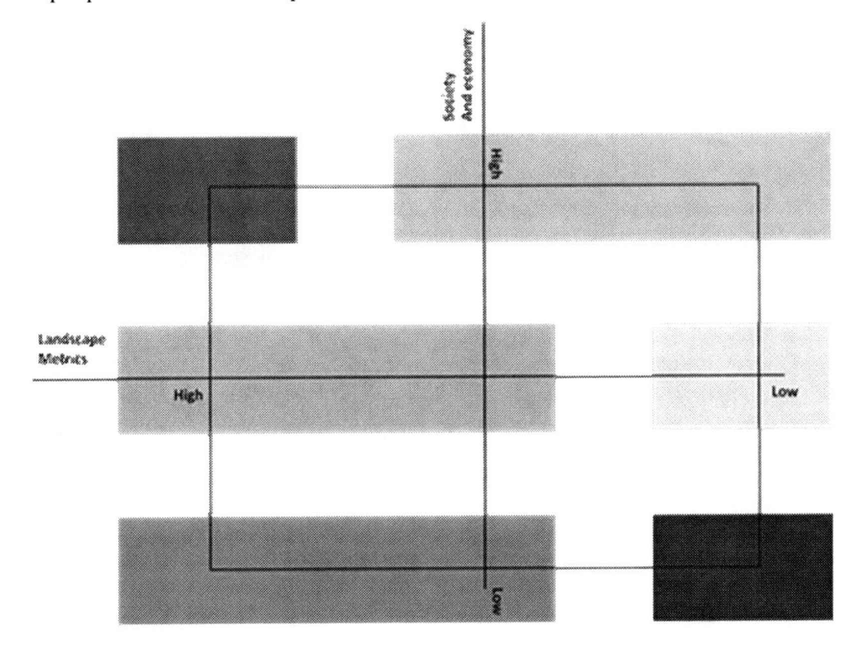

Fig. 4. Scatter plot representation of co-occurrences pairs according to the average values of each cluster; colours match those of spatial thematic

4 Interweaving Cluster and Volunteered Information Available on Google Earth

Throughout the definition of mixed clusters of landscape structural and socio-economic descriptors we have tried to shed lights on the possibility to discover underrated or concealed patterns. This represents a challenging issue that has been only partially met. Nevertheless, to an effective identification of the best suitable areas for the implementation of a RQD being, at the same time, able to capture all the nuances of the potential of living rurality, it is necessary to enrich the analysis with sound information on the perception of people living there, their culture and their local tradition.

To this purpose, the classification based on both spatial and human society descriptors has been converted into a suitable format (Kmz) for Google Earth® platform. This has allowed for the gathering of all voluntary information, e.g., photographs, land marks, labels of location of events, etc.. that people, living or visiting the places, have shared online.

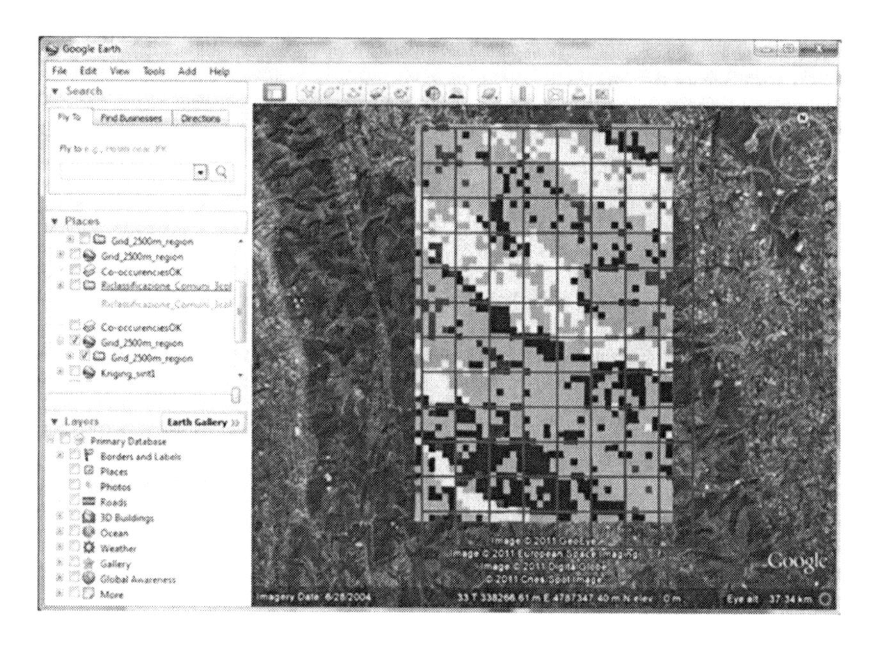

Fig. 5. Spatialisation of co-occurrence pairs in Kmz format and the grid used to account for the volunteered records classification

By overlaying the mixed clusters describing both the landscape structure and the social behave with a grid, of cells of 2.5 km on each side (Fig. 5), information available on GE have being classified, cell by cell, by a panel test analysis approach. At the current stage the experiment refers to the analysis of the photos layer, available on GE. Particularly, with regard to the subject of the shot, four classes have been defined: rural landscape; urban environment and built heritage; elements linked to the nature and biodiversity; and miscellaneous. To characterise the motivation of volunteers to share their pictures with other people, three classes of people have been assessed. The first class groups subjects whose intent was to communicate and share the beauty of the places (communicate outstanding aesthetics). The second class is composed of peoples wanting to expose or report negative facts. Finally, peoples whose purpose was not clearly assigned have been classified as neutral.

5 Results

Thanks to a meaningful description of landscapes characteristics the proposed classification can best serve as an effective tool to support decision makers on the identification of the best suitable RQD areas, if compared with the classification currently in force for the rural areas in Marche region [10]. In fact, the possibility to enrich our reading with integrated sets of heterogeneous and integrated socio-economic and spatial metrics sets may refuel the process of information production.

Furthermore, the overly with volunteered information has shown how despite local rural land managers depicts the territories comprise in the study area as typically dominated by rural landscapes, only the 15% of the sharing has been dedicated to that subject (Fig. 7 up). Most of the items in the Photos layer on GE are rather showing subjects related to the cultural and architectural heritage (32%) and to the biodiversity of the sites (25%). The remaining 27% is accountable to miscellaneous. The reasons why a such significant amount of light-motivated-shots should be further investigated to address what drives people to invest time to share trivial images in a web network.

Fig. 6. the panel test analysis. All records of available volunteered information appeared during the screening, have been accounted according to their contents. The spatial correspondence of each record with the relative clusters has also been taken into account

The analysis of motivation (Fig. 7 down) shows an even distribution between the willingness to communicate the beauties of places and neutral expressions, whilst the burden of shots to expose negative aspects represents not more than 7% of the selected records.

For what concerns the distribution of the sampled photos within the clusters, it is to be noticed how the photos referring the built heritage occur the most within the two large areas (blue and dark-blue pixels, Fig. 5) stretching across the syncline axes, as well as records dedicated to rural elements are evenly distributed in all clusters where landscape patterns have been classified as simplified. Finally, photos whose intent

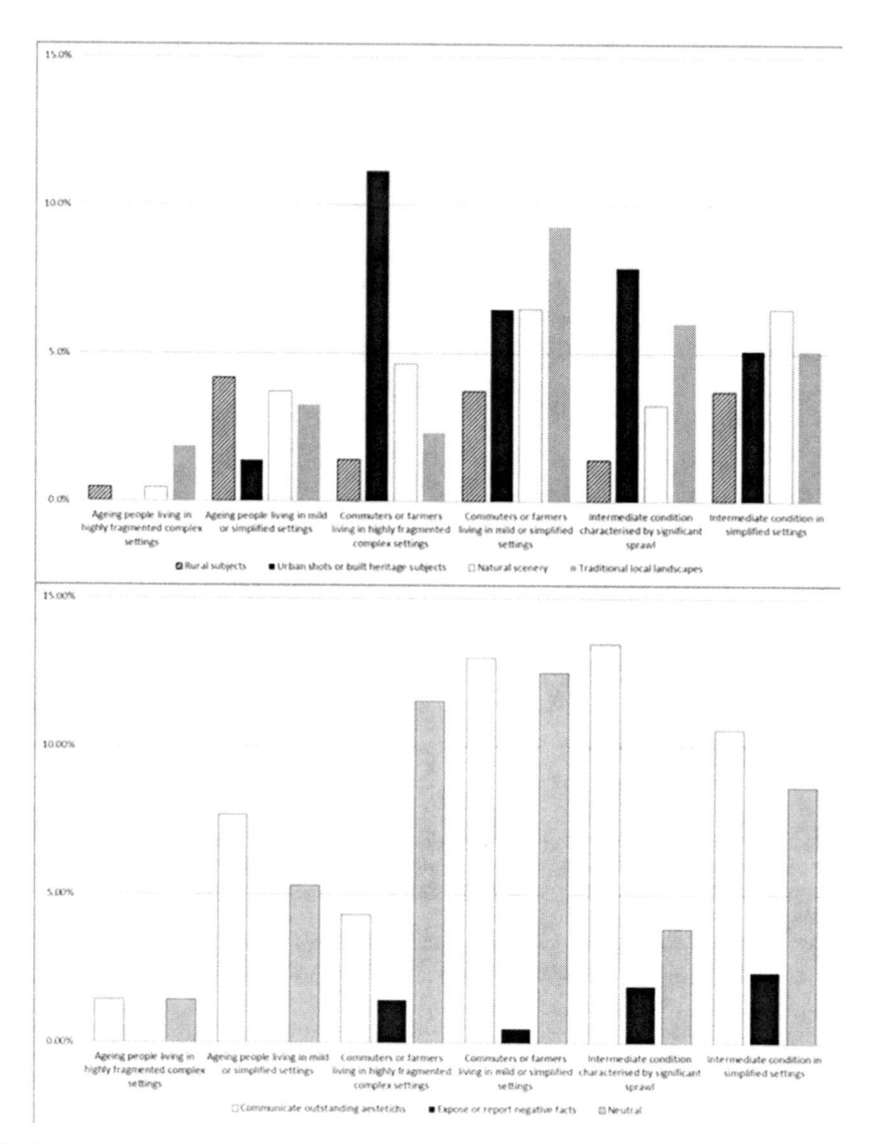

Fig. 7. The relative distribution of the subjects depicted (up) and the motivation of the picture's author (down) of each of the volunteered records within the clusters

was to expose or report degraded environments, are gathered the most within the clusters classified as intermediate situation with sprawl and or simplified landscape patterns.

6 Discussion

This paper represents the first step of an overall decisional process leading to the planning (forward-looking) and management (acting) of landscape. The discussion has aimed to contribute to the actual debate by proposing novel approaches to the characterisation of complex rural open spaces and aspires, in final synthesis, to offer a sound methodology suitable to land managers, planners and public authorities. On one hand, our purpose opens to challenge the analysis of the relationships of landscape metrics with indices or indexes referred to social aspects, as advocate by Uuemaa et al. [19]. On the other, our attempt of make a sound "re-use" of volunteered information is in concordance with the recalls from the international literature that assert the possibility of a better integration with aspects inherent relationships among people and the territories they live. In other words, to bridge the gap between "...the technical knowledge and methods of landscape planning and the cultural, economic, and political knowledge, perceptions, and practices of real people who ultimately determine the condition of real landscapes." [8].

Despite the research is still in an explorative stage, and the purpose designed for the case study –the definition of boundaries for specific QRD- has not yet completely been fulfilled, the first results allows for a reflection on the potential of new methodological approaches to soundly assess the degrees of rurality, as an important step towards improving the semantic and analytical toolbox of landscape analysis. In a landscape-oriented focus this means to assess the vitality of a rural system intended as a system that includes the soil and climate potentials, the land managerial know-how of the practiced agricultural uses, the economic behaviour, the ecological functioning, the social and cultural values of the local communities and the intrinsic qualities that may support not yet operational services. Living rurality is all but empty: it is filled with people and groups with specific relationships to their environment. In essence rurality is, and remains, a concept about the relationship of interested parties (locals, farmers, communities, etc.) to "their" open spaces, this is the unsealed soil, not ultimately threatened by urbanisation. Of course the specific expressions of rurality will change, e.g. the specific land use, but also the rural population. That means that measuring the degree of rurality is a matter of measuring the vitality of the rural system, whatever this system is composed of: it includes land use, communities, and their culture. We believe that the integration efforts among spatial and human society descriptors, as well as the gathering of sound volunteered information could be used as an outreach that could result in a proactive process to better meet the current needs of spatial planning authorities and land managers in all that complex settings where classical urban or nature protection normative and planning frameworks fail.

References

1. Abd-Elrahmana, A.H., et al.: A community-based urban forest inventory using online mapping services and consumer-grade digital images. International Journal of Applied Earth Observation and Geoinformation 12, 249–260 (2010)
2. Amichai-Hamburger, Y.: Potential and promise of online volunteering. Computers in Human Behavior 24, 544–562 (2008)
3. Cruickshank, J. A.: A play for rurality – Modernization versus local autonomy. Journal of Rural Studies 25, 98–107 (2009)

4. Klosterman, R.E.: The appropriateness of geographic information systems for regional planning in the developing world. In: Comput. Environ. and Urban Systems. European Landscape Convenction, ETS No 176, vol. 19(1), p. 13. Council of Europe, Florence (1995)
5. Cushman, S.A., McGarigal, K., Neel, M.C.: Parsimony in landscape metrics: Strength, universality, and consistency. Ecological Indicators 8, 691–703 (2008)
6. Gilg, A.: An Introduction to Rural Geography. Arnold, London (1985)
7. Leitão, A.B., et al.: Measuring Landscapes: A Planners Handbook. Island Press, Washington, D.C (2006)
8. Linehan, J., Meier, G.: Back to future, back to basic: the social ecology of landscape and the future of landscape planning. Landscape and Urban Planning 42, 207–223 (1998)
9. Marcheggiani, E., module, M., Colantonio R., V., Galli, A.: Integrated indicators in environmental planning: methodological considerations and applications. Ecological Indicators 6(1), 228–237 (2006)
10. Marcheggiani, E., Bomans, K., Galli, A., Gulinck, H.: New ways of landscape diagnosis. In Living Landscape: The European Landscape Convention in Research Perspective 9, 256–270 (2010) ISBN: 978-88-8341-458-9
11. Matthews, K.B., Sibbald, A.S., Craw, S.: Implementation of a spatial decision support system for rural land use planning: integrating geographic information system and environmental models with search and optimisation algorithms. Computers and Electronics in Agriculture 23, 9–26 (1999)
12. McGarigal, K., Marks, B.J.: FRAGSTATS: a spatial pattern analysis program for quantifying landscape structure. USDA Forest Service. GTR PNW 351 (1995)
13. McGarigal, K., Cushman, S.A., Stafford, S.G.: Multivariate Statistics for Wildlife and Ecology Research. Springer, New York (2000)
14. McGarigal, K., Cushman, S.A., Neel, M.C., Ene, E.: FRAGSTATS: Spatial Pattern Analysis Program for Categorical Maps. In: Computer software program produced by the authors at the University of Massachusetts, Amherst, Massachusetts, USA (2002), http://www.umass.edu/landeco/research/fragstats/fragstats.html.
15. Millenium Ecosystem Assessment, Ecosystems and Human Well-Being: A Framework for Assessment. Report of the Conceptual Framework Working Group of the Millennium Ecosystem Assessment. Island Press, Washington, p. 245 (2003)
16. Regione Marche, Programma di Sviluppo Rurale 2007-2013, Reg.CE 1698/05 (2008)
17. Mountrakis, G., AvRuskin, G.: Modeling Rurality using Spatial Indicators (2005), http://www.geocomputation.org/2005/
18. Nowak, D.J., Noble, M.H., Sisinni, S.M., Dwyer, J.F.: People & trees: assessing the US urban forest resource. Journal of Forestry, 37–42 (March 2001)
19. Riitters K.H., O'Neil R.V., Hunsaker C.T., Wickham J.D., Yankee D.H., Timmins S.P., Jones K.B., and Jackson B.L.: A factor analysis of landscape pattern and structure metrics, Landscape Ecology 10(1) 23-39,(1955)
20. Rizov, M.: Rural development and welfare implications of CAP reforms. Journal of Policy Modeling 26, 209–222 (2004)
21. Uuemaa, E., Antrop, M., Roosaare, J., Marja, R., Mander, U.: Landscape Metrics and Indices: An Overview of Their Use in Landscape Research. Living Rev. Landscape Res. 3, 1 (2009)

Conceptual Approach to Measure the Potential of Urban Heat Islands from Landuse Datasets and Landuse Projections

Christian Daneke, Benjamin Bechtel, Jürgen Böhner, Thomas Langkamp, and Jürgen Oßenbrügge

Institut für Geographie, Universität Hamburg, KlimaCampus. Bundesstraße 55, 20146 Hamburg, Germany
christian.daneke@uni-hamburg.de

Abstract. Urban morphology plays a crucial role in the alteration of the local climate, resulting in the formation of Urban Heat Islands. Regarding the steady growth of cities and the impact of global climate change, the risk of overheating is expected to increase. In order to reduce this risk and the resulting health problems, planning measures are needed to adapt to these severe events. Planners however need tools to quantify and evaluate different adaption strategies to judge its effectiveness. This paper presents conceptual thoughts towards the development of such a planning tool, as well as proposing a method to derive potential areas of intra Urban Heat Islands. A first calibration of the landscape metric for city of Hamburg region is presented, which showcases the method.

Keywords: Landscape Metrics, Urban Heat Island, Land Use Modeling, Hamburg.

1 Introduction

Cities and climate interact in two distinct ways. Firstly, they are responsible for the majority of greenhouse gas emissions. Secondly, they are one of the main receptors of global climate change induced impacts. Already today, most of the world's population is living in city regions and future projections indicate a steady increase of urban population. In this sense the vulnerability, the exposition to severe events, is considerably high in cities. Floods or heat waves can be named as two examples. Even in developed countries, heat is considered to have a considerable effect on mortality [1]. For example during the European Heat Wave in 2003, at least 35.000 heat-related deaths were recorded [2].

Climate in urban areas is considerably different to that of natural areas. Due to urbanization processes the local climate is altered in terms of wind or temperature. Further, due to constant increase of urban population the influences of urban fabric on the local urban climate will increase. Additionally, it is argued that global climate change will amplify effects of extreme weather events – including heat waves [3].

In order to cope with the risk of such extreme weather events, different mitigation and adaptation strategies have to be developed. Mitigation strategies revolve around

B. Murgante et al. (Eds.): ICCSA 2011, Part I, LNCS 6782, pp. 381–393, 2011.

the question how to reduce greenhouse gas emissions, while adaptation strategies aim to adapt to altered conditions. A simple adaption measure is to introduce green areas or prevent dense urban areas. However, for a planner it is difficult to quantify the impact and judge whether a certain adaption strategy is "better" or "worse".

This paper discusses a method to evaluate different land use projections and highlight potential areas of overheating. Firstly, a set of conceptual thoughts is discussed, expressing the requirements associated with the calculation of the overheating potential. Then, in the second part of this paper a metric is proposed and a first calibration using Hamburg as a test study is presented. In the end such a metric can be used to evaluate land use projections from urban land use models in terms of their UHI potential.

2 Urban Land Use as a Trigger of Urban Heat Islands

Urbanization processes have a measurable influence on the local climate, thereby producing the phenomena of the Urban Heat Island (UHI). The UHI is defined as the highest nocturnal air temperature difference at screen height [4]. Generally said, an UHI can be described as the temperature difference between urban areas and rural areas. The formation is caused by an agglomeration of different effects, which alter the energy balance in cities affecting the urban thermal environment [5]. A schematic overview of these interactions is illustrated in Figure 1, showing both small scale influence and large scale characteristics.

There are several influences that lead to the formation of an UHI. For example a positive correlation between UHI intensity and population size was found [7]. However, the largest influencing factor is related to the urban geometry and morphology. Buildings and other urban structures act as obstacles for the wind, making "Cities the roughest of all aerodynamic boundaries" [5]. In this process wind speeds in general

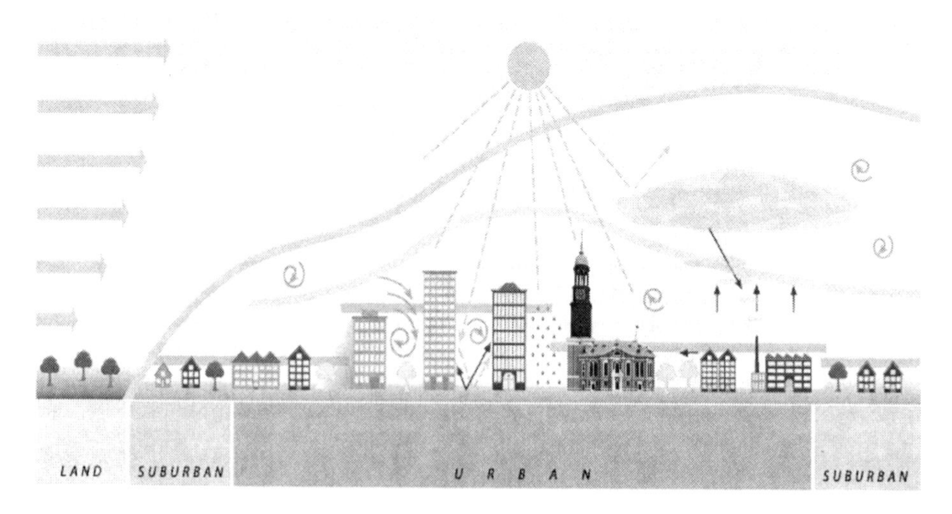

Fig. 1. Conceptual image showing the different impacts on the urban climate: enhanced turbulences, trapping of sunlight energy, anthropogenic heat based on different land use types [6]

are slowed down, decreasing energy transfer and thus intensifying the UHI. Especially the street orientation and building geometry are a crucial factor in the alteration of wind flow. This influence on the wind is defined by the roughness and can be classified [8][9]. Heat capacity is increased by the urban geometry and building materials by trapping incoming energy. As a result long wave radiation is emitted back at night leading to the effect that the UHI is strongest during nighttime. Not only is heat stored but also emitted by anthropogenic energy consumption. Thereby, heat is either emitted by heating in the winter or cooling during the summer. Additionally, the burning of fossil energy by driving cars releases heat into the surrounding. Evapotranspiration and heat fluxes are dramatically altered by the lack of vegetation [10] as well as the degree of soil sealing. But, the UHI intensity is not only related to the anthropogenic induced alteration but is also related to the synoptic weather conditions [11]. It could be found that wind speed and direction, as well as cloud cover both influence the intensity of the UHI.

UHIs are dependent on the spatial characteristics of the city [5]. Moreover, urban heat islands show a high spatial variability, which is related to the type of land use, vegetation, urban morphology and intensity of anthropogenic heat emission. For instance high density areas, such as the Central Business District (CBD) or industrial areas with little vegetation and dense building morphologies produce higher local temperatures. In return areas which are less dense and comprise wider streets produce a less intense UHI. Further, areas such as parks or large water bodies have a cooling effect on the surrounding [12]. In order to quantify the influence as well as the strength on the UHI formation local measurements are needed.

Saatoni et. al. for example found that for the city of Tel Aviv the heat island could be related to high building densities and heavy traffic. They also found cooler areas to be within the neighborhood of open, green areas [12]. A similar result can be found in the Portland region, where mobile vehicle measurements were used to quantify the intra UHI characteristics during warm summer days [13]. They found that the lowest temperatures could be found next to a large inner-city forest whereas the highest temperatures were correlated with industrial and commercial land use. A clear correlation between the intra UHI intensity and the land cover type is shown using mobile measurements for a Japanese city [14]. The influence of missing vegetation is shown in [10], where a linear correlation between the temperature and the amount of vegetation is found using Landsat imagery. The relationship between the urban heat island and urban morphology for a European city was tested in Aachen [15]. There, the highest temperatures were found in the medevil city centre too. Using Landsat imagery and temperature proxy data the relationship between urban fabric and temperature could be derived for Hamburg [16].

Such studies are of great importance to understand the interactions between urban morphology and the UHI. Further, these studies are the essence for the proposed metric in this paper. A distinct knowledge of these interactions is necessary in order to judge whether a certain urban development is causing an intensification of intra UHI or not.

3.1 Adaption Strategies and the Consequences on Urban Climate

Already in 1978 Oke emphasized the need for urban planners to consider the interaction between the urban areas and the atmosphere [7]. In order to plan adaptation strategies accurately, knowledge of the interaction between urban structures and the local climate are necessary.

Essentially, planning strategies aim to reduce negative effects while increasing positive effects at the same time. Hence, for new urban developments the urban geometry and building heights need to be considered. However, the redesign of street orientations or buildings in existing areas is nearly impossible. Introducing vegetation, by either planting new trees or developing new intra urban green areas is a highly effective measure. Parks act as cooling spots and increase the air quality, while trees create a shading effect within street canyons [17]. Other planning measures are aimed to improve building design and the materials [18].

Urban planners, however, have very little knowledge of the influences leading to the intra UHI. This emphasizes the need for scientists to communicate and create an awareness of the problem to planners so that it is incorporated into the planning process [11]. Further, the development of practical tools is necessary to present the results efficiently and effectively. Tools to assess different policy measures regarding the influence on UHI have been developed by different researchers [19]. However, all of the tools were designed to be used on a very high resolution, for example to assess the introduction of trees into a street canyon. A tool to assess land use datasets for a whole city, which enables to compare different scenarios of future land use change is still missing.

3 Requirements to Evaluate Urban Heat Island Potential from Policy Implementations

Ideally, before a new policy is being implemented, an urban planner is interested to see the possible outcomes. In the case of UHI formation it would be beneficial for a planner to have an indication, whether a specific land use trajectory increases the potential for intra UHI. A common example cited is the policy of the compact city to reduce greenhouse gas emissions as means of a mitigation strategy. However, in return compact structures, with a high fraction of impervious surfaces increase the potential of the UHI.

But, what measures are necessary in order to evaluate a land use policy decision? In this section, some conceptual requirements are highlighted and discussed.

Land Use Dataset as a representation of homogeneous urban climate

In the previous section it was discussed how different urban morphologies show different climatic responses. In order to assess these responses it is essential to have a general typology of urban structures. Such a typology should be based on geometric features, but also include some of the thermal properties and information on vegetation. Such typologies were proposed by many scholars [20] [21]. However one of the more recent typologies were proposed by Stewart and Oke, called the Thermal Climate Zones (TCZ) [4]. The aim of this typology is to integrate all existing typologies and form a consistent and objective approach [22]. For the Hamburg region it was attempted to create a land use dataset based on the TCZ. It was found however that some essential structures which are found in European cities could not be mapped with this US-focused approach. Thus a slightly modified urban classification scheme was proposed [23]. A land use dataset is needed which reflects the morphological characteristics as well as thermal properties.

Future projections of urban land use

In order to assess land use changes and create realistic projections different modeling approaches can be used, for example Cellular Automata [24]. Plus, Urban Land Use Models are a well-established method for scenario studies and act as Spatial Decision Support Systems [25]. By creating future projections of land use change based on different assumptions, planners and scientists get a picture of the possible developments before a plan is incorporated and can be adjusted accordingly.

3.1 Conceptual thoughts on the Implementation of a Landscape Metric

The proposed tool is intended to enable a comparison between different land use scenarios for one region and indicate which scenario yields a "better" potential to prevent the formation of an intra-urban UHI. For the calculation the following steps need to be defined based on several hypothesis concerning the interaction between the urban environment and the UHI:

1. Land use patterns have to be divided into "bad" ones like urban fabric and "good" ones which lower the UHI effect such as parks or water.
2. The influence of a land use type and the strength needs to be defined. As stated in section 1.1. the spatial variability of the UHI is strongly correlated with the existing land use.
3. Neighborhood effects play an important role. For example it was shown that areas close to water or green areas show lower temperatures. Further, the UHI potential is also dependent on the homogeneity of a specific area.
4. As the UHI effect can be described as the sum of the influences over a given area, the size and shape of a land use element has to be considered. It is hypothesized that large homogenous areas of urban cover which are compact result in a larger UHI potential.
5. It is hypothesized that the homogeneity and patch geometry play a role, but it can also be assumed that the location of a structure, in relation to the center is also very important. If this structure is placed close to the edge, then influences from other land use start to show. If a structure is found close to the center influences likely start to accumulate.

4 Method

A similar approach to the hypothesized metric can be found in [26], to measure habitat fragmentation. In this specific metric, classes are split into natural classes and classes which disturb natural habitats. Furthermore, neighborhood effects and the patch shape are accounted for. As a first start the metric by Klepper is used in order to investigate the usefulness for the proposed research question. An implementation of the Habitat Fragmentation Metric can be found in the Metronamica Modelling Framework [27] and is described shortly in this section.

Defining influences on class level:
As the first instance, land use classes are divided into positive and negative influences. Then, a fractional value is defined for each class, which indicates the amount of natural or in this case urban fraction. Lastly, weights are associated with each class, representing the resistance. Values for the weights are 1, 10, 100 and 1000, where increasing weights are harder to overcome, while 1 indicates no resistance.

Neighborhood effects and patch shape.
Based on a defined search radius r_0, given in pixel values, the neighborhood of a cell is analyzed. The algorithm looks for cells of the same type and calculates the equivalent area A' defined as:

$$A' = \sum w_i a_0 \tag{1}$$

Where w_i is the weight and a_0 is the area of a cell. A' will yield low values when the cell is surrounded by the same class. From A' the equivalent radius r' is calculated as:

$$r' = \sqrt{A'/\pi} \tag{2}$$

These two steps are done for every cell within a patch of the defined classes.

To account for the position of a cell within a patch, cells are weighted according to the distance to the centroid by:

$$v = \exp\left(-\frac{r'}{r_0}\right) \tag{3}$$

where r_0 is the search radius. The smaller r' is, the higher the weight v the chance of that cell being in center increases. By increasing the search radius it becomes more likely that more negative effects are included. When r' is high at a search radius of 1, then it is very likely that this cell will be on the edge.

Further the new equivalent radius s is given by the integral of the weight v, the fraction f and r':

$$s = \int_0^\infty v f dr \tag{4}$$

Computing s accounts for the compactness and size of the patch. If for example the patch is very compact and large, then values for v will be very high, resulting in a large equivalent radius. If however, the shape is stretched and thin, the sum of all values of v will likely be very low.

Finally, *KOV*, which indicates the chance of appearance, is calculated as:

$$KOV = \left(\frac{s}{r_0}\right)^z \tag{5}$$

The z parameter is defined by the user and controls the gradient towards the center of a patch. Low z values generally returns a shallow gradient and values of *KOV* are spread more evenly. Higher z values result in a steep gradient giving more emphasize on the center location. In the original setup of the metric, z refers to an empirical relationship between the area and the number of species. For the analysis of UHI potential such a relationship still needs to be defined.

5 Results of a First Calibration

In a first calibration process, the metric was used to analyze the urban heat potential using Hamburg as a test case.

5.1 Data

For the calculation a land use dataset for the greater Hamburg region is used, which is available for three time steps, 1960, 1990 and 2005. The dataset was created based on Aerial Photographs, Topographic Maps and 3D Building data at a resolution of 100 meters. Land use classes in the dataset are defined by morphological criteria, which reflect the influencing factors on the local climate. A description of the classification scheme can be found in [23] and is an adaption of the thermal climate zones [4]. For every class, morphological parameters like building height, building density and also preliminary figures for anthropogenic heat emissivity are known. The land use dataset for 2005 is shown in figure 2.

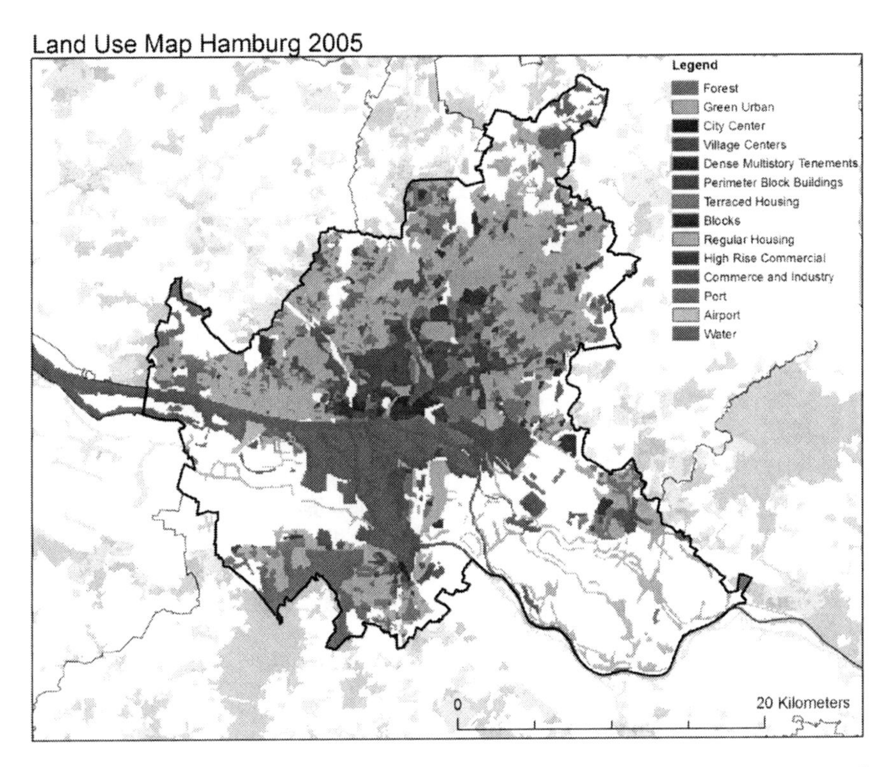

Fig. 2. Land Use dataset based on morphological characteristics for Hamburg and surrounding counties

5.2 Existing Assessments of the Urban Heat Island in Hamburg as Reference

To create a first estimation of calibration parameters, the produced results need to be compared to existing studies related to UHI alteration. Temperature changes within

Hamburg were measured using long term temperature recordings [28]. However, only few stations, mainly in rural areas, were used which don't give a good indication of the UHI variability. Temperature proxy data as presented in [16] are another valuable measure in fine tuning and will be implemented in the next iteration of calibration.

During the last summer in Hamburg an Urban Heat Island Map was published, indicating local hotspots and cold-spots. The map was produced by an interpolation of "school-stations" datasets within the city region[1]. Even though the accuracy has been criticized of these measurements, it is one of the few indications and visual representations of the UHI available. In a first iteration, the UHI-Map of Hamburg was used to generate a first fit and assess the parameters of the Habitat Fragmentation Metric.

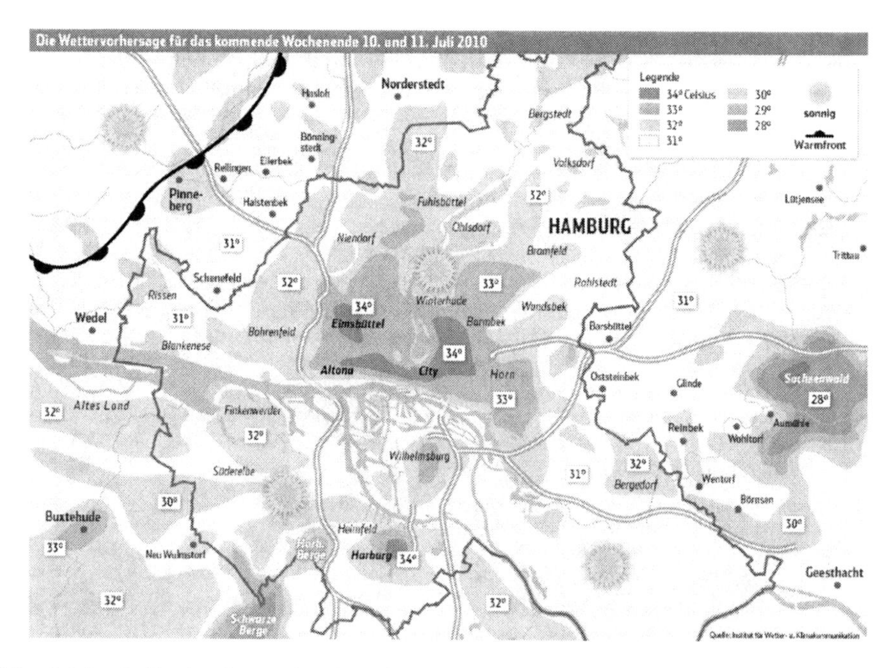

Fig. 3. Map indicating intra urban heat islands, as a wheather prediction during the last heat period in the summer 2010 (created at Institut für Wetter- und Klimakommunikation Hamburg)

5.3 Assigning Fractional Values and Weights to Each Class

The most difficult part of the calibration procedure is to define a fraction showing the influence on the local climate for every class. Because geographic location and characteristics of cities are different these fractional values have to be defined for each city individually. The fractional value *f* can be described as a function of all the geometric parameters available in a dataset. However, to derive the correct fractional values for every land use class, local measurements are essential as discussed in [15] for instance. For this test case, values for *f* are preliminarily set, based on assumptions derived from literature (Table 1). Regular housing, which refers to single family

[1] A project where weather stations are installed in various schools spread through the city of Hamburg. More information at: http://www.wetterspiegel.de

houses was given a weight of 10, which indicates a neutral behavior. Even though these areas are urbanized, they have very little influence on the urban heat island due to a high fraction of greening and the present sparse building geometry. The highest weighted classes are the urban core and dense multistory tenements. These classes are defined by high buildings, as well as dense building geometry. Additionally, the fraction of greening is very low.

Table 1. Table showing the parameter set up for the urban classes

Urban Classes		
Class	**Fraction**	**Weight**
Sealed area	0.6	1
Rail-Tracks	0.6	1
Urban Core	0.9	1
Village Core	0.8	1
Dense Multistory	0.9	1
Perimeter Blocks	0.7	1
Terraced housing	0.4	1
Blocks	0.5	1
Regular Housing	0.2	10
High Rise Commercial	0.6	1
Industry and Commerce	0.7	1
Port	0.7	1
Airport	0.5	10

Natural classes, which reduce the urban heat island formation are defined as:

Table 2. Fractions and weights of the „negative classes“

Natural Classes		
Class	**Fraction**	**Weight**
Arable Land	0	100
Pastures	0	100
Other Agricultural Areas	0	100
Forest	0	100
Shrub Lands	0	100
Wetlands	0	100
Green Urban Area	0	1000
Waterbody	0	1000

All natural classes were given a fraction 0 simply because no urban fabric can be found here. Green Urban Areas and Water bodies were given resistance values of 1000. Every other class, which mainly can be found in the rural areas, was given a value of 100.

5.4 Search Radius and the z Parameters

Search radius r_0 is given in cell space, and thus also dependent on the resolution of the existing dataset. It could be shown for example that the influence of a park can be measured 150 – 200 m into the neighborhood [14]. At a resolution of 100 meters in the underlying dataset a radius of 1.5 was chosen and tested.

The parameter z, which controls the gradient of values and is referred to an empirical relationship in ecology, was estimated to be around 0.5 and 0.6.

5.5 First Estimate of Urban Heat Island Potential in Hamburg

Using the above stated parameters, the following results can be presented for the land use map of 2005, Figure 5.

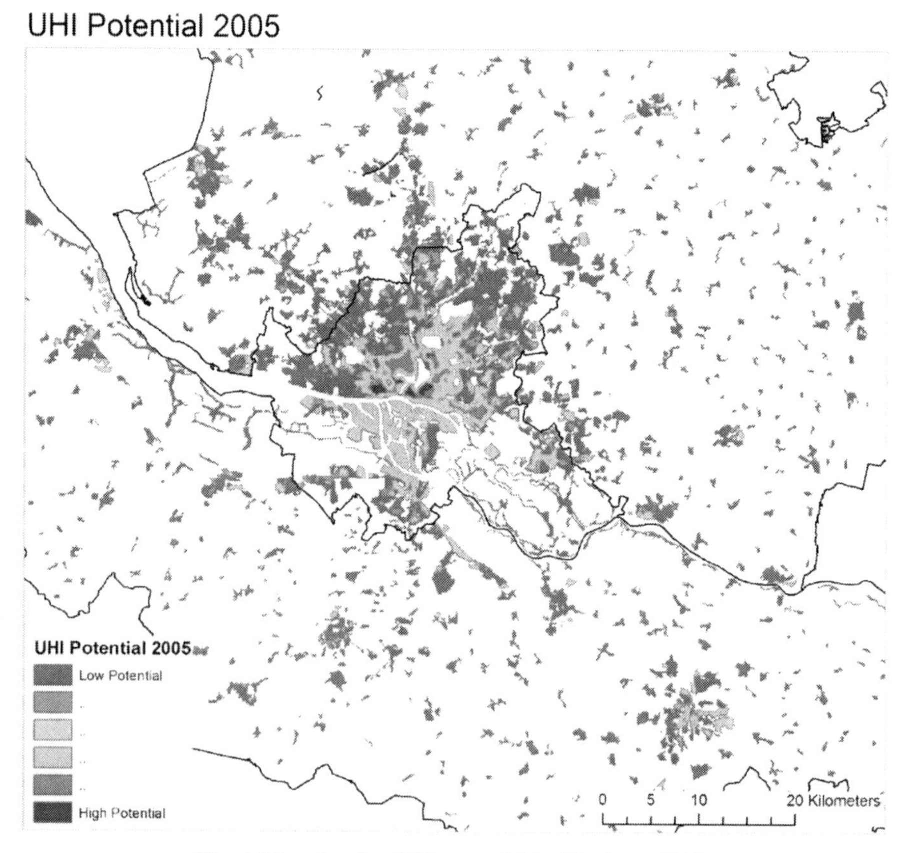

Fig. 4. Map showing UHI potential for Hamburg 2005

From the map it can be seen that the highest urban heat island potential can be found in the center and around the Alster Lake, which has the largest building densities in Hamburg. Also, the industrial areas in the east as well as of the medieval town centers in Hamburg show a high UHI potential. During the first iteration of the calibration it was the aim to achieve a fit close to Figure 5. Thereby, only the areas showing maxima were observed and compared. Transitions between areas and areas of medium potential have to be done during fine-tuning of the parameter.

Another example to present the UHI potential is through change, indicating where a certain development has caused an increase. The following figure shows the development of the UHI potential in Hamburg from 1960 to 2005.

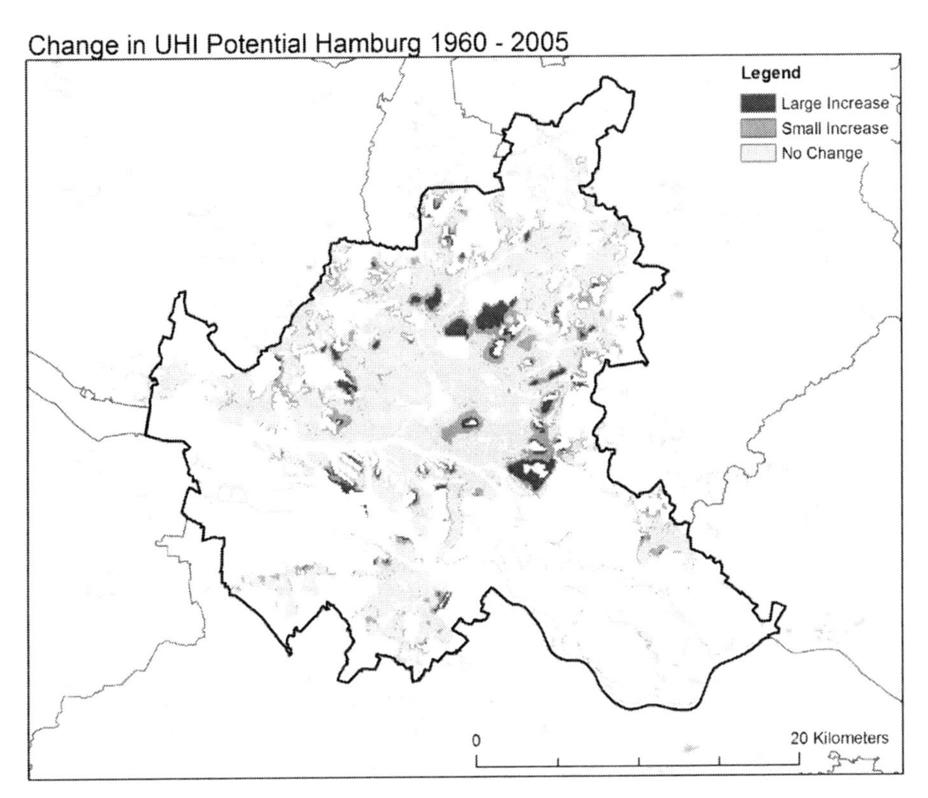

Fig. 5. Figure showing the increase of UHI potential for Hamburg from 1960 to 2005 based on land use changes in this period

6 Discussion and Concluding Remarks

A first calibration of the proposed Metric showed satisfactory fits, however only by a qualitative assessment. Calibration was done until the local maxima shown in the UHI map (Figure 2), the areas where the UHI is considered highest, were met and a qualitative comparison was satisfactory. Assumptions on the influences of the different morphologies are based on literature review and need further review. Therefore, the validity of the results needs to be tested using more precise local temperature measurements. In a first conceptual iteration however, it was shown how the Habitat Fragmentation can be used to evaluate land use change in terms of UHI potential. Furthermore, the need for a geocomputation-method to analyze future UHI distribution was highlighted.

Because of the ability to parameterize the Habitat Fragmentation Metric in order to reflect climatic responses, this method proves to be an excellent starting point. Plus, the proposed approach follows the call to create practical tools for planners to assess policy measures [11]. Also, from the change map produced, possible results were shown which can be used to transfer knowledge and impacts on UHI formation to the planners.

For the future, there remain a handful of tasks ahead in order to fully calibrate and validate the metric and each input parameter. Every parameter, especially the search radius and the power z need further investigation. In order to do so, more knowledge is needed how different urban classes behave in their influence of the intra urban heat island. Studies like [28] and [16] will be a first start to further calibrate and validate the metric. Still, a sufficient understanding of the influences can only be gained by conducting direct measurements within the street canyons, for example using mobile measurement methods as reported in [13][14]. For the Hamburg region it is planned to conduct mobile temperature measurements, using public Busses during the upcoming summer of 2011.

References

1. Poumadère, M., Mays, C., Le Mer, S., Blong, S.: The 2003 heat wave in France: dangerous climate change here and now. Risk Analysis 25, 1483–1494 (2005)
2. Haines, A., Kovats, R., Campbell-Lendrum, D., Corvalan, C.: Climate change and human health: impacts, vulnerability and public health. Public Health 120, 585–596 (2006)
3. Rosenzweig, C., Solecki, W., Parshall, L., Chooping, M., Pope, G., Goldberg, R.: Characterizing the urban heat island in current and future climates in New Jersey. Env. Hazards 6, 51–62 (2005)
4. Stewart, I.D., Oke, T.R.: Newly developed thermal climate zones for defining and measuring urban heat island magnitude in the canopy layer. In: Preprints, T. R. OkeSymp. & 8th Symp. onUrb. Env. Phoenix, January 11-15 (2009)
5. Oke, T.R.: Boundary Layer Climates, 2nd edn. p. 435 (1987)
6. Oßenbrügge, J., Bechtel, B.: Klimawandel und Stadt: Der Faktor Klima als neue Determinante der Stadtentwicklung. Hamburger Symposium Geographie, Band 2 (2010)
7. Oke, T.R.: City size and the urban heat island. Atmosphere and Environment 7, 769–779 (1973)
8. Davenport, A.G., Grimmond, C.S.B., Oke, T.R., Wieringa, J.: Estimating the roughness of cities and sheltered country. In: Preprint for the 12th AMS Conf. Appl.Clim. Asheville, N.C, pp. 96–99 (2000)
9. Wieringa, J., Davenport, A.G., Grimmond, C.S.B., Oke, T.R.: New revision of Davenport roughness classification. In: Proc. of the 3rd European & African Conf. on Wind Eng. vol. 8, Eindhoven, Netherlands (July 2001)

10. Weng, Q., Lu, D., Schubring, J.: Estimation of land surface temperature-vegetation abundance relationship for urban heat island studies. Remote Sensing of Environment 89, 467–483 (2004)
11. Yow, D.M.: Urban Heat Islands: Observations, Impacts, and Adaptation. Geography Compass 1(6), 1227–1251 (2007)
12. Saaroni, H., Ben-Dor, E., Bitan, A., Potchter, O.: Spatial distribution and microscale characteristics of the urban heat island in Tel-Aviv, Israel. Landscape Urban Plan 48, 1–18 (2000)
13. Hart, M.A., Sailor, D.J.: Quantifying the influence of land-use and surface characteristics on spatial variability in the urban heat island. Theor.Appl.Climatol. 95, 397–406 (2009)
14. Yokobori, T., Ohta, S.: Effect of land cover on air temperatures involved in the development of an intra-urban heat island. Climate Research 39, 61–73 (2009)
15. Buttstädt, M., Sachsen, T., Ketzler, G., Merbitz, H., Schneider, C.: Urban Temperature Distribution and Detection of Influencing Factors in Urban Structure. In: International Seminar on Urban Form, Hamburg (2010)
16. Bechtel, B.: Floristic mapping data as a new proxy for the mean urban heat island and comparison of predictors in Hamburg. Submitted to Climate Research (2011) (in review)
17. Spronken-Smith, R.A., Oke, T.R., Lowry, W.P.: Advection and the surface energy balance across an irrigated urban park. International Journal of Climatology 20, 1033–1047 (2000)
18. Kikegawa, Y., Genchi, Y., Kondo, H., Hanaki, K.: Impacts of city-block-scale countermeasures against urban heat-island phenomena upon a building's energy-consumption for air-conditioning. Applied Energy 83, 649–668 (2006)
19. Randall, T.A., Churchill, C.J., Baetz, B.W.: A GIS-based decision support system for neighbourhood greening. Environment and Planning B: Planning and Design 30, 541–563 (2003)
20. Auer, A.H.: Correlation of land use and cover with meteorological anomalies. J. Appl. Meteorol. 17, 636–643 (1978)
21. Baumüller, J., Reuter, U., Hoffmann, U., Esswein, H.: Klimaatlas Region Stuttgart. Schriftenr. In: Schriftenr. Verb. Reg. 26th edn. Verband Region Stuttgart, Stuttgart (2008)
22. Langkamp, T., Daneke, C., Bechtel, B.: Alteration of Urban Climate by Urban Morphology taking Wind and Temperature as examples. In: International Seminar on Urban Form, Hamburg (2010)
23. Daneke, C., Bechtel, B.: Classification scheme of urban structures based on climatic characteristics, designed for land use modeling applications. In: Daneke, C., Bechtel, B. (eds.) International Seminar on Urban Form, Hamburg (2010)
24. Batty, M., Yichun, X., Zhanli, S.: Modeling urban dynamics through GIS-based cellular automata. Computers, Environment and Urban Systems 23, 205–233 (1999)
25. Hurkens, J., Hahn, B.M., Van Delden, H.: Using the GEONAMICA® software environment for integrated dynamic spatial modelling. In: Sànchez-Marrè, M., Béjar, J., Comas, J., Rizzoli, A., Guariso, G. (eds.) Proceedings of the iEMSs Fourth Biennial Meeting: Integrating Sciences and Information Technology for Environmental Assessment and Decision Making. International Environmental Modelling and Software Society, Barcelona, Spain (2008)
26. Klepper, O.: Stapeling van Milieuthema's in termen van kans op voorkomen (ECO-notitie 97-01). In: RIVM, The Netherlands, pp. 97–91. Bilthoven (1997)
27. RIKS:Metronamica - Model descriptions. RIKS, Maastricht, the Netherlands (2009)
28. Schlünzen, K.H., Hoffmann, P., Rosenhagen, G., Riecke, W.: Long-term changes and regional differences in temperature and precipitation in the metropolitan area of Hamburg. Int. J. Climatology 30(8), 1121–1136 (2010)

Integration of Temporal and Semantic Components into the Geographic Information through Mark-Up Languages. Part I: Definition

Willington Siabato and Miguel-Angel Manso-Callejo

Universidad Politécnica de Madrid, Mercator Research Group. Autovía de Valencia Km. 7.5
ETSI Topografía, 28031, Madrid, Spain
{w.siabato,m.manso}@upm.es

Abstract. This paper raises the issue of a research work oriented to the storage, retrieval, representation and analysis of dynamic GI, taking into account the semantic, the temporal and the spatiotemporal components. We intend to define a set of methods, rules and restrictions for the adequate integration of these components into the primary elements of the GI: theme, location, time [62]. We intend to establish and incorporate three new structures into the core of data storage by using mark-up languages: a semantic-temporal structure, a geosemantic structure, and an incremental spatiotemporal structure. The ultimate objective is the modelling and representation of the dynamic nature of geographic features, establishing mechanisms to store geometries enriched with a temporal structure (regardless of space) and a set of semantic descriptors detailing and clarifying the nature of the represented features and their temporality. Thus, data would be provided with the capability of pinpointing and expressing their own basic and temporal characteristics, enabling them to interact with each other according to their context, and their time and meaning relationships that could be eventually established. All of this with the purpose of enriching GI storing and improving the spatial and temporal analyses.

Keywords: spatiotemporal, temporal, reasoning, GIS, time, geosemantic, dynamic storage, GIR.

1 Introduction

Time and semantics are two broad, general concepts applicable to a considerable number of scenarios (physics, geology, grammar and geography, among others). Time seems to be bound to every performed action; and every one of these actions has a meaning, a sense or a way of being interpreted that may be described through semantics.

In the realm of the Geographic Information Sciences these concepts have evolved following different paths. For over 25 years, time in GIS has been an active research line [51, 53, 54] with important theoretical and conceptual advances having been achieved. The time models developed for the temporal databases (see [63]) have chiefly influenced the trends followed for incorporation of temporal structures into GIS. Semantics in turn derives directly from the study of language and meaning; from a computational viewpoint, these concepts are framed within the Natural Language

B. Murgante et al. (Eds.): ICCSA 2011, Part I, LNCS 6782, pp. 394–409, 2011.

Processing (NLP) [40], a line of the Artificial Intelligence dealing with modelling and processing of human language in which a lot of work has been carried on for over 60 years [70]. The study of meaning in the realm of Geographic Information (GI) and the Geosciences is widely related to these elements, giving way to specific working areas such as Geosemantics or Geographic Information Retrieval (GIR) [32, 41, 12].

These constructs have come together leading to initiatives such as the computational processing of temporal expressions, a research area which has aroused the interest of the academic community, as witnessed by the related multiple academic events which have been held [5, 6, 11, 26, 28, 47]. In this regard relevant advances have been made: initiatives for the generation of temporal mark-up standards [24], annotation tools and systems [27, 28, 43], temporal annotation corpora [59, 69], annotation languages [66] and assessment methods [8], most of this work having been headed by Dr. James Pustejovsky.

The analysis of temporal expressions allows placing data, facts and events on timelines subjectively, correlating and arranging them chronologically. With a plain temporal description there would be sufficient available elements to solve elementary questions such as When do events occur? How often are they updated? Or which one has occurred before or afterwards? In order to provide data with the ability to describe themselves and relate to one another, taking into account temporality criteria and their representation meaning, it would be necessary to enrich them with a structure enabling identification of their temporal characteristics, their context and their meaning. Such a structure should be computationally usable and should be based on standards to ensure the interoperability of the data enriched with semantic and temporal elements.

This paper raises the issue of a research work oriented to the storage, retrieval and representation of dynamic GI, taking into account the semantic, the temporal and the spatiotemporal components. We intend to establish and incorporate three new structures into the core of data storage using mark-up languages: a semantic-temporal structure, a geosemantic structure, and an incremental spatiotemporal structure. The ultimate objective is the modelling and representation of the dynamics nature of geographic features (which are dynamic by definition), establishing mechanisms to store geometries enriched with a temporal structure and a set of semantic descriptors detailing and clarifying the nature of the represented features and their temporality.

A result of this proposal would be the definition of a new storing format which unlike the current model, would be made up of the three basic components (attributes, location and temporal reference) in addition to three layers to incorporate the semantic and temporal components, these layers are the core and innovation of this proposal (see Fig.1). Initially a file-type storing format is proposed; besides, the possibility of integrating the new concepts into a spatial database engine will be assessed according to the degree of progress of the research work, and we will analyse advantages, drawbacks and feasibility.

This piece of work is primarily based on concepts and studies related to space and time[1]; semantic and semantic interoperability, annotation of temporal expressions, work related to space and time labelling as well as GI retrieval[2].

---

[1] Langran [36-38], Worboys [75-78], Peuquet [52-58], Yuan [82-87], Wachowicz [73, 74], Galton [16-19], Hornsby [23, 81], Nixon [46], Koubarakis [33], Erwig [14], Güting [20, 21].

[2] Pustejovsky [27, 28, 59, 72], Verhagen [71], The MITRE Corporation [8, 66], Jones [30-32], Manning [40], Markowetz [41], Mennis [44, 45], Lemmens [39, 60].

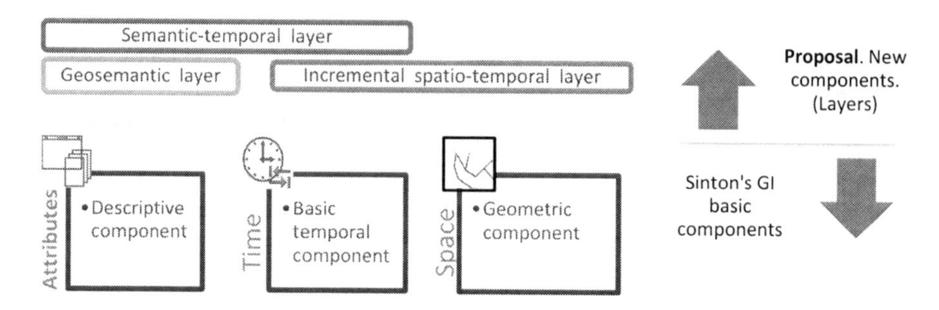

Fig. 1. Basic components of the GI and new structures proposed

This paper is structured as follows; part two presents basic and related concepts of time and geosemantic as well as the initial point of this research. Part three shows the problem that this research is intended to solve, it presents current models and some pros and cons. Part four describes the proposal and the research scope by giving the hypothesis, objectives, and the metamodel elements. It also shows the new components and how they will be included into de GI. Finally, part five enumerates the conclusion and the anticipated contributions to be gained from this research.

2 Relevant Concepts, Grounds and Starting Point

Both time and space appear to be indivisible or at least deeply related to one another; they make up an abstract universal frame, and everything surrounding us is contained therein. In this regard, every geographic feature and geographic entity belongs to that frame, therefore also its characteristics, properties and all possible analyses developed on it. In this context, Couclelis [9] describes what the spatiotemporal analysis and reasoning is:

"Geographic entities, like everything else in the world, exist in time as well as in space;...... Spatio-temporal reasoning is not reasoning about some abstract (x,y,z,t) framework: it is mainly reasoning about the appearance, change, and disappearance of things in space and over time."

For decades, Geography has emphasised the spatial component, leaving aside the temporal component. From its inception, the GIS development focused on the analysis of the geographic elements (space); this characteristic was inherited and the temporal aspect was consigned to the second place. By following the quantitative revolution of the 60's, Geography was catalogued as the "space science", so that the temporal component was limited to a simple attribute [51]. One of the first references to spatiotemporal data is to be found in the study published by Donna Peuquet [53], who by reflecting on temporal data series, identified the nature and the importance of the temporal dimension setting it apart from the spatial dimension.

Such as Ott and Swiaczny indicate [51], time is a relevant aspect in the subject field of GIScience, an element attracting ever so much attention by its importance, particularly during the last decade. This might be due to the large volume of data we now have at our disposal and to the potential that the historical series of data have generated[3], whereby the lack of effective methods and tools to deal with these data is shown; there are not methods to process and analyse temporal data appropriately. The time variable provides an additional component in GI management and it needs to be dealt with differently using new methods and models.

2.1 Concepts about Time

Regarding time, it is necessary to define it, to be aware of the type we are talking about and the model describing it: mechanical, mathematical, organic or psychological. How to perceive and model time is a mainstay of this research; to that end we are starting from the pre-established models, accepting their premises, theorems, restrictions and additional characteristics. In this research study, the mechanical and mathematical models will be taken on, with which all geographic features perceived by our senses (rivers, ways) or abstractly modelled (airways) may be described from a temporal viewpoint. With these two types of models it is possible to define elements such as measure, interval, dimension, modification, instant and position among others.

Regarding the form of relating space and time, two schools of thought have evolved from the ancient Greece [51]: (i) the absolute perspective where space and time are perceived as a single container within which everything exists, and (ii) the relativist perspective, where space and time are perceived as interrelated, dependent elements. In this research project the applied perspective will be the Newtonian one, where time is envisioned as an independent dimension though widely related and similar to the spatial dimension. This approach is used in view of the flexibility it provides for independent handling of the temporal component.

Castagneri [29] defines the temporal GIS (time-integrative GIS) as a means of storing and analysing spatial objects and changes in their attributes over time. This definition allows identifying the three types of elements (data) proposed by Peuquet as the basic elements of a GIS incorporating the time variable [56]: space, time, and attributes, basic components of the triadic model defined by Sinton [62] (see Fig. 2 and Fig. 3). The three main components of GI have been successfully and independently analyzed for over four decades; instances of these analyses and the definition of their primary elements are the studies of Berry [4] and Sinton. Peuquet's concept would be reinforced later on by Openshaw [50] through the analysis of patterns in the triad space-time-attribute. The triadic model surmises the representation of the attributes of an object with a specific position at a certain time, and like Openshaw's work, it shows the existing relationship between the where, the when and the what.

Nevertheless, it should be noted that these studies were oriented toward the structure and storage in the databases to be used by the GIS and that the concept of analysis and exploitation was still fuzzy; the interoperability concept was not taken into account either. The whole and integrated analysis of the main components is still a pending task; the proposed models so far are not entirely satisfactory and they do not provide a solution to the current needs for analysis and extraction of information.

---

[3] On this matter Peuquet [58] states that there is a large amount of spatiotemporal data at different scales, geographically taking up wide zones covered by satellite images captured by specialized sensors or random elements captured by mobile devices, and temporally including collections gathered during decades or datasets captured in real time.

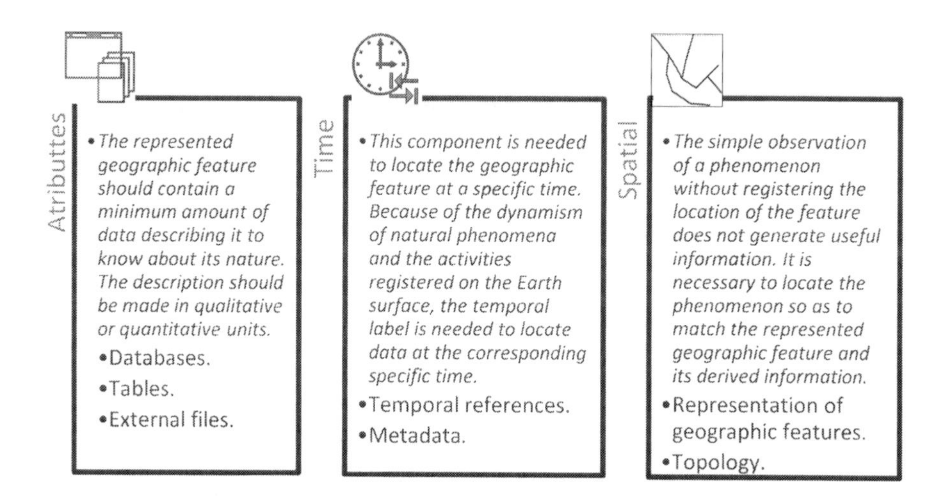

Atributtes
• The represented geographic feature should contain a minimum amount of data describing it to know about its nature. The description should be made in qualitative or quantitative units.
•Databases.
•Tables.
•External files.
Time
• This component is needed to locate the geographic feature at a specific time. Because of the dynamism of natural phenomena and the activities registered on the Earth surface, the temporal label is needed to locate data at the corresponding specific time.
•Temporal references.
•Metadata.
Spatial
• The simple observation of a phenomenon without registering the location of the feature does not generate useful information. It is necessary to locate the phenomenon so as to match the represented geographic feature and its derived information.
•Representation of geographic features.
•Topology.

Fig. 2. Description of the three basic components of Geographic Information defined by Sinton

Ott and Swiaczny [51] remark that the chief purpose of a temporal GIS is to reproduce temporal processes or sequences of events of the real world in a model, so as to make them accessible for the realisation of spatial queries, analysis and visualisation. However, in order to present and analyse the temporal and spatiotemporal information and their relationships, the modelling process requires an absolute accuracy in all the characteristics of the modelled, stored data: geometry, temporality, cartographic projection and attributes, among others.

The main issue in a historical or temporal GIS is the need to count on a high degree of reliability in their two basic components, spatial and temporal. It is possible to reach that degree of reliability only if the problem is tackled from the very beginning, at the start of the process, this is from the modelling of both the data and the storage. So far many applications have been carried out which solve specific problems; then they are used in generic scenarios with the purpose of creating frameworks for the treatment of spatiotemporal data. Nevertheless, when working with models intended for specific settings or when trying to develop dynamic tasks with tools designed for static scenarios and data, it becomes patently clear that those models cannot be applied to general aspects of the geographic features; they cannot be applied to the dynamic environment containing them either.

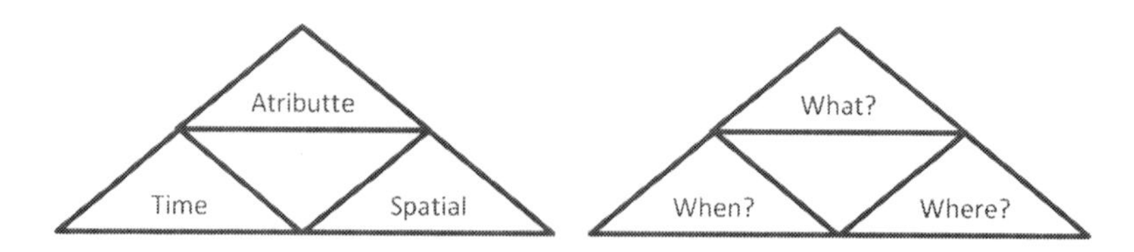

Fig. 3. Triadic model of space, time and attributes (Peuquet) [56]

Peuquet [54], Langran [35, 37, 38], Ott and Swiaczny [51], have stressed the need to work with space and time in GIS in a correlated fashion. By analysing the traditional way in which GIS are modelled by layers, it becomes apparent that the thematic and temporal levels are separated in different categories. This is due on the one hand to the above mentioned issue of the preference of space over time and on the other hand because the current models gear (almost compel) the user to work in a predetermined way with the existing spatial tools, having to forego the temporal component. Many free, open, and proprietary GI management software handle vector formats where the geometric component displaces the temporal component; this is due to the form in which data are structured and stored in the different types of formats and/or store methods; as an example, the old Shapefile format stores geometry and attributes representing and describing geometric entities, but temporality and evolution of the displayed geographic feature over time is omitted. More recent, advanced formats such as the Geodatabase, in addition to storing geometry and attributes, incorporate characteristics and topological relationships, spatial integrity, versioning and an option of historical file of data based on replication; however none of them carries these functions in an incremental manner and without geometry duplication.

2.2 Geosemantic and Temporal Analysis

Regarding the semantic and geosemantic components, multiple studies [3, 7, 30, 31] show how the temporal and spatial components play an important role by filtering, gathering together and prioritising information resources, thus motivating research in methods that would allow turning geotemporal references described by humans in structures and representations understandable by computers and digital information processing systems.

The automatic retrieval of geotemporal information is related to (i) the processes of identification and text analysis, (ii) the search of elements and geographic and temporal references and their accurate location on the space, and (iii) the combination of these references in meaningful semantic compilations such as the indicated search context [42, 61]. The analysed text may come from previously structured metadata, from the actual information contained in the geographic features or from expressions incorporated by the system's user, either as annotations or as texts for queries.

Although the issue has been handled with a degree of success in communities such as the Natural Language Processing and the Geographical Information Retrieval [2], the issue discussed in this research work takes in other different domains, since the knowledge developed by the reference investigators is neither reproducible as yet nor integrated into the GI or the GIS.

2.3 Starting Point

The work of many authors[1] has left a wide, robust theoretical base on the aspects to be taken into account for incorporation of time into information systems and the integration of the spatiotemporal variables into the GIS. Others[2] have established the necessary concepts for the extraction and treatment of the temporal expressions and the bases for semantic analysis of the GI. In spite of the significant advances carried out, both the issue of the space-time relationship and the incorporation of the semantic analysis into the GIS are still open research fields, and more importantly, none of the authors has raised the issue as the space-time-semantics triad. There are no previous studies on this matter going beyond modelling based on cognitive analysis [45], the frameworks for semantic interoperability in GIS applications [39, 67], or the use of general semantic elements (Web Service Modelling Ontology, Internet Reasoning Service) in specific geographic settings such as emergency management [68].

A landmark in the research line in which this proposal is framed is the work presented by Gail Langran [34, 35, 37, 38], who stated that a reasonable objective for the GIS would be to be capable of following up the changes occurring in a certain area by storing the historical data. She pointed out that the Information Systems in the 80's (both spatial and alphanumerical) tended since then to omit the historical record of the data and they did not store previous versions of the system. In order to resolve this weakness, she proposed a conceptual, logical and physical model to enrich systems with temporal capabilities. She introduced the necessary concepts for the changes in geographic features to be stored cumulatively without data duplication at the time of their storage. She later presented [36] a series of annotations applicable to the spatio-temporal systems with which the first general bases of this type of systems were settled. Although the incremental model presented by Langran [38] is conceptually similar to our proposal, we will explain later on this article (Section 4) how the spatiotemporal factor +/- δt and the semantic descriptor S determine that the proposed meta-model is different in concept, hence new and innovative. Following sections will explain in depth our proposal.

3 Identification of the Problem

At the present time the models and data storage systems of GI manage the time variable inefficiently in regard to users' needs, most particularly the possibility of carrying out analyses and a follow-up of the geographic features as a continuum. It is possible to register changes that occur in the reality that surround us (real world) and there are methods for implementation of basic temporal analyses allowing us to find out about the changes occurring within a certain period of time. Nonetheless, this task is carried out with a dataset captured at different times and states, what implies the utilisation of multiple *photos of the reality*. Every one of these photos is generated at a specific time t_n and registered independently within the system, ignoring the temporal and spatial correlation of the particular feature and the correlation of the features as a set[4]. This model known as Snapshot [1] was the first one to incorporate temporality into the spatial databases, and in spite of being so old and inefficient due to the constant duplication of data and attributes, their basic concepts still persist in the present-day systems. Thus, in a follow-up system one has an independently stored layer set that is kept as historical archives to which additional processes have to be applied in order to check the change zones and other characteristics inherent to the evolution of

[4] The correlation of the geographic features as a set was established by Waldo Tobler – First Geography Law.

the territory. From this viewpoint, an initial finite set representing *Current Reality (CR)* is derived. We define the set *CR* as follows:

$$CR = \{p_0, p_1, p_2, p_3, \ldots, p_n \;/\; p_0 \neq p_1 \ldots \neq \ldots, p_n \;\wedge\; p_0 \rightarrow t_0,\; p_1 \rightarrow t_1, \ldots,\; p_n \rightarrow t_n, \}$$

where: p_0 → Initial reality (geographic layer). (Photo *0)*
p_n → Reality in a subsequent time n (Photo *n*)
t_0 → Initial temporal reference
$t_0 < t_1 \ldots < t_n$

Each element $\boldsymbol{p_n}$ corresponds to a graphic output representing one or several levels of the modelled reality (layers) at a specific time. The fact that the set *CR* is defined by multiple elements and the number of these elements increases over time as the changes are registered implies many drawbacks. From the viewpoint of usage and capabilities provided by the system, it is possible to mention:

- Lack of a binding historical register of the represented features.
- Users do not achieve directly (or even indirectly in some cases) the desired answers from the spatiotemporal analyses carried out.
- Inability to develop real spatiotemporal analyses. Michael Worboys [77] indicates two cases that are not resolved naturally and are related to changes in space, time and attributes: (i) changes in population density in a certain district within a fixed time interval; in this case not only the variation in the number of inhabitants is solely taken into account, but also the changes in the legal and geographic boundary line of the particular zone; (ii) the evolution of the morbidity in a fast growing city in a period of two decades.
- Attribute→space→time relationships without one-to-one matching.
- Inability to find other related levels of information or associated geographic features.
- Time of query and processing longer than really needed in queries related to historical data or temporal characteristics.

From the viewpoint of the control and register of the information status, a possible solution to these deficiencies would be the handling of the versioning, however difficulties arise such as:

- The intrinsic need to manage versions. The historical registry of spatial data is carried out following the methods developed for conventional databases [63, 64].
- Information duplication (geometries and attributes) in zones not going through changes. A difficulty derived from versioning style with which the data are currently treated within the spatial databases e.g., the Esri® GeoDataBase model.
- Lack of data versions (releases). There is evidence of the registered versions but this does not imply identifying how many versions an entity has in store; there might be one or as many as existing versions in the database.

- Possible incompatibilities between the spatial and the descriptive components, taking into account a possible update of the attributes (alphanumerical data) but not of the spatial component (geometry). Example: the census.
- Unnecessary package traffic on the computer networks derived from data duplication.

In addition to these deficiencies, we have to mention the inability of the GI to be comprehended from a semantic perspective, i.e. its meaning. The data are not prepared to interact with users in natural language and they are not ready to define neither their own context nor their meaning. This is due to the fact that the natural flow of information treatment in the GIS is User→System←Data, the system being the one interpreting the user's commands (requests) to process subsequently the assigned user's commands taking the related data. In this scenario, everything revolves around the system by having the ability to interpret user commands and data structure; the user does not need to directly interact with the stored data, it just proposes tasks and processes. Under these conditions the user loses self-reliance, remains assigned to a second place and sees him(her)self constrained by the own capabilities of the system. If data were better fitted semantically, able to describe their context and inform who, what and how they are, defining the set they belong to, then, they could interact among them, relate to one another and set up natural subsets (e.g. buildings, rivers, ways, trees), independent of their storage, going beyond the data model controlling them. This would help the user with procedures since the system would not have to transform every order, and part of it could be comprehended directly by the data. In this case a flow of the type User←System←→Data would be generated.

4 Description of the Proposal

The above-mentioned deficiencies are indeed a problem requiring attention and needing solutions. To this effect, we have considered interesting to present a proposal improving the storage of the dynamic GI in the spatiotemporal (δt) and semantic (S) domains, with the purpose of optimising its retrieval, management, analysis, and general tasks based on spatial and spatiotemporal reasoning. Therefore, we propose the definition of a metamodel for storage of the dynamic GI –DGI– to be appropriate for different application domains through specific models and apt for materialisation through mark-up languages. As a result, a new robust, dynamic and flexible storage format is anticipated that integrates the spatial, the temporal and the semantic components. Based on the presented bibliographic reference framework and having exposed the issue, the starting hypothesis for definition of this research work is as follows.

The hypothesis on which this research is based on is the lack of the semantic and temporal components in the current structures of Geographic Information storage and which causes the spatiotemporal analyses to be deficient. The proposal of a new model incorporating an independent temporal structure and a semantic meaning would optimise such storage and would allow improving GI retrieval, processing and analysis capability. If this hypothesis is substantiated, the integration of the geographic, semantic and temporal components through standards would allow:

- exploiting efficiently the dynamic nature of GI;
- optimising the modelling, analysis and transfer of spatiotemporal GI;
- changing the current GI storage methods that do not appropriately fit the dynamic reality they represent;
- relying on well structured data to carry out geographic analysis taking into account the meaning of the data and the time variable;
- optimising response time in queries and spatiotemporal analyses.
- taking the first steps toward compatibility in the representation and analysis of data that include the time variable, hence moving forward toward the semantic and temporal interoperability of the spatiotemporal data.

In order to substantiate our hypothesis, we expect to propose and implement a metamodel for the enriched storage of GI involving temporal and semantic structure, also enabling the optimization of retrieval and dynamic representation of geographic features, and the interaction and exploitation of data in natural language. Broadly speaking, the objective of this work is to define a model of GI storage with the following characteristics:

- Integration of the dynamic nature of GI.
- Incorporation of semantic components describing the meaning and the temporal aspect of the stored data.
- Incorporation of semantic elements enabling the user to naturally interact with them (Natural language interaction).
- Integration of spatiotemporal structures that will allow registering incrementally the changes occurred in geographic features.
- Definition of rules for interaction of any stored dataset.
- Design to be implemented with mark-up languages.

This research work hopes to answer some open questions. Why has not been possible to implement the spatiotemporal models proposed in the last 10 years? What challenges are involved therein? Why have researchers not implemented them? Is it possible to improve spatiotemporal reasoning by changing the paradigm of storage of the dynamic geographic data and by adding temporal and semantic components? Does it make sense to implement a model envisaging the semantic enrichment of data beyond their description with metadata? Is it possible to assign semantic structure to geographic data so that the user might interact with them in natural language?

4.1 Proposal, Anticipated Advantages and Methodology

This work would optimise the register of geographic information through a new storing structure, incorporating the succeeding geometric and alphanumerical changes occurring over time on the geographic features, hence avoiding information duplication; furthermore, this work would permit to know the reality of a registered geographic feature at any time. We intend to provide data with semantic elements that should in turn describe them by using mainly available standards and specifications. We propose to generate a model describing a unitary set (singleton) for dynamic data storing, semantically enriched and called *Proposed Reality –PR–* defined as follows:

$PR_0 = p_0 {}^{+}/_{-} \delta t;$

$PR = \{PR_0, S\}; \rightarrow PR = \{(p_0 {}^{+}/_{-} \delta t) / S \in PR \wedge \exists \text{ one and only one } S\};$

where: $p_0 \rightarrow$ Initial reality
$\delta t \rightarrow$ Spatiotemporal change of reality (t_x)
Spatiotemporal factor

The spatiotemporal factor δt would make possible to know the reality (current state) of a geographic feature at a certain time and to represent the changes of the modelled object, hence of the stored feature. The changes in attributes and geometry may be registered and looked up at any point of the temporal scale in which the information has been registered; a discretisation of the model may therefore be inferred. The set contains a semantic descriptor (S) that will grant the stored data the necessary description so that data are self-describable and may relate according to their geographic and temporal nature. The following conditions are proposed for this set:

- $\forall p_x \in PR_0 \; \exists\rightarrow S \in PR$
- $PR_0 \subset S \quad v \quad S \in PR_0$
- $t_0 = dom(t)$
- $p_x = p_{x-1} + \delta t \quad \wedge \quad p_x = p_{x+1} - \delta t$
- $\sum_{X=0}^{n} p_x = p_n$
- $\sum_{X=0}^{n} \delta t_x = \delta t_n$

This proposal could optimise the current process through the creation of a model for GI storage in which the dynamic component of the represented reality is registered, thus reducing the problems exposed for the finite set *CR*. With the *PR* model, in addition to the above-mentioned aspects in the hypothesis, it is anticipated that the outcome of this research study allow:

- Optimising the volume of storage due to the elimination of coincident geometries, hence improving transfer time of GI belonging to time series or collections.
- The dynamic representation of the spatiotemporal changes registered (linear or ciclic) in geometry and attributes owing to the data intrinsic timeline.
- Optimising processes applied to spatial and temporal reasoning.
- Improving the interaction and relationship of data of the same nature through semantic descriptors.

4.2 Methodology

This research work will be mainly based on standard mark-up languages. Among the likely useful languages for implementation of the proposal, in addition to XML [22], the following stand out: Geography Mark-up Language –GML– [49], Keyhole Mark-up Language –KML– [48], SpatialML [66], TimeML [24], DARPA Agent Mark-up Language [10], Web Ontology Language –OWL– [79], Resource Description Framework (RDF) [80], SPARQL Query Language for RDF [13], in addition to other *de facto* or *ad hoc* standards related to semantic, temporal, and/or GI storage aspects. It is required to analyse different formalisms (XML, RDF, OWL, etc.) and studying which one(s) meet(s) at best the research's needs. To substantiate the defined hypothesis and to answer the above-mentioned open questions, the following activities are put forward:

- Finding an efficient way of fusing together and uniting space and time in the structures of storage with mark-up languages (e.g. GML and TimeML). Here the storing methods based on mark-ups and binaries, as well as the coincidences and differences between methods and time mark-up standards will be assessed.
- Evaluation of GML-based temporal models in other specific application fields. The analysis of the temporal elements of the Aeronautical Information eXchange Model (AIXM) [15], and of the models applied to the Geological Time Systems [65] and CHRONOS [25] are proposed.
- Evaluation of spatial and temporal database models. An important milestone regarding this area is the results of the CHOROCHRONOS project [33].
- Analysis of the appropriate form for incorporation of semantic, temporal, and descriptive annotations of data. The concepts applied here will be based on widely disseminated methodologies such as the GIR[12, 41, 32] and the NLP [40].
- Definition of the general concepts of the GI storing metamodel. Setting up the model layers (components) and their characteristics.
- In order to validate the proposal, the metamodel will be implemented through two specific models: aeronautical and meteorological. The implementation will be able to answer questions that have not been possible to solve so far. We will expose the deficiencies and weak points as well as the strengths and opportunities offered.

5 Conclusion and Expected Contributions

As a conclusion, it is possible to state that the semantic and temporal enrichment of the GI and its implementation through mark-up languages for its integration into the GI management systems is, we believe, the next natural step in the research carried out on the space-time issue in the Information Systems. It is necessary to carry on with the work of Peuquet, Langran, Armstrong, Snodgrass, Worboys, Yuan, Wachowicz, Galton, Hornsby, Frank, Jones, Shen, and all the researchers who have contributed some element to lay the foundations of the change in the paradigm toward the temporal analysis of GIS and the semantic interpretation of the GI. This proposed research work is warranted since it is anticipated to make progress, (i) setting up mark-up languages as an integrating element of the spatial, temporal and semantic elements; (ii) adding further descriptions to data so that they will be able to interact, to describe themselves and to provide the system with temporal and spatiotemporal dynamism as well as meaning. Temporal dynamism through a robust mark-up language; spatiotemporal dynamism by identifying incremental variations; enhancement of meaning by identifying the represented feature type; and (iii) developing concepts enabling progress in geosemantics.

The innovation of this research work lies in the proposal of a metamodel for representation, retrieval, reasoning and spatiotemporal and semantic analysis of GI. The anticipated contributions to be gained from this research are:

- Proposal of a metamodel enabling exploitation of the GI dynamic component.
- Definition of the method for incorporation of the semantic component as well as for integration of an independent temporal component into the GI storage structure.
- Proposal of a new format for the GI integrating the temporal, spatial, and attribute components, providing a semantic-temporal-spatial triadic support.

References

1. Armstrong, M.P. (ed.): Temporality in spatial databases The Urban and Regional Information Systems Association, Falls Church, USA (1988)
2. Association forComputing Machinery, http://www.sigir.org/forum/
3. Bates, M.J., Wilde, D.N., Siegfried, S.: An analysis of search terminology used by humanities scholars: the Getty Online Searching Project. The Library Quarterly 63(1), 1–39 (1993)
4. Berry, B.J.: Approaches to regional analysis: a synthesis. Annals of the Association of American Geographers 54(1), 2–11 (1964)
5. Brandeis University, http://www.timeml.org/acl2006time/
6. Brandeis University, http://www.dagstuhl.de/de/programm/kalender/semhp/?semnr=05151
7. Chen, Y.R., Fabbrizio, G.D., Gibbon, D., Jora, S., Renger, B., Wei, B.: Geotracker: geospatial and temporal RSS navigation. In: WWW 2007, pp. 41–50. ACM Press, New York (2007)
8. Corporation, T.M.: Time Expression Recognition and Normalization Evaluation. In: TERN-2004 Evaluation Workshop, MITRE (2004)
9. Couclelis, H.: Aristotelian Spatial Dynamics in the Age of Geographic Information Systems. In: Egenhofer, M.J., Colledge, R.G. (eds.) Spatial and temporal reasoning in geographic information systems, pp. 109–118. Oxford University Press, New York (1998)
10. DARPA's InformationExploitation Office, http://www.daml.org
11. Dipartimento diInformaticai Comunicazione, http://time.dico.unimi.it/TIME_Home.html
12. Ellen Voorhees, http://www.itl.nist.gov/iaui/894.02/
13. Eric Prud'hommeaux, http://www.w3.org/TR/2008/REC-rdf-sparql-query-20080115/
14. Erwig, M., Güting, R.H., et al.: Spatio-Temporal Data Types: An Approach to Modeling and Querying Moving Objects in Databases. GeoInformatica 3(3), 269–296 (1999)
15. EUROCONTROL, http://www.aixm.aero/public/subsite_homepage/homepage.html
16. Galton, A. (ed.): Qualitative spatial change. Oxford University Press, Oxford (2001)
17. Galton, A.: Desiderata for a Spatio-temporal Geo-ontology. In: Kuhn, W., Worboys, M.F., Timpf, S. (eds.) COSIT 2003. LNCS, vol. 2825, pp. 1–12. Springer, Heidelberg (2003)
18. Galton, A.: Fields and objects in space, time, and space-time. Spatial Cognition and Computation 4(1), 39–68 (2004)
19. Galton, A.: Space, time, and the representation of geographical reality. Topoi 20(2), 173–187 (2001)
20. Güting, R.H., Böhlen, M.H., Erwig, M., Jensen, C.S., Lorentzos, N.A., Schneider, M., Vazirgiannis, M.: A foundation for representing and querying moving objects. ACM Transactions on Database Systems (TODS) 25(1), 1–42 (2000)

21. Güting, R.H., Schneider, M.: Moving Objects Databases. Morgan Kaufmann, San Francisco (2005)
22. Thompson, H.S., Beech, D., Maloney, M., Mendelsohn, N.: http://www.w3.org/TR/2004/REC-xmlschema-1-20041028/
23. Hornsby, K.S., Yuan, M.: Understanding Dynamics of Geographic Domains, 1st edn. CRC Press, USA (2008)
24. ISO: Language resource management – Semantic Annotation Framework (SemAF) – Part1: Time and events. Technical Report (ISO/CD 24617-1). ISO, Geneva - Switzerland (2007)
25. Iowa State University and National Science Foundation, http://chronos.org/index.html
26. Pustejovsky, J.: http://www.timeml.org/site/terqas/index.html
27. Pustejovsky, J., Mani, I.: http://www.timeml.org/site/tarsqi/index.html
28. Pustejovsky, J., Mani, I.: http://www.timeml.org/site/tango/index.html
29. Jim Castagneri, http://www.gisworld.com/gw/1998/0998/998tmp.asp
30. Jones, C.B., Abdelmoty, A.I., Finch, D., Fu, G., Vaid, S.: The SPIRIT spatial search engine: Architecture, ontologies and spatial indexing. In: Egenhofer, M.J., Freksa, C., Miller, H.J. (eds.) GIScience 2004. LNCS, vol. 3234, pp. 125–139. Springer, Heidelberg (2004)
31. Jones, C.B., Alani, H., Tudhope, D.: Geographical Information Retrieval with Ontologies of Place. In: Montello, D.R. (ed.) COSIT 2001. LNCS, vol. 2205, pp. 322–335. Springer, Heidelberg (2001)
32. Jones, C.B., Purves, R.S.: Geographical Information Retrieval. Intl. J. of Geographical Information Science 22(3), 219–228 (2008)
33. Sellis, T.K., Koubarakis, M., Frank, A., Grumbach, S., Güting, R.H., Jensen, C., Lorentzos, N.A., Manolopoulos, Y., Nardelli, E., Pernici, B., Theodoulidis, B., Tryfona, N., Schek, H.-J., Scholl, M.O.: Spatio-Temporal Databases: The CHOROCHRONOS Approach. LNCS, vol. 2520. Springer, Heidelberg (2003)
34. Langran, G., Chrisman, N.: A framework for temporal geographic information. Cartographica 25(3), 1–14 (1988)
35. Langran, G.: A review of temporal database research and its use in GIS applications. Intl. J. of Geographical Information Systems 3(3), 215–232 (1989)
36. Langran, G.: Issues of implementing a spatiotemporal system. Intl. J. of Geographical Information Systems 7(4), 305–314 (1993)
37. Langran, G.: Temporal GIS design tradeoffs. URISA Journal 2(2), 16–25 (1990)
38. Langran, G.: Time in geographic information systems. Taylor & Francis, London (1992)
39. Lemmens, R., Wytzisk, A., By, R., Granell, C., et al.: Integrating Semantic and Syntactic Descriptions to Chain Geographic Services. IEEE Internet Computing 10(5), 42–52 (2006)
40. Manning, C., Schütze, H.: Foundations of statistical natural language processing, 6th edn. MIT Press, Cambridge (2003)
41. Markowetz, A., Brinkhoff, T., Seeger, B.: Geographic information retrieval. In:Agouris P, Croitoru A (eds.). In: Next generation geospatial information, pp. 5–17. Taylor & Francis, Abington (2005)
42. Martins, B., Manguinhas, H., Borbinha, J., Siabato, W.: A geo-temporal information extraction service for processing descriptive metadata in digital libraries. e-Perimetron 4(1), 25–37 (2009)
43. Mazur, P., Dale, R.: The DANTE temporal expression tagger. In: Vetulani, Z., Uszkoreit, H. (eds.) LTC 2007. LNCS, vol. 5603, pp. 245–257. Springer, Heidelberg (2009)
44. Mennis, J.L., Peuquet, D.J., Qian, L.: A conceptual framework for incorporating cognitive principles into geographical database representation. Intl. J. of Geographical Information Science 14(6), 501–520 (2000)

45. Mennis, J.L.: Derivation and implementation of a semantic GIS data model informed by principles of cognition. Computers, Environment and Urban Systems 27(5), 455–479 (2003)
46. Nixon, V., Hornsby, K.S.: Using geolifespans to model dynamic geographic domains. Intl J of Geographical Information Science 24(9), 1289–1308 (2010)
47. Office, D.A.: Proceedings of the 6th conference on Message understanding. In: MUC6 1995, Association for Computational Linguistics, Morristown (1995)
48. OGC OGC® KML. OGC Standard (OGC 07-147r2). Open GIS Consortium Inc. (2008)
49. OGC OpenGIS® Geography Markup Language (GML) Encoding Standard 3.2.1. OGC Standard (OGC 07-036). Open Geospatial Consortium Inc. (2007)
50. Openshaw, S.: Two exploratory space-time-attribute pattern analysers relevant to GIS. In: Fotheringham, S., Rogerson, P. (eds.) Spatial analysis and GIS, 1st edn., pp. 83–104. Taylor & Francis, London (1994)
51. Ott, T., Swiaczny, F.: Time-integrative Geographic Information Systems - Management and Analysis of Spatio-Temporal Data, 1st edn. Springer, Heidelberg (2001)
52. Peuquet, D.J., Duan, N.: An event-based spatiotemporal data model (ESTDM) for temporal analysis of geographical data. Intl J of Geographical Information Systems 9(1), 7–24 (1995)
53. Peuquet, D.J.: A conceptual framework and comparison of spatial data models. Cartographica 21(4), 66–113 (1984)
54. Peuquet, D.J.: It's About Time: A Conceptual Framework for the Representation of Temporal Dynamics in Geographic Information Systems. Annals of the Association of American Geographers 84(3), 441–461 (1994)
55. Peuquet, D.J.: Making space for time: Issues in space-time data representation. GeoInformatica 5(1), 11–32 (2001)
56. Peuquet, D.J.: Representations of geographic space: toward a conceptual synthesis. Annals of the Association of American Geographers 78(3), 375–394 (1988)
57. Peuquet, D.J.: Representations of space and time. The Guilford Press, London (2002)
58. Peuquet, D.J.: Theme section on advances in spatio-temporal analysis and representation. ISPRS Journal of Photogrammetry and Remote Sensing 60(1), 1–2 (2005)
59. Pustejovsky, J., Hanks, P., Saurí, R., et al.: The TIMEBANK corpus. In: Proceedings of the Corpus Linguistics 2003 conference, pp. 647–656 (2003)
60. Lemmens, R.: Semantic interoperability of distributed geo-services, vol.63, (Doctoral Dissertation), NCG, Delft - Netherlands (2006), `http://www.narcis.nl/publication/RecordID/oai:tudelft.nl:uuid:31b0eae6-c411-4bbd-a631-153498889671`
61. Siabato, W., Fernández-Wyttenbach, A., Martins, B., Bernabé, M.Á., Álvarez, M.: Análisis semántico del lenguaje natural para expresiones geotemporales. In: Jornadas Técnicas de la IDE de España -JIDEE 2008, Cartográfica de Canarias S.A., Tenerife - España (2008)
62. Sinton, D.F.: The inherent structure of information as a constraint to analysis: Mapped thematic data as a case study. In: First Intl Advanced Study Symposium on topological data structures for GIS, pp. 1–17. Harvard University LCGSA, Cambridge (1978)
63. Snodgrass, R.T.: Temporal databases status and research directions. ACM SIGMOD Record 19(4), 83–89 (1990)
64. Snodgrass, R.T.: Temporal databases. In: Frank, A.U., Formentini, U., Campari, I., et al. (eds.) GIS 1992. LNCS, vol. 639, pp. 22–64. Springer, Heidelberg (1992)

65. Solid Earth and Environment GRID,
`https://www.seegrid.csiro.au/twiki/bin/view/CGIModel/GeologicTime`
66. Spatio Temporal MITRE: SpatialML: Annotation Scheme for Marking Spatial Expressions in Natural Language 3.0. Technical Report ©The MITRE Corporation (2009)
67. Stoimenov, L., Dordevic-Kajan, S.: Framework for semantic GIS interoperability. Facta Universitatis (Series: Mathematics and Informatics) 17(1), 107–125 (2002)
68. Tanasescu, V., et al.: A Semantic Web GIS based emergency management system. In: Semantic Web for eGovernment 2006, pp. 1–12. Tech University of Athens, Greece (2006)
69. TimeML Working Group,
`http://www.timeml.org/site/timebank/timebank.html`
70. Turing, A.M.: Computing machinery and intelligence. MIND 59(236), 443–460 (1950)
71. Verhagen, M., Mani, I., et al.: Automating Temporal Annotation with TARSQI. In: Interactive Poster and Demonstration Sessions, pp. 81–84. Univ. of Michigan, USA (2005)
72. Verne J: Viaje al centro de la Tierra, vol.19. Casa Editorial El Tiempo, Bogota D.C (2001)
73. Wachowicz, M., Healey, R.G.: Towards temporality in GIS. In: Worboys, M. (ed.) Innovations in GIS: selected papers from the first National Conference on GIS Research UK. Innovations in GIS, vol. 1, pp. 105–115. CRC Press, London (1994)
74. Wachowicz, M., Owens, J.B.: Space-time representations of complex networks: What is next? GeoFocus 9(1), 1–8 (2009)
75. Worboys, M.: A generic model for spatio-bitemporal geographic information. In: Egenhofer, M.J., Colledge, R.G. (eds.) Spatial and Temporal Reasoning in Geographic Information Systems. Spatial Information Systems, pp. 25–39. Oxford University Press, New York (1998)
76. Worboys, M.: A model for spatio-temporal information. In: 5th Intl Symposium on Spatial Data Handling, pp. 602–611. University of South California, USA (1992)
77. Worboys, M.: A unified model for spatial and temporal information. The Computer Journal 37(1), 26–34 (1994)
78. Worboys, M.: Unifying the spatial and temporal components of geographic information. In: Sixth Intl Symp on Spatial Data Handling, pp. 505–517. Taylor & Francis, London (1994)
79. W3C, `http://www.w3.org/TR/2009/REC-owl2-overview-20091027/`
80. W3C, `http://www.w3.org/RDF/`
81. Yuan, M., Hornsby, K.S.: Computation and visualization for understanding dynamics in geographic domains: a research agenda, 1st edn. CRC Press, Boca Raton (2007)
82. Yuan, M., Mark, D.M., Egenhofer, M.J., Peuquet, D.J.: Extensions to Geographic Representation. In: McMaster, R.B., Usery, E.L. (eds.) A Research Agenda for Geographic Information Science, pp. 129–156. CRC Press, Boca Raton (2004)
83. Yuan, M.: Modeling semantical, temporal and spatial information in geographic information systems. In: Craglia, M., Couclelis, H. (eds.) Geographic Information Research: Bridging the Atlantic, vol. 1, pp. 334–347. Taylor & Francis, London - UK (1997)
84. Yuan, M.: Temporal GIS and spatio-temporal modeling. In: 3rd Intl Conf. on Integrating GIS and Environmental Modeling, pp. 21–26. Univ of California, Santa Barbara (1996)
85. Yuan, M.: Use of a Three-Domain Representation to Enhance GIS Support for Complex Spatiotemporal Queries. Transactions in GIS 3(2), 137–159 (1999)
86. Yuan, M.: Use of knowledge acquisition to build wildfire representation in Geographical Information Systems. Intl J. of Geographical Information Science 11(8), 723–746 (1997)
87. Yuan, M.: Wildfire conceptual modeling for building GIS space-time models. In: GIS/LIS 1994, pp. 860–889. ASPRS, Falls Church (1994)

Resilient City and Seismic Risk: A Spatial Multicriteria Approach

Lucia Tilio, Beniamino Murgante, Francesco Di Trani, Marco Vona, and Angelo Masi

Università degli Studi della Basilicata,
Viale dell'Ateneo Lucano, 10, 85100 Potenza, Italy
firstname.surname@unibas.it

Abstract. Nowadays, the most common approach to seismic risk mitigation is characterized only by strategies reducing building vulnerability, through structural interventions, and it does not consider the possibility to intervene at urban scale, reducing urban seismic vulnerability. This paper deals with the concept of urban seismic vulnerability, and introduces resilience, as the capacity of a system to adapt itself to new, generally negative, conditions, in order to re-establish normal conditions. Each city can express resilience, and the identification of its elements is the goal of our research. A spatial multi-criteria approach is here proposed.

Keywords: Resilient cities, Seismic Risk, Seismic Vulnerability, Urban Vulnerability, Spatial Multicriteria Analysis.

1 Introduction: Seismic Risk and Resilient Cities

Considering cities as complex systems, according to Salzano [1], we can recognize *urbs*, *civitas* and *polis*, respectively representing aspects related to physical environment, to the society living there and to governmental activities through which spaces are organized. These three main components – *urbs*, *civitas* and *polis* – interact each other in a continuous way, making complex governance and making it more complex when risks must be managed.

The first question which arises is what to do to face emergency conditions. Further questions concern how to manage crisis in order to limit damages caused by natural and other disasters and how to go over crisis and to guarantee the re-establishment of ordinary conditions.

At present, we are conscious that warding off occurrence of natural disasters is not always possible, even if we know that we can intervene in several ways; for instance, in Italy prevention could be more efficient, in particular if we consider hydro-geological disasters, which often could be avoided by a careful maintenance of hydrographic network. Anyway, considering the possibility of such disasters, we must work in order to face them, and to react, with as least as possible loss.

Considering natural events, risk assessment takes into account several components: generally, it is defined as a function of hazard, exposure and vulnerability [2]. Hazard

B. Murgante et al. (Eds.): ICCSA 2011, Part I, LNCS 6782, pp. 410–422, 2011.

concerns natural characteristics of a natural phenomenon; for instance, if we consider seismic risk, hazard depends on historical seismic characteristics, ground geological characteristics, geotectonic and seismic-genetic structure characteristics, which do not depend on human intervention, so that we are not able to control them.

Exposure, instead, concerns the human presence in a certain area. So, we could affirm that if hazard conditions are worrying (i.e. hazard is high), then we should not establish human settlements. Generally, this kind of decision depends on planning and urban tools. Sometimes, after a disaster, a law can intervene to forbid human settlements in a certain area: for instance, in Italy, Law 405 of 1907 ratified the displacement of many urban centres elsewhere in reason of dangerous landslides. This kind of decision, nevertheless, is hard to accept: historical centres, built over centuries, probably ignoring hazard conditions, because of limited knowledge, and now well-established, also in terms of urban shape and community identity, are hard to uproot and re-build in another place, even if this presents safer characteristics.

The only possibility is therefore to intervene on vulnerability of elements that are exposed to hazard. Generally, vulnerability is defined as the tendency of a certain element to be subjected to damages or corruptions, depending on its own physical and functional [3] characteristics. Generally, vulnerability is referred to the elements composing a settlement; in the case of seismic risk, for instance, vulnerability mainly refers to seismic vulnerability of buildings, and it is evaluated considering their structural characteristics. Such structural aspects, therefore, determine building behaviour in case of a seismic event.

In literature, however, the concept of urban seismic vulnerability has already been used (see for instance [4]): it has been recognized that global activity in a town can be compared to activity of a network system, where each edge, working at local level, contributes at global level. From this point of view, it becomes evident that physical damages are not only components of global damage. Moreover, it has been observed that earthquake effects are not limited to physical damages, but they have some ripples on economic, social and political activities, and they have a strong role onto city capacity to react.

That being so, risk prevention must be characterized by a new approach, that should go over building structural adjustment, and that recognizes single components working as a whole system: these components, that are not only physical ones (such as buildings and streets), but that refer to social, economic and political functions, strongly contribute to urban seismic vulnerability. New approaches must define tools able to mitigate such urban seismic vulnerability; therefore, it should forecast, before a seismic event, what kind of response the single components might show.

In other words, we affirm that an approach aiming at mitigating urban seismic vulnerability, must maximize system resilience, as the capacity of a certain system to adapt to new , generally negative conditions, to re-establish normality [5].

The paper is organized as follows: in the next paragraph, resilience concept and its relation with vulnerability will be deepened, and the idea of resilient city will be described, with some considerations about its relationship with both urban and emergency planning tools.

In the third paragraph, we will provide some theoretical considerations about spatial multicriteria approach, that is adopted in order to identify resilient city. The fourth paragraph contains a description of the study case, the identification of a resilient city in Marsicovetere, Southern Italy and finally, some opening questions and future directions of research.

2 What Is a Resilient City?

2.1 What Is Resilience?

In the last paragraph a resilience definition is provided, and an equivalence relation between urban seismic mitigation and resilience maximization is proposed. In this paragraph we want to deepen this relation, starting from the several facets showed by resilience concept and from several interpretations that nourish the debate.

As known, resilience concept has been developed in origin in the field of ecology. Holling [6] defines resilience as a property of a system that measures its ability to absorb changes of state variables, driving variables and parameters and still persist, and relates its concept to that of stability, intended as the ability of a system to return to an equilibrium state after temporary disturbance. In the last years, resilience became a usual term in the field of risk management. Pelling [7], for instance, affirms that resilience to natural hazards is the ability of an actor to cope with or to adapt to hazard stress. It is a product of the degree of planned preparation undertaken in the light of potential hazard, and spontaneous or premeditated adjustments made in response to felt hazard, including relief and reuse. Concept of seismic resilience considers also the social dimension: according to [8], community seismic resilience is defined as the ability of social units to mitigate hazards, to contain the effects of disasters when they occur, and to carry their recovery activities in ways that minimize social disruption and mitigate the effects of future earthquakes. This can be achieved both working on structural aspects and emergency response and strategies, involving institutions and organizations, and in particular those related to essential functions for community well-being, as acute-care hospitals.

Therefore, The International Strategy for Disaster Reduction, Hyogo Framework for Action 2005-2015, is called "Building the resilience of nations and communities to disasters", where resilience is defined as the capacity of a system, community or society potentially exposed to hazards to adapt, by resisting or changing in order to reach and maintain an acceptable level of functioning and structure, and is determined by the degree to which the social system is capable of organizing itself to increase its capacity of learning from past disasters for better future protection and to improve risk reduction measures [9].

These last considerations highlight that as resilience is considered a strategy to mitigate risks, communities recognize to be capable of coping a stress situation, where they must manage demands, challenges and changes, with available resources and competences [10]. Considering that a society flexible and able to shift rapidly, is also able to exploit any positive opportunity that might arise in an uncertain future [11], flexibility must be strongly enhanced.

Obviously, this is not a unique aspect to be considered. Considering natural disasters, several strategies could lead to enhance resilience in terms of augmented capacity of absorption and recovering from changes [11]. In particular, properties of resilience can be considered in order to identify strategies [8]: robustness, intended as the ability of elements to resist to a certain stress without suffering degradation or loss of functionality; redundancy, intended as the substitutability of elements in order to satisfy some requirements no more satisfied by a degraded element; resourcefulness, particularly concerning social systems, intended as the capacity of identifying problems and finding solutions depending on priorities and available resources; ability, defined as the capacity to meet priorities and to achieve goals as quickly as possible. Even if such properties seem to be abstract, they can find a concrete application. For instance, decentralization of decision making (i.e. creating several decision making centres in a town) or strategies about mobility, generally refer to redundancy, and so on.

2.2 What Is the Resilient City?

Paton et al. [10] define a resilient city as a sustainable network of physical systems and human communities, where the first ones include all kind of structures and infrastructures, "acting as the body of the city, its bones, arteries, and muscles", and the second ones represent the social and institutional components of the city, including all kind of associations and organizations and "acting as the brain of the city, directing its activities, responding to its needs, and learning from its experience". The metaphor makes clear that during and after a stress, both systems are determinant: if body collapses, the entire system collapses; if brain breaks down, the entire system breaks down.

Therefore, the most important aspect concerns how to define and apply strategies. In order to answer to the question, we started from the observation of the city; we recognized that, if we model city as a network system, it is characterized by a main trunk and some secondary branches, whose elements are hierarchically less important than the trunk ones. In terms of response to an earthquake, therefore, trunk elements must have a faster response, because they are charged of main activities of city, and moreover they represent place identity.

Adopting such an approach, it is required to define what elements of a city can represent the minimum set able to guarantee functionality.

2.3 How to Identify a Resilient City?

The minimum set of elements can be identified with reference to the four phases that Civil Protection indicates as the phases of disaster management. Considering forecasting and prevention (referred to peacetime) and emergency and post-emergency phases (in the aftermath), minimum set can be sketched as the nesting of four sub-sets, as in the following scheme:

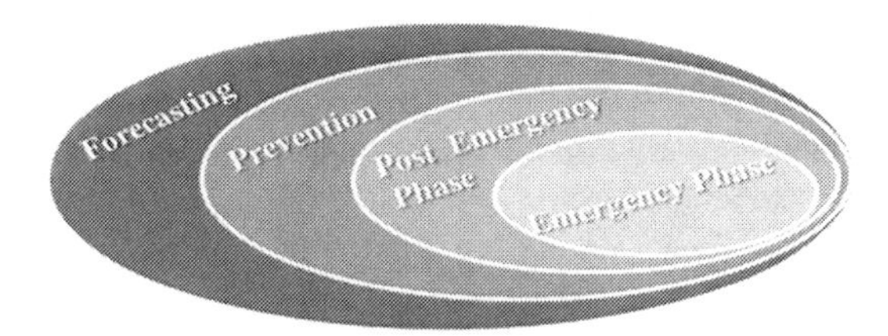

Fig. 1. Elements composing resilient city, sub-divided in four sub-sets referred to disaster management phases. Our elaboration

In peacetime, in particular during forecasting phase, once the expected seismic scenario is defined, it is needed to identify elements:

- requesting prevention, which do not satisfy acceptable vulnerability levels;
- needed in emergency phases, referred to the expected seismic scenario;
- needed to overpass the emergency phase, referred to the expected seismic scenario, and to re-establish normality.

Prevention phase will be characterized by all the actions aiming to bring elements composing sub-sets in acceptable vulnerability conditions.

All such elements are referred to the several aspects of city organization, such as accessibility, lifelines, etc.. In Figure 2, main systems of resilient city are showed, and their main components are synthetically listed. Their identification depends on functional, morphological and dimensional characteristics of the considered urban system [21]. A brief description of systems composing resilient city is presented below.

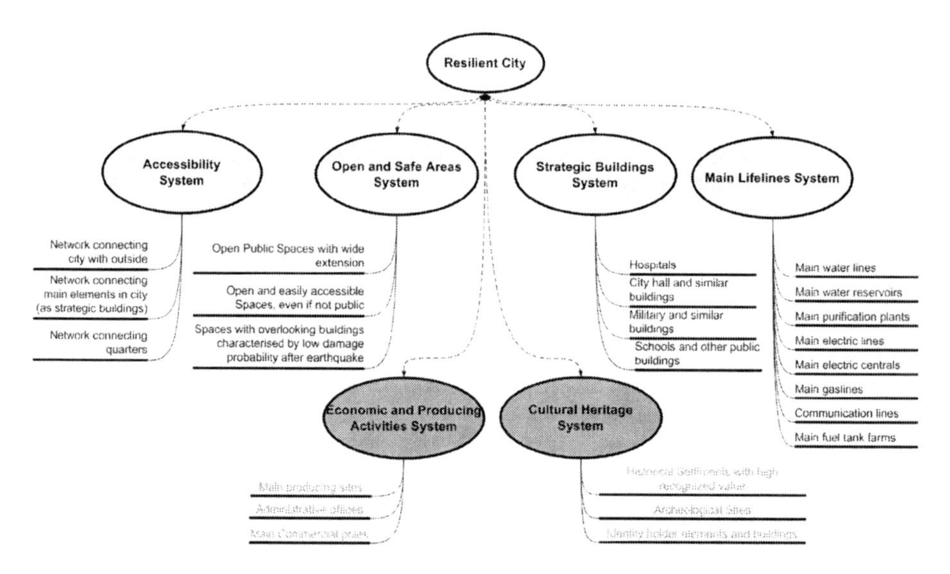

Fig. 2. Systems composing resilient city. Our elaboration.

- Accessibility system: in order to guarantee a minimum of normal cities functionalities, one aspect is related to accessibility: first, identification of main roads, useful as way of escape and allowing access to strategic buildings, as hospitals, and to shelter areas, then, roads connecting quarters and finally internal roads.
- Open and Safe areas Systems: at the same time, open spaces where gathering people, offering a recover, disposing a field hospital and so on, must have been identified, with strong guarantee of their safety.

- Strategic Building Systems: emergency activities need some buildings where decisions can be made, but firstly need hospitals, military buildings, in order to help hit people.
- Main Lifelines Systems: all these activities assume that main services work (water, gas, electricity distribution and communication must be efficient).
- Economic and producing activities and cultural heritage are presented in the figure, but represented in grey, in order to highlight their relative importance, depending on specific characteristic of the considered town.

2.4 Who Decides about Resilient City?

Recognition of resilient city is not so useful if it is not connected to a set of strategies aiming to reduce vulnerability and maintaining characteristics of resilience of the considered elements. At present, who has this responsibility? What is the relation among resilient city, government and planning tools? Is resilient city something with no-ordinary conditions, so that it is related only to emergency and disaster management or has it an ordinary component, and does it need to be introduced into ordinary management tools?

Considering the Italian situation, we notice that:

- Seismic Risk management is almost totally entrust to Civil Protection;
- each municipality should adopt an Emergency Plan, aiming at defining a possible risk scenario and the subsequent actions to manage the emergency;
- emergency plan does not consider the possibility of intervening to mitigate risk before seismic event occurrence;
- laws concerning spatial planning generally do not consider seismic risk as a crucial element influencing development strategies and policies. Considering, for instance, Basilicata region, law N.23/1999, even if the main part of region is classified as high hazard area, there are not specific directions in order to reduce risk.

So, we can affirm that at now resilient city does not yet represent a tool useful in risk mitigation. Civil Protection activities are quite different, and they seem very far from a prevention approach; at the same time, spatial planning tools do not consider seismic risk.

A reason of this situation can be found, probably, in the confusion related to the resilience concept. It is widely used, but it does not seem an operative concept, so that administrations are not able to acknowledge and take it into their instruments.

A possible solution can be an operative approach aiming at resilient city definition, as that one of spatial multicriteria approach, proposed in the next paragraphs.

3 Spatial Multicriteria Analysis

In order to identify resilient city, and define strategies to improve resilience and mitigate urban seismic vulnerability, in this paper we propose a spatial multicriteria approach to identify what parts of territory must resist to a seismic event and must rapidly re-establish their functions to guarantee a return to normal.

Considering multicriteria analysis as "a decision-aid and a mathematical tool allowing the comparison of different alternatives or scenarios according to many criteria, often conflicting, in order to guide the decision maker toward a judicious choice" [12], when alternatives and criteria have an explicit spatial dimension [13], models become "spatial" [14] and can benefit from using Geographic Information Systems; GIS, indeed, provides a powerful set of tools for manipulation and analysis of spatial information [15].

In this context of Spatial MCA, criteria are represented as map layers, and generally they are indicated by the term criterion maps. In particular, a criterion map represents the spatial distribution of an attribute that measures the degree to which its associated objective is achieved [16].

Considering that generally nature or human beings imposed some limitations that do not permit certain actions to be taken [17], in Spatial MCA it is necessary to model also constraint maps, representing restrictions and modelled as territory portions to subtract from criterion maps: constraints play as a hole in territorial extension, as showed in figure 3.

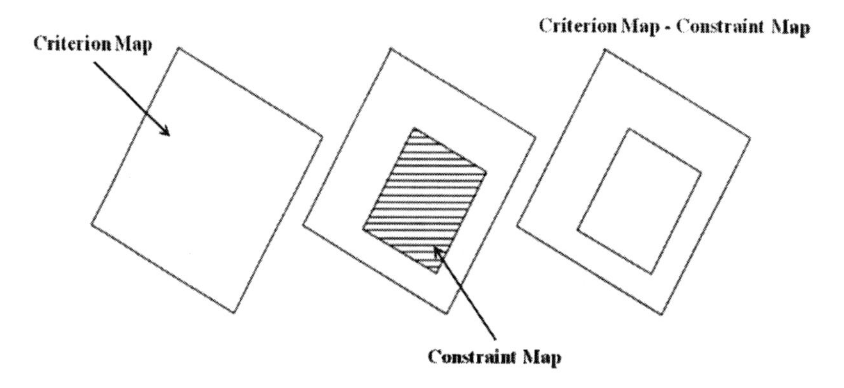

Fig. 3. Criterion maps and constraint maps

3.1 Spatial Multicriteria Analysis Model

Flowchart in figure 4 shows the process of modelling spatial multicriteria analysis.

The main phases are those of Intelligence, Design, Choice and Implementation; each one is characterized by several operations, deepen described below.

In order to compare criterion maps each others, various scales on which attributes are measured must be transformed to comparable units: this is a standardization process, showed in figure 4.

Considering deterministic maps (as in the study case, below described), transformation of input data can be made through several methods; here, linear scale transformation has been adopted, through maximum score procedure.

Linear scale transformation consists of a transformation of raw data into standardized criterion score, applying for each object (that can be a point, a line, a polygon if criterion map has vector data model, or grid cells if it has raster format) a

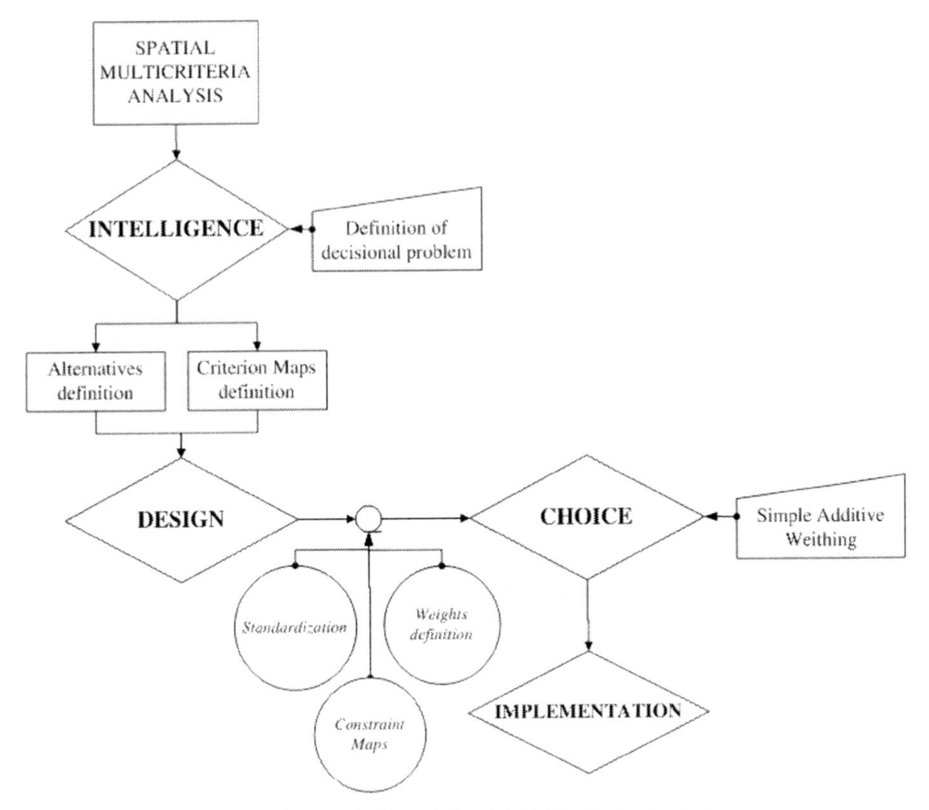

Fig. 4. Our modelling of Spatial Multicriteria Analysis

simple formula. In particular, the Maximum Score Procedure, that consists in a proportional transformation, is obtained through formula showed in (1),

$$x_{i,j}' = \frac{x_{i,j}}{x_j^{\max}} \tag{1}$$

where $x_{i,j}$ is the raw score for the ith object considering the jth attribute, $x_{i,j}'$ is the standardized score and x_j^{max} is the maximum score for the jth attribute on all objects; this formula is adopted when the criterion map represents a benefit, and it is to be maximized; in the case of minimization, when a criterion map represents a cost, formula (2) is adopted:

$$x_{i,j}' = 1 - \frac{x_{i,j}}{x_j^{\max}} \tag{2}$$

As a result, a criterion map ranges from 0 to 1. This procedure, anyway, does not guarantee that the lowest standardized value is zero, sometimes making the interpretation of criterion difficult.

Multi-criteria analysis allows to take into account several stakeholders playing a role into decisional process and which sometimes play a real role into decision. During the modelling phase, stakeholder value systems are to be considered, and introduced into analysis evaluating the relative importance of criterions for each stakeholder. This means identifying criterion weights (weights definition in figure 3), and it is the most subjective aspect of MCA, even if several methods help the analyst in weight identification. In the study case, a pair wise comparison method has been adopted, referring to the analytic hierarchy process by Saaty [18], and developed in subsequent steps, as showed in figure 5.

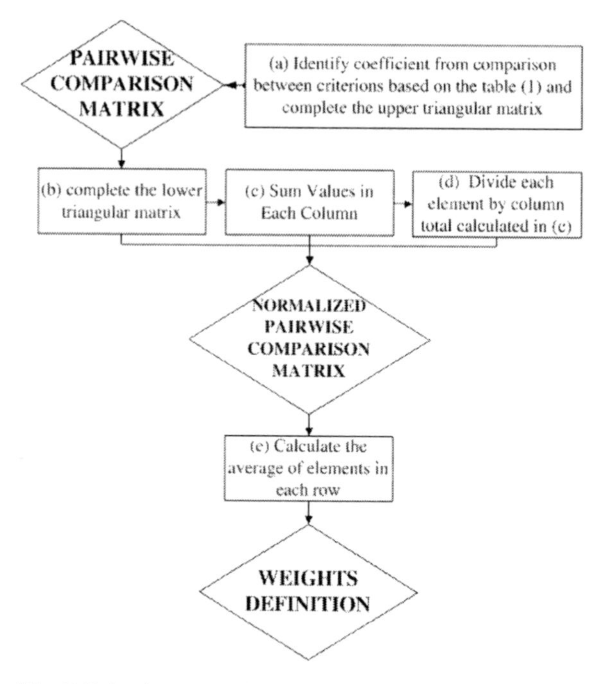

Fig. 5. Pair wise comparison method procedure, our scheme

After weight definition, next step in multicriteria analysis is the definition of decision rule, the procedure allowing ordering alternatives [19], in order to choose the best or the most preferred alternative. Among several decision rules which can be adopted, in the study case additive decision rules have been considered, and in particular the simplest method, simple additive weighting method [20], based on the concept of a weighted average, expressed in the (3)

$$A_i = \sum_j w_j x_{i,j} \tag{3}$$

where A_i represents the object-alternative (point, line, polygon or grid cell), $x_{i,j}$ is the score of the alternative on j^{th} criterion and w_j is the weight of the jth criterion.

4 Resilient City in Marsicovetere (South of Italy)

In order to validate the use of a spatial multicriteria approach to identify resilient city, Marsicovetere municipality (Southern Italy) has been chosen as a sample city, in reason of the importance that this municipality shows in a wider area, Val d'Agri. In last decades this area has been the scene of a development propulsion; main activities, administrative bureaus, health services, shopping centres, etc. are born in Marsicovetere area; development concentrates on valley area, named Villa d'Agri, where today the main part of inhabitants live, also thanks to road network configuration.

Moreover, Val d'Agri is a seismic area, classified in the higher risk class[1], and in the past has been hit by strong and devastating earthquakes.

Figure 6 shows the study area context. Directions of main road crossing territory are highlighted.

Fig. 6. Marsicovetere area, image from google maps

4.1 Data Acquisition and Criterion Maps Design

Considering the procedure described in figure 3, definition of decisional problems led to recognize that this particular problem is characterized by the absence of a priori defined alternatives: each grid cell is potentially an element of resilient city, and its suitability depends on several aspects, to evaluate through a multicriteria approach.

[1] Considering in force law, Ordinanza PCM 3274, 20.03.2003 and following.

Unfortunately, the design phase highlighted the important gap between informative layers. Acquired data do not allow to run analysis for all system listed in figure 1, so that some of them have been omitted, as lifelines.

Available information have been organized in order to define the following criterion maps:

- Accessibility: in order to define accessibility system, road network has been used to identify areas of territory with a certain degree of accessibility. This means that through Euclidean distance function, a criterion map has been built, where grid cells have increasing values with the increase of their distance from roads. In standardization step, formula (2) has been adopted, in order to obtain a high value for the areas close to roads. If a seismic event occurs, areas more accessible, closer than others to roads, are more suitable to receive people and/or to become shelter areas.
- Slope: after digital terrain model definition, a slope map has been calculated, and then, adopting formula (2) areas with soft slope have been considered more suitable to receive people and/or to become shelter areas.
- Urban centre proximity: this criterion highlights importance of proximity of shelter areas to urban centre, not only in reason of their greater accessibility, but also from a psychological point of view. After an earthquake, often, people do not want to leave their houses and places where their time is spent. Criterion map has been obtained applying Euclidean distance function to built-up areas, and then applying formula (2), similarly to accessibility criterion.
- Hydrographic network distance: in order to guarantee safe conditions, proximity of shelter areas to rivers and streams must be avoided, due to overflowing risk. This criterion has been obtained starting from hydrographic network, applying Euclidean distance function and then formula (1): grid cells with higher values are more suitable to become shelter areas.

Weight definition step, generally, should involve decision makers, in order to elicit their perception of criterion relative importance and to define numerical weights. Due to the experimental nature of the study case, a simulation led to weight definition. In particular, three weight sets have been exploited, considering (Set A) firstly criterions with same importance, then (Set B) stressing importance of criterions linked to functional aspects (accessibility and built-up areas proximity), and finally, (Set C) stressing importance of safety (slope and hydrographical network proximity).

The simple additive weighting procedure, then, has been iterated, obtaining three different evaluation scenarios.

At this point, an open and safe areas system can be identified, through subtraction of constraint maps. Areas physically occupied by buildings and roads are considered as constraints; moreover, areas where a hydro-geological constraint (in particular, areas with high landslides risk and flooding areas) is imposed by some territorial or urban plan, are considered as constraints.

Results are showed in figure 7. Darkest areas are the candidates to resilient city.

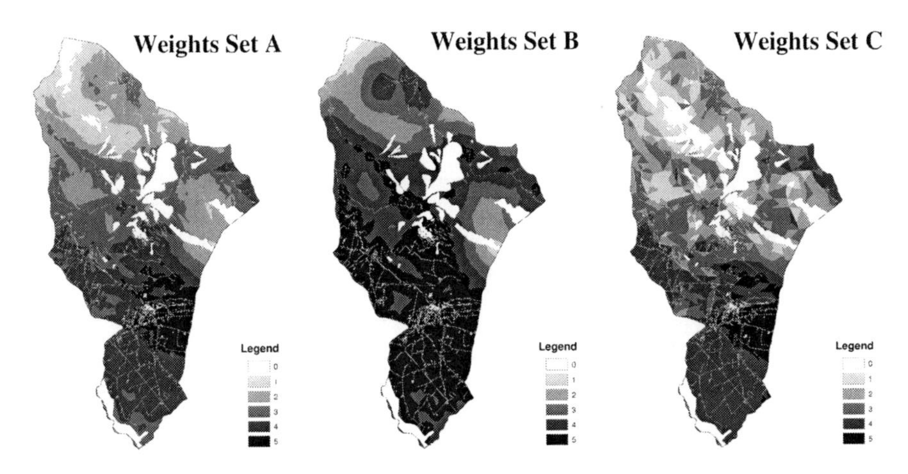

Fig. 7. Analysis results: darkest areas are candidates the resilient city, respectively considering the three sets of weights

5 Conclusion and Future Research

The applied methodology can be considered suitable to identify the resilient city, even if there are several critical aspects to consider.

As declared, simplest methods have been chosen, considering both the standardization phase and decision rule. More sophisticated methods probably would produce better results; choosing other multicriteria approach, moreover, might allow to go over the compensatory effect produced by methods based on average; another difficulty related to simple additive weighting methods is on the hypothesis of not additivity between criterions, not always guaranteed.

Another important aspect is the already declared lack of some information. The developed geographic information system is lacking in several kinds of informative layers: lifelines, but also information about main activities on territory, information about people, information about people who do not live in Marsicovetere, but work there and spend there main part of their day, information about seismic hazard and so on. In addition, at the same time, some available information have not yet been used, as buildings vulnerability. Such information require to be combined with a seismic scenario, in order to evaluate what can happen with the most probable earthquake.

Our last remarks concern the role of resilient city in government and the true contribution that its identification can produce in terms of seismic risk mitigation. At present, Civil Protection is demanded to manage activities of prevention and protection, but its role sometimes contrasts with the role of municipalities, which define urban plans, not always considering natural risks on their territory. Resilient city could become the link between Municipalities and Civil Protection, and its identification is only a first step towards risk mitigation: identified elements need a deeper analysis, a continuous monitoring and, if necessary, economic resources to guarantee their survival to disastrous events. According to Barnett [11], this means also defining a context where horizontal and vertical exchanges in social systems are encouraged to contribute to discussions about risks, enhancing theirs perception, and highlighting importance of prevention.

References

1. Salzano, E.: Memorie di un urbanista. L'Italia che ho vissuto. Corte del Fontego (2010)
2. UNDRO: Natural Disasters and Vulnerability Analysis, Report of Experts Group Meeting, Geneva (1980)
3. Fera, G.: La città antisismica, Gangemi (1997)
4. Manyena: The concept of resilience revisited. Disasters 30(4), 433–450 (2006)
5. Comfort, L.: Shared Risk: Complex systems in seismic response. Pergamon, New York (1999)
6. Holling, C.S.: Resilience and stability of ecological systems. Review of Ecology and Systematics 4, 1–23 (1973)
7. Pelling, M.: The vulnerability of cities: natural disasters and social resilience. Earthscan Publications LTD, London (2003)
8. Bruneau, M., Chang, S.E., Eguchi, R.T., Lee, G.C., O'Rourke, T.D., Reinhorn, A.M., Shinozuka, M., et al.: A Framework to Quantitatively Assess and Enhance the Seismic Resilience of Communities. Earthquake Spectra 19(4) (2006)
9. UN/ISDR, Report of the World Conference on Disaster Reduction, Hyogo, Japan (2005)
10. Paton, D., Millar, M., Johnston, D.: Community Resilience to Volcanic Hazard Consequences. Natural Hazards 24(2), 157–169 (2001)
11. Barnett, J.: Adapting to Climate Change in Pacific Island Countries: The Problem of Uncertainty. World Development 29(6), 977–993 (2001)
12. Roy, B.: Multicriteria methodology for decision aiding. Kluwer Academic Publishers, Dordrecht (1996)
13. Chakhar, S., Mousseau, V.: Spatial Multicriteria Decision Making. In: Shekhar, S., Xiong, H. (eds.) Encyclopedia of GIS, Springer, Heidelberg (2008)
14. Chakhar, S., Mousseau, V.: An algebra for multicriteria spatial modeling. Computers, Environment and Urban Systems 31(5), 572–596 (2007)
15. Carver, S.: Integrating multi-criteria evaluation with geographical information systems. International Journal of Geographical Information Science 5(3), 321–339 (1991), doi:10.1080/02693799108927858
16. Malczewski, J.: GIS and Multicriteria Decision Analysis. John Wiley & Sons, USA (1999)
17. Keeney, R.L.: Siting energy facilities. Academic Press, USA (1980)
18. Saaty, T.L.: The analytic hierarchy process. McGraw-Hill, USA (1980)
19. Starr, M.K., Zeleny, M.: MCDM: state and future of the arts. In: Starr, M.K., Zeleny, M. (eds.) Multiple criteria decision making, pp. 5–29. North-Holland, Amsterdam (1977)
20. Heywood, I., Oliver, J., Tomlinson, S.: Building an explanatory multi-criteria modeling environment for spatial decision support. In: Fischer, P. (ed.) Innovation in GIS 2, pp. 127–136. Taylor & Francis Group, London (1995)
21. Olivieri, M.: Dalla prevenzione edilizia alla prevenzione urbanistica. In: Olivieri, M. (ed): Regione Umbria. Vulnerabilità urbana e prevenzione urbanistica degli effetti del sisma: il caso di Nocera Umbra. INU Edizioni (2004)

Towards a Planning Decision Support System for Low-Carbon Urban Development

Ivan Blecic[1,4], Arnaldo Cecchini[1], Matthias Falk[2,4], Serena Marras[3,4], David R. Pyles[2,4], Donatella Spano[3,4], and Giuseppe A. Trunfio[1,4]

[1] DADU, Department of Architecture, Planning and Design
University of Sassari, Alghero, Italy
{ivan,cecchini,trunfio}@uniss.it
[2] LAWR, Land, Air and Water Resources,
University of California, Davis, CA, USA
{mfalk,rdpyles}@ucdavis.edu
[3] DESA, Dipartimento di Economia e Sistemi Arborei,
Università di Sassari, Italy
{serenam,spano}@uniss.it
[4] CMCC, Centro Euro-Mediterraneo per i Cambiamenti Climatici,
IAFENT-Sassari, Italy

Abstract. The flows of carbon and energy produced by urbanized areas represent one of the aspects of urban sustainability that can have an important impact on climate change. For this reason, in recent years the quantitative estimation of the so-called urban metabolism components has increasingly attracted the attention of researchers from different fields. On the other hand, it has been well recognized that the structure and design of future urban development can significantly affect the flows of material and energy exchanged by an urban area with its surroundings. In this context, the paper discusses a software framework able to estimate the carbon exchanges accounting for alternative scenarios which can influence urban development. The modelling system is based on four main components: (i) a Cellular Automata model for the simulation of the urban land-use dynamics; (ii) a transportation model, able to estimate the variation of the transportation network load and (iii) the ACASA (Advanced Canopy-Atmosphere-Soil Algorithm) model which was tightly coupled with the (iv) mesoscale weather model WRF for the estimation of the relevant urban metabolism components. An in-progress application to the city of Florence is presented and discussed.

Keywords: urban metabolism, urban sustainability, cellular automata, land-use dynamics.

1 Introduction

In recent years the topic of urban sustainability and its relation with the climate change has increasingly attracted the attention of researchers from different fields. In scientific literature we find two main research lines . On one hand, a large amount of

B. Murgante et al. (Eds.): ICCSA 2011, Part I, LNCS 6782, pp. 423–438, 2011.

effort has been devoted to quantitative estimates of components of the so-called urban metabolism [1,2], that is, the relevant energy and matter fluxes between city landscapes and the atmosphere. In particular, several advanced models, operating at different spatial and temporal scales, have been developed and used for this purpose. On the other hand, it has been recognised the need to develop suitable tools and quantitative indicators in order to effectively support urban planning and management with the goal of achieving a more sustainable metabolism in future cities.

In fact, it is well known that many urban landscape features can significantly influence both energy and material flows. For example, sprawled, low-density cities usually have higher per capita energy consumption for transportation than compact cities (e.g. different rates of motor vehicle use) and, ultimately, different scenarios of carbon emissions. In addition, other anthropogenic sources, such as those related to heating and hence to the characteristics of the buildings, play an important role in the urban metabolic balance. Also, different extension and position of urban green areas can influence the balance of carbon and heat fluxes. Further urban characteristics which might also impact its metabolism are the age of the city, its overall infrastructure and its degree of industrial development.

Besides factors strictly related to the urban fabric and to anthropic activities, climate also has an impact on the urban metabolism. For example, a city with continental climate usually consumes more energy for winter heating and summer cooling than does a city with a more temperate climate. Also, regardless of their natural or anthropogenic origin, the involved fluxes of matter and energy interact in a complex, nonlinear way with the local weather conditions.

Another aspect, which should be taken into account to support a urban planning and management oriented to achieve sustainable metabolism, is related to the nature of the complexity that characterises urban systems. In fact, urban planners intervene in a spatial system inhabited by *free* human agents in which the detailed location of activities is controlled by spatial interaction between them (usually competing heterogeneous agents) and between different (attractive and repulsive) land uses and functions. As a result, given a temporal horizon, decisions about a planning alternative (e.g. to allocate a specific area to one or more land uses, to build specific infrastructures or to make available public services) usually do not lead to easily predictable configurations of the urban land uses (intended as a map of the allocation of the relevant urban activities/functions).

Given the above considerations, the development of an integrated modelling system able to link urban planning decisions to the indicators of sustainable urban metabolism estimates , is a nontrivial task necessarily involving an interdisciplinary modelling effort. A contribution in such direction is provided by the study object of this paper, which is part of the research that is being undertaken within the European project BRIDGE and aims to set up a software framework for estimating carbon exchanges alternative scenarios of urban development.

The paper is organised as follows. Section 2 outlines the framework architecture and illustrates some of the characteristics of its main components. Section 3 presents and discusses an in-progress application on the City of Florence. The paper ends with section 4 in which some strengths and limitations of the proposed methodology are addressed, and planned and potential future development of the framework are outlined.

2 The Framework Components

The proposed modelling framework is based on four main components: (*i*) a Cellular Automata model for the simulation of the urban land-use dynamics [3,4,5]; (*ii*) a transportation model for estimating how different land-use scenarios impact the transportation network load [6,7]; (*iii*) a Soil-Vegetation-Atmosphere Transfer model (SVAT) [8-14], which was tightly coupled with the (*iv*) mesoscale weather model WRF [15] for the accurate simulation of carbon fluxes accounting for the local meteorology.

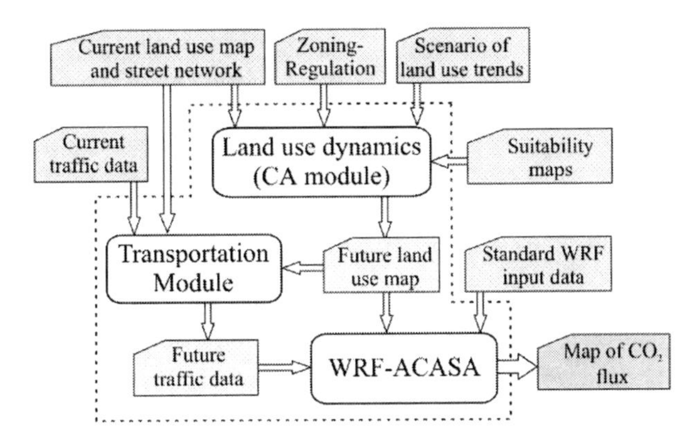

Fig. 1. Outline of the modelling framework highlighting the most relevant data exchange between the involved components

As shown in Fig. 1, the land-use dynamics simulation module takes as main input the current map of land uses, the street network, the constraints related to the zoning regulation, the suitabilities of the cells to support the modelled land uses and the hypothesis on the future land-use trends. The latter may come from a demographic study and/or from assumption on the development of specific economic sectors. The results produced by the land-use dynamics module consist of a map of future land uses, which represent a spatial distribution of the aggregate land-use demand consistent with the main rules governing the functioning of an urban system.

Such future land use map, together with the street network including the current traffic data, are used by the transportation module for estimating future traffic data coherent with the assumed land uses trends.

As the final step of the modelling workflow, the future scenario of land use and traffic data, together with other relevant input data, are used by the coupled model WRF-ACASA for estimating future maps of CO_2 fluxes in the urban area under consideration.

Below, the main components of the modelling system outlined in Fig. 1 are described in more detail.

2.1 The CA-Based Land-Use Dynamics Model

The adopted CA model for the simulation of land use dynamics is an adaptation, with some modifications, of the well known Constrained Cellular Automata approach (CCA) [3,4]. In particular, the CA module has been designed in order to satisfy different objectives, namely: *(i)* the ability to operate at a reasonably high spatial resolution; *(ii)* the inclusion of an adequate simulation of the spatial processes that determine the land use patterns; *(iii)* the capability of processing a suitable representation of relevant landscape features and legal and planning restrictions on land use.

The model enables to incorporate the dynamics caused by large-scale processes (e.g. the demography or the development of specific economic sectors) through the linkage of more traditional dynamic models (i.e. a-spatial demographic, economic and environmental models) or even through the use of simple trends (e.g. extrapolations based on historical data or representing scenarios of development) [3,4]. In particular, the allocation of land uses depends both on an aggregate model (or trend) exogenous to the CA, and on the local CA-based interactions (i.e. on the basis of the local transition rules and of the cell characteristics). In this sense, the CCA can be viewed as a way to determine the spatial distribution of an aggregate land-use demand, taking into account for the local interaction between different land uses as well as the physical, environmental and institutional factors and other relevant characteristics typical of each cell. Thus, the CCA model can easily account for the planning decisions whose broader effects in terms of a spatial distribution of land-uses have to be evaluated.

In the CCA model, each cell has a set of properties representing all relevant physical, environmental, social and economical characteristics, as well as cell accessibility depending on the transportation network, and the imposed legislative constraints (i.e. the zoning status for each land-use). This allows the model to be linked both conceptually and practically with GIS [3,16,17], and indeed the cellular space on which the CA operates can be easily obtained from the layers of a raster GIS.

The cellular space consists of a rectangular grid of square cells, corresponding to the resolution of the data which are used as the source of land cover. Clearly, the size of the grid can vary according to the map of the city being modelled. Each cell is characterised by:

- a *suitability factor* $S_j \in [0,1]$ for each dynamic's land use. The suitabilities represent the "propensity" of a cell to support a particular activity or land use (e.g. can be computed as a normalised weighted sum or product of relevant physical and environmental factors characterising each cell). The suitabilities can be either pre-calculated in a GIS environment, and in this case remain constant during the simulation, or dynamically computed by the CCA model itself during the simulation;
- an *accessibility factor* $A_j \in [0,1]$ for each land use, reflecting the importance of access to the transportation networks for the various land uses or activities (e.g. commerce generally requires better accessibility than residence). These quantities are computed by the CCA module itself before starting the simulation using the street network provides as input;

- a value $Z_j \in [0, 1]$, defining the degree of legal or planning permissibility of the *j*-th land use (for example due to zoning regulations).
- its *current land use/function*. In general, the model includes both static land uses (i.e. not changing during the simulation but influencing the dynamics in the cell neighbourhood in terms of attractive or repulsive effect) and dynamic land uses. The specific land uses included in the current version of the model reflects at least the details of the CORINE land cover data and can be improved in case of more detailed data availability . Broadly speaking, static land uses include: road and rail networks, subways, airports, vegetated areas, water bodies, agricultural areas, forests etc. The actively modelled (i.e. dynamic) land uses in the present application include: *continuous urban fabric, discontinuous urban fabric, industrial areas, and commercial areas*. As mentioned before, the growth of dynamic land uses is defined by exogenous demands.

The state of a cell also includes the *transition potential* P_j for each land use *j*, which is computed by the CCA transition function and expresses the land propensity level to acquire the *j*-th use.

As in every CA, each cell is characterised by a neighbourhood, namely the set of cells the state of which can influence the dynamics of the cell itself. In other words, the change of a cell state at each time-step depends on the states of its neighbouring cells. In the model adopted here the neighbourhood is defined as the circular region around the cell with the radius of 1km which is considered sufficient to allow local-scale spatial processes to be captured in the CA transition rules.

The first phase of the transition function, executed at each step by all cells, consists of the computation of the transition potentials (one for each actively modelled land use) on the basis of the suitabilities, accessibilities, zoning, and states of the cells in the neighbourhood. In particular, the following equation was used:

$$P_j = A_j S_j Z_j N_j \tag{1}$$

where N_j is the so called *neighbourhood effect*, which represents the sum of all the attractive and repulsive effects of land uses and land covers within the neighbourhood, on the *j*-th land use which the current cell may assume. Since, in general, more distant cells in the neighbourhood have smaller influence, in our version of the model the factor N_j is computed as:

$$N_j = I_k + \sum_{c \in V} f_{ij}(d_c) \tag{2}$$

where the summation is extended to all the cells of the neighbourhood V (which does not include the owner cell itself) and: i denotes the current land use of the cell $c \in V$, d_c is the distance between the central cell and the neighbouring cell c, and the function $f_{ij}(d)$ is a parameterised function expressing the influence of the i-th land-use at the distance d on the potential land use j. In addition, the positive term I_k, where k denotes the current land use of the cell, accounts for the effect of the cell on itself (zero-distance effect) and represents an inertia effect due to the transformation costs from one land use to another.

The second phase of the transition function takes place on a non-local basis and consists of transforming each cell into the state with the highest potential, given the constraint of the overall number of cells in each state imposed by the exogenous trend for that iteration. Details of the procedure can be found in [3,4].

Since, as mentioned above, the transition function depends on parameters, the model needs a preliminary calibration phase which can be based on available historical spatial data of the area under study.

The relevant output from the CCA model are maps showing the predicted evolution of land uses in the area of interest over a predefined period of time.By varying the inputs into the CCA model (e.g. zoning status, transport networks, presence of facilities and services), the model can be used to explore the future urban development of the area of interest under alternative spatial planning and policy scenarios.

The CCA model, together with the Transportation model described below, was implemented using the MAGI C++ class library described in [18].

2.2 The Transportation Model

The adopted transportation model consists of a dynamic formulation of the well-known gravity model of trips distribution [6,7,18] and aims at capturing the long-term average vehicle load on the road network.In particular, the current version of the model provides an estimate of the load variation related to the future land uses scenarios. For this purpose it uses an origin-destination (OD) trip matrix, expressing the distribution of trip demand, which is computed on the basis of relevant land uses (i.e. the ones modelled in the CA model described above) considered as trip sources and attractors .

In the model, the street network is composed of a set $\mathcal{N}$ of nodes and a set $\mathcal{L}$ of directed links. Also, the entire urban area is partitioned in a set K of zones, each including nodes $n \in \mathcal{N}$ which represent sources and/or destination of vehicular trips. In particular, an OD matrix $\boldsymbol{X}$ provides, for each couple of zones $(k_i, k_j) \in \mathcal{K} \times \mathcal{K}$ the number of trips x_{ij} for the time period under consideration. In the model, the matrix $\boldsymbol{X}$ is estimated on the basis of the mix of land uses in both the source and destination zones.

Each of the x_{ij} trips may use different network paths connecting nodes belonging to the zones k_i and k_j. The dynamic assignment of the path to each trip is based on a simulation procedure similar to that proposed in [6,7].

In a preliminary phase of the algorithm, a source node $n_i \in \mathcal{N}$ and a destination node $n_j \in \mathcal{N}$ are assigned to the x_{ij} trips. Currently, n_i and n_j are chosen as the nodes in $\mathcal{N}$ closest to the centroids of k_i and k_j, respectively.

The dynamic phase of the adopted algorithm is composed of m steps, which can be thought as corresponding to m time intervals. At each time step and for each couple $(k_i, k_j) \in \mathcal{K} \times \mathcal{K}$, a path is assigned to the corresponding fraction $\delta_{ij} = x_{ij}/m$ of trips. At the end of each step, the traffic load of each link $l \in \mathcal{L}$ is updated on the basis of the amounts of trips the route of which includes the link l itself.

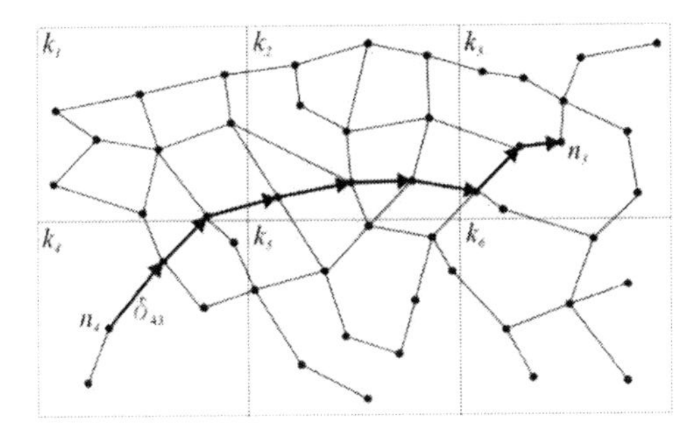

Fig. 2. Partition of the urban area into sectors and allocation of a packet of trips between two sectors during a step of the algorithm

More in detail, at each step each packet δ_{ij} of travellers departing from a particular node $n_i \in \mathcal{N}$ chooses the shortest path to the destination node, based on the travel time perceived according to the current estimate of the traffic conditions [23]. To this end, a procedure based on a variation of the well-known Dijkstra algorithm, calculates the path between a couple (n_i,n_j) having the minimum travel cost on the basis of the current traffic load of each involved link.

In particular, the network is considered as congestible, that is, every traveller imposes a certain amount of delay on every other traveller. This congestion delay is accounted for by assigning, at the end of the *r-th* step, to each link $l_k \in \mathcal{L}$ a generalised cost $c_k^{(r)}$ expressed as:

$$c_k^{(r)} = \frac{| l_k |}{\sigma_k(q_k^{(r)})} \tag{3}$$

where $| l_k |$ is the length of l_k, and σ_k is a function, which depends on the link capacity and provides the vehicles' velocity given the current load $q_k^{(r)}$ of the link.

At the end of the *m* steps, the total amount of trips $\sum_i \sum_j x_{ij}$ are allocated and the final estimate $q_k^{(m)}$ of the load is available for each link l_k. Running the model for both the current land use scenario and for future scenarios under consideration provides estimate of possible future street network load variations.

Thanks to its simplicity, the adopted approach allows for the simulation of real networks through reasonable computational resources. In addition, it requires a relatively small amount of parameters to be estimated.

2.2 The WRF-ACASA Coupled Model

SVAT (Soil-Vegetation-Atmosphere Transfer) models are used to accurately describe how soil, vegetation, and water surfaces exchange energy (heat), moisture, and trace gases with the atmosphere. Traditionally, such models have been embedded into meteorological simulations because they provide essential information on the heat and moisture input to the atmosphere from the Earth's surface (e.g., soil, vegetation, water bodies) and on how the surface extracts momentum and kinetic energy from the atmosphere. SVAT models are used for producing and updating surface-air forcing boundary conditions needed by atmospheric circulation models such as WRF. In addition, biophysicists and ecologists use SVAT models to determine how plants and plant communities respond to environmental conditions.In the context of climate change, simulating future climate and land use scenarios using SVAT models that incorporate CO_2 exchange processes can play an important role in helping policy makers design the most efficient climate protection strategies.

The SVAT model adopted in the present study is the Advanced Canopy-Atmosphere-Soil Algorithm (ACASA), a multilayer model which was originally developed by the University of California Davis (UCD) to simulate the exchange of heat, water vapour and CO_2 within and above a canopy [8,9]. ACASA has been widely applied as a stand-alone (*in-situ*) model over natural and agricultural ecosystems [10,13,14] and the model ability to adequately reproduce turbulent fluxes in different environmental conditions has been evaluated as obtaining reliable flux simulations. To properly work in urban environments, the model was recently modified to account for the anthropogenic contribution to heat exchange and carbon production [11].

ACASA was recently coupled with the well known Weather Research and Forecasting model (WRF) [15]. WRF is a mesoscale numerical weather prediction system which, thanks to its flexibility and computational efficiency, is largely used for both operational forecasting and atmospheric research. In particular, the ACASA model was adopted as an alternative to the existing suite of surface-layer schemes available in WRF due to the need to establish more flexible and realistic representations of surface-layer physics and physiology [12]. The WRF model, in our case driven by North American Regional Reanalysis data (NCAR-NCEP), is run down to its planetary boundary layer, where ACASA is called (see Fig 3).

In ACASA, the canopy is represented as a horizontally homogeneous medium with all leaves and branches arranged with spherical symmetry. This includes both plant and urban (building) canopies. The surface-layer domain simulated by ACASA represents average conditions and fluxes for a 30-minute time intervals at each spatial point in WRF. Twenty equally spaced atmospheric layers represented by a steady-state 3rd order turbulence extend to twice the canopy height [19,20], withthe canopy occupying the lowermost 10 layers. All sources of heat and mass fluxes occur within the canopy and at the soil- (or snowpack)-atmosphere interface only, while constant-flux assumptions prevail in the 10 layers above the canopy. Within the soil horizon, the model includes thermal and hydrologic diffusion among an adjustable number and depth of soil layers; varying by soil type. . Short-wave (solar) radiative transfer within the canopy initially occurs among 100 layers to suit the numerical needs of the radiative transfer equations, which are solved for both beam and diffuse PAR (0.36-0.72 μm) and NIR (0.7 – 2.0 μm) wavelength bands. The short-wave radiation flux streams needed by the rest of the model are then interpolated to 10 layers. The latter are used to help drive the calculations of surface temperatures, terrestrial-infrared radiative transfer, physiological conditions, and associated fluxes needed for the turbulence calculations; all of which come into iterative numerical convergence. Once accomplished, ACASA provides the 30-minute surface flux and boundary conditions needed by WRF to operate.

The required set of driving meteorological data from the lowest WRF atmospheric layer, applied to ACASA on a half-hourly basis, are: : precipitation rate and form (kg m^{-2} $timestep^{-1}$), specific humidity (kg kg^{-1}), wind speed (m s^{-1}), downwelling short-wave radiation (W m^{-2}), downwelling long-wave radiation (W m^{-2}), air temperature (K), air pressure (hPa), and carbon dioxide concentration (ppm). Furthermore, initial soil temperature (K) and moisture (volumetric) profiles for the first model timestep are provided by the WRF initialization routines. Surface morphological parameters that drive the physiological responses also have to be specified, which vary by WRF land use type (including CCA and/or satellite-derived data wherever possible). The set of key morphological parameters includes: total (green) leaf area index (m^2 m^{-2}), maximum canopy height (m), leaf-scale ideal photosynthetic potential (μmol m^{-2} s^{-1}), human population density (# people km^{-2}, currently keyed in by WRF urban land use type), and eventually vehicle flux density (# vehicles m^{-2}).. Morphological parameters not represented in the WRF land-use parameter suite, quantities such as mean leaf diameter and basal respiration rates for plant tissues, are specified with constant near-cardinal values for all land points. The key set was chosen for a focus in this study, due both to model sensitivity, cogency with CCA architecture, and data availability. Staudt *et al* [24] provide additional background information on morphological parameters and model sensitivities to each.

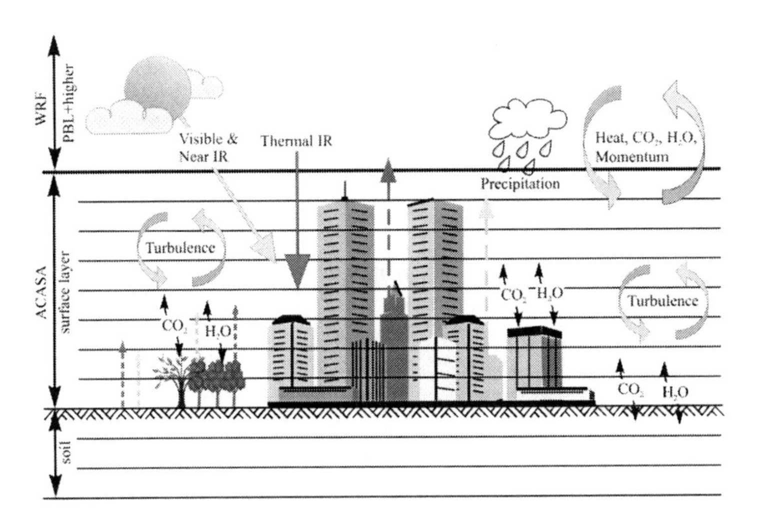

Fig. 3. Scheme of the processes modelled by the coupled model WRF-ACASA

The ACASA model output that feed information back to the WRF-simulated atmosphere include half-hourly vertical fluxes of heat (Wm^{-2}), water vapour (kg m^{-2} s^{-1}), and CO_2 (μmol m^{-2} s^{-1}) momentum flux density (as friction velocity, m s^{-1}), turbulence kinetic energy (m^2 s^{-2}). In addition, snowpack and/or soil and canopy thermal and hydrological states, needed for adequate simulation of the surface-layer, are updated at all land points and are stored between timesteps.

The initial conditions for a WRF runs are pre-processed through a separate package called the WRF Preprocessing System (WPS). Among others, the input to WRF-ACASA from WPS contains the vegetation/land-use type coded according to the U.S. Geological Survey (USGS) standard and to the Urban Canopy Model (UCM) [21]. Other relevant input are represented by the CO_2 anthropogenic emissions which can be directly related to the estimated traffic load of the road network, to the prevalent land use of each cell and to the local population density.

In order to simulate energy and carbon fluxes in urban environment, the coupled WRF-ACASA model needs information about the land use and traffic scenarios produced respectively by the CCA and transportation simulation modules. The current configuration includes three levels of urban intensity (CCA→WRF) corresponding to three levels of population density (WRF→ACASA), represented in Figure 4. For this purpose, the CCA module is able to export the future land use projections into the binary format accepted by the WPS. The coupling was based on a suitable table of conversion between CORINE codes, which are used by the CCA, and the USGS + UCM codes used by WRF.

According to recent validated results [12], WRF-ACASA produces reliable patterns of heat and CO_2 fluxes at high spatial resolutions over urban areas.

3 An In-Progress Application Example

After the model design and set-up phase, an exploratory application on the city of Florence is now in progress in order to test the limits and potential of the framework.

The city of Florence was chosen because it is one of the case studies of the European project BRIDGE, in the context of which this research is being conducted. In particular, for this city, the WRF-ACASA model has already been calibrated through sampled fluxes, and traffic data as well as zoning regulation.

As for the CCA module, the first phase of the setup was dedicated to the selection of the proper input data sources. In particular, after a preliminary analysis of the available spatial data, it was decided to adopt the 100 *m* resolution CORINE land cover (CLC) as the input land-use layer (see Fig. 4). The choice was supported by the fact that, although not very accurate in urban areas, CLC data are widely available in Europe. Nevertheless, the CCA model was designed in order to allow for the use of a more accurate, CORINE-like, raster when and where available.

In order to effectively incorporate the zoning regulation data into the CCA module, a semiautomatic pre-processing of the Florence urban masterplan ("Piano Regolatore Generale", PRG) was carried out. In particular, as shown in Fig. 5, the masterplan prescriptions were reduced to land-uses permissions and prohibitions relevant for the CA simulation, both in terms of types of land-uses simulated by the CA (high and low density residential, industrial, commercial, agriculture), and in terms of its spatial scale of operation (pixel lattice representing 100×100 meters square). This zoning data layer was subsequently imported into the CCA model, assigning to each zone type a "permissibility factor" of new urban development for each land-use type. In fact, this factor is an important component of the CA transition rules outlined in section 2.1.

Another essential input used by both the CCA and Transportation modules, is the road network of the urban area. In order to improve the framework usability, the transportation module was designed to exploit the XML data from the OpenStreetMap collaborative project [22], which provides free street network data with a noticeable detail and geographic coverage.

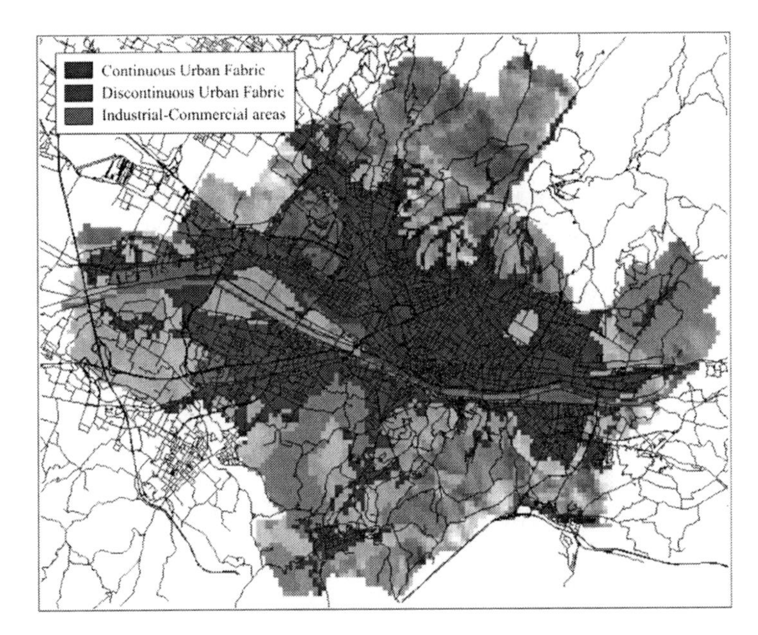

Fig. 4. Current land cover within the administrative boundary of Florence, according to CORINE (CLC2000) inventory, and street network imported from the OpenStreetMap project

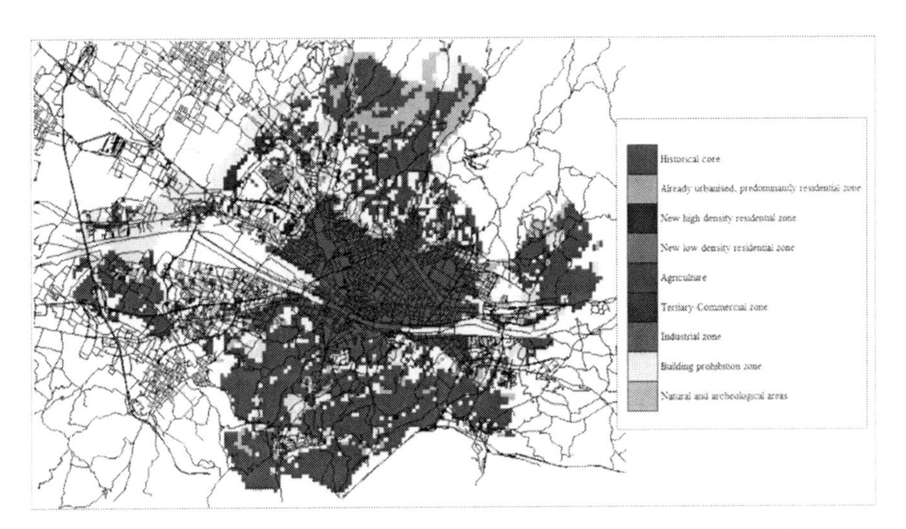

Fig. 5. Schematic regulation for Florence derived from the urban masterplan

In the preliminary application on the city of Florence, using the current street network represented in Fig. 4, a pre-processing phase for the calculation of the cells accessibilities (see section 2.1) was carried out by the CCA module. During the same pre-processing phase, a simple formula was used for deriving the suitabilities from a Digital Elevation Model of the urban area under consideration. In addition, a parameter calibration phase was preliminarily carried out exploiting the CLC1990 dataset, the map of Florence was provided by the Urban ATLAS EU project, and additional spatial data were provided by the municipality of Florence to the BRIDGE project consortium.

Then, using the land cover and (Fig. 4) the planning regulation map (Fig. 5.) as further input information for the CCA model, several future land-use scenarios were generated. For example, the future land use projection represented in Fig. 6 corresponds to a 20-year evolution of the urban area. Fig. 6 also includes the assumed evolution over time of the actively modelled land uses.

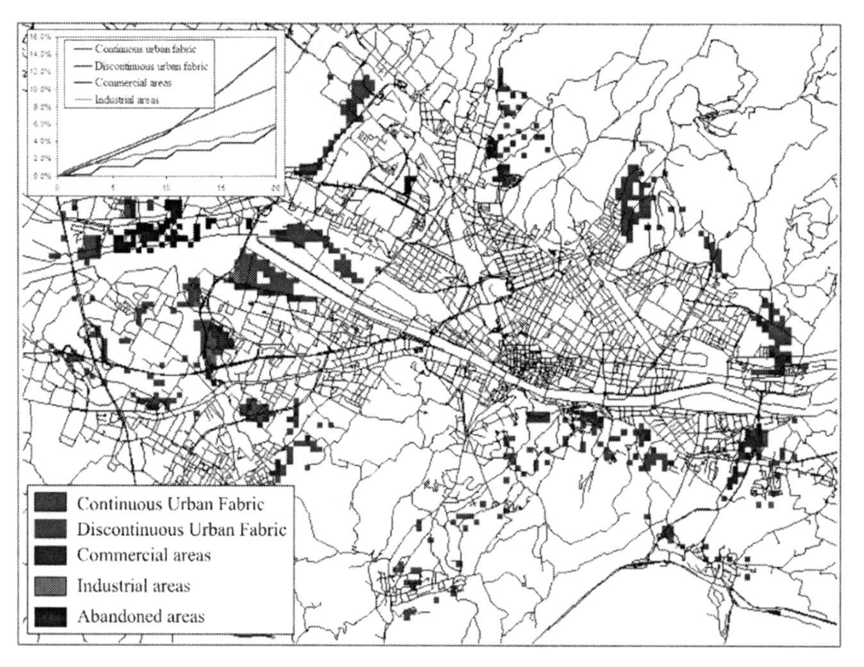

Fig. 6. Future land use projection obtained by the CCA module in 20 step of simulation. Only the cells where the land use was changed by the simulation are depicted. The embedded graph shows the extensions evolution of the actively modelled land uses.

Based on such future land-use scenarios, and after a preliminary phase of calibration based on the available traffic data, the transportation model is able to estimate the expected variation of the load of vehicles on the road network. For example, projecting on the road network the estimates of the traffic demand, driven by the future land-use scenario in Fig. 6, the transportation model generated the map of load variations for each road segment of the network depicted in Fig. 7.

Finally, as explained above, the WRF-ACASA model is able to include the land-use maps and the traffic load information in its simulations to generate maps of CO_2 and other meteorologically significant fluxes.

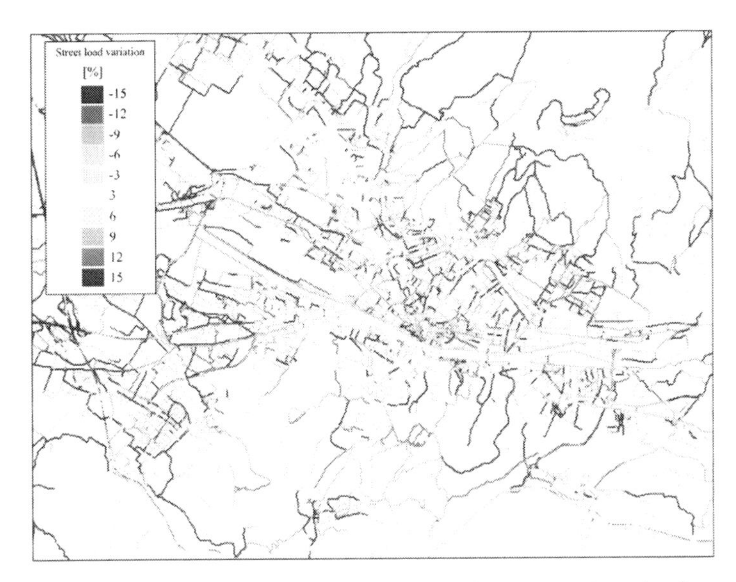

Fig. 7. Variation map of the street network loads which correspond to the future land use projection depicted in Fig. 6

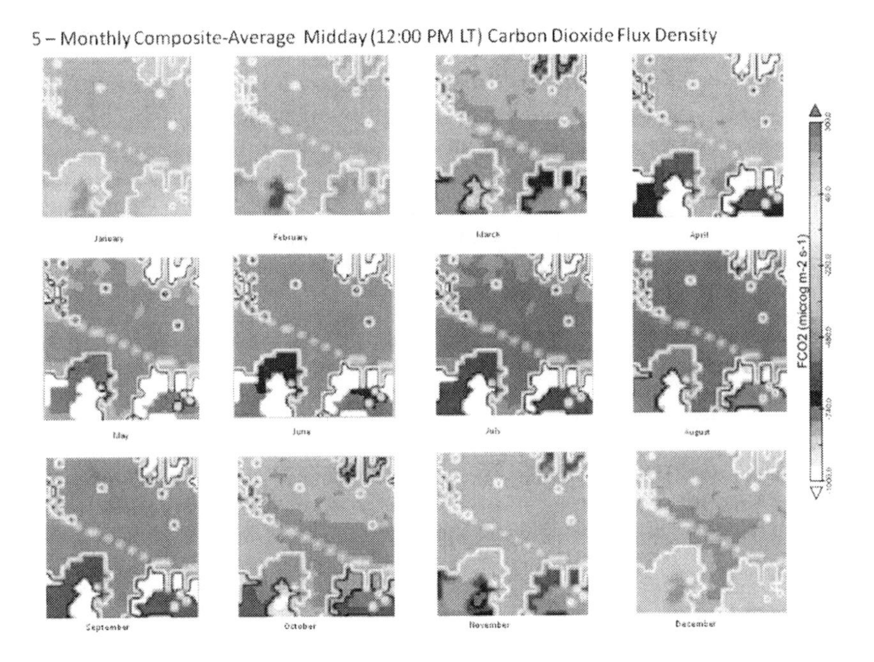

Fig. 8. Monthly-averaged values of midday (12:00 PM, Local time) carbon dioxide flux density obtained through WRF-ACASA for all of 2008 using the current land uses and traffic data for Florence case study

An example of such a set of maps, obtained for the current land use and traffic data, using the meteorology of the year 2008 is shown in Fig. 9.

Currently, the WRF-ACASA module is simulating the whole year 2008 in correspondence to the scenario of future land use in Fig. 6 and using the traffic data corresponding to the variations depicted in Fig. 7.

4 Conclusions and Future Work

The complexity typical of urban systems, the interactions between emissions and the influence of meteorological factors, make it difficult to effectively support urban planning and management with the goal of achieving a more sustainable metabolism in future cities.

In this context, we have presented an effective integration between different models, with the purpose of linking urban planning decisions to the estimates of CO_2 fluxes in urban environment. An application example on the City of Florence, even while it is still being carried out, has also been illustrated. This enabled to show in concrete terms what type of issues (i.e. data acquisition, calibration of the models) are involved for applying the proposed modelling system and what type of results can be expected.

The proposed framework allows, for example, to obtain realistic estimates on the impact of future planning decisions on CO_2 emissions in terms of its potential reduction from mitigation strategies. In more detail, information on zoning regulation and assumptions about the demography or development of specific economic sectors, as well as future changes in land uses are needed to obtain CO_2 flux estimates in urban area. On the other hand, as it is often the case, model calibration represents a crucial point for the application of such type of models. In particular, the CCA and the transportation models require historical land use data and traffic data, respectively. Also ACASA requires a preliminary calibration phase via collection of real CO_2 concentration values and other input parameters.

Ongoing work includes the development of a methodology and the definition of a reliable protocol calibrating the involved models, as well as the application of the proposed modelling approach to different urban areas.

Acknowledgments

This research was funded by the Euro-Mediterranean Centre for Climate Change (CMCC) (Agreement CMCC-UNISS DESA n° 070115) and by the FP7 European Project BRIDGE (Grant agreement n° 211345).

References

1. Wolman, A.: The metabolism of cities. Scientific American 213 (1965)
2. Newman, P.: Sustainability and cities: extending the metabolism model. Landscape and urban planning 44 (1999)

3. White, R., Engelen, G., Uljee, I.: The use of constrained cellular automata for high resolution modelling of urban land use dynamics. Environment and Planning B 24, 323–343 (1997)
4. White, R., Engelen, G.: High-resolution integrated modelling of the spatial dynamics of urban and regional systems. Computers, Environment and Urban Systems 2824, 383–400 (2000)
5. Blecic, I., Cecchini, A., Trunfio, G.A.: A General-Purpose Geosimulation Infrastructure for Spatial Decision Support. Transactions on Computational Science 6, 200–218 (2009)
6. Tsekeris, T., Stathopoulos, A.: Gravity models for dynamic transport planning: Development and implementation in urban networks. Journal of Transport Geography 14(2), 152–160 (2006)
7. Tsekeris, T., Stathopoulos, A.: Real-time dynamic Origin-Destination matrix adjustment with simulated and actual link flows in urban networks. Transportation Research Record. Journal of the Transportation Research Board 1857, 117–127 (2003)
8. Pyles, R.D., Weare, B.C., Paw, U.K.T.: The UCD Advanced Canopy-Atmosphere-Soil Algorithm: Comparison with observations from different climate and vegetation regimes. Q.J.R. Meteorol. Soc. 126, 2951–2980 (2000)
9. Pyles, R.D., Weare, B.C., Paw, U.K.T., Gustafson, W.: Coupling between the University of California, Davis, Advanced Canopy-Atmosphere-Soil Algorithm (ACASA) and MM5: Preliminary Results for for Western –North America. J Appl. Meteorol. 42, 557–569 (2003)
10. Staudt, K., Falge, E., Pyles, R.D., Paw, U.K.T., Foken, T.: Sensitivity and predictive uncertainty of the ACASA model at a spruce forest site. Biogeosciences 7, 3685–3705 (2010)
11. Marras, S., Spano, D., Pyles, R.D., Falk, M., Snyder, R.L., Paw, U.K.T.: Application of the ACASA model in urban environments: two case studies. In: AMS Conference on Ninth Symposium on the Urban Environment, Keystone, Colorado, USA, August 2-6 (2010)
12. Falk, M., Pyles, R.D., Marras, S., Spano, D., Snyder, R.L., Paw, U.K.T.: A Regional Study of Urban Fluxes from a Coupled WRF-ACASA Model. American Geosciences Union Fall Meeting. San Francisco, California (2010)
13. Staudt, K., Serafimovich, A., Siebicke, L., Pyles, R.D., Falge, E.: Vertical structure of evapotranspiration at a forest site (a case study). Agric. Forest Meteorol. 151, 709–729 (2011)
14. Marras, S., Pyles, R.D., Sirca, C., Paw, U.K.T., Snyder, R.L., Duce, P., Spano, D.: Evaluation of the Advanced Canopy-Atmosphere-Soil Algorithm (ACASA) model performance over Mediterranean maquis ecosystem. Agric. Forest Meteorol. 151, 730–745
15. Skamarock, W.C., Klemp, J.B., Dudhia, J., Gill, D.O., Barker, D.M., Duda, M.G., Huang, X.Y., Wang, W., Powers, J.G.: A Description of the Advanced Research WRF Version 3. NCAR/TN–475 + STR - NCAR TECHNICAL NOTE (2008)
16. Batty, M., Xie, Y.: From cells to cities. Environment and Planning B, 31–48 (1994)
17. Clarke, K.C., Gaydos, L.J.: Loose-coupling a cellular automaton model and GIS: long-term urban growth predictions for San Francisco and Baltimore. International Journal of Geographic Information Science, 699–714 (1998)
18. Erlander, S., Stewart, N.F.: The Gravity Model in Transportation Analysis: Theory and Extensions. VSP, Utrecht (1990)
19. Meyers, T.P., Paw, U.K.T.: Testing of a higher-order closure model for modelling airflow within and above plant canopies. Bound. Lay. Meteorol. 31, 297–311 (1986)
20. Meyers, T.P., Paw, U.K.T.: Modelling the plant canopy micrometeorology with higher-order closure principles. Agric. For. Meteorol. 41, 143–163 (1987)

21. Kusaka, H., Kondo, H., Kikegawa, Y., Kimura, F.: A simple single-layer urban canopy model for atmospheric models: Comparison with multi-layer and slab models. Bound-Layer Meteorol 101, 329–358 (2001)
22. OpenStreetMap, http://www.openstreetmap.org/
23. Ran, B., Boyce, D.: Modeling Dynamic Transportation Networks: An Intelligent Transportation System Oriented Approach, 2nd edn. Springer, Berlin (1996)
24. Staudt, K., Falge, E., Pyles, R.D., PawU, K.T., Foken, T.: Sensitivity and predictive uncertainty of the ACASA model at a spruce forest site. Biogeosciences 7, 3685–3705 (2010), http://www.biogeosciences.net/7/3685/2010/

Quantitative Analysis of Pollutant Emissions in the Context of Demand Responsive Transport

Julie Prud'homme, Didier Josselin, and Jagannath Aryal

UMR ESPACE – CNRS 6012, University of Avignon, France
julie.prud-homme@etd.univ-avignon.fr

Abstract. Nowadays, transport contributes significantly towards environmental problems (about 50% of the total CO and NOx). To date, most environmental issues due to transport have focused on general transportation methods. On the other hand, a popular approach of transport – Demand Responsive Transport (DRT) – has been studied for various aspects but rarely from the perspective of environmental issues. In this paper, we investigate the impacts of DRTs on pollutant emissions. For this purpose, we adapt a method established by European research co-operation - Methodologies for Estimation of Emissions from Transport (MEET). We create a specific model to estimate the pollution of a DRT system (GREEN-DRT) adapted from the MEET. We simulate DRT operation on three overlapping territories in France. The results show that optimising the DRT induces a significant decrease of pollutant emission due to the reduction of vehicles and travelled distances.

Keywords: Pollutant Emission Model, GREEN-DRT, Demand Responsive Transport, Territories and Scales, DARP Optimization, MEET.

1 Introduction

Gas and substances sent in atmosphere during fuel combustion have a global impact, with greenhouse problem, and a local one, with direct effects on public health. Road transports are directly concerned by the second aspect. Their environmental impacts, principally pollutant emissions, are henceforth well known with several European projects, mainly ARTEMIS (Assessment and Reliability of Transport Emission Models and Inventory Systems) [2] and COPERT (Computer Program to Calculate Emissions from Road Transport) program (1989 until now) [16], with Methodologies to Estimate Emissions from Transport (MEET) project (1996 – 1998) and COST Actions on the Estimation of Emissions from Transport (1993 – 1998). Main results are related in [11].

Demand Responsive Transport systems, which support grouping people in vehicles, are proposed as a relevant solution to reduce road transports economical costs and environmental impacts [14], [1], [3], [5]. DRTs have been studied for approximately two decades but the research had rarely concerned specifically on environmental impacts. Therefore, we bring some elements of DRT environmental performance analysis. We investigate the pollutant relationship in the DRT context

B. Murgante et al. (Eds.): ICCSA 2011, Part I, LNCS 6782, pp. 439–453, 2011.

(Geographical Reasoning on Emission Estimations based on road Network applied to Demand Responsive Transports – GREEN-DRT) by adapting an existing methodology *(MEET)* and highlight the associated parameters.

This research is developed around two main questions:

- How can we estimate the pollutant emissions of vehicles, according to the DRT operational efficiency (service flexibility, routes optimised 'on the fly', vehicle depots disseminated on the whole territory)?
- Does a high passenger grouping rate in vehicles significantly reduce emissions?

2 Evaluation of Atmospheric Pollution from Transport Systems

Since the 1970's, with a renewal in the 1990's and the vision of the sustainability concept, many researches have been conducted to evaluate atmospheric pollution emitted from transport systems. Researchers have been interested to understand and to identify parameters which influence fuel consumption, therefore emissions of pollutants from car traffic [9], [15].

There exist different kinds of environmental impacts from car traffic that we present in this paper. Similarly, we present parameters that we could exploit to estimate transport pollutant emissions. At the end of the paper, some simulations illustrate the global impact of DRT optimisation on pollutant emission.

2.1 Different Kinds of Environmental Impacts

Car traffic has various types of environmental impacts. The light density of car flow causes noisy and visual impacts to the neighbourhoods. Road infrastructures cause a territorial fragmentation and a social segregation which make pressure on residential and natural protected areas. The simple fact of using a car produces also a lot of pollutant emissions. Indeed, the fuel combustion by cars, lorries and all road vehicles creates pollutants gas and substances. The main pollutants are carbon dioxide (CO_2), carbon monoxide (CO), hydrocarbon (HC), oxides of nitrogen (NO_x) and particulate matter (PM).

In this paper, we present a study of a particular transport system, a Demand Responsive Transport (DRT), which uses available road infra-structures and tends to reduce the amount of car traffic due to passengers grouping. Further, we develop the aspect of pollutant emissions from traffic fumes.

2.2 Estimate Pollutant Emissions from Road Transports

The current research is based on the Methodologies for Estimation of Emissions from Transport (MEET) project, in conjunction with the COST Action 319, methodology for calculating emissions and energy consumption prepared for the European Commission DGVII by Transport Research Laboratory (1999). After exploiting the method, we present parameters linked to the MEET and a synthesis of pollutant estimation method that will be partly applied in simulations.

Pollutant emissions are directly related to the cars fuel consumption. Emission performances vary depending on vehicle types. To characterise a vehicle type, we would need to consider several criteria like:

- motor vehicle purpose (passenger car, light duty vehicle, lorry, bus, coach) ;
- vehicle size (cylinder) ;
- vehicle age and associated emission control level. EU legislation fixes standards of emissions to validate the sale of vehicles. It informs on pollutant emissions from the vehicle ;
- fuel (mainly petrol, diesel). Depending on the fuel types and their combustion, produced pollutants are different.

2.3 Calculation of Pollutant Emissions, according to the MEET Methodology

We base our DRT emission model (GREEN-DRT) on MEET's one, related in [11]. The main sources of emission from road vehicles are the exhaust gases and hydrocarbons produced by evaporation of the fuel. When an engine is started below its normal operating temperature, it uses fuel inefficiently, and the amount of pollution produced is higher than when it is hot. These observations lead to the first basic relationship used in the calculation method:

$$E = E_{hot} + E_{start} + E_{evaporative} \quad (1)$$

where:

E	is the total emissions
E_{hot}	is the emission produced when the engine is hot
E_{start}	is the emission when the engine is cold
$E_{evaporative}$	is the emission by evaporation (only for Volatile Organic Compound)

Hot emissions are produced when the engine and the pollution control systems of the vehicle (*e.g.* catalyst) have reached their normal operating temperature. For hot emissions, the amount of operation related emission factor is primarily expressed as a function of the average speed of the vehicle. It is carried at a particular average speed, on roads with a certain gradient, for vehicles with a certain load. Calculation of emissions is "*stressed that distance is distributed over different types of roads. A part of the distance is travelled in urban areas, a part in rural areas and the rest on highways, each type of road having a different average speed and affecting the emission factors*" [11].

Taking into account the different vehicle categories, the general equation for hot emission estimation, depending on pollutant type, can be derived:

$$E_k = \sum_i n_i l_i \sum_j p_{ij} e_{ijk} \quad (2)$$

where:

k	identifies the pollutant
i	is the vehicle category
j	is the type of road
n_i	is the number of vehicles in category i
l_i	is the average annual distance travelled by the vehicles of category i
p_{ij}	is the percentage of the annual distance travelled on road type j by vehicle category i
e_{ijk}	is the emission factor of pollutant k corresponding to the average speed on road type j, for vehicle category i

The method proposed by MEET for estimating start-related emissions was developed empirically, using data assembled from many European test programmes. According to the MEET, for each pollutant and vehicle type, a reference value is defined for the excess emission as the value corresponding to a start temperature of $20^{\circ C}$ and for an average trip speed of 20 km/h.

Start emissions, because they occur only during the early part of a journey, are expressed as an amount of pollutants produced per trip, and not over the total travelled distance. The emission factor e_{ijk} is calculated as a function of the average vehicle speed, the engine temperature, the length of the trip and the length of the cold part of the trip which is about the first 6 km [12],[13]. More engine heats up, less start emissions are large.

Cold start emissions are estimated as constants (excess emissions per cold start).

$$E_{start} = \omega\left[f(V) + g(T) - 1\right]h(d) \tag{3}$$

where:

E_{start}	for a trip is expressed in gram
V	is the mean speed in km/h during the cold period
T	is the temperature in °C (ambient temperature for cold start, engine start temperature for starts at an intermediate temperature)
d	is the travelled distance
ω	is the reference excess emission (at $20^{\circ C}$ and 20 km/h)
$f(), g(), h()$	are functions

According to [12], evaporative emissions occur in a number of different ways. Fuel vapour is expelled from the tank each time it is refilled, the daily increase in temperature causes fuel fumes to expand and be released from the fuel tank. Vapour is created wherever fuel may be released to the air, especially when the vehicle is hot during or after use. There are therefore different emission factors depending on the type of evaporative emissions. Generally, these factors are a function of the ambient temperature and the fuel volatility. Similarly, a number of activity data are also needed, including total travelled distance and number of trips according to the temperature of the engine at the end of the trip.

Thus a very accurate study of air polluting from DRT vehicles should take into account all these various criteria. However, we are currently not able to consider this whole set of parameters in our study. It is due, on the one hand, to the data availability and some material aspects, and, on the other hand, to the peculiarities of DRTs. We are going now to explain how we propose to estimate pollutant emissions using a simplified and adapted method.

3 Application to Demand Responsive Transport Systems: The GREEN-DRT Model

We apply pollutant emission estimations in the special case of a flexible transport system, a Demand Responsive Transport (DRT). At first, we present DRT systems. Then, we present how to adjust the MEET model to provide the dedicated GREEN-DRT model, *Geographical Reasoning on Emission Estimations based on road Network shape adapted to DRT system.*

3.1 Demand Responsive Transport Systems

According to [4], DRT is a kind of public transport which combines the advantages of collective transport and individual vehicles. It is often considered as a versatile transport because it offers more flexibility for the users, and provides some saving for the carrier and for the local transportation authorities. It generally serves an extended territory at various times from a lot of destinations, and sometimes with a door-to-place or door-to-door service, depending on the kind of DRT. DRT states as a sharing transport which aims to group the passengers from different individual demands within a same or similar trip. This includes important travelled distance reduction, depending on the quantity of clients and their origin and destination points. This feature can be significant for environmental impacts of DRT systems, in comparison with other transport systems.

Environmental impacts can also be restrained because DRT vehicles run only when it is used (the system is activated by a demand at least from a single client) and because the system succeeds in grouping clients, and therefore reduces the number of vehicles on the road.

A simplified version of DRT pollutant emissions pattern is presented hereafter in Figure 1. In practice, a DRT journey implies one great start emission (at the vehicle station), as many smaller start emissions as the number of stops fixed by clients, a part of the travel is in cold conditions for fuel consumption, before a warm distance. This information is very difficult to get. We will see later in simulations how we nevertheless can make a general trend analysis using a simplified process, without considering such an accurate granularity of spatial information.

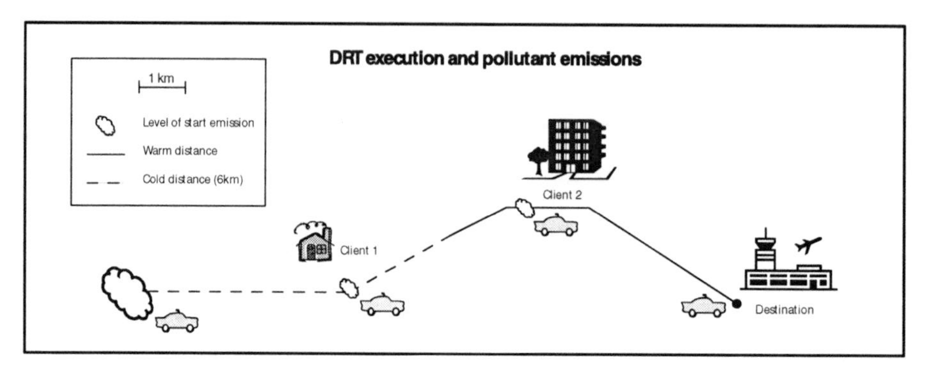

Fig. 1. DRT operation and pollutant emissions

In the schematic diagram above (Figure 1), we consider two people who want to go to the airport to take a similar time flight. Despite DRT execution imposes some stops, car engine remains still hot. We consider that in few minutes, the engine temperature does not significantly change.

3.2 Adaptation of MEET for Pollutant Emissions Estimation

To adapt the MEET model to DRT system (GREEN-DRT model), we should take into account the network shape, which may have some implications on the driving behaviour. For example, turns provoke decelerations and accelerations, as well as road intersections. We are aware that in the emission calculation the shape of each road section has a major impact which we consider in the future work and hence not presented in this paper. First, as mentioned before, we cannot consider evaporative emissions because of the complexity in getting the input data. Indeed, we don't know about the really used vehicles fleet and the way those are driven. It is also complicated to assess the stop time duration, because it depends on the number of clients and the type of crossing. Moreover, using GIS, we only have a simple topology of the road network, with section impedance and length. These data will not be used in the presented results.

The MEET model is applied to all road vehicles in a territory. This vehicle fleet is composed of various kinds of vehicles. This model is based on estimation of vehicle amount in each category. Mileage is obtained from the vehicle category mileage average. Moreover, in order to avoid too much imprecision, the reasoning is made at the year scale.

In DRT simulation conditions, we exactly know the components of the fleet and the mileage of each vehicle. Equations from MEET model need to be adapted to this context of use. From previous equations, we can derive the following equation:

$$E_k = \sum_t D_{v,t,j} . e_{v,j,k} \tag{4}$$

where:

E_k is the quantity of each pollutant k emitted, expressed in gram

k identifies the pollutant

t is the type of road

$D_{v,t,j}$ is the distance travelled on road type j by the vehicle V

$e_{v,j,k}$ is the emission factor of pollutant k corresponding to the average speed on road type j, for the vehicle V

and

$$E_{hot} = \sum E_k \tag{5}$$

where:

E_{hot} is the emissions during the warm distance by a vehicle

Start emission calculation (eq. 3) still the same than in MEET model, where $d = 6\ km$.

Finally, this is the general equation for emission estimation in GREEN-DRT:

$$E_V = E_{hot} + E_{start} \tag{6}$$

Let us notice that this formula is simplified for the estimations provided in this paper, because we only know the sequence of stops for each vehicle, but not the real final path.

3.3 Comparison of MEET and GREEN-DRT Models

We presented MEET and GREEN-DRT models in the equations above. The comparison of these two models is presented in a summary table below.

Table 1. Comparison of MEET and GREEN-DRT Models

	MEET	**GREEN-DRT**
Context	All vehicles are on a defined area	DRT vehicles
Hypothesis	Unknown vehicle fleet	Known vehicle fleet
Temporal granularity	Year estimate	Calculation of routes
Network shape in the calculation of emissions	Rate of each road type used by similar vehicles	Details of parameters for every road sections

4 Methodology for Simulations

Equations set in previous section enable us to evaluate DRT environmental successes and to identify and quantify DRT leading characteristics in their performance, according to different territorial structures [8]. With purpose of a routing optimisation kernel, we can test several parameters.

4.1 Optimisation Kernel

The used algorithm [10] solves the Dial A Ride Problem with Time Windows and many vehicles (N-TW-DARP) [6], [7]. It generates a good solution using a heuristic approach, due to a very high complexity. It is a classical insertion algorithm. It iterates and mixes candidate sets of routes that converge to a final solution, which is optimal but approximate. This means we cannot ensure to find the 'best' solution, but we are certain to obtain a very good one. Over a certain quantity of demands, an exact method would not be able to produce any solution in a reasonable time.

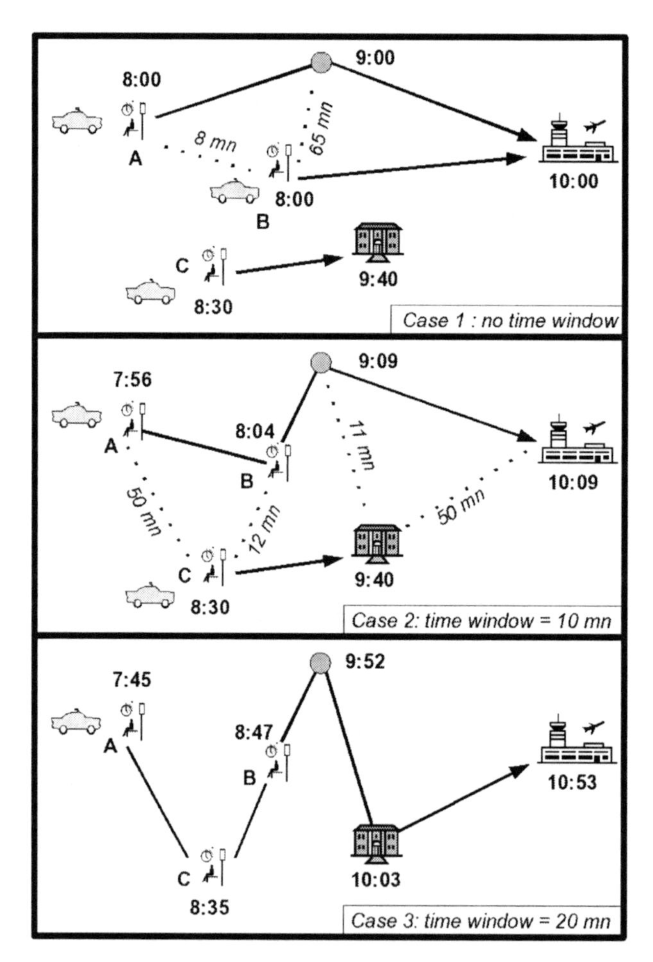

Fig. 2. Insertion process and time windows to increase the grouping rate and decrease the number of vehicles

The objective function aims at reducing the global cost, that is to say decreasing the number of vehicles and the travelled distances. The grouping objective is pursued using possible deviations (time windows). The complete operating service, from the passenger reservation to the kernel execution, is possible on-line, using a web service that also allows to locate the vehicle depots, the stops and the clients (their address and their travel from origin to destination). This tool, firstly developed by University of Avignon, has been adapted and is now improved and maintained by the private firm Prorentsoft.

The optimisation kernel has to deal with two constraints in contradiction. On the one hand, it must find routing solutions within a given time window for vehicle detours. On the other hand, the service must respect the client mobility practice and their wish. In Figure 2, we draw three different cases that show the influence of increasing the time window and give some examples of the insertion process principle. When there is no time window (Case 1), it is necessary, as taxis usually make, to assign one vehicle per client (A, B and C). At the opposite (Case 3), a large time window allows large detours but those can imply very important time loss, which are probably unbearable (client A). Sometimes, the regular bus lines can have such an impact on public transport use. We can notice in this case that a single vehicle is necessary: this would have a great impact on the environment, even if the travelled distances are longer. Between these two cases, we find the Case 2 where the time window is sufficient to provide reasonable grouping rates (two vehicles) and acceptable travel durations for the clients. This is where Demand Responsive Transport can be the most useful.

4.2 Workflow of GREEN-DRT Model

The following figure (Fig.3) presents the flow of GREEN-DRT model execution. The execution follows five successive steps. A Geographical Information System (GIS) can be used at different steps to perform the spatial aspect of emission calculation and to focus on their spatial distribution.

Currently, there are two different ways to apply GREEN-DRT model. The first one requires an on-line GIS based on Google maps and networks, dedicated to DRT implementation. Using this software, it is possible to fix a set of stops, a set of depots and a set of client demands (origins and destinations). In a second stage, a matrix of shortest paths is processed. According to the vehicle location and availability, we launch the optimisation kernel, which gives a sheet of parameters about the routes of all the vehicles used. From these data, we calculate the emissions. This is what is executed in this paper. Thus, the estimation of pollution is based on aggregated information. For instance, we cannot precisely estimate the hot/cold emissions and the speed effects. For each vehicle, we finally obtain its route, *i.e.* a complete sequence of stops. So we know its start, the travelled distances and how many stops compose the route until the end.

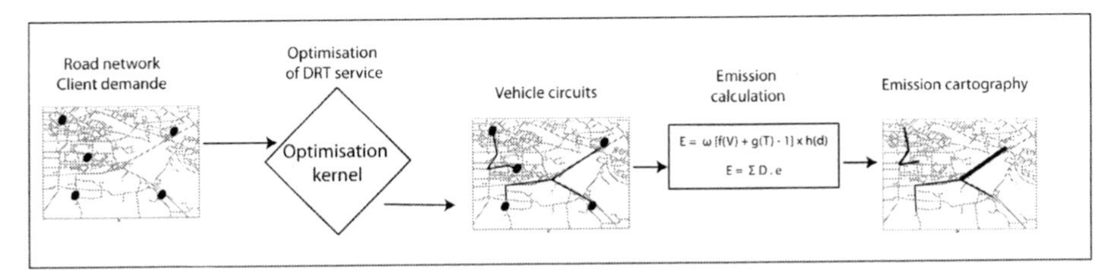

Fig. 3. Steps of GREEN-DRT workflow

The second possible process depicted in the figure 3 consists in detailing the real paths of all the vehicles on the network. This requires to use a classical GIS and to precisely redraw the circuits on the network. This work is not presented in this paper. It will allow us to get a more accurate assessment of pollutant emissions, taking into account the type of roads and the average of allowed speed, the type of emission (start, cold/warm) and any other effects due to the network structure (windings). Now we only present a simulation of a DRT using the online software 'Pronto' from Prorentsoft. Although this tool is already used in other locations with concrete clients and real vehicles from 2006 ('Doubs Central', in north-east of France), we here simulate a virtual DRT in Brittany, a French western region, in collaboration with our partners, the local transport authorities involved in the 'Modulobus' research program funded by the French Research Agency. This may prefigure what could be an optimised DRT on this territory and its environmental impact, thanks to different simulations we proposed.

4.3 Parameters of Simulations

To analyse DRT environmental performances, we run many simulations. We arbitrarily located DRT stops on the territory of Brittany. Then, we vary several parameters depending on different simulation cases. We acted on the deviation time allowed (5, 10, 15 and 20 minutes) which has been applied on different DRT operation territories (Fig. 4). On the one hand, the variation of deviation time must act on the grouping rate: as variation time increases, grouping rate increases too. On the other hand, as transport systems have a spatial dimension, it is relevant to consider the DRT system deployment scale. Our study is based on two territories, having various sizes (Fig. 4). In the Region of Brittany (level 1), there is the level 2 territory corresponding to Agglomeration of Rennes (Fig. 5). The level 3 is the City of Rennes (Fig. 6). The level results are not relevant in this paper and are not presented.

All the simulations are made with 50 client demands and various numbers of stops and vehicles. Two types of vehicles in equal proportion namely petrol and diesel are considered. The emission is the total quantity of the different emitted pollutants (CO CO_2, NO_x etc.). We do not take into account the types of road section, the speed and as well the shape of the network.

In the simulations, the set of stops and time deviation are predefined. Trip times, origins and destinations are chosen randomly. As we cannot know the real (even theoretical) speed for each road section, an average speed is given all over the network. However, the travelled distances are accurate and include the distances from the depots to the clients and between the ordered stops. So, these simulations provide interesting global results to study the relation between several criteria: distance, grouping rate, emissions, time window inducing route deviations, space extent and scale.

We located stops for each overlapping level. At Level 2, we selected 33 stops, one in each city of agglomeration of Rennes (608 km²) (Fig. 5). We obtain a stop density of 0.054 stop per km². At Level 3, we selected 79 stops in the city of Rennes (50 km²) (Fig. 6). The stops are located on some places of interest, mostly downtown. We obtain a stop density of 1.58 stop per km². Therefore, they are closed from each others (hundreds of metres).

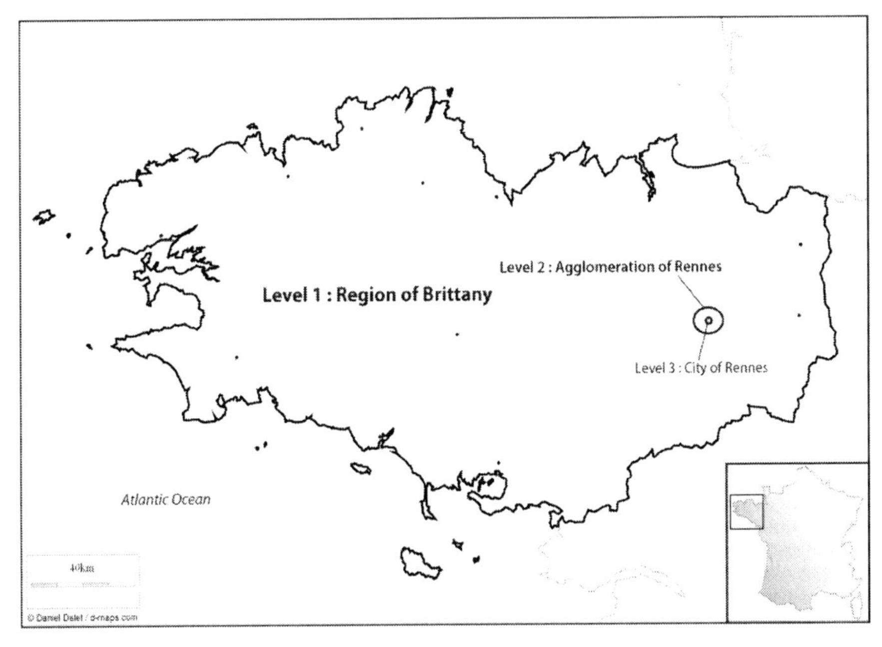

Fig. 4. Scale levels of study in the Brittany territory (level 1)

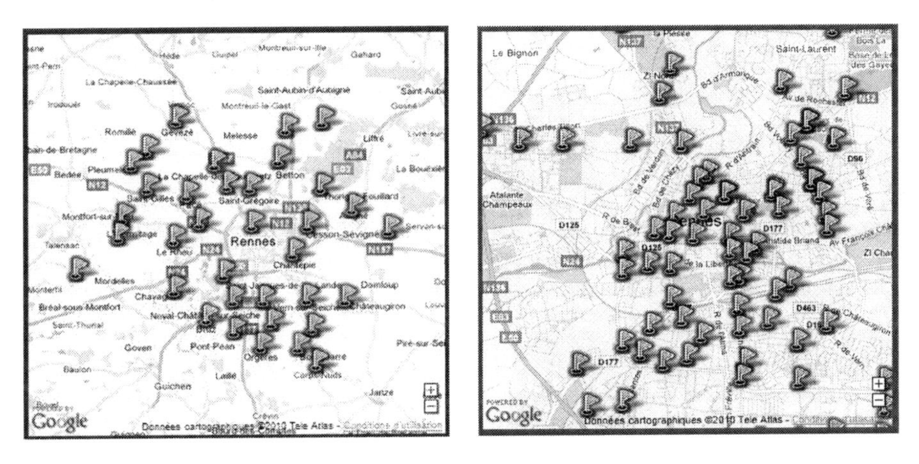

Fig. 5. DRT stops on Level 2 (scale = 1:800,000)

Fig. 6. DRT stops on Level 3 (scale = 1:130,000)

5 Experimentations and Results

These experiments logically show that grouping clients in the vehicles decrease the number of vehicles used. Subsequently, we observe a decrease of pollutant emissions, with an increase of grouping rate.

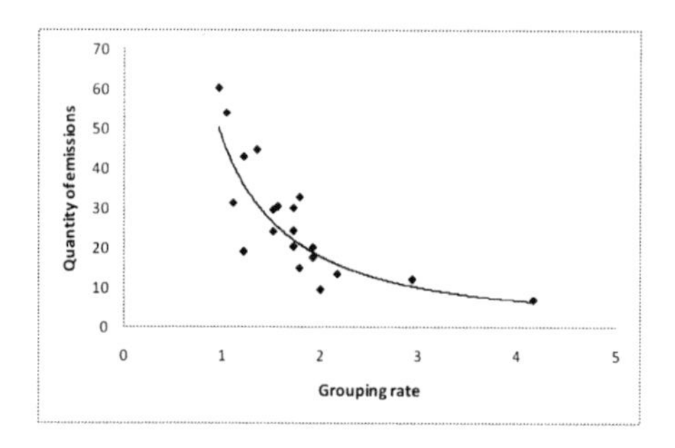

Fig. 7. Relation between grouping rate of emissions emitted

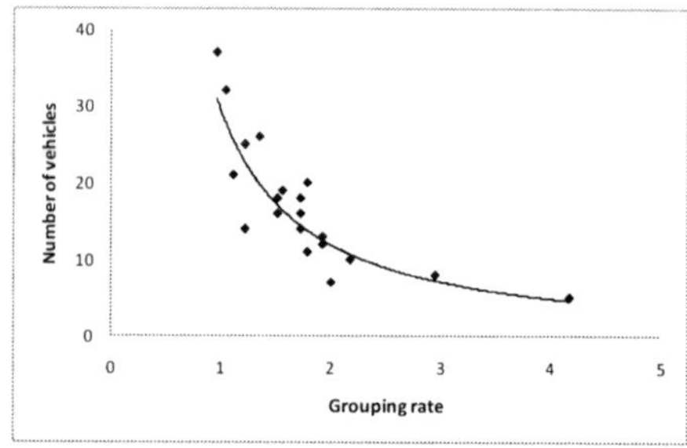

Fig. 8. Relation between grouping rate and quantity and number of vehicles used

The fitted curves in Figure 7 and 8 show the relationship between these variables and attests the above statement.

The general trend of the observed relationship corresponds to the statement: emissions decrease when grouping rate increases, but there are many irregular points. It is partially due to the heuristic that finds good solutions but not always the best ones. However, we need to explore additional variable correlations to understand these phenomena. Let us now focus on the factors influencing the DRT pollutant emissions: the scale level and the deviation time allowed, for instance.

Due to the different scales, the curves must be analysed at each scale, in terms of shapes and breakdown points. Between scale levels, mileage and emission dimensions can be different, because of territory size.

The simulation results for the Level 2 (Table 2, Figs. 9, 10) are as follows:

Table 2. Results for Level 2

Deviation time (mn)	Grouping rate (people/vehicle)	Pollutant emissions (kg)	Mileage (km)
5	1.72	20.32	1992
10	1.92	17.68	1969
15	1.92	17.60	1896
20	2.94	12.17	1704

At a medium scale level (Level 2), deviation time and grouping rate vary linearly (Table 2). When the first increases, the second too. Relationship between grouping rate and pollutant emission is very clear (Fig. 9), such as the link between emissions and mileage (Fig. 10).

At this scale level, the main statement is verified. There is a threshold of deviation time (15 minutes) over which the relation is distinctly different.

Simulation results for the Level 3 are the following (Table 3, Figs. 11, 12).

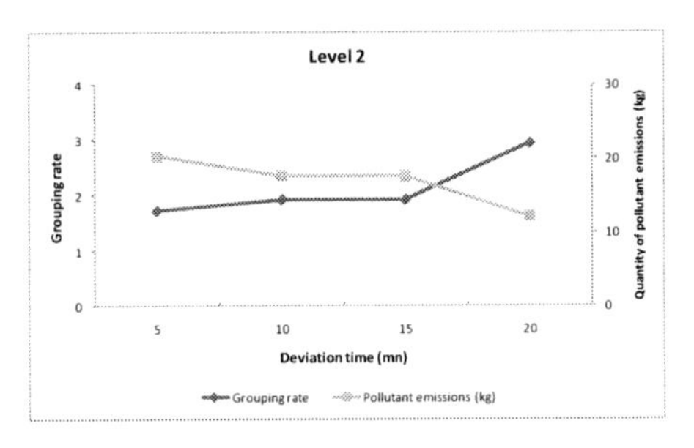

Fig. 9. Link between grouping rate and quantity of pollutant emissions depending deviation time at Level 2

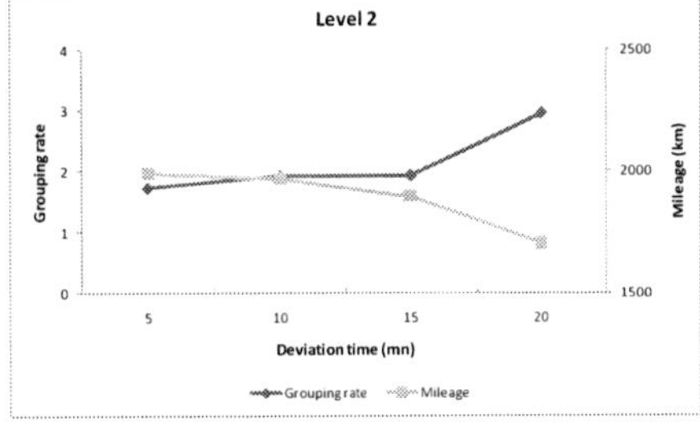

Fig. 10. Link between grouping rate and mileage depending on deviation time at on Level 2

Table 3. Results for Level 3

Deviation time (mn)	Grouping rate (people/vehicle)	Pollutant emissions (kg)	Mileage (km)
5	2.17	13.45	362
10	2.00	9.59	337
15	4.17	6.92	311

At this scale level, we have three observations. We obtain results which indicate a close situation to the Level 2. Curves in Figure 11 and Figure 12 show an increase of grouping rate while emissions decrease. In this case, the change is visible for 10 minutes of possible detours, due to shortest distances between stops.

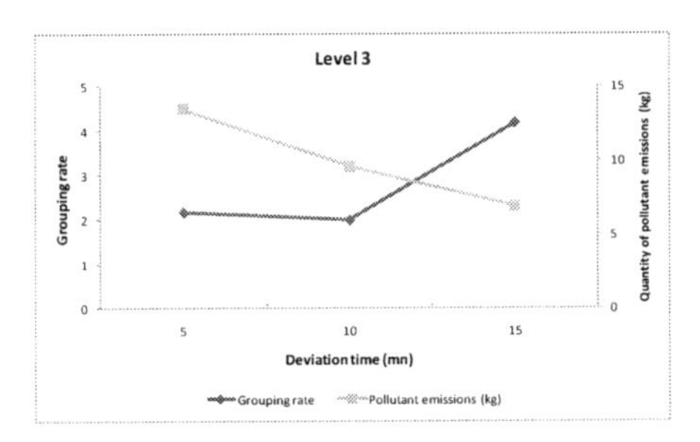

Fig. 11. Link between grouping rate and quantity of pollutant emissions depending on deviation time at Level 3

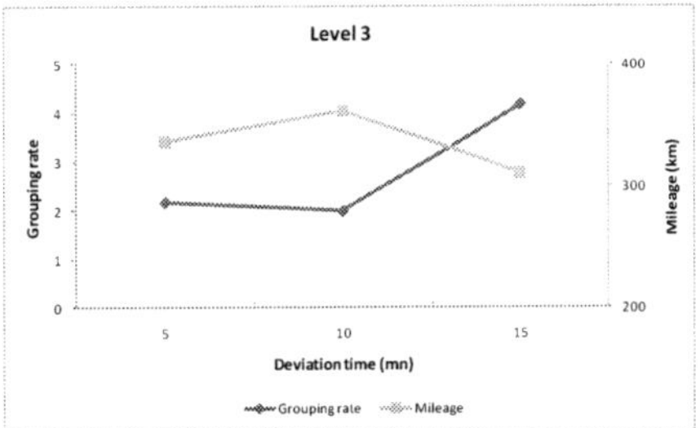

Fig. 12. Link between grouping rate and mileage depending on deviation time at Level 3

We showed that the results for Level 2 and 3 are perfectly matching with the general statement of relationship between mileage and emitted pollutants. Globally, increase in the grouping rate provides, on the one hand, more important deviation time, on the other hand, lower mileage and emissions. In other terms, the detours are not favourable to an emission decrease, but the effect of the grouping rate is so strong that the pollution per capita or per vehicle is significantly reduced, although this favourable impact is progressively minimized itself for high grouping rates. So the grouping causes more benefits than detours inducing more travelled distances. This is in favour to optimisation. In the next section, we discuss about results and research prospects.

6 Discussion and Conclusion

6.1 Effect of the Routing Optimisation on Transport Efficiency and Pollutant Emission

We showed that the increase of the deviation time supports the grouping rate, decreases the vehicle quantity and the pollutant emissions. The simulation results confirm this statement. It also shows that the grouping rate can increase quickly over a certain threshold, depending on the stops density and the distance apart.

Moreover, it underlines that the increase of deviation time enlarges the possible area to be visited by the vehicles. When there is a few stop density, there is a higher probability of using the maximum limit of deviation time. With more deviation time, it is possible to visit farther to pick up the client, that increases mileage. This is fortunately compensated by the reduction in the number of vehicles used.

The simulations do not exactly represent the real functioning of DRT systems. Generally, clients use them in some time slots along the day or the week. This means that the simulations probably underestimate the efficiency of the service (grouping rate). In real situation, we would expect a highly likely efficient service with a better grouping rate.

We addressed the pollutant emissions due to DRT for the first time by adapting an established model and adjusting the associated equations. We applied computational simulations to produce parameters at different scales, corresponding to some French territories. The relationship between some of these parameters has been clarified. But there remains some work to improve the model.

6.2 Further Works

Indeed, we noticed that the above parameters are not sufficient to represent the DRT pollutant emissions. For the proper representation and estimation of DRT pollutant emissions, some territorial aspects should be added. We conclude that more spatial aspects have to be considered in the simulations to explore the details of DRT operation environmental impacts.

For instance, we remarked that, due to client grouping, there are fewer vehicles, less start emissions, less mileage and less hot emissions. This reasoning is limited by the scale of DRT territory extent and stop density: while stop density decreases, distances increase. We could link stop density to the road network characteristics to explain a part of DRT systems performances. The challenge will be to evaluate contributions of each factor and to determine the importance of the road network neighbourhood context (*e.g.* urban *versus* rural areas) in mileage.

At the moment, we cannot precisely state which factor is the most important. To figure out this, we have to observe the results with more parameters added in the computational simulation. In further works, we shall test more simulation parameters along with sophisticated material facilities to better understand the DRT pollutant emissions, using a GIS. Especially, we shall re-build the shortest path of each vehicle from the complete and already known sequence of stops, in order to get a more accurate and realistic operating of the vehicles.

References

1. Ambrosino, G., Nelson, J.D., Romanazzo, M.: Demand Responsive Transport Services: towards the Flexible Mobility Agency. ENEA edn. (2004)
2. Boulter, P., McCrae, I.: ARTEMIS: Assessment and reliability of transport emission models and inventory systems - Final Report (2007)
3. Castex, E.: Le Transport À la Demande (TAD) en France: de l'état des lieux à l'anticipation. Modélisation des caractéristiques fonctionnelles des TAD pour développer les modes flexibles de demain. Ph.D. thesis, Avignon, France (2007)
4. Castex, E., Houzet, S., Josselin, D.: Prospective research in the technological and mobile society: new Demand Responsive Transports for new territories to serve. AGILE (2004)
5. CERTU: Le transport à la demande en 140 questions. Éditions du CERTU (2010)
6. Cordeau, J.F., Laporte, G., Potvin, J.Y., Savelsbergh, M.W.P.: Transportation on Demand. Octobre 2004, Edition CRT, Québec (2004)
7. Diana, M., Dessouky, M.: A new regret insertion heuristic for solving large-scale dial-a-ride problems with time windows. J. Transportation Research Part B 38, 539–557 (2004)
8. Foltête, J.C., Genre-Grandpierre, C., Josselin, D.: Impact of Road networks on Urban Mobility, in Modelling Urban Dynamics. Editions Theriault M & Des Rosiers F, Chap. 5 (2010)
9. Gallez, C.: Indicateurs d'évaluation de scénarios d'évolution de la mobilité urbaine. Rapport sur convention DTT-INRETS (2000)
10. Garaix, T., Josselin, D., Feillet, D., Artigues, C., Castex, E.: Transport à la demande points à points en zone peu dense. In: Proposition de méthode d'optimisation de tournées (2005); Cybergeo, article selected from SAGEO 2005, Avignon 20-23 juin 2005, France, http://www.cybergeo.eu/index2650.html
11. Hickman, J., Hassel, D., Joumard, R., Samaras, Z., Sorenson, S.: Environment, energy and transport. Written material extended version by EU-PORTAL, Transport teaching material, based on MEET and COST results (2003)
12. Joumard, R., Sérié, E.: Modelling of cold start emissions for passenger cars. INRETS report. LTE 9931 (1999)
13. Joumard, R., Vidon, R., Paturel, L., Pruvost, C., Tassel, P., de Soete, G., Saber, A.: Evolution des émissions de polluants des voitures particulières lors du départ à froid. J. The Science of the Total Environment 169, 185–193 (1995)
14. Lebreton, E.: Le transport à la demande comme innovation institutionnelle. J. Flux 43, 58–69 (2001)
15. Nicolas, J.P., Pochet, P., Poimboeuf, H.: Indicateurs de mobilité durable – Application à l'agglomération de Lyon. Report, Association pour les pratiques du developpement durable (APDD) et Laboratoire d'Economie des Transports (LTE), Lyon, France (2001)
16. Ntziachristos, L., Samaras, Z.: Exhaust emissions from road transport. EMEP/EEA emission inventory Guidebook – COPERT program (2009)

Individual Movements and Geographical Data Mining. Clustering Algorithms for Highlighting Hotspots in Personal Navigation Routes

Gabriella Schoier* and Giuseppe Borruso

DEAMS, University of Trieste,
Piazzale Europa 1, 34127 Trieste, Italy
{gabriella.schoier,giuseppe.borruso}@econ.units.it

Abstract. The rapid developments in the availability and access to spatially referenced information in a variety of areas, has induced the need for better analysis techniques to understand the various phenomena. In particular our analysis represents a first insight into a wealth of geographical data collected by individuals as activity dairy data.The attention is drawn on point datasets corresponding to GPS traces driven along a same route in different days. Our aim here is to explore the presence of clusters along the route, trying to understand the origins and motivations behind that in order to better understand the road network structure in terms of 'dense' spaces along the network. In this paper the attention is therefore focused on methods to highlight such clusters and see their impact on the network structure. Spatial clustering algorithms are examined (DBSCAN) and a comparison with other non-parametric density based algorithm (Kernel Density Estimation) is performed. A test is performed over the urban area of Trieste (Italy).

Keywords: DBSCAN, Kernel Density Estimation, GPS traces, activity dairy data.

1 Introduction

The rapid developments in the availability and access to spatially referenced information in a variety of areas has induced the need for better analysis techniques to understand the various phenomena. In particular spatial clustering algorithms, which groups similar spatial objects into classes, can be used for the identification of areas sharing common characteristics. The aim of this paper is to present a density based algorithm for the discovery of clusters of units in large spatial data sets. In particular the analysis represents a first insight into a wealth of geographical data collected by individuals as activity dairy data, these representing the routes a set of individuals drive in their daily movements for reasons connected to work, children picking, free time, rest. In this analysis

* The paper derives from joint reflections of the authors. Paragraphs 1, 4, 5.1 and 7 are realized by Giuseppe Borruso and paragraphs 2, 3 and 6 by Gabriella Schoier while paragraphs 5.2 and 5.3 are developed in common.

B. Murgante et al. (Eds.): ICCSA 2011, Part I, LNCS 6782, pp. 454–465, 2011.

the attention is drawn on point data sets corresponding to GPS traces driven along a same route in different days. Our aim here is to explore the presence of clusters along the route, trying to understand the origins and motivations behind that in order to better understand the road network structure in terms of 'dense' spaces along the network, these representing road congestion, rather than the presence of junctions or other factors affecting individual mobility. In this paper the attention is therefore focused on methods to highlight such clusters and see their impact on the network structure. Spatial clustering algorithms are examined (DBSCAN) and a comparison with other non-parametric density based algorithm (Kernel Density Estimation) is performed. A test is performed over the urban area of Trieste (Italy).

2 Geographical Data Mining

In recent years geographic data collection devices linked to location-aware technologies such as the global positioning system allow researchers to collect huge amounts of data. Other devices such as cell phones, in-vehicle navigation systems and wireless Internet clients can capture data on individual movement patterns. The process of extracting information and knowledge from these massive geo-referenced databases is known as Geographic Knowledge Discovery (GKD) or Geographical Data Mining.

The nature of geographic entities, relationships and data means that standard Knowledge Discovery in Databases (KDD) or Data Mining techniques are not sufficient ([14]). Specific reasons include the nature of geographic space, the complexity of spatial objects and relationships as well as their transformations over time, the heterogeneous and sometimes ill-structured nature of geo-referenced data, and the nature of geographic knowledge.

Spatial objects are embedded in a continuous space that serves as a measurement framework for all other attributes. This framework generates a wide spectrum of implicit distance, directional, and topological relationships, particularly if the objects are greater than one dimension (such as lines, polygons and volumes). Moreover geographic data often exhibits the properties of spatial dependency and spatial heterogeneity. Spatial dependency is the tendency of observations that are more proximal in geographic space to exhibit greater degrees of similarity or dissimilarity depending on the observed data. Proximity can be defined in highly general terms, including distance, direction and/or topology.

Spatial heterogeneity or the non-stationarity of the process with respect to location is often evident since many geographic processes are local.

Real-time environmental monitoring systems such as intelligent transportation systems and location-based services are generating geo-referenced data in the form of dynamic flows and space-time trajectories.

The complexity of spatial objects and relationships in geo-referenced data, as well as the computationally intensity of many spatial algorithms, means that geographic background knowledge can play an important role in managing the GKD process.

All these aspects have taken to the development of models such as spatial associations rules which show dependency relationships between attributes in a database, spatial clustering algorithms which includes techniques for classifying data objects into similar groups,geographic visualization algorithms which support the different phases of the GKD process such as data preprocessing, the selection of Data Mining tasks and techniques, interpretation and integration with existing knowledge.

3 Spatial Clustering Algorithms

Spatial clustering is the process of grouping a set of objects into clusters so that objects belonging to a cluster have high similarity in comparison to one another but low similarity with respect to objects in others clusters, taking into account spatial characteristics of the data. Spatial clustering algorithm need minimal requirements of domain knowledge to determine the input parameters, can discover clusters of arbitrary shape, have to have good efficiency on large database[13].

Among the different types of clustering algorithms, density-based methods regard clusters as dense regions of objects in the data space which are separated by regions of low density (representing noise). They can be used to filter out noise(outliers) and discover clusters of arbitrary shape.

3.1 DBSCAN

DBSCAN Density-Based Spatial Clustering of Applications with Noise [10]and its extensions (see e.g. [10], [17], [23], [22], [18], [19], etc.) judges the density around the neighborhood of an object to be sufficiently dense if the number of points within a distance *Epscoord* of an object is greater than *MinPts* number of points.

It classifies regions with sufficiently high density into clusters, and discovers clusters of arbitrary shape. The density based notion of clustering states that within each cluster the density of the points has to be significantly higher than the density of points outside the cluster. The algorithm uses two parameters *Epscoord* and *MinPts* to control the density of the cluster.

The neighborhood within a radius *Epscoord* of a point is called *Epscoord - neighborhood* of that point. The *Epscoord - neighborhood* of a point is defined by

$$N_{Epscoord}(p) = \{q \in D \mid dist(p,q) < Epscoord\}. \tag{1}$$

where D is a data set of points.

The distance function determines the shape of the neighborhood. *MinPts* is the minimum number of points that must be contained in the neighborhood of that point in the cluster.

In the following lines we give some definitions.

A point with at least *MinPts* number of points within its *Epscoord - neighborhood* is a *core* point.

A point p is *density reachable* from a point q if there is a chain of points $p_1, ..., p_{n-1}, p_n$ where $p_1 = p$ and $p_n = q$ such that p_i is *direct density reachable* from p_{i+1}.

A point p is *directly density reachable* from a point q if p belongs to the neighborhood of q and q is a *core* point.

A point p is density connected to a point q if there is a point o such that both, p and q are density reachable from o.

The clustering formed from DBSCAN follows the rules below:

1 A point can only belong to a cluster if and only if it lies within the *Epscoord - neighborhood* of some *core* point in the cluster.

2 A *core* point o within the *Epscoord - neighborhood* of another *core* point p must belong to the same cluster as p.

3 A *border* point r within the *Epscoord - neighborhood* of some *core* point $p_1, ..., p_{n-1}, p_n$ must belong to the same cluster to at least one of the *core* points.

4. A *border* point which does not lie within the *Epscoord- neighborhood* of any *core* point is considered to be *noise.*

This algorithm is based on two concepts: density reachability and density connection. Density-based Cluster Definition Let D be a database of points.

A cluster C ,wrt. *Epscoord* and *MinPts*, is a non empty subset of D satisfying the following conditions: 1) For $\forall p, q$: if $p \in C$ and if q is density reachable from p then $q \in C$. 2) For $\forall p, q \in C : p$ is density connected to q.

In order to discover clusters DBSCAN checks the *Epscoord- neighborhood* of each point in the database. If it contains more than *MinPts* a new cluster is created with p as a core point.

Remark 1. Note that both the notion of clusters and the algorithm DBSCAN, apply as well to $2D$ or $3D$ Euclidean space as to some higher dimensional feature space.

4 Kernel Density Estimation

We use Kernel Density Estimation to compare the results obtained using the DBSCAN algorithm over the point dataset. The basic elements of GPS traces were considered, that implying using the original GPS point datasets as the starting point for the analysis. GPS points belonging to the track-log files were used as events in space as the observed locations in a distribution As the simple observation of events in space provides just a first insight on the structure of the distribution, more refined techniques can be used to identify clusters or regularity in the distribution relative to a model, (i.e., complete spatial randomness: CSR. [12], [1]). A Kernel Density Estimation (KDE) can be used to transform point events in space in a continuous density function over a study region considered, creating a three dimensional surface expressing the variation of density of events over space. KDE applications spans from epidemiology to earth science, with more recent applications also in social and human sciences. Recent applications of KDE deals with seismic risk ([7]), crime analysis ([5]) as well as point events over network spaces, in many cases implying also a modification of the kernel searching function to considering the network structure of space as a subset of the uniform, Euclidean space ([3], [2], [15], [21]).

The kernel consists of a family of 'moving three dimensional functions that weight events within its sphere of influence according to their distance from the point at which the intensity is being estimated' ([12]). The general form of a kernel estimator is:

$$\hat{\lambda}(s) = \sum_{i=1}^{n} \left[\frac{1}{\tau^2} k\left(\frac{s - s_i}{\tau}\right) \right] . \tag{2}$$

where $\hat{\lambda}(s)$ is the estimate of the density of the spatial point pattern measured at location s, s_i the observed i^{th} event, $k()$ represents the kernel weighting function and τ is the bandwidth, or searching radius, centred in location s, within which events s_i are counted and will contribute to the density function ([12]). By modifying τ, the main arbitrary variable defining the width of the searching function, it is possible to obtain either a more spiky or smoothed representation of the phenomenon's density. From a more operational point of view, the procedure involve 'discretizing' a continuous study region using a fine grid and therefore considering the cells as approximations of continuous point locations in space. The algorithms computes the distances between each of the reference cells and the event's locations, evaluates the kernel function for each measured distance and sums the results for each reference cell. The result is an estimate in every point of space - or each cell in the study region - of the density of events observed within a distance expressed by the bandwidth and weighted according a distance - decay function from the point of estimate. Cell's density values form a continuous density surface, which 'peaks and valleys' resembles the different patterns drawn by the events' distribution. As bandwidth is the main parameter to be set, one can either define a fixed radius or an adaptive bandwidth ([20], [4]). In this latter case a number of events to be included into the searching function is set, and therefore the bandwidth varies according to the proximity of the events. Most authors emphasize the fact that a bandwidth's choice is more important than choosing the weighting function, as the statistical results are not significantly affected by the various kernel functions ([9]). In this example we consider a quartic function, weighting events closer to the estimate point more than those farther apart.

5 Some Comparisons between DBSCAN and KDE

In this section we consider a comparison between a Density-Based Spatial clustering algorithm (DBSCAN) and a non-parametric density based algorithm

(Kernel Density Estimation). A test is performed over the urban area of Trieste (Italy).

5.1 The Data and the Study Area

The analysis was performed over an activity dairy dataset as registered by individual users relying on a commercial GPS receiver for personal car navigation. The full dataset consists of journeys registering home-work, home-school, work-free time locations (gym, theatres, etc.) collected over a 6 months period (July 2010 - January 2011) on a nearly daily basis and mainly based on car usage. The data were mainly collected in the Province of Trieste (Italy) although travels in the region, abroad or in other Italian cities were recordered.

In Fig. 1 one can see the route.

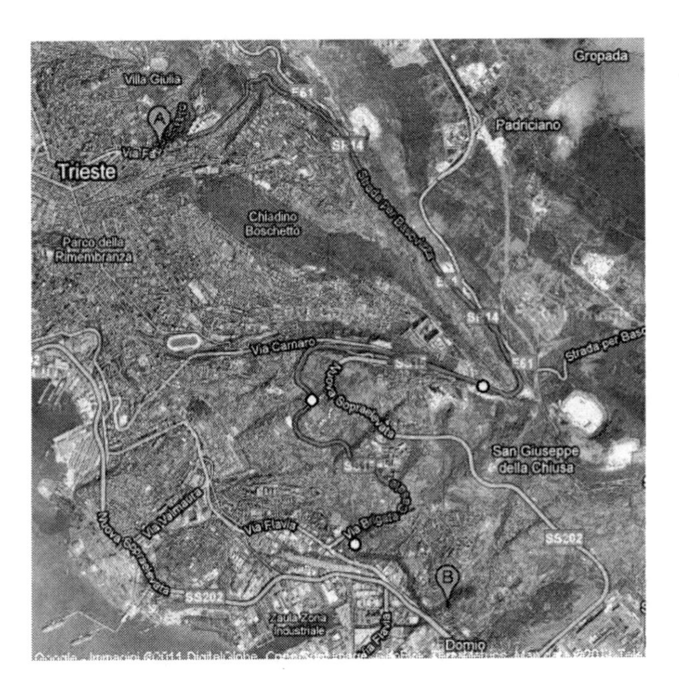

Fig. 1. The route

In this initial stage of the research, the attention is dedicated to highlighting hot-spots in a linear dataset, as that represented by GPS traces. Hotspots and can be of interest for highlighting braking areas along a route and therefore those critical nodes and junctions in an urban road network characterized by traffic or other forms of congestion. Also, the aim is that of allowing the realization of maps of fluidity or accessibility given the users travel time in certain areas.

Without focusing the attention on the full database, here we decided to focus on just one route registered on September 1st 2010 and portraying a car user travelling from home address to child's school. Data were collected by a receiver and converted into a standard *.gpx* format, readable by most of the commercial and free GIS packages and then converted into a standard GIS format. The data were then opened and visualized for a first visual analysis of the phenomenon.

A single file consisting of 1 059 point events, these being generally spaced of 1 second in time, of a home - school journey for a total length of 13 575 m was used. Such uniformity and regularity in the timetable of recording allow us to map interestingly places of slower traffic, because of congestion or simple presence of other cars. The file used here was also chosen as it presented different streets and road environment over which a wealth of situations can be found: urban street network, extra urban ones, urban motorways, state roads, important junctions between motorways and other streets. That allowed to experiment in one journey several situations where a point cluster can found that implying a slowing down of a car when crossing a road or at junctions or in the traffic.

5.2 The Analysis

The analysis on the dataset was performed both on the DBSCAN algorithm and on KDE.

As known the input parameters for the DBSCAN are the radius *Epscoord* and *MinPts* that is the minimum number of points that must be contained in the neighborhood of that point in the cluster. In the application *Epscoord*=0.0004 as the bandwidth is 44 m. while *MinPts*= 15 in order to have meaningful clusters.

The analysis was also performed using a quartic kernel function, with a 5 m. cell resolution and a 44 m. bandwidth. Tests with a narrower width band of 25 m. as well as with a wider one at 100 one were also done. The former (25 m.) proved to be interesting as well as the 44 m. one, mainly in highlighting hotspots, while the latter (100 m) over smoothed too much the data therefore diluting the phenomenon over an excessive extension of the searching algorithm. The 44 m. bandwidth was considered as an intermediate measure, furthermore comparable with the DBSCAN algorithm.

In this stage of the research we did not consider applying a network-based density estimation, as in the literature cited, that would probably allow more refined results, as a route can be considered as a subset of space, in this case belonging to a network space. However, we realized that given the thickness and proximity of events along the route and the choice of a narrow bandwidth to highlight hotspots, a standard, Euclidean kernel density estimation function would have suited the analysis, particularly at the present, initial stage of our research. Furthermore, our aim consisted of a direct comparison of the two algorithms as the DBSCAN and KDE over a linear route, and therefore the searching function should have been changed in both algorithms for a more precise comparison.

In Tab.1 are reported some summary results regarding the application of DBSCAN on the route registered on September 1st 2010 and portraying a car user travelling from home address to school. The total number of points are 1 059. As one can see there have been some slow-downs which form the six clusters described below.

Table 1. Route registered on September 1st 2010 (DBSCAN)

	border	clus. 1	clus. 2	clus. 3	clus. 4	clus. 5	clus. 6
seed	0	27	14	34	23	6	19
border	884	8	7	5	11	13	8
total	884	35	21	39	34	19	27

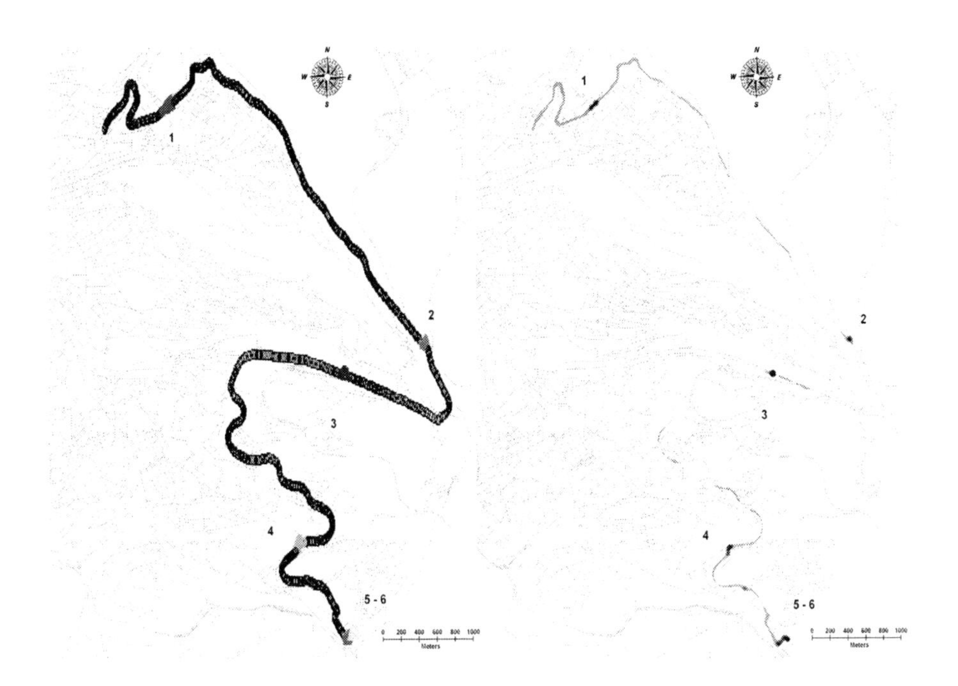

Fig. 2. Results from DBSCAN (right)and KDE (left)

5.3 The Results

The results obtained through DBSCAN are displayed in Fig. 2 (right) while those obtained through the KDE algorithm are displayed in Fig. 2 (left).

The results from the DBSCAN are six clusters visible from the numbers. They are represented with different colors: cluster 1 with red, cluster 2 green, cluster 3 blue, cluster 4 light blue, cluster 5 pink and cluster 6 yellow.

The results from the density analysis are visible as darker areas from light grey to black in the map. Clusters here visible as hotspots can be seen as black areas, while denser dark grey areas can be interpreted as slower segments of the journey. In this sense the results from the KDE analysis can suggest twofold comments.

On one side we can draw a more direct comparison between the KDE hotspots and the DBSCAN clusters. Fig. 2 portrays the results from the application of both algorithms. Clusters are numbered from 1 to 6 starting from the north of the map, with Cluster 1 being close to the starting point of the journey and 6 being

Fig. 3. Cluster 1

located at the end of the journey. The journey follows approximately a North-east heading in its first part, followed by a South-east one, this latter comprehending some wide bends as deviations from the main bearing. The elongated Cluster 1 in DBSCAN (left) is visible as a double-peaked hotspot in KDE (right) using the same searching radius. In the journey the area is located in proximity of the main bus stop close to the Universityś main building and driven at a peak hour time (before classes start) and therefore a stop and go can be justified by the presence of people hoping off busses and crossing the road to reach the Universityś buildings (Fig. 3).

Clusters 2, 3 and 4 are related to junctions between major roads and therefore areas where speed decreases because of the need to give way to other cars travelling in other directions. As in Cluster 4, the shape in KDE is elongated as a double junction is met and therefore a driver is forced to slow down twice in crossing different routes. A minor hotspot is visible between Clusters 4 and 5, not highlighted by DBSCAN, located at a minor junction where a car stop is not always necessary but just a slowing down of the car. Clusters 5 and 6 are located in the final part of the journey and to the stopping area. Therefore they resemble slowing down when approaching the final junctions that lead to the parking area (Fig. 4).

On the other side we can say something about non-peak areas deriving from KDE analysis, as the dark grey areas along the journey. These can be interpreted as slow parts of the journey when compared to the rest of it. In fact they can noticed in close proximity of the hotspots, when speed is slower than in the normal

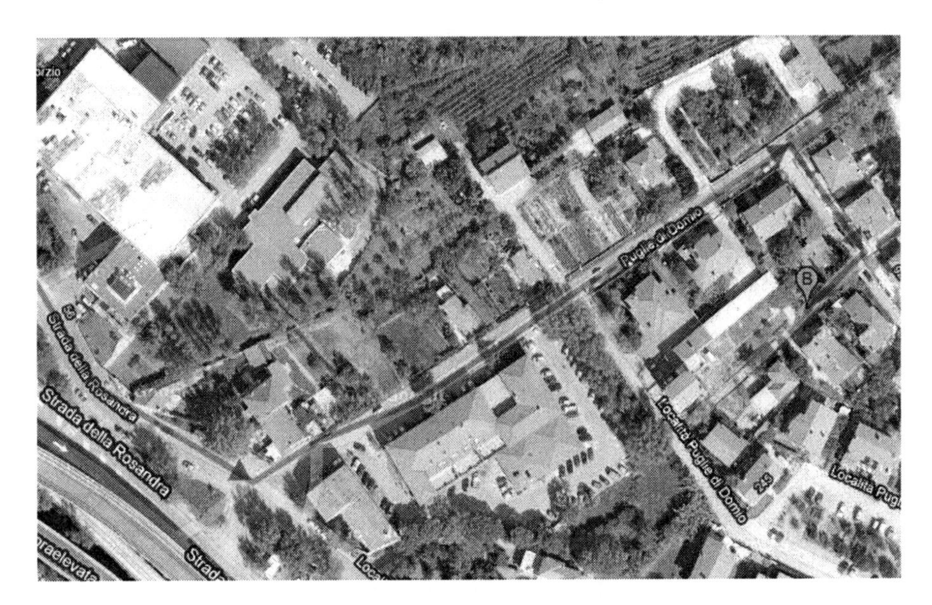

Fig. 4. Clusters 5 and 6

journey, and also in other parts of the route, in urbanized areas (beginning and ending of the journey) as well as in out of town road segments in correspondence to major bends.

6 Conclusions

In this paper the attention is drawn on point datasets corresponding to GPS traces driven along a same route in different days. Our aim here is to explore the presence of clusters along the route, trying to understand the origins and motivations behind that in order to better understand the road network structure in terms of 'dense' spaces along the network, these representing road congestion, rather than the presence of junctions or other factors affecting individual mobility. The attention is therefore focused on methods to highlight such clusters and see their impact on the network structure. Spatial clustering algorithms are examined (DBSCAN) and a comparison with other non-parametric density based algorithm (Kernel Density Estimation) is performed. A test is performed over the urban area of Trieste (Italy). The results from the application of both the DBSCAN algorithm and the KDE one, using the same searching radius,are comparable. They show the hotspots in the same places when the speed is slower than in the normal journey, in urbanized areas (beginning and ending of the journey) as well as in out of town road segments in correspondence to major bends. This may encouraged in go on in this conjoint use of the two approaches.

7 Further Research Directions

This paper represents a first analysis on GPS-collected personal navigation data, therefore focusing on activity dairy data and human behavior through space and time. The database collected so far is quite huge and concerns personal movements by car in a wealth of origins and destinations in an urban environment and involving different individuals. The data collected span from same journeys driven in different days and in same hours, to same journeys in different times of a day and also different journeys in a same timeframe. Such variety of data could allow, in our opinion, a wealth of analysis to be performed in order to understand pattern of movements of a set of individuals in a same day, as well as differences in 'city uses' in different times of the day, and therefore highlight a city's 'variable geometry' in time and space. Also, the presence of hotspots replicated on different journeys can allow highlighting the different speeds in an urban environment and also their directions, therefore allowing to understand urban metrics and shapes in a different way than that usually allowed by more simplified data sets. Such opportunities are just examples of the possibilities allowed by a different availability of geographical data, 'unofficially' collected but based on real journeys and uses of a city. Point pattern analysis in its various declinations, and particularly those based on network constraints of space, seem to be logical evolutions of the research carried on so far. Therefore an application of modified DBSCAN and KDE in a network environment represents the next step that will be added to this line research, as well as the consideration of more differentiated journeys in the urban area.

Acknowledgments. GIS analysis was performed using Intergraph GeoMedia Professional and GRID 6.1 under the Registered Research Laboratory (RRL) agreement between Intergraph and the Department of Education Sciences and Cultural Processes of the University of Trieste.

References

1. Bailey, T.C., Gatrell, A.C.: Interactive Spatial Data Analysis. Addison Wesley Longman, Edinburgh (1996)
2. Borruso, G.: Network Density Estimation: Analysis of Point Patterns over a Network. In: Gervasi, O., Gavrilova, M.L., Kumar, V., Laganá, A., Lee, H.P., Mun, Y., Taniar, D., Tan, C.J.K., et al. (eds.) ICCSA 2005. LNCS, vol. 3482, pp. 126–132. Springer, Heidelberg (2005)
3. Borruso, G.: Network Density Estimation: A GIS Approach for Analysing Point Patterns in a Network Space. Transactions in GIS 12(3), 377–402 (2008)
4. Brunsdon, C.: Analysis of Univariate Census Data. In: Openshaw, S. (ed.) Census Users Handbook. GeoInformation International, Cambridge, pp. 213–238 (1995)
5. Chainey, S., Reid, S., Stuart, N.: When is a hotspot a hotspot? A procedure for creating statistically robust hotspot maps of crime. In: Kidner, D., Higgs, G., White, S. (eds.) Socio-Economic Applications of Geographic Information Science Innovations in GIS 9, Taylor & Francis, Abington (2002)
6. Cressie, N.A.C.: Statistics for spatial data. John Wiley & Sons, London (1993)

7. Danese, M., Lazzari, M., Murgante, B.: Kernel Density Estimation Methods for a Geostatistical Approach in Seismic Risk Analysis: the Case Study of Potenza Hilltop Town (southern Italy). In: Gervasi, O., Murgante, B., Laganà, A., Taniar, D., Mun, Y., Gavrilova, M.L. (eds.) ICCSA 2008, Part I. LNCS, vol. 5072, pp. 415–429. Springer, Heidelberg (2008)
8. El-Sonbaty, Y., Ismail, M.A., Farouk, M.: An Efficient Density-Based Clustering Algorithm for large Databases. In: Proceedings of the 16th IEEE International Conference on Tods with Artificial Intelligence, ICTAI (2004)
9. Epanechnikov, V.A.: Nonparametric estimation of a multivariate probability density. Theory of probability and its applications 14, 153–158 (1969)
10. Ester, M., Kriegel, H.P., Sander, J., Xiaowei, X.: A Density-Based Algorithm for Discovering Clusters in Large Spatial Databases with Noise. In: Proceeding of the 2nd International Confererence on Knowledge Discovery and Data Mining, pp. 94–99 (1996)
11. Gatrell, A., Bailey, T., Diggle, P., Rowlingson, B.: Spatial Point Pattern Analysis and its Application in Geographical Epidemiology. Transactions of the Institute of British Geographers 2, 1256–1274 (1996)
12. Gatrell, A.: Density Estimation and the Visualisation of Point Patterns. In: Hearnshaw, H.M., Unwin, D. (eds.) Visualisation in Geographical Information Systems, pp. 65–75. John Wiley, Chichester & Sons (1994)
13. Han, J., Kamber, M., Tung, A.K.H.: Spatial Clutering Methods in Data Mining: A Survey (2001), ftp://ftp.fas.sfu.ca/pub/cs/han/pdf/gkdbk01.pdf
14. Koperski, K., Han, J., Adhikary, J.: Mining Knowledge in Geographical Data (1998), ftp://ftp.fas.sfu.ca/pubcs/han/pdf/geo_survey98.pdf
15. Okabe, A., Satoh, T.: Uniform Transformation for Points Pattern Analysis on a Non-uniform Network. Journal of Geographical Systems 8, 25–37 (2006)
16. O'Sullivan, D., Unwin, P.J.: Geographic Information Analysis. Wiley, Chichester (2003)
17. Sander, J., Ester, M., Kriegel, H.P., Xiaowei, X.: Density-Based Clustering in Spatial Databases: The Algorithm GDBSCAN and its applications (1999), http://www.dbs.informatik.uni-muenchen.de/Publikationen/
18. Schoier, G., Bato, B.: A modification of the DBSCAN Algorithm in a Spatial Data Mining Approach. In: Meeting of the Classification and Data Analysis Group of the SIS: CLADAG 2007, Macerata, pp. 395–398 (2007)
19. Schoier, G., Borruso, G.: A Clustering Method for Large Spatial Databases. In: Laganá, A., Gavrilova, M.L., Kumar, V., Mun, Y., Tan, C.J.K., Gervasi, O. (eds.) ICCSA 2004. LNCS, vol. 3044, pp. 1089–1095. Springer, Heidelberg (2004)
20. Silverman, B.W.: Density Estimation for Statistics and Data Analysis. Chapman Hall, London (1986)
21. Yamada, I., Thill, J.: Comparison of Planar and Network K-functions in Traffic Accident analysis. Journal of Transport Geography 12, 149–158 (2004)
22. Yu, X., Zhou, D., Zhou, Y.: A New Clustering Algorithm Based on Distance and Density, pp. 1016–1021. IEEE, Los Alamitos (2005)
23. Yue, S., Li, P., Gou, J., Zhou, S.: Using greedy Algorithm: DBSCAN revisited II. Journal of Zhejiang University SCIENCE, 1404–1412 (2004)

How Measure Estate Value in the Big Apple? Walking along Districts of Manhattan

Carmelo M. Torre

Department of Architecture and Urban Planning, Polytechnic of Bari,
Via Orabona 4, 70125 Bari, Italy
torre@poliba.it

Abstract. The paper explains a procedure to discover the coherence of the relationship between physical distance and real estate value variation.

Many author consider (both in the past and in the recent time) the possibility that real estate value can depend on distance from some central point.

Such convintion lead to the use of geostatistical approaches based on kriging techniques. In the same time literature teach that the market shows a higher value where several amenities are coexisting.

But in those urban realities where the number of central points and the number of amenities is high, the complexity does not support the construction of models, and this complexity leads to a different concept of identity as synthesis of distance, borders and concentration.

The use of fuzzy cluster can support the analysis. The paper gives a brief example about how this works in the case of New York core.

Keywords: Metropolis, Real estate market, Fuzzy clustering, Urban Identity.

1 Introduction

Real estate appraisal has founded its main approach on multiple regression analysis for a long time. Any kind of parameter has been investigated in the main urban reality of the World.

Furthermore, property value represents a major indicator of quality of life and services. In the Seventies, the concept of hedonic pricing has been pointed, to define the relationship between the presence of the so-called *Amenities* (environment, urban services, cultural heritage) and the level of housing estate prices [1].

In the recent years, anyway, a new rise on the fore of the relationship distance-estate value, put the attention on *kriging* techniques to make real estate value varying by the distance [2].

Therefore, we can doubtless think that it is possible, and it is legitimized by scientific literature the search for a model based on distance and/or based on contiguity of settlements to some centrality.

But some limits of such models are identifiable with the aspects that will be reminded as it follows.

The first is that in urban, complex, realities, distance can be measured towards/from a number of reference points, all potentially affecting real estate value.

B. Murgante et al. (Eds.): ICCSA 2011, Part I, LNCS 6782, pp. 466–476, 2011.

The second limitation regards the co-presence of many central amenities; that makes difficult to identify the contribution of each one of the same amenities to the variation of value. In simple words the social complexity affects the value with a non-linear rule.

Respect this way of reasoning we can consider a good proof the metropolitan reality of New York City; its reality represents the conjunction between an the old metropolis of the Thirty's and a the new one after 2001, the coexistence of Harlem with Chinatown and the cross road between the Fifth avenue and the Broadway so well described by Luis Mumford [3].

The less fascinating, but considerably relevant aspect is that the variability of the urban context in New York is surely not gradually changing.

A sudden difference can arise between one avenue to another, in the high density, in the dynamic of buildings' substitution, decay, restoring, in a high density tissue; some buildings, some public spaces denote the immediate boundary between different neighbours that have a trade-marked identity.

If we refer to the models of above (based on spatial density of amenities and distance), the high density, in the same time, denotes high concentration of possible amenities affecting real estate value; and the existence of physical boundary put on evidence a criticism for the idea of a gradual variation of values.

In other works we have already consider some criticism related with the assumption that given a unique segment of market, *ceteris paribus*, the only variable affecting the real estate value, is the position.

In this paper we assume that it is possible to define with a fuzzy clustering a similarity function, in order to compare the districts inside the quarter of Manhattan, New York, and in order to give a validation/confutation of the toblerian relationship value-distance [4].

In the first part, after a brief description of the area of study, a resume of fuzzy clustering method is given, and more in detail, the FCM set by Bezdeck [5].

In the second part the FCM is utilised for a classification of similarity among the different considered districts, where the measure of dissimilarity/similarity derives from a peculiar definition of fuzzy distance [6].

The results will be discussed, in order to underline that in a complex space, as the way as Manhattan peninsula can be defined, the relationship value-distance is a controversial function, in a fuzzy environment [7].

2 Fuzzy Clustering Methods

2.1 Measuring Values of Housing Property in New York

2.1.1 General Data of the Case of Study

New York City accounts more less 8.400.000 inhabitants. Population is spread in five different boroughs: Manhattan, Brooklyn, The Bronx, Queens and Staten Island.

Manhattan and Bronx are divided from Brooklyn and Queens by the East River, crossed by some of the most famous urban bridges in the world.

Despite to the physical continuity, the character of the city changes fast, especially when pedestrians pass from south Manhattan to Harlem along the Central Park, or from the Greenwich Village to the East side (Figures 1 and 2).

Fig. 1. New York Quarter of Harlem, nearby the north east corner of Central Park

Fig. 2. The core area of Manhattan , nearby the south west corner of Central Park

The city is facing with the greatest estate market crisis of the last two decades. Property prices are now at the same level of fifteen years ago [8].

The refurbishment process shows some stops going around, and it is possible to discover some abandoned building also not so far from the seat of Wall Street.

The high price of buildings even if degraded, is due not only to the fabric, but also to the land; it happens therefore that the process of substitution/renewal, is not immediately starting (Figure 3).

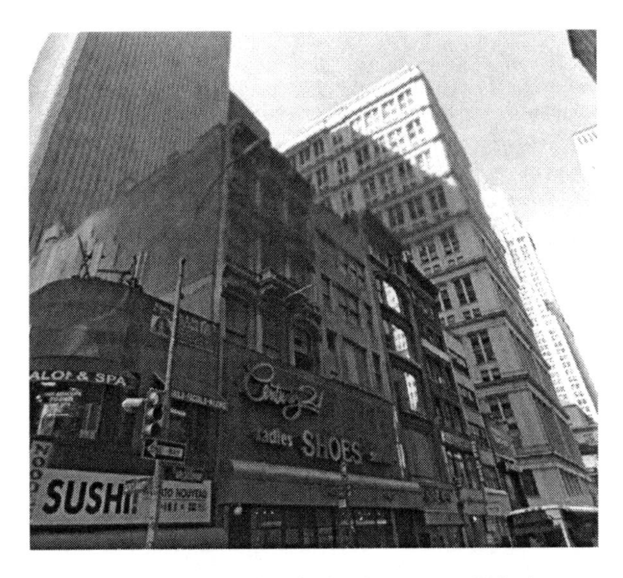

Fig. 3. Degraded buildings in the downtown of Manhattan

All the city is divided in more less sixties administrative contexts, the so called Districts.

Manhattan accounts 12 districts, Brooklyn 18, Queens 13 and Bronx 12. In addition to the previous Quartiers, should be taken into account the Staten Island, subdivided in three Districts that only with some approximation could be defined metropolitan.

The investigation has been carried out by looking at a restricted group of Districts surrounding the core area of Manhattan, adjunct to the facing District of the same Peninsula (Figure 4).

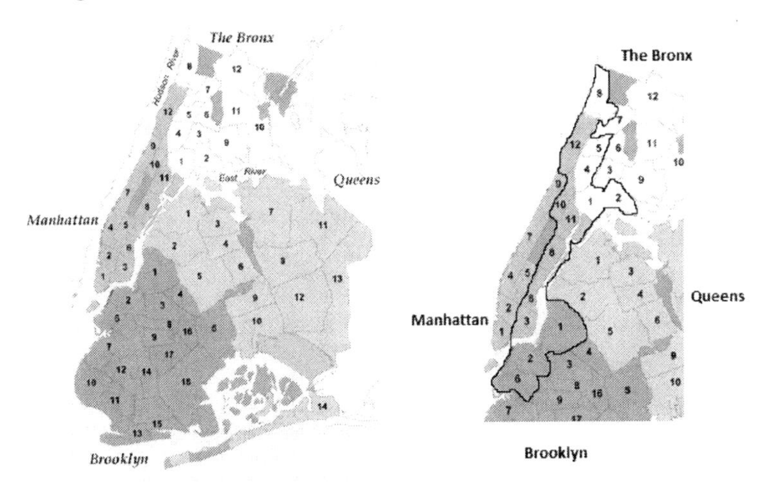

Fig. 4. New York metropolitan area subdivided by Districts (on the left), and the area of study (on the right)

Moving from the southern to the northern area of the Neighbour, all Districts belonging to the Eastern part of Manhattan have been considered, (in the paper identified by the acronyms: Mn3, Mn6, Mn8, Mn10, Mn11, Mn12).

In addition, those district that represent the enclosure of the Manhattan: in the north side a piece of the Bronx's Districts; in the South area Some District of Brooklyn.

The Bronx's Districts (in the paper identified by the acronyms: Bx1, Bx 2, Bx 4, Bx 5, Bx 8) represent a physical area of continuity with Columbia and Harlem.

The considered Brooklyn's Districts (in the paper identified by the acronyms: Bx1, Bx 2, Bx 4, Bx 5, Bx 8) are joining the Skyscrapers' Peninsula by bridges.

All the area of study is posed along a south-north axis, for around seven kilometres.

Table 1 shows the main data regarding price values and rent values of Districts in the case study.

Note that some districts of Brooklyn show a median value of prices which is not dissimilar from the price of Manhattan area.

Table 1. Sale price and monthly gross rent of Housing property. Median values for Districts

	Bx 1+2	Bx 4	Bx 5	Bx 8	Bk 1	Bk 2	Bx 6	Mn 3	Mn 6	Mn 8	Mn 10	Mn 11	Mn 12
Price	343,7	153,1	419,2	318,4	662,5	656,1	941,6	611,7	698,8	1000	696	543,6	380,7
Rent	654	870	890	1023	964	1063	1399	782	1746	1745	720	624	921

Source: Elaboration from New York City Department of City Planning Survey, November 2008

2.2 The Use of Fuzzy Clustering

Every economic condition is characterized by an information set composed by descriptive data, represented by financial metrics (e.g. in our case rent or prices) and geographic areas' that describe the way to uses the descriptive information for describing the spatial variation of estate market.

The assumption that all character of richness of an area can be resumed by the rule high price = high social class or social-spatial identity does not fit with the real world. Since, generally real-world problems are not direct win-lose situations (simple games), but a certain degree of compromise is needed, two main cases can be distinguished [9]:

1) broad commonality of attributes (i.e. the downtown correspond mostly with highest property value and rents)

2) direct conflict of attributes (i.e. controversial between high rents and low incomes, or high prices and low quality of settlements).

The following main assumptions are made:

1) only a set of well defined measures has to be taken into account;

2) the areas are often aggregation of small geographic reference sub-areas too, but it is taken for grant that it is possible to represent their attributes independently from the way they are derived from the sub areas.

On an axiomatic basis, cluster analysis can be distinguished in

a) deterministic,
b) stochastic and
c) fuzzy.

By taking into consideration the "clustering criteria", the following distinction exists [5] [10]:

i. hierarchical methods,
ii. graph theoretic methods,
iii. objective functional methods.

The hierarchical clustering approach FCM, in particular, allows an evolutionary view of the aggregation process and can easily be dealt within fuzzy terms [munda rietveld nijkamp].

Fuzzy c-means (FCM) is a method of clustering which considers possible that a number of data can belong to two or more clusters [ ] [ ]. The approach starts with the minimization of the objective function (1):

$$J_m = \sum_{i=1}^{N}\sum_{j=1}^{C} u_{ij}^{m}\left\|x_i - c_j\right\|^2 . \tag{1}$$

where m is any real number greater than 1, u_{ij} is the degree of ownership of x_i to the cluster j, x_i is the ith of multi-dimensional measured data, c_j is the multi-dimension center of the cluster, and the expression inside || || is any kind of rule forming the similarity between given data and the center.

Lets' given a controversial indicator, a fuzzy cluster algorithm can be used in order to have an idea of the spatial aggregation (minimizing such indicator) that are "possible".

Whit this assumption it is possible to evaluate the similarity among areas when an assessment of their different geo-economic data is given.

By using the semantic distance s described in paragraph 3 as distance indicator, a similarity matrix (achieved by means of the simple rule s=1/1+d) for all possible pairs of areas can be obtained, so that the following clustering procedure is meaningful.

Fuzzy partitioning is carried out for a certain number of steps (let's be k steps) through an iteration of the objective function (above), with the upgrade of membership u_{ij} and the cluster centers c_j by:

$$u_{ij} = \frac{1}{\sum_{k=1}^{C}\left(\frac{\left\|x_i - c_j\right\|}{\left\|x_i - c_k\right\|}\right)^{\frac{2}{m-1}}} \quad \text{and} \quad c_j = \frac{\sum_{i=1}^{N} u_{ij}^{m}\cdot x_i}{\sum_{i=1}^{N} u_{ij}^{m}} \tag{2}$$

This iteration will stop when:

$$\max_{ij}\left\{\left|u_{ij}^{(k+1)} - u_{ij}^{(k)}\right|\right\} < \varepsilon \tag{3}$$

where ε is a test criterion between 0 and 1.

That is to say that after k iterations, if at the k +1 the difference of membership is sufficiently small (less than ε), the process is completed and the value of u_{ij} converges to a local minimum or a saddle point of J_m.

However, in a fuzzy spatial environment a relevant question still survives, i.e. the relation between the concepts of partition and equivalence class.

In our case of study the equivalence class is represented by the number of range of variation of rents and prices, but partition consider the median value.

In a crisp environment, the choice of treatment of data in terms of partitions or equivalence relations is a matter of convenience, since the two models are fully equivalent (philosophically and mathematically).

The probabilistic approach in fact considers equivalent a given distribution (i.e. a symmetric and non-negative Gauss distribution) and its expected value (equal to the median in a Gauss distribution).

On the contrary, fuzzy equivalence relations and partitions are philosophically similar, but their mathematical structures are not isomorphic.

The semantic distance s, try to solve the question as explained in the next paragraph.

2.3 Munda's Semantic Distance Applied in FCM

The utilized fuzzy similarity rule was set by Munda [5]: the difference of value between elements of a cluster is measured on the basis of semantic distance, [11] [12].

Relation 1 shows its formulation.

Let's give a j-th quantitative attribute of a set of two element X and Y; let's suppose that $f_j(X)$ and $g_j(Y)$ represent the value functions of the fuzzy attribute for X and Y.

The function f_j and g_j can be crisp numbers (this means that the function give a certain result), probabilistic values (this means that f_j and g_j represent expected values), or fuzzy numbers (this means that f_j and g_j represent ownership function).

The function has been shaped by the use of data showed in Table 2.

In this last case the "Semantic Distance", is represented by the sum of two double integral:

$$S_d(f_j(x), g_j(y)) = \int_{-\infty}^{+\infty}\int_{X}^{+\infty} |Y - X| g_j(Y), f_j(X) dY dX + \int_{-\infty}^{+\infty}\int_{-\infty}^{X} |X - Y| f_j(X), g_j(Y) dY dX \tag{4}$$

In the clustering the *Similarity index* has been calculated. The index aggregates the group on the basis of two indicator: price value and rent value.

Table 3 shows the matrix of similarity among all couples of districts.

The dendrogram in figure 5 shows the relevant similarities. In the case of study some relevant clusters have been identified as significant of toblerian condition of contiguity of values: the couple of the 11th and 10th District of Manhattan represent a Grey area where Manhattan reality is melt with the Bronx in correspondence with the poorest part of Harlem (similarity equal to 0.89).

Table 2. Variation of Real Estate (Prices and rents) Values for Districts

GROSS Monthly RENT ($)	Bronx 1+2	Bronx 4	Bronx 5	Bronx 8	Brook 1	Brook 2	Brook 6	Manh 3	Manh 6	Manh 8	Manh 10	Manh 11	Manh 12
Less than 200	3083	827	1088	331	953	1019	541	2084	393	538	206	261	1059
200 to 299	5661	2958	248	707	3047	2325	1263	6026	669	1022	4333	6149	2517
300 to 499	6604	3114	2955	121	4275	2384	1368	7532	1181	1225	5799	6366	4204
500 to 749	81	7185	5798	3724	6831	4706	2842	13684	2416	4356	10756	9314	10462
750 to 999	8297	1512	13879	8047	6529	441	3371	8687	3624	506	8827	5541	18583
1000 to 1,499	6701	12067	10923	11902	10619	6217	6459	9475	13348	14822	6912	574	19592
1,500 or more	1577	142	1731	3652	9169	10274	13455	1407	332	45629	4565	4515	6077
Prices ($)	**Bronx 1+2**	**Bronx 4**	**Bronx 5**	**Bronx 8**	**Brook 1**	**Brook 2**	**Brook 6**	**Manh 3**	**Manh 6**	**Manh 8**	**Manh 10**	**Manh 11**	**Manh 12**
Less than 50.000	164	533	144	1301	416	387	46	675	164	333	433	338	245
50.000 to 999.99	66	402	84	499	202	402	50	229	66	270	148	44	113
100.000 to 149.999	302	241	0	872	183	140	73	125	164	43	188	126	229
150.000 to 199.999	191	169	29	1323	338	173	111	102	267	257	52	19	191
200.000 to 299.999	552	205	115	2298	287	1451	555	665	910	1212	444	276	880
300.000 to 499.999	1478	569	931	329	1485	3362	1999	174	5697	5455	854	736	2498
500.000 to 999.999	420	267	379	2558	5598	5583	658	3581	1087	13387	2038	903	1563
1.000.000 or more	31	0	0	1058	1391	4653	8136	1939	7937	22232	2113	815	237

Source: Elaboration from New York City Department of City Planning Survey, November 2008

Table 3. Similarity matrix among Districts

	Bx 1+2	Bx 4	Bx 5	Bx 8	Bk 1	Bk 2	Bx 6	Mn 3	Mn 6	Mn 8	Mn 10	Mn 11	Mn 12
Bx 1+2	1	.6790	.8123	.7731	.8061	.7687	.7279	.8061	.6752	.6439	.8683	.8683	.8123
Bx 4		1	.6993	.6968	.6630	.6611	.6000	.6630	.6416	.5731	.6505	.6505	.6993
Bx 5			1	.8591	.8591	.8496	.7789	.8591	.7789	.7176	.8068	.8068	.8695
Bx 8				1	.8522	.8683	.7882	.8522	.8286	.7450	.7410	.7410	.8618
Bk 1					1	.8683	.8286	.8800	.7882	.7554	.8187	.8187	.8618
Bk 2						1	.8435	.8714	.8435	.7943	.7477	.7477	.8547
Bx 6							1	.8299	.8105	.8540	.7230	.7230	.7796
Mn 3								1	.7882	.7554	.8187	.8187	.8618

Table 3. (*continued*)

Mn 6	1	.8307	.6789	.6789	.7796
Mn 8		1	.6639	.6639	.7172
Mn 10			1	.8919	.8078
Mn 11				1	.8078
Mn 12					1

Another relevant group is represented by some district of Brooklin connected with the corresponding district of Manhattan from one side to another of the east river. These realities represent the richest part of the "skyscrapers area". The districts are Bk1 and Bk2 of Brooklyn and the Mn3 of Manhattan (similarity 0.88).

The richest part of Manhattan (District Mn8) is near of Guggenheim Museum, towards the east side, moving from the Fifth Avenue and Central Park.

Other group look less spatially related. In front of such part we find the richest district of Brooklyn (District Bk6), that is shifted more towards south.

The similarity decreases (0.85).

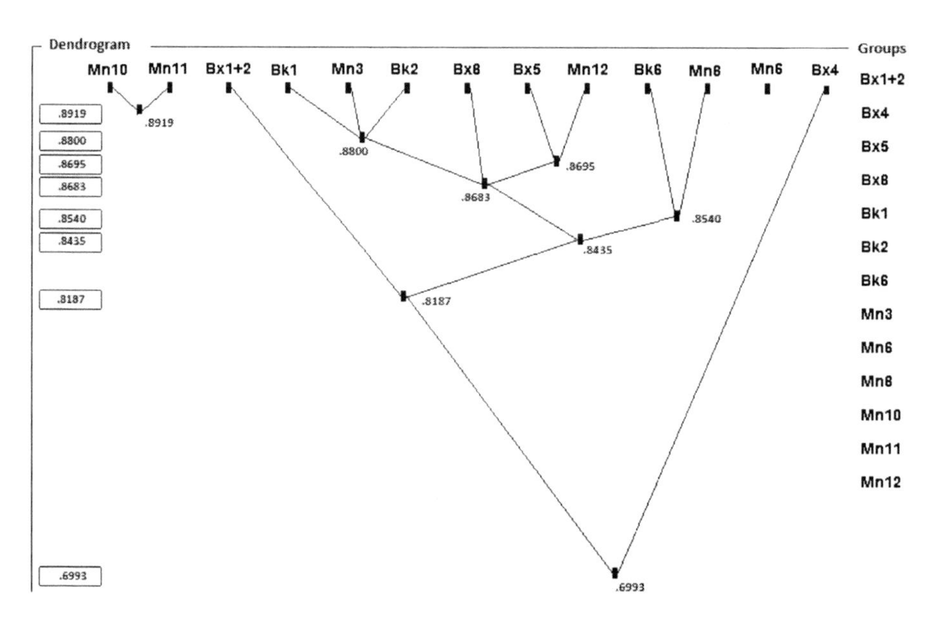

Fig. 5. Dendrogram of similarity for New York Districts

3 Concluding Remarks

The use of fuzzy cluster gives back a idea of relationship between values, and identity of quarters in the New York reality. The idea of contiguity, instead, appear more strongly than the concept of spatial correlation based on distance [13] [14].

In fact the similarity identifies immediately some areas that are on the borderline of neighbors and that show a proper identity (as the case of the poorer Harlem, or the richer South-Est Side).

It is possible to better consider at the local scale, inside each part of the neighbors the influence of distance.

The reality of the city shows a step-up variation that is well identified with the physical/perceptual modern/historical barriers, that still sign the urban tissue.

Even if these part of New York have changed their social character, and face a continuous dynamic substitution of some ethnic group with another, the structural and dimensional character of property seems to survive as evidence of a different economic status of residents in some area.

Such estate values, in its negative and positive peaks, is often corresponding to a “status symbol”, while in the intermediate conditions maintains a character of fuzzyness.

A proof that anyway the toblerian rule (nearly condition are more similar than far ones), is that when distance make sense (i.e. bridges that joins Bronx and Manhattan represents a spatial continuity more than a barrier), the social status and the estate values decrease more sweetly.

Finally the clustering appears useful to put on a better evidence the identicalness of some well characterized areas, evidence to the walker, assign of a social complex value [15]. In that case, and only for identity, distance loses its significance, and the urban history prevails making difference (Chinatown vs core Manhattan or Broadway, Harlem vs the fifth avenue nearby Central Park).

References

1. Rosen, S.: Hedonic Prices and Implicit Markets: Product Differentiation in Pure Competition. Journal of Political Economy 82(1), 34–55 (1974)
2. Torre, C.M., Mariano, C.: Analysis of Fuzzyness in Spatial Variation of Real Estate Market: Some Italian Case Studies. In: Phillips-Wren, G., Jain, L.C., Nakamatsu, K., Howlett, R.J. (eds.) Advances in Intelligent Decision Technologies. Springer, Berlin (2010)
3. Wojtowicz, R. (ed.): Sidewalk Critic: Lewis Mumford’s Writings on New York. Princeton Architectural Press (2000)
4. Tobler, W.R.: A computer movie simulating urban growth in the Detroit region. Economic Geography 46(2), 234–240 (1970)
5. Bezdek, J.C.: Pattern Recognition with Fuzzy Objective Function Algoritms. Plenum Press, New York (1981)
6. Munda, G.: Social Multicriterial evaluation for a sustainable economy. Springer, Berlin (2008)
7. Hui-yong, Y., Gittelsohn, J.: U.S. Home Values to Drop by $1.7 Trillion This Year. Dec 9, 11:08 (2010), http://www.bloomberg.com

8. Zadeh, L.A.: Fuzzy sets. Information and Control 8(3), 338–353 (1965)
9. Hartigan, J.: Clustering algorithms. John Wiley and Sons, West Sussex, England (1975)
10. Dunn, J.C.: A Fuzzy Relative of the ISODATA Process and Its Use in Detecting Compact Well-Separated Clusters. Journal of Cybernetics 3, 32–57 (1973)
11. Takahashi, K., Tango, T.: A flexibly shaped spatial scan statistic for detecting clusters. International Journal of Health Geographics 4, 11–13 (2005)
12. Munda, G., Nijkamp, P., Rietveld, P.: Fuzzy multigroup conflict resolution for environmental management". Working paper, Free University Amsterdam, Faculty of Economics Sciences, Business Administration and Econometrics (1992)
13. Kerry, R., Haining, R.P., Oliver, M.A.: Geostatistical Methods in Geography: Applications in Human Geography. Geographical Analysis 42/1, 5–6 (2010)
14. Miyamoto, S.: Fuzzv sets in information retrieval and cluster analvsis. Kluwer Academic Publishers, Dordrecht (1990)
15. Fusco Girard, L.: The complex social value of the architectural heritage. Icomos Information 1, 19–22 (1986)

Modelling Proximal Space in Urban Cellular Automata

Ivan Blecic, Arnaldo Cecchini, and Giuseppe A. Trunfio

Department of Architecture and Planning - University of Sassari,
Palazzo Pou Salit, Piazza Duomo 6, 07041 Alghero, Italy
{ivan,cecchini,trunfio}@uniss.it
http://www.lampnet.org

Abstract. In the great majority of urban models based on Cellular Automata (CA), the concept of proximity is assumed to reflect two fundamental sources of spatial interaction: (1) accessibility and (2) vicinity in Euclidean sense. While the geographical space defined by the latter clearly has an Euclidean representation, the former, based on the accessibility, does not admit such a regular representation. Very little operational efforts have been undertaken in CA-based urban modelling to investigate and provide a more coherent and cogent treatment of such irregular geometries, which indeed are a fundamental feature of any urban geography. In this paper, we suggest an operational approach – entirely based on cellular automata techniques – to model the complex topology of proximities arising from urban geography, and to entangle such proximity topology with a CA model of spatial interactions.

Keywords: urban cellular automata, land-use dynamics, proximal space, irregular neighbourhood, informational signal propagation, informational field.

1 Introduction

The idea of spatial interaction in cellular automata (CA) models is strongly related to that of proximity. Indeed, a CA transition rule is always, by definition, a function describing the relation between a cell and its neighbouring, *viz.* proximal cells. And the nature of proximity depends fundamentally on the kind of geometry used to describe the underlying geographical space.

When modelling urban dynamics based on spatial interactions, we are always implicitly or explicitly making assumptions about the nature of proximities within an urban geography, and are hence seeking for its suitable geometrical representation.

In the great majority of CA urban models, the "proximities" considered relevant from the theoretical standpoint are related to two sources of spatial interaction: (1) accessibility and (2) vicinity. By latter we mean a *regular* spatial distance, most commonly Euclidean, or more generally a special case of the Minkowski distance.

However, in the case of accessibility the situation is profoundly different. The accessibility refers here to *a measure* of how (how much, how easy, how quickly) places are mutually accessible (i.e. reachable) to human beings. Such accessibility may clearly be further subdivided by means of transportation (pedestrian, bicycle, automobile, heavy vehicle, railway, and so on) as they all give rise to different

B. Murgante et al. (Eds.): ICCSA 2011, Part I, LNCS 6782, pp. 477–491, 2011.

accessibilities of places. Anyhow, in all these cases, the accessibility itself is deeply determined by the relevant underlying urban geography. The web of pedestrian, road, railway and underground transportation networks shape such geography, bringing about a highly *irregular* geometry of the accessibility-type proximity. This type of proximity manifestly does not admit a regular representation, let alone Euclidean. Indeed, an Euclidean representation of the geometry of the accessibility-type proximity would be to a large extent inappropriate, if not fundamentally flawed.

Considering these what seem to us rather straightforward observations, it is quite surprising how little operational efforts (e.g. [1]) have been undertaken in CA-based urban modelling to investigate and provide a more consistent treatment of such irregular geometries, which indeed are an important feature of every urban geography.

This lack of treatment can for instance be easily acknowledged in the two families of urban CA models which in derived, extended, specialised or inspired-by forms have often been proposed for CA-based urban simulation: the so called Constrained Cellular Automata (CCA) [2-4] and the CA models based on the SLEUTH approach [5-7]. In both these families of models, the CA does not employ a strictly local neighbourhoods, and therefore does indeed simulate spatial interactions over greater distances, but the distance is intended exclusively in the Euclidean sense. This same general approach is taken for granted in various attempts to comprehensively present and discuss the theory and applications of urban CA (see for example [8,9]).

Such being the landscape of CA urban applications, to be fair there have been invitations from theoretical standpoints to develop a more appropriate understanding of the concept of nearness, to give it a deeper geographical meaning, in a way, to enrich it with a thicker geographical theory. This line of reasoning may, for instance, almost directly be derived from the notion of *proximal* space, developed within the research on 'cellular geography' [10] which set the basis for the so called *geo-algebra* approach proposed by Takeyama and Coucleis [11]. In that paper, the homogeneity of cells' neighbourhoods has been questioned by Takeyama and Coucleis precisely on the ground that every cell may have a different neighbourhood, defined by its specific relations of "nearness" with other spatial entities, where "nearness" can means both topological relation or a more generic functional influence.

In this paper, we take on the task to suggest an operational approach – entirely based on cellular automata techniques – to model the complex topology of proximities arising from urban geography, and to entangle such proximity topology with a CA model of spatial interactions.

2 Proximal Spaces as Informational Fields

The approach we follow to model the irregular geometry of proximities begins by having each cell emit informational signals propagating throughout the cellular space. However, depending on the signal type, its propagation is not uniform, but is conditioned by the "propagation medium" the signal encounters. This means that, starting from the emitting cell, the signal is diffused in all directions, but the decay of the signal's intensity depends on the state (e.g. land use) of the cells crossed by the signal. As a consequence of such signal propagation, each cell generates an *informational field* around itself, whose shape and intensity at every cell of the cellular space depends on the states of the cells along all the paths the signal propagated through.

To see how these general concepts may relate and be applied to urban context, we can for example think of a model in which the above described signals propagate better (i.e. with a lower rate of decay) along the roads, and that they more easily spill over to the cells surrounding the roads. Another example could be a railway transportation network. Here, the signal would propagate smoothly along the railway, but by model design would not be allowed to spill over to the surrounding cells, except at cells representing railway stations.

To sum up, the beforehand suggested method allows us to generate an irregular geography-based "informational fields" around each cell. In other words, seen from the opposite perspective, every cell receives a set of signals of different type and intensity from other (potentially all) cells. Once generated, the informational fields are used in the CA transition rules, stated in a way to combine the received signals as the input information.

The suggested strategy of modelling proximities by mean of information signals propagating throughout cellular space is similar in spirit to the "at-a-distance interaction fields" proposed in [12]. The specific contribution of our proposal should therefore be seen in its attempt to apply and embed these concepts into specifically *urban* CAs, and to conceive a particular operational simulation approach for that purpose.

3 An Application to a CA Model

The experimental setting for demonstrating the above ideas is a bi-dimensional CA composed of square cells providing a raster representation of a geographical area. The state of every cell represents its land use, of one of the following seven types: *residential, industrial, commercial, undeveloped land, road, railways, railway station.* The latter three types are considered as static and thus cannot change nor be transformed endogenously during the simulation. Starting from a given initial configuration, the automaton evolves in discrete steps simulating the land-use dynamics of the area.

At each simulation step, the execution of the CA model is divided into two distinct phases: (1) *informational fields propagation phase* and (2) *land-use dynamics phase.*

3.1 Informational Fields Propagation Phase

This phase serves to generate the informational fields around each cell. Each signal carries the following information: (1) the ID of the *source* cell, (2) the *land use* type of the source cell, (3) the *propagation rule*, and (4) the signal's intensity. During the subsequent steps, every signal held by a cell is transmitted to its Moore-neighbouring cells, provided that the signal's intensity is above a predefined threshold.

In particular, at the first step of this phase, the cells having residential, industrial and commercial land use are made to emit two informational signals, to account for two relevant modes of spatial interaction discussed in the introduction (see section 1): (1) *vicinity signal* and (2) *accessibility signal.*

The *vicinity signal* propagates regularly, in the sense that the signal decay from cell to cell is uniform in all directions and does not depend on the land uses crossed by the signal. More specifically, each source cell emits a vicinity signal of an initial intensity of σ_{max}. This intensity is subject to a constant decay when passing from one cell to another. In our experimental setting, $\sigma_{max} = 1000$ and the decay constant is 100.

The *accessibility signal* uses an irregular propagation, by which the signal decay from cell to cell depends on the land uses of the cells being crossed by the signal. Again, the source cell emits the signal with an initial intensity of σ_{max}, but the propagation rule, expressed in terms of decay of intensity on cell-to-cell basis, depends on the combination of land uses of both sender and receiver cell. This propagation rule is therefore based on a "sender-receiver land-use matrix". In our experimental setting, we use $\sigma_{max} = 1000$ and in Table 1 we report the values used for the sender-receiver land-use matrix. The decay values in this table reflect the intuitions on the accessibility signal propagation exemplified above in section 2.

Table 1. Sender-receiver land-use matrix used for propagation decay of accessibility information signal (N/A = "propagation not allowed")

Receiver cell → ↓Sender cell	**R**	**C**	**I**	**U**	**Ro**	**Rw**	**RwS**
R (Residential)	100 (1)	100 (1)	100 (1)	100 (1)	50 (3)	N/A (5)	5 (7)
C (Commercial)		100 (1)	100 (1)	100 (1)	50 (3)	N/A (5)	5 (7)
I (Industrial)			100 (1)	100 (1)	50 (3)	N/A (5)	5 (7)
U (Undeveloped)				150 (2)	50 (3)	N/A (5)	5 (7)
Ro (Road)					20 (4)	N/A (5)	5 (7)
Rw (Railway)						5 (6)	5 (7)
RwS (Railway station)							5 (7)

The values in this table reflect some intuitions on the nature of the accessibility signal discussed above in subsection 2. The decay of intensity is rather high (=100) when the signal propagates from one to another of the three dynamic land uses (residential, commercial and industrial) (1). This is to account for the fact that in that case the effective accessibility among places is achieved using local low-speed/capacity road network (not explicitly represented on the map by road cells). The decay of the signal is even greater (=150) if propagating from one to another undeveloped cell (2), for we assume there to be even less road infrastructure. The propagation from dynamic land uses to road cells (3) are suffering somewhat minor decay (=50), while the decay along the road (4) is even smaller (=20), for obvious reasons. The railway station cells are used as access interfaces to railway transportation infrastructure. While the signal propagates with small decay (=5) along the railways (6), from there it cannot propagate directly to the cells surrounding the railways (5: = N/A) but needs to pass through a railway station (7: = 5) in order to be propagated further.

For a visual comparison of the two types of signal and propagation modes, see Fig. 1.

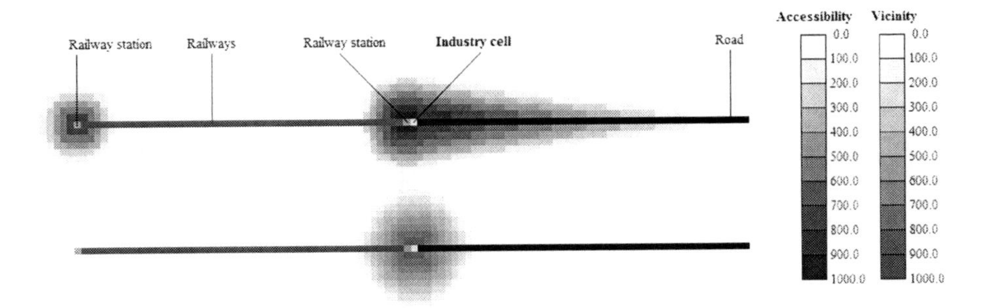

Fig. 1. The propagation of the two types of signal originating from the same cell (industry cell in the middle). *Above*: accessibility signal, with lower decay rate along roads and railways. *Below:* vicinity signal with uniform decay in all direction.

The phase of informational fields propagation ends when no signal needs to be further propagated throughout the CA. This condition is satisfied when every cell has already emitted all its signals and all the intensities of received signals are below a predefined propagation threshold.

3.2 Land-Use Dynamics Phase

At the end of an informational fields propagation phase, every cell has received a series of signals from other cells. This information is used as an input in the land-use dynamics phase.

This phase is based on the computation of the so-called *transition potentials* expressing the propensity of the land to be transformed into possible land uses. In [2-4], where the hereby employed concepts of transitional potentials have been developed, the cell neighbourhood is a circular region of a given radius around the cell. Therefore, we adapted the thereby presented rules to our circumstance of irregular neighbourhood patterns shaped by informational fields. Hopefully, we maintain the spirit of the spatial interaction principles inherent in the original rules.

In the canonical constrained model (see [2]), the transition potential of a cell depends on three factors: physical (geological, orographical) suitability, zoning regulation and the so called neighbourhood effect. Only the neighbourhood effect is driven by intrinsically cellular-automata logic. The former two are, of course, essential for practical application of the constrained cellular automata model in specific urban contexts, for they may pose relevant constraints on the development of a specific urban area. However, given the scope of this paper to make a comparison of constrained models with and without taking proximal space into account, we assume isotropic space and no planning regulation, which make physical suitability and zoning regulation factors unnecessary.

The neighbourhood effect $N_j \in \mathbb{R}$ is the combination of all the relevant attractive and repulsive influences of the neighbouring cells on the land use j, which a cell may assume. In the present model, and critically different from [2-4], N_j is computed using all the informational signals received by the cell:

$$N_j = \sum_{t}^{T} \sum_{i}^{I_t} f_{tj}(\sigma_{ti}) \quad (1)$$

where:

- T is the number of signal types;
- I_t is the number of signals of type t received by the cell;
- σ_{ti} is the intensity of the single signal i of type t.
- and $f_{tj}(\cdot)$ is a function yielding the component of the neighbourhood effect on the land use j due to a signal of type t;

The function $f_{tj}(\cdot)$ may in principle be of various forms. For the sake of simplicity, we use the linear form:

$$f_{tj}(\sigma) = b_{tj}\sigma \quad (2)$$

where b_{tj} is the interaction factor, giving the direction (positive or negative) and the intensity of the influence of a signal of type i on the land use j. In Table 2 we report the values of interaction factors b_{tj} used by our model and provide a brief rationale for those values in the table footnote.

Table 2. Values of interaction factors (b_{tj}) among land uses used in the experiments

Signal type	Land use	R	C	I
R	vicinity	-0.3 (1)	0 (2)	-1.5 (3)
	accessibility	0 (4)	1 (5)	0.5 (6)
C	vicinity	0 (7)	-0.3 (8)	-1 (9)
	accessibility	1 (10)	0 (11)	0.5 (12)
I	vicinity	-1.5 (13)	-1 (14)	0 (15)
	accessibility	0.5 (16)	0.5 (17)	0.2 (18)

We assume that, due to adversity for excessive local crowdedness, the vicinity among residential cells (1) has a small negative effect (b = -0.3) on residential potential. The mutual accessibility among residential cells (4) is assumed to have no direct influence (b=0). Residential potential is insensitive to the vicinity of commercial cells (2: b = 0), but is highly influenced by their accessibility (5: b = 1), due to commercial cells offering retail and services for residential areas as well as providing jobs. These relations between residential and commercial are symmetric (7, 10). The relations among industrial and residential uses are divergent: there is a strong mutual repulsive influence for vicinity (3, 13: b = -1.5) due to pollution, noise and other environmental incompatibilities among the two uses; however, their mutual accessibility (6, 16) does enforce each other's potential (b = 0.5), because the residential cells provide industrial workforce. The effect of vicinity of industry on commercial potential (14) and vice versa (9) is negative due to environmental incompatibilities, somewhat less strong (b = -1) than in the case of residential-industrial repulsion. Nonetheless, commercial potential gains (b=0.5) from greater accessibility of industrial cells (17) and vice versa (12), for it is assumed that the goods manufactured by industries are transported to commercial areas for retail, as well as commercial areas providing services to industrial activities. Finally, there is a slight self-enforcing relation among industrial uses (18: b = 0.2), accounting for possible economies of agglomeration.

It is important to note that, in spite of the simple linear form of functions $f_{tj}(\sigma)$, the combination of all contributions given by Eq. (1) can on the aggregate level reproduce a variety of spatial interaction patterns. An example is given in Fig. 2 where we illustrate the effect of a single industrial cell (see Fig. 1 above for its information fields) on residential potential of all the cells on the map.

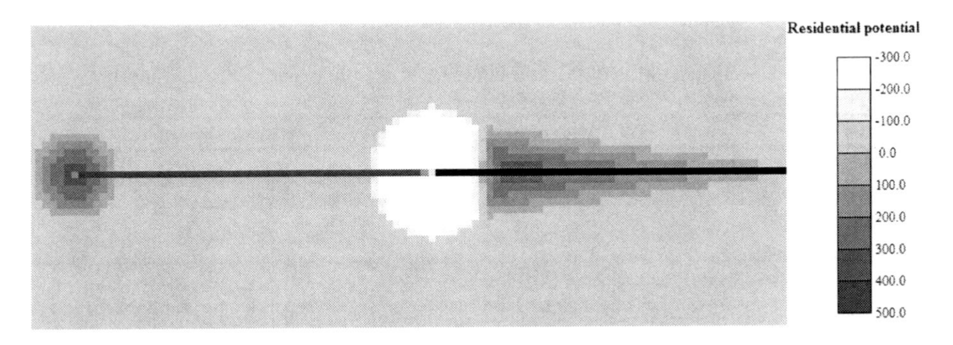

Fig. 2. Residential potential due to an industrial cell (cfr. Fig. 1 above). Nota bene, the colour scale goes from -300 to 500, so white and the two brightest shades represent negative values; the predominant background shade is used for cells with potential 0.

When all transition potential are computed, the following transition rule is applied to all cells of the automaton:

Transition rule: a cell of the land use k is transformed into the land use j, iff P_j is the highest of all the cell's transitions potentials, provided that $P_j - P_k > I_k$, where I_k can be thought of as a kind of inertial of the current land use, a minimum difference threshold for a cell to be transformed into another land use.

The rationale behind this transition rule is straightforward. A cell has a transition potential to be transformed into other dynamic land uses. Of all the possible land uses, it will transform into the one having the highest transition potential, provided that it is strong enough to "overturn" the current land-use k (therefore it must hold $P_j - P_k > I_k$). This latter threshold (I_k) accounts therefore for the inherent inertia or sunk costs of urban transformations.

4 Example Runs

In this section we present few results of a simulation exercise, offering a qualitative comparison of two CA models: one applying concepts of proximal space as shaped by the effective accessibility of places, and the other using conventional neighbourhoods with purely Euclidean distances. For brevity, we call the former the "wave model" and the latter the "classical model".

But beforehand, we want to emphasise that qualitative comparisons of this kind are not meant to, nor could they, deliver a conclusive argument in favour of one or other modelling approach. The primarily illustrative purpose of this exercise is therefore to discuss some implications, in terms of qualitative spatial patterns, of our modelling approach. A richer and more attentive to accessibility a description of the proximal space seems to us theoretically sound and a promising modelling innovation. That notwithstanding, eventually a more decisive case may arise only through a deeper and more rigorous exploration of its potential explicative power and possible empirical corroborations. This is something we are not able to do, here and yet.

The comparative specification of the two models is reported in Table 3.

Table 3. Specification of the two CA models used in experimental setting

Parameters	Wave model	Classical model
Signal intensity at origin (σ_{max})	1000	1000
Cell-to-cell decay of vicinity signal	100	100
Cell-to-cell decay of accessibility signal	As per Table 1 (depends on land uses)	30 (independent on land uses)
Dynamic land uses interaction (b_{tj} in Eq. (2))	As per Table 2.	As per Table 2.
Direct influence of roads and railway stations on dynamic land-use potentials?	No, the influence is indirect, because roads and railway stations are used only to propagate dynamic land-use signals	Yes, the intensity of the signal from *the most accessible* road (if present) and *the most accessible* railway station (if present) are used in the computation of dynamic land-use potentials with following b_{tj}: j=R j=C j=I t=Road 200 200 200 t=Rw. station 50 50 50
Residential inertia ($I_{k=res.}$)	60.000	120.000
Commercial inertia ($I_{k=com.}$)	40.000	90.000
Industrial inertia ($I_{k=ind.}$)	40.000	90.000
Dynamic land uses added at each CA step	40 residential 40 commercial 40 industrial	40 residential 40 commercial 40 industrial

As can be noted in Table 3., the distinctive generative logic of the two models resides in the differences between their respective ways to model accessibility, that is, to make use of the transportation network. In the case of the wave model, the only thing that counts for future transformation potential of cells is their proximity to the existing dynamic uses. Proximity is intended as an interplay of vicinity and accessibility, the latter being shaped in a nontrivial way by available transportation network. Being close to a road or a railway station *is not in itself an independent advantage*, but rather a mean to access other cells. Therefore, the wave model uses transportation networks indirectly but, so to say, in a more natural way.

Instead, in the classical model, the accessibility component of proximity of dynamic land uses is calculated simply as Euclidean distance. The closer in Euclidean sense, the more mutually accessible. When modelled in that way, the accessibility remains uninfluenced by the existing transportation networks. That is the reason why, to model the indisputable importance of transportation networks for urban development, the accessibility of transportation network is made to enter into the equation directly, but in a trivial and unnatural manner. Indeed, in the classical model, being close to a road or a railway station cell *is an indepentent advantage*, no matter how close or far away that road or railway station cell is to other land uses, no matter even if the road connects to somewhere or it is just a road to nowhere.

Cellular automata are complex systems, and their results may be highly variable even with small changes in initial parameters. Our two models are no different in that regard, and this has another important implication for the kind of comparison we are presenting here. We kept many parameters equal between the two models, and tried as much as possible to reconstruct the classical model in a way to make it more comparable to the wave model. Also, for a "fair comparison", we set the parameters of the classical model to have it produce spatial patterns similar to those generated by the wave model. Even had we undertaken and performed to perfection all the necessary steps to make the two models as comparable as possible (which we surely hadn't), the intrinsic underlying complexity is such to invite us to refrain from a "literal" comparison of spatial patterns. What seems to us a more productive attitude is to call attention to distinctive workings of different generative principles in observable patterns. Hopefully, this should shed some light on the modelling potential of our approach and foster some inspiration for further investigation.

In Fig. 3 we present a side-by-side comparison of the two models starting from the simplest of all the test scenarios. A road connects two initial nuclei at both ends (1), each comprising two residential, two commercial and two industrial cells. After 5 simulation steps, in the wave model the two nuclei have expanded (2), while in the classical model the expansion has already reached the middle point of the road (3). This difference, which may not seem too relevant, is however intuitively interpretable keeping in mind the distinct generative logic between the models.

After 10 steps, the new developments in the wave model have reached the "centre" (4). We see the stabilisation of the observed patterns after 20 steps. In the wave model, the two initial areas of growth have maintained their dominance in size (5), which is a plausible result, considering that all the successive development was driven by new land use cells seeking to be accessible to what was already there. The three distinct and quite homogeneous industrial areas have arisen due to repulsive forces between industry on one side, and residential and commercial uses on other side, but also due to self-attractive forces among industry cells. The classical model has produced a different pattern. The two original nuclei have not developed much (7), and the majority of urban mass is clustered around the centre. The reason for this is that during the initial steps, when there were still little cells occupied, the vicinity to the road itself was a weighty factor for transformation potential. That is why, after few initial steps, all the cells around the road get occupied (3) much quicker that in the wave model. Once the cells surrounding the road were occupied, the centre played

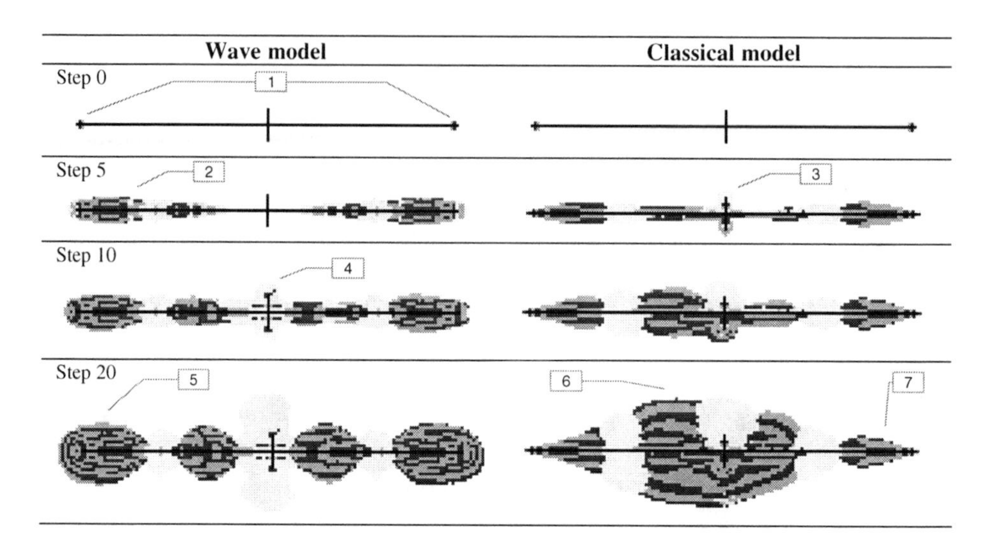

Fig. 3 "Simple linear"

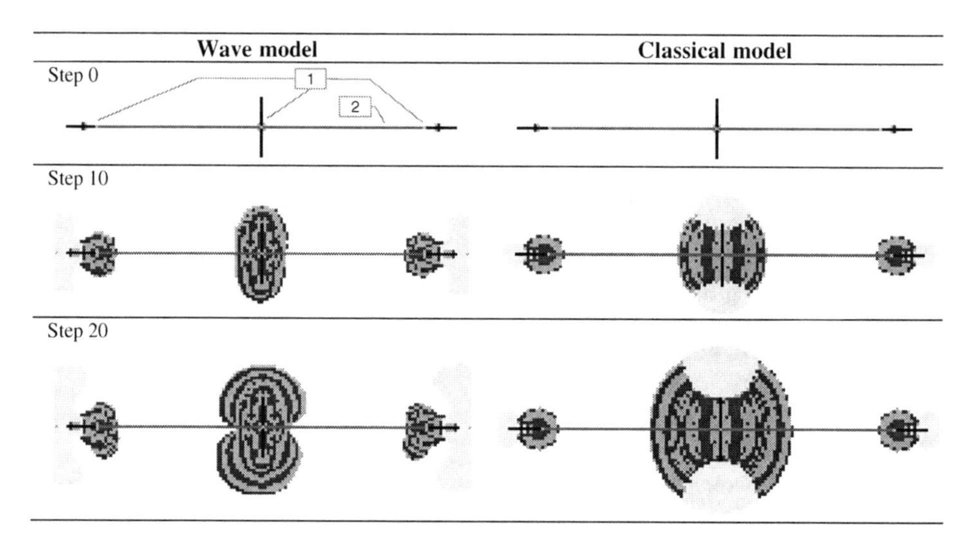

Fig. 4. "Linear with railways"

a dominant attractive role because the cells there are on average closer to other occupied cells, and therefore progressively gained higher transformation potential.

As in the previous Fig. 3, the initial scenario used in Fig. 4 have the same two original nuclei at both ends of the road. However, the centre is connected by a high-speed railway (orange: railway station (1), red: railways (2)). The two patterns after 10 steps bear some resemblances, but the causes for them are different. In the wave model, the centre gets occupied soon because it is highly accessible to the two initial areas of growth. The latter two did not so much develop residential and commercial uses because the industrial developments at the two extremes of the road, formed up in few early steps of the simulation starting from the two initial industry cells, exerted negative influence on local residential and commercial transformation potentials. On the other side, the occupation of the centre in the classical model is due to it being close to road and railway station which played important attractive role, especially during early steps with there being little land use cells to exercise strong enough attractive influence to anchor the future development around the original nuclei.

For both models, we can notice the unfolding of more or less similar pattern trends until the step 20. One relevant observation to be made is that in the classical model from step 10 to 20 the development of the two original areas was nearly arrested, being them unable to withstanding the dominance, and hence attractive force, of the central urban mass. Also, note how in the wave model the industrial agglomerates are better separated from the mixed residential-commercial masses. This is due to their mutual accessibility, an *equally* favourable factor in both models, being accomplished differently than in the classical model. That was then conducive to a different "settlement" of the interplay between the repulsive effect of the vicinity and the attractive effect of the accessibility between industry and residential/commercial uses.

In the scenario in Fig. 5, the connection to the centre is asymmetric, with western arm served by railway and eastern by road. Similar observations as for the fig. 4 can be made. In the wave model, the two original nuclei maintain their dominance, while in the classical model the eastern area remain completely undeveloped and the western nuclei stops growing after step 10.

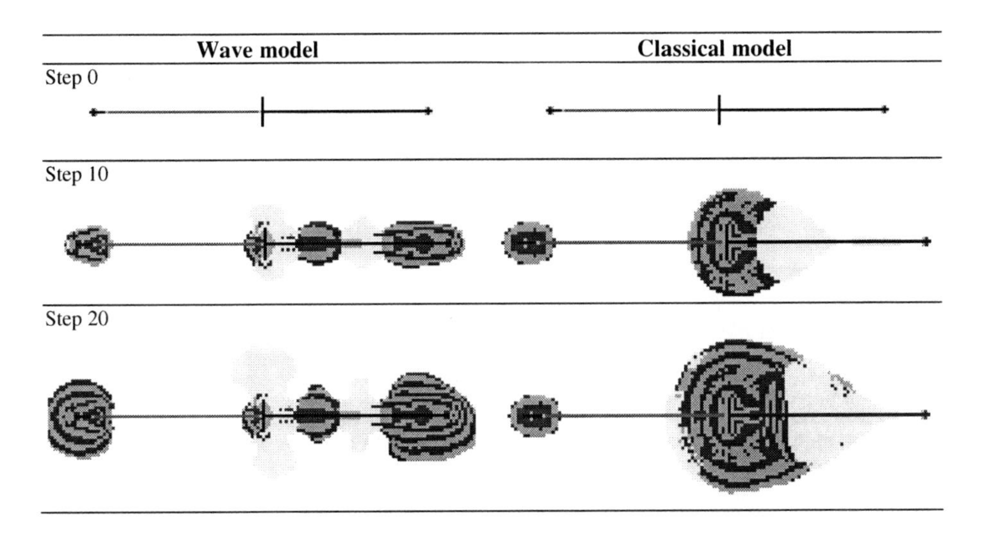

Fig. 5. "Linear asymmetric"

Fig. 6 starts from a more articulated scenario, offering better possibility to appreciate the differences in spatial patterns between the two models. The initial situation is given by 4 residential, 4 commercial and 4 industrial cells located in the centre (1). The evolution of the wave model after 10 steps have produced a strong development in the centre, but also in the north (2). The industrial areas have

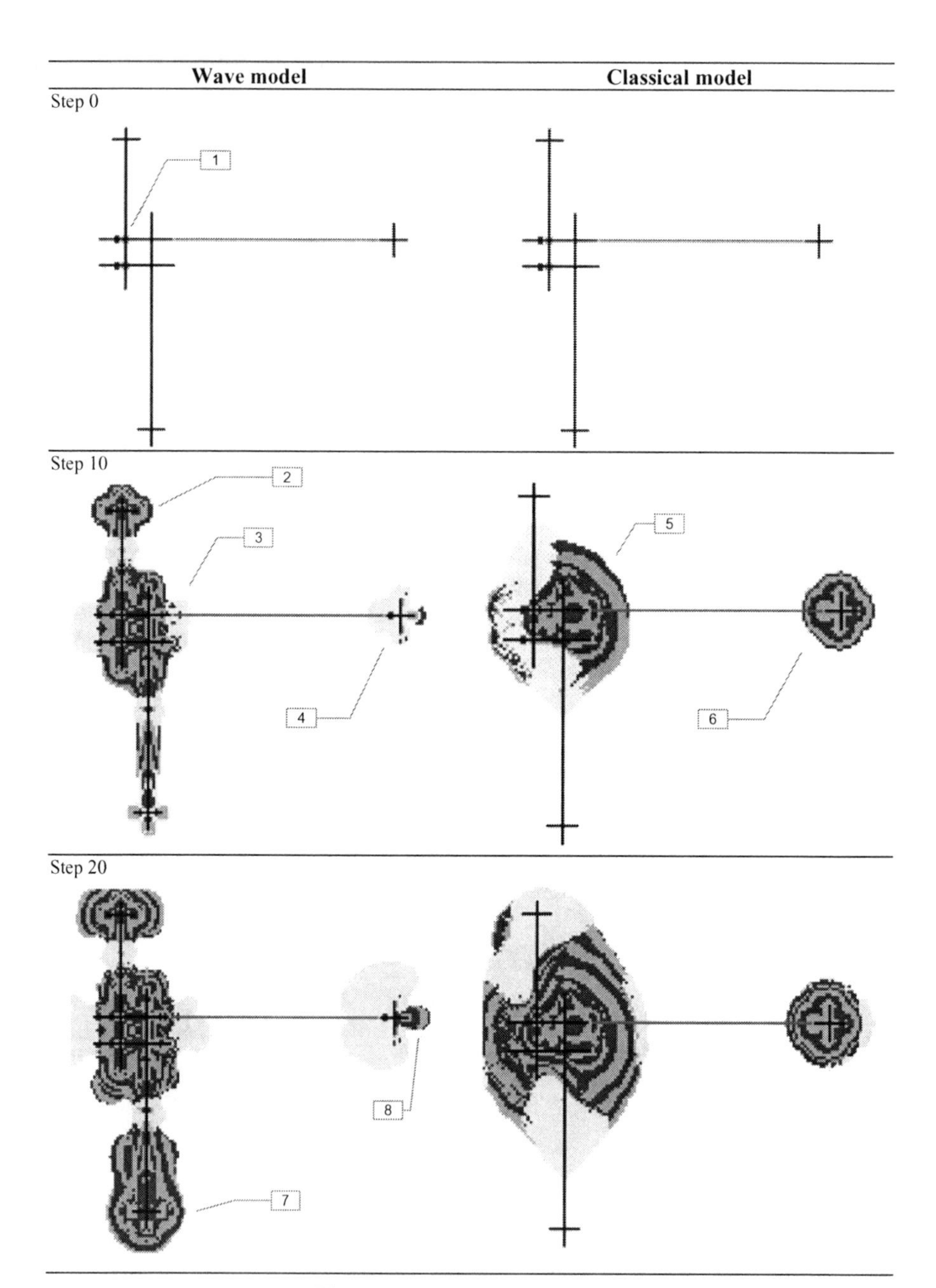

Fig. 6. "Centre and peripheries"

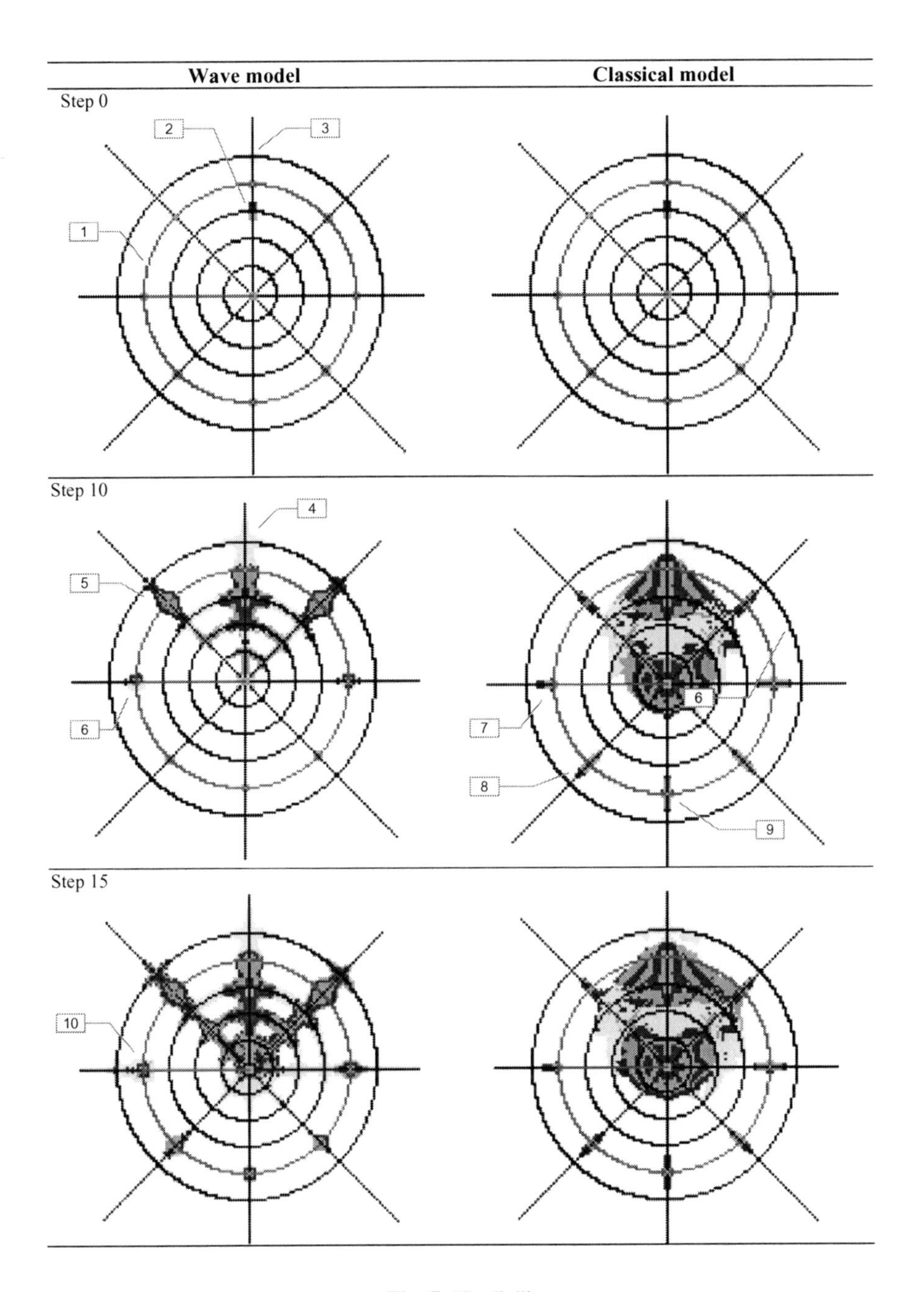
Wave model
Classical model
Step 0
1
2
3
Step 10
4
5
6
6
7
8
9
Step 15
10

Fig. 7. "Radial"

developed mainly at the margins of the central development (e.g. 3) but also in the eastern sector which is well connected to the centre by railway. At the same step 10, the classical model has evolved in a more concentric manner. The large residential/commercial areas in the centre (5) and in the east (6) are due to the fact that at both ends of the railway tracks there are both road and railway station cells, which exercised *per se* a strong attractive force in the early steps of the simulation.

After 20 steps, we see that the classical model has evolved more or less along the same trends observable after 10 steps. On the other hand, the wave model has developed a new clearly distinguishable residential/commercial area at the southern end of the road (7), and a smaller one in the east (8), which has started to crowd out and push the large industrial area more towards the west.

In the last scenario, shown in Fig. 7., we have drawn a set of concentric roads rings and one railway ring (1). The east-west symmetry is broken by the railway connection with the centre available only in the west. The initial land uses were assigned to 4 residential and 4 commercial cells in the centre-north (2) and 4 industrial cells further to the north (3). After step 10, the wave model has developed a large industrial area (4) around the initial industrial zone, and have grown a residential/commercial area in the north-west (5) and an approximately specular area in the north-east. Note also that some development has begun in the west (6) and east near the railway stations, due to good accessibility of the main urban masses. The classical model evolved a significantly different and more compact spatial pattern. The developments around the railway stations along the railway ring were driven not as much by them being near the existing urban mass, but rather for the fact that railway stations and roads exercised an independent attractive force. This fact is recognisable by the approximately equal sizes of the developments around railway stations (7, 8, 9, and their symmetric counterparts) no matter the distance/accessibility from the main existing urban mass. At step 15, we notice in the classical model that the attractive force of the progressively growing predominant mass has arrested the developments around the far-away railway stations. In the wave model, we can notice the beginning of the break of east-west symmetry (10) due to the presence of the eastern railway connection to the centre.

In spite of simplicity and abstractness of the test scenarios, the results of the above presented comparisons indicate that including an improved description of the proximities arising in an urban geography can lead to different, and possibly more realistic, urban patterns.

5 Conclusions

In this paper, our primary aim was to suggest a possible modelling of the notion of proximity and proximal space arising in urban geography, and to employ it in urban CA. The point of departure was the idea that the proximity usually held to be relevant for spatial interactions cannot be assumed to exist in and as a regular Euclidean space, since the "distortions" of urban geography bring about highly irregular and complex topology of proximities. To account for this complexity, we have thereafter suggested a description of the proximal space as a set of informational signals propagating through the space, hence generating informational fields around each cell. Every such field exhibits different strength (intensity) at different points (cells) in space, since the irregularity of its geometry is inherently dependant on the different "propagation media" (road, railways, residential, commercial or any other area) crossed by the informational signals.

A cell is then able to "know" its proximity to another cell by knowing the intensity of the signal – the "radiation", so to say carrying on with the wave metaphor – it receives from that cell. Finally, the combination of all the received radiations is the information input to the CA transition rules through which to model the land-use dynamics.

In the examples presented, our focus was not so much on the plausibility and theoretical foundation of the transition rules and of the proximity-based land-use urban dynamics. Rather, it was a quite expedient (and probably somewhat coarse) exemplification of how even existing models based on assumptions of spatial interaction may in a reasonably convenient manner be adapted to employ our more generalised, and more attentive to actual urban geography, notion and description of the proximal space.

References

1. Batty, M.: Distance in space syntax. CASA Working Papers (80). Centre for Advanced Spatial Analysis (UCL), London, UK (2004)
2. White, R., Engelen, G.: Cellular automata and fractal urban form: A cellular modeling approach to the evolution of urban land-use patterns. Environment and Planning, 1175–1199 (1993)
3. White, R., Engelen, G., Uljee, I.: The use of constrained cellular automata for high-resolution modelling of urban land use dynamics. Environment and Planning B 24, 323–343 (1997)
4. White, R., Engelen, G.: High-resolution integrated modelling of the spatial dynamics of urban and regional systems. Computers, Environment and Urban Systems 28(24), 383–400 (2000)
5. Clarke, K., Hoppen, S., Gaydos, L.: A self-modifying cellular automaton model of historical urbanization in the san francisco bay area. Environment and Planning B 24, 247–261 (1997)
6. Clarke, K.C., Gaydos, L.J.: Loose-coupling a cellular automaton model and GIS: long-term urban growth predictions for San Francisco and Baltimore. International Journal of Geographic Information Science, 699–714 (1998)
7. Project Gigalopolis, NCGIA (2003), `http://www.ncgia.ucsb.edu/projects/gig/`
8. Benenson, I., Torrens, P.M.: Geosimulation: object-based modeling of urban phenomena. Computers, Environment and Urban Systems 28(1-2), 1–8 (2004)
9. Torrens, P.M., Benenson, I.: Geographic Automata Systems. International Journal of Geographical Information Science 19(4), 385–412 (2005)
10. Tobler, W.: Cellular geography, in S. Gale & G. Olsson. In: Philosophy in Geography, pp. 379–386. Reidel, Dordrecht (1979)
11. Takeyama, M., Couclelis, H.: Map dynamics: integrating cellular automata and GIS through Geo-Algebra. Intern. Journ. of Geogr. Inf. Science 11, 73–91 (1997)
12. Bandini, S., Mauri, G., Vizzari, G.: Supporting Action-at-a-distance in Situated Cellular Agents. Fundam. Inform 69(3), 251–271 (2006)

Improvement of Spatial Data Quality Using the Data Conflation

Silvija Stankutė and Hartmut Asche

University of Potsdam, Department of Geography,
Karl-Liebknecht-Strasse 24/25, 14476 Potsdam, Germany
{silvija.stankute,gislab}@uni-potsdam.de
www.geographie.uni-potsdam.de/geoinformatik

Abstract. After the introduction of digital mapping techniques in the 1960s and then GIS shortly afterwards, researchers realized that error and uncertainty in digital spatial data had the potential to cause problems that had not been experienced with paper maps [1]. Spatial data quality is a very active domain in geographic information research. This paper describes the significance of data quality and how is data quality defined. Data conflation can help to increase the amount of suitable for usage data. This paper analyzes results of spatial data conflation.

Keywords: data fusion, data conflation, spatial data quality, edge matching, vector data, homogenization, heterogeneous spatial data, data mining.

1 Introduction

With the introduction of digital mapping techniques in the 1960s and then GIS shortly afterwards, researchers realized that error and uncertainty in digital spatial data had the potential to cause problems that had not been experienced with paper maps. An international trend started in the early-1980s to design and implement data transfer standards which would include data quality information that had disappeared from the margins of paper maps with the transformation to digital data products. The standard that clearly led the way in documenting data quality was the U.S. Spatial Data Transfer Standard [1]. Spatial data quality research results to international conventions and standards for the description of spatial data (e.g. ISO 19107, ISO 19111, ISO 19112, ISO 19115, ISO 19118, ISO 19119, ISO 19126).

Spatial data sets are used in different branches of industry or academic fields, e.g. landscape planning, environmental protection, traffic, urbanism, economy, tourism et cetera [2]. They can now be downloaded on the Internet (e.g. OpenStreetMap). Geographical data are represented at various levels of abstraction [3]. It could be vague, inaccurate or incomplete. These divergences between reality and representation could be acceptable within the framework of certain applications [4].

The main intention of this paper is to present the data conflation as one of the options for improvement of spatial data quality.

B. Murgante et al. (Eds.): ICCSA 2011, Part I, LNCS 6782, pp. 492–500, 2011.

2 Problems of Data Quality

Many organizations create data for the best applicability for their own special applications. Producers of spatial data have developed different methods for data acquisition. This leads to the accruement of different data types, data formats and different semantic information. The acquisition of spatial data for the particular goals results in multiplicity of spatial data. However these spatial data cannot be circulated and used multiply. Also, they can result in inefficient solutions. For this reason the integration or conflation of spatial data is very expedient. A conflation allows the use of already available spatial data and improvement of its quality.

Generating integrated and semantically accurate spatial dataset demands the correction of the data. The input datasets have different accuracy, because they are collected in different ways in different time-frames and independent of each other. The data can also have large differences in topicality. Depending on the size of the data correction, one can make the statement about the quality of the data.

2.1 Indicators of Spatial Data Quality

In a number of fields, the approach to quality evolved into a definition based on *fitness-for-use* [5]. ISO 8402 defines the quality as the "'totality of characteristics of a product that bear on its ability to satisfy stated or implied needs"'. This means that to define the quality two informations are needed: the information on the data being used and on the users needs [6]. Spatial data is defined to be *fitness-for-use* if it meets requirements of the target application [7].

Data quality is defined by one or more quality dimensions [5]. Quality dimensions for geographic data are called spatial data quality elements. They include completeness, logical consistency, positional accuracy, temporal accuracy (the accuracy of reporting time associated with the data [1]) and thematical/semantical or attribute accuracy [7]. Typically, metadata for spatial data include descriptions of data quality and include information about these elements [8].

Spatial data are different in the data aquisitionacquisition methods. Depending on the data aquisitionacquisition method a spatial data will can be classified into two groups: original data and derived data. The original spatial data will beis collected by means of the direct aquisitionacquisition of the object. The derivative or secondary data are arisedarises in a sources, which are as athe result of already existent data with scale and theme [9].

2.2 Multiple Representations

In order to transmit the real world into the language understandable for the computer, it should be modeled according to specific rules in a simplified form. Such data models represent the objects of reality as points, lines or areas (polygons). Each of these objects is are provided with the x-and y-coordinates and contains an information on the spatial reference.

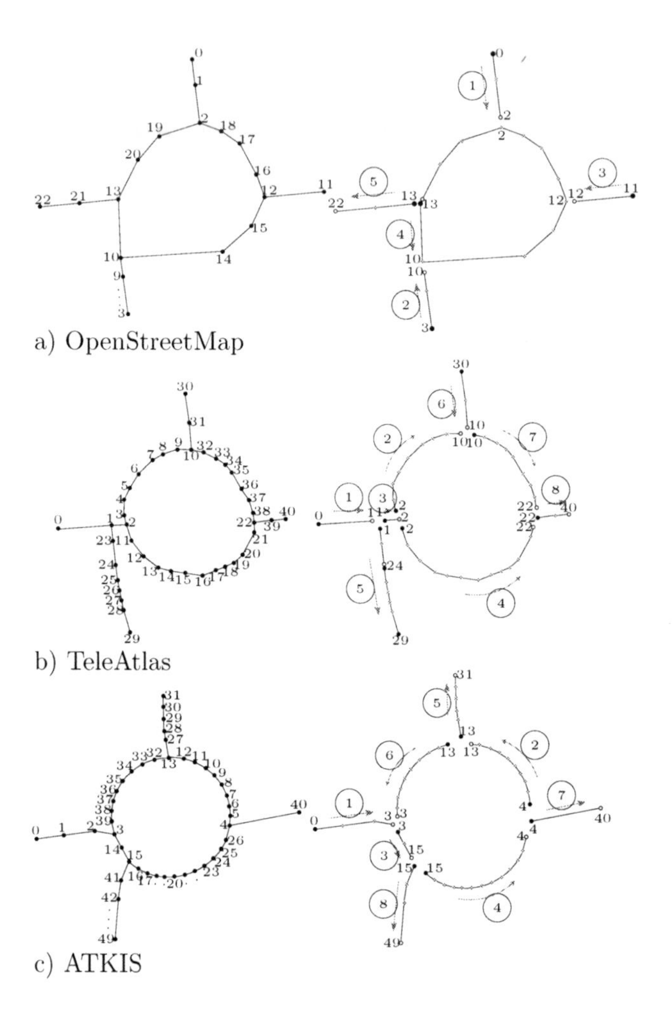

Fig. 1. Spatial data formats of the same object created by different providers

In other words, the spatial reference is the description of a place or object. This description may have a direct or an indirect form. The direct description of the object is a coordinate or address. The indirect form describes the position of an object relative to other objects [10].

The different producers of spatial data detected depict the same object of the real world differently. There are no uniform rules for acquisition of spatial data. According to this the different abstract representations of one and the same object of the real world may arise [11]. Figure 2 shows an example of alternative geometric representations of the same real world object. Each representation was generated by different spatial data providers.

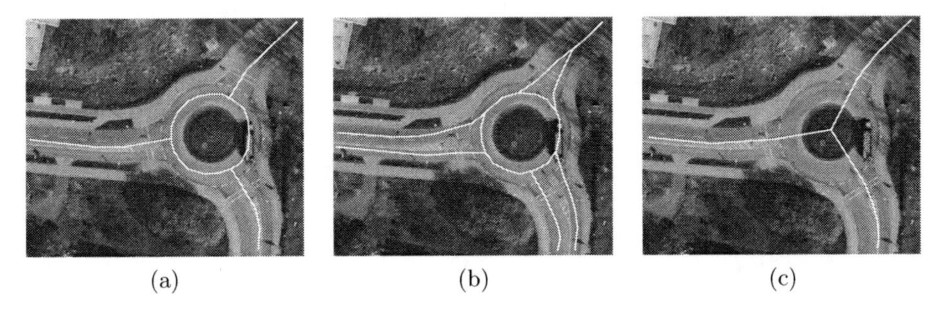

(a) (b) (c)

Fig. 2. Multiple representations of the same real world object

Another example shows the differences of dataformats of the same object (see Fig. 1). This fact makes the data conflation complex. As a result of this many non-trivial problems arise that require a more complex integration process. It makes the implementation of specific preprocessing necessary.

3 Improved Data Quality

During conflation process the following quality elements of spatial datasets should be improved: completeness, logical consistency, positional/geometrical accuracy, temporal accuracy, attribute accuracy. The difference between the quantity of real world objects in a given area and the quantity of available objects in the dataset is called the completeness of this dataset. In principle, the geometrical and thematic accuracy together provides information about completeness of a dataset.

During the conflation process information from the source input dataset (*SDS*) and the target input dataset (*TDS*) have to be assigned to each other. The *SDS* is defined as the dataset from where the geospatial information is taken (e.g. thematic information) and the *TDS* is defined as the dataset to which the geospatial information taken from the *SDS* is being transferred, i.e. the expanded dataset [12].

After merge process of two or more datasets, the completeness of input data is always increased. One condition must be fulfilled - one of the input datasets must have more information than the other. Not all new geometrical object of the end dataset include information about attributes. The completeness of the end dataset can never be complete in terms of thematic information. Datasets generated by conflation can be complete only in terms of geometrical information. The following example illustrates this problem.

The figure 3 shows an example of two datasets. The first dataset (source dataset) includes the information about 6 buildings. However in the real world total number of buildings is 8, so two objects in this dataset are not provided. The source dataset includes thematic information about type of use of these buildings. The second dataset (target dataset) includes geometrical information about 5

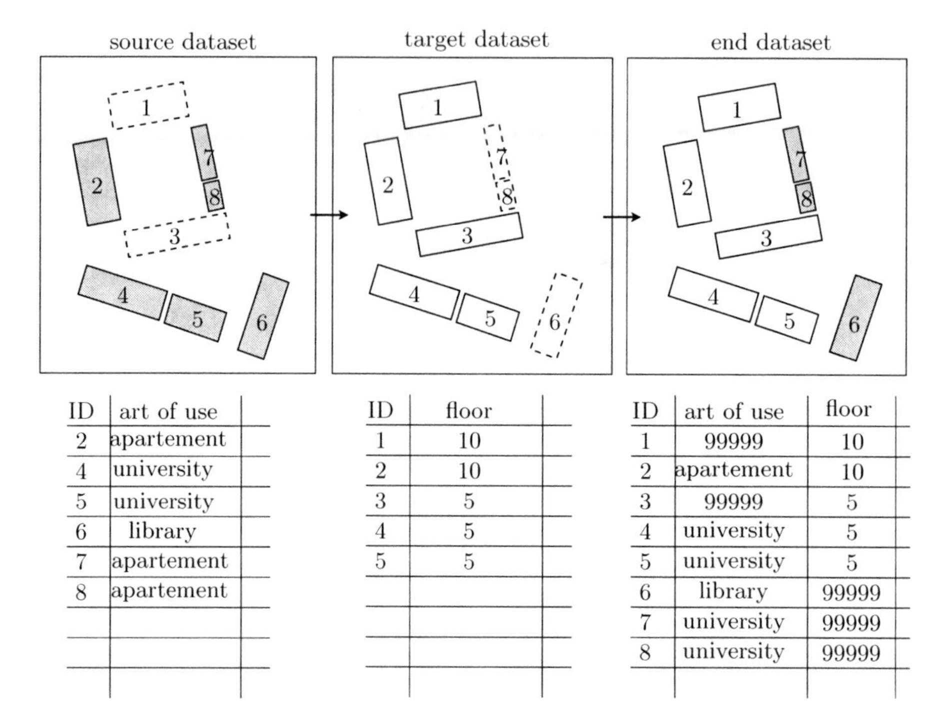

ID	art of use
2	apartement
4	university
5	university
6	library
7	apartement
8	apartement

ID	floor
1	10
2	10
3	5
4	5
5	5

ID	art of use	floor
1	99999	10
2	apartement	10
3	99999	5
4	university	5
5	university	5
6	library	99999
7	university	99999
8	university	99999

Fig. 3. An example for improvement of spatial data by means of data fusion

objects. The information about existence of the buildings number 6, 7 and 8 is not available. Unlike source dataset, target data have information about quantity of floors. This information in the first dataset is missing. The end dataset in the figure 3 shows the complete dataset in terms of geometric information. The table under it shows increment of attributes. Geometrical objects, which are available in both input datasets, have 100% thematical completeness. The missing objects have thematic information of only one input dataset. This applies to all data types: polygons, lines, points.

Conflation approaches allow the improvement of positional and temporal accuracy as well. Positional accuracy of a dataset can be increased with the information given by another input dataset. If both datasets have the major variance from real world, the arithmetic average of all input datasets can increase this quality element. The temporal accuracy will be improved if metadata provide information about actuality of spatial data.

The second section introduces the data conflation method for improvement of road networks.

3.1 Methodology

The approach presented here improves the quality of spatial data. This method illustrates how to increase the geometrical completeness of the road networks

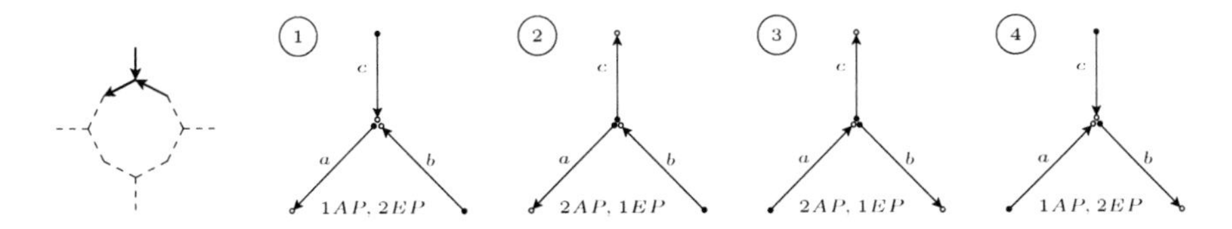

Fig. 4. Define the roundabout in the both input datasets

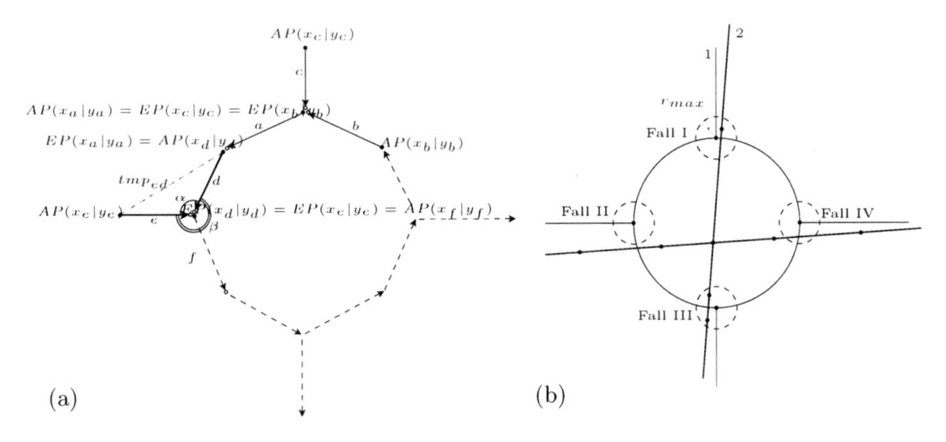

Fig. 5. Inserting missing roundabouts: (a) Edge tracing for the identification of a roundabout, (b) search for roundabout access or exits in the second input dataset (2)

data. In the source dataset available objects such roundabouts must be found in the target dataset and assigned to the new amended dataset.

The problem is, that crossroads, which are roundabouts, in the dataset are saved as simple crossroad. At first a position of all available crossroads in the both datasets have to be found. A roundabout is find, if minimum three edges of the road network have the same start- and endpoint (see figure 4). If there are three edges, wich have the same node, regardless of that is start or end point of each edge, then this intersection is a part of the roundabout. Whether it really a roundabout is will be tested as follows. The edges subsequented to the edges a and b have to come to a same point (see Figure 5 a)) and thus form a circle. If they not meets, it is a normal crossroad. A roundabout consists of several access and exits. At these points in the dataset, the tracing of the edge curve (starting from the edges a and b) can assumes the wrong way. Therefore, each point must be considered, whether it occurs only twice (Fig. 5 a)). If it occurs thrice (e.g. $EP(x_d|y_d) = EP(x_e|y_e) = AP(x_f|y_f)$), so the further run of the curve is defined with the calculation of the interior angle α 1

$$\alpha = \arccos\left(\frac{d_d^2 + d_e^2 - d_{tmp_{ed}}{}^2}{2 \cdot d_d \cdot d_e}\right),$$

wobei

$$\begin{aligned} d_d &= \sqrt{(EPx_d - APx_d)^2 + (EPy_d - APy_d)^2} \\ d_e &= \sqrt{(APx_e - EPx_d)^2 + (APy_e - EPy_d)^2} \\ d_{tmp_{ed}} &= \sqrt{(APx_e - APx_d)^2 + (APy_e - APy_d)^2}. \end{aligned} \tag{1}$$

It will verifyed whether the distance d_e between the points $AP(x_e|y_e)$ and $EP(x_d|y_d)$

$$d_e = \sqrt{(APx_e - EPx_d)^2 + (APy_e - EPy_d)^2} \tag{2}$$

larger than the maximum permissible distance d_{max} (this parameter is user-defined)

$$d_e > d_{max} \qquad \text{ist.} \tag{3}$$

In this case the angle β

$$\beta = 360° - \alpha \tag{4}$$

will be take for the following run of the curve. If the angle β fulfils the condition

$$\beta_{min} \leq \beta \leq 180°, \tag{5}$$

then the edge continues at the point $EP(x_d|y_d) = EP(x_e|y_e) = AP(x_f|y_f)$ the right path. The process will be accomplished successively. If the current point is again as the startpoint, so is the traffic node a crossroad.

In this way every crossroad of the dataset is verifyed. If a roundabout is defined, than at the second step the adequate crossroad is surched in the second dataset. Therefor the points are used, which are valid as traffic access or exits.

For example in the figure 5 a) it would be the points $AP(x_a|y_a)$ and $EP(x_d|y_d)$. The points of traffic nodes in the second dataset will be searched in the circle with the maximum Radius r_{max} (user-defined, dependending on data quality). There the different cases are possible. They are illustrate in the figure 5 b). In case *I* the point is find inside of specified circle. If any point will be found, so it is the case *II* and the case *IV*. The case *III* is then positive, if two poins are found within cicle. At the case *II* the algorithm will be run with the rest of the points.

All access or exits of roundabout are found in the first input dataset. The corresponding edges in the second input dataset are also found. Now the geometrical information about new objects can be assigned. There are three possibilities. Radius of roundabout in the first input dataset will be defined. Only the polylines, which form the circle are used. Polyline includes nodes. Two pair of nodes are extracted. The first pair with minimum an maximum y-coordinate value and the second pair with minimum and maximum x-coordinate value. The x-values and analogical y-values of both pairs are not considered. With the help of these points the average diameter $d = (d_1 + d_2)/2$ and the radius r of the circle are

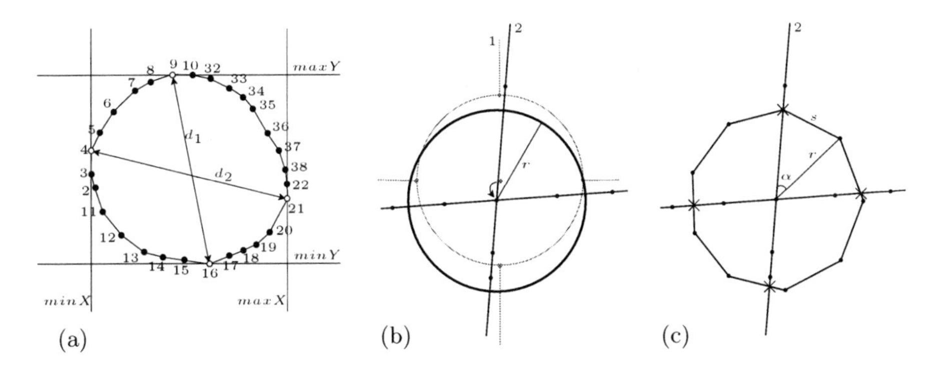

Fig. 6. Identification and transmission of the new roundabouts: (a) the calculation of the radius r for the new roundabout, (b) transferring the center of the circle, (c) circle as a polyline with the newly calculated traffic access and exit points. 1 - *source*-dataset, 2 - *target*-dataset

calculated (see fig. 6 a)). If the node of a crossroad in the target dataset is known, then this node is a centre point of the new circle with the defined radius r. The new circle is for the first time temporary saved in the separated file as a polyline. Each segment of polyline corresponds chord s, which length dependents on the size of radius

$$s = 2r \cdot sin(\frac{\alpha}{2}) \tag{6}$$

(see Figure 6 c)).

After the accurate assignment of all source and target datasets among each other the information about roundabout was transferred. This approach improves spatial data quality of network data. The results are identical with the results in section 3.

4 Summary and Concluding Remarks

This paper gives an overview of problems of spatial data quality. The data conflaton allows to increase the available datasets. In this paper only one method for merging of data and improvement of data quality was presented. It shows that the conflaton or fusion of different spatial datasets is the complex and not trivial technique. Therefore it makes a generation of improved spatial data possible. In future work the presented method of data fusion will be extended and proved by analysing more complex application examples.

References

1. Hunter, G.J., Bregt, A.K., Heuvelink, G.B.M., De Bruin, S., Virrantaus, K.: Spatial Data Quality: Problems and Prospects. In: Navratil, G. (ed.) Research Trends in Geographic Informations Science. Lecture Notes in Geoinformation and Cartography, pp. 101–121. Springer, Heidelberg (2009)

2. Stankutė, S., Asche, H.: An Integrative Approach to Geospatial Data Fusion. In: Gervasi, O., Taniar, D., Murgante, B., Laganà, A., Mun, Y., Gavrilova, M.L. (eds.) ICCSA 2009. LNCS, vol. 5592, pp. 490–504. Springer, Heidelberg (2009)
3. Veregin, H.: Data Quality Parameters. In: Longley, P.A., Goodchild, M.F. (eds.) Geographical Information Systems, Principles and Technical Issues, pp. 177–189. John Wiley & Sons, Chichester (1999)
4. Olteanu, A.-M.: A Multi-Citeria Fusion Approach For Geographical Data Matching. In: Proceedings of the 5th International Symposium Spatial Data Quality 2007, Enschede, June 13-15 (2007)
5. Chrisman, N.: Development in the Treatment of Spatial Data Quality. In: Devillers, R., Jeansoulin, R. (eds.) Fundamentals of Spatial Data Quality, pp. 21–30. Iste, London (2006)
6. Devillers, R., Gervais, M., Bedard, Y., Jeansoulin, R.: Spatial Data Quality: From Metadata to Quality Indicators and Contextual End-User Manual. In: Proceedings of the OEEPE/ISPRS Joint Workshop on Spatial Data Quality Management, Istanbul, March 21-22, 2002, pp. 45–55 (2002)
7. Onchaga, R., Morales, J., Widya, I., Lambert, J.M.: An Ontology Framework for Quality of Geographical Information Services. In: Proceedings of the 16th International Symposium on Advances in Geographic Information Systems, ACM GIS 2008, Irvine, California, USA, November 5-7 (2008)
8. Comber, A.J., Fisher, P.F., Wadsworth, R.A.: Semantics, Metadata, Geographical Information and Users. Transactions in GIS 12(3), 287–291 (2008)
9. Hake, G., Grünreich, D., Meng, L.: Kartographie. Visualisierung raum-zeitlicher Informationen. Berlin, New York (2002)
10. Albrecht, J.: Key Concepts & Techniques in GIS. Sage, London (2007)
11. Balley, S., Parent, C., Spaccapietra, S.: Modelling geographic data with multiple representations. International Journal of Geographical Information Science 18(4), 327–352 (2004)
12. Stankute, S., Asche, H.: Geometrical DCC-Algorithm for Merging Polygonal Geospatial Data. In: Taniar, D., Gervasi, O., Murgante, B., Pardede, E., Apduhan, B.O. (eds.) ICCSA 2010. LNCS, vol. 6016, pp. 515–527. Springer, Heidelberg (2010)
13. Chrisman, N.R.: The Error Component in Spatial Data. In: Longley, P.A., Goodchild, M.F., Rhind, D.J. (eds.) Geographical Information Systems: Principles and Applications, vol. 1, pp. 165–174. Longman, London (1990)
14. Haunert, J.-H.: Link based Conflation of Geographic Datasets. In: Proceedings of the 8th ICA WORKSHOP on Generalisation and Multiple Representation, A Coruña, July 7-8 (2005)

Territories of Digital Communities. Representing the Social Landscape of Web Relationships

Letizia Bollini

Psychology Department, University of Milano-Bicocca
Piazza dell'Ateneo Nuovo 1, 20126 Milano, Italy
letizia.bollini@unimib.it

Abstract. People establish connections with the territory that live in and the territory is modelled by their presence. The sedimentation of the places, by chance or planned, get *inhabited* people according to models and connections that are deeply linked to the individual and social experience. The digital territories as *social-network* seem to set up similar connections as the traditional social dynamics do with the physical territory.

But how are they able to (re)produce significant experiences for the different *social-tribes* that in the territory interact and communicate?

These questions are investigated by this study carried out on the field and focused on Milano-Bicocca district.

The study consists of two parts: a qualitative/ hybrid research about the *urban-island* to understand the connections that the social groups establish with the space and between themselves. The second – *design-oriented* – proposes some models of representation and tools for planning web communication shaped on the *social-tribes* and their relational experience.

Keywords: Digital networks, social tribes & territories, digital territories, design of social communities, interaction design based on social communities.

1 Introduction

The people that live and inhabit a place create with it a connection that drives its roots in the social and personal dimension of the places' history.

The places – in turn – are modelled by the historical sedimentation of their functional transformations and by the lived experience of peoples that inhabit them, that change them, that adapt them, beyond of their mere spatial functionality. The places are not just portions of space with a mainly urban function, but rather a set of cultural meanings, physical environments which are assigned symbolic and social meanings.

The spaces as consequence represent the social or *ethnics* groups that inhabit them, returning an image of the constructions, of the trends and of the social relations of the groups.

Historically this conception of the space – mainly the urban one – as an *anthropomorphous* system was used by different and opposite design and town-planning approaches mainly in modern time.

B. Murgante et al. (Eds.): ICCSA 2011, Part I, LNCS 6782, pp. 501–511, 2011.

After the fading of the utopia of the Italian Renaissance's *ideal towns* and after the return to the past of the Neoclassicism, during the XIX century the town becomes place of the contrast between the planned and the lived experience. The *artificial* interventions that work out new plans for the large European capitals – like *Plan Cerdà* in Barcelona or the Haussmann's boulevard in Paris according to an *a priori* model that considers the town as a scenographic stage for the self-portrait of the new society, of its power and of its rites – are evolving toward reflections more focused on the relationship between humankind and space.

The new century opens the urban reflection proposing again ideal towns, collective tenement houses as far to the experimental districts of the modern Rationalism – Siedlung as Weißenhof in Stuttgart – a solution is tentatively proposed to give a reply – even if quantitative – to the phenomenon of the industrial urbanization, planning the urban district, planning the services and the individual residential units according to a model of social organisation and to architectural *top-down* approach that risk to transform the town as an alienating place for people that inhabit it, *a machine pour habiter* [1].

Only in the 70's with the so-called *participated architecture* the attention of architects and city-planners has been focussed mainly on people as active subject in defining the space, its physical function and its social meaning in living experience.

The design process seems to give again more centrality to the symbolic, cultural and relational appearance in an attempt – perhaps utopian – to redefine the project according to a *bottom-up* approach.

This attention for the resident or what we could define – according to the new theories of the *user-centered design* – the *user* seems be even more urgent as the urban landscape becomes, as at the present, more a stage for the representation than a place of the daily life.

2 From Social Territories to Virtual Communities

Starting from Egon Brunswik's studies [2], the connection between people and environment takes again centrality also in psychology. Following a sort of inversion in the "connection between shape and background" the physical context is carried in front and is getting studied and conceptually defined with deeper detail than the individuals and the groups.

The approaches to the subject are very different. Following the psychological perspective different interpretations of the connection *environment-individual* are proposed: the first one attributes to the environment the role of independent variable that – by means of the actual stimulations – produces effects on the individual behaviour; in a second perspective the persons are interpret of the environment according to the specific peculiarities; finally an hypothesis assumes that people and the environment due to the mutual interactions give rise to reciprocal influences. The first point of view – that finds application in the *architectural psychology* – proposes to individualize the features of the physical environment that obstruct or facilitate the behaviour of the persons taking into account physical and quantitative issues. The second one – applied in the field of research and studies of perception-environmental knowledge, whose main item is the *behavioral geography* – focuses on the individuals, on its knowledge and its environment evaluations.

The hypothesis that presupposes a context of interactions gives greater attention to the variable of socio-cultural nature following the lines of search proposed by Ittelson [3]. But the point is surely the work of Kevin Lynch [4]: at the end of the 60's *The city image and its elements* has been published. A revolutionary outlook is proposed, suggesting to plan a town starting from the image that residents have of it. The model is widely accepted in the 60's and has further extended by the work of Kaplan & Kaplan [5] based on the effective variables which influence the emotional evaluations of the places: *coherence, complexity, legibility* and *mystery*.

Finally, the geographers have investigated some relation *people-territory* applying the discipline and the methods of the *behaviorism* and using projective test, completion, verbal association and expressive techniques, representation test – producing maps of different areas from data recollected from memory is the method used extensively in this study – and expressive methodologies.

2.1 Landscape as Social Representation

Inside the contemporary culture it is born therefore – according to the definition of Michael Jacob – the *omnipaysage*: "A landscape is, a landscape is, a landscape..." The concept that generates the question about the meaning of the contemporary landscape and the paradox of its representation. The landscape is a complex cultural construction: "The landscape is the artificial, not natural result, of a culture, that redefines perpetually its connection with the nature. [...] The experience of the landscape is, in general and in the first place, an experience of the himself." [6]

According to Lynch, the *public image* is the mental framework that shared by the majority of a town population; the occurrence alone that the people live or enjoy of the same physical reality produces the possibility to share the same *image* of the town. It was supposed that the people that live in the same district and share a common culture have also a common image of the town and this can differ from the public image of other citizens. It seems indeed that "for every town it exists a public image that is the superimposition of a lot of individual images" [4]. These individual images are indispensable to be able to live in the actual environment and to collaborate with the other people. The feature of the urban landscape that produces the identification of these elements is the *legibility*: how easy is to recognize its parts and how they can be organized in a consistent system. A representable and readable environment facilitates the motion across the environment and avoids the anguish that the chaotic town produces and – even more important – gives the possibility to the observer to find out and to emphasize the useful and significant elements that operate as a system of reference.

If the territory is representation – this means – social landscape of the daily experience of peoples who lives it, and if this experience is not limited to single individual, but also commonly shared inside the social groups – or paraphrasing Michel Maffesoli [7] – *socio-urban tribes*.

2.2 User Personas vs. Digital Communities

In contemporary culture the dual relationship between group and territory – or rather between individual and real – have included a variety of digital tools based primarily on the Internet, which reflect and reproduce the dynamics of the relationship between individuals and individual-community-individual in a territory, translating into an area that could be defined generically as *digital territory*. If the web is a universal interconnection system – now literally the most complete form of globalization – the people have shown a tendency to organize themselves – as happens in real social landscape – according to aggregation dynamic that assume a common interest as catalyst. We see – as regards the normal web design approach – a divergent phenomenon. Traditional methods – typically the *user-centered design* [8] – seek to prefigure homogeneous aggregations of potential users who share *psycho-demographic* profile, computer skills, dynamic exploration, expectations and use of information – according to the methods of *user personas* and *user scenario* – but the web 2.0 requires – conversely – to organize information and user experience more according to a social logic. The design focus is on interaction between people instead between people and digital technologies.

The classic web design methodology – that seeks to identify characteristics of individuals and generalized to users sharing specific answers in terms of experience, enjoyment – is based on the idea of homogeneous groups about practices. The emerging phenomenons – like the so-called Web 2.0 and specifically social networks – seem to largely ignore this dynamic challenging the previous paradigms. The digital social groups seem to reorganize themselves in terms of *communities of interest* in which the reason aggregator becomes the true focus of collective identity. Facebook fans of *I love tiramisù* fan-page or the Zynga-FarmVille *neighbors* who exchange daily favours, gifts and cooperation, do not share common demographic characteristics – like *user personas* presumes - but in the babel of their real or *fake* identities, their experiential and daily lives, are strongly interconnected by a common interest, although probably they don't share any other characteristic or social demographics. In this case, the social group that in the aggregate and virtual *community of interest* go beyond a typical parameter that is the *shared repertory*. [9]

3 An Hybrid Methodological Approach to Web Communities Design

According to this evolution, the research project tries to explore the creation of communities both in physical and virtual environment – using the privileged relationship that people establish with a given territory real or/and virtual – to design their everyday social digital experience.

The research tries to test methodologies and disciplinary approaches to implement a hybrid reproducible practice guidelines and directions for a *social-centered design* less abstract – world wide – and more *g-local*. This approach assumes the idea, and within the scope of this conjecture is checked for its validity, to analyse and translate the connection that people establish with their territory and between them, in the *web-world*. The aim is to reintroduce the social dynamics and identification with the land returning the core issues – real and symbolic – of its representation to create experiential privileged locations where – depending on the socio-urban tribe membership – users can identify themselves.

The experimental research project proposes an interpretive and methodological model to support the design hypothesis relevant to web design – portals, blogs or social network – that have a strong connection with the territory (For further information about this issue please refer to the previous research & publication: *Knowledge sharing and management for local community: logical and visual georeferenced information access*) [10].

The current analytic and design hypothesis has been already tested during a precedent study on a micro-area scale (the University Campus of Milano-Biccoca in 2007: the results have been recently published) [11]. The hypothesis is at the present tested on urban scale with the grant of a national research project PRIN 2007-09 (*Territorio e rappresentazione. Paesaggi urbani. Paesaggi sociali. Paesaggi digitali. Rimini e l'altro Mediterraneo.* In press).

The case study is focused on Milano-Bicocca district located in the extreme northeast suburbs of Milan at the boundary with the suburbs. The area is an ex-dismissed industrial area that in the last 10 years has been recovered with a cooperative public-private project using the master plan of the italian architect Vittorio Gregotti. Different residential urban settlements, the second public University of Milano (one of the seven Milano Universities) and a cluster of buildings dedicated to tertiary & services and to high-tech activities.

The district is limited on two sides from physical boundaries (an overpass to the district of Milano-Greco and the railroad) and on the other two boundaries respectively with two municipalities of the suburbs and is delimited from an urban high way and the work in progress of the fourth subway line. For its geographical conformation, for the peculiarity of its oversized architectures and for the missed connections with the surrounding urban settlements, during the study was defined as the *urban island*.

3.1 Phase 1: Identifying Urban-Social Tribes

The project is divided into two phases began with the exploration of the social *texture* through qualitative research methodologies in the field and a traditional user-centered approach. The analytical phase was conducted according to the following process/ methodologies:

a. Others Identification of the physical space and parts of the territory through significant architectural reconnaissance survey.
b. Identification and definition of *urban-social tribes* through a field research conducted with a mixed approach: qualitative interviews based on the model of story-telling and a search based on lifestyle approach, user personas & scenarios.
c. Cognitive tests to verify the links with the territory, the mental model, and the similarities/differences between different tribes: task-based user test and expressive techniques (drawing from memory a map of the district).

d. Reconnaissance on the digital reality in web & social-network relate to the social-subject divided into institutional sites, blogs and forums, Linden Second-Life, Web 2.0 social networking platforms (particularly Facebook).

3.2 Phase 2: Designing the Social User Experience

The second phase used collected data and interpretation to suggest the possible models of information architecture of a social online community linked to the district. From this last phase of the project ideas are born – developed in the course of Master Degree in *Theory and Communication Technology* – that illustrate this research.

The planning phase has been divided into two main activities:

a. Identification and reorganization of information architecture that could provide an adequate user-experience and for each tribe.
b. Development of different concepts and interaction systems between the portal and users to create identification (social brand) and sense of belonging (community).
c. Development of concepts prototypes based on different metaphors (time, space, 3d, travel, interior space, etc.) for user-test scenario/task-based.

4 Milano-Bicocca District: A Case Study

During the research were identified four social-urban tribes that use the space according to day-time, routes and perceptions of the environment, different, but similar within the same social group.

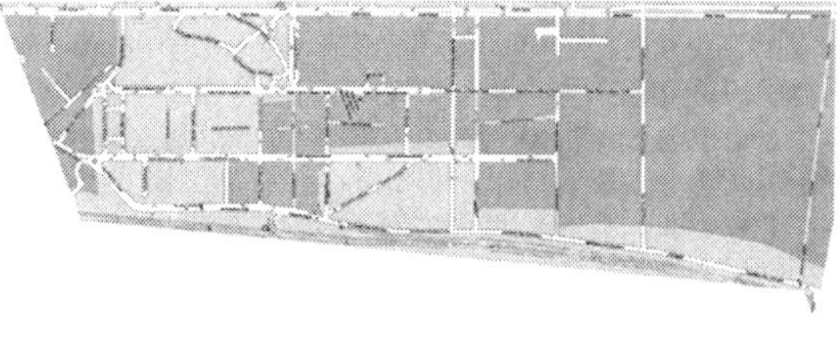

Fig. 1. Perspective of the University Campus and district map

The identified *social-urban tribes* are:

- **Resident:** usually young couples with children
- **Students:** who once arrived in the district spend the day in the same building using services generally concentrated in the central square
- **Employees and workers:** concentrated at the extreme boundary of the district, don't have a significant link with it
- **City-users:** *use* the district and services during the free time: gym, cinema and theatre, (Teatro degli Arcimboldi replaced during the renovation La Scala di Milano)

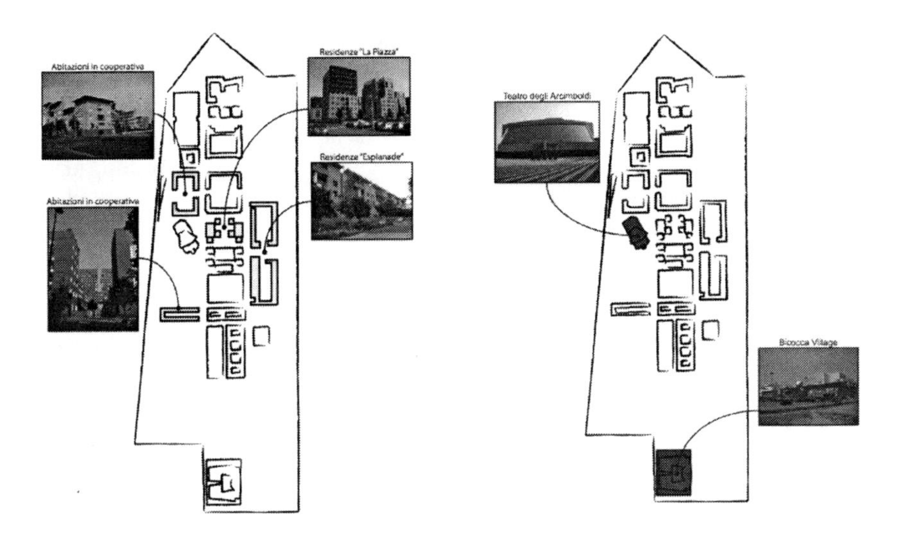

Fig. 2. The places of *social-urban tribes*: residents and city-users [Student: Tommaso Rossi]

Fig. 3. Maps sketched by: residents, students, workers (on the left the main street: Viale Sarca)

Figures 2 and 3 show some of the research results: the division of spaces attended by the various tribes (phase 1) and the representation of some interviewed (phase 1.d). The analytical phase revealed significant guidelines for further conceptualization steps: the various tribes – even in the internal variability subjective – identify specific recurring area of the district by prevalent use.

4.1 Residents: Socialise

Residents recall and represent locations on the map of daily life such as supermarket, pharmacy, church, schools, the central square and the cherry-hill they represent the only gathering places, including urban morphology of the neighbourhood.

The existing web-sites dedicated to this audience are very institutional - the local committee, etc. - and low popular, some even haven't been updated until 2004!

The requirement that residents express is primarily to have real and virtual places to socialize. In a new urban, suburban living and the presence of very different functions on a relational social networking emerges as one of the important functions of the interconnection that social-network could provide. The requirement that emerges from interviews is twofold: a place of gathering and sharing information, often ignored by the residents themselves and the use the web as a knowledge-sharing and socialization channel.

4.2 Students: Share

For students, the time spent in Bicocca is a significant part of a day although the district doesn't seem to encourage other forms of attraction or involvement of this tribe who spend their social time elsewhere. The web-sites are popular and represented the university buildings – in more detail depending on the Faculty – the food-services, concentrated in the central square underground and two university libraries. If transport – train and tram – are clearly identified, other places seem totally lacking in their experience. This is the case of the disproportionate presence of Teatro degli Arcimboldi that with its anomalous location compared to the rigid regularity of the grid construction completely disappears in the students' drawings.

Students are the tribe that mostly uses the Internet – many of them are already digital natives – so that their participation is particularly strong (students' official forum, FaceBook groups & fan pages official or spontaneously created).

The need is clearly to have structured and autonomous social sites that allow a/synchronous interaction very focused on the exchange, peers. The mood is micro-blogging, conversation or chatting.

4.3 Workers: Inform

This tribe spends the day in the discritc and often remains – unlike students –using services both for residents and city-users. It is a tribe with blurred and multiform boundaries composed by employees and managers of multinational companies and workers of the shops and services.

The spaces' knowledge is superficial: they recognize the strong presences – university buildings, theatres – but they tend to attend only the central space of collective and multi-purpose square (Piazza Trivulziana). The web needs are informative or finalized the discovery of places and services. More difficult to use the social-network often blocked by corporate IT policies.

4.4 City Users: Evaluate

This *cross*-tribe have a limited and *adulterated* experience of the district – usually in the evening or during the week end – *using* artificial commerce-places and entertainment services. The vision of the whole place is almost poor. The neighbourhood is perceived as a place of transit to theatre, cinema or mall, rather than as a living place with its own specific identity.

Conversely they are strong web-users to get information and to assess opportunities. Especially appreciated the chance to read and share comments and ratings – typical *user-generated* content – on restaurants, exhibitions and movies. These are often influencing the decision whether or not to move to these places.

5 Designing the Spatial and Social Metaphors

According to this analysis were produced several concept design in phase 2. In particular from a two-dimensional matrix in which were placed the 4 tribes (vertical path) and macros informative areas/functions of the district was done to achieve a multifaceted content classification that can intercept both the need a specific tribes and content in the system. The exploration of this conceptual and experiential organization has produced several proposals in the following are the most interesting.

A first model has focused its attention on the temporal overlap between the various tribes within the same urban area – work/free time - and has produced a structure interaction based on the presence synchronicity of users as a reason for aggregation.

Fig. 4. The *time* metaphor [Students: Donetti, Lacarbonara & Ostuni]

A second has instead used the concept of flow, by analogy with the navigation within a site, such as interpretative key.

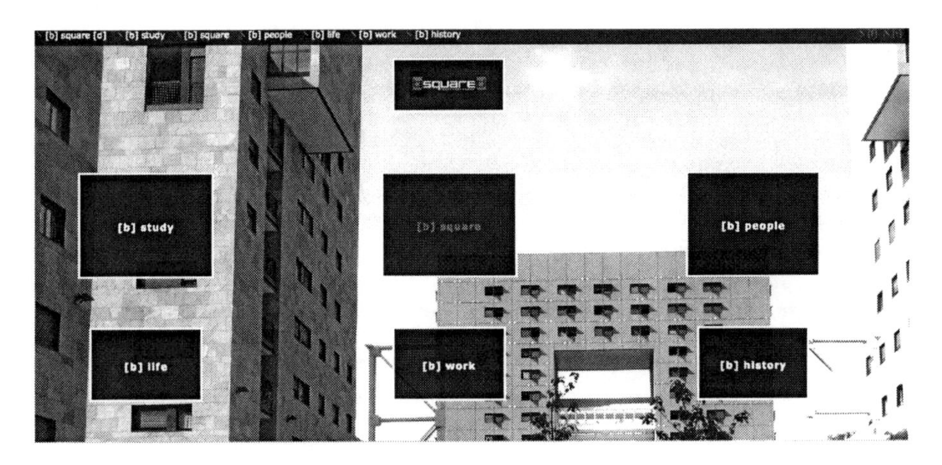

Fig. 5. The *square* metaphor [Students: Ciccarelli, Falcone & Montoli]

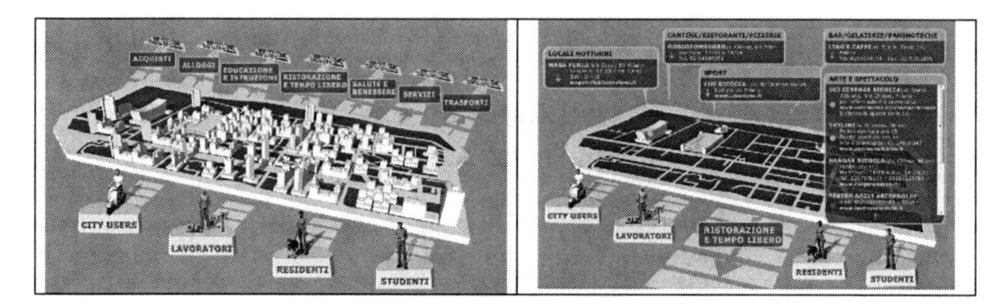

Fig. 6. Functions (in background) and social-urban tribes & interaction [Students: Merighi, Marcon & De Santis]

The most interesting proposal takes – even against the flatness of existing social life – the spatial metaphor as a key three-dimensional interpretation and identification of the place/tribe (see Figure 6). The 3D map is a sedimentation of places and relationships that can be scanned vertically – within the tribe – within or across the functions – or a combination of both. According to the possible paths the architectural objects are illuminated or faded in function of the social pertinence – derived from the perceptual model delineated in the survey – of places.

The system proposes a cognitive interface model to facilitate the use of the site or, to use Lynch's conjecture, the perceived *readability* of public image of urban land.

The information presented in pop-up instead comes from the interaction of users according to a bottom-up selection process. Are users, and considering voting for individual services to determine their presence in the first position, and therefore most clearly visible on the map. The community plays a role not only relational but also build collaborative & shared content.

6 Conclusions

The experience of the proposed research from a cultural, theoretical and methodological tools try to compose different disciplines and tries to articulated them in a hybrid approach for the design of Web 2.0. exceeded architectural hierarchical models, quantitative usability, the new Internet generation needs tools that are able to grasp the technology, communications, and above all relational.

Although applied to a specific case study – but already tested on a lesser scale of intervention and now revived on a large and complex project – the research model and design process have the potential to become a methodological approach reproducible and applicable to similar contexts.

Acknowledgments. Thanks to all the students of *Visual & Interface Design* 2009-10 – *Theory and Technology of Communication Master Degree*, University of Milano-Bicocca and in particular to Donetti, Lacarbonara & Ostuni; Ciccarelli, Falcone & Montoli; Merighi, Marcon & De Santis authors of the presented projects and to Tommaso Rossi: The Urban Island (2009), final dissertation in *Communication Science Degree*, University of Milano-Bicocca.

References

1. Choay, F.: Pour une Anthropologie de l'Espace. Seuil, Paris (2006)
2. Brunswick, E.: The Conceptual Framework of Psychology. In: Bechtel, R.B., Ts'erts'man, A., Churchman, A. (eds.) Handbook of environmental psychology, John Wiley and Sons, New-York (2002)
3. Ittelson, W.H., Proshansky, H., Rivlin, L., Winkel, G.: An Introduction to Environmental Psychology. Holt, Rinehart and Winston, New York (1974)
4. Lynch, K.: The City Image and its Elements. Prentice-Hall, New Jersey (1969)
5. Kaplan, R., Kaplan, R., Wendt, J.S.: Rated Preference and Complexity for Natural and Urban Visual Material. Perceptionand Psychophysics 12, 354–356 (1972)
6. Jacob, M.: Il Paesaggio. Il Mulino, Bologna (2009)
7. Maffesoli, M.: The Time of Tribes: The Decline of Individualism in Mass Society. Sage Publications, London (1996)
8. Cato, K.: User-Cenered Web Design. Addison-Wesley, Reading (2001)
9. Wenger, E.: Communities of Practice: Learning, Meaning, and Identity. Cambridge University Press, Cambridge (1998)
10. Bollini, L., Cerletti, V.: Knowledge Sharing and Management for Local Community: Logical and Visual Georeferenced Information Access. In: Granville, B., Majkic, Z., Li, C. (eds.) ISRST 2009 Proceedings, Orlando, pp. 92–99 (2009)
11. Bollini, L.: Learning by Doing: a User-Centered Approach to Signage Design. Milano-Bicocca a Case Study. In: EduLearn 2010, IADE, Barcelona (2010)

Decentralized Distributed Computing System for Privacy-Preserving Combined Classifiers – Modeling and Optimization

Krzysztof Walkowiak, Szymon Sztajer, and Michał Woźniak

Department of Systems and Computer Networks, Faculty of Electronics, Wroclaw
Wrocław University of Technology, Wybrzeze Wyspianskiego 27, 50-370 Wroclaw, Poland
Krzysztof.Walkowiak@pwr.wroc.pl

Abstract. The growing amount of various kinds of information triggers the need to develop efficient network computing systems, as single machines in many cases are not able to provide effective processing and analysis. One of the very promising approaches of distributed data analysis is combined classification, which could be relatively easily implemented in distributed computing systems. In this paper we address problem of decentralized distributed computing system for mentioned above classification method. We focus on the system fairness. The performance metric is defined as a maximum response time, i.e., the computing system should be designed to minimize the response time of each client using the system. We assume that the system is decentralized and each request is sent by the client directly to computing nodes without assistance of a central service. An ILP (Integer Linear Programming) model is formulated and applied to obtain optimal results provided by branch-and-cut algorithm included in the CPLEX solver. Widespread simulations are performed to evaluate properties of the computing system in terms of several parameters describing the system.

Keywords: distributed computing, grid computing, privacy-preserving combined classifiers, ILP modeling, optimization.

1 Introduction

Fast development of telecommunication and information technologies prompts to create distributed systems, which offer very large computational and processing capabilities that cannot be supported in traditional architectures. As the most popular examples of distributed systems we can list grids, public-resource computing systems (known also as global computing or peer-to-peer computing) and cloud computing. A wide range of applications have been adopted to be executed in distributed environments, e.g., collaborative visualization of large scientific databases, financial modeling, bioinformatics, experimental data acquisition, earthquake simulation, medical data analysis, climate/weather modeling, astrophysics and many others [1-8].

In recent years, progress in IT sector causes production and collecting of huge amounts of data. It has become necessary to start looking for new efficient data

B. Murgante et al. (Eds.): ICCSA 2011, Part I, LNCS 6782, pp. 512–525, 2011.

analysis methods. One of the most promising directions of that research is data mining, which is widely used in computer security (e.g. designing IDS/IPS), medicine (e.g. patient diagnosis), finance (e.g. credit approval), or trade.

For practical reasons, it is usually impossible to gather large amounts of data in a single dataset. Therefore, many institutions are gathering data in separated facilities. Considering the need of developing distributed data analysis tools seems inevitable, one of the great advantages is that data mining methods can be, pretty easily, implemented in distributed computing systems. On the other hand, one should consider that distributed data mining incurs some serious difficulties like a privacy preserving problem. One of the most popular issues of data mining are pattern recognition methods, which are usually used for important and confidential data which cannot be subject to the risk of disclosure. Therefore, research connected with Secure Multi-party Computation is a necessity.

In this paper we focus on the application of distributed computing systems to combined classifiers, so-called Multiple Classifier Systems (MCSs), which are one of the most promising directions of pattern recognition. They group methods, which exploit strengths of an ensemble of individual classifiers and allow improving accuracy or response time of the simple recognition methods. This can be achieved if we assume diversity in the sense that the individual classifiers have to be different. The difference can be understood in the meaning of different datasets, for example learning on the partitions of the original dataset, or even using a different recognition method for each of the individual classifiers. The classifier ensemble is a group of N elementary classifiers and each of them makes a decision about object classification. The final decision of the combined classifier is made on the basis of the mentioned above decisions by a fusing algorithm. Besides the aforementioned advantages of MCSs, they are also – in comparison to many other recognition algorithms – much easier to be implemented in distributed environments. For that reason, they can be used for databases which are partitioned, e.g., for privacy reasons.

The main contributions of the paper are as follows. (i) Detailed discussion on Multiple Classifier Systems in the context of privacy requirements. (ii) Architecture of a decentralized distributed computing system dedicated for destined privacy-preserving combined classifiers. (iii) Formulation of an ILP model of the system. (iv) Extensive numerical experiments run to examine performance of the system in terms of several parameters describing the system.

The remainder of the paper is organized in the following way. In the next section we present a review of recognition algorithms. Section 3 introduces the architecture of a distributed computing system and an ILP (Integer Linear Programming) model of the system. Section 4 reports results of numerical experiments. Finally, the last section concludes this work.

2 Model of Combined Classification Task

The pattern recognition groups together useful data analysis tools, which could be applied to several practical applications as computer network security, computer aided medical diagnosis or fraud detection to enumerate only a few. This approach tries to classify a given object to one of the predefined categories on the basis of selected features. Usually mentioned above interconnection is trained on the basis of collected dataset or it is given by experienced experts in the form of a set of rules. There are numerous models of classifiers and methods how to train them [9]. In recent years the hybrid approaches like combined pattern recognition become more and more popular. One of the main reason why they are attractive from the practical point of view is that nowadays enterprises collect their data usually in distributed storages and combined classifiers are able to extract and use knowledge which is hidden is distributed databases [10, 11, 12].

Such an implementation of MCS seems incredibly tempting with its speed and computing capabilities, but it could be also very hazardous. In every distributed environment there is a big chance that the privacy of analyzed data can be compromised. During the last couple of years several researches have been made in the field of privacy-preserving data mining [13, 14]. Of course, also some specific studies are being made on secure multi-party computation [15].

2.1 Models of Multiple Classifier Systems

The general idea of Multiple Classifier System is as follows. A simple classifier is replaced by a more sophisticated functional block, i.e., a single classifier is replaced by an ensemble of elementary classifiers and each of them is built on a separate dataset. The final decision of this ensemble is made by a fuser, which takes decisions made by simple classifiers into consideration. The idea of a Multiple Classifier System is depicted in Fig. 1.

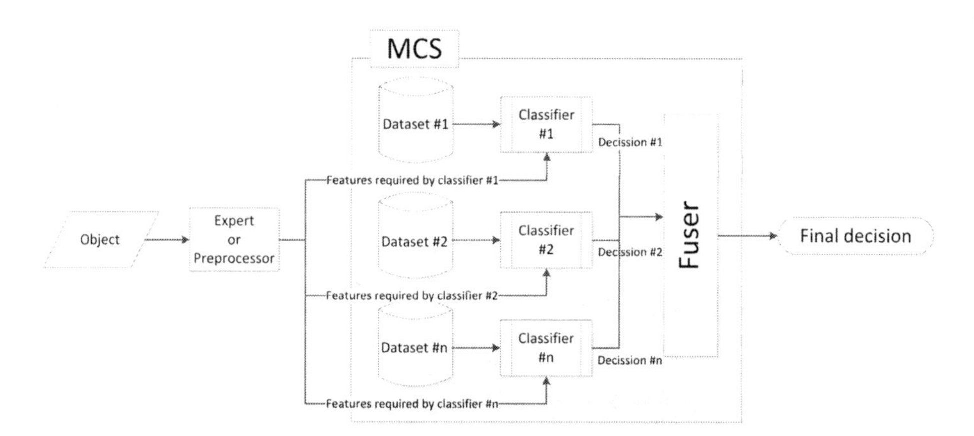

Fig. 1. Scheme of a Multiple Classifier System

The main effort of Multiple Classifier Systems focuses on combining knowledge coming from the set of elementary classifiers. One of the greatest advantages of MCSs is that each individual classifier can be computed independently e.g., in a separated node of a computing network, which makes them very easy to be implemented in distributed computing environment. Another motivation to use MCSs is the possibility for improving classification accuracy or response time by using partitions of the original dataset for the individual classifiers. It is proven that using combined classifiers can give smaller classification error than the best of the individual algorithms [10]. It also avoids choosing the worst classifier from the pool of available elementary classifiers.

There are several methods of combining elementary classifiers. The first group consists of methods for classifier fusion at the level of their responses [16]. The second one, gathers classifier fusion methods based on discriminant analysis, especially the posterior probability estimators, associated with probabilistic models of a given pattern recognition task [17-18].

In this paper we are going to focus on majority voting, which is a traditional representative for the group of algorithms making fusion of classifiers on the level of their discrete outputs - decisions. Majority voting is a simple yet effective method. It gives us also a possibility to easily extend the system with a weighted voting method in the future. However, as yet, researches have not shown much classification efficiency improvement coming from replacing majority voting by weighted voting [10].

There is also another group of methods, which make the classifier fusion at the level of their responses. In this group, the decision is formed by the classifier fusion on the level of their continuous outputs – support functions, the main form of which are the *posterior* probability estimators, referring to the probabilistic model of a pattern recognition task [18]. The aggregating methods, which do not require learning, perform fusion with the help of simple operators, such as the maximum, minimum, mean or product, but they are typically relevant in specific, clearly defined conditions [19]. However, we shall not turn our interest onto this group of methods for the reasons, which will be presented in the following subsection.

2.2 Secure Multi-party Computation

One of the most important issues in the field of distributed pattern recognition is the case of privacy preserving. There are many situations where many sides want to receive knowledge from the entire data set without compromising the privacy of the individual data sets within the different participants.

According to [13], the key to the definition of privacy in the context of distributed data mining is that nothing is learned beyond what is inherent in the result. Basing on that, we can also assume that the less redundant information comes with the result the better.

Let us consider a hypothetical distributed medical diagnosis environment comprised of V medical institutions. Each of them holds a computing node with one or more individual classifiers. The database is horizontally partitioned between all of the institutions. Each of the classifiers is connected with a database (there can be one per institution or more, e.g., a specific DB for each of the institutions units). Databases contain personal medical data, which obviously are confidential. The goal is to allow other fellow participants (medical institutions) to use knowledge gathered in databases for diagnosis of their own patients without compromising any specific data from it. After gathering answers from the computing nodes, client (one of the medical institutions) combines them using a chosen fuser.

The selection of the combining algorithm is very relevant. We decided that simple majority voting will be a good choice. This is a good moment to present the reasons for such a selection, mentioned in the previous subsection. As it was also mentioned, minimizing the amount of information given by the result to as little as necessary is crucial. Majority voting needs only a straight answer from the elementary classifiers stating: "this object belongs to class *i*". For example, if we choose a mean combiner we would need an answer containing *posterior* probabilities of the fact that object x belongs to class i. Such an answer can give a lot more information about how our database is constructed to an untrusted adversary when he e.g. intensively queries our computing node. This solution should provide quite a high level of privacy, which can be also improved by using some cryptographic solutions in order to secure the data during transport through the network.

3 Optimization Model

In this section we present an Integer Programming formulation of an optimization problem related to a decentralized distributed computing system designed for privacy-preserving combined classifiers. There are R computational projects indexed $r = 1,2,\ldots,R$. Each project denotes a database used by elementary classifiers. Each project is divided into units of the same size including a particular number of individual training samples. Let n_r denote the number of uniform units in project $r = 1,2,\ldots,R$. We assume that the distributed computing system contains V computing nodes indexed $v = 1,2,\ldots,V$. Nodes are connected by a computer network (e.g., Internet). Each computing node represents a single machine or a cluster located in the same physical location. There is a limit on the maximum number of units that each node can store denoted by c_v, i.e., the number of units of all possible projects assigned to node v can not exceed c_v. This limit includes capacity constraints of each computing node related to storage space, link capacity and others. For each node we are given processing rate p_v given in units/millisecond. This limit denotes the number of project units that node v can process in one millisecond. To make the model more realistic, we assume that each project can be split to maximum S computing nodes, i.e., the number of computing nodes involved in a computational project cannot exceed S.

In the network, there is a set of demands (clients) indexed $d = 1,2,\ldots,D$. Each client generates requests related to one (or more) of computational projects r and wants to receive the decision as fast as possible. Constant b_{rd} is 1, if demand d generates requests related to project r and 0 otherwise. It is assumed that each demand d knows, which computing nodes are involved in project r, if $b_{rd} = 1$. Note that this information is provided by a special indexing service for each requesting demand. Moreover, the assignment of computing nodes to projects is relatively stable. When the system is redesigned, i.e., the allocation of projects to computing nodes is changed, this information is delivered to all interested clients (demands). The request of demand d related to a particular project r is sent to each computing node involved in project r. We are given network delays between for each computing node v and each client d denoted as t_{vd} and given in milliseconds. This delay can be estimated by special networking techniques, e.g., using the ICMP protocol.

The objective of the optimization is to minimize the response time of the computing system, which includes (i) the overall time required to send all requests and replies through the network and (ii) the processing time. The main decision variable is x_{rv} that denotes the number of project r units located in node v. Consequently, the processing time of project r in node v is x_{rv} / p_v. Moreover, we introduce an auxiliary binary variable y_{rv} which is 1, if at least one unit of project r is located at node v; 0 otherwise.

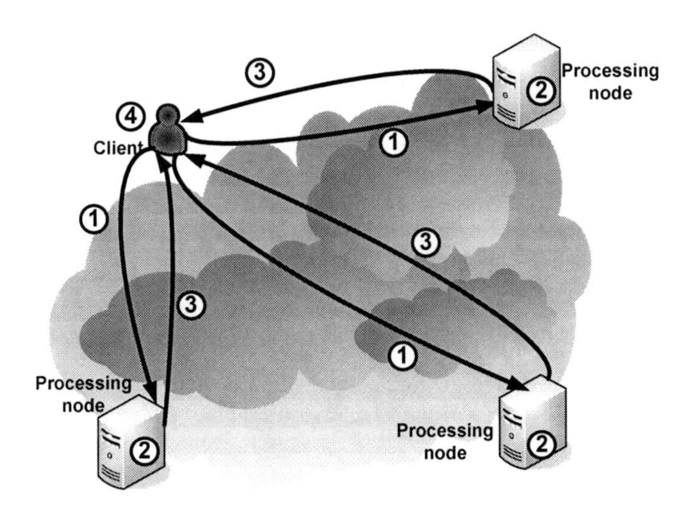

Fig. 2. Model of a distributed computing system

The workflow of the system is shown in Fig. 2. We assume that 3 computing nodes are involved in the considered computing project. Step 1 (number of steps are shown on the figure in circles) is issued by the client of demand d, which sends a query to each computing node v related the considered project. The delay of this operation is t_{vd} for each computing node v. Subsequently, each computing node processes the query (step 2) what takes x_{rv} / p_v and returns to the client node the decision (step 3 and delay t_{vd}). When, the demand node collects all answers from computing nodes, it makes the final processing in a very small time which is a constant, so it is not considered in the model (step 4). Note that in our previous paper [20], we presented a similar model, however a central server processing all requests was assumed. In contrast, this work assumes that the computing system is decentralized and each client directly sends requests to computing nodes.

Using variables x_{rv} and y_{rv} we can define $z_{rvd} = 2y_{rv}t_{vd} + x_{rv} / p_v$ – the response time related to project r, computing node v and demand d. Notice that z_{rvd} includes: (i) the network delay between the demand node d and computing node v (and in the opposite direction); (ii) processing time required in node v and project r. The overall decision time related to demand d and project r (considering requests to all computing nodes participated in project r) denoted as z_{rd} is defined as the maximum value of z_{rvd} over all nodes $v = 1,2,\ldots,V$. This follows from the fact that the final decision can be made only when the client collects all responses. The objective is to minimize the maximum response time of the computing system – denoted as z. We take into account all projects and demands, i.e., we want to minimize the maximum value of z_{rd} over all $r = 1,2,\ldots,R$ and $d = 1,2,\ldots,D$. Note that in this work we do not address the problems of tasks' scheduling. We make an assumption that the computing system is dimensioned accordingly to the predicted load. More precisely, each arriving request is processed almost immediately without the need to queue the request. Thus, the system objective (maximum response time) does not include any queuing delay. This can be achieved by parallel processing of requests and overprovisioning of processing units at each computing node according to forecasted arrival rate of requests. However, resources related to storage capacity of each node are included in the model, since databases used in privacy-preserving combined classifiers methods in many cases are of large size.

To formulate the ILP model we use notation as in [21].

Indices

$v = 1,2,\ldots,V$ computing (processing) nodes
$d = 1,2,\ldots,D$ demands (clients).
$r = 1,2,\ldots,R$ projects

Constants

c_v capacity of node v
p_v processing rate of node v - the number of units that v can process in 1 ms
t_{vd} network delay between computing node v and end node of demand d (ms)
b_{rd} =1, if demand d generates requests related to project r; 0, otherwise
n_r size of the project r (number of database units)
S split, i.e., the maximum number of computing nodes involved in one project
M large number

Variables

x_{rv} the part of project r (number of units) located on node v (integer)
y_{rv} =1, if the part of project r is located on node v; 0, otherwise (binary)
z_{rvd} overall response time related to project r, computing node v and demand d
z_{rd} decision time for project r and demand v
z maximum response time of the system

Objective

It is to find allocation of computational projects to computing nodes satisfying the node capacity, split value and minimizing the maximum response time:

$$F = z. \tag{1}$$

Constraints

a) All units of each project $r = 1,2,\ldots,R$ must be allocated for processing to computing nodes:

$$\textstyle\sum_v x_{rv} = n_r \tag{2}$$

for each $r = 1,2,\ldots,R$.

b) The number of units assigned to computing node v cannot exceed the node capacity:

$$\textstyle\sum_r x_{rv} \le c_v \tag{3}$$

for each $v = 1,2,\ldots,V$.

c) Binary variable y_{rv} (denoting if node v is used to process project r) is 1, only if at least one unit of project r is assigned to node v ($x_{rv} > 0$) and 0 otherwise ($x_{rv} = 0$):

$$y_{rv} \leq x_{rv} \tag{4}$$

$$x_{rv} \leq My_{rv} \tag{5}$$

for each $r = 1,2,\ldots,R$, $v = 1,2,\ldots,V$.

d) The number of computing nodes involved in every project r cannot exceed the split S:

$$\sum_v y_{rv} \leq S \tag{6}$$

for each $r = 1,2,\ldots,R$.

e) Definition of z_{rvd} denoting the response time related to project r, computing node v and demand d in the case when demand d participates in project r ($b_{rd} = 1$)

$$z_{rvd} = 2y_{rv}t_{vd} + x_{rv} / p_v \tag{7}$$

for each $r = 1,2,\ldots,R$, $v = 1,2,\ldots,V$, $d = 1,2,\ldots,D$, $b_{rd} = 1$.

f) If demand d does not participate in project r ($b_{rd} = 0$), then z_{rvd} must be 0:

$$z_{rvd} = 0 \tag{8}$$

for each $r = 1,2,\ldots,R$, $v = 1,2,\ldots,V$, $d = 1,2,\ldots,D$, $b_{rd} = 0$.

g) Definition of z_{rd} denoting the overall response time related to project r, and demand d:

$$z_{rvd} \leq z_{rd} \tag{9}$$

for each $r = 1,2,\ldots,R$, $v = 1,2,\ldots,V$, $d = 1,2,\ldots,D$.

h) Definition of z denoting the overall maximum response time:

$$z_{rd} \leq z \tag{10}$$

for each $r = 1,2,\ldots,R$, $d = 1,2,\ldots,D$.

The presented model (1)-(10) is strongly NP-hard problems since it is equivalent to the Multidimensional Knapsack Problem [22]. Note that the problem (1)-(10) is constructed in the context of Multiple Classifier Systems, however it can be applied to model a wide range of computational tasks that can be processed in a distributed manner.

4 Results

In this section we report results of numerical experiments. Since the optimization model formulated in the previous section belongs to a class of ILPs (Integer Linear Programs), we applied branch-and-cut algorithm included in CPLEX 11.0 solver [23]. All results reported below are optimal. Note that to obtain these optimal results, in reasonable time, relatively small problem instances were examined.

The first goal of experiments was to evaluate influence of the split S. Recall that the split value denotes the maximum number of computing nodes involved in one project. The value of the split should be a tradeoff between two requirements:

minimization of the maximum response time and management issues. Notice that if the split has a relatively small value, each project can be split to few computing nodes what increases the response time. However, management of such projects is simpler. In a case when the split is increased, the response time should be lowered, but the management of the project becomes more challenging. To make numerical experiments several sets of input data were randomly generated. The computing system includes 10 nodes and is described by randomly selected parameters: c_v denoting the capacity of node v (in range 200-400) and p_v denoting the processing rate of node v (in range 0.2-0.5). Each project set contains 10 computational projects – the project size (parameter n_r) was chosen at random in range 100-300. Finally, each demand set includes 100 demands (clients) with the following parameters selected at random. Parameter t_{vd} denoting network delay between computing node v and end node of demand d is in range 10-200. There are two scenarios related to parameter b_{rd}, i.e., in scenario A each demand is connected to only one project, in scenario B each demand is assigned on average to 25% existing projects.

Figs 3 and 4 show the maximum response time as a function of the split value for scenarios A and B, respectively. Each figure includes three curves showing performance of three selected tests. The results are in harmony with our intuition, i.e., increasing of the split leads to reduction of the maximum response time. However, the largest gain is between S = 1 and S = 2. Starting from S = 6, the reduction is declining.

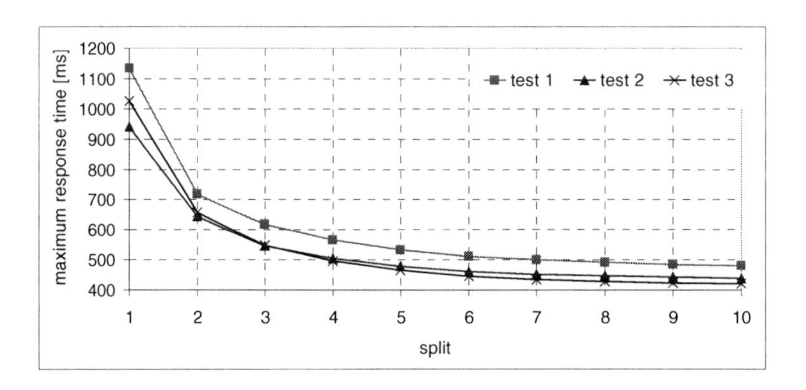

Fig. 3. Maximum response time as a function of split – scenario A

Fig. 5 reports the average execution time of CPLEX (MIP solver) required to solve the model in optimal way. All experiments were made on a PC with Intel Dual Core Processor T7500, 2.2 GHz, 3GB RAM, Windows Vista. Each presented curve relates to one of considered simulation scenarios. We can easily notice that the in the case of S = 2 the execution time is the smallest (about 22 seconds). For larger values of the split, the solution time grows, however the detailed performance depends on the scenario. This can be explained by the fact that increasing of the split increases the solution space and consequently the branch-and-cut method implemented in the CPLEX solver becomes less efficient.

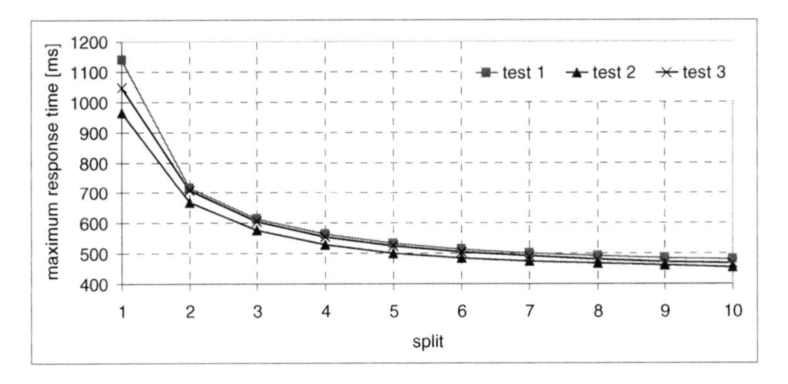

Fig. 4. Maximum response time as a function of split – scenario B

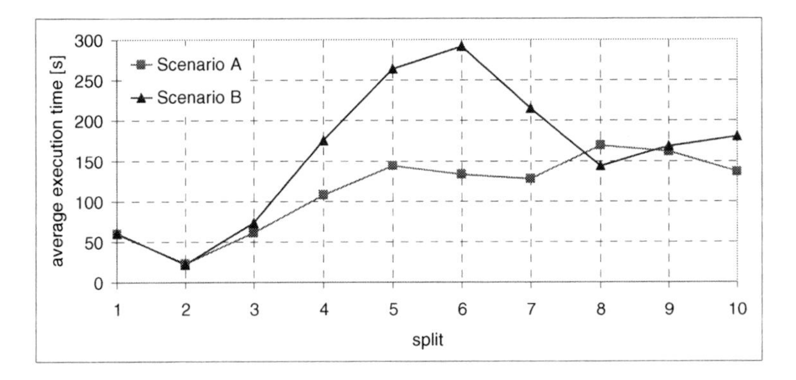

Fig. 5. CPLEX average execution time as a function of split

In the second experiment we examined how the objective function (maximum response time) depends on the number of computing nodes available in the computing system. The project set contains 10 projects (parameter n_r denoting the size is in the range 100-300). The demand set includes 30 clients and each client is assigned on average to 25% existing projects. Moreover, in tests we used 10 various computing systems including from 6 to 15 nodes with parameters randomly selected in the same ranges as in the previous case. In Fig. 6 we present six curves related to different values of the split parameter. Note that $S = \max$ means that for a given computing system (axis x), we select the maximum possible split, e.g., in the case of 15 nodes, $S = 15$. The main lesson learned from Fig. 6 is that, whatever the split value is, the response time decreases with the size of the computing system (number of nodes). However, the largest gain is observed in cases of $S = 1$ and $S = \max$. In the former case the difference between a system including 6 nodes and a system with 15 nodes is 12.23%. In the latter case ($S = \max$) the corresponding value is 14.20%. For other values of the split, the difference between 6 and 15 nodes is always below 7%. Consequently, we can conclude that the performance of the considered computing system defined as the maximum response time does significantly depend on the computing system size.

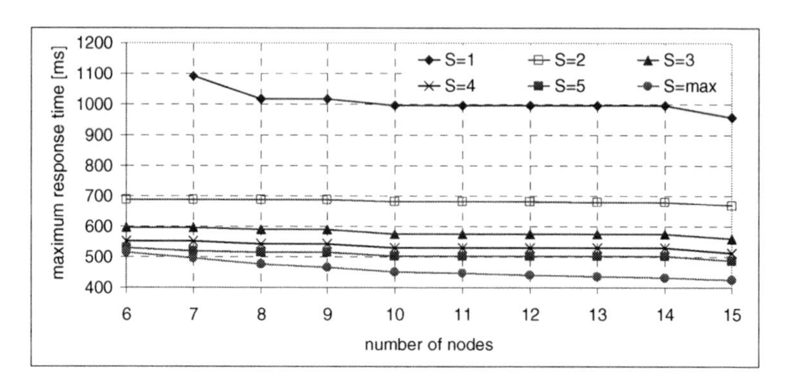

Fig. 6. Maximum response time as a function of number of nodes and split

The next goal of simulations was to verify if the system capacity influences the considered performance metric. The system capacity is defined as $\sum_v c_v$ (sum of nodes' capacity). We examine the system capacity in relation to overall project demand defined as $\sum_r n_r$ (sum of all projects' size). As in previous cases, we generated at random computing systems including 6 nodes, project sets containing 6 projects and a demand set including 30 clients connected to a one randomly selected project. In the first experiment, we fixed the overall project demand and tested 10 sets of nodes with different values of the system capacity. The ratio (system capacity)/(overall project size) was in range 100%-103%. Fig. 7 reports the maximum response time as a function of this ratio. We present three curves representing various values of the split.

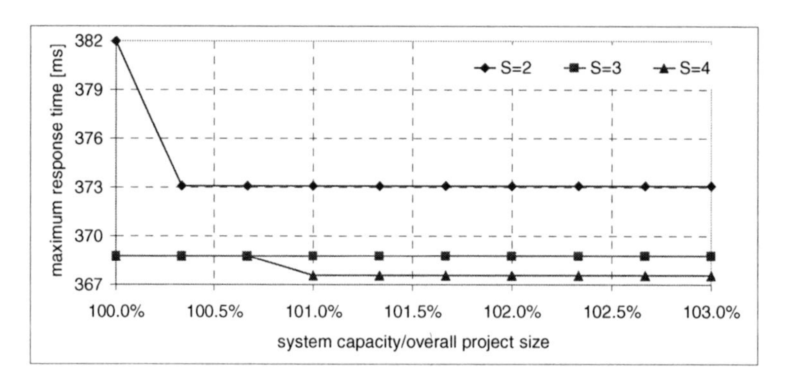

Fig. 7. Maximum response time as a function of (system capacity)/(overall project size) ratio and split

In the next experiment (Fig. 8) we considered a reversed situation, i.e., the system capacity was fixed and we changed the overall project size. In this case the (overall project size)/(system capacity) ratio was in range 95%-100%. The general trend observable on both figures is that the maximum response time of the system is not sensitive to the system capacity, i.e., increasing of the system capacity to enable more flexible allocation of projects to computing nodes does not have a strong influence on the maximum response time.

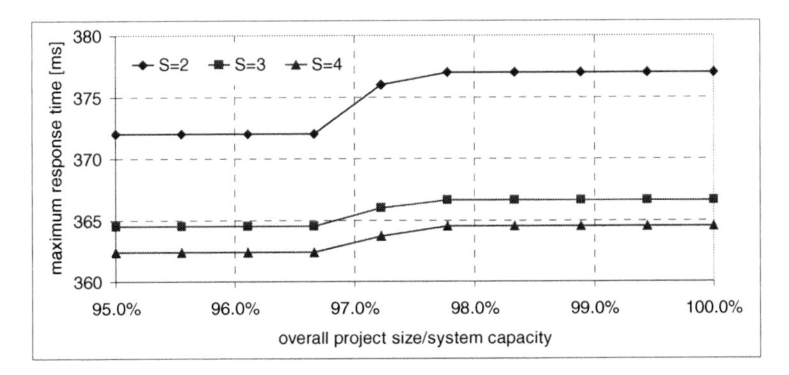

Fig. 8. Maximum response time as a function of (overall project size)/(system capacity) ratio and split.

In the last experiment, we focused on the parameter b_{rd}, which denotes if demand d generates requests to project r. We tested computing systems with 8 nodes, 8 projects and 50 demands (with randomly selected values of parameters using the same ranges as above). Values of b_{rd} parameter were selected to obtain the demand/project ratio in range 10%-100%. Notice that demand/project ratio equals to 10% means that on average each demand is assigned to 10% of existing projects. In Fig. 9 we report the maximum response time as a function of demand/project ratio and split. We can observe that with the increase of the demand/project ratio, the maximum response time slowly grows when $S > 8$. However, the difference between 10% and 100% is only about 6%.

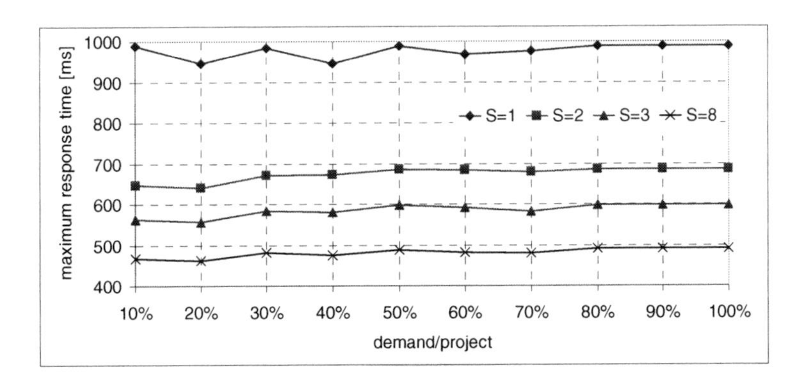

Fig. 9. Maximum response time as a function of demands/projects ratio and split

5 Concluding Remarks

In this paper we focused on the decentralized distributed computing system modeling. We introduced a simple and intuitive solution for preserving privacy in a Multiple Classifier System. It guarantees that the analyzed data are stored and computed in secure isolation from the untrusted adversaries, which can only obtain knowledge about the result of the computation, and not about the shape of the database. Since the proposed distributed system is aimed to be applied in the context of combined classifiers, the objective was defined as the maximum response time. We have formulated an ILP model of the system, which was next used to obtain optimal results provided by the CPLEX solver. The properties of the computing system were extensively evaluated in terms of several parameters, including split, system size (number of computing nodes), system capacity, project size, demand/project ratio. The reported results show that the largest impact on the system performance (i.e., maximum response time) has the split factor. Other examined parameters (system size, system capacity, project size, demand/project ratio) do not significantly influence the maximum response time.

In future work we would like to develop effective heuristic algorithms for the presented ILP model to obtain results for larger problem instances. Moreover, we plan to consider distributed computing systems with augmented reliability requirements, when the system is designed to protected operation against selected kinds of failures (e.g., computing node failure or network failure).

Acknowledgments. This work was supported in part by The Polish Ministry of Science and Higher Education.

References

1. Anderson, D.: BOINC: A System for Public-Resource Computing and Storage. In: Proc. of the Fifth IEEE/ACM International Workshop on Grid Computing, pp. 4–10 (2004)
2. Buford, J., Yu, H., Lua, E.: P2P Networking and Applications. Morgan Kaufmann, San Francisco (2009)
3. Milojicic, D., et al.: Peer to Peer computing. HP Laboratories Palo Alto, Technical Report HPL-2002-57 (2002)
4. Nabrzyski, J., Schopf, J., Węglarz, J. (eds.): Grid resource management:state of the art and future trends. Kluwer Academic Publishers, Boston (2004)
5. Shen, X., Yu, H., Buford, J., Akon, M. (eds.): Handbook of Peer-to-Peer Networking. Springer, Heidelberg (2009)
6. Tarkoma, S.: Overlay Networks: Toward Information Networking. Auerbach Publications (2010)
7. Taylor, I.: From P2P to Web services and grids: peers in a client/server world. Springer, Heidelberg (2005)
8. Travostino, F., Mambretti, J., Karmous Edwards, G.: Grid Networks Enabling grids with advanced communication technology. Wiley, Chichester (2006)
9. Jain, A.K., Duin, P.W., Mao, J.: Statistical Pattern Recognition: A Review. IEEE Trans. on PAMI 22(1), 4–37 (2000)
10. Kuncheva, L.I.: Combining Pattern Classifiers: Methods and Algorithms. New Jersey (2004)
11. Miller, D.J., Zhang, Y., Kesidis, G.: Decision Aggregation in Distributed Classification by a Transductive Extension of Maximum Entropy/Improved Iterative Scaling, Hindawi Publishing Corporation. EURASIP Journal on Advances in Signal Processing Volume (2008)

12. Kacprzak, T., Walkowiak, K., Woźniak, M.: Optimization of Overlay Distributed Computing Systems for Multiple Classifier System – Heuristic Approach. Logic Journal of IGPL (in press, 2011)
13. Vaidya, J., Clifton, C.W., Zhu, Y.M.: Privacy Preserving Data Mining. Springer, New York (2006)
14. Aggrawal, C.C., Yu, P.S.: Privacy-Preserving Data Mining: Models and Algorithms. Springer, New York (2008)
15. Lindell, Y., Pinkas, B.: Secure Multiparty Computation for Privacy-Preserving Data Mining. The Journal of Privacy and Confidentiality 1(1), 59–98 (2009)
16. Kuncheva, L.I., Whitaker, C.J., Shipp, C.A., Duin, R.P.W.: Limits on the Majority Vote Accuracy in Classier Fusion. Pattern Analysis and Applications 6, 22–31 (2003)
17. Alexandre, L.A., Campilho, A.C., Kamel, M.: Combining Independent and Unbiased Classifiers Using Weighted Average. In: Proc. of the 15th ICPR, vol. 2, pp. 495–498 (2000)
18. Biggio, B., Fumera, G., Roli, F.: Bayesian Analysis of Linear Combiners. In: Haindl, M., Kittler, J., Roli, F. (eds.) MCS 2007. LNCS, vol. 4472, pp. 292–301. Springer, Heidelberg (2007)
19. Duin, R.P.W.: The Combining Classifier: to Train or Not to Train? In: Proc. of the ICPR 2002, Quebec City (2002)
20. Przewoźniczek, M., Walkowiak, K., Woźniak, M.: Optimizing distributed computing systems for k-nearest neighbors classifiers - evolutionary approach. Logic Journal of IGPL (2011), doi:10.1093/jigpal/jzq034
21. Pioro, M., Medhi, D.: Routing, Flow, and Capacity Design in Communication and Computer Networks. Morgan Kaufmann, San Francisco (2004)
22. Puchinger, J., Raidl, G., Pferschy, U.: The Multidimensional Knapsack Problem: Structure and Algorithms. Informs Journal on Computing (2009), doi:10.1287/ijoc.1090.0344
23. ILOG CPLEX, 12.0 User's Manual, France (2007)

A New Adaptive Framework for Classifier Ensemble in Multiclass Large Data

Hamid Parvin, Behrouz Minaei, and Hosein Alizadeh

School of Computer Engineering, Iran University of Science and Technology (IUST), Tehran, Iran
{parvin,b_minaei,halizadeh}@iust.ac.ir

Abstract. This paper proposes an innovative combinational algorithm to improve the performance of multiclass problems. Because the more accurate classifier the better performance of classification, so researchers have been tended to improve the accuracies of classifiers. Although obtaining the more accurate classifier is often targeted, there is an alternative way to reach for it. Indeed one can use many inaccurate classifiers each of which is specialized for a few dataitems in the problem space and then s/he can consider their consensus vote as the classification. This paper proposes a new ensembles methodology that uses ensemble of classifiers as elements of ensemble. These ensembles of classifiers jointly work using majority weighted voting. The results of these ensembles are in weighted manner combined to decide the final vote of the classification. In empirical result, these weights in final classifier are determined with using a series of genetic algorithms. We evaluate the proposed framework on a very large scale Persian digit handwritten dataset and the results show effectiveness of the algorithm.

Keywords: Genetic Algorithm, Optical Character Recognition, Pairwise Classifier, Multiclass Classification.

1 Introduction

In practice, there may be problems that one single classifier can not deliver a satisfactory performance [13]. In such situations, employing ensemble of classifying learners instead of single classifier can lead to a better learning [11]. Although obtaining the more accurate classifier is often targeted, there is an alternative way to obtain it. Indeed one can use many inaccurate classifiers each of which is specialized for a few dataitems in the problem space and then employ their consensus vote as the classification. This can lead to better performance due to reinforcement of the classifier in error-prone problem spaces.

In General, it is ever-true sentence that "combining the diverse classifiers which are better than random results in a better classification performance" [5], [11] and [15]. Diversity is always considered as a very important concept in classifier ensemble methodology. It refers to being as much different as possible for a typical ensemble. Assume an example dataset with two classes. Indeed the diversity concept for an ensemble of two classifiers refers to the probability that they produce dissimilar

B. Murgante et al. (Eds.): ICCSA 2011, Part I, LNCS 6782, pp. 526–536, 2011.

results for an arbitrary input sample. The diversity concept for an ensemble of three classifiers refers to the probability that one of them produces dissimilar result from the two others for an arbitrary input sample. It is worthy to mention that the diversity can converge to 0.5 and 0.66 in the ensembles of two and three classifiers respectively. Although reaching the more diverse ensemble of classifiers is generally handful, it is harmful in boundary limit. It is very important dilemma in classifier ensemble field: the ensemble of accurate-diverse classifiers can be the best. It means that although the more diverse classifiers, the better ensemble, it is provided that the classifiers are better than random.

Evolutionary computations are considered universal optimizers or problem solvers. It is common to take it as an optimizer in large fields of science. The most well-known of them is considered to be Genetic Algorithm (GA). John Holland first introduced GA [6].

GA like other machine learning algorithms is based loosely on mechanism of biological evolution. It is applied in the wide problem spaces [11] in two ways: their direct usage as classifiers [8], and their usage as optimizing tools for determining parameters of classifiers. In [2], the GA is used to find decision boundaries in feature space. Another application of the GA is optimization of parameters in classification process. Many researchers also use GA in feature subset selection [1], [4], [10], [14] and [16]. Combination of classifiers is another field that GA has a hand as an optimization tool. Indeed, GA has also been used for feature selection in classifier ensemble [9] and [12].

An Artificial Neural Network (ANN) is a model which is to be configured to be able to produce the desired set of outputs, given an arbitrary set of inputs. An ANN generally composed of two basic elements: (a) neurons and (b) connections. Indeed each ANN is a set of neurons with some connections between them. From another perspective an ANN contains two distinct views: (a) topology and (b) learning. The topology of an ANN is about the existence or nonexistence of a connection. The learning in an ANN is to determine the strengths of the topology connections. One of the most representatives of ANNs is MultiLayer Perceptron. Various methods of setting the strength of connections in an MLP exist. One way is to set the weights explicitly, using a prior knowledge. Another way is to 'train' the MLP, feeding it by teaching patterns and then letting it change its weights according to some learning rule. In this paper the MLP is used as one of the base classifiers.

Decision Tree (DT) is considered as one of the most versatile classifiers in the machine learning field. DT is considered as one of unstable classifiers. It means that it can converge to different solutions in successive trainings on same dataset with same initializations. It uses a tree-like graph or model of decisions. The kind of its knowledge representation is appropriate for experts to understand what it does [17].

Its intrinsic instability can be employed as a source of the diversity which is needed in classifier ensemble. The ensemble of a number of DTs is a well-known algorithm called Random Forest (RF) which is considered as one of the most powerful ensemble algorithms. The algorithm of RF was first developed by Breiman [3].

This paper proposes a framework to develop combinational classifiers. In this new paradigm, a multiclass classifier in addition to a few pairwise classifiers creates a classifier ensemble. At last, to produce final consensus vote, different votes (or outputs) are gathered, after that the weighted majority voting algorithm is employed to aggregate them. The weights are determined by universal optimizer problem solvers like genetic algorithm.

This paper focuses on Persian handwritten digit recognition (PHDR), especially Hoda dataset [7]. Although there are well works on PHDR, it is not rational to compare them with each other, because there was no standard dataset in the PHDR field until 2006 [7]. The contribution is only compared with those used the same dataset used in this paper, i.e. Hoda dataset.

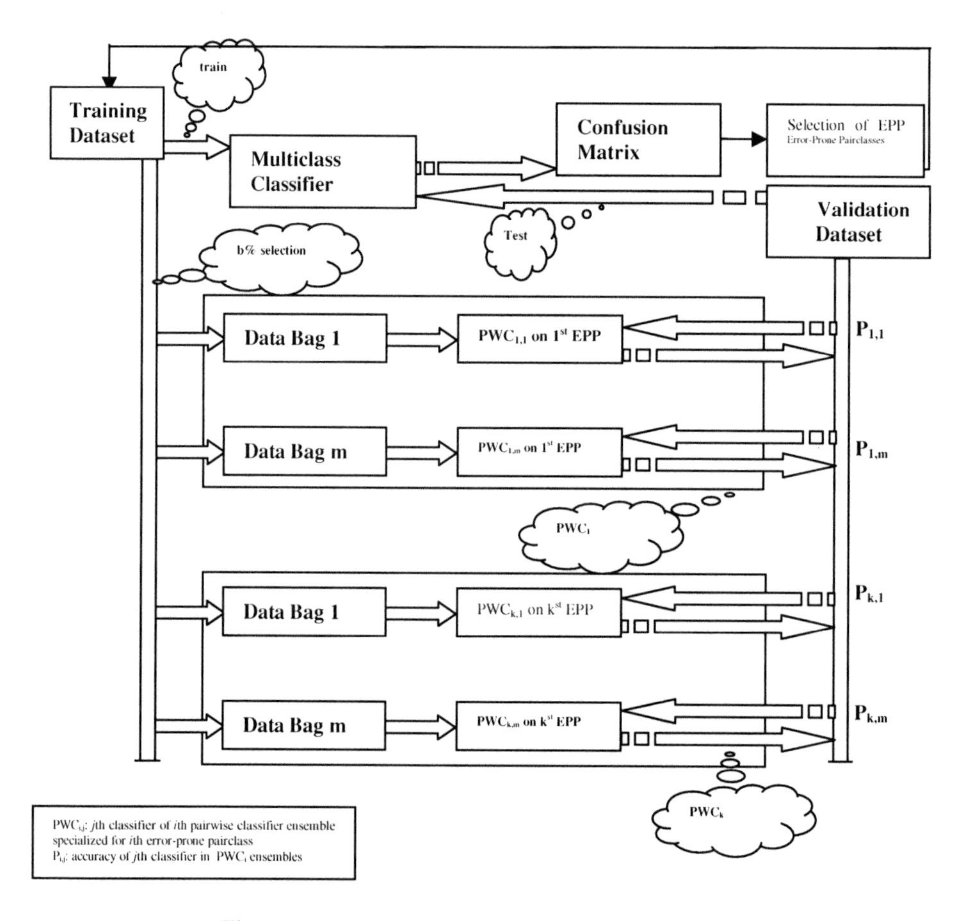

Fig. 1. The first training phase of the proposed method

2 Proposed Algorithm

The main idea behind the proposed method is to use a number of pairwise classifiers to reinforce the main classifier in error-prone regions of problem space. Figure 1 depicts the training phase of the proposed method schematically.

In the proposed algorithm, a multiclass classifier is first trained. Its duty is to obtain confusion matrix over validation set. Note that this classifier is trained over the total train set. At next step, the pair-classes which are mostly confused with each other and are also mostly error-prone are detected. After that, a number of pairwise classifiers are employed to reinforce the drawbacks of the main classifier in those error-prone regions. A set of distinct classifiers is used for each class as an ensemble which is to learn that class. Considering the outputs of the main multiclass classifier and ones of the pairwise classifiers totally as a new space, GA is finally used to determine the weight of each classifier to vote in the ensemble. So, GA is run as many as the number of classes. It means GA is utilized as an aggregator in an ensemble detecting a class. Assume that the number of classes is denoted by c. So GA is run c times, each of them is denoted by GA_1, GA_2, ..., GA_c. GA_i means the running of GA to detect i-th class or equivalently (i-1)-th digit.

2.1 Determining Erroneous Pair-Classes

At the first step, a multiclass classifier is trained on all train data. Then, using results of this classifier on the evaluation data, confusion matrix is obtained. This matrix contains important information about the functionalities of classifiers in the dataset localities. The close and Error-Prone Pair-Classes (EPPS) can be detected using this matrix. Indeed, confusion matrix determines the between-class error distributions. Assume that this matrix is denoted by a. Item a_{ij} of this matrix determines how many instances of class c_j have been misclassified as class c_i.

Table 1 shows the confusion matrix obtained from the base multiclass classifier. As you can see, digit 5 (or equivalently class 6) is incorrectly recognized as digit 0 fifteen times (or equivalently class 1), and also digit 0 is incorrectly recognized as digit 5 fourteen times. It means 29 misclassifications have totally occurred in recognition of these two digits (classes). The mostly erroneous pair-classes are respectively (2, 3), (0, 5), (3, 4), (1, 4), (6, 9) and so on according to this matrix. Assume that the i-th mostly EPPC is denoted by $EPPC_i$. So $EPPC_1$ will be (2, 3). Also assume that the number of selected EPPC is denoted by k.

2.2 Training of Pairwise Classifiers

After determining the mostly erroneous pair-classes, or EPPCs, a set of m binary classifiers is to be trained to jointly, as an ensemble of binary classifiers, reinforce the main multiclass classifier in the region of each EPPC. So as it can be inferred, it is necessary to train k ensembles of m binary classifiers. Assume that the ensemble which is to reinforce the main multiclass classifier in the region of $EPPC_i$ is denoted by PWC_i. Each binary classifier contained in PWC_i, is trained over a bag of train data like RF. The bags of train data contain only b percept of the randomly selected of train data. It is worthy to be mentioned that pairwise classifiers which are to participate in PWC_i are trained only on those instances which belongs to $EPPC_i$. Assume that the j-th classifier binary classifier of PWC_i is denoted by $PWC_{i,j}$. Because there exists m classifiers in each of PWC_i and also there exists k EPPC, so there will be $k*m$ binary classifiers totally. For example in the Table 1 the EPPC (2, 3) can be considered as an erroneous pair-class. So a classifier is necessary to be trained for that EPPC using those dataitems of train data that belongs to class 2 or class 3. As mentioned before, this method is flexible, so we can add arbitrary number of PWC_i to the base primary classifiers. It is expected that the performance of the proposed framework outperforms the primary base classifier.

It is worthy to note that the accuracies of $PWC_{i,j}$ can easily be approximated using the train set. Because $PWC_{i,j}$ is trained only on *b* percept of the train set with labels belong to $EPPC_i$, provided that *b* is very small rate, then the accuracy of $PWC_{i,j}$ on the train set with labels belong to $EPPC_i$ can be considered as its approximated accuracy. Assume that the mentioned approximated accuracy of $PWC_{i,j}$ is denoted by $P_{i,j}$.

It is important to note that each of PWC_i acts as a binary classifier. As it mentioned each PWC_i contains *m* binary classifiers with an accuracy vector, P_i. It means of these binary ensemble can take a decision with weighed sum algorithm illustrated in [9]. So we can combine their results according to weighs computed by the below equation.

$$w_{i,j} = \log(\frac{p_{i,j}}{1-p_{i,j}}) \tag{1}$$

where $w_{i,j}$ is the accuracy of *j*-th classifier in the *i*-th binary ensemble. It is proved that the weights obtained according to the equation 1 are optimal weights in theory. Now the two outputs of each PWC_i are computed as equation2.

$$PWC_i(x \mid h) = \sum_{j=1}^{m} w_{i,j} * PWC_{i,j}(x \mid h) \quad , \quad h \in EPPC_i \tag{2}$$

where *x* is a test data.

2.3 Fusion of Pairwise Classifiers

The last step of the proposed framework is to combine the results of the main multiclass classifier and those of PWC_i. It is worthy to note that there are $2*k$ outputs from the binary ensembles plus *c* outputs of the main multiclass classifier. So the problem is to map a $2*k+c$ intermediate space to a *c* space each of which corresponds to a class. The results of all these classifiers are fed as inputs for the aggregators. Note that there are *c* aggregators, one per each class. The Output of aggregator *i* is the final joint output for class *i*. Here, the aggregation is done using a special weighting method. The problem here is how one can optimally determine these weights. In this paper, GA is employed to find these weights.

Because of the capability of the GA in passing local optimums, it is expected that the accuracy of this method outperforms a simple MLP or unweighted ensemble. Figure 1 along with Figure 2 and Figure 3 depicts the structure of the ensemble framework.

As it is shown in Figure 2, in the proposed framework, the number of times that GA is invoked is equal to *c*, which is the number of digits (classes). This GA-based algorithm is overall illustrated by Figure 2.

In fact, each GA creates an ensemble to detect one digit (class), by considering the $2*k+c$ intermediate space obtained by the multiclass classifier plus the binary classifier ensembles as new feature space. Each GA uses one hyper-line in this new intermediate feature space, by assigning a weight to each dimension. The chromosome representation of GA_i is a vector of real numbers. The function of GA_i is calculated as equation 3.

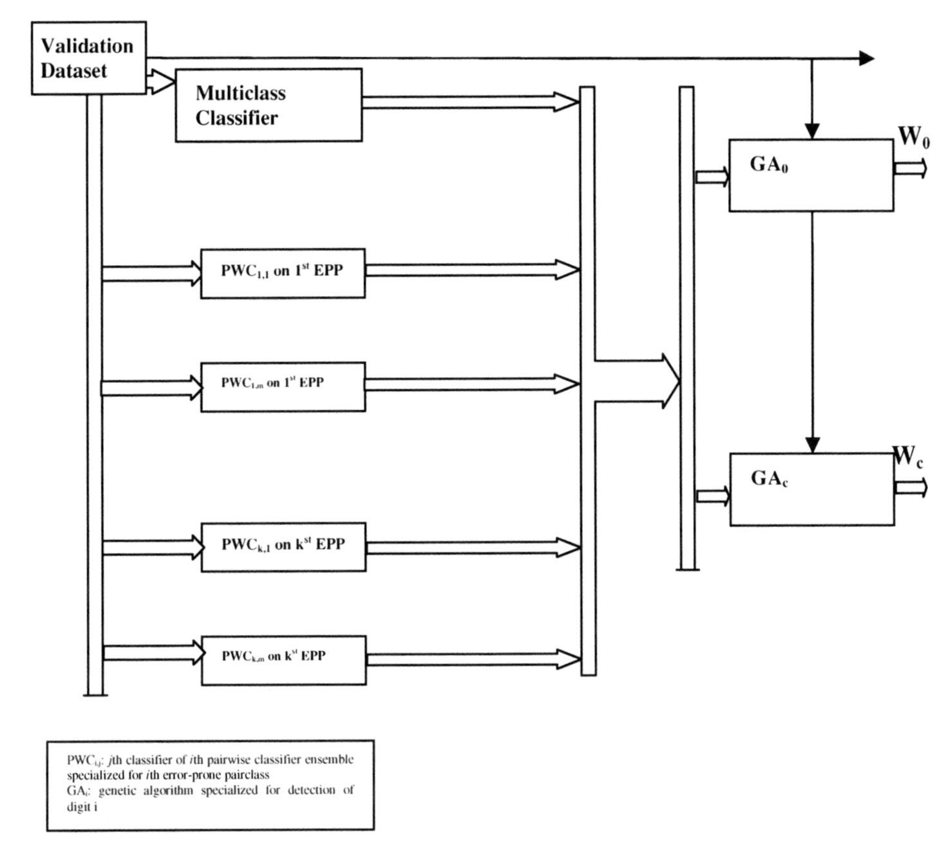

Fig. 2. The second training phase of the proposed method based on GA

$$Fitness(W_i, ValSet) = \sum_{x \in ValSet} f(x, W_i) \tag{3}$$

where the function $f(x, W_i)$ is computed as equation 4.

$$f(x, W_i) = sign(x, i) * \left(BinOuts(x, W_i) + MultiOuts(x, W_i)\right) \tag{4}$$

where the function $sing(x, i)$ is computed as equation 5.

$$sign(x, i) = \begin{cases} 1 & label(x) = i \\ -1 & label(x) \neq i \end{cases} \tag{5}$$

and *ValSet* in equation 4 is validation set. In the equation 4, *BinOuts* is the weighted sum of the outputs of the binary ensembles, given an input sample x, which is computed as equation 6, and *MultiOuts* is the weighted sum of the outputs of the main multiclass classifier, given an input sample x.

$$BinOuts(x,W_i) = \sum_{j=1}^{k} \sum_{h=1}^{2} W_i(s) * PWC_{i,j}(x \mid EPPC^{h}_{\lceil j/2 \rceil}) \tag{6}$$

where s is computed as equation 7.

$$s = (j-1) * 2 \tag{7}$$

and *MultiOuts* is also computed as equation 8:

$$MultiOuts(x,W_i) = \sum_{j=1}^{c} W_i(2*k+j) * MCC(x \mid j) \tag{8}$$

Indeed GA_i try to better discriminate the class i from other classes. Finally, the most voted class is selected as final decision of the framework as depicted in the Figure3. This is simply done using a max function as it is obvious from the Figure 3. It means that the final decision is taken by equation 9.

$$FinalDecision(x) = \arg\max_{i} f(x,W_i) \tag{9}$$

3 Experimental Results

This section evaluates the results of applying the proposed framework on a Persian handwritten digit dataset named Hoda [7]. This dataset contains 102,364 instances of digits 0-9. Dataset is divided into 3 parts: train, evaluation and test sets. Train set contains 60,000 instances. Evaluation and test datasets are contained 20,000 and 22,364 instances. The 106 features from each of them have been extracted which are described in [7]. Some instances of this dataset are depicted in Figure 4.

3.1 Parameter Setting

In this paper, MLP and DT are used as base primary classifier. We use an MLPs with 2 hidden layers including respectively 10 and 5 neurons in the hidden layer 1 and 2, as the base Multiclass classifier. Confusion matrix is obtained from its output. Also DT's measure of decision is taken as Gini measure. The classifiers' parameters are kept fixed during all of their experiments. It is important to take a note that all classifiers in the algorithm are kept unchanged. It means that all classifiers are considered as MLP in the first experiments. After that the same experiments are taken by substituting all MLPs whit DTs.

The parameter k is set to 11. So, the number of pairwise ensembles of binary classifiers added equals to 11 in the experiments. The parameter m is also set to 9. So, the number of binary classifiers per each EPPC equals to 9 in the experiments. It means that 99 binary classifiers are trained for the pair-classes that have considerable error rates. Assume that the error number of each pair-class is available. For choosing the most erroneous pair-classes, it is sufficient to sort error numbers of pair-classes. Then we can select an arbitrary number of them. This arbitrary number can be determined by try and error which it is set to 11 in the experiments.

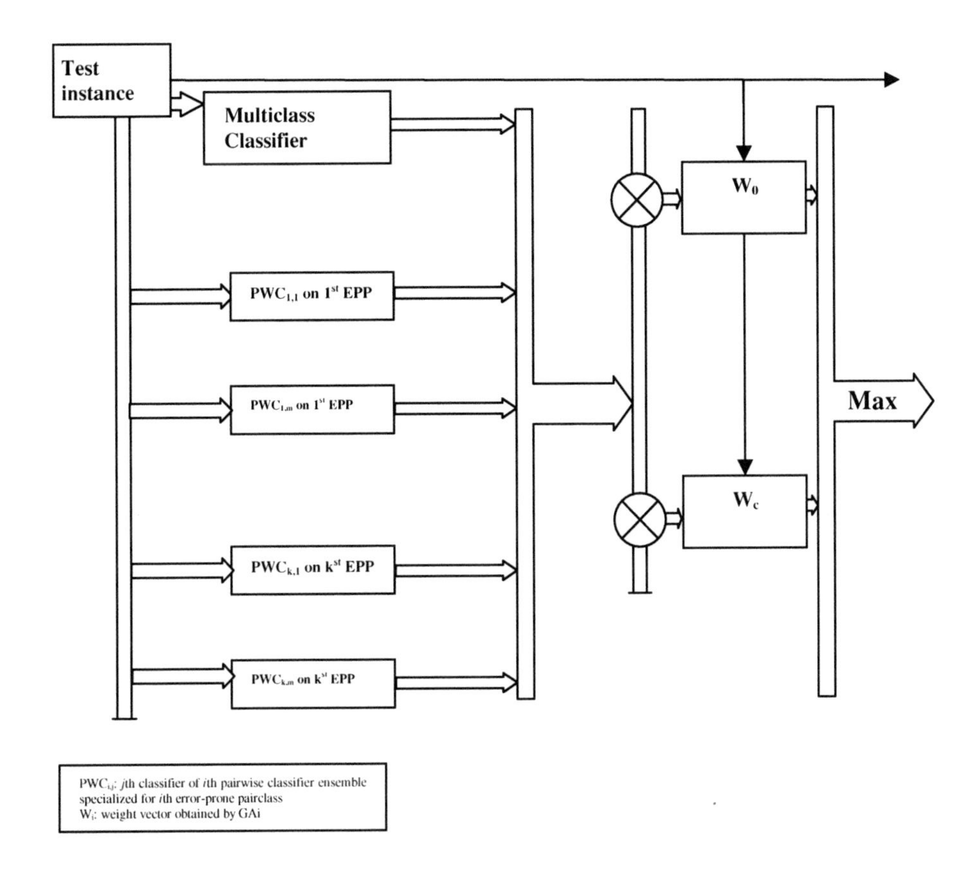

Fig. 3. Test phase of the proposed method based on GA

Standard English	0 1 2 3 4 5 6 7 8 9
Standard Farsi-1	۰ ۱ ۲ ۳ ۴ ۵ ۶ ۷ ۸ ۹
Standard Farsi-2	٠ ١ ٢ ٣ ٤ ٥ ٦ ٧ ٨ ٩
Hand-Written Farsi	

Fig. 4. Some instances of Persian OCR data set, with different qualities

As mentioned 9*11=110 pairwise classifiers are added to main multiclass classifier. As the parameter b is selected 20, so each of these classifiers is trained on only b precepts of corresponding train data. It means each of them is trained over 20 percept of the train set with the corresponding classes. The cardinality of this set is calculated by equation 10.

$$Car = \|train\| * 2 * b / c = 60000 * 2 * 0.2 / 10 = 2400 \qquad (10)$$

It means that each binary classifier is trained on 2400 datapoints with 2 class labels.

Table 1. Unsoft confusion matrix pertaining to the Persian handwritten OCR using an MLP

	0	**1**	**2**	**3**	**4**	**5**	**6**	**7**	**8**	**9**
0	969	0	0	4	1	14	2	0	0	1
1	4	992	1	0	2	4	1	1	1	15
2	1	1	974	18	9	1	4	4	0	1
3	0	0	13	957	12	0	3	2	0	1
4	5	0	3	17	973	3	2	2	0	3
5	15	0	0	0	0	977	1	0	0	0
6	2	6	2	1	3	0	974	5	1	3
7	3	0	3	1	0	1	1	986	0	0
8	0	1	0	1	0	0	2	0	995	0
9	1	0	4	1	0	0	10	0	3	976

The results of primary multiclass classifier and those of pairwise binary classifier ensembles are given to 10 GAs as inputs. Therefore, each chromosome contains 32 (22 for outputs of pairwise classifiers ensemble and a more 10 for outputs of the primary multiclass classifier) genes per each class. The number of GAs equals to the number of labels. Gaussian and Scattered operators are respectively used for mutation and recombination. Also, population size is 500. P_{mut} is set to 0.01 and the mutation is considered bitwise. $P_{crossover}$ is set to 0.8.

Termination condition is passing of 200 generations. Fitness function for GA is as mentioned in equation 3. The output of each GA is certainty of GA to select its corresponding class. Max function selects the most certain decision as final joint decision. Table 2 shows the accuracies of the different methods.

Table 2. Accuracies of different settings of the proposed framework

Methods	Ensemble of DTs	Ensemble of MLPs
A simple multiclass classifier	96.57	97.83
Weighed fusion with GA	98.99	99.04

4 Conclusion

In this paper, a new method is proposed to improve the performance of multiclass classification system. An arbitrary number of binary classifiers are added to main classifier to increase its accuracy. Then results of all these classifier are given to a set of GAs. The final results can competently obtain by certain weighting approach.

Usage of confusion matrix make proposed method a flexible one. The number of all possible pairwise classifiers is *c*(c-1)/2* that it is *O(c^2)*. Using this method without giving up a considerable accuracy, we decrease its order to *O(1)*. This feature of our proposed method makes it applicable for problems with a large number of classes. The experiments show the effectiveness of this method. Also we reached to very good results in Persian handwritten digit recognition.

References

1. Bala, J., De Jong, K., Huang, J., Vafaie, H., Wechsler, H.: Using learning to facilitate the evolution of features for recognizing visual concepts. Evolutionary Computation, Special Issue on Evolution, Learning, and Instinct: 100 years of the Baldwin Effect 4(3) (1997)
2. Bandyopadhyay, S., Muthy, C.A.: Pattern Classification Using Genetic Algorithms. Pattern Recognition Letters 16, 801–808 (1995)
3. Breiman, L.: Bagging Predictors. Journal of Machine Learning 24(2), 123–140 (1996)
4. Guerra-Salcedo, C., Whitley, D.: Feature Selection mechanisms for ensemble creation: a genetic search perspective. In: Freitas, A.A. (ed.) Data Mining with Evolutionary Algorithms: Research Directions – Papers from the AAAI Workshop, Technical Report WS-99-06, pp. 13–17. AAAI Press, Menlo Park (1999)
5. Gunter, S., Bunke, H.: Creation of classifier ensembles for handwritten word recognition using feature selection algorithms. In: IWFHR 2002, January 15 (2002)
6. Holland, J.: Adaptive in Natural and Artificial Systems. MIT Press, Cambridge (1992) (1st edn. The University of Michigan Press, Ann Arbor (1975))
7. Khosravi, H., Kabir, E.: Introducing a very large dataset of handwritten Farsi digits and a study on the variety of handwriting styles. Pattern Recognition Letters 28(10), 1133–1141 (2007)
8. Kuncheva, L.I., Jain, L.C.: Designing Classifier Fusion Systems by Genetic Algorithms. IEEE Transaction on Evolutionary Computation 33, 351–373 (2000)
9. Kuncheva, L.I.: Combining Pattern Classifiers, Methods and Algorithms. Wiley, New York (2005)
10. Martin-Bautista, M.J., Vila, M.A.: A survey of genetic feature selection in mining issues. In: Proceeding Congress on Evolutionary Computation (CEC 1999), Washington DC, pp. 1314–1321 (1999)
11. Minaei-Bidgoli, B., Punch, W.F.: Using Genetic Algorithms for Data Mining Optimization in an Educational Web-based System. In: Cantú-Paz, E., Foster, J.A., Deb, K., Davis, L., Roy, R., O'Reilly, U.-M., Beyer, H.-G., Kendall, G., Wilson, S.W., Harman, M., Wegener, J., Dasgupta, D., Potter, M.A., Schultz, A., Dowsland, K.A., Jonoska, N., Miller, J., Standish, R.K. (eds.) GECCO 2003. LNCS, vol. 2723, Springer, Heidelberg (2003)
12. Minaei-Bidgoli, B., Kortemeyer, G., Punch, W.F.: Mining Feature Importance: Applying Evolutionary Algorithms within a Web-Based Educational System. In: Proc. of the Int. Conf. on Cybernetics and Information Technologies, Systems and Applications (2004)

13. Parvin, H., Alizadeh, H., Fathi, M., Minaei-Bidgoli, B.: Improved Face Detection Using Spatial Histogram Features. In: Improved Face Detection Using Spatial Histogram Features. The 2008 Int. Conf. on Image Processing, Computer Vision, and Pattern Recognition (IPCV 2008), Las Vegas, Nevada, USA, July 14-17 (2008)
14. Punch, W.F., Pei, M., Chia-Shun, L., Goodman, E.D., Hovland, P., Enbody, R.: Further research on Feature Selection and Classification Using Genetic Algorithms. In: 5th International Conference on Genetic Algorithm, Champaign IL, pp. 557–564 (1993)
15. Saberi, A., Vahidi, M., Minaei-Bidgoli, B.: Learn to Detect Phishing Scams Using Learning and Ensemble Methods. In: IEEE/WIC/ACM International Conference on Intelligent Agent Technology, Workshops (IAT 2007), Silicon Valley, USA, November 2-5, vol. 5, pp. 311–314 (2007)
16. Vafaie, H., De Jong, K.: Robust feature Selection algorithms. In: Proceeding 1993 IEEE Int. Conf on Tools with AI, Boston, Mass., USA, pp. 356–363 (1993)
17. Yang, T.: Computational Verb Decision Trees. International Journal of Computational Cognition, 34–46 (2006)

An Efficient Hash-Based Load Balancing Scheme to Support Parallel NIDS

Nam-Uk Kim, Sung-Min Jung, and Tai-Myoung Chung

Internet Management Technology Laboratory,
Department of Computer Engineering,
School of Information and Communication Engineering,
Sungkyunkwan University,
300 Cheoncheon-dong, Jangan-gu,
Suwon-si, Gyeonggi-do, 440-746, Republic of Korea
Tel.: +82-31-290-7222; Fax: +82-31-299-6673
{nukim,smjung}@imtl.skku.ac.kr, tmchung@ece.skku.ac.kr

Abstract. Today, as the scale of network grows up, a standalone NIDS with only one intrusion detection node is not enough to inspect all traffic. One of the most widely considered solutions to address this problem is to configure parallel NIDS in which multiple intrusion detection nodes work together. A load balancing mechanism enables this configuration by distributing traffic load to several nodes. In the frequently changing environment of today's network, it is an important issue for load balancing mechanism to distributing traffic equally to each node. Meanwhile, several studies have been made on the load balancing scheme, but they do not satisfy the requirements of load balancing for parallel NIDS. Thus we proposed HLPN (Hash-based Load balancing scheme suitable for Parallel NIDS) which satisfies these requirements. As a result of the performance evaluation, HLPN represented 58% better performance in terms of the fairness of the traffic distribution than static hash-based scheme, and gave almost equal, or rather better, performance to that of DHFV.

Keywords: NIDS; Load Balancing; Hash-based Load balancing.

1 Introduction

Over the past years, information technology has been developed rapidly. So national or industrial major facilities increasingly rely on computing systems to produce, process, and preserve information. But side effects such as threats against secret information and stability of information processing systems also have been increased. This situation has become remarkably worse since computing systems have been connected each other through network.

A Network Intrusion Detection System (NIDS) is a system for preventing network of limited size from various threats. The researches for NIDS have been progressed since 1990s, and this system has been one of the most important

B. Murgante et al. (Eds.): ICCSA 2011, Part I, LNCS 6782, pp. 537–549, 2011.

equipment for providing security when organizing local network in one's company. An NIDS detects attacks arising from inside, as well as outside of network by monitoring and analysing network traffic.

Recently, the advance in computing performance of network equipments, transfer rate of transmission media network and so on, enabled high-speed and large-scale data transmission. So several studies have been made on design of NIDS in future network environment, and they include NetSTAT[6], EMERALD[7], Prelude[8]. In large-scale network, a standalone NIDS with only one detecting node is not enough to inspect all traffic for detecting attack. One of the most widely considered schemes to solve this problem is to configure parallel NIDS in which multiple detecting nodes work together. In parallel NIDS, incoming traffic should be divided into manageable size and distributed to multiple network intrusion detection nodes through load balancing mechanism. Furthermore, because all major NIDSs keep significant per-flow state to facilitate reassembling TCP byte streams, it is necessary for load balancing in parallel NIDS to preserve flows to each detecting node.

Flow based load balancing scheme is one of the load balancing scheme. In this scheme, the flow groups to be forwarded toward each node are determined, and traffics are distributed based on these groups. One of the most powerful schemes for flow based load balancing is hash based load balancing. SPANIDS[1], DHFV[2] are such hash based schemes which are our primary concern. They are efficient method for general distributed system in network, but have some problems to be implemented in parallel NIDS system. In this paper, we present HLPN which solves these problems and show performance evaluation of our proposed system.

This paper is organized as follows. The chapter 2 introduces hash based load balancing scheme and related researches, and describe about the problems of the proposed schemes. In this chapter, we also describe about characteristics of network traffic which have been used for our scheme. The chapter 3 presents essential mechanism of HLPN. The chapter 4 shows performance evaluation of HLPN. Finally, we conclude this paper in the chapter 4.

2 Related Work

2.1 Hash Based Load Balancing Scheme

As mentioned in previous chapter, hash based load balancing is powerful scheme for flow based load balancing. In this scheme, a load balancer has a lookup table in which each entry has mapping information between hash bin value and node ID number. For an incoming packet, a flow is identified by IP address pair and port number pair of source and destination. A hash bin value is generated by hashing this flow identification information. A node ID number indicates the NIDS node to which the flow identified by a hash bin value will be delivered. Hash based approach for load balancing is appropriate for flow based distribution because of two reasons. One is that a hash bin identifies a group of flows, and the other one is that hash value for arbitrary input is almost equally distributed.

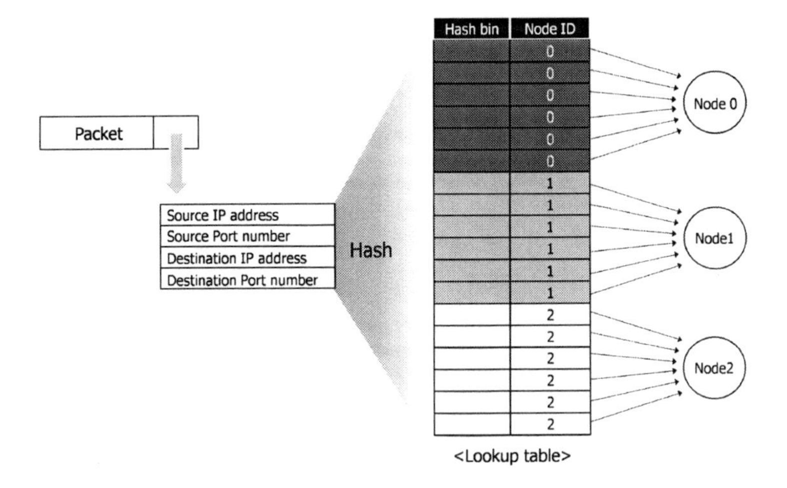

Fig. 1. Static hash based load balancing scheme

Fig. 1 introduces how hash based load balancing works. There are two kind of hash based load balancing which are static and dynamic hash based scheme. In static scheme, if lookup table entries are once generated, they are never changed. So breaking flow streams never occurs, but also, controlling flow is never available. However in dynamic scheme, the values inside of lookup table entries can be changed so that mapping between hash bin values and NIDS nodes to be altered. It means that if traffic loads are unbalanced between nodes, the dynamic scheme can control the traffics to be balanced in some measure. We will call the changing the mapping between hash bin values and NIDS nodes as rearrangement[1].

2.2 Related Studies of Dynamic Hash Based Load Balancing Scheme

SPANIDS is an architecture for parallel NIDS designed in 2005. SPANIDS uses a dynamic hash-based packet distribution scheme to address the challenges of a high-speed distributed NIDS. The rearrangement of lookup table occurs when an NIDS node issues flow control message to the load balancer when the input buffer of the node reaches a certain threshold. The hash table size scales in powers of two, proportional to the number of nodes[1].

Dynamic hashing with flow volume (DHFV) is one of the dynamic hash based load balancing scheme that uses flow volume for rearrangement. Flow volume of each flow is calculated by sum up the size of incoming packet which is available to get from the packet header[2].

Fig. 2 describes how the DHFV algorithm works. In the DHFV scheme, hash table size is set to be large enough to identify each active flow by unique bin value. Load balancing system observes outgoing buffer of each sublinks to node at every fixed period and check if there are buffers reached to a certain threshold. When overloaded buffers are found, it sorts the flows by volume size and get a

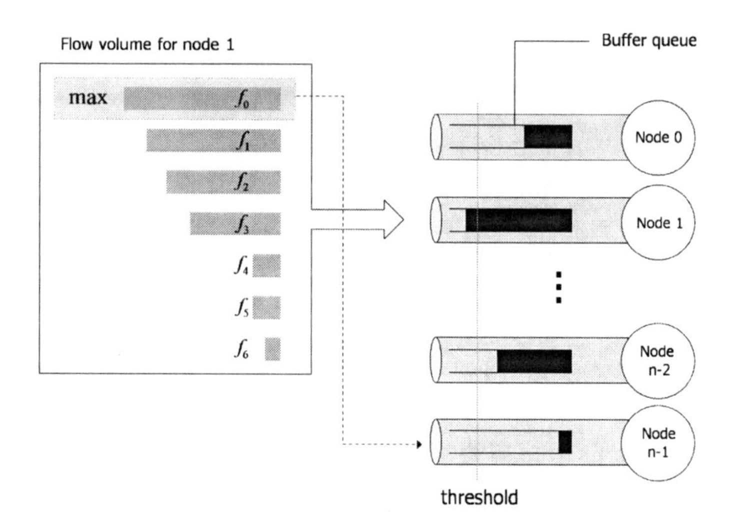

Fig. 2. Load balancing algorithm of DHFV

flow which has largest volume. Then reassigns the chosen flow to the node that shows minimum occupancy rate of buffer.

2.3 Problems of Existing Load Balancing Schemes When Applied in Parallel NIDS

Distributing packets using the schemes we've introduced has some problems when applied in parallel NIDS. Static hash based scheme is the best choice if we consider only accuracy in intrusion detection with unbroken flow. However, it cannot address load fluctuation at all. SPANIDS and DHFV perform rearrangement in lookup table to manage the situation when traffics are poorly distributed. However, their rearranging algorithm cannot prevent breaking flow streams. SPANIDS has so few number of hash bins, therefore a hash bin indicates so many flows. Thus if a rearrangement occurs, the direction of lots of flow streams are changed. In DHFV algorithm, the flow of largest volume is reassigned to other node when the outgoing buffer to a node reached to a certain threshold. Because the possibility of sustaining its flow stream grows when the flow volume size gets larger, It could be said that DHFV algorithm also changes the direction of flow streams.

In this paper, we present the system which provides the mechanism to rearrange the mapping table without breaking flow maintenance. In our proposed scheme, packets are distributed by preceded calculation that considers various aspects of network traffic and each node.

2.4 The Properties of Network Traffic Considerable for Load Balancing in NIDS

In 2001, the principle properties of network traffic was organized by Carey Williamson[3]. He analyzed Internet traffic and deduced some general properties of network traffic. Of course, it has been nearly 10 years since his study was published, but the result still be applicable because the main protocols used in communication in network are not changed at all. The properties of network traffic considerable for load balancing in NIDS are as follows:

1. Packet sizes are bimodally distributed. Approximately 50% of packets in network traffic have the size of Maximum Transmission Unit (MTU). On the other hand, nearly 40% them have the Minimum size, because TCP acknowledgment packets are frequently generated and transmitted.
2. Packet traffic is non-uniformly distributed. The source and destination addresses in packets are highly non-uniformly distributed. It means that the communicating frequency of each host in the Internet is greatly different each other.
3. Network traffic exhibits "locality" properties. In a short period, packets belongs to a flow tend to be sustainedly flowed if the volume of them is near to the size of MTU.

3 Proposed Load Balancing Scheme

In this chapter, we present our proposed load balancing system. First of all, we will introduce overall structure of load balancing system, and then describe principle mechanism. We will also show an example scenario at the last of this chapter.

Fig. 3 shows the overall structure of our proposed load balancing system. The structure was designed based on Gero Dittman's load balancer, adding table rearrangement unit which performs calculations for rearrangement at every fixed period[4].

Traffic distribution unit extracts flow identification information from incoming packet and get a hash value of it. Then the node to which the packet should be forwarded is decided by searching the lookup table.

Table rearrangement unit operates as a independent unit, seperated from the traffic distribution unit. It receives packet size, flow identification information and its hash value as input from the traffic distribution unit. With these values, it performs calculation that we will describe in detail later and updates lookup table at each fixed period.

As shown in Fig. 4 below, there are three types of period for our proposed scheme. Period for observation is only for simulation, in which the traffic volume for each NIDS node is calculated and shown as simulation result. In every period for rearrangement, table rearrangement unit performs calculation for learning

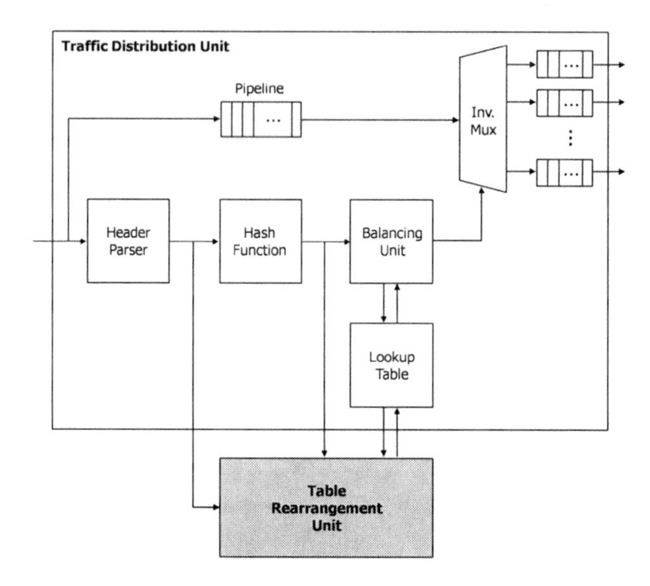

Fig. 3. The structure of load balancer

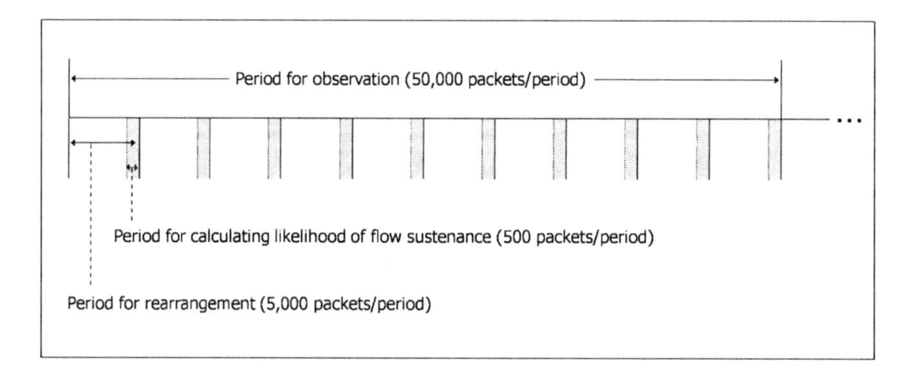

Fig. 4. Three types of period for proposed scheme

exact amount of traffic forwarded to each NIDS node during this period. Period for calculating likelihood of flow sustenance means the period in which the possibility of each flow's maintenance is judged by its volume size.

Table 1 shows terminologies and variables used for explaining operating mechanism of the table rearrangement unit. The table rearrangement unit get each flow f_i 's volume by adding packet size indicated in IP header, for two different periods. One is s_i, that is the flow volume measured during comparatively short period P_s, and it is used for considering temporal locality of a flow. The other is l_i, that is the flow volume measured during comparatively long period P_l ($= t \cdot P_s$) and it is used for sum up the amount of data forwarded to each node during fixed period.

Table 1. Terms and Variables

Term/Variable	Description
N	The number of NIDS nodes
m	The range of hash values (The hash values are distributed from 0 to $m-1$)
f_i	Flow i
$node_i$	Node i
P_{ob}	Observation period
P_s	Short period (Period for calculating likelihood of flow sustenance)
P_l	Long period (Rearrangement period)
t	P_l is an integer multiple of P_s ($P_l = t \cdot P_s$)
l_i	Long period flow volume of f_i
$l_{node(i)}$	Sum of all flows belong to $node_i$
l_{all}	Sum of all $l_{node0} \sim l_{node(N-1)}$
s_i	Short period flow volume of f_i
$W_{node(i)}$	Weight value that represents the relative performance of $node_i$
$T_{node(i)}$	Weight value that represents the relative amount of traffic forwarded to $node_i$
W	Sum of all $W_{node0} \sim W_{node(N-1)}$ or Sum of all $T_{node0} \sim T_{node(N-1)}$

Table update occurs at the end of every period P_l, and just before that, the calculation proceeds as follows. First of all, get $l_{node(j)}$ by adding up all the l_i s of each node. Then the traffic proportion $T_{node(i)}$ of each node is calculated using $l_{node0} \sim l_{node(N-1)}$ and l_{all}. For each node, $T_{node(i)}$ is compared $W_{node(i)}$ and if $T_{node(i)}$ is greater than $W_{node(i)}$, suitable amount of flows mapped to the node are rearranged to the node in which $T_{node(i)}$ is smaller than $W_{node(i)}$. Each flow's s_i is used for deciding what flows should be rearranged.

As mentioned in earlier chapter, Internet traffic has some temporal locality. Thus, if the period is short enough, as the flow volume measured in one period is longer, the possibility of maintaining the flow with similar size is higher. It means that, to accomplish flow based distribution, it is better to rearrange the flows that have shorter flow volume if a rearrangement is necessary. Therefore in our scheme, all flows in each node are sorted by s_i and the flows that have minimum s_i are decided to be rearranged.

Finally, the mapping between each flow and node is completely decided by some calculation that using l_i of each flow to be rearranged and difference between $W_{node(i)}$, $T_{node(i)}$ of each node. The overall rearrangement algorithm which should be implemented at the end of every period P_l is described below in detail.

Fig. 5 shows an example scenario of operation implemented in table rearrangement unit. In this example, we suppose that the number of NIDS nodes N is 4, the entire weight value for the system performance W is 20, and W_{node0}, W_{node1} for the $node_0$, $node_1$ are same as 5.

Rearrangement Algorithm

```
for i = 0 to m - 1
    l_all += l[i]

unit_w = round(l_all/w)

for i = 0 to N - 1
    for all flow j s in node i
        l_node[i] += l[j]
    T_node[i] = round(l_node[i]/unit_w)
    if T_node[i] - W_node[i] > 1
        E_node[i] = 0, l_sum = 0, j = 0
        sort flows in node i with s[] by ascending order and
            insert them into fsorted[]
        while l_sum < ( l_node[i] - W_node[i] * unit_w)
            l_sum += l[fsorted[j]]
            insert fsorted[j] into queue
            j++
    else if T_node[i] - W_node[i] < -1
        E_node[i] = W_node[i] * unit_w - l_node[i]
    else E_node[i] = 0

while queue is empty
    dequeue flow j from queue
    for i = 0 to N - 1
        if E_node[i] > l[j]
            rearrange flow j to node i
            E_node[i] -= l[j]
```

On the upper side in Fig. 5, there exists l_i and s_i values for each flow calculated during certain period. We suppose that l_{node0} is 550, l_{node1} is 200 and l_all is 1600. Then the system can learn that the proportion of traffic forwarded to $node_0$ is higher than the value of W_{node0}/W and to $node_1$ is lower than the value of W_{node0}/W. Therefore the flows belong to $node_0$ are sorted by s_i value and rearrangement occurs. The lower side in Fig. 5 shows the mapping status of the flows and nodes after the rearrangement is done.

4 Performance Evaluation

In this chapter, we present performance evaluation of our proposed scheme. The simulation was focused on distribution of traffic sizes forwarded to each node at every observation period. We generated 1,000,000 packets, and the same sequence of packets is used for simulating packet distribution with the schemes which are static hash based scheme, DHFV and HLPN. We assume that there are 4 NIDS nodes in parallel NIDS system, and each node has the same capacity for performing intrusion detection.

Before rearrangement

Hash bin value	Long period flow volume	Short period flow volume	
11	l_{11}	s_{11}	max
0	l_0	s_0	
2	l_2	s_2	
3	l_3	s_3	
7	l_7	s_7	
16	l_{16}	s_{16}	
21	l_{21}	s_{21}	min
30	l_{30}	s_{30}	max
1	l_1	s_1	
9	l_9	s_9	
28	l_{28}	s_{28}	
17	l_{17}	s_{17}	
32	l_{32}	s_{32}	min

Node 0
Node 1
Node i

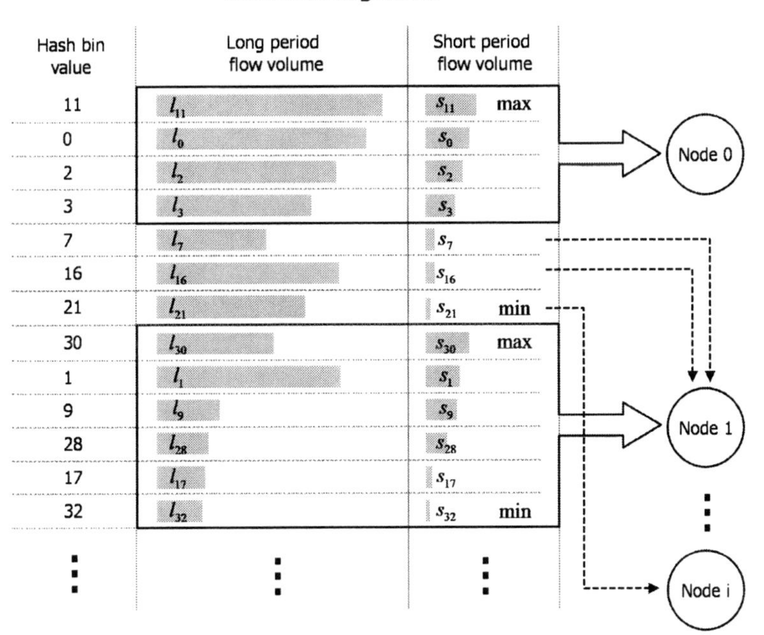

Fig. 5. Before & After rearrangement in certain period

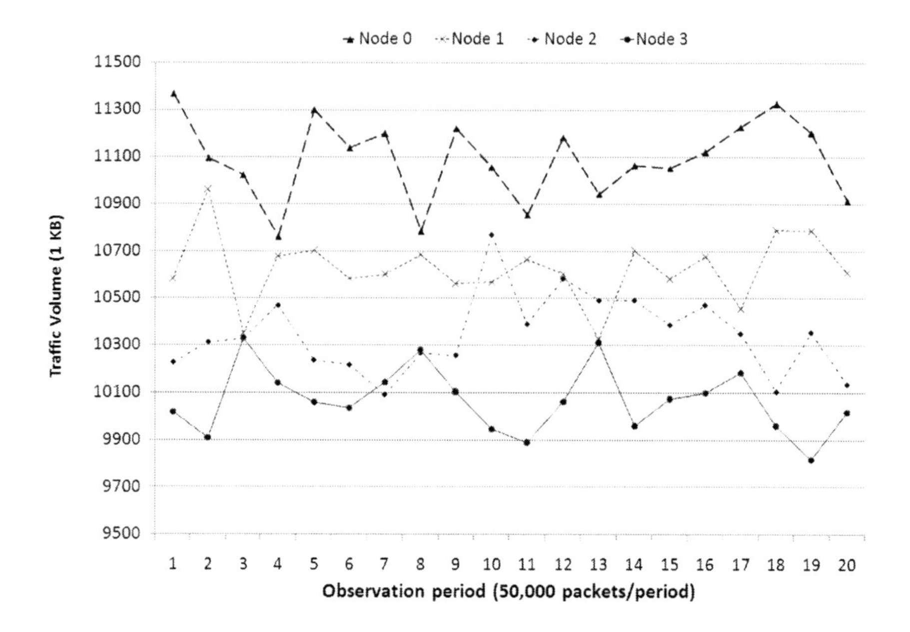

Fig. 6. Distribution of traffic loads by static hash-based scheme

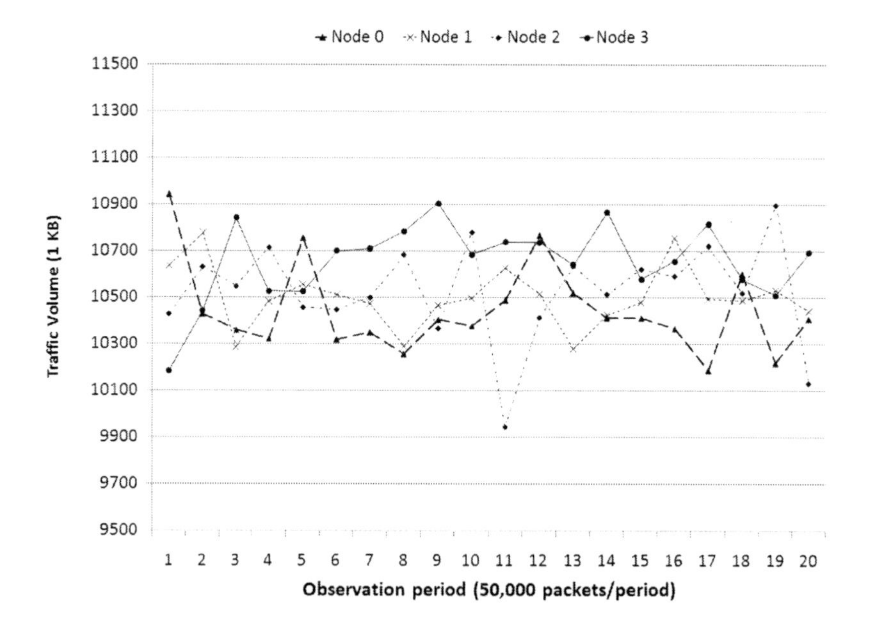

Fig. 7. Distribution of traffic loads by DHFV scheme

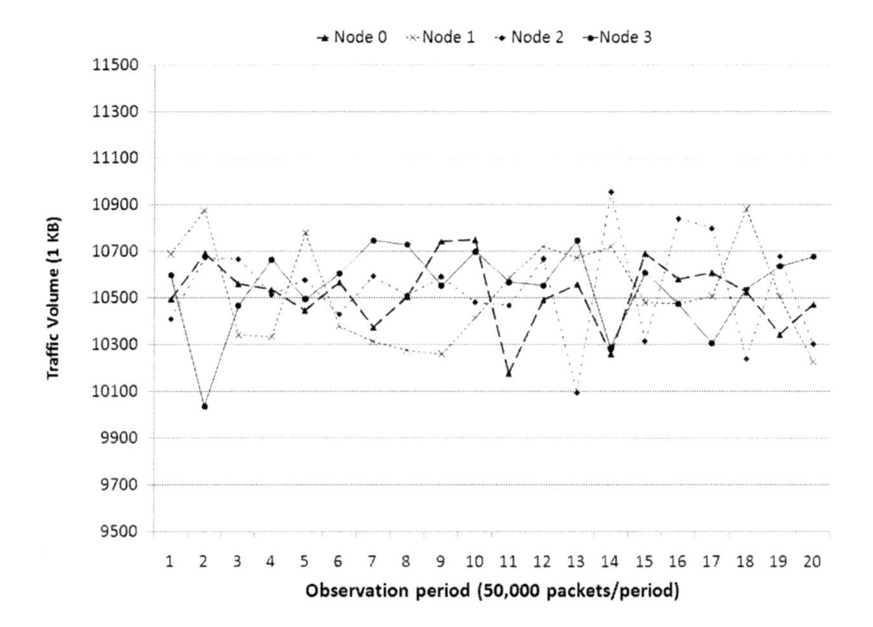

Fig. 8. Distribution of traffic loads by HLPN scheme

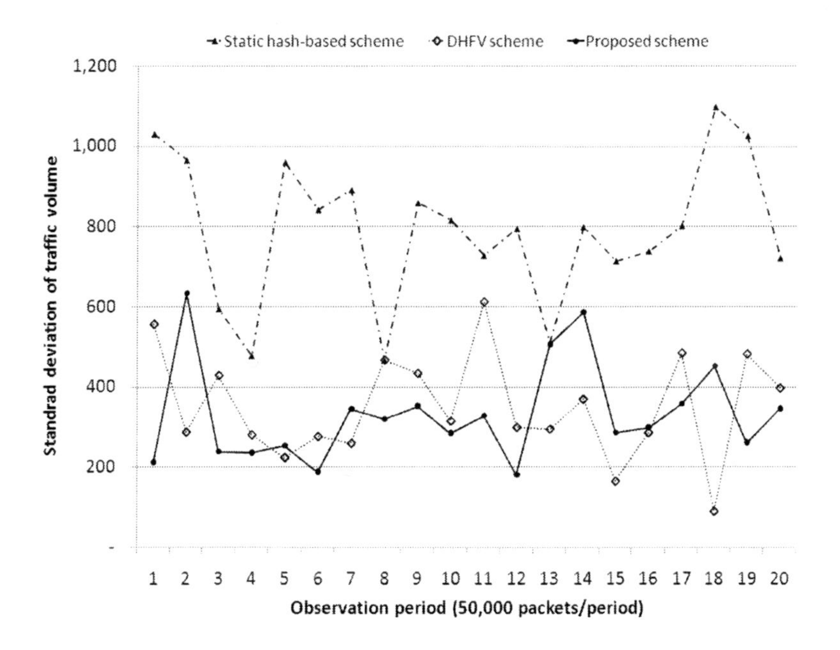

Fig. 9. Standard deviation of traffic loads for each node

Fig. 6 - 8 illustrate the distribution of traffic loads for each node at every observation period when implementing static hash-based scheme, DHFV and HLPN. The simulation result shows that DHFV and HLPN distribute traffic much more evenly than static hash-based scheme. It also says that HLPN has similar performance to DHFV in terms of evenness of traffic distribution.

For the three schemes, we also calculated σ (standard deviation) value of traffic volume for each node to numerically evaluate the fairness of the traffic distribution, as shown in Fig. 9. In static hash-based scheme, the maximum σ value was 1,098 and the minimum was 466. In DHFV and HLPN, the maximum σ value was 611 and 633, and the minimum was 90 and 180, respectively. The mean of σ was 791.79, 349.97 and 333.04 for hash-based scheme, DHFV and HLPN respectively. This result states that HLPN gives significant improvement of about 58% in terms of fairness of the traffic distribution, relative to the case of implementing static hash based scheme. Meanwhile HLPN gives performance almost equal to, or rather better than, that of DHFV in terms of fairness of the traffic distribution.

5 Conclusion

In this paper, we proposed the load balancing system in which packets are distributed by preceded calculation that consider various aspects of network traffic and each node's performance. HLPN provides the mechanism to rearrange the lookup table without breaking flow sustenance. Therefore, it is much more suitable for implementing in parallel NIDS than any other hash-based load balancing methods. Moreover, as shown in performance evaluation, HLPN also gives 58% better performance in terms of the fairness of the traffic distribution than static hash-based scheme, and gives almost equal, or rather better, performance to that of DHFV.

In the future work, we will perform various experiments on HLPN scheme changing the number of NIDS nodes, the range of hash bin value, P_s, P_l, and so on. From that, we will find appropriate values for variables mentioned above in HLPN. After all, we expect that the HLPN would become more realistic and sophisticated load balancing scheme suitable for parallel NIDS.

References

1. Schaelicke, L., Wheeler, K., Freeland, C.: SPANIDS: a Scalable Network Intrusion Detecion Loadbalancer. In: 2nd Conference on Computing Frontiers, pp. 315–322. ACM, New York (2005)
2. Jo, J.-Y., Kim, Y.-H., Chao, H.J., Merat, F.: Internet Traffic Load Balancing using Dynamic Hashing with Flow Volume. In: Proceeding of SPIE ITCom, Boston, vol. 4865, pp. 154–165 (2002)
3. Williamson, C.: Internet Traffic Measurement. IEEE Internet Computing 5, 70–74 (2001)
4. Dittmann, G., Herkersdorf, A.: Network Processor Load Balancing for High-Speed Links. In: SPECTS 2002, San Diego, pp. 727–735 (2002)

5. Vallentin, M., Sommer, R., Lee, J., Leres, C., Paxson, V., Tierney, B.: The NIDS Cluster: Scalable, Stateful Network Intrusion Detection on Commodity Hardware. In: Kruegel, C., Lippmann, R., Clark, A. (eds.) RAID 2007. LNCS, vol. 4637, pp. 107–126. Springer, Heidelberg (2007)
6. Vigna, G., Kemmerer, R.: NetSTAT: A Network-based Intrusion Detection System. Journal of Computer Security 7, IOS Press
7. Porras, P., Neumann, P.: EMERALD: Event Monitoring Enabled Response to Anomalous ive Disturbances. In: Proceeding of the 20th National Information Systems Security Conference, pp. 353–365 (1997)
8. Blanc, M., Oudot, L., Glaume, V.: Global Intrusion Detection: Prelude Hybrid IDS. Technical report (2003)
9. Cao, Z., Wang, Z., Zegura, E.W.: Performance of Hashing-based Schemes for Internet Load Balancing. In: IEEE INFOCOM 2000, Israel, vol. 1, pp. 332–341 (2000)
10. Martin, R., Menth, M., Hemmkeppler, M.: Accuracy and Dynamics of Hash-based Load Balancing Algorithms for Multipath Internet Routing. In: IEEE International Conference on Broadband Communication, Networks and Systems (BROADNETS), San Jose (2006)
11. Shi, W., MacGregor, M.H., Gburzynski, P.: Load Balancing for Parallel Forwarding. IEEE/ACM Transactions on Networking (TON) 13, 790–801 (2005)

Penalty Functions for Genetic Programming Algorithms

José L. Montaña[1], César L. Alonso[2],
Cruz Enrique Borges[1], and Javier de la Dehesa[1]

[1] Departamento de Matemáticas, Estadística y Computación,
Universidad de Cantabria, 39005 Santander, Spain
{montanjl,borgesce}@unican.es

[2] Centro de Inteligencia Artificial, Universidad de Oviedo
Campus de Viesques, 33271 Gijón, Spain
calonso@aic.uniovi.es

Abstract. Very often symbolic regression, as addressed in Genetic Programming (GP), is equivalent to approximate interpolation. This means that, in general, GP algorithms try to fit the sample as better as possible but no notion of generalization error is considered. As a consequence, overfitting, code-bloat and noisy data are problems which are not satisfactorily solved under this approach. Motivated by this situation we review the problem of Symbolic Regression under the perspective of Machine Learning, a well founded mathematical toolbox for predictive learning. We perform empirical comparisons between classical statistical methods (AIC and BIC) and methods based on Vapnik-Chrevonenkis (VC) theory for regression problems under genetic training. Empirical comparisons of the different methods suggest practical advantages of VC-based model selection. We conclude that VC theory provides methodological framework for complexity control in Genetic Programming even when its technical results seems not be directly applicable. As main practical advantage, precise penalty functions founded on the notion of generalization error are proposed for evolving GP-trees.

Keywords: Genetic Programming, Symbolic Regression, Inductive Learning, Regression Model selection, genetic programming, symbolic regression.

1 Introduction

In the last years Genetic Programming (GP) has been applied to a range of complex learning problems, including that of symbolic regression in a variety of fields like quantum computing, electronic design, sorting, searching, game playing, etc. For dealing with these problems GP evolves a population composed by symbolic expressions built from a set of functionals $F = \{f_1, \ldots, f_k\}$ and a set of terminals $T = \{x_1, \ldots, c_1, \ldots, c_t\}$ (including the variables and the constants). Once the functionals and the terminals have been selected, the regression task can be thought as a supervised learning problem where the hypothesis class $\mathcal{H}$

B. Murgante et al. (Eds.): ICCSA 2011, Part I, LNCS 6782, pp. 550–562, 2011.

is the tree structured search space described from the set of leaves T and the set of nodes F. Analogously, the GP algorithm evolving symbolic expressions representing the concepts of class $\mathcal{H}$ can be regarded as a supervised learning algorithm that selects the best model inside the class $\mathcal{H}$.

Regarding this consideration of GP as a supervised learning task, we use tools from Statistical Learning Theory (SLT) ([10]) with the purpose of model selection in Genetic Programming. This point of view has been previously proposed in [9] (see also [2]). In that works the core of Vapnik-Chervonenkys theory is translated into the GP domain with the aim of addressing the code-bloat problem. A further development of this point of view, including some experimental discussion, can be found in [7].

In the present paper we focus our attention on problems presenting noisy data. We try to identify the shape of good penalty terms in order to minimize the error of generalization, that is, the error over unseen points. Usually, analytic model selection criteria like AIC (Akaike Information Criterium) and BIC (Bayesian Information Criterium) estimate the generalization error as a function of the empirical error with a penalization term related with some measure of model complexity. Then this function is minimized in the class of concepts $\mathcal{H}$. Since most model selection criteria, in particular analytic model selection and Structural Risk Minimization based on VC theory, are based on certain assumptions, mainly linearity and exact computation of the classification capacity of the class of concepts $\mathcal{H}$, it is important to perform empirical comparisons in order to understand their practical usefulness in settings when these assumptions may not hold, which is the case of Genetic Programming.

The paper is organized as follows. Section 2 is devoted to present some useful tools from statistical learning theory. Section 3 describes classical model selection criteria (AIC and BIC) and the structural risk minimization approach with a new measure of model complexity founded in VC analysis of GP. Section 4 describes the experimental setting and results. Finally, Section 5 contains some conclusions.

2 Statistical Learning Theory

In the seventies the work by Vapnik and Chervonenkis ([12], [10], [11]) provided a remarkable family of bounds relating the performance of a learning machine. The Vapnik- Chervonenkis dimension (VC-dimension) is a measure of the capacity of a family of functions (or learning machines) $f \in \mathcal{H}$ as classifiers.

In a binary classification problem, an instance x is classified by a label $y \in \{-1, 1\}$. Given a vector of n instances, $(x_1, \ldots, x_n)$, there are 2^n possible classification tuples $(y_1, \ldots, y_n)$, with $y_i \in \{-1, 1\}$. If for each classification tuple $(y_1, \ldots, y_n)$ there is a classifier $f \in \mathcal{H}$ with $f(x_i) = y_i$, for $1 \leq i \leq n$, we say that $(x_1, \ldots, x_n)$ is shattered by the class $\mathcal{H}$. The VC dimension of a class $\mathcal{H}$ is defined as the maximum number of points that can be shattered by $\mathcal{H}$. If VD dimension is h this means that there exists at least some set of h points which can be shattered. For instance, VC dimension of lines in the plane is 3, more generally, VC dimension of hyperplanes in $\mathbb{R}^n$ is $n + 1$.

In general, the error, $\varepsilon(f)$, of a learning machine or classifier f is written as

$$\varepsilon(f) = \int Q(x, f; y) d\mu, \tag{1}$$

where Q measures some notion of loss between $f(x)$ and y, and μ is the distribution from which examples (x, y) are drawn to the learner, usually x is called the instance and y the label. For example, for classification problems, the error of misclassification is given taking $Q(x, f; y) = |y - f(x)|$. Similarly, for regression tasks one takes $Q(x, f; y) = (y - f(x))^2$ (mean square error). Many of the classic applications of learning machines can be explained inside this formalism. The starting point of statistical learning theory is that we might not know μ. At this point one replace theoretical error $\varepsilon(f)$ by empirical error that is estimated from a finite sample $\{x_i, y_i)\}_{i=1}^n$ as:

$$\varepsilon_n(f) = \frac{1}{n} \sum_{i=1}^{n} Q(x_i, f; y_i)). \tag{2}$$

Now, the results by Vapnik state that the error $\varepsilon(f)$ can be estimated independent of the distribution of $\mu(x, y)$ due to the following formula.

$$\varepsilon(f) \leq \varepsilon_n(f) + \sqrt{\frac{h(log(2n/h) + 1) - log(\eta/4)}{n}}, \tag{3}$$

where η is the probability that bound is violated and h is the VC dimension of the family of classifiers $\mathcal{H}$ from which function f is selected. The second term of the right hand side is called the VC confidence (as example, for $h = 200$, $n = 100000$ and $\eta = 0.95$ the VC confidence is 0.12. While the existence of the bounds in Equation 3 is impressive, very often these bounds remain meaningless. For instance, VC dimension of the family of Support Vector Machines embedded in m dimensions with polynomial kernels is infinite, however if we also bound the degree of the polynomials, then VC dimension is finite and depends of dimension m and degree d, indeed, it is bounded by $(4ed)^m$. Note that this quantity grows very quickly which again makes the bound in Equation 3 useless.

2.1 VC Dimension of GP

The VC dimension h depends on the class of classifiers, equivalently on a fully specified learning machine. Hence, it does not make sense to calculate VC dimension for GP in general, however it makes sense if we choose a particular class of computer programs as classifiers (i.e. a particular genotype). For the simplified genotype that only uses algebraic analytic operators of bounded degree (polynomials, square roots and, in general, power series with fractional exponents), some chosen computer program structure and a bound on the non-scalar height of the program, VC dimension remains polynomial in the non-scalar height of the program and in the number of parameters of the learning machine.

To make clear the notion of non-scalar height it is enough to define the notion of non-scalar (or non-linear) node. We say that a node of a GP-tree is non-scalar if it is not a linear combination of its sons. With this notion we can estate the following general result whose proof is a refinement of the techniques introduced in [6], but having into account that degree only increases at non-scalar nodes.

Theorem 1. *Let $\mathcal{C}_{k,n}$ be the concept class whose elements are GP-trees $\mathcal{T}_{k,n}$ having $k+n$ terminals (k constants and n real variables) and non-scalar height $h = h(k,n)$. Assume that the GP-tree $\mathcal{T}_{k,n}$ has at most q analytic algebraic nodes of degree bounded by $D \geq 2$ and number of sons bounded by β. Then, the VC dimension of $\mathcal{C}_{k,n}$ is in the class*

$$O((log_2 D + log_2\ max\{\beta, 2\})\ k(n+k+\beta q)^2 h^2).$$

Hence, GP approach with analytic algebraic functionals, and "short" programs (of height polynomial in the dimension of the space of events) has small VC dimension. For a class of models $\mathcal{H}$ with finite complexity (for instance –in the case of GP– trees with bounded size or height), the model can be chosen minimizing the empirical error:

$$\varepsilon_n(f) = \frac{1}{n}\sum_{i=1}^{n} Q(x_i, f, y_i) \tag{4}$$

The problem of model selection –also called complexity control– arises when a class of models consists of models of varying complexity (for instance –in the case of Genetic Programming– trees with varying size or height). Then the problem of regression estimation requires optimal selection of model complexity (i.e., the size or the height) in addition to model estimation via minimization of empirical risk as defined in Equation 4.

3 Penalty Functions

In general, analytical estimates of error (Equation 1) as a function of empirical error (Equation 4) take one of the following forms:

$$\varepsilon(f) = \varepsilon_n(f).pen(h, n) \tag{5}$$

$$\varepsilon(f) = \varepsilon_n(f) + pen(h/n, \sigma^2), \tag{6}$$

where f is the model, *pen* is called the penalization factor, h is the model complexity, n is the size of the training set and σ, when used, is the standard deviation of the additive noise (Equation 6).

The first two analytical estimates of error that we shall use in this work are the following well known representative statistical methods:

- Akaike Information Criterium (AIC) which is as follows (see [1]):

$$\varepsilon(f) = \varepsilon_n(f) + \frac{2h}{n}\sigma^2 \tag{7}$$

- Bayesian Information Criterium (BIC) (see [3]):

$$\varepsilon(f) = \varepsilon_n(f) + (ln\ n)\frac{h}{n}\sigma^2 \quad (8)$$

As it is described in [4], when using a linear estimator with parameters, one first estimates the noise variance from the training data (x_i, y_i) as:

$$\sigma^2 = \frac{n}{n-h}\frac{1}{n}\sum_{1\leq i\leq n}(y_i - \hat{y}_i)^2 \quad (9)$$

$\hat{y}_i$ is the estimation of value y_i by model f, i.e. $\hat{y}_i = f(x_i)$. Then one can use Equation 9 in conjunction with AIC or BIC for each (fixed) model complexity. The estimation of the model complexity h for both methods is the number of free parameters of the model f.

The third model selection method used in this paper is based on the Structural Risk Minimization (SRM) (see [10])

$$\varepsilon(f) = \varepsilon_n(f).\left(1 - \sqrt{p - p\ ln\ p + \frac{ln\ n}{2n}}\right)^{-1}, \quad (10)$$

where $p = \frac{h}{n}$, and h stands for the Vapnik-Chervonenkis (VC) dimension as a measure of model complexity. Note that under SRM approach, it is not necessary to estimate noise variance. However an estimation of the VC dimension is required.

3.1 The Genetic Programming Approach

The above model selection criteria are used in the framework of linear estimators and the model complexity h, in this case, is the number of free parameters of the model (for instance, in the familiar case where the models are polynomials, h is the degree of the polynomial).

In our attempt to carry the above methods to GP, we will use GP-trees as the evolving structures that represent symbolic expressions or models. The internal nodes of every GP-tree are labeled by functionals from a set $F = \{f_1, \ldots, f_k\}$ and the leaves of the GP-tree are labeled by terminals from $T = \{x_1, \ldots, c_1, \ldots, c_t\}$. As a GP-tree represents some symbolic expression f, we will use the equations 7, 8 and 10 as the different fitness functions for f in our study. For our GP version of AIC and BIC we will maintain as model complexity the number of free parameters of the model f represented by the tree. On the other hand, in the case of SRM we will introduce a new estimator for the VC dimension of GP-trees. This estimator consists in the number of non-scalar nodes of the tree, that is, nodes which are not labeled with $\{+, -\}$ operators. This is a measure of the non-linearity of the considered model and can be seen as a generalization of the notion of degree to the case of GP-trees. This notion is related with the VC dimension of the set of models given by GP-trees using a bounded number of non-scalar operators. The exact relationship between non-scalar size of a GP-tree (more generally, a computer program) and its VC dimension is showed in Theorem 1 stated in Section 2. The recombination operators used are the well known standard crossover and mutation operators for trees in Genetic Programming (see [5]). Then an extensive experimentation has been done in order to compare the performance of these three model selection criteria in the Genetic Programming framework.

4 Experimentation

4.1 Experimental Settings

We consider instances of symbolic regression problem for our experimentation. We have executed the algorithms over two groups of target functions. The first group includes the following three functions that also were used in [4] for experimentation:

Discontinuous piecewise polynomial function:

$$\begin{array}{ll} g_1(x) = 4(x^2(3-4x) & x \in [0, 0.5] \\ g_1(x) = (4/3)x(4x^2 - 10x + 7) - 3/2 & x \in (0.5, 0.75] \\ g_1(x) = (16/3)x(x-1)^2 & x \in (0.75, 1]] \end{array} \quad (11)$$

Sine-square function:

$$g_2(x) = sin^2(2\pi x),\ x \in [0, 1] \quad (12)$$

Two-dimensional *sin* function:

$$g_3(x) = \frac{sin\sqrt{x_1^2 + x_2^2}}{x_1^2 + x_2^2},\ x_1,\ x_2 \in [-5, 5] \quad (13)$$

The second group of functions is constituted by five functions of several classes: trigonometric functions, polynomial functions and one exponential function. These functions are the following:

$$\begin{array}{ll} f_1(x) = x^4 + x^3 + x^2 + x & x \in [-5, 5] \\ f_2(x) = e^{-sin\ 3x + 2x} & x \in [-\frac{\pi}{2}, \frac{\pi}{2}] \\ f_3(x) = e\, x^2 + \pi\, x & x \in [-\pi, \pi] \\ f_4(x) = cos(2x) & x \in [-\pi, \pi] \\ f_5(x) = min\{\frac{2}{x}, sin(x) + 1\} & x \in [0, 15] \end{array} \quad (14)$$

For the first group of target functions we use the following set of functionals $F = \{+, -, *, //\}$, incremented with the *sign* operator for the target function g_1. In the above set F, "//" indicates the protected division, i.e. $x//y$ returns x/y if $y \neq 0$ and 1 otherwise. The terminal set T consists of the variables of the corresponding function and includes the set of constants $\{0, 1\}$.

For the second group of functions, the basic set of operations F is incremented with other operators. This aspect for each function is showed in table 1.

Table 1. Function set for the second group of target functions

Function	Function set
f_1	$F \cup \{sqrt\}$
f_2	$F \cup \{sqrt, sin, cos, exp\}$
f_3	$F \cup \{sin, cos\}$
f_4	$F \cup \{sqrt, sin\}$
f_5	$F \cup \{sin, cos\}$

We use GP-trees with height bounded by 8. As it was mentioned above, the model complexity h is measured by the number of non-scalar nodes of the tree for the SRM method (fitness equation 10) and by the number of free parameters of the model f represented by the tree, for AIC and BIC methods (fitness equations 7 and 8 respectively). The rest of the parameters for the genetic training process are the following: population size $M = 100$; maximum number of generations $G = 1000$; probability of crossover $p_c = 0,9$ and probability of mutation $p_m = 0,1$. Tournament selection and the standard operators of crossover and mutation for tree-like structures are used. For all the executions, the genetic training process finishes after 10^7 operations have been computed. Observe that the number of computed operations equals the internal nodes of the trees that are visited during the process. Training sets of $n = 30$ examples are generated where the x- values follow from uniform distribution in the input domain. For the computation of the y-values, the equation **??** is used in order to corrupt the values with noise. The noise variance σ was fixed to 0.2.

The experimentation scheme is as follows: For each model selection criterium (AIC, BIC, SRM) and each target function, we use a simple competitive co-evolution strategy where 10 populations of 100 individuals evolve independently, considering same training set. Then, we select the model proposed by the best of these 10 executions. The above process completes one experiment. We have performed 100 experiments and for each one a different random realization of 30 training samples was considered. Hence for each target function, we have executed each algorithm 1000 times and finally 100 models have been selected. These models are the best ones of each group of 10 populations related to same training set.

4.2 Experimental Results

When the competitive genetic training process finishes, the best individual is selected as the proposed model for the corresponding target function. In order to measure the quality of the selected model, it makes sense to consider a new set of points generated without noise from the target function. This new set of examples is known as the test set or validation set. So, let $(x_i, y_i)_{1 \leq i \leq n_{test}}$ a validation set for the target function $g(x)$ (i.e. $y_i = g(x_i)$) and let $f(x)$ be the model estimated from the training data. Then the prediction risk $\varepsilon_{n_{test}}$ is defined by the mean square error (MSE) between the values of f and the true values of the target function g over the validation set:

$$\varepsilon_{n_{test}} = \frac{1}{n_{test}} \sum_{i=1}^{n_{test}} (f(x_i) - y_i)^2 \tag{15}$$

Frequently, when different Genetic Programming strategies solving symbolic regression instances are compared, the quality of the selected model is evaluated by means of its corresponding fitness value over the training set. But with this quality measure, a fitness value close to zero does not necessary imply a good model for the target function. It is not possible to distinguish between good executions and overfiting executions. In fact when the training set is corrupted with noise (which is our case), final selected models with very low fitness values are probably not so good models as they seem.

Figures 1 and 2 show comparison results for AIC, BIC and SRM, after that the 100 experiments were completed. The empirical distribution of the prediction risk for each model selection is displayed using standard box plot notation with marks at 25%, 50% and 75% of that empirical distribution. The first and last mark in each case, stands for the prediction risk of the best and the worst selected model respectively. In all cases the size n_{test} of the validation set is 200. Note that the scaling is different for each function.

A first analysis of the figures concludes that SRM model selection performs clearly better than AIC and BIC. Note that for all of the studied target functions, the SRM strategy obtains prediction risk values for the best execution that are lower than those obtained by AIC and BIC strategies. This situation also happens for the most part of the target functions, if we consider the 25%, 50% or 75% of the selected models from the empirical distribution. On the other hand, AIC and BIC perform quite similar. This could be because both fitness functions (equations 7 and 8) take the same structure (equation 6) with very similar additive penalization terms.

In table 2 the best prediction risk value for each target function and model selection criterium is displayed and tables 3 and 4 show, respectively, the mean of the prediction risk values considering the 5 and the 25 best experiments.

Table 2. Prediction risk values of the best obtained model for each target function

Function	AIC	BIC	SRM
f_1	2.44E-07	2.40E-07	**2.13E-07**
f_2	3.47E+00	1.33E+00	**8.29E-02**
f_3	1.26E+00	1.26E+00	**6.71E-01**
f_4	4.44E-01	4.44E-01	**3.87E-01**
f_5	3.84E-01	2.15E-01	**2.64E-02**
g_1	2.94E-02	3.07E-02	**1.33E-02**
g_2	1.26E-01	1.15E-01	**1.04E-01**
g_3	3.21E-02	3.21E-02	**1.85E-02**

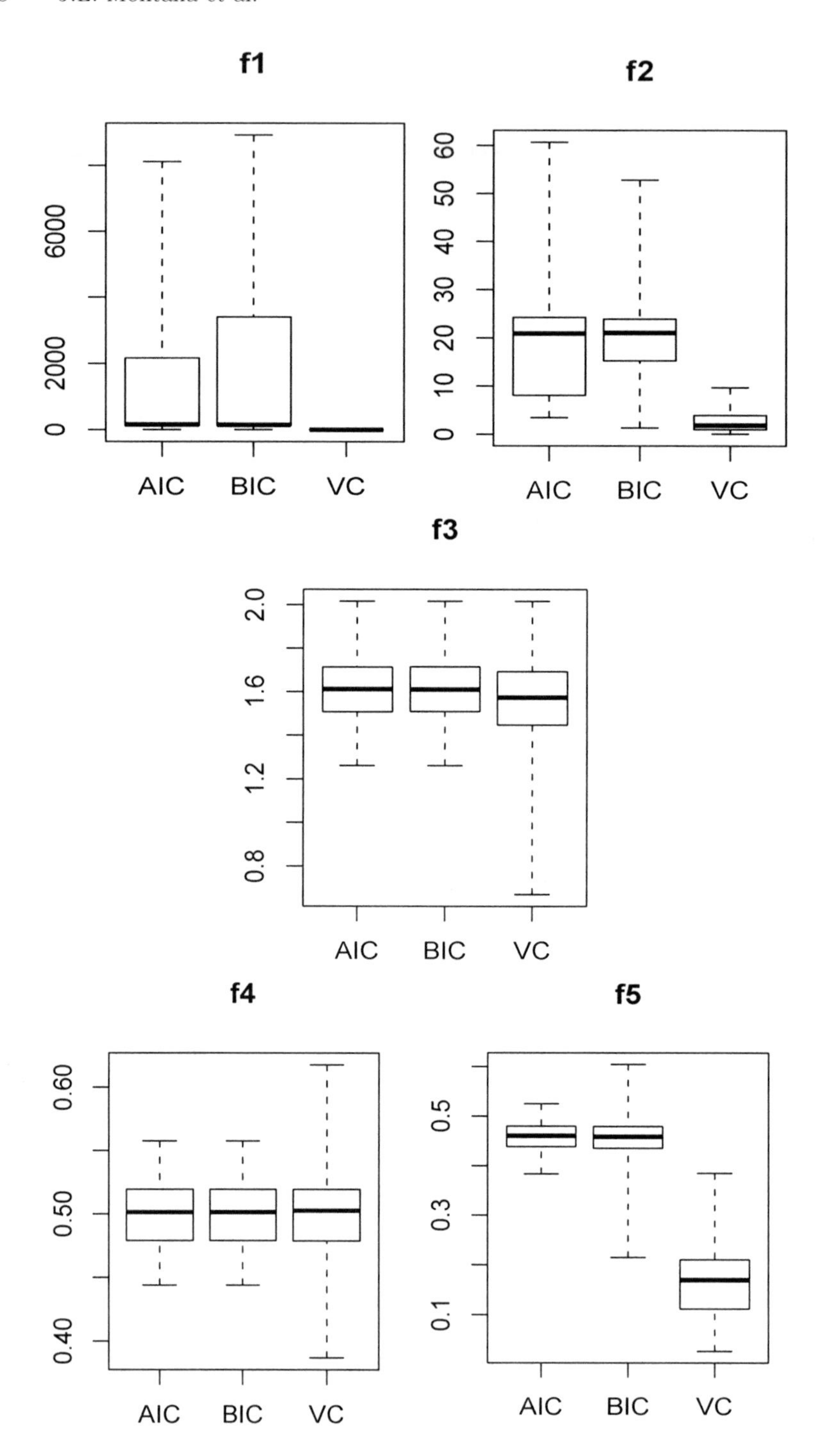

Fig. 1. Empirical distribution of the prediction risk $\varepsilon_{n_{test}}$ for target functions f_1 to f_5, over 100 experiments

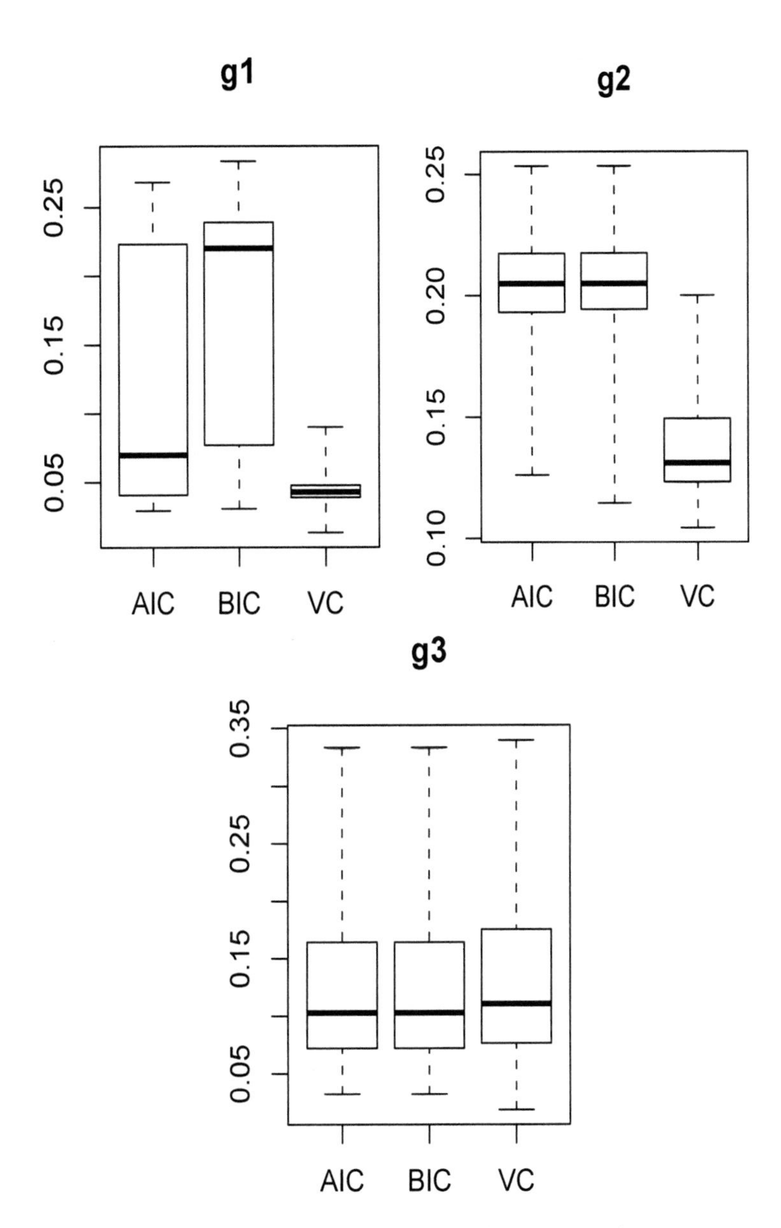

Fig. 2. Empirical distribution of the prediction risk for target functions g_1 to g_3, over 100 experiments

Table 3. Mean prediction risk of the best 5%

Function	AIC	BIC	SRM
f_1	3.01E-07	2.74E-07	**2.42E-07**
f_2	4.57E+00	3.48E+00	**5.06E-01**
f_3	1.31E+00	1.31E+00	**6.91E-01**
f_4	4.50E-01	4.50E-01	**4.36E-01**
f_5	4.05E-01	3.34E-01	**3.84E-02**
g_1	3.13E-02	3.33E-02	**2.30E-02**
g_2	1.33E-01	1.55E-01	**1.12E-01**
g_3	4.24E-02	4.24E-02	**3.67E-02**

Table 4. Mean prediction risk of the best 25% of the models, for each target function

Function	AIC	BIC	SRM
f_1	1.74E+01	3.24E+01	**2.89E-07**
f_2	6.37E+00	9.76E+00	**8.27E-01**
f_3	1.42E+00	1.43E+00	**1.04E+00**
f_4	4.66E-01	4.66E-01	**4.64E-01**
f_5	4.24E-01	4.10E-01	**7.47E-02**
g_1	3.67E-02	4.00E-02	**3.35E-02**
g_2	1.71E-01	1.81E-01	**1.18E-01**
g_3	**5.64E-02**	**5.64E-02**	5.82E-02

Table 5. Mean prediction risk of the 100 executions for each pair strategy-target function

Function	AIC	BIC	SRM
f_1	1.62E+03	1.57E+03	**3.58E-07**
f_2	1.93E+01	2.11E+01	**2.63E+00**
f_3	1.62E+00	1.62E+00	**1.50E+00**
f_4	**5.00E-01**	**5.00E-01**	5.03E-01
f_5	4.60E-01	4.55E-01	**1.66E-01**
g_1	1.23E-01	1.79E-01	**4.61E-02**
g_2	2.03E-01	2.05E-01	**1.36E-01**
g_3	**1.25E-01**	**1.25E-01**	1.33E-01

Results showed in table 2 confirm that SRM produces the best obtained model for all target functions. This is more clear for the functions f_2, f_3 and f_5. From the results displayed in table 3 and table 4 it can be deduced that considering a reasonable set of executions of the studied model selection strategies for each target function, the mean quality of the strategy SRM is the best one.

Finally, in table 5 we present the mean prediction risk, considering all the performed executions. Cause we have 10 executions for each experiment and 100 experiments were completed, each result presented in this table corresponds to a mean value over 1000 runs.

The results in table 5 confirm again that SRM presents the best results for the most part of the target functions. There is a remarkable difference in terms of performance for the polynomial function f_1, that can be also seen in table 4.

Taking into account the above experimentation we can conclude that Structural Risk Minimization method (based on VC-theory) as a model selection criterium when using genetic training from noisy data, combined with our model complexity measure that counts the number of non-scalar nodes of the tree, clearly outperforms classical statistical methods as AIC or BIC. In general, SRM obtains better solutions than AIC or BIC in almost all studied cases.

5 Conclusions

In this paper we have presented an empirical comparative study of three model selection criteria for learning problems with GP-trees. The first two methods (AIC and BIC) are classical statistical methods and the third one (SRM) is a selection criterium based on VC-theory with a new model complexity measure. The strategy used for the selection of the model was a genetic training method over a finite set of noisy examples. For measuring the quality of the selected model after the training process, a validation set of noise free examples was generated and a mean square error fitness of the model over the validation set was computed. An extensive experimentation over several symbolic regression problem instances, suggests that SRM selection criterium performs better than the other two considered methods. However AIC and BIC methods, that usually are employed in combination with least-squares fitting techniques, also perform quite well when using genetic training.

As final remark we note that it is impossible to draw any conclusions based on empirical comparisons unless one is sure that model selection criteria use accurate estimates of model complexity. There exist experimental methods for measuring the VC-dimension of an estimator ([13], [8]); however they are difficult to apply for general practitioners. In this paper we have used a new complexity measure for GP-trees. Essentially, under this approach we combine the known analytical form of a model selection criterium, with appropriately tuned measure of model complexity taken as a function of (some) complexity parameter (i.e. the value h that measures the non-linearity of the considered tree). This alternative practical approach is essentially to come up with empirical 'common-sense' estimates of model complexity to be used in model selection criteria.

References

1. Akaike, H.: Statistical prediction information. Ann. Inst. Statistic. Math. 22, 203–217 (1970)
2. Amil, N.M., Bredeche, N., Gagné, C., Gelly, S., Schoenauer, M., Teytaud, O.: A statistical learning perspective of genetic programming. In: Vanneschi, L., Gustafson, S., Moraglio, A., De Falco, I., Ebner, M. (eds.) EuroGP 2009. LNCS, vol. 5481, pp. 327–338. Springer, Heidelberg (2009)

3. Bernardo, J., Smith, A.: Bayesian theory. John Wiley & Sons, Chichester (1994)
4. Cherkassky, V., Yunkian, M.: Comparison of Model Selection for Regression. Neural Computation 15, 1691–1714 (2003)
5. Koza, J.: Genetic Programming: On the Programming of Computers by Means of Natural Selection. MIT Press, Cambridge (1992)
6. Montaña, J.L.: Vcd bounds for some gp genotypes. In: ECAI, pp. 167–171 (2008)
7. Montaña, J.L., Alonso, C.L., Borges, C.E., Crespo, J.L.: Adaptation, performance and vapnik-chervonenkis dimension of straight line programs. In: Vanneschi, L., Gustafson, S., Moraglio, A., De Falco, I., Ebner, M. (eds.) EuroGP 2009. LNCS, vol. 5481, pp. 315–326. Springer, Heidelberg (2009)
8. Shao, X., Cherkassky, V., Li, W.: Measuring the VC-dimension using optimized experimental design. Neural Computation 12, 1969–1986 (2000)
9. Teytaud, O., Gelly, S., Bredeche, N., Schoenauer, M.: Statistical Learning Theory Approach of Bloat. In: Proceedings of the 2005 conference on Genetic and Evolutionary Computation, pp. 1784–1785 (2005)
10. Vapnik, V.: Statistical learning theory. John Wiley & Sons, Chichester (1998)
11. Vapnik, V., Chervonenkis, A.: On the uniform convergence of relative frequencies of events to their probabilities. Theory of Probability and its Applications 16, 264–280 (1971)
12. Vapnik, V., Chervonenkis, A.: Ordered risk minimization. Automation and Remote Control 34, 1226–1235 (1974)
13. Vapnik, V., Levin, E., Cun, Y.: Measuring the VC-dimension of a learning machine. Neural Computation 6, 851–876 (1994)

Generation of Pseudorandom Binary Sequences with Controllable Cryptographic Parameters*

Amparo Fúster-Sabater

Institute of Applied Physics, C.S.I.C.
Serrano 144, 28006 Madrid, Spain
amparo@iec.csic.es

Abstract. In this paper, a procedure of decomposition of nonlinearly filtered sequences in primary characteristic sequences has been introduced. Such a procedure allows one to analyze structural properties of the filtered sequences e.g. period and linear complexity, which are essential features for their possible application in cryptography. As a consequence of the previous decomposition, a simple constructive method that enlarges the number of known filtered sequences with guaranteed cryptographic parameters has been developed. The procedure here introduced does not impose any constraint on the characteristics of the nonlinear filter.

Keywords: stream cipher, pseudorandom sequence, nonlinear filter, linear complexity, period, cryptography.

1 Introduction

Pseudorandom binary sequences are typically used in a wide range of applications such as: spread spectrum communication systems, multiterminal system identification, global positioning systems, software testing, error-correcting codes and cryptography. This work deals specifically with this last application.

In fact, confidentiality in sensitive information is a crucial feature in such types of systems. This quality makes use of an encryption function currently called *cipher* that converts the original message (*plaintext*) into the ciphered message (*ciphertext*). Symmetric cryptography or secret-key cryptography is split into two large classes [20]: block ciphers and stream ciphers depending on whether the encryption function is applied to a bit block or to each individual bit, respectively.

At the present moment, stream ciphers are the fastest among the encryption procedures so they are implemented in many technological applications e.g. algorithms A5 in GSM communications [13], the encryption function E0 in Bluetooth specifications [2] or the stream cipher RC4 for encrypting Internet traffic [18].

* This work was supported in part by CDTI (Spain) and the companies INDRA, Unión Fenosa, Tecnobit, Visual Tools, Brainstorm, SAC and Technosafe under Project Cenit-HESPERIA; by Ministry of Science and Innovation and European FEDER Fund under Project TIN2008-02236/TSI.

B. Murgante et al. (Eds.): ICCSA 2011, Part I, LNCS 6782, pp. 563–572, 2011.

Stream ciphers are designed to generate from a short seed, the *key*, a long sequence of pseudorandom bits, the *keystream sequence*. Such a sequence is XORed with the plaintext (in emission) in order to obtain the ciphertext or with the ciphertext (in reception) in order to recover the plaintext. Security of a stream cipher resides in the characteristics of the keystream sequence:

- Long period
- Good statistical properties, (see [5] and [17])
- High linear complexity related to the amount of known sequence necessary to recover the whole sequence, (see [7] and [10]).

All of them are necessary conditions that every keystream sequence must satisfied. The central requirement for the keystream bits should be that they appear like a random sequence to an attacker as proposed in [6] and [19].

Traditionally, keystream generators make use of maximal-length Linear Feedback Shift Registers (LFSRs) [12] whose output sequences (the PN-sequences) are combined in a nonlinear way in order to produce the desired keystream sequences, see as well known examples of keystream generators [4], [11], [14], [22] and [23]. One of the most popular pseudorandom sequence generators is the nonlinear filter of PN-sequences, which is a very common procedure to generate keystream sequences in stream ciphers [1]. Generally speaking, sequences obtained in this way are supposed to satisfy the above mentioned characteristics. In [9], it is proved that the probability of choosing a nonlinear filter of PN-sequences with optimal properties tends asymptotically to 1 as far as the LFSR length increases. However, there are only a few design principles for such filters [21]. The present paper tackles this problem and introduces a simple design strategy for nonlinear filters with cryptographic application. The method is based on the decomposition of the output sequence of a nonlinear filter, the *filtered sequence*, in *primary characteristic sequences* what allows us to analyze the cryptographic properties (period and linear complexity) of the filtered sequence. As a natural consequence of this decomposition, an easy method of generating keystream sequences with application in cryptography is developed.

The paper is organized as follows. In Section 2, some basic concepts and definitions are introduced. The decomposition of the filtered sequence in primary characteristic sequences is given in section 3. The study of cryptographic parameters in terms of the previous decomposition is considered in Section 4. The strategy of generation of new filtered sequences is defined and analyzed in Section 5. Finally, conclusions in Section 6 end the paper.

2 Basic Concepts and Definitions

Specific notation and different basic concepts are introduced as follows:

Let $\{a_n\}$ be the binary sequence generated by a maximal-length LFSR of L stages [20] and [21]. That is to say, a LFSR whose feedback polynomial [12] is primitive of degree L and whose output sequence is a PN-sequence of period $2^L - 1$. In the sequel and without loss of generality we assume that $\{a_n\}$ is in its characteristic phase: it means that the initial state has been chosen in such a way that the generic element a_n can be written as:

$$a_n = \alpha^n + \alpha^{2n} + \ldots + \alpha^{2^{L-1}n}, \quad (1)$$

$\alpha \in GF(2^L)$ being a root of the LFSR characteristic polynomial and L the length of the LFSR.

$F : GF(2)^L \longrightarrow GF(2)$ denotes a m-th order nonlinear filter applied to the L stages of the previous LFSR. That is, F includes at least a term product of m distinct elements of the sequence $\{a_n\}$ of the form $a_{n+t_1}\, a_{n+t_2}\, \ldots\, a_{n+t_m}$, where the concatenation of variables denotes the logic function AND and the indexes t_j $(j = 1, 2, \ldots, m)$ are integers satisfying the inequalities:

$$0 \leq t_1 < t_2, < \ldots < t_m < 2^L - 1.$$

The binary sequence $\{z_n\}$ is the sequence obtained at the output of the nonlinear filter F, that is the filtered sequence.

Let Z_{2^L-1} denote the set of integers $[1, 2, \ldots, 2^L - 1]$. We consider the following equivalence relation R defined on its elements: $q_1\ R\ q_2$ with $q_1, q_2 \in Z_{2^L-1}$ if there exists a j, $0 \leq j \leq L - 1$, such that:

$$2^j \cdot q_1 = q_2 \ mod\ 2^L - 1. \quad (2)$$

The resultant equivalence classes into which Z_{2^L-1} is partitioned are called [16] the cyclotomic cosets module $2^L - 1$. All the elements q_i of a cyclotomic coset have the same Hamming weight, that is, the same number of 1's in their binary representation.

The element leader, E, of every cyclotomic coset E, is the smallest integer in such an equivalence class. If L is a prime number, then the cardinal of every cyclotomic coset is L (except for coset 0 whose cardinal is always 1). If L is a composite number, then the cardinal of a cyclotomic coset E may be L or a proper divisor of L.

For a maximal-length LFSR, we can give the following definitions:

Definition 1. *The characteristic polynomial of a cyclotomic coset E is a polynomial $P_E(x)$ defined by:*

$$P_E(x) = (x + \alpha^E) \cdot (x + \alpha^{2E}) \cdot \ldots \cdot (x + \alpha^{2^{(r-1)}E}), \quad (3)$$

where the degree r $(r \leq L)$ of $P_E(x)$ equals the cardinal of the cyclotomic coset E.

Definition 2. *The primary characteristic sequence of a cyclotomic coset E is a binary sequence, denoted by $\{S_n^E\}$, that satisfies the expression:*

$$\{S_n^E\} = \{\alpha^{En} + \alpha^{2En} + \ldots + \alpha^{2^{(r-1)}En}\},\ \ n \geq 0. \quad (4)$$

Recall that the sequence $\{S_n^E\}$ is in its characteristic phase and satisfies the linear recurrence relation given by $P_E(x)$. Moreover, according to [7], $\{S_n^E\}$ is a decimation of $\{a_n\}$ made out from such a sequence by taking one out of E terms.

Primary characteristic sequences of cyclotomic cosets will be the fundamental tools used in this paper.

If coset E is a proper coset [12], then $P_E(x)$ is a primitive polynomial of degree L and its characteristic sequence $\{S_n^E\}$ is a PN-sequence. If coset E is an improper coset [12], then $P_E(x)$ is either a primitive of degree $r < L$ or an irreducible polynomial of degree L, consequently the period of its characteristic sequence, notated $T(\{S_n^E\})$, will be a proper divisor of $2^L - 1$. For more details concerning the cyclotomic cosets the interested reader is referred to [[12], chapter 4] and [3].

In brief, every cyclotomic coset E can be characterized by its leader element E or its characteristic polynomial $P_E(x)$ or its characteristic sequence $\{S_n^E\}$.

3 Decomposition of Filtered Sequences in Primary Characteristic Sequences

According to equation (1), the generic element z_n of the filtered sequence $\{z_n\}$ can be written as:

$$\begin{array}{c} z_n = F(a_n, a_{n+1}, \ldots, a_{n+L-1}) = \\ C_1\alpha^{E_1 n} + (C_1\alpha^{E_1 n})^2 + \ldots + (C_1\alpha^{E_1 n})^{2^{(r_1-1)}} + \\ C_2\alpha^{E_2 n} + (C_2\alpha^{E_2 n})^2 + \ldots + (C_2\alpha^{E_2 n})^{2^{(r_2-1)}} + \\ \vdots \\ C_i\alpha^{E_i n} + (C_i\alpha^{E_i n})^2 + \ldots + (C_i\alpha^{E_i n})^{2^{(r_i-1)}} + \\ \vdots \\ C_{N_m}\alpha^{E_{N_m} n} + (C_{N_m}\alpha^{E_{N_m} n})^2 + \ldots + (C_{N_m}\alpha^{E_{N_m} n})^{2^{(r_{N_m}-1)}} \end{array} \quad (5)$$

r_i being the cardinal of coset E_i with $1 \leq i \leq N_m$, (N_m being the number of cosets E_i with binary weight $\leq m$) and C_i constant coefficients in $GF(2^L)$.

If L is prime (which is the most common case in stream ciphers), $r_i = L \ \ \forall i$ and N_m is:

$$N_m = 1/L \cdot \sum_{i=1}^{m} \binom{L}{i}. \quad (6)$$

At this point, different features can be pointed out:

- Recall that every row in (5) corresponds to the n-th element of the characteristic sequence $\{C_i\alpha^{E_i n} + \ldots + (C_i\alpha^{E_i n})^{2^{(r_i-1)}}\}$. The coefficient C_i determines the starting point of such a sequence. If $C_i = 1$, then the above sequence is in its characteristic phase and equals the generic characteristic sequence $\{S_n^E\}$ described in equation (4).
- It can be proved [15] that every coefficient $C_i \in GF(2^{r_i})$, so that as long as C_i ranges in its corresponding subfield we move along the sequence $\{S_n^{E_i}\}$.

- If $C_i = 0$, then $\{S_n^{E_i}\}$ would not contribute to the filtered sequence $\{z_n\}$. In that case the cyclotomic coset E_i would be degenerate. On the other hand, if $C_i \neq 0$, then $\{S_n^{E_i}\}$ would contribute to the filtered sequence $\{z_n\}$ and the cyclotomic coset E_i would be nondegenerate.

Now equation (5) can be rewritten in a more compact way as:

$$\{z_n\} = \{F(a_n, a_{n+1}, \ldots, a_{n+L-1})\} = \sum_{E_i} C_{E_i}\{S_n^{E_i}\}, \tag{7}$$

where $C_{E_i}\{S_n^{E_i}\}$ denotes the primary characteristic sequence

$$\{C_{E_i}\alpha^{E_i n} + \ldots + (C_{E_i}\alpha^{E_i n})^{2^{(r_{E_i}-1)}}\}$$

and the sum is extended to the leaders E_i of the cosets of weight $\leq m$.

Thus, the output sequence of an arbitrary m-th order nonlinear filter can be decomposed as the term-wise sum of the primary characteristic sequences associated with the cyclotomic cosets of weight less or equal than m. Such a decomposition is called the decomposition in primary characteristic sequences. According to this representation, different cryptographic parameters can be analyzed and computed:

- The number of different m-th order nonlinear filters that is related to the range of possible values for the coefficients C_{E_i}. In fact, there is a one-to-one correspondence between every N_m-tuple of coefficients $(C_{E_1}, C_{E_2}, \ldots, C_{E_{N_m}})$ and every m-th order nonlinear filter.
- The possible periods of the filtered sequences that are related to the periods of the characteristic sequences $\{S_n^{E_i}\}$ as the period of the filtered sequence is the $l.c.m.(T(\{S_n^{E_i}\}))$.
- The possible linear complexities of the filtered sequences that are related to the number of coefficients C_{E_i} different from zero as the contribution to the linear complexity of any nondegenerate coset equals the cardinal of such a coset.

Now, let us see a simple example in order to clarify the previous concepts.

Example 1. For the pair $(L, m) = (4, 2)$, that is a nonlinear filter of second order applied to the stages of a LFSR of length 4, we have three different cosets of binary weight less or equal than 2: coset 1= $\{1, 2, 4, 8\}$, coset 3 = $\{3, 6, 12, 9\}$ and coset 5 = $\{5, 10\}$. In this case:

$$\begin{aligned}\{z_n\} &= \{F(a_n, a_{n+1}, a_{n+2}, a_{n+3})\}\\ &= \{C_1\alpha^n + (C_1)^2\alpha^{2n} + (C_1)^4\alpha^{4n} + (C_1)^8\alpha^{8n} +\\ &\quad C_3\alpha^{3n} + (C_3)^2\alpha^{6n} + (C_3)^4\alpha^{12n} + (C_3)^8\alpha^{9n} +\\ &\quad C_5\alpha^{5n} + (C_5)^2\alpha^{10n}\}\\ &= C_1\{S_n^1\} + C_3\{S_n^3\} + C_5\{S_n^5\}\,.\end{aligned}$$

with $C_1, C_3 \in GF(2^4)$ and $C_5 \in GF(2^2)$.

Thus $\{z_n\}$ can be decomposed as the term-wise sum of three primary characteristic sequences $\{S_n^1\}$, $\{S_n^3\}$, $\{S_n^5\}$.

Every second order nonlinear filter F determines a possible choice of the triplet (C_1, C_3, C_5) as well as every coefficient C_i determines the starting point of the corresponding characteristic sequence $\{S_n^{E_i}\}$. In addition, as F is a second order nonlinear filter, at least one of the coefficient C_3 or C_5 corresponding to the cosets of weight 2 must be different from zero.

4 Analysis of Cryptographic Parameters in Terms of the Primary Characteristic Sequences

According to the decomposition in primary characteristic sequences, every m-th order nonlinear filter $F(a_n, a_{n+1}, \ldots, a_{n+L-1})$ applied to the L stages of a maximal-length LFSR can be uniquely characterized by a N_m-tuple of coefficients $(C_{E_1}, C_{E_2}, \ldots, C_{E_{N_m}})$, in brief (C_{E_i}) with $(1 \leq i \leq N_m)$. In the sequel, both characterizations will be used indistinctly.

Let A be the set of the m-th order nonlinear filters. We can group all the elements of A that produce the same filtered sequence $\{z_n\}$ or a shifted version notated $\{z_n\}^*$. From equation (5), it is easy to see that if C_{E_i} is substituted by $C_{E_i} \cdot \alpha^{E_i}$ $\forall i$, then the sequence $\{z_{n+1}\}$ is obtained. In general,

$$C_{E_i} \rightarrow C_{E_i} \cdot \alpha^{jE_i} \;\; \forall i \Rightarrow \{z_n\} \rightarrow \{z_{n+j}\} \, . \tag{8}$$

This fact enables us to define a new equivalence relation, notated $\sim$, on the set A in such a way that: $F_1 \sim F_2$ with $F_1, F_2 \in A$ if

$$\{F_1(a_n, a_{n+1}, \ldots, a_{n+L-1})\} = \{F_2(a_n, a_{n+1}, \ldots, a_{n+L-1})\}^* \, . \tag{9}$$

Therefore, two different nonlinear filters of the same order F_1, F_2 in the same equivalence class will produce shifted versions of the same filtered sequence. In addition, it is easy to see that the relation above defined is an equivalence relation and that the coefficients associated with F_1, F_2, notated $(C^1_{E_i}), (C^2_{E_i})$ respectively, verify:

$$C^2_{E_i} = C^1_{E_i} \cdot \alpha^{jE_i} \, . \tag{10}$$

Clearly, the number of elements in every equivalence class equals the period of the filtered sequence, notated T, so that in equation (10) the index j ranges in the interval $[1, \ldots, T-1]$.

According to the previous expressions, we can summarize the study of the cryptographic parameters for the m-th order nonlinear filters such as follows:

Remark 1: The number of m-th order nonlinear filters (cardinal of A) is:

$$Cardinal(A) = (2^{\binom{L}{m}} - 1) \cdot 2^{\binom{L}{m-1}} \cdot 2^{\binom{L}{m-2}} \ldots 2^{\binom{L}{1}} \, . \tag{11}$$

Remark 2: The period T of the filtered sequence $\{z_n\}$ obtained from a m-th order nonlinear filter is the $l.c.m.(T\{S_n^{E_i}\})$, that is the least common multiple of the periods of the characteristic sequences $\{S_n^{E_i}\}$ that appear in the decomposition of primary characteristic sequences as $\{z_n\}$ is the term-wise sum of sequences with different periods.

Remark 3: The linear complexity LC of the filtered sequence $\{z_n\}$ obtained from a m-th order nonlinear filter is:

$$LC = \sum_{E_i} r_{E_i} \ , \tag{12}$$

where r_{E_i} is the cardinal of coset E_i as the contribution of a nondegenerate coset to the linear complexity of $\{z_n\}$ equals its cardinal.

Let us see an illustrative example of the above remarks.

Example 2. For the pair $(L, m) = (4, 2)$ and the LFSR of length 4 whose characteristic polynomial is $P(x) = x^4 + x + 1$, we have:

$$\{z_n\} = C_1\{S_n^1\} + C_3\{S_n^3\} + C_5\{S_n^5\} \ .$$

If $C_1 = 1$, $C_3 = \alpha^3$ and $C_5 = \alpha^{10}$, then

$$\begin{aligned} \{z_n\} &= 1\{S_n^1\} + \alpha^3\{S_n^3\} + \alpha^{10}\{S_n^5\} \\ &= \{0,0,0,1,0,0,1,1,0,1,0,1,1,1,1,\dots\} \\ &+ \{1,1,1,1,0,1,1,1,1,1,0,1,1,1,1,\dots\} \\ &+ \{1,0,1,1,0,1,1,0,1,1,0,1,1,0,1,\dots\} \end{aligned}$$

where the first sequence is in its characteristic phase, the second is shifted one position regarding the characteristic sequence $\{S_n^3\}$ and the last sequence appears shifted two positions regarding $\{S_n^5\}$. Different features of $\{z_n\}$ are:

- $T(\{z_n\}) = l.c.m.(15, 15, 3) = 15$
- $LC(\{z_n\}) = \sum_{E_i} r_{E_i} = 4 + 4 + 2 = 10$
- The minimal polynomial of $\{z_n\}$ is:

$$P(x) = P_1(x) \cdot P_3(x) \cdot P_5(x) = \\ x^{10} + x^8 + x^5 + x^4 + x^2 + x + 1 \ .$$

- The second order nonlinear that generates $\{z_n\}$ is:

$$F(a_n, a_{n+1}, \dots, a_{n+L-1}) = \\ a_n a_{n+2} + a_n a_{n+3} + a_{n+1} a_{n+2} + a_{n+2} a_{n+3} + a_{n+2} \ .$$

When C_1, C_3 range in $GF(2^4)$ and C_5 in $GF(2^2)$, we have $(2^4-1)(2^4-1)(2^2-1) = 675$ different second order nonlinear filters generating sequences with the same period, linear complexity and minimal polynomial as those of $\{z_n\}$.

5 A Practical Method of Constructing Nonlinear Filters with Controllable Cryptographic Parameters

In this section, a practical method of constructing good cryptographic nonlinear filters for stream cipher is given. In fact, we start from a nonlinear filter f whose number of coset nondegenerate is known to be large, see [8]. In addition, f is the m-th order function applied to the stages of a LFSR of length L, that is:

$$f = a_{n+t_1} \cdot a_{n+t_2} \cdot \ldots \cdot a_{n+t_m},$$

and f' is a different nonlinear filter related to f in the sense that:

$$f' = a_{n+t'_1} \cdot a_{n+t'_2} \cdot \ldots \cdot a_{n+t'_m}$$

with

$$t'_i = 2^k \cdot t_i \ mod\ 2^L - 1 \quad \forall i, \quad k \in N \, .$$

According to the decomposition in primary characteristic sequences, the sequence $\{z_n\}$ generated by the filter f can be written as:

$$\{z_n\} = \{ \prod_{i=1}^{m} (\sum_{j=0}^{L-1} \alpha^{2^j t_i} \cdot \alpha^{2^j n}) \} = \sum_{i=1}^{N_m} C_i \, \{S_n^{E_i}\} \, , \tag{13}$$

where N_m is defined as before, $C_i \in GF(2^L)$ is a constant coefficient and $\{S_n^{E_i}\}$ is the characteristic sequence of coset E_i.

By analogy, the sequence $\{z'_n\}$ generated by the filter f' can be represented as:

$$\{z'_n\} = \{ \prod_{i=1}^{m} (\sum_{j=0}^{L-1} \alpha^{2^j t_i} \cdot \alpha^{2^j n}) \} = \sum_{i=1}^{N_m} C'_i \{S_n^{E}\} \, , \tag{14}$$

where the coefficients C_i and C'_i in (13) and (14) are related through the expression:

$$C'_i = (C_i)^{2^k} \ \forall i \, , \tag{15}$$

since in $GF(2^L)$ and $\forall k$ the following expression is verified:

$$(\alpha^{a \cdot 2^k} + \ldots + \alpha^{p \cdot 2^k}) = (\alpha^a + \ldots + \alpha^p)^{2^k} \, . \tag{16}$$

Consequently, if some C_i is zero in equation (13), then the corresponding C'_i in (14) will also be zero and viceversa. Therefore, both filters have the same number of nondegenerate cosets, e.g. the same linear complexity.

In addition, the period of both filters is the same since they contain the same characteristic sequences except for a phase shift.

Finally, this results is valid for all maximal-length LFSR since equation (16) is independent of $GF(2^L)$. Thus, the constructing method can be stated as follows:

Step 1: Start from a nonlinear filter whose number of coset nondegenerate is known to be large (e.g. a unique maximum order term with equidistant phases [8]).
Step 2: From f determine the corresponding filter f' with phases separated a distance of value 2^k.
Step 3: Repeat step 2 for different values of k with $k \in N$.

In this way, we get a family of nonlinear filters to be applied at a LFSR of length L preserving the same period and linear complexity as those of the initial filter. The procedure is general and can be applied to the generation of nonlinear filters in the range of cryptographic application.

6 Conclusions

According to the decomposition of filtered sequences in primary characteristic sequences, an equivalence relation has been defined on the set of m-th order nonlinear filters. Such nonlinear filters can be classified according to their period, linear complexity and characteristic polynomial of the generated sequences. Thanks to this decomposition, the structural properties of the filtered sequences can be easily analyzed. In particular, their cryptographic parameters.

The work concludes with a constructive method, based on the design of new filters from a given one, that allows us to generate filtered sequences having maximum period and a guaranteed large linear complexity. Consequently, they can be used for cryptographic purposes.

References

1. Awad, W.: Bias in the Nonlinear Filter Generator Output Sequence. Information Technology Journal 7(3), 541–544 (2008)
2. Bluetooth, Specifications of the Bluetooth system, Version 1.1, `http://www.bluetooth.com/`
3. Biggs, N.: Discrete Mathematics, 2nd edn. Oxford University Press, New York (2002)
4. Coppersmith, D., Krawczyk, H., Mansour, Y.: The shrinking generator. In: Stinson, D.R. (ed.) CRYPTO 1993. LNCS, vol. 773, pp. 22–39. Springer, Heidelberg (1994)
5. Diehard Battery of Tests of Randomness (1995), `http://i.cs.hku.hk/~diehard/`
6. eSTREAM-The ECRYPT Stream Cipher Project (2007), `http://www.ecrypt.eu.org/stream/`
7. Fúster-Sabater, A., Caballero-Gil, P.: On the linear complexity of nonlinearly filtered PN-sequences. In: Safavi-Naini, R., Pieprzyk, J.P. (eds.) ASIACRYPT 1994. LNCS, vol. 917, pp. 80–90. Springer, Heidelberg (1995)
8. Fúster-Sabater, A., Garcia, J.: An efficient algorithm to generate binary sequences for cryptographic purposes. Theoretical Computer Science 259, 679–688 (2001)
9. Fúster-Sabater, A., Caballero-Gil, P.: Strategic Attack on the Shrinking Generator. Theoretical Computer Science 409(3), 530–536 (2008)
10. Fúster-Sabater, A., Caballero-Gil, P.: Synthesis of Cryptographic Interleaved Sequences by Means of Linear Cellular Automata. Applied Mathematics Letters 22(10), 1518–1524 (2009)
11. Fuster-Sabater, A., Delgado-Mohatar, O., Brankovic, L.: On the Linearity of Cryptographic Sequence Generators. In: Taniar, D., Gervasi, O., Murgante, B., Pardede, E., Apduhan, B.O. (eds.) ICCSA 2010. LNCS, vol. 6017, pp. 586–596. Springer, Heidelberg (2010)
12. Golomb, S.: Shift-Register Sequences. Aegean Park Press, Laguna Hill California (1982)

13. GSM, Global Systems for Mobile Communications, http://cryptome.org/gsm-a512.htm
14. Hu, Y., Xiao, G.: Generalized Self-Shrinking Generator. IEEE Trans. Inform. Theory 50, 714–719 (2004)
15. Lidl, R., Niederreiter, H.: Introduction to Finite Fields and Their Applications. Cambridge University Press, Cambridge, England (1986)
16. Limniotis, K., Kolokotronis, N., Kalouptsidis, N.: Nonlinear Complexity of Binary Sequences and Connections with Lempel-Ziv Compression. In: Gong, G., Helleseth, T., Song, H.-Y., Yang, K. (eds.) SETA 2006. LNCS, vol. 4086, pp. 168–179. Springer, Heidelberg (2006)
17. NIST Test suite for random numbers, http://csrc.nist.gov/rng/
18. Rivest, R.: The RC4 Encryption Algorithm. RSA Data Sec., Inc. (March 1998), http://www.rsasecurity.com
19. Robshaw, M.J.B., Billet, O. (eds.): New Stream Cipher Designs. LNCS, vol. 4986. Springer, Heidelberg (2008)
20. Rueppel, R.: Analysis and Design of Stream Ciphers. Springer, New York (1986)
21. Simmons, G.: Contemporary Cryptology, The Science of Information, pp. 65–134. IEEE Press, Los Alamitos (1992)
22. Tan, S.K., Guan, S.: Evolving cellular automata to generate nonlinear sequences with desirable properties. Applied Soft Computing 7(3), 1131–1134 (2007)
23. Teo, S.G., Simpson, L., Dawson, E.: Bias in the Nonlinear Filter Generator Output Sequence. International Journal of Cryptology Research 2(1), 27–37 (2010)

Mobility Adaptive CSMA/CA MAC for Wireless Sensor Networks

Bilal Muhammad Khan and Falah H. Ali

Communication research group, School of Engineering and Design,
University of Sussex, UK
bmk21@sussex.ac.uk, f.h.ali@sussex.ac.uk

Abstract. In this paper we propose high throughput low collision, mobility adaptive and energy efficient medium access protocol (MAC) called Mobility Adaptive (MA-CSMA/CA) for wireless sensor networks. MA-CSMA/CA ensures that transmissions incur less collision, and allows nodes to undergo sleep mode whenever they are not transmitting or receiving. It uses contention based as well as contention free period efficiently together to minimise the number of collisions cause by the mobile node entering and leaving the clusters. It also allows nodes to determine when they can switch to sleep mode during operation. MA-CSMA/CA for mobile nodes provides fast association between the mobile node and the cluster coordinator. The performance of MA-CSMA/CA is evaluated through extensive simulation, analysis and compared with the existing IEEE 802.15.4 industrial standard. The results show that MA-CSMA/CA outperforms significantly the existing CSMA/CA protocol including throughput, latency and energy consumption.

Keywords: Mobile CSMA/CA, Mobile MAC, WSN.

1 Introduction

Recent improvements in affordable and efficient integrated electronic devices have a considerable impact on advancing the state of the art of wireless sensor networks (WSNs). It constitutes a platform of a broad range of applications related to security, surveillance, military, health care, environmental monitoring, inventory tracking and industrial controls [1,2]. Handling such a diverse range of application will hardly be possible with any single realization of a WSN.

The IEEE 802.15.4 wireless personal area network (WPAN) [3], which has been designed to have low data rate, short transmission distance and low power consumption is a strong candidate for WSN.

The medium access control (MAC) protocol plays major role in determining the throughput, latency, bandwidth utilization and energy consumption of the network. Therefore it is of a paramount importance to design and choose the MAC protocol to provide the required quality of service QoS for a given application. There are several MAC protocols available for multihop wireless networks [4-9] which can be topology dependent or independent [10-13]. These protocols are broadly classified as schedule and contention based MAC.

B. Murgante et al. (Eds.): ICCSA 2011, Part I, LNCS 6782, pp. 573–587, 2011.

In [14] IEEE 802.15.4 industrial standard CSMA/CA MAC protocol is presented. CSMA/CA uses random backoff values as collision resolution algorithm. Since WSN is resource constrained therefore the entire backoff phenomenon is performed blindly without the knowledge of the channel condition, this factor contributes in high number of collision especially when the number of active nodes are high. CSMA/CA is also used in mobile applications however the QoS is very poor. The protocol incurs high latency over the mobile node as it requires long association process for the node leaving from one cluster to another cluster. If the node miss the beacon from the new cluster head consecutively for four times then the node will go in orphan realignment process hence inducing more delays. In order for the node to join the new cluster and not to miss the beacon from the new cluster head it tends to wake up for long duration which in turns significantly increases the energy consumption of the individual node. Moreover as the mobile node succeeds in migrating from one cluster to another it has to compete with the already available active nodes in the cluster which contributes in significant number of collision and packet drop, resulting in overall degradation of network QoS.

In [15] MOB-MAC is presented which uses an adaptive frame size predictor to significantly reduce the energy consumption. A smaller frame size is predicted when the signal characteristics becomes poor. However the protocol incurs heavy delays due to the variable size in the frame and is not suitable for mobile real time applications. AM-MAC is presented in [16]; it is the modification of S-MAC to make it more useful in mobile applications. In AM-MAC as the mobile node reaches the border node of the second cluster copies and hold the schedule of the approaching virtual cluster as well as the current virtual cluster. By adopting this phenomenon the protocol provides fast connection for the mobile node moving from one cluster and entering the other. The main drawback however is the node has to wakeup according to both the schedule but cannot transmit neither receive data packet during the wakeup schedule other than the current cluster. This contributes to significant delay and loss of energy due to idle wakeup. In [17] another variation of S-MAC is presented by the name of MS-MAC. It uses signal strength mechanism to facilitate fast connection between the mobile node and new cluster. If the node experience change in received signal strength then it assumes that the transmitting node is mobile. In this case the sender node not only sends the schedule but also the mobility information in the synchronous message. This information is used by the neighboring node to form an active zone around the mobile node so that whenever the node reaches this active zone it may be able to update the schedule according to the new cluster. The down side of this protocol is idle listening and loss of energy. Moreover nodes in the so called active zone spend most of the time receiving synchronous messages rather than actual data packets thus resulting in low throughput and increase latency. In [18] another real time mobile sensor protocol DS-MAC is presented. In this protocol according to the energy consumption and latency requirement the duty cycle is doubled or half. If the value of energy consumption is lower the threshold value (*Te*) than the protocol double its wakeup duty cycle to transfer more data, thus increasing the throughput and decreasing the latency of the network. However the protocol requires overheads and during the process of doubling of wakeup duty cycle if the value crosses *Te* the protocol continues to transmit data using double duty cycle resulting in loss of energy. MD-MAC is proposed in [19] which is the extension of DS-MAC and MS-MAC. The protocol enforces *Te* value and during any time if the value of energy consumption is doubled then it halves the duty cycle. Moreover the mobile node only undergoes neighborhood discovery rather than other neighboring node which forms an active zone as in the case of MS-MAC. MD-MAC is complex and requires high overheads.

In this paper we fundamentally address the above mentioned limitations by employing MA-CSMA/CA. The protocol enables existing CSMA/CA wireless standard used in WSN to be used in mobile applications efficiently. The proposed MAC is energy efficient while maintaining high goodput, low latency and fairness among the nodes compared to the existing mobile sensor standard and protocol. The energy efficiency is achieved by (i) minimizing the number of collisions and by (ii) transmitting the sleep schedule every time so that the node does not waste unnecessary energy to perform backoff and sensing of the channel as the channel is being occupied by some other contending nodes. High goodput and fairness among the nodes is achieved by reallocating the GTS slots to the nodes according to their priorities as well as data transmission frequency ensuring that every node will be able to access the shared communication link on fair bases.

Once the node joins the network it contend normally as per CSMA/CA standard for the transmission of data, however in case of the mobile node leaving one cluster and joining another cluster, the new cluster head allocates a GTS slot for the incoming node. This in turns minimizes the collisions in the new cluster because of the entrance of the new node and also by pre-allocating GTS slot to the mobile node the protocol save valuable association and re-association time for the nodes which in turns improves the latency of the network and save valuable energy of the mobile unlike to the current CSMA/CA standard.

The basic design principle of the proposed protocol is given in section 2. In section 3 working principle of MA-CSMA/CA protocol is presented. Section 4 presents the analytical performance analysis of MA-CSMA/CA protocol. Section 5 presents extensive simulation results and comparisons. Section 6 gives the conclusions and future work.

2 Proposed MA-CSMA/CA Protocol Design

In wireless sensor networks majority of the contention based protocols use random backoff mechanism to avoid collision. The main drawback of using such protocols is that they rely on allocating random backoff delays in order to resolve the collisions but as the number of nodes increases the probability of nodes selecting the same backoff increases thus resulting in access collision. In case of mobile sensor network this problem become more serious, as the mobile node enter from one cluster to another and try to utilise the available resources which are already limited for the nodes in the cluster. This creates congestion, high latency and low throughput in the network. In this paper we are presenting a novel MA-CSMA/CA MAC which enables the existing IEEE 802.15.4 CSMA/CA industrial standard to work in mobile scenario while maintaining an acceptable quality of service. The protocol minimises effect of collision and topology change due to mobile node. Furthermore MA-CSMA/CA decreases the association time for the nodes which are moving from one cluster to another considerably without incurring energy losses and computational complexities.

This results in solving one of the key issues of mobile sensor MAC protocols as well as significant improvement in throughput as well as reduction in energy due to the significant reduction in the number of retransmissions.

2.1 Network Model for MA/CSMA/CA

The standard supports three types of topology: 1) star; 2) peer to peer and; 3) cluster tree topology. Each WPAN has one coordinator function as the central controller to organize the WPAN and coordinate the other components of WPAN [7]. There are two different types of devices in an IEEE 802.15.4 network; a full function device (FFD) and a reduce function device (RFD). The FFD can operate in three modes serving as personal area network (PAN) coordinator, a coordinator or a device. An FFD can communicate to RFD and other FFD; however an RFD can only communicate to FFD. RFD is intended to simple applications like sensing. The coordinator in any kind of network topology serves the role of sink. The MA-CSMA/CA network is in beacon enable mode, in this mode constant beacon are transmitted from the coordinator towards the node before the start of the data transmission frame. Due to the beacon enable mode the nodes are synchronized with the coordinator. Fig. 1 shows the network model for MA-CSMA/CA. It can be seen from the figure that the network topology is multi-cluster topology. There is one central PAN coordinator and the information is relayed to it by other intermediate nodes. These intermediate nodes sometimes becomes coordinator to other WPAN, referred to as bridges [8].The PAN coordinator may instruct a FFD to become the coordinator of the new cluster, this new coordinator will serve as bridge node to the main central coordinator. The two coordinators of different clusters being the FFD can communicate directly and share the information of each other cluster especially when the mobile node is moving from one cluster to another, these cluster heads (coordinators) communicate the information and gives the new delay value to the mobile node as it enters new cluster. The advantage of multi-cluster topology is increased coverage area.

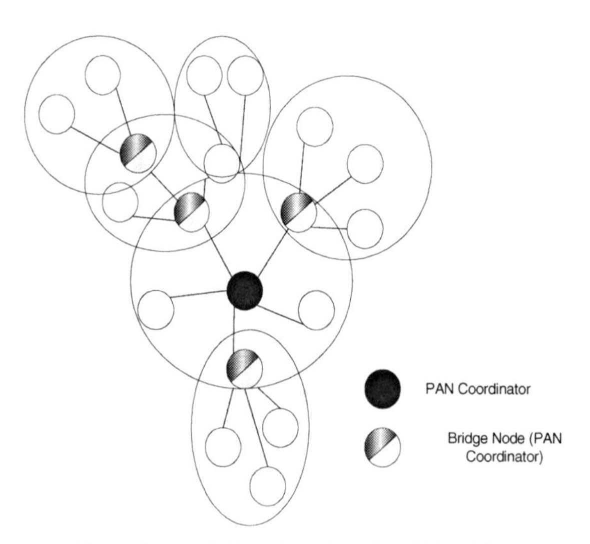

Fig. 1. Network Topology for MA-CSMA/CA

2.2 Transmission Stages of MA-CSMA/CA

The transmission of each successful frame in MA-CSMA/CA MAC undergoes five main stages:

- Stage1: The contending node acquires backoff delay as per standard CSMA/CA protocol. In case of mobile node moving from one cluster to another, it has highest priority. The node constantly monitors the beacon strength received from different coordinator apart from the parent coordinator. At any stage if the beacon strength from one particular coordinator increases, the node assumes that it is moving towards the new coordinator and sends the request to the current cluster coordinator to allocate a guaranteed time slot (GTS) in the cluster towards which the node is moving. In order to maintain fairness among the nodes the GTS resources are reserved once for the mobile node as they are about to enter the new cluster. After the first transaction the node has to compete in the cluster like the other nodes. This GTS allocation before entering the new cluster significantly reduces the association time for the new which in normal circumstances any node using the standard CSMA/CA protocol undergoes. Moreover as the node enters the new cluster since it is pre-allocated with GTS slot then any ongoing transmission within the cluster will not be interrupted. This in turns made MA-CSMA/CA more suitable for mobile scenario.
- Stage2: Nodes having data packet in the buffer undergoes backoff delay before the transmission of packet.
- Stage3: Nodes performs clear channel access (CCA). IF the channel is busy then increment the backoff value.
- Stage4: If the channel is available after the second CCA then the node transmit the data towards the intended receiver.
- Stage5: If the node is mobile and enters in the new cluster then after the first transmission GTS is de-allocated in order to maintain fairness in the cluster.

MA-CSMA/CA protocol is illustrated in Fig.2 for two nodes A and B. Node A is mobile therefore it will continue to receive beacon from other nearby coordinators. As the signal strength of a particular coordinator increases in consistent manner then the node assumes that it is moving in the direction of that coordinator and requests the current coordinator to send request for GTS allocation in the adjacent cluster for the mobile node. The coordinator upon receiving the request establishes the link with the adjacent cluster coordinator and request for GTS for the mobile node.

In this way the time of association is reduce considerably and the node will be able to transmit and receive the data packets normally. It is assumed that all the nodes A and B have data packets in their buffers. Therefore according to the allocated delays these nodes will perform the backoff. After the backoff period, node A performs CCA. After the two successful CCA the node A sends data towards coordinator 1, in the mean time Node B also start performing CCA. As the channel was preoccupied by node A; node B increases its backoff value, and repeats the whole procedure again until it transmit the buffered data packet.

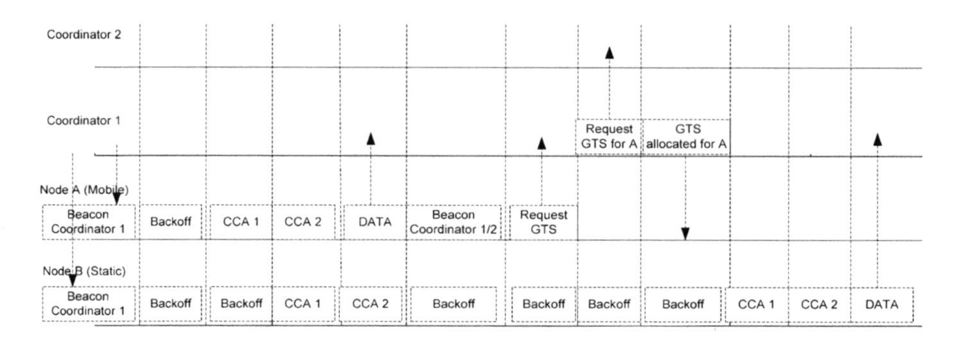

Fig. 2. Illustration of MA-CSMA/CA MAC protocol

3 Working Principle of MA-CSMA/CA

The working of the proposed protocol is shown in Fig. 3. As the node joins the cluster it receives beacon from the potential cluster heads. The node decides on the basis of the signal strength and chose the cluster head having the strongest signal. In case of mobile node since it is constantly on the move the proposed protocol uses the measure of signal strength from different cluster head to its advantage. The node monitors the beacon signal strength from different cluster heads, as it reaches to point where the energy level of received beacon from the cluster head keeps on increasing consistently the node assumes that it is moving towards the direction of this particular cluster head. The protocol allows the node to send a request to gain GTS in the upcoming cluster head before even joining the new cluster through its parent cluster head. Upon receiving the request of the mobile node the current cluster head establishes the link with the potential new cluster head for the mobile node and receive the GTS for the node. This whole process is not in the conventional CSMA/CA WSN standard. Once the node leaves the cluster it performs the association process, the process takes long time and if the nodes misses beacon frame from the new coordinator four times then it would go in orphan alignment process. This whole process of association going into new cluster increases latency of the system. Moreover after association process the new incoming node has to compete with the other nodes previously available in the cluster which increase the probability of collisions, low throughput, high latency and overall degradation of quality of service of the network.

By allocating the GTS to the node before even it enters the new cluster via the old cluster head, MA-CSMA/CA significantly reduces the association time for the mobile node, moreover due to GTS the node will be able to successfully transmit the data in the new cluster without competing with the local nodes and not contributing to the number of collisions of the new cluster. In order to maintain fairness among the node the GTS is only allocated once for every mobile node entering the cluster after that the node has to contend and perform the same procedure which other nodes follow in order to transmit the data towards destination

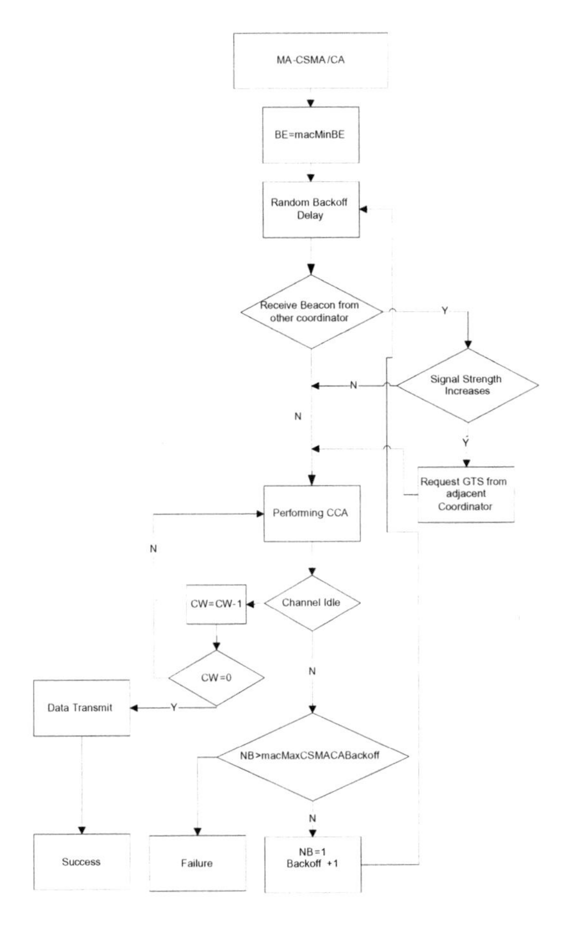

Fig. 3. Block diagram representation of MA-CSMA/CA protocol

4 Performance Analysis

Latency, Goodput and Energy performance of MA-CSMA/CA protocol are analysed in this section. We assume that the channel is error free and the frame arrival rate of each node follow a Poisson distribution. The probability $P(i,t)$ that exactly i number of frames arrive at a node during the time interval t is given by

$$p(i,t) = \frac{(rt)^i}{i!} e^{-rt}, i = 0,1,2,3, \dots \infty. \quad (1)$$

Where t is the unit backoff period and r is the data payload arrival rate. Hence the total traffic load (L) of the network that consist of (N) number of nodes will be $L=N.r.\ell$ where ℓ is the mean length of data payload. Therefore the probability that no frames arrive at the network within the time interval t is given by

$$P(0,t) = e^{-Nrt} \quad (2)$$

Each successful transmission in the contention access period for proposed protocol protocol constitutes three states: 1) Delay Backoff *(DB)* state, 2) Data transmission state *(DTx)*, and 3) ACK state. For the jth successful data frame transmission in the kth superframe these states are denoted as $\tau_j^k DB$, $\tau_j^k DT_x$, $\tau_j^k ACK$, respectively. The value of $\tau_j^k DT_x$ depends upon the network load and the frame size. Hence the total successful transmission $\left(\tau_j^k ST\right)$ time for the jth data frame in the kth superframe is given as

$$\tau_j^k ST = \tau_j^k DB + \tau_j^k DT_x + \tau_j^k ACK \tag{3}$$

The values of $\tau_j^k DB$ and $\tau_j^k DT_x$ majorly contribute towards the value of $\tau_j^k ST$. The value of $\tau_j^k DB$ depends upon the number of active nodes present in the network as well as the priority of the node itself. $\tau_j^k DT_x$ is the summation of all the time required for the successful as well as the unsuccessful transmission of data over the shared communication link. Hence the total time period for data transmission is given by

$$\tau_j^k DT_x = \tau_j^k SDT_x + \tau_j^k FDT_x \tag{4}$$

Where $\tau_j^k SDT_x$and $\tau_j^k FDT_x$ are the successful and failed data transmission times respectively, which depends on the number of retransmission *(RT)* and setup time *(SP)* for the retransmission.

If the number of active nodes changes from one superframe to another then the network traffic rate during the particular superframe will also be changed, hence the traffic arrival rate on variable active number of nodes in a superframe can be computed as

$$p(i, \propto, t) = \frac{(\propto rt)^i}{i!} e^{-\propto rt} \tag{5}$$

Where $\propto$ is the number of active nodes in any particular superframe. Let τCCA be the total time taken to perform two clear channel accesses (CCA). Each CCA requires one unit backoff period. Upon the completion of the backoff procedure the node activates the receiver and performs channel sensing. Any node will be aware if the channel is busy due to the ongoing transmission from a neighbouring node or any node within its radio range also known as non hidden node only if the node starts the transmission in the period $\tau DT_x + \tau ACK + \tau CCA - 1$ ahead of the end of its carrier sensing. Let $P_j^k CHA$ be the probability that the channel is available for transmission then we have

$$P_j^k CHA = P\left(0, \left(N_j^k AN - 1\right). \left(\tau DT_x + \tau ACK + \tau CCA - 1\right)\right) \tag{6}$$

Where $N_j^k AN$ is the total number of active nodes desired to transmit in a particular subperiod or slot of the respective superframe. In order to calculate the goodput *(G)* of the proposed scheme let DFT^k be the total number of data frames that are transmitted during the contention access period of kth superframe. DFT^k depends on three major factors, the time length of contention access period, total time of each successful transmission and the number of active nodes in a specific superframe. Hence the goodput can be defined as the ratio of the total number of data transmission over the time period and can be calculated from the following

$$G = \frac{L \sum_{k=1}^{x} DFT^{K}}{(X)(UBP)(SD)} \tag{7}$$

Where L is the length of data payload, X is the total number of superframes, UBP is the unit backoff period and SD is the superframe duration.

5 Simulation Results

Matlab is used as the simulation platform for the implementation of the new MA-CSMA/CA and the existing WSN industrial standard CSMA/CA. It is assumed that there are no retransmissions. Multi-hop network is considered and all nodes can send maximum of 1K of data which is divided into a number of packets having maximum of 143 data bits per packet. As slotted CSMA/CA is considered therefore the size of beacon interval *(BI)* and super frame duration *(SD)* is calculated as follows:

$$\begin{aligned} &BI = (aBaseSuperframeDuration)2^{BO} \\ &0 \leq BO \leq 14 \end{aligned} \tag{9}$$

$$\begin{aligned} &SD = (aBaseSuperframeDuration)2^{SO} \\ &0 \leq SO \leq BO \leq 14 \end{aligned} \tag{10}$$

In Eq. (9) and (10), *BI* and *SD* are dependent on two parameters beacon order *(BO)* and super frame order *(SO)*. The *aBaseSuperframeDuration* denotes the minimum duration of super frame, corresponding to $SO = 0$. This duration is fixed to 960 symbols where every symbol corresponds to four bits and the total duration of the frame equals to 15.36 msec, assuming 250kbps in the 2.4GHz frequency band, hence each time slot has duration of 0.96 msec. In the simulation analysis the above mentioned conditions are considered and performance analysis are carried out by assigning same values to *SO* and *BO*. Moreover in order to support mobility Random Waypoint model [20] is used. Rest of the simulation parameters are presented in Table1.

The major performance matrices are simulated and the results of the proposed MA-CSMA/CA and existing CSMA/CA MAC protocol are compared in the following sections.

5.1 Setup Time

Setup time can also be referred to as response time. It is the time required by any node in the network when the packet arrives in its buffer to the time the node about to transmit the packet. The setup time is mainly dependent on the MAC layer and its performance. The more efficient MAC protocol is the less setup time will be required by the contending node to access the shared communication link. The setup time is

Table 1. Simulation Parameters

ABaseSlotDuration	60symbol
AMaxSIFSFrameSize	18 octets
ANumSuperframeSlots	16
AUnitBackoffPeriod	20symbol
MacMinSIFSPeriod	12symbol
CW	2
MacMinBE	0 to 3
MaxCSMABackoff	5
Radio Turnaround Time	12symbol
CCA detection time	8symbol
E_{TX}	0.0100224mJ
E_{RX}	0.0113472mJ
E_{CCA}	0.0113472mJ
E_{idle}	0.000056736mJ

dependent on many parameters including; control signal time (τ_{CS}), backoff exponential time (τ_{BE})in case of CSMA/CA, association time (τ_{AS}) for MA-CSMA/CA, radio turnaround time(τ_{RT}). The average setup time (τ_{sp}) for MA-CSMA/CA and CSMA/CA can be calculated from Eq (11) and (12) as follows:

$$\tau_{sp} = \sum_{m=1}^{M} \tau_{AS(m)} + \tau_{CS(m)} + \tau_{RT} \tag{11}$$

$$\tau_{sp} = \sum_{m=1}^{M} \tau_{BE(m)} + \tau_{CS(m)} + \tau_{RT} \tag{12}$$

The value of aforementioned backoff exponent time is given by:

$$\tau_{BE} = (2^{BE} - 1)\tau_{st};\ 0 \leq BE \leq 5 \tag{13}$$

$$\tau_{CS} = \left\lceil \frac{Tnb}{Tcs} \right\rceil \tau_{st} \tag{14}$$

Where (τ_{st})is the symbol time, (Tnb)is total number of control bits and (Tcs) is total number of bits per symbol.

$$\tau_{CS} = (ns)\tau_{st} \tag{15}$$

Where (ns) is total number of symbols for which timer is activated.

Fig. 4 shows the simulation results of average setup time for different number of active nodes *N* and traffic generated per node. The simulation results are obtained by choosing the values of backoff exponent *(BE)* equals to 3 and 2 respectively for CSMA/CA. In case of CSMA/CA for *(BE=3)* the setup time is much higher, this is due to the reason that the node before starting to send the packet undergoes a backoff delay, after the backoff delay it performs channel sensing, if the node fond that the channel is busy than it undergoes a random delay. Due to this reason the latency increases. In case of *(BE=2)* the value of setup time is less than *(BE=3)* but still considerably greater than the other techniques. In case of mobile nodes CSMA/CA apart from initial backoff delay exhibit more delays due to initialization and orphan process. In case of CSMA/CA if the node goes out of the range of the coordinator than it starts the process of association. During this process any packets send towards the node are dropped. If the node gets acknowledgement during association process then it waits for the beacon from the corresponding coordinator. In case the association process fails than the node can go in orphan realignment procedure or perform the association procedure again. This whole procedure contributes to the latency of the network. The MA-CSMA/CA incurs low delays. The mechanism of requesting the GTS slot from the coordinator of other cluster before being disconnected from the parent coordinator saves considerable amount of time which in the aforementioned protocols is wasted due to association process. Under variable and different number of active nodes conditions the proposed protocol shows significant reduction in latency as compared to other protocols which can be seen from the results.

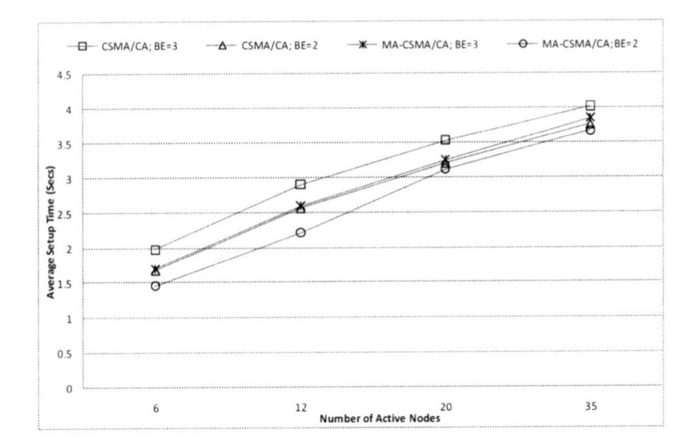

Fig. 4. Comparison of setup time for MA-CSMA/CA and CSMA/CA

5.2 Idle Channel Time

Idle channel time is one of the key factors in determining the efficiency of the MAC protocol. In WSNs the resources especially the bandwidth is also scarce and hence the efficient utilization of the bandwidth becomes priority for any MAC protocol. Fig. 5 shows the comparison of the existing industrial standard MAC protocol and the proposed protocol which supports mobility. In case of CSMA/CA it is clear from the

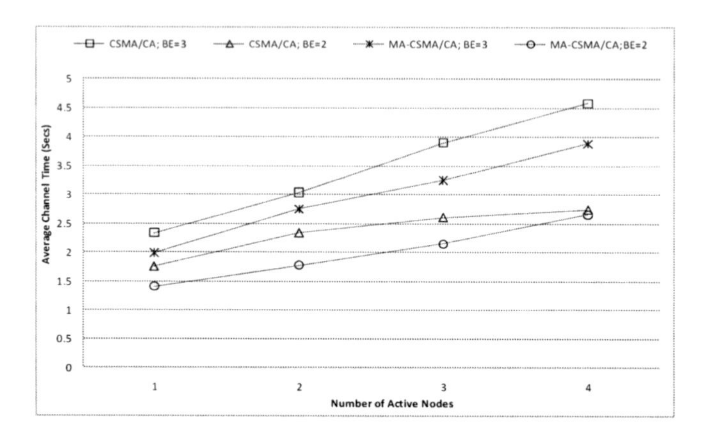

Fig. 5. Comparison of Idle Channel time for MA-CSMA/CA and CSMA/CA

results that the protocol shows worst bandwidth utilization. This is due to the fact that as the nodes collide or senses any activity over the channel they backoff randomly and this random backoff increases exponentially thus increasing the delay and decreasing the bandwidth utilization. It is because of this nature of CSMA/CA protocol the nodes undergo long delays before transmission of data. In case of proposed MA- CSMA/CA MAC nodes undergoes initial backoff values which is according to the standard, moreover before the start of transmission a feedback is provided to all the potential contending nodes about the duration of the current transmission. Therefore the node freezes their backoff values and goes to sleep mode, as they wake up the nodes start the remaining backoff count and access the channel. Hence instead of providing a long random backoff delay which results in excessive wait state with the probability that the channel may have been available during all the excessive random backoff duration, the proposed protocol provides initial backoff values along with the feedback of transmission so that nodes will know exactly when the channel is available for the transmission of data. It can be seen clearly from the results that the proposed MAC protocol significantly improves the channel utilization as compared IEEE 802.15.4 industrial standard CSMA/CA in mobile application scenario.

5.3 Goodput

Goodput is the total number of bits received correctly at the destination without retransmissions and control signals. Fig.6 shows the comparison of the goodput of the proposed protocol and CSMA/CA in mobile application. The result shows significant degradation of goodput in case of CSMA/CA. There are several factors behind the degradation of the goodput for the CSMA/CA. As from the previous section it is clear that the channel utilization is very poor and the latency in terms of setup time is very high. The other influencing factor is the presence of random backoff and problem of collision. As the number of nodes in the network increases it also increases the number of potential contending nodes which causes higher collision more random backoff and more retransmission of data. Hence more and more packets are collided

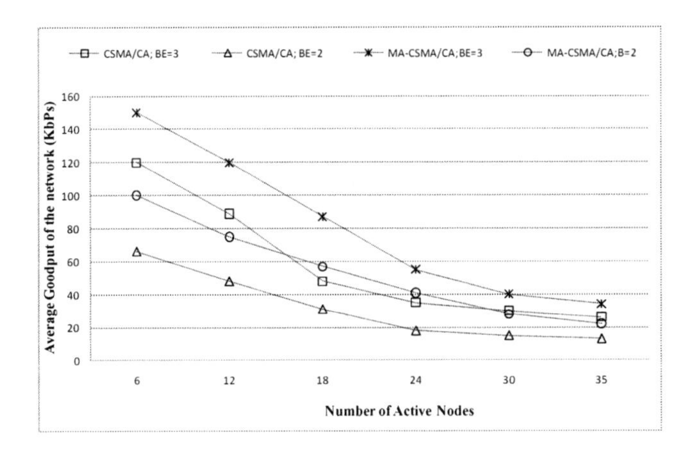

Fig. 6. Comparison of Goodput for MA-CSMA/CA and CSMA/CA

and the overall quality of service falls drastically. It can be seen that at lower value of *(BE)* the goodput of the network is decreased. This is due to the fact that more number of nodes will be ready to transmit the data with lower backoff delay causing considerable rise in collision and thus contributing to lower goodput results. In case of MA-CSMA/CA collision is minimized by allocating GTS slot for the mobile node entering the new cluster. Moreover the average setup time for the proposed protocol is very low and the channel utilization is also on the high side. Hence the goodput of the proposed scheme increases not only under increase number of active nodes but also on variable data traffic.

5.4 Energy

Energy is also one of the key factors in wireless sensor networks. Since sensors are tiny devices they are required to save energy. Fig. 7 shows the comparison of energy for MA-CSMA/CA and CSMA/CA. It can be seen from the result that as the number of nodes increases the energy consumption of CSMA/CA increases. The reason behind this is that CSMA/CA suffers heavily from collisions due to hidden nodes as well as contention collisions. As the packets of the nodes are collided they have to be retransmitted. This retransmission of data packet along with the normal transmission of the packets over the network results in more collisions which can be referred to as collision chain reaction. Due to retransmission and performing additional carrier sensing, nodes waste considerable amount of energy. Since the proposed protocol minimizes collision and all the nodes receive feedback from the coordinator about the ongoing transmission over the link, thus the nodes dose not waste unnecessary time for the channel to become idle and goes to sleep mode. Moreover due to the introduction of mechanism by virtue of which mobile node can request slot to transmit the data in upcoming cluster before even joining the cluster saves significant time required for association, thus saving considerable amount of energy as compared to the conventional standards.

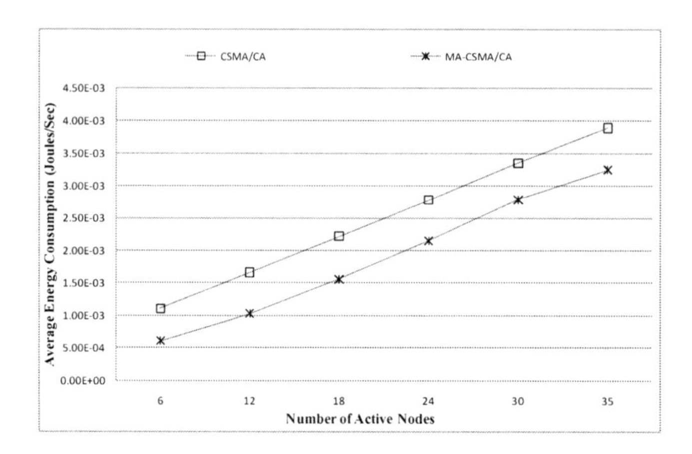

Fig. 7. Comparison of Energy for MA-CSMA/CA and CSMA/CA

6 Conclusion

In this paper a novel MA-CSMA/CA MAC protocol which is to the best of author's knowledge the first of its type is presented. The proposed MAC provides a contention based collision minimize transmission by employing mechanism of GTS allocation to the mobile nodes joining the new cluster which resolves the major issue of control message collisions as well. MA-CSMA/CA MAC solves the major problem of association for mobile node by introducing the request mechanism before entering a new cluster. Thus in this way the association time is almost negligible. MA-CSMA/CA works well for both static and mobile scenario without compromising the performance requirements. The proposed MAC is extensively compared with the current state of the art industrial standard CSMA/CA for IEEE 802.15.4. The results clearly show that it outperforms the existing industrial standard of CSMA/CA protocol in all the performance aspects. There is considerable gain in the reduction of latency of about 5 times and the channel utilization is about 6 times better. A significant improvement of about 70% to 80% in goodput and consumption of energy is reduced by 2 folds. The results also show the superiority of the proposed protocol MA-CSMA/CA not only at low traffic load but at high network load. As number of active and mobile nodes in the network increases the proposed protocol accomplishes far better performance than the existing IEEE 802.15.4 standard.

References

[1] Lee, J.H., Hashimoto, H.: Controlling mobile robots in distributed intelligent sensor network. IEEE Trans. Ind. Electron 50(5), 890–902 (2001)

[2] Ray, S., Starobinski, D., Trachtenberg, A., Ungrangsi, R.: Robust location detection with sensor networks. IEEE Journal of selected areas in Communication 22(6), 1016–1025 (2004)

[3] IEEE 802 Working Group, Standard for part 15.4: Wireless Medium Access Control (MAC) and Physical Layer (PHY) Specifications for Low Rate Wireless Personal Area Networks (LR-WPAN), ANSI/IEEE std. 802.15.4 (September 2006)
[4] Misic, J., Fung, C.J., Misic, V.B.: On node population in a multilevel 802.15.4 sensor network. In: Proc. GLOBECOM, pp. 1–6 (November 2006)
[5] Chlamtac, I., Lerner, A.: Fair algorithms for maximal link activation in multihop radio networks. IEEE Transactions on Communications 35(7), 739–746 (1987)
[6] Cidon, I., Sidi, M.: Distributed assignment algorithms for multihop packet radio networks. IEEE Transactions on Computers 38(10), 1236–1361 (1989)
[7] Ephremides, A., Truong, T.: Scheduling broadcasts in multihop radio networks. IEEE Transactions on Communications 38(4), 456–460 (1990)
[8] Kleirock, L., Tobagi, F.: Packet switching in radio channels, part 1: Carrier sense multiple-access models and their throughput delay Characteristics. IEEE Transactions on Communications 23(12), 1400–1416 (1975)
[9] Kleirock, L., Tobagi, F.: Packet switching in radio channels, part 2: Hidden-terminal problem in carrier sense multiple access and the busytone solution. IEEE Transactions on Communications 23(12), 1417–1433 (1975)
[10] Lam, S.: A carrier sense multiple access protocol for local networks. Computer Networks 4, 21–32 (1980)
[11] Bao, L., Garcia-Luna-Aceves, J.J.: A new approach to channel access scheduling for Ad Hoc networks. In: Seventh Annual International Conference on Mobile Computing and Networking 2001, pp. 210–221 (2001)
[12] Chlamtac, I., Farago, A.: Making transmission schedules immune to topology changes in multi-Hop packet radio networks. IEEE/ACM Transactions on Networking 2(1), 23–29 (1994)
[13] Ju, J., Li, V.: An optimal topology-transparent scheduling method in multihop packet radio networks. IEEE/ACM Transactions on Networking 6(3), 298–306 (1998)
[14] Ramanathan, S.: A unified framework and algorithm for channel assignment in wireless networks,Wireless Networks. Springer, Wireless Networks 5(2), 81–94 (1999)
[15] Raviraj, P., Sharif, H., Hempel, M., Ci, S.: MOBMAC- An Energy Efficient and Low latency MAC for Mobile Wireless Sensor Networks. IEEE Systems Communications 370–375 (August 14-17, 2005)
[16] Choi, S.-C., Lee, J.-W., Kim, Y.: An Adaptive Mobility-Supporting MAC protocol for Mobile Sensor Networks. In: IEEE Vehicular Technology Conference, pp. 168–172 (2008)
[17] Pham, H., Jha, S.: An adaptive mobility-aware MAC protocol for sensor networks (MS-MAC). In: Proceedings of the IEEE International Conference on Mobile Ad-hoc and Sensor Systems (MASS), pp. 214–226 (2004)
[18] Lin, P., Qiao, C., Wang, X.: Medium access control with a dynamic duty cycle for sensor networks. In: Proceedings of the IEEE Wireless Communications and Networking Conference (WCNC), vol. 3, pp. 1534–1539 (2004)
[19] Hameed, S.A., Shaaban, E.M., Faheem, H.M., Ghoniemy, M.S.: Mobility-Aware MAC protocol for Delay Sensitive Wireless Sensor Networks. In: IEEE Ultra Modern Telecommunications & Workshops, October 2009, pp. 1–8 (2009)
[20] Bettstetter, C., Resta, G., Santi, P.: The node distribution of the random waypoint mobility model for wireless ad hoc networks. IEEE Transactions on Mobile Computing 2(3), 257–269 (2003)

A Rough Set Approach Aim to Space Weather and Solar Storms Prediction

Reza Mahini[2], Caro Lucas[1], Masoud Mirmomeni[1,4], and Hassan Rezazadeh[3]

[1] Control and Intelligent Processing Center of Excellence, School of Electrical and Computer Engineering, University of Tehran, Iran
lucas@ipm.ir, mirmomen@msu.edu
[2] Electrical, Computer and IT Eng. Department, Payame Noor University of Tabriz, Iran
r_mahini@pnu.ac.ir
[3] Industrial Engineering Department, University of Tabriz, Iran
h-rezazadeh@tabrizu.ac.ir
[4] Departmentment of Computer Science and Engineering, Michigan State University, USA

Abstract. This paper illustrates using Rough set theory as a data mining method for modeling Alert systems. A data-driven approach is applied to design a reliable alert system for prediction of different situations and setting off of the alerts for various critical parts of human industry sections. In this system preprocessing and reduction of data with data mining methods is performed. Rough set learning method is used to attain the regular and reduced knowledge from the system behaviors. Finally, using the produced and reduced rules extracted from rough set reduction algorithms, the obtained knowledge is applied to reach this purpose. This method, as demonstrated with successful realistic applications, makes the present approach effective in handling real world problems. Our experiments indicate that the proposed model can handle different groups of uncertainties and impreciseness accuracy and get a suitable predictive performance when we have several certain features set for representing the knowledge.

Keywords: Data Mining, Rough Sets, Data-Driven Modeling, Solar Activities, Alert Systems.

1 Introduction

Studying the features of the solar activities is of prime importance, not only for its effect on the climatological parameters, but also for practical needs such as telecommunications, power lines, geophysical exploration, long-range planning of satellite orbital trajectories and space missions planned by space organizations. The Sun has an obvious, direct and perceptible impact on the Earth's environment and life. The thermal influence of the solar energy on the earth environment is modulated on easily identified daily and seasonal time scales. These are essentially Sun-as-a-star global effects, in which solar variability does not play any role. In fact, solar activity comprises of all transient phenomena occurring in the solar atmosphere, such as sunspots, active regions, prominences, flares and coronal mass ejections.

B. Murgante et al. (Eds.): ICCSA 2011, Part I, LNCS 6782, pp. 588–601, 2011.

Space weather forecast service must be available in real-time to moderate the effects for the users. The service must also be useful and understandable to the user. Space weather deals with real-world problems, i.e. conditions and processes that most often are described as nonlinear and chaotic. Since real-world data are noisy and huge, there is a need to utilize suitable and efficient methods of using them in alert and prediction systems. Traditionally Artificial Intelligence (AI) represented the symbolic approach to knowledge processing and coding. Recently, however AI (the new AI) also includes soft computing methods, a consortium of methodologies that works synergistically and provides, in one form or another, flexible information processing capability for handling real life ambiguous situations, is used. Soft computing methodologies include fuzzy sets, neural networks, genetic algorithms, rough sets, and their hybridizations, have recently been used to solve data mining problems. Neurocomputing techniques have been successful in modeling and forecasting space weather conditions and effects, simply because they can describe non-linear chaotic dynamic systems. They are also robust and still work despite data problems [1].

Moreover, expert systems, genetic algorithms, and hybrid systems such as neurofuzzy systems and combinations of neural networks [2], [3], [4], [5], [6], [7] have been used. Many similar studies of the solar activities and prediction over indices such as: SSA methods in [8], [9], prediction of solar activity with BELBIC in [10], [11], Hybrid Predictor method in [12], Alert System Based on Knowledge management [13], Integrated methods such as Knowledge-Based Neurocomputing [14], are also used.

Current studies aim at modeling and predicting space weather using the knowledge-based Rough set method for alerting regarding the different situation in space weather. Different alert systems and models are presented for predicting and alerting to communication and satellite system which support specific aims and use different approaches and ways such as: MHD [17], WINDM [18], MSFM [19] and LSWM[20], [21]. These models choose different indices as inputs and description of conclusion is used for prediction and setting off alerts.

The rest of this paper is organized as follows: Section 2 discusses the basics of rough set theory. Section 3 describes a Data-Driven Modeling Approach for the designed system. Section 4 describes the research case study. Section 5 concludes the paper and proposes further works.

2 Rough Set Methods

During twenty years of following Pawlak's [22] rough set theory and its applications it has reached a certain degree of maturity. In recent years, a rapid growth of interest in rough set theory and its application is seen worldwide. The theory has attracted attention of many researchers and practitioners all over the world who contributed fundamentally to its development and applications. From logical point of view rough set theory is a new approach to uncertainties. From practical point of view rough set theory seems to be of fundamental significance to AI and cognitive sciences, especially to machine learning, knowledge discovery, decision analysis, inductive reasoning and pattern recognition. It seems also important to decision support systems and data mining. In fact it is a new mathematical approach to data analysis. Rough set theory is based on sound mathematical foundation [23], [24].

Data reduction consists of eliminating of superfluous data from the information system in such a way that basic approximation properties of the system remain intact. Finally certain and possible decision rules are defined, which form a logical language to describe the lower and the upper approximation. Later, decision rules are used to describe patterns in the data. The main advantage of rough set theory is that it does not need any preliminary or additional information about data like probability in statistics, or basic probability assignment in Dempster-Shafer theory and grade of membership or the value of possibility in fuzzy set theory [23]. Some of the Rough set theory actions such as following: Characterizing of set of objects in terms of attribute values, Finding dependencies (total or partial) between attributes, Reduction of superfluous attributes (data), Finding significance attributes, Decision rule generation and others[23].

Rough set theory as one of the most important data mining method which is presented with the concept of an approximation space, which is a pair $\langle U, R \rangle$, where U is a non-empty set (the universe of discourse) and R an equivalence relation on U, i.e., R is reflexive, symmetric, and transitive. The relation R decomposes the set U into disjoint classes in such a way that two elements x, y are in the same class if $(x, y) \in R$. If two elements x, y in U belong to the same equivalence class, we say that x and y are indistinguishable. For $X \in 2^U$, in general it may not be possible to describe X precisely in $\langle U, R \rangle$ One may then characterize X by a pair of lower and upper approximations defined as follows [22].

$$\underline{RX} = \{x \in U \mid [x]_R \subseteq X\}; \quad \overline{RX} = \{x \in U \mid [x]_R \cap X \neq \phi\} \tag{1}$$

where $[x]_R$ stands for the equivalence class of x by R. The pair $(\underline{R}X, \overline{R}X)$ is the representation of an ordinary set X in the approximation space $\langle U, R \rangle$ or simply called the rough set of X.

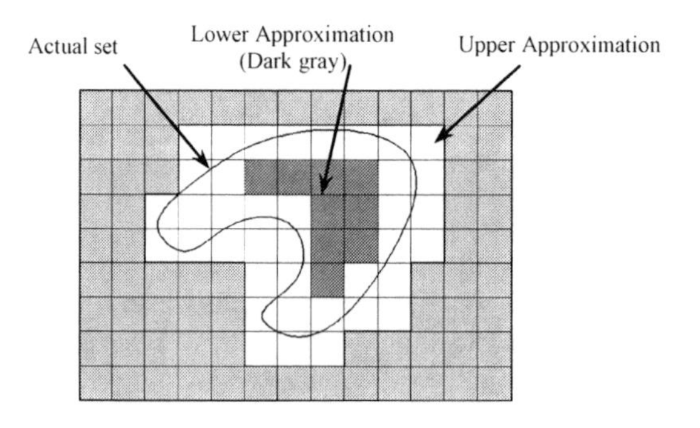

Fig. 1. Properties of rough sets approximation

An information system I is a pair $I = \langle U, A \rangle$, where U is a set of objects, A is a set of attributes, and each attribute $a \in A$ associated with the set of attribute values V_a is understood as a mapping $a: U \rightarrow V_a$. An information system is called a decision system if assuming that the set of attributes $A = C \bigcup D$ and $C \bigcap D \neq \phi$, where C is the set of conditional attributes and D is the set of decision attributes. Given an information system *I*, each subset *P* of the attribute set *A* induces an equivalence relation *IND (P)* called *P*-indiscernibility relation as follows:

$$IND(P) = \{(x, y) \in U^2 \mid a(x) = a(y), \, for \; all \; a \in P\}, \tag{2}$$

and

$$IND(P) = \bigcap_{a \in P} IND(\{a\}) \cdot \tag{3}$$

If $(x, y) \in IND(P)$ we then say that objects x and y are indiscernible with respect to attributes in *P*. In other words, we cannot distinguish *x* from *y*, and vice versa, in terms of attributes in *P*. Note that the partition of *U* generated by *IND(P)*, denoted by *U/IND(P)*, can be calculated in terms of those partitions generated by single attributes in *P* as follows [25]:

$$IND(P) = \otimes_{a \in P} Y/IND(\{\alpha\}), \tag{4}$$

where

$$A \otimes B = \{X \bigcap Y : \forall X \in A; \forall Y \in B; X \bigcap Y = \phi\}. \tag{5}$$

In [23], Pawlak firstly introduces two numerical characterizations of imprecision of a subset *X* in the approximation space $\langle U, P \rangle$ accuracy and roughness. Accuracy of *X*, denoted by $\alpha_P(x)$ is simply the ratio of the number of objects in its lower approximation to that in its upper approximation;

$$\alpha_P(x) = \left| \frac{\underline{P}(X)}{\overline{P}(X)} \right|, \tag{6}$$

Where $|\,.\,|$ denotes the cardinality of a set. It is possible to measure roughness and accuracy of approximation quality on the relations over *U* [25].

The attributes in decision table must be reduced with an efficient attributes reduction algorithm. For conciseness, this algorithm is summarized in pseudo code (see Fig.2).

In reduction of knowledge the basic role is played, in the proposed approach by two fundamental concepts of *a* reduct and the core. Intuitively, *a* reduct knowledge is its essential part, which suffices to define all basic concepts accruing in the considered knowledge, whereas the core is in a certain sense it's most important part. The following is an important property establishing the relationship between the cores and reductions.

Algorithm: *Attributes reduction (C, D)*
Input: *C, the set of criteria attributes; D, the set of class attribute (decision).*
Output: *R, the attribute reduct,* $R \subseteq C$

- $R \leftarrow \{\}$
- *do*
- $T \leftarrow R$
- *for each* $x \in (C - R)$
- *if* $\gamma_{R \cup \{x\}}(D) > \gamma_T(D)$
- $T \leftarrow R \cup \{x\}$
- $R \leftarrow T$
- *until* $\gamma_R(D) = \gamma_C(D)$
- *return* R

Fig. 2. Attributes Reduct algorithm [27]

$$CORE(P) = \bigcap RED(P), \tag{7}$$

Where $RED(P)$ is the family of all reductions of P and P is represented knowledge.

As mentioned in $\varphi \rightarrow \psi$ decision, attribute α is dispensable if:

$$\varphi \rightarrow \psi \Rightarrow \varphi/(P - \{a\}) \rightarrow \psi, \tag{8}$$

Where the set of all indispensable attributes in $\varphi \rightarrow \psi$ is called the core of $\varphi \rightarrow \psi$, and is denoted by $CORE(P \rightarrow Q)$ and

$$CORE(P \rightarrow Q) = \bigcap RED(P \rightarrow Q), \tag{9}$$

Where $RED(P \rightarrow Q)$ is the set of all reductions of $(P \rightarrow Q)$ [23].

3 Data-Driven Modeling Approach

This section of paper describes the designed architecture for prediction and setting off of the suitable alerts to the mentioned systems by using Rough set method. There are two basic approaches to prediction: model-based approach and nonparametric method (data-driven). Model-based approach assumes that sufficient prior information is available with which one can construct an accurate mathematical model for prediction. Nonparametric approach, on the other hand, directly attempts to analyze a sequence of observations produced by a system to predict its future behavior. Though nonparametric approaches often cannot represent full complexity of real systems, many contemporary prediction theories are developed based on the nonparametric approach because of difficulty in constructing accurate mathematical models. The most motivated idea in model analysis is to use a data-driven model to get a deeper insight to the underlying real world system.

According to the data-driven characteristics of the system and available data in this case (several years' solar activities data) after the system obtaining the previous behavior data of the solar activities obtained from authentic world data centers the system reduces the decision rules by using rough set theory reduction algorithms . For a better performance of rough set theory reduction algorithms and reducing noisy and missing values of data, an initial preprocessing phase on the raw data is performed.

In order to apply our approach, it is necessary to select reduce the set of attributes precisely (for attaining better result). It is also needed to normalize the data in standard data set form features. In this stage for having reliable results from the selected features, rough set feature reduction (RSFR) algorithm is used [27]. Finally, prior to this step it has a number of rules in decision table which leads to the behavior of the system, they are not neat and tidy however. As it was mentioned the rough set learning algorithm reduces these rules efficiently i.e. with using combination and removing repeated or inconsistent rules (caused by noise). In conclusion with combining of the decision rules for each class of the decision concept, the learning procedure from the off-line data was done. In the third phase, the model could function as simple as a gradient that updates with newly arrived on-line data, or an approximation model linearized around an operating point using historical data for the nominal part and on-line data for updating and re-tuning. The model architecture is described in figure 4. In order to guarantee the model's stability, new events can be considered as new observation or test data for checking system performance. The algorithm contains the following steps:

Initialize phase

› Getting observation from information centers.

› Normalization input data with determined range of data.

› Building standard decision table.

Learning phase

Data mining with Rough Set

› Checking out of decision table consistency and making it consistent.

› Removing redundant row of table.

› Finding Core and Reductions for any decision rule.

› Removing replicate rows to get the simple table.

› Achieving decision rules for decision table.

› Combining the decision rules of each class.

Reasoning phase and retuning

› Considering the result and setting off of suitable alerts.

› Continuing to Training designed system from observation data and previous results (section two).

Fig. 3. An overview of designed algorithm for predicting solar activities

As reflected in figure 4, the creation of descriptive production rules from given feature patterns is central to the present work.

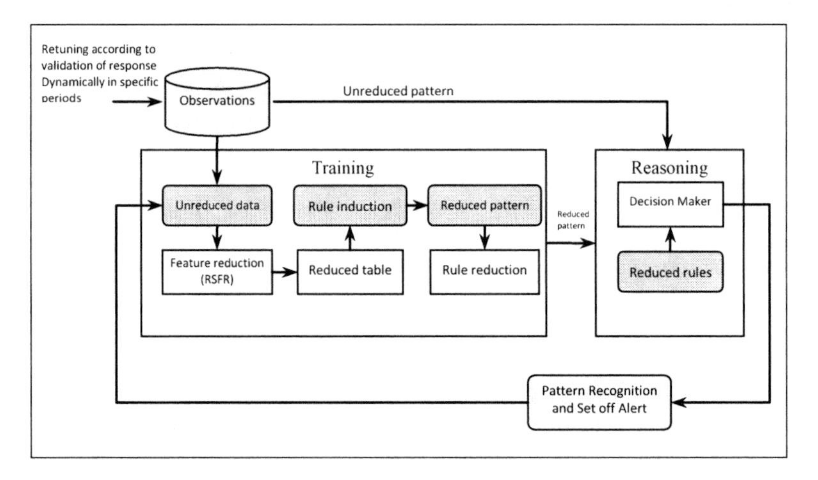

Fig. 4. Architecture of the designed system for setting off of alerts

Due to the fact that designed system may need to train again and have low efficiency at a point of time e.g. in an 11-year sunspot cycles of solar activities, the system will begin to learn from its previous results and also observations as new observations. So performance of the system will not be reduced and the system will have dynamic performance.

4 Case Study

The Sun exerts profound control over a natural hazard - space weather - that poses a risk to modern technology. This is a hazard we knew little of until the Space Age. The Sun's control of space weather is exercised through the charged particles and magnetic fields that are carried by the solar wind as it buffets the Earth's magnetic field. Our data and expertise help to develop scientific understanding of the evolution of the solar activities and space environments. In order to modeling and monitoring these activities, Rough sets are used as a tool. Specifically, it provides a mechanism to represent the approximations of concepts in terms of overlapping concepts [16].

4.1 Data Description

There are many important solar indices which researchers use for forecasting and designing the alert systems. In this paper, we selected several most important and commonest prediction and alert indices according to (the datasets of which are available from authentic information services from the world) to set off the alerts. The selected indices for this study are: Sunspot number (SN), Solar Radio Flux (SRF), Kp and Dst that we will illustrated them in the section below.

Sunspot Numbers (SN). Sunspots are darker than the rest of the visible solar surface because they are cooler. Sunspots appear often in groups, and the sunspot number R is defined as $R = k\ (f + 10\ g)$, where f is the total number of spots visible to the observer, g is the number of disturbed regions (single spots or groups of spots), and k is a constant for the observatory related to the sensitivity of the observing equipment. This index is a suitable criterion for the solar activities and source of many events of space weather.

Solar flux (SF). One of the major indicators of solar activity is known as the solar flux. It provides an indication of the level of radiation that is being received from the Sun. The level of ionising radiation that is received from the Sun is approximately proportional to the Solar Flux.

Geomagnetic Index (Kp). The Kp index is obtained from a number of magnetometer stations at mid-latitudes. Also this index is a suitable criterion for the alert systems.

Storm Time Index (Dst). The hourly Dst index is obtained from magnetometer stations near the equator but not so close that the E-region equatorial electrojet dominates the magnetic perturbations seen on the ground and uses in alert systems.

4.2 Experimental Results

This section first provides previous results and alerts from web sites [29], [30] and appropriates them as inputs for the designed system. It must be declared that providing these data in outright framework is difficult because different prediction models exploit different combinations of these parameters such as SN, Kp, Dst, Solar Wind, Solar Flux, AE, etc. Although the selected attributes were based on other real trusty alert systems, it must be noted that experimental result of algorithm in fig.2 [27] would also obtain the same indices as the important attributes. Also missing value in observation data, uncertainty and accuracy of measurement systems leads to unreliable predictions. Hence regarding described algorithm, preprocessing on data is done and the solar activity indices. The data sets are normalized in several specific separated classes. The chart of observation data for years 1996 – 2006 are shown in figure 7. The general objective is to obtain the partition that, for a fixed number of classes. Next, the indices data are shown with class numbers such as 1-2-3-4.

In next step, the algorithm begins cleaning data for obtaining some primary rules. The result of rule reduction step is illustrated in table 1. After reduction of the rules, the algorithm reduces redundant rules obtained from algorithm's step 5 of learning phase.

For implementation of the system, we selected 3000 out of 4000 decision rules for training phase and the system was expected to response to new event. In training phase, the algorithm reached 73 efficient rules to make decision which 10 of them are presented in table 1. In fact, we have just reduced the primary rules without any combining. Combination of the rules to reach minimal number of them is done with rough set algorithms. In the next section of the algorithm for attainment of efficient and valid rules for decision making, support and confidence of the rules are computed. Hence the number of rules for using decision system is 36 decision rules which are presented in table 2.

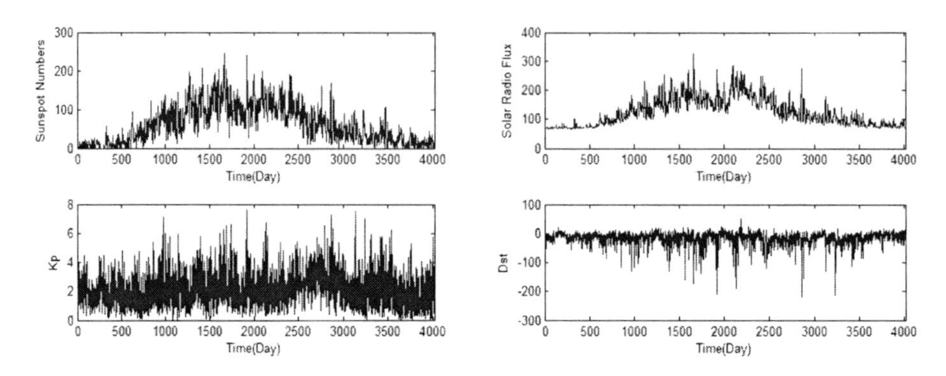

Fig. 7. The real data charts from solar indices (SN, SF, Kp, Dst)

Table 1. Results of decision rules reduction algorithm (DS stands for system decision)

	Antecedent				consequent
IF	SN is 1	------	& Kp is 1	& Dst is 1	THEN DS is 1
IF	SN is 2	------	------	------	THEN DS is 2
IF	SN is 1	------	& Kp is 1	& Dst is 3	THEN DS is 2
IF	SN is 2	------	& Kp is 3	& Dst is 2	THEN DS is 3
IF	------	& SF is 2	& Kp is 3	& Dst is 2	THEN DS is 3
IF	------	& SF is 2	& Kp is 3	& Dst is 3	THEN DS is 3
IF	SN is 3	& SF is 3	& Kp is 4	& Dst is 4	THEN DS is 4
IF	SN is 2	------	------	& Dst is 4	THEN DS is 4
IF	SN is 2	& SF is 2	------	& Dst is 4	THEN DS is 4
IF	SN is 2	& SF is 2	& Kp is 4	& Dst is 4	THEN DS is 4

Table 2. The combination of decision rules results (final rules)

Rows	The reduced rules
1	$SN_1\ Kp_1\ Dst_1 \rightarrow DS_1$
2	$SN_1\ SF_1\ Kp_1\ Dst_1 \rightarrow DS_1$
3	$SN_1 \vee Kp_1 \rightarrow DS_2$
4	$(SN_1 \vee SF_1) Kp_1\ Dst_3 \rightarrow DS_2$
5	$SN_1\ Kp_2\ (SF_1 \vee Dst_2) \rightarrow DS_2$
6	$SN_2\ Kp_1\ (SF_1 \vee Dst_2) \rightarrow DS_2$
7	$(SN_2 \vee Dst_3)\ SF_1\ Kp_2 \rightarrow DS_2$
8	$SN_2\ SF_2\ (Kp_2 \vee Dst_1) \rightarrow DS_2$
9	$SN_2\ Kp_1\ (SF_3 \vee Dst_3 \vee Dst_1) \rightarrow DS_2$
10	$SN_1\ Kp_2\ (S\ F_2 \vee Dst_3 \vee Dst_1) \rightarrow DS_2$
11	$Kp_3 \rightarrow DS_3$
12	$SN_3\ SF_1\ (Kp_1 \vee Dst_1) \rightarrow DS_3$

Table 2. (Continued)

13	$SN_3\ SF_2\ (Kp_1 \vee Dst_1 \vee Dst_2) \rightarrow DS_3$
14	$SN_3\ SF_1\ (Kp_2 \vee Dst_2) \rightarrow DS_3$
15	$Kp_3\ Dst_2\ (SN_2 \vee SF_2) \rightarrow DS_3$
16	$Kp_3\ Dst_3\ (SN_2 \vee SF_2) \rightarrow DS_3$
17	$SN_3\ SF_3\ (Dst_1 \vee Dst_2 \vee Dst_3 \vee Kp_1) \rightarrow DS_3$
18	$SN_2\ SF_3\ (Kp_3 \vee Kp_2 \vee Dst_3 \vee Dst_1) \rightarrow DS_3$
19	$SN_4 \vee SF_4 \vee Kp_4 \vee Dst_4 \rightarrow DS_4$
20	$SN_4\ Dst_2\ (SF_1 \vee Kp_1 \vee Kp_2 \vee Kp_3) \rightarrow DS_4$
21	$SN_4\ SF_4\ Dst_1\ (Kp_1 \vee Kp_2) \rightarrow DS_4$
22	$SN_4\ Dst_4 \rightarrow DS_4$
23	$SN_4\ Dst_1 \rightarrow DS_4$
24	$SN_4\ Dst_3 \rightarrow DS_4$
25	$SN_4\ SF_4\ Dst_3\ (Kp_2 \vee Kp_3) \rightarrow DS_4$
26	$SN_4\ SF_4\ Kp_4\ (Dst_3 \vee Dst_4) \rightarrow DS_4$
27	$SN_4\ SF_4\ (Kp_1 \vee Dst_3 \vee Dst_4) \rightarrow DS_4$
28	$SN_4\ SF_3\ (Dst_3 \vee Dst_4) \rightarrow DS_4$
29	$SN_4\ SF_3\ Kp_1 Dst_4 \rightarrow DS_4$
30	$SN_4\ SF_3\ Kp_4\ Dst_3 \rightarrow DS_4$
21	$SN_4\ SF_4\ Dst_4\ (Kp_2 \vee Kp_3 \vee Kp_4) \rightarrow DS_4$
32	$SN_3\ Dst_4 \rightarrow DS_4$
33	$SN_3\ SF_3\ Dst_4 \rightarrow DS_4$
34	$SN_2\ Dst_4 \rightarrow DS_4$
35	$SN_2\ SF_2\ Kp_4\ Dst_4 \rightarrow DS_4$
36	$SN_2\ SF_2\ Dst_4 \rightarrow DS_4$

For examination of this algorithm 1000 data from the whole 4000 were selected. As it is shown, the alert system experimental result is performed over the test data. The training data is obtained from [28], [29] and the experimental results of the system are shown in table 4 compared with the real alert in [3]. Comparing two different systems which use different indices and method for prediction and alert might be not easy because in real world we have a limit number of intelligent alert system, in other words, systems which have special algorithms and methods of performing predictions and alerts.

Table 3. The system performance based on Kp index

Kp Warning	Kp <= 3	3 <Kp<= 5	Kp> 5
Detection	41.40%	97.26%	68.75%
Miss	58.59%	2.73%	31.25%
False Alarm	0.00%	1.05%	11.99%

In table 3 we the system performance with common alert systems alerts based on Kp index values is presented. due to the fact that, the system works with 4 criteria especially when the values of the Kp is low the system has rather low performance but against high value of Kp index the alert of the system is appropriate. The real alerts are described with these linguistic terms: low activity (LA), active (A), high activity (HA), very high activity (VHA).

Table 4. Comparison of system results with the expert detection[28]

SN	SF	Kp	Dst	Rough set Output	System Alert	Expert Detection
0	72.6	1.2	-5.5	1	quiet	LA
13	75.1	1.312	-62.12	3	danger	A
76	128	2.325	-24.04	3	danger	HA
144	210.6	3.037	-13.41	4	H.danger	VHA
151	210.3	1.237	1.25	4	H.danger	VHA
107	174.4	3.675	-46.04	3	danger	HA
50	112.7	2.412	-14.37	2	active	A
22	77.9	0.612	-8.208	1	quiet	LA
16	90.7	1.662	-6.416	1	quiet	LA

Table 5. Comparison of 10 results of system detection and real alerts [30]

YY	MM	DD	SSN	SF	Kp	Dst	Syste	Real Alert
2004	3	19	58	111.3	1.625	-10.12	3	risk
2004	3	27	88	127.2	3.362	-27.25	3	risk
2004	4	29	24	89.8	1.037	-7.916	2	prudence
2004	7	20	91	180.8	2.012	-7.041	3	risk
2004	7	21	88	177.7	0.837	-0.875	3	risk
2004	7	23	74	170.4	4.737	-19.12	3	risk
2004	7	24	69	151.8	4.175	-118.3	4	High risk
2004	7	25	57	143.9	7.275	-66.54	4	High risk
2004	7	26	64	132	3.625	-132.4	4	High risk
2004	7	27	55	121.8	7.575	-83.79	4	High risk
2004	12	31	22	95.2	1.8	-20.2	2	prudence
2005	1	17	64	133.1	5.2	-73.58	3	risk
2005	1	19	45	128.3	5.037	-47.16	3	risk
2005	1	27	20	84.3	0.487	-8.583	1	safe
2005	3	3	9	75.7	1.075	-6.125	1	safe
2005	3	25	34	81.7	3.35	-18.83	2	prudence
2006	2	21	0	74.2	3.337	-17.37	2	prudence
2006	2	22	0	74.4	2.912	-11.45	2	prudence
2006	4	14	36	79.4	5.325	-46.12	3	risk
2006	11	21	0	75.6	0.175	11.333	1	safe

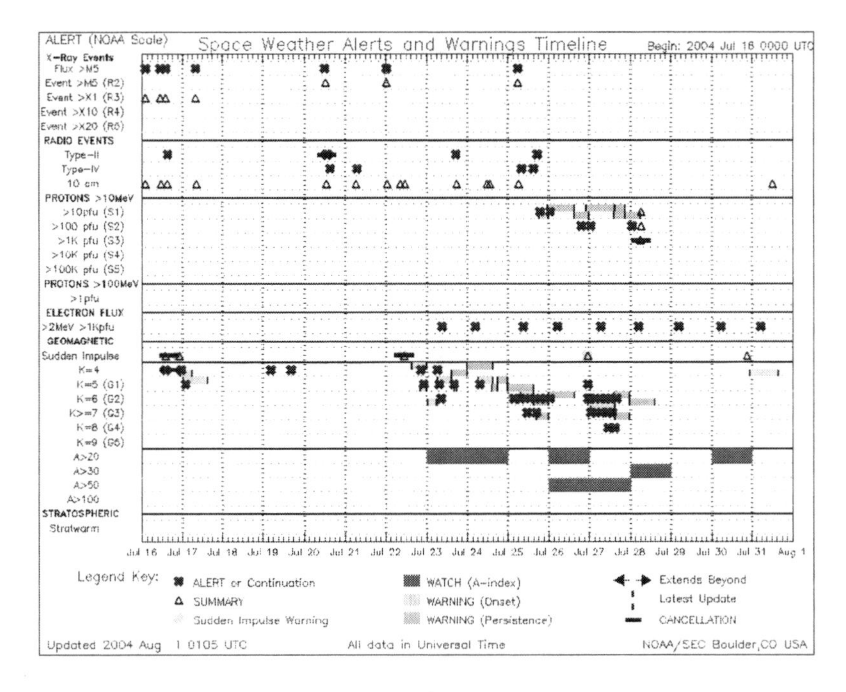

a)

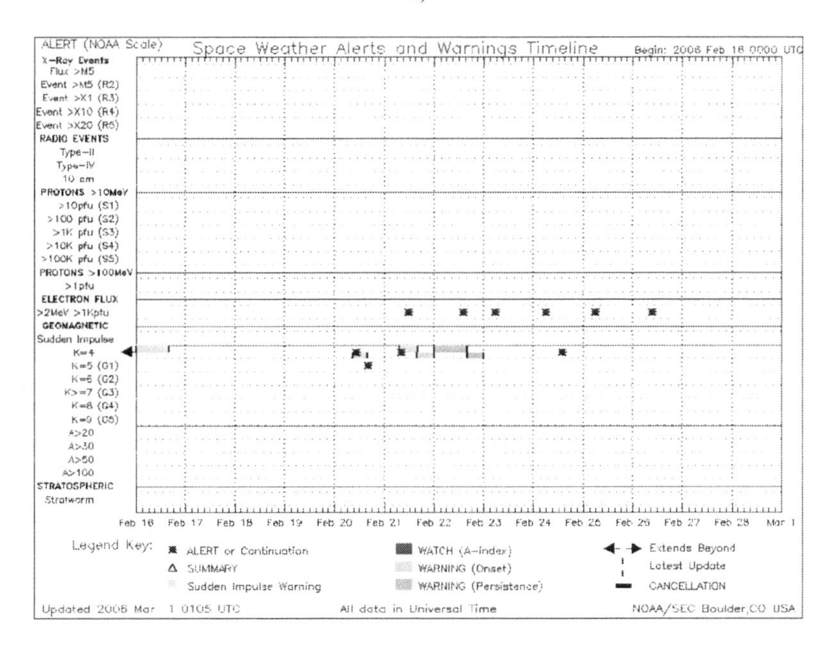

b)

Fig. 8. The alerts from the alert system in NOAA in a) Jul 2004 , b) Feb 2006

5 Conclusion and Future Works

Developing alert and prediction systems using noisy data in huge amount is one of the most important research topics. Regarding the importance of application of these systems in expensive satellite and communications and power systems in recent years, using data mining and suitable combinations of their algorithms can be useful in simplifying and solving the problems. As it was seen, in this research an initial normalization of data for converting them to practical information was utilized. Since in data mining based on rough set theory, there is no need to an expert or prior knowledge and this system learns from data or solar events, for discovering and formulating knowledge a data driven model was used. So all occurred patterns can be members of their class such as quiet class and etc. Finally, it is possible to conclude that the designed system has a suitable capability in reducing the effects of noises in decision making. Moreover, having high reliability rate is one of its advantages. In future works, in order to arrive at more accuracy and better performance in prediction and alerting, other collective combinations of rough set theory can be used with other methods which yet have other approaches toward the issue.

Acknowledgements. This paper is dedicated to the memory of the deceased professor Dr. Caro Lucas.

References

1. Mitra, S., Pal, S.K., Mitra, P.: Data mining in Soft Computing Framework: A Survey. IEEE transactions on neural networks 13(1) (January 2002)
2. Lundstedt, H.: Neural networks and prediction of solar terrestrial effects. Planet Space Science 40, 457–464 (1992)
3. Boberg, F., Wintoft, P., Lundstedt, H.: Real time Kp predictions from solar wind data using neural networks. Phys. Chem. Earth. 25(4), 275–280 (2000)
4. Gholipour, A., Abbaspour, A., Araabi, B.N., Lucas, C.: Enhancements in the prediction of solar activity by locally linear model tree. In: Proc. of MIC 2003: 22nd Int. Conf. on Modeling, Identification and Control, Innsbruck, Austria, pp. 158–161 (2003)
5. Gholipour, A., Lucas, C., Araabi, B.N., Mirmomeni, M., Shafiee, M.: Extracting the main patterns of natural time series for long-term neurofuzzy prediction. Neural Comput. & Applic. Springer, London (2006), doi:10.1007/s00521-006-0062-x
6. Attia, A.F., Hamed, R.A., Quassim, M.: Prediction of Solar Activity Based on Neuro-Fuzzy Modeling. Springer Solar Physics 227, 177–191 (2005)
7. Gholipour, A., Araabi, B.N., Lucas, C.: Predicting Chaotic Time Series Using Neural and Neurofuzzy Models: A Comparative Study. Springer Neural Processing Letters 24, 217–239 (2006)
8. Gholipour, A., Lucas, C., Araabi, B.N., Shafiee, M.: Solar activity forecast: Spectral analysis and neurofuzzy prediction. Elsevier Journal of Atmospheric and Solar-Terrestrial Physics 67, 595–603 (2005)
9. Prestes, A., Rigozo, N.R., Echera, E., Vieira, L.E.A.: Spectral analysis of sunspot number and geomagnetic indices (1868–2001). Journal of Atmospheric and Solar-Terrestrial Physics 68, 182–190 (2006)

10. Lucas, C., Abbaspour, A., Gholipour, A., Araabi, B.N., Fatourechi, M.: Enhancing the Performance of Neurofuzzy Predictors by Emotional Learning Algorithm. Informatica 27(2), 165–174 (2003)
11. Babaie, T., Karimizandi, R., Lucas, C.: Prediction of solar conditions by emotional learning. Intelligent Data Analysis 9, 1–15 (2006)
12. Chen, Y.P., Wu, S.N., Wang, J.S.: A Hybrid Predictor for Time Series Prediction. IEEE, Los Alamitos (2004), 0-7803–8359
13. Mahini, R., Lucas, C., Mirmomeni, M.: Designing a New Alert System Based on KM in Fuzzy Expert System. In: 2nd Int. Conf. knowledge management. KMCM (2008)
14. Cloete, I., Zurada, J.M.: Knowledge-Based Neurocomputing. MIT Press, Cambridge (2000)
15. Li, R.F., Wang, X.Z.: Dimension reduction of process dynamic trends using independent component analysis. Computers and chemical engineering 26, 467–473 (2002)
16. Chimphlee, S., Salim, N., Ngadiman, M.S.B., Chimphlee, W., Srinoy, S.: Independent component analysis and rough fuzzy based approach to web usage mining. In: Proceeding of 24th IASTED international Multi-Conference Artificial Intelligence and applications, February 13–16 (2006)
17. De Zeeuw, D.L., Gombosi, T.I., Groth, C.P., Powell, K.G., Stout, F.: An adaptive MHD method for global space weather simulations. IEEE Tran. on Plasma Science 28(6), 1956–1965 (2002)
18. Horton, W., Doxas, I.: A low dimensional dynamical model for the solar wind driven geotail-ionosphere system. Journal of Geophysical Research 103, 4561–4572 (1998)
19. Freeman, J., Nagai, A., Reiff, P., Denig, W., Gussenhoven, S.S., Heinermann, M., Rich, F., Hairston, M.: The use of neural networks to predict magnetospheric parameters for input to a magnetospheric forecast model. In: Joselyn, J., Lundstedt, H., Trollinger (eds.) Artificial Intelligence Applications in Solar Terrestrial Physics, Boulder, Colorado, vol. 167, Natl. Oceanic and Atmos. Admin, Boulder, Colorado (1994)
20. Gleisner, H., Lundstedt, H., Wintoft, P.: Predicting geomagnetic storms from solar wind data using time delay neural networks. Annales Geophysicae 14, 679–686 (1996)
21. Gleisner, H.: Solar wind and Geomagnetic activity: predictions using neural networks. PhD thesis, Lund University, Lund, Sweden (2000)
22. Pawlak, Z.: Rough sets. International Journal of Computer and Information Sciences 11, 341–356 (1982)
23. Pawlak, Z.: Rough Sets, Theoretical Aspects of Reasoning about Data. Kluwer Academic Publishers, Dordrecht (1991)
24. Ziarko, W. (ed.): Rough Sets, Fuzzy Sets and Knowledge Discovery. Proceeding of the International Workshop on Rough Sets and Knowledge Discovery (RSKD 1993), Banff, Alberta, Canada, October 12–15. Springer, Heidelberg (1993)
25. Huynh, V., Ho, T., Nakamori, Y.: An Overview on the Approximation Quality Based on Rough-Fuzzy Hybrids. Studies in Fuzziness and Soft Computing (2008)
26. Shen, Q., Chouchoulas, A.: A rough-fuzzy approach for generating classification rules. Pattern Recognition Society, pp. 31–3203. Elsevier Science Ltd., Amsterdam (2002)
27. Jensen, R., Shen, Q.: Fuzzy–rough attribute reduction with application to web categorization. Fuzzy Sets and Systems 141, 469–485 (2004)
28. The Space Physics Interactive Data Resource, `http://spidr.ngdc.noaa.gov/spidr/index.jsp`
29. NOAA's National Geophysical Data Center, `http://www.nesdis.noaa.gov`
30. Space Weather Alerts Archives, `http://www.swpc.noaa.gov/alerts/archive`

A Discrete Flow Simulation Model for Urban Road Networks, with Application to Combined Car and Single-File Bicycle Traffic

Jelena Vasic and Heather J. Ruskin

School of Computing, Dublin City University, Dublin 9, Ireland
{jvasic,hruskin}@computing.dcu.ie

Abstract. A model, discrete in terms of time, geometrical space and velocity, is defined for a mix of car and bicycle traffic. Although based on cellular automata (CA) and interchangeable with a CA model in some special cases, the spatial aspect of the model presented here includes some characteristics that set it apart from CA models, such as overlapping cells and extended stochasticity. These characteristics allow easy incorporation of a variety of network elements into a spatial network model. The behaviour model includes rules for movement along stretches of road, as well as rules of behaviour at decision and conflict points on the road. Agent based simulations are run for three simple scenarios and results of these simulations are presented.

Keywords: cellular automata, heterogeneous traffic flow, bicycles, urban roads, simulation.

1 Introduction

Since the publication of the seminal papers by Nagel and Schreckenberg [17] and Biham et al. [3], the former proposing a one-dimensional and the latter a two-dimensional model, cellular automaton (CA) models have figured prominently in the area of traffic flow modelling. Characteristics of urban traffic have been studied using these models for a variety of network scenarios [8,5,21,19], including unsignalised individual network elements with either homogeneous [18,22,23,12,24,13] or heterogeneous traffic [7,6,9,14] and signalised elements [4,2]. Mixed motorised and non-motorised traffic has been modelled using cellular automata also: Gundaliya et al. [10] and Mallikarjuna and Rao [16] developed such models based on multiple cell occupancy. The CA approach has been used to some extent for additional modelling of bicycle flows and interactions with motorised traffic [11,14].

Herein we present a three-way (time, space, velocity) discrete flow model for heterogeneous vehicular traffic on a network. The spatial component of the model accommodates road sharing by vehicles of different sizes and maximal velocities, through use of heterogeneous cellular automaton systems. It addresses the spatial aspect of complex manoeuvres that take place at intersections - and other

B. Murgante et al. (Eds.): ICCSA 2011, Part I, LNCS 6782, pp. 602–614, 2011.

points of conflict and decision making in the network - by defining rules for the transposition of geometrically natural space representations into abstract discrete models. The behaviour model is based on the Nagel-Schreckenberg cellular automaton model for road traffic [17], which was considerably modified to incorporate lateral interaction between vehicles, as well as the complex behaviour that arises from network structure.

While the model is general with respect to vehicle and infrastructure element types, it has been applied here to the case of mixed car and pedal-bicycle urban traffic and the elemental topologies of straight road, left turn and right turn.

This paper is organised as follows. Section 2 presents the model. Section 3 describes the simulation scenarios that were implemented, while Sect. 4 presents the results obtained from the simulations. Section 5 concludes the paper by discussing the merits of the model, summarising the results and indicating related planned future work.

2 Model

The spatial and behaviour model are presented separately.

2.1 Spatial Model

One of the principal characteristics of the spatial model construction, introduced in [20], is that different vehicles move on separate, potentially overlapping cellular systems. Thus a larger vehicle moves on a lattice consisting of larger cells, while a smaller vehicle moves on a different lattice, consisting of proportionally smaller cells. Network features where routes diverge or converge are modelled using overlapping cells, in that the cells of two convergent or divergent routes will overlap somewhat. Intersections and other complex network elements are modelled using geometrically natural routes (to determine the number of cells and overlapping of routes within an element) rather than represented using a rectangular cell lattice, as is typically the practice in cellular automata models.

Some terms need to be defined at this point:

- a **track** is a single-width sequence of cells along which a vehicle moves forward
- the cells represent vehicle **positions** on the track and are assigned whole numbers in ascending order, from 0 to N - 1, where N is the number of cells in the track; they may overlap with cells in the same track or with cells in other tracks
- a **decision point** is a point at which a track diverges to become more than one track: a vehicle must make a decision as to which of the possible directions it will take, before it reaches the decision point
- a **conflict point** is a position at which a track is in conflict with another track because in some part the two tracks overlap; generic examples of such conflicts are found at intersections in urban traffic or entry-ramps on motorways

Three infrastructure elements are constructed using the described model. Their geometric representation is shown in Fig. 1.

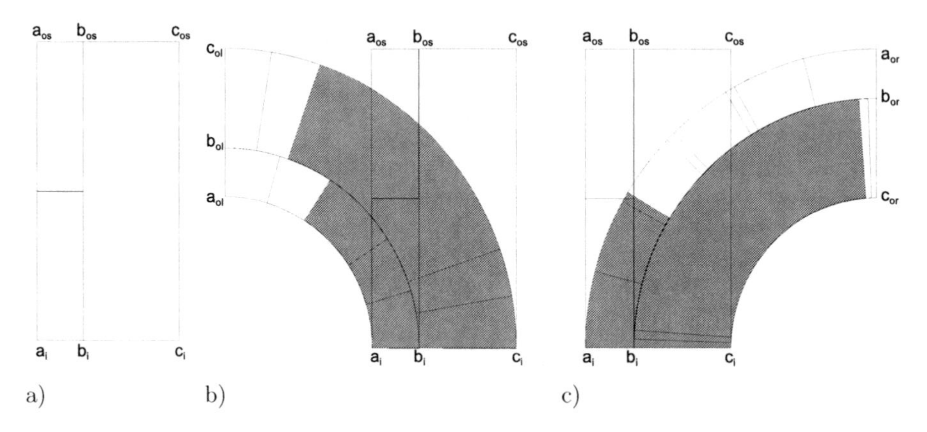

Fig. 1. Spatial model elements (model representations of real infrastructure elements): (a) cell of a road shared by cars and bicycles, (b) the spatial model of a left turn in the same type of road and (c) the spatial model of a right turn in that same type of road. The shaded areas show the size of a cell in each track.

Figure 1a represents a cell of a straight road where bicycles and cars move side-by-side but do not enter each other's space. Bicycles move on track $a_i b_i b_{os} a_{os}$, which contains two cells, while cars move on road section $b_i c_i c_{os} b_{os}$, which consists of a single cell. No cells overlap. This element is used for situations of "positional discipline", where bicycles keep to the left[1] side of the lane, while cars keep to the right. Figure 1b represents a left turn. The car and bicycle tracks for the straight ahead direction are identical to the straight road element. The turning tracks, i.e. left in Fig. 1b and right in Fig. 1c, are the same in number and relative positions as those of the straight tracks, but, as can be seen from the figure, cells within a route overlap. The purpose of this is to allow for space occupation geometry to be reflected in the model (for example, a car that fits exactly into track $b_i c_i c_{os} b_{os}$ in Fig. 1b cannot be accommodated exactly in track $b_i c_i c_{ol} b_{ol}$ and is allowed to "edge" forward in this track) and to force delay in a section of the road that slows vehicles down (in this case, the turn). Fine-tuning of this delay is introduced by placing a presence probability on cells. The presence probability expresses the difficulty of traversing the track (e.g. the difficulty of turning), where a greater probability represents a higher difficulty of turning. When a vehicle is passing over the cell, it is counted into the traversed length if it is present and is not counted if it is not present. The cell's

[1] The use of directions "left" and "right" is with reference to left-hand-side driving, as in Ireland or the UK. However, the model has its "mirror image", applicable to right-hand side driving, in what would result from an across-model exchange of "left" for "right" and vice versa. It is, therefore, generalisable.

presence is re-calculated, using the presence probability value, after a vehicle traverses it. It has the same value for all the cells in a track. Three overlapping cells constitute each of tracks $\mathrm{a_i b_i b_{ol} a_{ol}}$ and $\mathrm{b_i c_i c_{ol} b_{ol}}$ in Fig. 1b. Similarly, five and three cells constitute, respectively, tracks $\mathrm{a_i b_i b_{or} a_{or}}$ and $\mathrm{b_i c_i c_{or} b_{or}}$ in Fig. 1c. The straight road section, in Fig. 1a, has no conflict or decision points. The left turn element, Fig. 1b, has a single conflict, between bicycle track $\mathrm{a_i b_i b_{os} a_{os}}$ and car track $\mathrm{b_i c_i c_{ol} b_{ol}}$; thus bicycles and cars travelling on those tracks are presented with conflict points $\mathrm{a_i b_i}$ and $\mathrm{b_i c_i}$, respectively. The left turn element has a single decision point for each type of vehicle, at $a_i b_i$ for bicycles and at $b_i c_i$ for cars. The right turn element, Fig. 1c, has a single conflict, between car track $\mathrm{b_i c_i c_{os} b_{os}}$ and bicycle track $\mathrm{a_i b_i b_{or} a_{or}}$. Bicycles entering the element are presented with the conflict immediately, at $a_i b_i$, similarly to the cars, which see a conflict point at $b_i c_i$. A decision point in this element exists at $\mathrm{a_i b_i}$ for bicycles and at $\mathrm{b_i c_i}$ for cars.

The relationship between the cells of the model geometric space is expressed through (i) the sequence in a route, such as $\mathrm{a_i b_i b_{os} a_{os}}$, which consists of two cells, (ii) sequencing of infrastructure elements, where connected, and (iii) overlapping. The overlapping of cells can be expressed through a table that we call the impingement table, which indicates all instances of overlap between cells within an infrastructure element. For example, Table 1 is the impingement table for the left turn shown in Fig. 1. The binary representation of overlap, when transferred into code during implementation, facilitates fast run-time inspection of cell availability for occupation.

2.2 Behaviour Model

The behaviour model is built upon the original Nagel-Schreckenberg [17] CA model. Even though this basic model does not reproduce all the properties of traffic flow on freeways, it is sufficient as a base for the purpose of modelling movement in an urban network [15].

Table 1. Impingement table for left turn shown in Fig. 1b. Column and row headings refer to cells. For example, BL2 is the second cell in the bicycle left-turn track.

	BS1	**BS2**	**CS1**	**BL1**	**BL2**	**BL3**	**CL1**	**CL2**	**CL3**
BS1	1	0	0	1	1	0	1	1	1
BS2	0	1	0	0	0	0	1	1	1
CS1	0	0	1	0	0	0	1	1	1
BL1	1	0	0	1	1	1	0	0	0
BL2	1	0	0	1	1	1	0	0	0
BL3	0	0	0	1	1	1	0	0	0
CL1	1	1	1	0	0	0	1	1	1
CL2	1	1	1	0	0	0	1	1	1
CL3	1	1	1	0	0	0	1	1	1

In a cellular automaton model, in general, a set of rules is followed towards determining the velocity of vehicles, i.e. the number of cells that a vehicle will advance in the next time step, which then controls movement in any iteration. This position update can be performed (i) in parallel for each time step, where first the velocities for all the vehicles in the system are determined, then all the vehicles are moved or (ii) in sequence, where each vehicle moves immediately upon determining its velocity for the time step. We chose the more commonly used parallel update, because it is "less safe" from the point of view of conflict resolution, and thus more closely models the problems of interest here.

The following list contains the variables and constants used in the behaviour rule descriptions, with their definitions:

- $v_{\mathrm{LIM}i}(t+1)$ is the limiting value for i^{th} vehicle's velocity at time step $t+1$
- v_{MAX} is the maximal velocity
- $d_{\mathrm{I0}i}$ is the distance from the ith vehicle to the nearest impinged cell i.e. the number of unimpinged cells ahead of the vehicle; a cell is *impinged* if it is not available for occupation, i.e. if it is occupied or if any overlapping cells are occupied
- d_{di} is the distance to the dth decision point for vehicle i
- d_{WD} is the warning distance for a decision point
- $v_{\mathrm{LD}}(x)$ is the decision point imposed velocity limit at distance x to a decision point
- d_{ci} is the distance to the cth conflict point for vehicle i
- d_{WC} is the warning distance for a conflict point
- $v_{\mathrm{LC}}(x)$ is the conflict point imposed velocity limit at distance x to a conflict point
- $d_{\mathrm{BL0}i}$ is the distance to the nearest bicycle in the adjoining bicycle track to the left, for car i
- d_{WB} is the warning distance for a bicycle in the adjoining bicycle track to the left
- $v_{\mathrm{LB}}(x)$ is the velocity limit imposed by a bicycle in the adjoining bicycle track to the left at distance x
- $v_i(t)$ is the velocity of the ith vehicle at time step t
- $v_i(t+1)$ is the currently calculated velocity for the ith vehicle at time step $t+1$
- p_R is the randomisation parameter, with which stochasticity is introduced into the behaviour model
- C0 is the conflict nearest ahead to the ith vehicle
- $d_{\mathrm{C0}i}$ is the distance to the nearest conflict for vehicle i
- v_{C0O0} is the velocity of the first approaching vehicle on the conflicting track in conflict C0
- d_{C0O0} is the distance to C0 of the first approaching vehicle on the conflicting track in conflict C0

The rules that define the behaviour of vehicles in this model are as follows:

1. **Building a list of decision and conflict points within "warning distance"**

The *warning distance* is the greatest distance at which a vehicle needs to be aware of a decision point or an intersection it is approaching. Both the decision point and the conflict point warning distances are constant for a vehicle type and are denoted d_{WD} and d_{WC}, respectively. The positions at and within warning distance before a decision or conflict point are associated with velocity limits. This defines the warning distance as the farthest distance from a decision or conflict at which the velocity limit is lower than the maximal velocity for a vehicle type.

The velocity limits within warning distance can be chosen to specify different behaviours. The specific set of limits used here can be defined as those that allow a vehicle to reach the velocity of 1 before arriving at the decision or conflict point, while decelerating, at most, by 1. Table 2 shows the values resulting from that rule, at maximal velocities of 3 and 2 for cars and bicycles, respectively. From the table it can be seen that, for example, the warning distance for a bicycle approaching an intersection or conflict is 2, while the warning distance for a car approaching one of these entities is 5. A car is also warned if there is a bicycle alongside it or alongside its route within warning distance.

Table 2. Velocity limits at positions on approach, if v_{MAX} for cars is 3 and for bicycles is 2. *Notes:* (1) The limit of 1 is also imposed in the case of a bicycle alongside a car. In the given numbering context, this corresponds to an index of -1.

Zero-based index of position ahead	4	3	2	1	0
To intersection, by bicycle (if turning)	-	-	-	1	1
To intersection, by car (if turning)	2	2	2	1	1
To conflict, by bicycle	-	-	-	1	1
To conflict, by car	2	2	2	1	1
To bicycle on left side[1], by car	2	2	2	1	1

2. **Making any decisions presented in rule 1**
   Making the decisions involves intelligence on the part of the agent representing the vehicle-driver unit. A decision in the model consists of choosing a direction if more than one direction of movement is possible for the vehicle. This, for example, occurs at the left turn shown in Fig. 1b. Upon reaching the entry line into the turn, $a_i c_i$, a vehicle has the option of moving forward or moving left, and a decision as to route choice must be made.

   Decisions are made using a probability of turning, separately assigned to each vehicle type (car, bicycle). This approach is possible since the scenarios implemented here include only two-way decisions ("straight ahead or left?" and "straight ahead or right?").
3. **Determine limiting value for velocity**
   Determining the maximum value that a particular vehicle's velocity can take in the next time step is a multi-step task in itself:

(a) $v_{\text{LIM}i}(t+1) = v_{\text{MAX}}$
(b) if $d_{\text{I}0i} < v_{\text{LIM}i}(t+1)$ then $v_{\text{LIM}i}(t+1) = d_{\text{I}0i}$
(c) for each decision made in step 1: if $d_{di} < d_{\text{WD}}$ and vehicle i is not going straight through and $v_{\text{LD}}(d_{di}) < v_{\text{LIM}i}(t+1)$ then $v_{\text{LIM}i}(t+1) = v_{\text{LD}}(d_{di})$
(d) for each conflict encountered in step 1: if $d_{ci} < d_{\text{WC}}$ and vehicle i does not have priority and $v_{\text{LC}}(d_{ci}) < v_{\text{LIM}i}(t+1)$ then $v_{\text{LIM}i}(t+1) = v_{\text{LC}}(d_{ci})$
(e) if vehicle i is a car and $d_{\text{BL}0i} < d_{\text{WB}}$ and $v_{\text{LB}}(d_{\text{BL}0i}) < v_{\text{LIM}i}(t+1)$ then $v_{\text{LIM}i}(t+1) = v_{\text{LB}}(d_{\text{BL}0i})$

4. **Acceleration (equivalent to Nagel-Schreckenberg rule 1 [17])**
$v_i(t+1) = v_i(t) + 1$, if $v_i(t) < v_{\text{LIM}i}(t)$
5. **Slowing down based on the limiting velocity value (equivalent to Nagel-Schreckenberg rule 2 [17])**
if $v_{\text{LIM}i}(t+1) < v_i(t+1)$, then $v_i(t+1) = v_{\text{LIM}i}(t+1)$
6. **Randomisation (equivalent to Nagel-Schreckenberg rule 3 [17])**
if $v_i(t+1) > 0$ then, with probability p_R the velocity is re-calculated as $v_i(t+1) = v_i(t+1) - 1$
7. **Checking for unresolved conflicts**
if $v_i(t+1) > d_{\text{C}0i}$ and conflict C0 is unresolved, then $v_i(t+1) = d_{\text{C}0i}$
The particular conflict resolution rule employed in the simulations is that a conflict is unresolved if there is a vehicle approaching the conflict point on the conflicting track and if the current velocity of that vehicle will in any way allow it to cross into the conflict area, i.e. cross the conflict point, in the next time step. This corresponds to $v_{\text{C}0\text{O}0} < d_{\text{C}0\text{O}0} \Rightarrow$ conflict resolved, otherwise conflict not resolved. Note that at $d_{\text{C}0\text{O}0} = 0$ the velocity condition is $v_{\text{C}0\text{O}0} < 0$ and since a velocity cannot be negative in the model, no vehicles are permitted just at the conflict point if the conflict is to be considered resolved.
8. **Vehicle motion (equivalent to Nagel-Schreckenberg rule 4 [17])**
The ith vehicle moves ahead by $v_i(t+1)$.

It should be noted that, in the rules defined above, v_{MAX}, d_{WD}, d_{WC}, v_{LD} and v_{LC} values are different for different vehicle types. Application of the rules to a particular vehicle type implies the use of appropriate vehicle type-correspondent values of those variables.

3 Simulated Scenarios

Three scenarios were simulated: a stretch of straight road, a left turn and a right turn.

The straight road stretch was constructed from 100 elements of the type shown in Fig. 1a. The spatial model for this configuration is shown in Fig. 2. The left turn scenario is constructed from three groups identical to that for the straight road scenario and one left turn element of the type shown in Fig. 1b. The spatial configuration used for the left turn scenario is shown in Fig. 3. The right turn scenario is constructed analogously (using a right turn element from Fig. 1c instead of the left turn element).

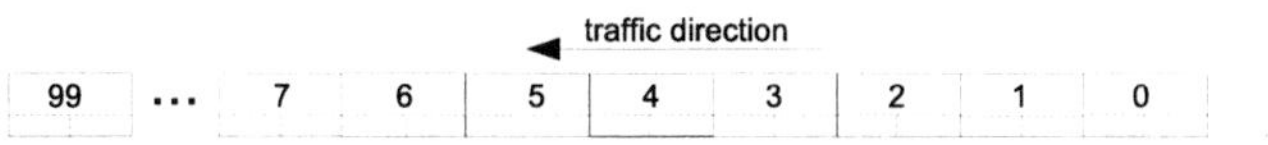

traffic direction

99 ... 7 6 5 4 3 2 1 0

Fig. 2. Spatial model for the straight road simulation scenario

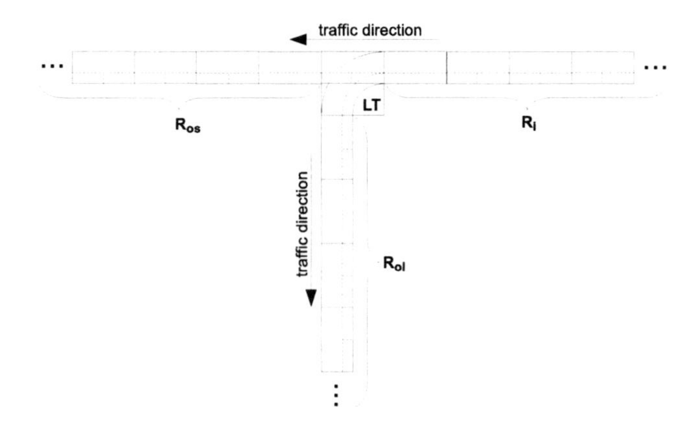

Fig. 3. Spatial model for the left turn road simulation scenario. Road sections $\mathrm{R_i}$, $\mathrm{R_{os}}$ and $\mathrm{R_{ol}}$ are each exactly the same as the straight road stretch shown in Fig. 2. LT is the element shown in Fig. 1b.

The following settings apply to the simulations:

- In all scenarios the constants used are: for cars $v_{\mathrm{MAX}} = 3$, $p_{\mathrm{R}} = 0.3$ and for bicycles $v_{\mathrm{MAX}} = 2$, $p_{\mathrm{R}} = 0.3$.
- In the left turn scenario the probability of turning, p_{T}, is set to 0 for bicycles and varied for cars: $p_{\mathrm{T}} \in \{0.0, 0.5, 1.0\}$; the cell presence probability for car track $\mathrm{b_i c_i c_{ol} b_{ol}}$ is varied as $p_{\mathrm{CP}} \in \{0.33, 6.67, 1.0\}$.
- In the right turn scenario the probability of turning, p_{T}, is set to 0 for cars and varied for bicycles: $p_{\mathrm{T}} \in \{0.0, 0.5, 1.0\}$; the cell presence probability for bicycle track $\mathrm{a_i b_i b_{or} a_{or}}$ is varied as $p_{\mathrm{CP}} \in \{0.2, 0.6, 1.0\}$.
- Each combination of parameters for each scenario was run for $t = 100000$ time-steps.
- A vehicle insertion attempt onto each suitable track takes place at each time step, before rule application, with probability p_{IB} and p_{IC} and initial velocity 1 and 2, for bicycles and cars, respectively. The vehicle is placed on the track at position $v_{\mathrm{MAX}} - 1$ or the farthest unimpinged cell, whichever is lesser. If position 0 is impinged, the insertion does not take place. In the straight road simulation scenario, both the bicycle and car track of the road stretch are "suitable". In the left and right turn scenarios, the only "suitable" tracks are the bicycle and car track of road section $\mathrm{R_i}$, since the initial cells of all other tracks immediately follow tracks of connected infrastructure elements.

– Vehicles move off the last cell of an open-ended stretch of road as if the road extends infinitely and has no vehicles on it beyond the last cell of the open-ended stretch.

4 Results

Each parameter combination for each simulation scenario was applied with car insertion probabilities taking all values $0 \leq p_{\mathrm{IC}} \leq 1$, with step 0.02, and bicycle insertion probabilities taking all values $0 \leq p_{\mathrm{IB}} \leq 1$, with step 0.02. A flow diagram, as a function of the two insertion probabilities, for all simulation scenarios and vehicle types, is shown in Fig. 4. The effect of the yield rule for cars and bicycles, respectively, is visible in Fig. 4b,f. The cases where priority is granted are illustrated in Fig. 4c,e, respectively, for cars and bicycles. Here the maximal flow capacities are almost maintained, relative to the straight stretch of road results in Fig. 4a,d, in spite of the increasing probability of insertion for the other type of vehicle. The small difference in maximal flow capacities between Fig. 4a and Fig. 4c for cars and between Fig. 4d and Fig. 4e for bicycles arises from the delay to vehicles on the higher priority track caused by those few vehicles on the lower priority track that manage to enter the intersection before an effective conflict

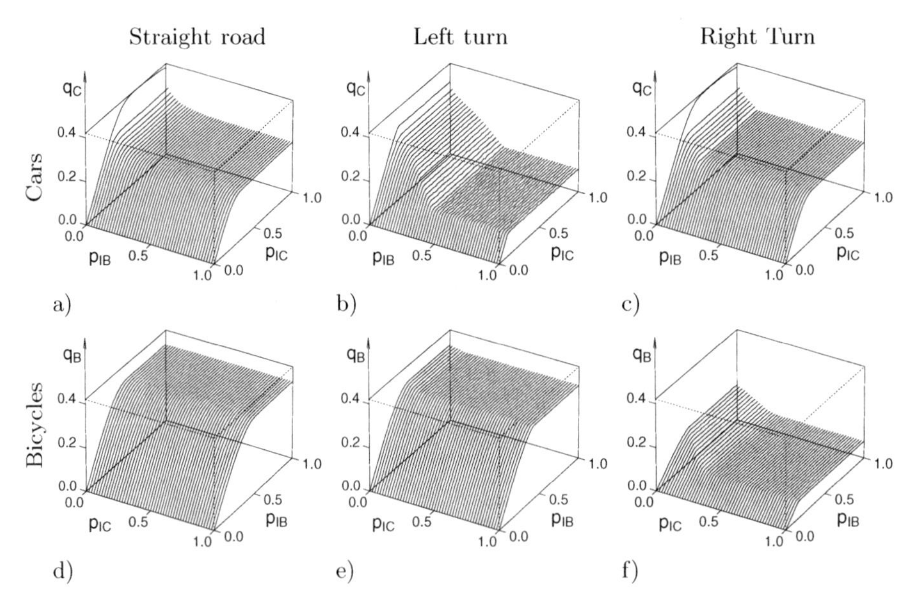

Fig. 4. Flow as a function of bicycle insert probability and car insert probability measured in **straight road** simulation for cars (a) and for bicycles (d); in **left turn** simulation with $p_{\mathrm{T}} = 0.5$, $p_{\mathrm{CP}} = 0.67$ for cars (b) and bicycles (e); and in **right turn** simulation with $p_{\mathrm{T}} = 0.5$, $p_{\mathrm{CP}} = 0.6$ for cars (c) and bicycles (f). The measurements were taken at the last cell of the initial stretch of road: $x = 99$ for cars and $x = 199$ for bicycles.

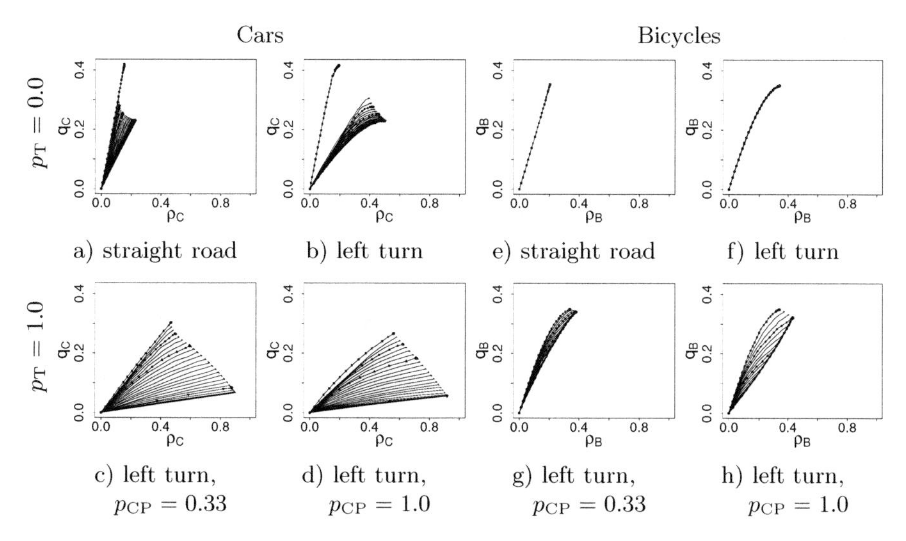

Fig. 5. Fundamental diagrams (density, flow) measured in **straight road** simulation for cars (a) and bicycles (e) and in **left turn** simulation with $p_T = 0.0$ for cars (b) and bicycles (f); with $p_T = 1.0$, $p_{CP} = 0.33$ for cars (c) and bicycles (g); and with $p_T = 1.0$, $p_{CP} = 1.0$ for cars (d) and bicycles (h). The measurements were taken at the last cell of the initial stretch of road: $x = 99$ for cars and $x = 199$ for bicycles; and for insertion probabilities $0 \leq p_{IC} \leq 1$, with step 0.02, and $0 \leq p_{IB} \leq 1$, with step 0.02. The car flow lines each correspond to a value of p_{IB}, while the bicycle flow lines each correspond to a value of p_{IC}. Increasing $p_{IB}(p_{IC})$, up to a certain value, result in fundamental diagram lines with lower flow for cars(bicycles). For any $p_{IB}(p_{IC})$ equal to or above that value, the fundamental diagram is on the lower limit line.

occurs and are slow to leave it. Also, in all the car flow diagrams, Fig. 4a,b,c, the negative impact of the presence of bicycles on the flow of cars is visible.

In Figs. 5 and 6 more detail can be seen on how the flows and densities develop as insertion probabilities of similar and other vehicle type change. These two figures contain fundamental diagrams for the two scenarios, across different vehicle types, over a range of values for turning probability and difficulty of turning. In both scenarios and for both vehicle types, the fundamental diagrams that are unaffected by flows of the other vehicle type exhibit flows equal to those of the straight road case, with higher corresponding vehicle densities. This is because of the difference in measurement position (in the middle of the road for left and right turn scenarios, as opposed to at the end of the road for the straight road scenario). This is the case with the highest flow lines in Figs. 5b,f,g,h and 6b,c,d,f. In Figs. 5g,h and 6c,d, the fundamental diagram lines show lower flows for higher insertion probabilities of other vehicle types. This is caused by the turning vehicles of lower priority that manage to enter the intersection (before a conflict arises) but linger long enough to affect the flow of the priority stream. This effect, however, is limited. In Figs. 5c,d and 6g,h, the fundamental diagram lines "fan out" much further towards the 0 flow line, which is an expected effect

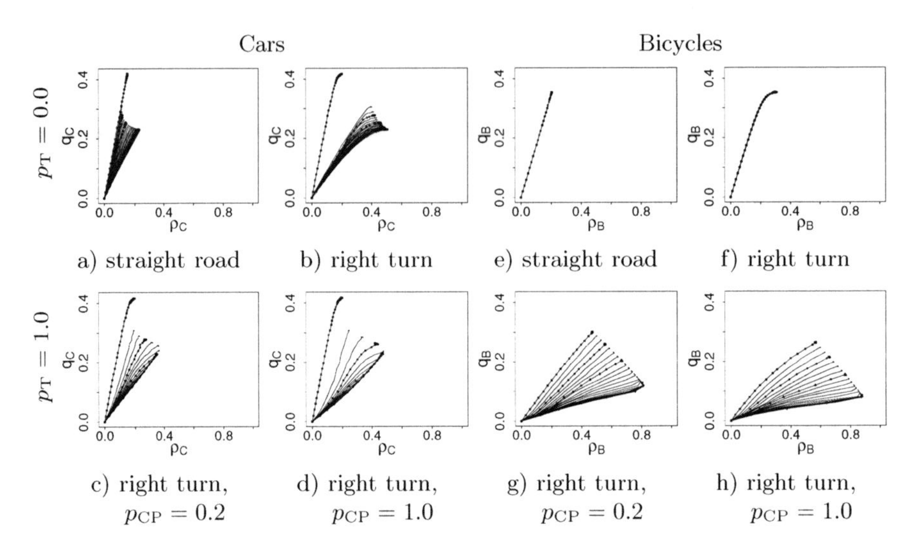

Fig. 6. Fundamental diagrams (density, flow) measured in **straight road** simulation for cars (a) and bicycles (e) and in **right turn** simulation with $p_T = 0.0$ for cars (b) and bicycles (f); with $p_T = 1.0$, $p_{CP} = 0.2$ for cars (c) and bicycles (g); and with $p_T = 1.0$, $p_{CP} = 1.0$ for cars (d) and bicycles (h). The measurements were taken at the last cell of the initial stretch of road: $x = 99$ for cars and $x = 199$ for bicycles; and for insertion probabilities $0 \leq p_{IC} \leq 1$, with step 0.02, and $0 \leq p_{IB} \leq 1$, with step 0.02. The car flow lines are each for a constant value of p_{IB}, while the bicycle flow lines are each for a constant value of p_{IC}. Increasing $p_{IB}(p_{IC})$, up to a certain value, result in fundamental diagram lines with lower flow for cars(bicycles). For any $p_{IB}(p_{IC})$ equal to or above that value, the fundamental diagram is on the lower limit line.

for the low-priority stream. However, the flows do not reach 0, because of the upper limit imposed on densities by the simulation model itself when used with open boundary conditions (cf. [1]). The decrease in flow that is present even at insertion rate of 0 for other vehicle type, in Figs. 5c,d and 6g,h, is a result of vehicles slowing down to turn. The additional "fanning out" of car fundamental diagram lines in all cases is caused by cars slowing down in the presence of bicycles at close proximity. Increasing proportions of turning vehicles and higher difficulty of turning values both negatively affect flows of each vehicle type.

5 Conclusion

The model, presented in this paper, builds on earlier cellular automaton modelling and simulation, of vehicle movement, to define a general and easily extensible model for heterogeneous traffic on urban traffic networks. Simulations were performed of three simple scenarios, in order to demonstrate the applicability of the model. Results show the expected capacity reduction for flows with lower priority, and variation of flow with changes in turning probability and difficulty of turning. Also, reduction in flow of cars, effective in the presence of bicycles in close proximity on the road, is successfully incorporated into the model.

Future efforts, building on the work presented here, will involve application of the model to more complicated scenarios, including additional control and rule based management techniques for urban networks, particularly those relevant to motorised/non-motorised heterogeneous traffic.

Acknowledgement. This work is funded by the Irish Research Council for Science, Engineering and Technology (IRCSET), through an "Embark Initiative" postgraduate scholarship, addressing the "greening" of city transport.

References

1. Barlovic, R., Huisinga, T., Schadschneider, A., Schreckenberg, M.: Open boundaries in a cellular automaton model for traffic flow with metastable states. Physical Review E 66, 46113 (2002)
2. Belbasi, S., Foulaadvand, M.E.: Simulation of traffic flow at a signalized intersection. Journal of Statistical Mechanics: Theory and Experiment 2008, P07021 (2008)
3. Biham, O., Middleton, A.A., Levine, D.: Self-organization and a dynamical transition in traffic-flow models. Physical Review A 46, R6124–R6127 (1992)
4. Brockfeld, E., Barlovic, R., Schadschneider, A., Schreckenberg, M.: Optimizing traffic lights in a cellular automaton model for city traffic. Physical Review E 64, 56132 (2001)
5. Chowdhury, D., Schadschneider, A.: Self-organization of traffic jams in cities: Effects of stochastic dynamics and signal periods. Physical Review E 59, R1311–R1314 (1999)
6. Deo, P., Ruskin, H.J.: Comparison of homogeneous and heterogeneous motorised traffic at signalised and two-way stop control single lane intersection. In: Gavrilova, M.L., Gervasi, O., Kumar, V., Tan, C.J.K., Taniar, D., Laganá, A., Mun, Y., Choo, H. (eds.) ICCSA 2006. LNCS, vol. 3980, pp. 622–632. Springer, Heidelberg (2006)
7. Deo, P., Ruskin, H.J.: Simulation of heterogeneous motorised traffic at a signalised intersection. In: El Yacoubi, S., Chopard, B., Bandini, S. (eds.) ACRI 2006. LNCS, vol. 4173, pp. 522–531. Springer, Heidelberg (2006)
8. Esser, J., Schreckenberg, M.: Microscopic simulation of urban traffic based on cellular automata. International Journal Of Modern Physics C 8, 1025–1036 (1997)
9. Feng, Y., Liu, Y., Deo, P., Ruskin, H.J.: Heterogeneous traffic flow model for a two-lane roundabout and controlled intersection. International Journal of Modern Physics C: Computational Physics & Physical Computation 18, 107–117 (2007)
10. Gundaliya, P.J., Mathew, T.V., Dhingra, S.L.: Heterogeneous traffic flow modelling for an arterial using grid based approach. Journal of Advanced Transportation 42, 467–491 (2008)
11. Jiang, R., Jia, B., Wu, Q.S.: Stochastic multi-value cellular automata models for bicycle flow. Journal of Physics A - Mathematical and General 37, 2063 (2004)
12. Li, X.B., Jiang, R., Wu, Q.S.: Cellular automaton model simulating traffic flow at an uncontrolled t-shaped intersection. International Journal of Modern Physics B 18, 2703–2707 (2004)

13. Li, X.G., Gao, Z.Y., Jia, B., Zhao, X.M.: Cellular automata model for unsignalized t-shaped intersection. International Journal of Modern Physics C: Computational Physics & Physical Computation 20, 501–512 (2009)
14. Li, X.G., Gao, Z.Y., Jia, B., Zhao, X.M.: Modeling the interaction between motorized vehicle and bicycle by using cellular automata model. International Journal of Modern Physics C: Computational Physics & Physical Computation 20, 209–222 (2009)
15. Maerivoet, S., De Moor, B.: Cellular automata models of road traffic. Physics Reports 419, 1–64 (2005)
16. Mallikarjuna, C., Rao, K.R.: Cellular automata model for heterogeneous traffic. Journal of Advanced Transportation 43, 321–345 (2009)
17. Nagel, K., Schreckenberg, M.: A cellular automaton model for freeway traffic. J. Phys. I 2, 2221–2229 (1992)
18. Ruskin, H.J., Wang, R.L.: Modelling traffic flow at an urban unsignalised intersection. In: Sloot, P.M.A., Tan, C.J.K., Dongarra, J., Hoekstra, A.G. (eds.) ICCS-ComputSci 2002. LNCS, vol. 2329, pp. 381–390. Springer, Heidelberg (2002)
19. Tonguz, O.K., Viriyasitavat, W., Bai, F.: Modeling urban traffic: A cellular automata approach. IEEE Communications Magazine 47, 142–150 (2009)
20. Vasić, J., Ruskin, H.J.: Cellular automaton simulation of traffic including cars and bicycles. Physica A (submitted)
21. Wahle, J., Schreckenberg, M.: A multi-agent system for on-line simulations based on real-world traffic data. Hawaii International Conference on System Sciences 3, 3037 (2001)
22. Wang, R.L., Ruskin, H.J.: Modeling traffic flow at a single-lane urban roundabout. Computer Physics Communications 147, 570–576 (2002)
23. Wang, R.L., Ruskin, H.J.: Modelling traffic flow at a multilane intersection. In: Kumar, V., Gavrilova, M.L., Tan, C.J.K., L'Ecuyer, P. (eds.) ICCSA 2003. LNCS, vol. 2667, pp. 577–586. Springer, Heidelberg (2003)
24. Wang, R.L., Ruskin, H.J.: Modelling traffic flow at multi-lane urban roundabouts. International Journal of Modern Physics C: Computational Physics & Physical Computation 17, 693–710 (2006)

A GPU-Based Implementation for Range Queries on Spaghettis Data Structure

Roberto Uribe-Paredes[1,2], Pedro Valero-Lara[3], Enrique Arias[4], José L. Sánchez[4], and Diego Cazorla[4]

[1] Computer Engineering Department, University of Magallanes, UMAG, Punta Arenas, Chile

[2] Database Group - UART, National University of Patagonia Austral, Río Turbio, Santa Cruz, Argentina

[3] Albacete Research Institute of Informatics, University of Castilla-La Mancha, Albacete, España

[4] Computing Systems Dept, University of Castilla-La Mancha, Albacete, España

roberto.uribeparedes@gmail.com

Abstract. Similarity search in a large collection of stored objects in a metric database has become a most interesting problem. The *Spaghettis* is an efficient metric data structure to index metric spaces. However, for real applications processing large volumes of generated data, query response times can be high enough. In these cases, it is necessary to apply mechanisms in order to significantly reduce the average query time. In this sense, the parallelization of metric structures is an interesting field of research. The recent appearance of *GPU*s for general purpose computing platforms offers powerful parallel processing capabilities. In this paper we propose a *GPU*-based implementation for *Spaghettis* metric structure. Firstly, we have adapted *Spaghettis* structure to *GPU*-based platform. Afterwards, we have compared both sequential and *GPU*-based implementation to analyse the performance, showing significant improvements in terms of time reduction, obtaining values of speed-up close to 10.

Keywords: Databases, similarity search, metric spaces, algorithms, data structures, parallel processing, GPU, CUDA.

1 Introduction

In the last decade, the search of similar objects in a large collection of stored objects in a metric database has become a most interesting problem. This kind of search can be found in different applications such as voice and image recognition, data mining, plagiarism and many others. A typical query for these applications is the *range search* which consists in obtaining all the objects that are at a definite distance from the consulted object.

1.1 Similarity Search in Metric Spaces

Similarity is modeled in many interesting cases through metric spaces and the search of similar objects through range search or nearest neighbour. A metric

B. Murgante et al. (Eds.): ICCSA 2011, Part I, LNCS 6782, pp. 615–629, 2011.

space $(\mathbb{X}, d)$ is a set $\mathbb{X}$ and a distance function $d : \mathbb{X}^2 \rightarrow \mathbb{R}$, so that $\forall x, y, z \in \mathbb{X}$; then there must be properties of positiveness $(d(x,y) \geq 0$ *and* $d(x,y) = 0)$ iff $(x = y)$, symmetry $(d(x,y) = d(y,x))$ and triangle inequality $(d(x,y) + d(y,z) \geq (d(x,z))$.

In a metric space $(\mathbb{X},d)$ given, a finite data set $\mathbb{Y} \subseteq \mathbb{X}$, a series of queries can be made. The basic query is the *range query*, a query being $x \in \mathbb{X}$ and a range $r \in \mathbb{R}$. The range query around x with range r is the set of objects $y \in \mathbb{Y}$ such that $d(x,y) \leq r$. A second type of query that can be built using the range query is k *nearest neighbour*, the query being $x \in \mathbb{X}$ and object k. Neighbors k nearest to x are a subset $\mathbb{A}$ of objects $\mathbb{Y}$, such that if $|\mathbb{A}| = k$ and an object $y \in \mathbb{A}$ does not exist an object $z \notin A$ such that $d(z,x) \leq d(y,x)$.

Metric access methods, *metric space indexes* or *metric data structures* are different names for data structures built over a set of objects. The objective of these methods is to minimize the amount of distance evaluations made to solve the query. Searching methods for metric spaces are mainly based on dividing the space using the distance to one or more selected objects. As they do not use particular characteristics of the application, these methods work with any type of objects [1].

Among other important characteristics of metric structures, we can mention that some methods may work only with discrete distances, while others also accept continuous distances. Some methods are static, since the data collection cannot grow once the index has been built. Others accept insertions after construction. Some dynamic methods allow insertions and deletions once the index has been generated.

Metric space data structures can be grouped in two classes [1], *clustering*-based and *pivots*-based methods.

The *clustering*-based structures divide the space into areas, where each area has a so-called center. Some data is stored in each area, which allows easy discarding the whole area by just comparing the query with its center. Algorithms based on clustering are better suited for high-dimensional metric spaces, which is the most difficult problem in practice. Some clustering-based indexes are *BST* [2], *GHT* [3], *M-Tree* [4], *GNAT* [5], *EGNAT* [6], and *SAT* [7].

There exist two criteria to define the areas in clustering-based structures: *hyperplanes* and *covering radius*. The former divides the space in *Voronoi* partitions and determines the hyper plane the query belongs to according to the corresponding center. The covering radius criterion divides the space in spheres that can be intersected and one query can belong to one or more spheres.

The *Voronoi diagram* is defined as the plane subdivision in n areas, one per each center c_i of the set $\{c_1, c_2, \dots, c_n\}$ (centers) so that a query $q \in c_i$ area if and only if the Euclidean distance $d(q, c_i) < d(q, c_j)$ for every c_j, with $j \neq i$.

In the *pivots*-based methods, a set of pivots are selected and the distances between the pivots and database elements are precalculated. When a query is made, the query distance to the pivots is calculated and the triangle inequality is used to discard the candidates. Its objective is to filter objects during a request through the use of a triangular inequality, without really measure the distance between the object under request and the discarded object.

An abstract view of this kind of algorithms is the following:

- A set of k pivots ($\{p_1, p_2, \ldots, p_k\} \in \mathbb{X}$) are selected. During indexing time, for each object x from the database $\mathbb{Y}$, the distance to the k pivots is calculated and stored $(d(x, p_1), \ldots, d(x, p_k))$.
- Given a query (q, r), the result $d(p_i, x) \leq d(p_i, q) + d(q, x)$ is obtained by triangular inequality, with $x \in \mathbb{X}$. In the same way, $d(p_i, q) \leq d(p_i, x) + d(q, x)$ is obtained. From these inequations, it is possible to obtain a lower bound for the distance between q and x given by $d(q, x) \geq |d(p_i, x) - d(p_i, q)|$. Thus, the objects x are the objects that accomplish with $d(q, x) \leq r$, and then the rest of objects that do not accomplish with $|d(q, p_i) - d(x, p_i)| \leq r$ can be excluded.

Many indexes are trees, and, the children of each node define areas of space. Range queries traverse the tree, entering into all the children whose areas cannot be proved to be disjoint with the query region. Other metric structures are arrays; in this case, the array usually contains all the objects of the database and maintains the distances to the pivots.

The increased size of databases and the emergence of new types of data, where exact queries are not needed, creates the need to raise new structures to similarity search. Moreover, real applications require that these structures allow them to be stored in secondary memory efficiently, consequently optimized methods for reducing the cost of disk accesses are needed.

Finally, the need to process large volumes of generated data requires to increase processing capacity and so to reduce the average query times. In this context, the study is relevant in terms of parallelization of algorithms and distribution of the database.

1.2 Parallelization of Metric Structures

Currently, there are many parallel platforms for the implementation of metric structures. In this context, basic research has focused on technologies for distributed memory applications, using high level libraries for message passing as MPI [8] or PVM [9], and shared memory, using the language or directives of OpenMP [10].

In [11] and [12] we can find information about testing done on the *MTree*; in this case, the authors focus their efforts on optimizing the structure to properly distribute the nodes on a platform of multiple disks and multiple processors.

Some studies have focused on different structures parallelized on distributed memory platforms using MPI or BSP. In [13] several methods to parallelize the algorithms of construction and search on *EGNAT*, analyzing strategies for distribution of local and/or global data within the cluster, are presented. In [14] the problem of distributing a metric-space search index based on clustering into a set of distributed memory processors, using *List of Clusters* like base structure, is presented.

In terms of shared memory, [15] proposes a strategy to organize metric-space query processing in multi-core search nodes as understood in the context of search engines running on clusters of computers. The strategy is applied in each search node to process all active queries visiting the node as part of their solution which, in general, for each query is computed from the contribution of each search node. Besides, this work proposes mechanisms to address different levels of query traffic on a search engine.

Most of the previous and current works developed in this area are carried out considering classical distributed or shared memory platforms. However, new computing platforms are gaining in significance and popularity within the scientific computing community. Hybrid platforms based on *Graphics Processing Units* (GPU) is an example.

In the present work we show a version of the pivot-based metric structure called *Spaghettis* [16] implemented on a GPU-based platform. There are very little work in metric spaces developed in this kind of platforms. In Section 2.2 we show related work in this area.

2 Graphics Processing Units

The era of single-threaded processor performance increases has come to an end. Programs will only increase in performance if they utilize parallelism. However, there are different kinds of parallelism. For instance, multicore CPUs provide task-level parallelism. On the other hand, Graphics Processing Units (GPUs) provide data-level parallelism.

Current *GPU*s consist of a high number (up to 512 in current devices) of computing cores and high memory bandwidth. Thus, GPUs offer a new opportunity to obtain short execution times. They can offer 10x higher main memory bandwidth and use data parallelism to achieve up to 10x more floating point throughput than the CPUs [17].

GPUs are traditionally used for interactive applications, and are designed to achieve high rasterization performance. However, their characteristics have led to the opportunity to other more general applications to be accelerated in GPU-based platforms. This trend is called General Purpose Computing on GPU (GPGPU) [18]. These general applications must have parallel characteristics and an intense computational load to obtain a good performance.

To assist in the programming tasks of these devices, the GPU manufacturers, like NVIDIA or AMD/ATI, have proposed new languages or even extensions for the most common used high level programming languages. As example, NVIDIA proposes CUDA [19], which is a software platform for massively parallel high-performance computing on the company powerful GPUs.

In CUDA, the calculations are distributed in a mesh or grid of thread blocks, each with the same size (number of threads). These threads run the GPU code, known as kernel. The dimensions of the mesh and thread blocks should be carefully chosen for maximum performance based on the specific problem being treated.

Current GPUs are being used for solving different problems like data mining, robotics, visual inspection, video conferencing, video-on-demand, image databases, data visualization, medical imaging, etc and it is increasingly the number of applications that are being parallelized for GPUs.

2.1 CUDA Programming Model

The *NVIDIA*'s *CUDA* Programming Model ([19]) considers the GPU as a computational device capable to execute a high number of parallel threads. CUDA includes *C/C++* software development tools, function libraries, and a hardware abstraction mechanism that hides the GPU hardware to the developers by means of an Application Programming Interface (API). Among the main tasks to be done in CUDA are the following: allocate data on the GPU, transfer data between the GPU and the CPU and launch kernels.

A CUDA kernel executes a sequential code in a large number of threads in parallel. The threads within a block can work together efficiently exchanging data via a local shared memory and synchronize low-latency execution through synchronization barriers (where threads in a block are suspended until they all reach the synchronization point). By contrast, the threads of different blocks in the same grid can only coordinate their implementation through a high-latency accesses to global memory (the graphic board memory). Within limits, the programmer specifies how many blocks and the number of threads per block that are allocated to the implementation of a given kernel.

2.2 GPUs and Metric Spaces

As far as we know, the solutions considered till now developed on GPUs are based on *kNN* queries without using data structures. This means that GPUs are basically applied to exploit its parallelism only for exhaustive search (brute force) [20,21,22].

In [20] both elements (A) and queries (B) matrices are divided on fixed size submatrices. In this way, the resultant submatrix C is computed by a block of threads. Once the whole submatrix has been processed, *CUDA-based Radix Sort* [23] is applied over the complete matrix in order to sort it and obtain the first k elements as a final result.

In [21] a brute force algorithm is implemented where each thread computes the distance between an element of a database and a query. Afterwards, it is necessary to sort the resultant array by means of a variant of the *insertion sort* algorithm.

As a conclusion, in these works the parallelization is applied in two stages. The first one consists in building the distance matrix, and the second one consists in sorting this distance matrix in order to obtain the final result.

A particular variant of the above proposed algorithms is presented in [24] where the search is structured into three steps. In the first step each block solves

Algorithm 1 *Spaghettis*: Construction Algorithm

1: {Let $\mathbb{X}$ be the metric space}
2: {Let $\mathbb{Y} \subseteq \mathbb{X}$ be the database}
3: {Let P be the set of pivots $p_1, \ldots, p_k \in \mathbb{X}$}
4: {Let S_i be the table of distances associated p_i}
5: {Let Spaghettis be $\cup S_i$}
6: **for all** $p_i \in P$ **do**
7: $\quad S_i \leftarrow d(p_i, \mathbb{Y})$
8: **end for**
9: **for all** S_i **do**
10: $\quad$ Order(S_i)
11: **end for**
12: Each element within S_i stores its position in the next table (S_{i+1})

a query. Each thread keeps a heap where stores the kNN nearest elements proccessed by this thread. Secondly, a reduction operation is applied to obtain a final heap. Finally, the first k elements of this final heap are taken as a result of the query.

3 *Spaghettis* Data Structure

Spaghettis [16] is a variant of data structure *LAESA* [25] based on pivots. The method tries to reduce the CPU time needed to carry out a query by using a data structure where the distance to the pivots is sorted independently. As a result there is an array associated to each pivot allowing a binary search in a given range.

For each pivot set $S_i = \{x : |d(x, p_i) - d(q, p_i)| \leq r\}, i = 1, ..., k$, is obtained, where q is a query and r is a range, and a list of candidates will be formed by intersection of the whole sets.

3.1 Construction

During the construction of the spaghettis structure, a random set of pivots $p_1, ..., p_k$ is selected. These pivots could belong or not to the database to be indexed. The algorithm 1 shows in detail the construction process. Each position on table S_i represents an object of the database which has a link to its position on the next table. The last table links the object to its position on the database. Figure 1 shows an example considering 17 elements.

3.2 Searching

During the searching process, given a query q and a range r, a range search on an *spaghettis* follows the following steps:

1. The distance between q and all pivots $p_1, \ldots, p_k$ is calculated in order to obtain k intervals in the form $[a_1, b_1], \ldots, [a_k, b_k]$, where $a_i = d(p_i, q)$ - r and $b_i = d(p_i, q) + r$.
2. The objects in the intersection of all intervals are considered as candidates to the query q.
3. For each candidate object y, the distance $d(q, y)$ is calculated and if $d(q, y) \leq r$, then the object y is a solution to the query.

Implementation details are shown in algorithm 2. In this algorithm, S_{ij} represents the distance between the object y_i to the pivot p_j.

Figure 1 represents the data structure *spaghettis* in its original form. This structure is built using 4 pivots to index a database of 17 objects. The searching process is as follows. Assuming a query q, the distance to the pivots $\{8, 7, 4, 6\}$, and a searching range $r = 2$, Figure 1 shows in dark gray the intervals $\{(6, 10), (5, 9), (2, 6), (4, 8)\}$ over which the searching is going to be carried out. Also, in this figure it is possible to see all the objects that belong to the intersection of all the intervals and then they are considered as candidates. Finally, the distance between the candidates and query has to be calculated in order to determine a solution from the candidates. The solution is given if the distance is lower than a searching range.

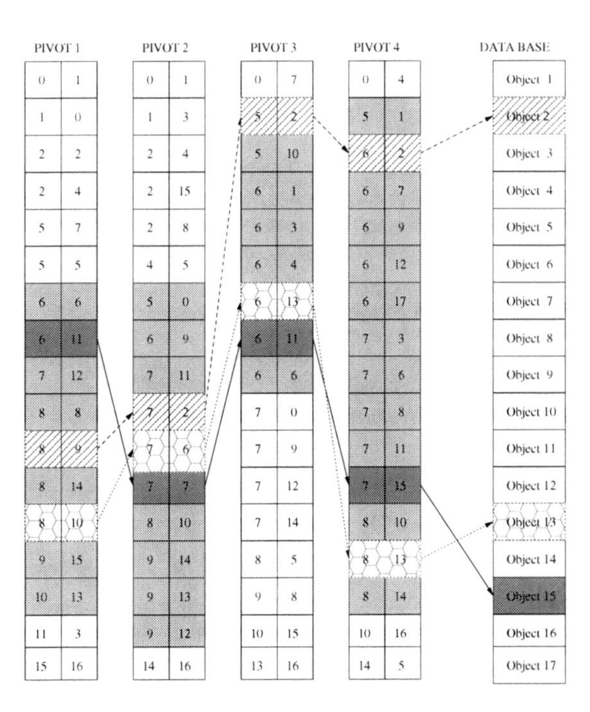

Fig. 1. *Spaghettis*: Construction and search. Example for query q with ranges $\{(6, 10), (5, 9), (2, 6), (4, 8)\}$ to pivots.

Algorithm 2 *Spaghettis*: Search Algorithm

rangesearch(query q, range r)
1: {Let $\mathbb{Y} \subseteq \mathbb{X}$ be the database}
2: {Let P be set of pivots $p_1, \ldots, p_k \in \mathbb{X}$}
3: {Let D be the table of distances associated q}
4: {Let S be Spaghettis}
5: **for all** $p_i \in P$ **do**
6: $D_i \leftarrow d(q, p_i)$
7: **end for**
8: **for all** $y_i \in \mathbb{Y}$ **do**
9: $discarded \leftarrow false$
10: **for all** $p_j \in P$ **do**
11: **if** $D_j - r > S_{ij} \,||\, D_j + r < S_{ij}$ **then**
12: $discarded \leftarrow true$
13: break;
14: **end if**
15: **end for**
16: **if** $!discarded$ **then**
17: **if** $d(y_i, q) \leq r$ **then**
18: add to result
19: **end if**
20: **end if**
21: **end for**

4 GPU-Based Implementation

The main goal of this paper is to develop a GPU-based implementation of the range query algorithms.

This type of process intrinsically has a high data-level parallelism with a high computing requirements. For that reason, GPU computing is very useful in order to accelerate this process due to the fact that GPUs exploit in an efficient way data-level parallelism. Moreover, these devices provide the best cost-per-performance parallel architecture for implementing such algorithms.

This section is divided in two different parts. First, we show the exhaustive search GPU-based implementation, and next we present the spaghettis GPU-based implementation.

4.1 Exhaustive Search GPU-Based Implementation

This implementation is an iterative process where in each iteration one kernel is executed, which calculates the distances between one particular query and every elements of the database. It is not possible to calculate all distances for all queries in only one kernel due to the GPU limitations (number of threads and memory capacity). In this kernel as many threads as number of elements in the database are launched. Each thread calculates the distance between one data of dataset and one particular query, and next, determines if this data is or not a valid solution.

4.2 Spaghettis GPU-Based Implementation

In order to obtain better performance on GPU, we have made some changes on the original *Spaghettis* structure. We adapt the structure for that it is very similar to an array, which is more efficient in GPU computing. In this implementation, each row is associated with an object of dataset and each column to a pivot. Therefore, each cell contains the distance between the object and the pivot. Moreover, unlike the original version, the array is sorted by the first pivot. Thus, the cells of the same row is associated with the same object.

The parallelization of the searching algorithm has been splitted into three parts, which are the most computationally expensive parts of this algorithm. These parts correspond to the three steps presented in Subsection 3.2.

The first part consists in computing the distances between the set of queries, Q, and the set of pivots, P. In order to exploit the advantages of using a GPU platform is necessary a data structure which stores all distances. Therefore, this structure is implemented as a $Q \times P$ matrix which allows us to compute all distances at the same time in a single call to kernel. This part is implemented in one kernel with as many threads as number of queries. In fact, each thread solves independently the distance from a query to all pivots. The algorithm 3 shows a general pseudocode of this kernel.

Algorithm 3 Distance generator kernel

__global__ KDistances(queries Q, pivots P, distances D)

1: {Let D be the table of distances associated to q}
2: {Let i be thread Id }
3: **for all** $p_j \in P$ **do**
4: $D_{ij} \leftarrow d(q_i, p_j)$
5: **end for**

The second part of the parallel implementation consists in determining if each element of the database is or not a candidate for every query. This part has been implemented as an iterative process. In each iteration the candidates for a particular query are computed in one kernel. As we have described above, it is not possible to calculate all candidates for every queries in only one kernel due to the GPU limitations. In this kernel as many threads as number of elements of the database are launched. Each thread of this kernel determines, for a given data (y_i) of the dataset, if this data is candidate or not. Thus, this kernel returns a list of candidates for a given query. Finally, when this process finishes we obtain one list of candidates for each query. This task is carried out by a kernel called *KCandidates* (see algorithm 4).

The kernel *KSolution* (see algorithm 5) correspond with the third part, and computes if each candidate is really a solution. In this kernel, the number of threads correponds to the number of candidates for each query. Each thread calculates the distance between one candidate and one query, and determines if

Algorithm 4 CUDA Search Algorithm

```
__global__ KCandidates(range r, Spaghettis S, distances D, pivots P, candidates C)
1: {Let P be set of pivots p_1, ..., p_2 ∈ X}
2: {Let D be the table of distances associated q}
3: {Let C be list of candidates for q}
4: {Let i be thread Id }
5: discarded ← false
6: for all p_j ∈ P do
7:    if D_j − r > S_ij || D_j + r < S_ij then
8:       discarded ← true
9:       break;
10:   end if
11: end for
12: if !discarded then
13:    add to C (candidates)
14: end if
```

this candidate is or not solution. Finally, as result we obtain one list of solutions for each query.

In the three kernels, threads belonging to the same thread block operate over contiguous components of the arrays. Therefore, more efficient memory accesses are allowed. This is due to the abovementioned changes in the spaghettis structure.

Algorithm 5 CUDA final solutions for query q

```
__global__ KSolution(range r, database Y, candidates C, query q, solutions R)
1: {Let Y ⊆ X be the database}
2: {Let C be list of candidates for q}
3: {Let R be list of solutions for q}
4: {Let i be thread Id }
5: if d(c_i, q) ≤ r then
6:    add to R (solutions)
7: end if
```

5 Experimental Evaluation

This section presents the experimental results obtained for the previous algorithms considering the Spanish dictionary as database. For this case study the generated *spaghettis* data structure is completely stored on the global memory of the GPU.

5.1 Experimental Environment

Tests made in one metric space from the Metric Spaces Library[1] were selected for this paper. This is a Spanish dictionary with 86,061 words, where the *edit distance* is used. This distance is defined as the minimum number of insertions, deletions or substitutions of characters needed to make one of the words equal to the other. We create the structure with the 90% of the dataset, and reserve the rest for queries. We have chosen this experimental environment because is the usual environment used to evaluate this type of algorithms.

Hardware platform used was a PC with the following main components:

- CPU: Intel Core 2 Quad at 2.66GHz and 4GB of main memory.
- GPU: GTX 285 with 240 cores and a main memory of 1 GB.

5.2 Experimental Results

The results presented in this section belong to a set of experiments with the following features:

- The selection of pivots ware made randomly.
- The *spaghettis* structure was built considering 4, 8, 16, and 32 pivots.
- For each experiment, 8,606 queries were given over an *spaghettis* with 77,455 objects.
- For each query, a range search between 1 and 4 was considered.
- The execution time shown in this paper is the total time of all the processes for both versions, parallel and sequential. Therefore, in the case of parallel version, the execution time also includes the data transfer time between the main memory (CPU) and global device memory (GPU).

Figure 2(a) shows the execution time spent by the sequential and GPU implementation for *Spaghetttis* structure. Notice that the parallel version based on CUDA reduces dramatically the execution time, increasing the performance. Figure 2(b) shows in detail the time spent by the CUDA implementation. As reference, the execution time spent by the sequential and GPU implementation for the exhaustive search (Seq. and GPU Brute Force) is included in both figures (2(a) and 2(b)).

According to experimental results, it is interesting to discuss the following topics:

- As can be observed, the use of Spaghettis structure allows us to decrease the number of distance evaluations, due to that to compute the distance between all the database objects is avoided. In Figure 2 we can deduce that:
  - When the number of pivots increases the performance of search algorithm is much better in sequential and GPU versions.
  - The use of GPU decreases considerably the execution time in both versions, exhaustive search and emphSpaghettis structure.

[1] `www.sisap.org`.

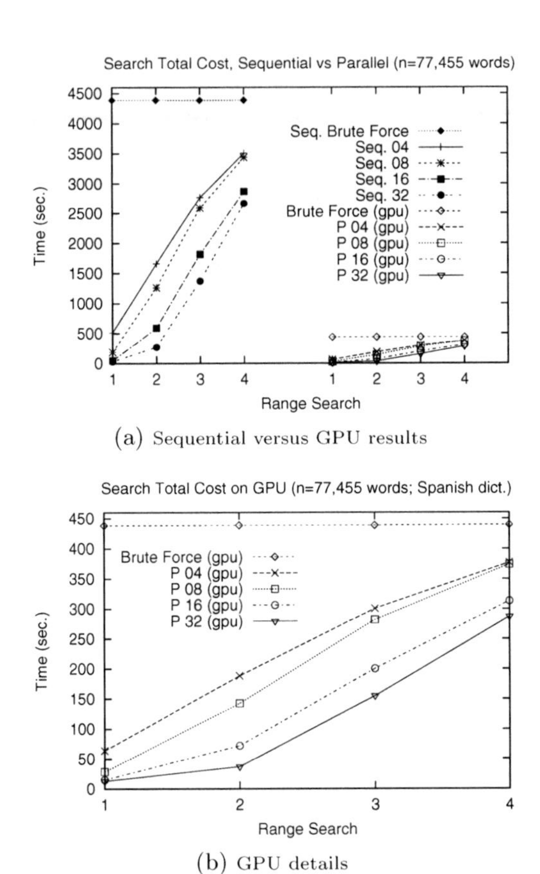

(a) Sequential versus GPU results

(b) GPU details

Fig. 2. Comparative results of search costs for the space of words for *Spaghettis* metric structure (Spanish Dictionary). Number of pivots 4, 8, 16 and 32, and range search from 1 to 4.

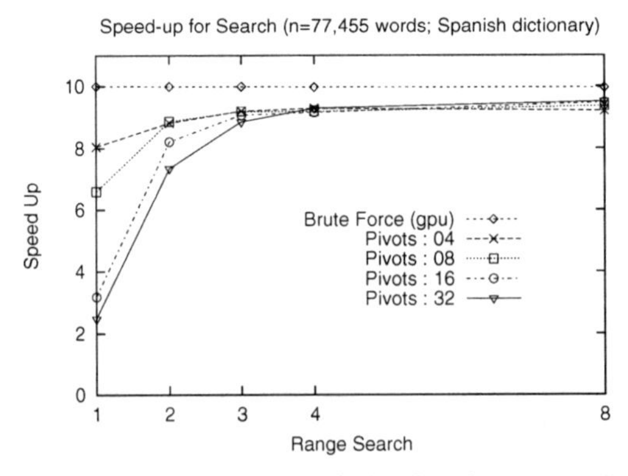

Fig. 3. Speed-up graphics to the space of words for *Spaghettis* metric structure (Spanish Dictionary)

- As can be observed in Figure 3 (range 1 and 2), the speed-up is smaller when the number of pivots is higher. Due to this fact, more number of pivots more workload for the threads. Moreover, when the range is higher (range 3 and 4) the speed-up increases, because the behaviour approaches to exhaustive search.
- There is an asymptotic speed-up around 9.5 (see Figure 3). It is possible to observe that this behaviour is shown when the range search is 4. But, in order to ensure this assertion, a proof considering a range search equal to 8 has been carried out.

6 Conclusions and Future Work

In this work, a parallel approach based on GPU has been carried out in order to reduce the execution time spent on the searching process of a query in a dataset using *Spaghettis* data structure.

This implementation has provided good results in terms of speed-up when considering suitable values for the input parameters as number of pivots and range search. In this case, a speed-up of 9.5 has been obtained.

To be able to continue with the study of this work in order to obtain more efficient implementations, and as future work, we have planned the following topics:

- To test the GPU-based implementation presented in this paper considering a different database.
- Moreover, we would like analyse the impact that different distance functions have on the global performance of this kind of algorithms, and on the acceleration obtained with parallel platforms. There are distance functions with a great computational load, like that presented in this paper, and others with minimum computational requirements. In these cases, hiding the overhead due to data transferences will be a challenge.
- In order to be able of executing the algorithms presented here on different GPU vendor platforms, OpenCL implementations will be carried out.
- To compare with other parallel platforms in terms of performance, energy consumption and economic cost. As a consequence, it is necessary to implement the work carried out here using MPI or OpenMP (or both) according to the target platform.

Acknowledgments

This work has been partially funded by research programs PR-F1-02IC-08, University of Magallanes, Chile and 29/C035-1, Academic Unit of Río Turbio, National University of Patagonia Austral, Argentina, and the Spanish project SAT-SIM (ref: CGL2010-20787-C02-02).

References

1. Chávez, E., Navarro, G., Baeza-Yates, R., Marroquín, J.L.: Searching in metric spaces. ACM Computing Surveys 33(3), 273–321 (2001)
2. Kalantari, I., McDonald, G.: A data structure and an algorithm for the nearest point problem. IEEE Transactions on Software Engineering 9(5) (1983)
3. Uhlmann, J.: Satisfying general proximity/similarity queries with metric trees. Information Processing Letters 40, 175–179 (1991)
4. Ciaccia, P., Patella, M., Zezula, P.: M-tree: An efficient access method for similarity search in metric spaces. In: The 23st International Conference on VLDB, pp. 426–435 (1997)
5. Brin, S.: Near neighbor search in large metric spaces. In: The 21st VLDB Conference, pp. 574–584. Morgan Kaufmann Publishers, San Francisco (1995)
6. Uribe, R., Navarro, G.: Egnat: A fully dynamic metric access method for secondary memory. In: Proc. 2nd International Workshop on Similarity Search and Applications (SISAP), pp. 57–64. IEEE CS Press, Los Alamitos (2009)
7. Navarro, G.: Searching in metric spaces by spatial approximation. The Very Large Databases Journal (VLDBJ) 11(1), 28–46 (2002)
8. Gropp, W., Lusk, E., Skelljum, A.: Using MPI:Portable Parallel Programming with the Message Passing Interface. Scientific and Engineering computation Series. MIT Press, Cambridge (1994)
9. Geist, A., Beguelin, A., Dongarra, J., Jiang, W., Manchek, B., Sunderam, V.: PVM: Parallel Virtual Machine – A User's Guide and Tutorial for Network Parallel Computing. MIT Press, Cambridge (1994)
10. Dagum, L., Menon, R.: OpenMP: An industry-standard API for shared-memory programming. IEEE Computational Science and Engineering 5(1), 46–55 (1998)
11. Zezula, P., Savino, P., Rabitti, F., Amato, G., Ciaccia, P.: Processing m-trees with parallel resources. In: RIDE 1998: Proceedings of the Workshop on Research Issues in Database Engineering, p. 147. IEEE Computer Society, Washington, DC (1998)
12. Alpkocak, A., Danisman, T., Ulker, T.: A parallel similarity search in high dimensional metric space using M-tree. In: Grigoras, D., Nicolau, A., Toursel, B., Folliot, B. (eds.) IWCC 2001. LNCS, vol. 2326, pp. 166–252. Springer, Heidelberg (2002)
13. Marin, M., Uribe, R., Barrientos, R.J.: Searching and updating metric space databases using the parallel EGNAT. In: Shi, Y., van Albada, G.D., Dongarra, J., Sloot, P.M.A. (eds.) ICCS 2007. LNCS, vol. 4487, pp. 229–236. Springer, Heidelberg (2007)
14. Gil-Costa, V., Marin, M., Reyes, N.: Parallel query processing on distributed clustering indexes. Journal of Discrete Algorithms 7(1), 3–17 (2009)
15. Gil-Costa, V., Barrientos, R., Marin, M., Bonacic, C.: Scheduling metric-space queries processing on multi-core processors. In: Euromicro Conference on Parallel, Distributed, and Network-Based Processing, pp. 187–194 (2010)
16. Chávez, E., Marroquín, J., Baeza-Yates, R.: Spaghettis: An array based algorithm for similarity queries in metric spaces. In: 6th International Symposium on String Processing and Information Retrieval (SPIRE 1999), pp. 38–46. IEEE CS Press, Los Alamitos (1999)
17. Wu-Feng, Manocha, D.: High-performance computing using accelerators. Parallel Computing 33, 645–647 (2007)
18. GPGPU, general-purpose computation using graphics hardware, `http://www.gpgpu.org`

19. NVIDIA CUDA Compute Unified Device Architecture-Programming Guide, Version 2.3. NVIDIA (2009), http://developer.nvidia.com/object/gpucomputing.html
20. Kuang, Q., Zhao, L.: A practical GPU based kNN algorithm. In: International Symposium on Computer Science and Computational Technology (ISCSCT), pp. 151–155 (2009)
21. Garcia, V., Debreuve, E., Barlaud, M.: Fast k nearest neighbor search using GPU. In: Computer Vision and Pattern Recognition Workshop, pp. 1–6 (2008)
22. Bustos, B., Deussen, O., Hiller, S., Keim, D.A.: A graphics hardware accelerated algorithm for nearest neighbor search. In: Alexandrov, V.N., van Albada, G.D., Sloot, P.M.A., Dongarra, J. (eds.) ICCS 2006. LNCS, vol. 3994, pp. 196–199. Springer, Heidelberg (2006)
23. Satish, N., Harris, M., Garland, M.: Designing efficient sorting algorithms for many-core GPUs. In: Parallel and Distributed Processing Symposium, International, pp. 1–10 (2009)
24. Barrientos, R.J., Gómez, J.I., Tenllado, C., Prieto, M.: Heap based k-nearest neighbor search on GPUs. In: Congreso Español de Informática (CEDI), Valencia, Septiembre (2010)
25. Micó, L., Oncina, J., Vidal, E.: A new version of the nearest-neighbor approximating and eliminating search (AESA) with linear preprocessing-time and memory requirements. Pattern Recognition Letters 15, 9–17 (1994)

A Concurrent Object-Oriented Approach to the Eigenproblem Treatment in Shared Memory Multicore Environments

Alfonso Niño, Camelia Muñoz-Caro, and Sebastián Reyes

QCyCAR research group, Escuela Superior de Informática, Universidad de Castilla-La Mancha, Paseo de la Universidad 4, 13071 Ciudad Real, Spain
{Alfonso.Nino, Camelia.Munoz, Sebastian.Reyes}@uclm.es

Abstract. This work presents an object-oriented approach to the concurrent computation of eigenvalues and eigenvectors in real symmetric and Hermitian matrices on present memory shared multicore systems. This can be considered the lower level step in a general framework for dealing with large size eigenproblems, where the matrices are factorized to a small enough size. The results show that the proposed parallelization achieves a good speedup in actual systems with up to four cores. Also, it is observed that the limiting performance factor is the number of threads rather than the size of the matrix. We also find that a reasonable upper limit for a "small" dense matrix to be treated in actual processors is in the interval 10000-30000.

Keywords: Eigenproblem, Parallel Programming, Object-Orientation, Multicore processors.

1 Introduction

The eigenproblem plays an important role in both science and engineering. Thus, it appears in problems such as the quantum mechanical treatment of time independent systems [1], in the principal components analysis (PCA) [2], in the specific application of PCA to face recognition (eigenfaces) or in the computation of the eigenvalues of a graph in spectral graph theory applied to complex networks [3]. The eigenproblem implies, in practical terms, the computation of the eigenvalues and eigenvectors of a matrix [4]. Very often the matrix is a symmetric, real, one. However, in the general case, we deal with Hermitian, complex matrices; see for instance [5]. An interesting fact is the quadratic dependence of the matrix with the problem size. This leads easily to large (or very large) matrices, for instance, in the variational treatment of quantum systems with several degrees of freedom. The capability of dealing with large eigenproblems (matrices of size about 10^6) is of great interest for tackling realistic problems in different fields of science and engineering.

The eigenvalue problem is a central topic in numerical linear algebra. The standard approach for the numerical solution of the eigenproblem is to reduce the matrix to some simpler form that yields the eigenvalues and eigenvectors directly [6]. The first method of this kind dates back to 1846 when Jacobi proposed to reduce a real

B. Murgante et al. (Eds.): ICCSA 2011, Part I, LNCS 6782, pp. 630–642, 2011.

symmetric matrix to diagonal form by a series of plane rotations [7]. From then, the field has experienced a huge development, especially since the availability of the modern computer in the early1950s [6]. A milestone of the field was due to Givens in 1954 [8]. In this work, Givens proposed to use a finite number of orthogonal transformations to reduce a matrix to a form easier to handle, such as a tridiagonal form. In addition, in 1958 Householder showed how to zeroing the elements outside the tridiagonal in a matrix row and column without spoiling any previous similar transformation [9]. The Householder method became the standard reduction method of matrices to tridiagonal form on serial computers [6]. The method is described in detail in any text dealing with the eigenvalue problem, see for instance [10-13]. From the 1960s the way to compute selected eigenvalues and eigenvectors of a tridiagonal matrix involves locating the eigenvalues using a Sturm sequence and obtaining the eigenvectors by inverse iteration [14]. This approach is well presented in the classical Wilkinson's book [10]. On the other hand, for computing the whole set of eigenvalues and eigenvectors the QR technique is more efficient [13]. A different standpoint is represented by the divide and conquer approach initially proposed by Cuppen in 1981 [15]. In this approach, the matrix is reduced to tridiagonal form, splitting this last in two blocks plus a rank-one update. The procedure can be recursively applied until the matrices are small enough. Then, we can treat the resulting blocks by other method such as QR. The "modern", stable implementation of the method was proposed in 1995 by Gu and Eisenstat [16]. The divide and conquer approach is the fastest way to obtain all the eigenvalues and eigenvectors of a symmetric matrix [6]. Besides, the method is well suited for parallel implementation.

Different available software packages allow treating the eigenvalue problem. Most of them implement descendants of the algorithms presented in the classical Wilkinson and Reinsch book [17]. In particular, many of these algorithms were codified in Fortran, in the 1970s, in the LINPACK (for numerical linear algebra) and EISPACK (for eigenproblems) packages [18, 19]. LINPACK and EISPACK give rise to LAPACK (also in Fortran) in the 1990s [20]. The last descendant in this family is ScaLAPACK, which provides a parallel implementation of a subset of LAPACK using a distributed memory parallel programming approach [21]. In this context, it is interesting to mention ARPACK (ARnoldi PACKage), which is a FORTRAN 77 numerical software library for solving large scale eigenvalue problems [22]. ARPACK is designed to compute a few eigenvalues, and its corresponding eigenvectors, of a general n by n matrix A. It is especially well suited for large sparse or structured matrices. The package is based on an algorithmic variant of the Arnoldi process called the Implicitly Restarted Arnoldi Method [23]. A parallel ARPACK (PARPACK) is available for distributed memory architectures [24]. Another package worth mentioning is SLEPc [25]. SLEPc (Scalable Library for Eigenvalue Problem Computations) is a software library for the treatment of large sparse eigenproblems on parallel computers with a distributed memory architecture.

These packages have been essentially implemented, or are intended to be used, under a traditional imperative programming model. However, to ease the modeling of complex application problems, it would be interesting to use an object-oriented approximation. In this context, we have proposed recently an object-oriented approach for the uniform treatment of eigenproblems in real symmetric and Hermitian matrices [26]. This approach has shown to yield speedups up to three over the standard LAPACK routines in small matrices (i.e. for maximum sizes of 10^4) [26]. On the other hand, it is clear that to deal with larger problems (say for matrices of size 10^5-10^6), we must resort to parallel programming.

When considering the parallel computing landscape over the last few years, we find an interesting evolution. Microprocessors based on a single processing unit (CPU) drove performance increases and cost reductions in computer applications for over two decades. However, this process reached a limit point around 2003 due to heat dissipation and energy consumption issues [27]. These problems limit the increase of CPU clock frequency and the number of tasks that can be performed within each clock period. The solution adopted by processor developers was to switch to a model where the microprocessor had multiple processing units known as cores [28]. Nowadays, we can speak of two approaches [28]. The first, multicore approach, integrates a few cores (at present between two and eight) into a single microprocessor, seeking to keep the execution speed of sequential programs. Actual laptops and desktops incorporate this kind of processors. The second, many-core approach, uses a large amount of cores (at present as many as several hundred) and are specially oriented to the execution throughput of parallel programs. This approach is exemplified by the Graphical Processing Units (GPUs) available today. Thus, parallel capabilities are available in the commodity machines we find everywhere. Clearly, this change of paradigm has had (and will have) a huge impact on the software developing community [29].

Traditionally, parallel systems, or architectures, fall into two broad categories: shared memory and distributed memory [30]. In shared memory architectures we have a single memory address space accessible to all the processors. Shared memory machines have existed for a long time in the servers and high-end workstations segment. On the other hand, in distributed memory architectures there is not global address space, but each processor owns its own memory space. This is a popular architectural model encountered in networked or distributed environments such as clusters or Grids of computers. As a consequence of the popularity of computer clusters in the past years, today's most used parallel approach in scientific computing is message passing. However, it is interesting to realize that, at present, we can exploit the shared memory nature of multicore microprocessors in the individual computer nodes of any cluster.

The conventional parallel programming practice involves a pure shared memory model [30], usually using the OpenMP API [31], in shared memory architectures, or a pure message passing model [30], using the MPI API [32], on distributed memory systems. Accordingly to this hybrid architecture, different parallel programming models can be mixed in what is called hybrid parallel programming. A wise implementation of hybrid parallel programs can generate massive speedups in the otherwise pure MPI or pure OpenMP implementations [33]. The same can be applied to hybrid programming involving GPUs and distributed architectures [34, 35].

As a needed step in the treatment of large eigenproblems, we present in this work an extension of our previous object-oriented approach to the parallel treatment of eigenproblems on the memory shared architecture of present multicore systems. Our aim is to obtain a solution with acceptable scaling for a few (say less than 10) concurrent execution threads. As an initial case, we apply the proposed solution to the computation of eigenvalues and eigenvectors for real symmetric matrices.

2 Proposed Object-Oriented Approach to the Eigenproblem

As commented above, it is of interest treating the eigenproblem in large matrices. For that, we can resort to some variant of the divide and conquer approach [15, 16]. In this form, we can factorize recursively the large matrix in smaller ones until the resulting matrices are small enough to be solved individually. This process can be implemented concurrently in a distributed memory system such as a computer cluster. However, the treatment of the small matrices is done on individual cluster nodes. Here, we can use a shared memory parallel model to take advantage of the multicore architecture of the nodes. In addition, this use of parallelism allows for increasing the size these "small" matrices can have. Therefore, the number of times the divide and conquer process must be applied can be reduced. The result would be a reduction of the overall computational effort.

To parallelize the computation of eigenvalues and eigenvectors on a shared memory system, we propose an extension of the object-oriented approach previously developed [26]. The corresponding UML class diagram is shown in Fig. 1.

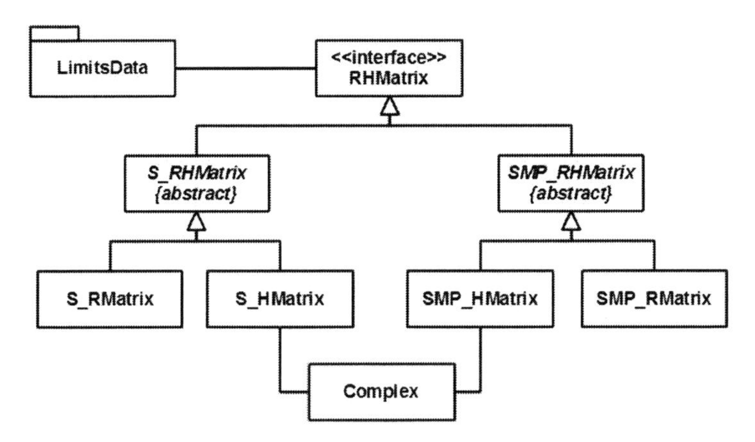

Fig. 1. Class diagram for the sequential and concurrent proposed treatment of real and Hermitian eigenproblems in multicore systems

Fig. 1 shows at the top of the diagram an interface defining (but not implementing) the functional behavior of every operative class. The system is organized in two branches corresponding to the Sequential (S_) and Shared Memory Parallel (SMP_) cases. At the lower end of the diagram we have the specialized classes that deal with the real (R) and Hermitian (H) cases. The use of the inheritance relationship along the diagram allows for the use of polymorphic references of the base class (RHMatrix) for referring to objects of any concrete class at the bottom. In this way, the same code can be used to invoke the processing of real or Hermitian problems either sequentially or concurrently. The diagram also shows that the RHMatrix interface class exhibits an association relationship with a package (LimitsData). This is used to define the different numerical limits needed in the algorithms. Finally, the classes dealing with Hermitian matrices have an association relationship with a "Complex" class, which is used to represent the behavior of complex numbers.

3 Sequential Algorithm

To compute eigenvalues and eigenvectors of real symmetric and Hermitian matrices, we use the classical procedure described in the introduction [14]. First, we reduce the matrix to tridiagonal form using a series of Householder reflections. Here, we introduce the method of Shukuzawa et al. [36] to transform the Hermitian matrices to real tridiagonal matrices by using modified Houselholder reflections [26]. In this form, real symmetric as well as Hermitian matrices yield a real tridiagonal matrix. Second, we compute the desired eigenvalues of the real tridiagonal matrix using a Sturm sequence and the bisection method [10-13]. Third, we compute the eigenvectors for each previous eigenvalue using inverse iteration [10-13]. The pseudocode for the algorithm used is shown in Algorithm 1. Here, too low level details are not shown for the sake of clarity. For specific details consult [26].

Begin_algorithm
  // Tridiagonalization of the n x n A matrix
  for *i←0 to n-2 (real symmetric matrix) or n-1 (Hermitian matrix)*
    Compute $s = \left(\sum_{j=i+1}^{n-1} |a_{j,i}|^2 \right)^{1/2}$
    Compute vector $\mathbf{x}$ *with* $x_j=0, \forall j \leq i;\ x_i=a_{i,\,i+1} \pm s;\ x_j=a_{i,\,j}, \forall j > i$
    Compute vector $\mathbf{u}=\mathbf{x} / |\mathbf{x}|$
    Compute $\beta=2$ *(real case) or*
        $\beta=1+(s+a^{*}_{i,\,i+1})/(s+a^{*}_{i,\,i+1})$ *(Hermitian case)*
    Compute vector $\mathbf{p}=\beta^{+}\cdot\mathbf{A}\cdot\mathbf{u}$
    Compute $K=2\cdot\mathbf{u}^{T}\cdot\mathbf{A}\cdot\mathbf{u}$ *(real case) or*
        $K=[1-\mathrm{real}(\kappa)]\cdot\mathbf{u}^{+}\cdot\mathbf{A}\cdot\mathbf{u}$ *(Hermitian case)*
    Compute vector $\mathbf{q}=\mathbf{p}-K\cdot\mathbf{u}$
    Update matrix $\mathbf{A}=\mathbf{A}-\mathbf{q}\cdot\mathbf{u}^{+}-\mathbf{u}\cdot\mathbf{q}^{+}$
  end_for

  // Computing m of the n possible eigenvalues (m ≤ n)
  Obtain whole eigenvalues interval using Gershgorin theorem
  for *i←m-1 to 0*
    while *error in eigenvalue* e_i *larger than a given limit*
      Bracket eigenvalue e_i *using bisection and a Sturm sequence*
    end_while
    Make last non-degenerate e_i *the new upper limit of the eigenvalues interval*
  end_for

  //Computing the m eigenvectors associated to the eigenvalues (optional)
  for *i←0 to m-1*
    Generate guess vector $\mathbf{v}_i$ *randomly*
    Perform LU decomposition of tridiagonal matrix
    while *error in eigenvector* $\mathbf{v}_i$ *larger than a given limit*
      Get $\mathbf{v}_i$ *using inverse iteration*
    end_while
    if *i>0 and* $\mathbf{v}_i$ *is degenerate* ***then***

```
        j←i-1
         while e_j = e_i
            Orthonormalize e_i respect to e_j
            j←j-1
         end_while
      end_if
      Rotate vector back to the original A matrix  v_i = v_i - β_j · u_j^+ · [u_j · v_i]
   end_for
End_algorithm
```

Algorithm 1. Sequential algorithm for the computation of eigenvalues and eigenvectors of real symmetric and Hermitian matrices.

Taking into account the different vector-vector and matrix-vector products, the pseudocode above shows that the tridiagonalization has $O(n^3)$ complexity. In addition, computation of the m eigenvalues exhibits $O(m \cdot n)$ complexity. Finally, the eigenvectors computation is $O(m \cdot n^2)$. The question now is how to use concurrency to lower the complexity of each of these processes.

4 Concurrent Algorithm

Different data oriented approaches have been proposed for parallelizing the Householder tridiagonalization. However, these approaches rely either in a distributed memory model [37, 38] or in the availability of a perfect square number of processors [39]. On shared memory systems it is interesting to mention the work of Honecker and Schüle [40]. These authors present an OpenMP version of the Householder algorithm for Hermitian matrices. However, they do not consider the subsequent problem of parallelizing the computation of eigenvalues and eigenvectors.

In the present work, we use a memory shared approach considering a limited number of cores. Here, Algorithm 1 is used as a basis for the concurrent computation of eigenvalues and eigenvectors. The algorithm shows that a task based parallelizing approach offers little room for improvement, since very few different tasks can be performed concurrently. Therefore we resort to a data (domain) decomposition to achieve a scalable solution. In our case the whole process is based in matrix and vector manipulations. Thus, a data decomposition approach is obtained by considering independent rows or columns. That depends on the most appropriate strategy for profiting from cache data locality in the used language. In practice, this is achieved by performing concurrently the iterations of the different loops presented in Algorithm 1.

Analyzing Algorithm 1 several facts are clear. First, in the tridiagonalization step, we have an update of the **A** matrix within the outer loop. Therefore, the different iterations of the outer loop are coupled together. So, they cannot be performed concurrently without incurring in a data race condition. Second, the computation of eigenvalues does not involve mixing eigenvalues. Thus, an independent eigenvalues computation is possible. For P concurrent execution threads, the complexity of this step would be reduce to $O(m \cdot n/P)$. Finally, calculation of each eigenvector is independent of the others, except if degeneracy is present. However, placing the degenerate eigenvectors orthonormalization outside the main loop solves the problem. Taking into account that in the general case the amount of degeneracy is small, if any, the complexity would be reduced to $O(m \cdot n^2/P)$.

Algorithm 2 shows that the main source of imbalance in the concurrent algorithm is due to the tridiagonalization part. Here, the overload due to the opening and closing of the concurrent threads is within the main loop. It is also the sequential, single thread, computation of the β factor (and some steps of the K factor), which introduces a sequential limit on the grounds of Amdahl´s law. Therefore, we can expect a decrease of performance with the size of the matrix and the number of concurrent threads.

With the previous considerations a concurrent algorithm can be devised as the one shown in Algorithm 2.

Begin_algorithm
  // Tridiagonalization of the n x n A matrix
  for *i←0 to n-2 (real symmetric matrix) or n-1 (Hermitian matrix)*
    open *P concurrent threads*
      do *among the P threads*
        Compute $s = \left(\sum_{j=i+1}^{n-1} |a_{j,i}|^2 \right)^{1/2}$
        Compute vector $\mathbf{x}$ *with* $x_j=0, \forall j \le i;\ x_i=a_{i,i+1} \pm s;\ x_j=a_{i,j}, \forall j > i$
        Compute vector $\mathbf{u}=\mathbf{x} / |\mathbf{x}|$
      end_do_threads
      do *single thread*
        Compute $\beta=2$ *(real case) or*
        $\beta=1+(s+a^*_{i,i+1})/(s+a^*_{i,i+1})$ *(Hermitian case)*
      end_single_thread
      do *among the P threads*
        Compute vector $\mathbf{p}=\beta^+ \cdot \mathbf{A} \cdot \mathbf{u}$
      end_do_threads
      do *among the P threads*
        Compute $K=2 \cdot \mathbf{u}^T \cdot \mathbf{A} \cdot \mathbf{u}$ *(real case) or*
        $K=[1-\mathrm{real}(\kappa)] \cdot \mathbf{u}^+ \cdot \mathbf{A} \cdot \mathbf{u}$ *(Hermitian case)*
      end_do_threads
      do *among the P threads*
        Compute vector $\mathbf{q}=\mathbf{p}-K \cdot \mathbf{u}$
      end_do_threads
      do *among the P threads*
        Update matrix $\mathbf{A}=\mathbf{A} - \mathbf{q} \cdot \mathbf{u}^+ - \mathbf{u} \cdot \mathbf{q}^+$
      end_do_threads
    close *concurrent threads*
  end_for
  // Computing m of the n possible eigenvalues (m ≤ n)
  Obtain whole eigenvalues interval using Gershgorin theorem
  open *P concurrent threads*

```
      do among the P threads
         for i←m-1 to 0
            while error in eigenvalue e_i larger than a given limit
               Bracket eigenvalue e_i using bisection and a Sturm sequence
            end_while
            Make last non-degenerate e_i the new upper limit of the eigenvalues interval
         end_for
      end_do_threads
   close concurrent threads


   //Computing the m eigenvectors associated to the eigenvalues (optional)
   open P concurrent threads
      do among the P threads
         for i←0 to m-1
            Generate guess vector v_i randomly
            Perform LU decomposition of tridiagonal matrix
            while error in eigenvector v_i larger than a given limit
               Get v_i using inverse iteration
            end_while
            Rotate vector back to the original A matrix  v_i = v_i - β_j · u_j^+ · [u_j · v_i]
         end_for
      end_do_threads
   close concurrent threads

   if i>0 and v_i is degenerate then
      j←i-1
      while e_j = e_i
         open P concurrent threads
            do among the P threads
               Orthonormalize e_i respect to e_j
               j←j-1
            end_do_threads
         close concurrent threads
      end_while
   end_if
End_algorithm
```

Algorithm 2. Concurrent algorithm for the computation of eigenvalues and eigenvectors of real symmetric and Hermitian matrices.

There are two implicit considerations not shown in Algorithm 2. The first is how to balance the workload among the different concurrent threads. The second is how to profit from the cache. These questions are implementation related and are considered in the next section.

5 Results and Discussion

As test case, the classes in Figure 1 and the Algorithm 2 are implemented for real symmetric matrices in C++ using OpenMP for shared memory parallelization. To account for a good workload balancing, the different loop indexes are associated to the concurrent threads using the guided self-scheduling algorithm [41] available in OpenMP. For the class Complex, see Figure 1, we use the C++ standard complex class. In addition, we profit from the symmetry in the matrix to store it in packed upper triangular form. To take advantage of the cache, we use row-major order. Compilation is carried out under the Linux operating system using the Solaris CC compiler with the –O5 compiling option, *i.e.*, the higher level of code optimization. The compiler version 5.11 is used. Performance of the implementation is tested considering matrix sizes of 5000 to 10000 in increments of 1000. The matrices are dense matrices, filled with real numbers generated randomly in the [0, 99] interval using the C language `rand()` function. The system used is a Quad-Core AMD Opteron™ 2376 HE with 4x512 KB and 6MB of L2 and L3 cache, respectively, and 8 GB of main memory. Assuming we use 64 bits for the real data type, the 512KB of L2 cache are enough to store rows up to 40000 elements. As shown later, this cache size is larger than the practical limit for a matrix to be considered "small". Therefore, for the matrices the present approach is intended for, there is no need to block the rows in smaller pieces in order to fit the cache. Finally, to use the worst case, we consider computation of 100% of the eigenvalues and eigenvectors in the series of matrices.

First of all, we compare the sequential results with the one-thread results using the concurrent algorithm. We find that the concurrent version is slightly faster by just a 0.7% in the worst case (matrix with size 7000 and sequential computation time of 1512 seconds). Therefore, the concurrent version can be considered as efficient as the sequential one.

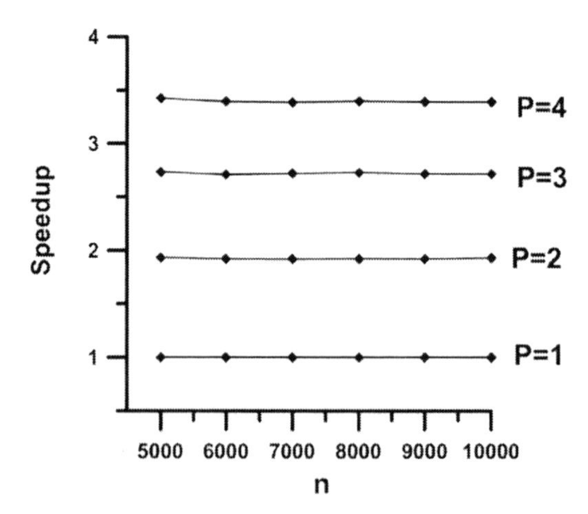

Fig. 2. Speedup for the computation of the whole set of eigenvalues and eigenvectors in real symmetric matrices as a function of the matrix size (n) and the number of cores (P)

The next question to answer is the performance of the concurrent version as a function of the matrix size and the number of cores. Thus, we compute the speedup for each matrix as a function of the number of cores. For the different matrix sizes, the speedup is defined as the quotient of the one core case with respect to each computing time. The result is shown in Figure 2. We observe that, for a given number of cores, the speedup is fairly independent of the matrix size. The computing time, using the most consuming case, n=10000, ranges from 4425 seconds (for P=1) to 1304 seconds (for P=4). The speedup reaches almost 3.5 in all the P=4 cases.

The variation rate of the speedup with the number of cores, P, is shown in Figure 3 for the worst case, n=10000. By fitting the speedup to P, we observe a quadratic variation with a good correlation coefficient, R=0.999. As shown in Figure 3, the variation is suboptimal. The maximum, obtained by extrapolating the regression curve, is predicted to be a speedup of 5 for P=9 processors. Anyway, the present results suggest that the limiting performance factor is the number of concurrent threads rather than the size of the matrix.

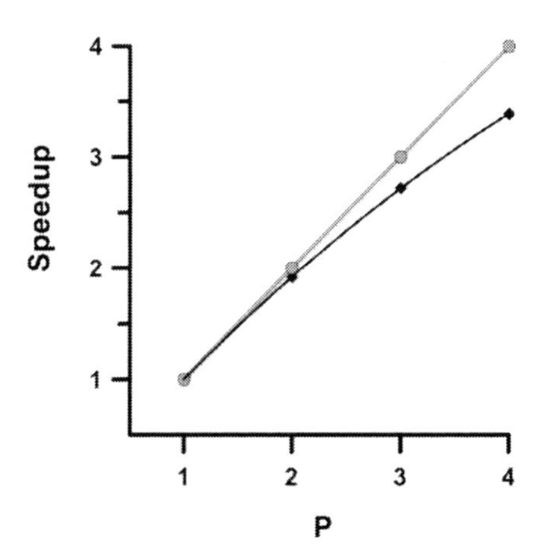

Fig. 3. Variation of the Speedup for the computation of the whole set of eigenvalues and eigenvectors in real symmetric matrices of n=10000 as a function of the number of cores (P). The continuous grey line represents the ideal scaling case. The diamonds represent the experimental data for a matrix size n=10000. The black line corresponds to the quadratic fitting of the data.

Since in the worst case, n=10000, we compute with P=4 the whole set of eigenvalues and eigenvectors in 1304 seconds, the question arises as what is the largest size we can treat in a reasonable time. In other words, what a "small" eigenproblem means in actual architectures. Thus, we have extended the computation from n=10000 to n=40000 in 5000 steps. The results are shown in Figure 4.

We observe that the worst case, n=40000, can be handled in less than 21 hours. Fitting the data to a third degree polynomial the correlation coefficient is very good, R=1.000, as expected from the algorithmic complexity. In addition, this uniform behavior shows that no degradation of the cache use does exist. So, what is the limit

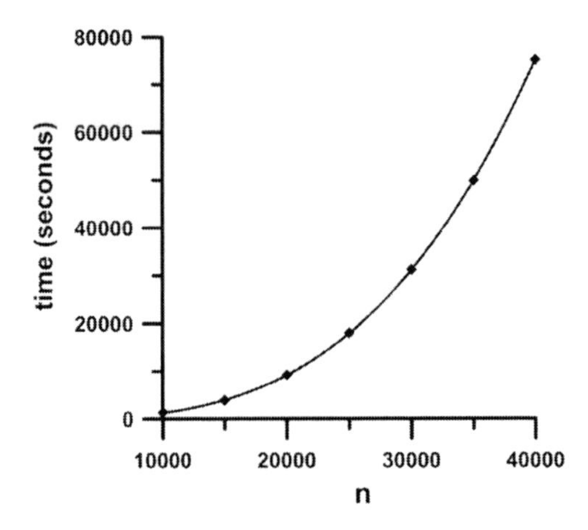

Fig. 4. Computation time, in seconds, as a function of the matrix size, n, for the P=4 case. The diamonds represent the experimental data. The black line corresponds to the cubic fitting of the data.

for a matrix to be considered small? Actually, Figure 4 shows that even the largest size here considered is affordable in a reasonable time. However, limiting ourselves to sizes processed in less than half a day the limit is about n=30000, which is processed in 8.7 hours.

Acknowledgments. This work has been co-financed by FEDER funds and the Consejería de Educación y Ciencia de la Junta de Comunidades de Castilla-La Mancha (grant # PBI08-0008). The Ministerio de Educación y Ciencia (grant # AYA 2008-00446) is also acknowledged.

References

1. Levine, I.N.: Quantum Chemistry, 5th edn. Prentice-Hall, Englewood Cliffs (1999)
2. Johnson, R.A., Wichern, D.W.: Applied Multivariate Statistical Analysis. Fifth Edition. Prentice-Hall, Englewood Cliffs (2002)
3. Mieghem, P.N.: Graph Spectra for Complex Networks. Cambridge University Press, Cambridge (2011)
4. Golub, G.H., van Loan, C.F.: Matrix Computations, 3rd edn. Johns Hopkins University Press, Baltimore, MD (1996)
5. Castro, M.E., Niño, A., Muñoz-Caro, C.: Evaluation and Optimal Computation of Angular Momentum Matrix Elements: An Information Theory Approach. WSEAS Trans. on Inf. Sci. and Appl. 7:2, 263–272 (2010)
6. Golub, G.H., van der Vorst, H.A.: Eigenvalue Computation in the 20th Century. J. Comput. and Appl. Math. 123, 35–65 (2000)
7. Jacobi, C.G.J.: Ueber ein Leichtes Verfahren, die in der Theorie der Säcularstörungen Vorkommenden GleichungenNnumerisch Auflösen. J. Reine Angew. Math. 30, 51–94 (1846)

8. Givens, W.: Numerical Computation of the Characteristic Values of a Real Symmetric Matrix. Oak Ridge Report Number ORNL 1574, physics (1954)
9. Householder, A.S.: Unitary Triangularization of a Nonsymmetric Matrix. J. ACM. 5, 339–342 (1958)
10. Wilkinson, J.H.: The Algebraic Eigenvalue Problem. Clarendon Press, Oxford (1965)
11. Parlett, B.N.: The symmetric Eigenvalue Problem. In: SIAM, Philadelphia. Republication of the original work, Prentice-Hall, Englewood Cliffs (1998)
12. Golub, G.H., van Loan, C.F.: Matrix Computations, 3rd edn. Johns Hopkins University Press, Baltimore, MD (1996)
13. Press, W.H., Flannery, B.P., Teukolsky, S.A., Vetterling, W.T.: Numerical Recipes. In: The Art of Scientific Computing, Cambridge University Press, Cambridge (2007)
14. Ortega, J.M.: Mathematics for Digital Computers. In: Ralston, Wilf (eds.), vol. 2, p. 94. John Wiley & Sons, Chichester (1967)
15. Cuppen, J.J.M.: A Divide and Conquer Method for the Symmetric Tridiagonal Eigenproblem. Numer. Math. 36, 177–195 (1981)
16. Gu, M., Eisenstat, S.C.: A Divide-and-Conquer Algorithm for the Symmetric Tridiagonal Eigenproblem. SIAM J. Matrix Anal. Appl. 16, 172–191 (1995)
17. Wilkinson, J.H., Reinsch, C.: Handbook for Automatic Computation. In: Linear Algebra, vol. 2, Springer, Heidelberg (1971)
18. Dongarra, J.J., Moler, C.B., Bunch, J.R., Stewart, G.W.: LINPACK Users' Guide; LINPACK (1979), http://www.netlib.org/lapack
19. EISPACK Last access (December 2010), http://www.netlib.org/eispack/
20. Anderson, E., et al.: LAPACK Users' Guide. In: SIAM, July 22, 2010 (1999), http://www.netlib.org/lapack/
21. Blackford, L.S., et al.: ScaLAPACK Users' Guide, December, 2010. SIAM, Philadelphia (1997), http://www.netlib.org/scalapack/
22. ARPACK Last access (December 2010), http://www.caam.rice.edu/software/ARPACK/
23. Lehoucq, R.B., Sorensen, D.C.: Deflation Techniques for an Implicitly Restarted Arnoldi Iteration, SIAM. J. Matrix Anal. & Appl. 17, 789–821 (1996)
24. PARPACK Last access (December 2010), http://www.caam.rice.edu/~kristyn/parpack_home.html
25. SLEPc Last access (December 2010), http://www.grycap.upv.es/slepc/
26. Castro, M.E., Díaz, J., Muñoz-Caro, C., Niño, A.: A Uniform Object-Oriented Solution to the Eigenvalue Problem for Real Symmetric and Hermitian Matrices. Comput. Phys. Comun. Comput. Phys. Comun. doi:10.1016/j.cpc.2010.11.022(in press)
27. Kirk, D., Hwu, W.: Programming Massively Parallel Processors: A Hands-on Approach. Morgan Kaufmann, San Francisco (2010)
28. Hwu, W., Keutzer, K., Mattson, T.G.: The Concurrency Challenge. IEEE Design and Test of Computers 25, 312–320 (2008)
29. Sutter, H., Larus, J.: Software and the Concurrency Revolution. ACM Queue 3, 54–62 (2005)
30. Sottile, M.J., Mattson, T.G., Rasmussen, C.E.: Introduction to Concurrency in Programming Languages. CRC Press, Boca Raton (2010)
31. OpenMP, A.P.I.: Specification for Parallel Programming. Last access (December 2010), http://openmp.org
32. Gropp, W., et al. (eds.): MPI: The Complete Reference, vol. 2. The MIT Press, Redmond, Washington (1998)

33. Kedia, K.: Hybrid Programming with OpenMP and MPI, Technical Report 18.337J, Massachusetts Institute of Technology (2009)
34. Jacobsen, D.A., Thibaulty, J.C., Senocak, I.: An MPI-CUDA Implementation for Massively Parallel Incompressible Flow Computations on Multi-GPU Clusters. In: 48th AIAA Aerospace Sciences Meeting and Exhibit, Florida (2010)
35. Jang, H., Park, A., Jung, K.: Neural Network Implementation using CUDA and OpenMP. In: Proc. of the 2008 Digital Image Computing: Techniques and Applications, Canberra, pp. 155–161 (2008)
36. Shukuzawa, O., Suzuki, T., Yokota, I.: Real tridiagonalization of Hermitian matrices by modified Householder transformation. Proc. Japan. Acad. Ser. A 72, 102–103 (1996)
37. Bischof, C., Marques, M., Sun, X.: Parallel Bandreduction and Tridiagonalization. Proceedings. In: Proc. Sixth SIAM Conference on Parallel Processing for Scientific Computing, pp. 383–390. SIAM, Philadelphia (1993)
38. Smith, C., Hendrickson, B., Jessup, E.: A Parallel Algorithm for Householder Tridiagonalization. In: Proc. 5th SIAM Conf. Appl. Lin. Alg. Lin. Alg. SIAM, Philadelphia (1994)
39. Chang, H.Y., Utku, S., Salama, M., Rapp, D.: A Parallel Householder Tridiagonalization Stratagem Using Scattered Square Decomposition. Parallel Computing 6, 297–311 (1988)
40. Honecker, A., Schüle, J.: OpenMP Implementation of the Householder Reduction for Large Complex Hermitian Eigenvalue Problems. In: Bischof, C., et al. (eds.) Parallel Computing: Architectures, Algorithms and Applications. NIC Series, vol. 38, pp. 271–278. John von Neumann Institute for Computing (2007)
41. Polychronopoulos, C.D., Kuck, D.: Guided Self-Scheduling: a Practical Scheduling Scheme for Parallel Supercomputers. IEEE Trans. on Computers 36, 1425–1439 (1987)

Geospatial Orchestration Framework for Resolving Complex User Query

Sudeep Singh Walia[1], Arindam Dasgupta[2], and S.K. Ghosh[2]

[1] Department of Computer Science & Engineering
[2] School of Information Technology
Indian Institute of Technology, Kharagpur, West Bengal, India 721302
sswalia@cse.iitkgp.ernet.in, adgkgp@gmail.com, skg@iitkgp.ac.in

Abstract. With the development of web based technology and availability of spatial data infrastructure, the demand for accessing geospatial information over web has increased significantly. The data sets are being accessed by standard geospatial web services. However, for complex user queries involving multiple web services, it is required to discover those services and logically compose them to deliver the intended information. In this paper, a geospatial orchestration engine has been proposed for composition geospatial web services. A rule repository has been designed within the orchestration engine by considering the basic geospatial web services. A framework has been presented to develop a complex information service which can be achieved by chaining of already existing services. The efficacy of the framework has been demonstrated through a case study.

Keywords: Geospatial Web Services, Service chaining, Service orchestration, Service Composition.

1 Introduction

With the tremendous progress of information technology, the demand of online access of geospatial information from the distributed data sources is increasing. During the past decade many research works have carried out for integration distributed geospatial repositories to retrieve spatial data from heterogeneous sources. However only accessing the geospatial information may not be sufficient to provide essential information for decision making. The complex geospatial queries need integration and processing of the information.

Many private and government organizations collect/maintain domain specific geospatial for their organizational needs, often in the proprietary format by using vendor specific database software. Further, these data repositories are too large in volume and highly heterogeneous, making it difficult to move and into the uniform format. On the other hand, sharing of these geospatial data are extensively useful to answer various different complex quires in many applications like environmental monitoring, disaster management, land-use mapping, transportation mapping and analysis, urban development planning, and natural resource assessment, etc.

B. Murgante et al. (Eds.): ICCSA 2011, Part I, LNCS 6782, pp. 643–651, 2011.

Enterprise Geographical Information Systems (E-GIS) framework is a service driven approach for integration and sharing of heterogeneous data repositories. The Open Geospatial Consortium (OGC) provides several standards for spatial web service for accessing and sharing of geospatial information. The OGC provides standards for several spatial services, namely, Web Feature Service (WFS), Web Coverage Service (WCS), Web Map Service (WMS) etc. Further, the Web Processing Service (WPS) provides the standard for the processing of geospatial information. However, it may not be possible to resolve complex geospatial queries through basic OGC services. Most of the recently available spatial data infrastructure emphasizes on simple retrieval and visualization, rather than providing geospatial information by processing of those data. In order to handle complex queries, usual practice is to download the relevant geospatial data in the local system and process the same with in-house GIS package.

The aim of orchestration is to facilitate collaboration of web services across the enterprise boundary to access composite information. In particular, orchestration is the description of interactions and messages flow between services in the context of a business process [3][4]. Most recently in the internet technology domain, web services have achieved a wide acceptance. T. Andrews et al. [5] identified that by using service composition technology with the use of BPEL (Business Process Execution Language) an advanced architectural models of web services could be developed. A similar process can be used in the mobile based software domain to access composite web services. Brauner et al. [6] propose to use the BPEL in combination with WSDL to execute such workflows. In geospatial domain M Gone et al. [7] analyzed the use of BPEL in comparison to Web Services. It states that current implementations of OGC services are some kind of hybrid *representational state transfer* (REST) based services. Since BPEL requires SOAP services, the OGC services, which do not provide SOAP interfaces, need a wrapper, which acts as a proxy to the OGC services. The suitability of the *Web Service Orchestration* (WSO) technology as a possible solution for disaster management scenarios has been evaluated by A Weiser [8]. In the paper [9] a framework has been designed to provide user specific geospatial information in mobile devices environment through the orchestration of geo-services. In this work an orchestration engine has been proposed to use predefined business logic for the composition of geospatial services. To establish a service chain of geospatial services three different approaches has been described in [10].

In order to create and provide user specific geospatial information for mobile device environment, there is a need of an Enterprise GIS platform which could resolve complex geospatial query. Researchers have attempted to integrate of spatial data repositories to provide a platform of spatial data infrastructure [1][2]. These types of spatial data infrastructures do not always provide essential information according to need of users especially in emergency situations. Most of time, for acquisition of geospatial information is done through the application of monolithic complex software which could be handled by professionals. Such types of applications are not suitable for geo processing through the utilization of web services due to following reasons.

- There is no OGC standard for chaining of geospatial web services for acquisition of complex information.
- The chaining of geospatial web services in rigid way cannot be utilized in many situations.
- Implementation of business logic to resolve complex information is difficult.

Let us consider following user query:

"Find the nearest k (k=1, 2, 3 ...) hospitals along the road network within a buffer area of d (d=1, 2, 3 ...) km from user's location"

To resolve the above query, following geospatial services are needed:

a. Web Feature Service (say, "WFS1") for retrieving the location of Hospital feature.
b. Web Feature Service (say, "WFS2") for Road feature.
c. Web Processing Service (say, "WPS1") to generate buffer from the user's position.
d. Web Processing Service (say, "WPS2") to generate the intersection between buffer (generated by WPS1) and the Hospital geometry features.
e. Finally, another Web Processing Service (say, "WPS3") is needed to locate the *k* nearest hospitals, along with the path to reach the hospitals.

Thus, it can be observed that in order to resolve the above query, it is required to orchestrate several geospatial web services.

In this paper, a geospatial orchestration engine framework has been proposed to resolve complex queries. The orchestration engine also handles the long running transaction between the geospatial web services. The problem of utilizing *Business Process Description Language* (BPEL) in chaining of geospatial web services for orchestration has been has been introduced. Then an alternative approach for orchestration of geospatial web services for providing the complex information in the mobile device has been presented. A case study has been presented to validate the framework.

2 Orchestration of Geospatial Web Services

The orchestration of geospatial web services refers to the procedure for integration of relevant web services by applying business logic to generate required geospatial information. The required geospatial services are selected and coordinated by an orchestration engine (OE). This architecture typically describes the way of selecting relevant data services to retrieve desired geospatial information through the use of catalog services. In order to generate information from heterogeneous dataset, it is required to integrate the various sources of distributed data into an Enterprise GIS platform. The main aim of proposed framework is to provide user specific information by discovering and coordinating relevant geospatial services by exploiting OGC based catalog services (CSW). In this framework the catalog sever not only used as geospatial web service registrar but also used as registrar of business logic for orchestration of web services. The business provides the integration logic activities that takes one or more types of input and creates an output according to user requirements. The user is able to define their required business logic to retrieve relevant geospatial information.

Most of the SDIs (spatial data Infrastructure) are supported by catalog services to publish geospatial services, discover relevant services and, retrieve data for further processing or displaying map for visualization. The Web Feature Service (WFS) is used to retrieve spatial feature data from the SDI in Geographic Markup Language (GML) format. To provide some specific information, some SDIs use fixed service chaining technique by compositing relevant geospatial web services. However, these types information retrieval techniques are not flexible and in most cases, unable to fulfill the actual requirements of the user. To process the geospatial data on the fly, the OGC defines *Web Processing Service* (WPS) standard to produce different types of information. WPS consists of libraries of relevant geospatial algorithms and can be accessed through the geospatial web services.

The identification of relevant services to capture the main goal of user is another task of the framework. To identify the required services, the service interface should provide meaningful service descriptions to enhance the services chaining process. However, in many situations, only the service description does not identify the proper services due to semantic heterogeneity problem. Finding the relevant services the semantic heterogeneity plays a vital role. The utilization of catalog services will be useful is used to discover the other geo services by analyzing the metadata of service descriptions. However, the metadata based search will lead to semantic heterogeneity problem. To overcome the semantic heterogeneity, an ontology enabled catalog service is required. The concept of ontology has been referred to the same meaning with several terms. In order to make the ontology machine-readable, it has to be formalized in some representation language. An explicit context model for ontology is used to resolve semantic heterogeneity of the geospatial services to identify the relevant services for service chaining.

3 Framework for Orchestration Engine

The main aim of the proposed framework is to provide processing of geospatial information through utilizing geospatial services. It provides uniform access point for accessing of complex geospatial information by composing different web services.. This architecture describes the way of selecting relevant web services according to user's query.

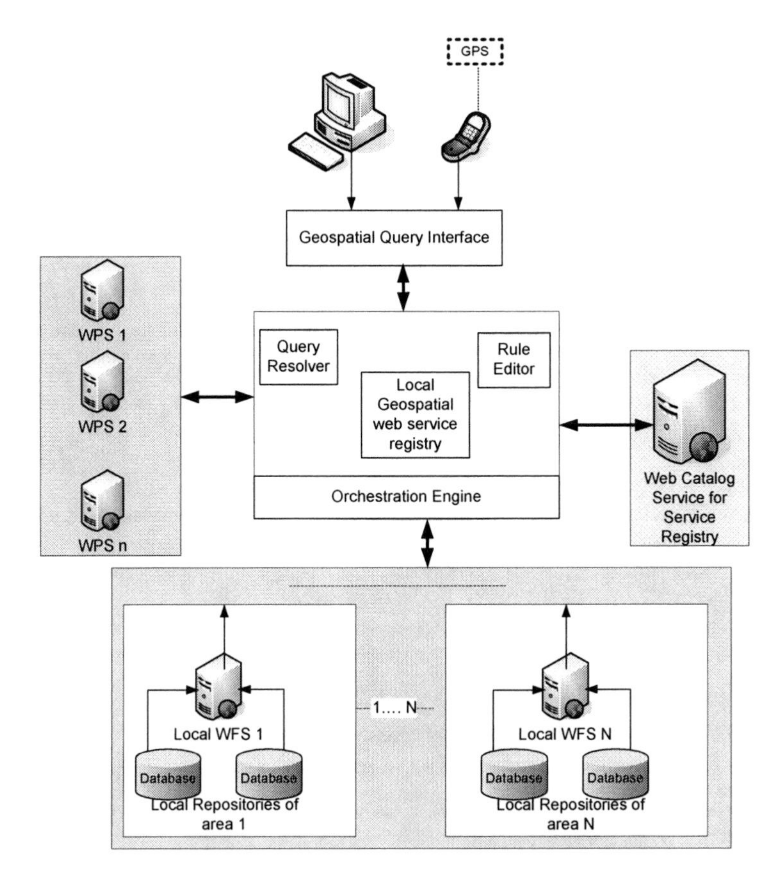

Fig. 1. Proposed Orchestration Architecture

Fig.1 shows the proposed framework for composing geospatial web services through service chaining and orchestration. It can access geospatial information from heterogeneous sources (through corresponding WFSs) and process the retrieved data using chained geospatial web services.

3.1 Components of Orchestration Engine

The orchestration engine consists of following components

Geospatial query interface

The system accepts the geospatial query in an extended SQL type format. It is assumed that orchestration engine gets query (from the user interface module) in this form along with the data sources to be accessed. The query resolver provides with the name of the feature information along with input parameters and output format to the query interface. After validating the query format the system parses the query to extract input parameters, output format and the feature information required for processing the query.

Rule editor
The rule editor is used to add or remove the geospatial services from the local service registry of the orchestration engine. It can be used to compose complex geospatial web services. The complex geospatial web services definition is stored as an entity, namely "complex web service", in the rule repository (explained later). As each entity in the rule repository in defined in terms of inputs it takes and output it produces, it is possible to chain the web services (both basic/atomic and complex services).

Local geospatial web service registry or rule repository
Local Geospatial service registry is a local registry service where different geospatial web feature services and web processing services are registered. Each service registered with the framework provides the description of the task it achieves, the input parameters required and the output parameters. The standards of OGC are used in defining the input and output. The basic four types of geospatial web services in the Local service registry are

- Web Feature services
- Web Processing services
- Chained Web Processing services
- Composite Web Feature and Processing services

Web Feature service and Web processing service are the standard services. Chained Web Processing services are formed by the orchestration of existing web processing services. Composite Web Feature and Processing services are formed by the orchestration of existing web processing services and web feature services. The order for execution of atomic web services in complex web services is defined in the local service registry itself.

Each geospatial web service is defined in terms of the input parameters it takes, the output parameters it offers, a description of the tasks it completes and an ordering set. An ordering set for each complex query provides with the order in which atomic web services are used to answer the complex query.

4 Case Study

Consider the following user query, ***Q***:
"Find the nearest k (k=1, 2, 3 ...) hospitals along the road network within a buffer area of d (d=1, 2, 3 ...) km from user's location"

The user query is parsed by the query resolver to extract the feature names and the input parameters to be fed to the orchestration engine. For the given query, the *Hospital* and *Road* databases need to be accessed to find out the position of the user and radius of interest.

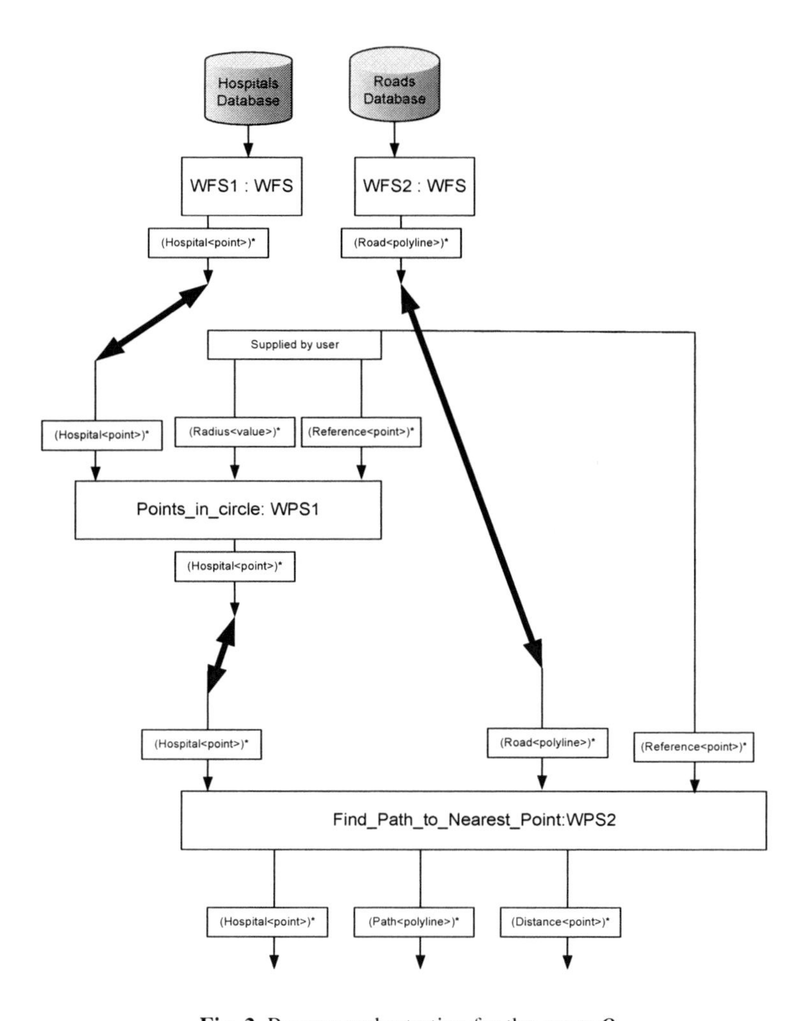

Fig. 2. Process orchestration for the query $\boldsymbol{Q}$

The rule repository contains a composite web feature and the processing service, "*Nearest*", to resolve the user query. The "*Nearest*" web service is achieved through orchestration of atomic web services. To resolve the query, $\boldsymbol{Q}$, following atomic services are required to be executed.

a. A web Feature service "WFS1", to get location of hospital features.
b. A web Feature service "WFS2", to get road network.
c. A web processing service "WPS1", to generate buffer centered at user's position.
d. A web processing service "WPS2", to get intersection between buffer generated and Hospital geometry features.
e. A web processing service "WPS3", to locate nearest hospitals, along with the path to hospital sorted by there distance from the user's location among the hospitals within the buffer area.

The above mentioned services are executed in an order which is defined in the service orchestration engine corresponding to the "*Nearest*" rule in the rule repository.

The proposed framework has been implemented in the mobile platform. The following results (refer Fig. 3) are obtained for query $\boldsymbol{Q}$.

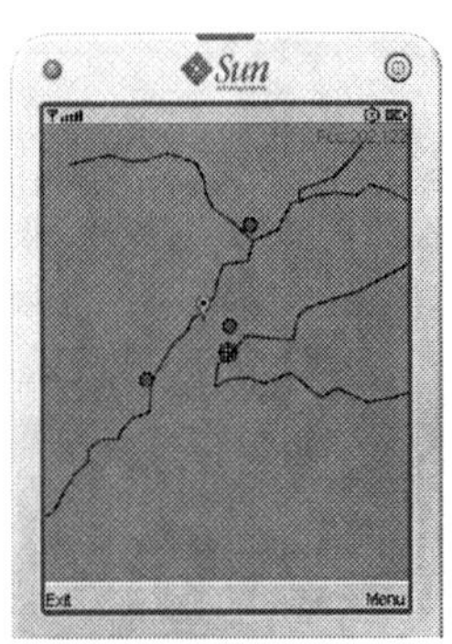

User position (reference point), along with road network and hospitals

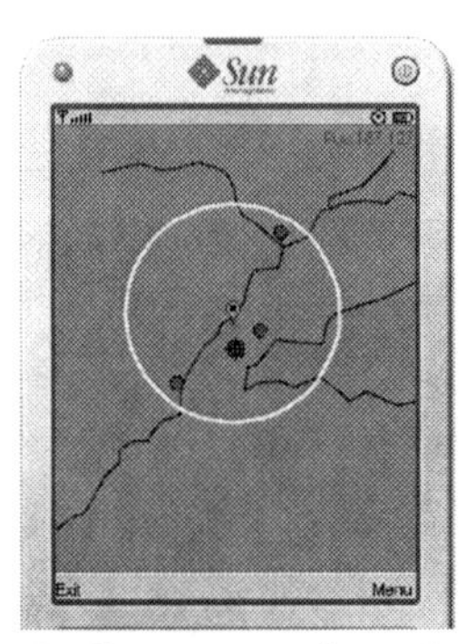

Buffer area (radius of interest) to locate hospitals

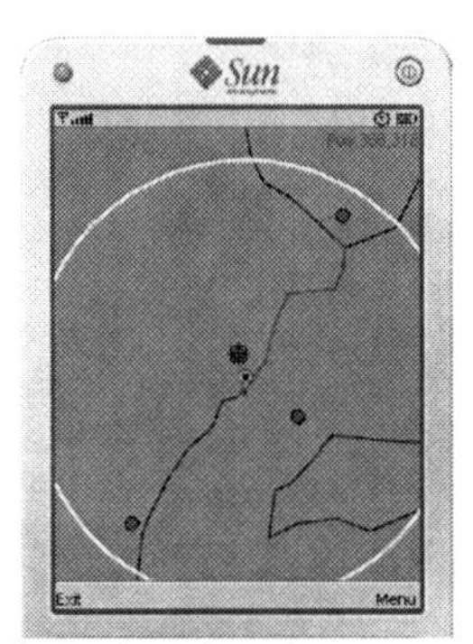

Path showing the nearest hospital from reference point

Fig. 3. Snapshot of Implementation of the CASE STUDY

5 Conclusion

In this paper a framework work has been proposed for orchestration of geospatial web services. The framework answers complex geospatial queries on a set of heterogeneous data sources. An orchestration engine has been developed on the top of the spatial data infrastructure. It utilizes the different geospatial services to implement chain of web services to answer complex queries. In this way the logical flow of different partial information are controlled to form the complete information against a user query. The major advantage of the proposed framework for orchestration of web services is that complex geospatial web services can further chained with other complex geospatial web services to further complete more complex tasks. A case study has been presented to show the efficacy of the proposed framework.

References

1. Masser, I., Rajabifard, A., Williamson, I.: Spatially enabling governments through sdi implementation. International Journal of Geographical Information Science 22, 5–20 (2008)
2. Sha, Z., Xie, Y.: Building spatial information systems based on geospatial web services. In: WCSE 2009, vol. 4, IEEE Press, New York (2009)

3. Peltz, C.: Web Services Orchestration and Choreography. IEEE Computer 36(10), 46–52 (October 2003)
4. Mazzara, M., Govoni, S.: A Case Study of Web Services Orchestration. In: Jacquet, J.-M., Picco, G.P. (eds.) COORDINATION 2005. LNCS, vol. 3454, pp. 1–16. Springer, Heidelberg (2005)
5. Andrews, T., Trickovic, I.: Business process execution language for web services version 1.1, Technical report, IBM (May 2003)
6. Brauner, J., Schae®er, B.: Integration of grass functionality in web based sdi service chains. In: Academic paper on the FOSS4G Conference, Cape Town, South Africa, 29 September - 3 October (2008)
7. Gone, M., Schade, S.: S Schade. Towards semantic composition of geospatial web services using wsmo in comparison to bpel. In: Proceedings of GI-Days, Munster, Germany, pp. 43–63 (2007)
8. Weiser, A., Zipf, A.: Web service orchestration (wso) of ogc web services (ows) for disaster management. In: Weiser, A., Zipf, A. (eds.) Proceedings of the 3rd International Symposium on Geoinformation for Disaster Management, Toronto, Canada, pp. 33–41 (2007)
9. Dasgupta, A., Ghosh, S.K.: A Framework for Ubiquitous Geospatial Information Integration on Mobile Device Using Orchestration of Geoservices. International Journal Of UbiComp (IJU) 1(3), 69–88 (2010)
10. Meng, X., Bian, F., Xie, Y.: Geospatial Services Chaining with Web Processing Service. In: Proceedings of the International Symposium on Intelligent Information Systems and Applications (IISA 2009), Qingdao, P. R. China, pp. 7–10 (2009)

A Bio-Inspired Approach for Risk Analysis of ICT Systems

Aurelio La Corte, Marialisa Scatá, and Evelina Giacchi

Department of Electrical, Electronics and Computer Science Engineering,
Faculty of Engineering, University of Catania,
viale A. Doria 6,95125, Catania-Italy
{lacorte, lisa.scata, evelina.giacchi}@diit.unict.it
http://www.unict.it

Abstract. In recent years, information and communication technology (ICT) has been characterised by several evolving trends and new challenges. The process towards the convergence has been developed to take into account new realities and new perspectives. Along with many positive benefits, there are several security concerns and ensuring privacy is extremely difficult. New security issues make it necessary to rewrite the safety requirements and to know what the risks are and what can be lost. With this paper we want to propose a bio-inspired approach as a result of a comparison between biological models and information security. The risk analysis proposed aims to address technical, human and economical aspects of the security to strategically guide security investments. This analysis requires knowledge of the failure time distribution to assess the degree of system security and analyse the existing countermeasures to decrease the risk, minimise the losses, and successfully manage the security.

Keywords: ICT; VoIP; NGN; Risk Analysis; Security; Failure Time Distribution.

1 Introduction

ICT technologies have pervaded all critical infrastructures. Thus, in many processes, the security risk has recently gained in significance. How to apply ICT in various fields has become essential to understanding how to protect what led to its introduction. The adoption of ICT by enterprises is a phenomenon that grows continuously and that allows the implementation of innovative solutions, exploiting emerging technologies and supporting decisions and management processes that are otherwise enormously complex. Innovation in the ICT thus brings countless benefits to enterprises in different areas of interest, and it ultimately enables the design of security solutions in many areas of study [1][5]. The complexity of the systems handling the information and communications, is changing. The networks are evolving towards convergence, and the aim is to form a single complex, dynamic and communicative body, capable of providing

B. Murgante et al. (Eds.): ICCSA 2011, Part I, LNCS 6782, pp. 652–666, 2011.

services, applications, dynamic models around each of users. With the increase of size, interconnectivity and convergence, the networks have become vulnerable to various form of threats[1][3][5]. The evolution of the communication implies the need of the new security requirements, new security mechanisms and efficient countermeasures. Biological organisms are also complex interconnected systems with many points of access. Using the concepts of biological systems and models we can inspire information and communication security. Thus, in examing some of common structures that will characterise the future of the ICT, we can find some striking similarities to biological systems[18][19]. This paper has been organised as follow. After a brief introduction and the related work, presented in Section I and Section II, in Section III and IV we discuss on Information and Communication Technology Security issues and security requirements related to the new challeges and the new tendency towards the convergence of the next future. An overview of Biological Models, Bio-Inspired Security and similarities with Information Security are presented in Section V. In Section VI we present a risk analysis model based on bio-inspired approach. Some consideration about strategic decisions and business investments on risk are presented in Section VII and finally we conclude with Conclusions and Future Work.

2 Related Work

Despite the growing interest in the research and standardisation of communities regarding the security of computer science and communication systems, a general consolidated study on the degree of security of an ICT system is still lacking [1][5]. Through standardisation activities is ongoing attempt to solve the security problems of ICT systems by promoting use of many different models of study and analysis. Additional studies have investigated ICT applications in various fields. Relatively little research has been conducted concerning mobile/wireless ICTs, software-as-service, RFID, storage infrastructure, social computing networks and VoIP. There is a general tendency to treat the issue of security in communications through taxonomies of vulnerabilities and threats [3][4][6]. Risk, instead, is rarely mentioned in quantitative terms about ICT Security. Recognising what is already described by other studies [11][12][13][15] on risk analysis and management for information system security, and recognising what is already studying and analysing abour statistical methods [7][8][16] this paper aims to propose a comparison of different disciplines. The aim is to learn from the study for the analysis of risk in the broader field of biology and demonstrate that there is a strong correlation between the study of Epidemiology and the study of security in communication systems, to find an appropriate model risk analysis for ICT systems.

3 Information and Communication Security Issues

As explained in [9][10] information can be defined as an important business asset that can exist in many forms.In [1][5], information security is defined as a range of controls that is needed for most situations in which information systems are used. In the past, information was stored using methods aimed at finding it, over the years, in the same format with the highest integrity. The main problem was that part of it was lost, because of the nature of the data and the breach and errors accidental. The data violation has acquired in time varied connotations because of the different security requirements, and the issue of data breach remains the same over the years. To strategically design the system is important to limit the damage, analysing aspects related to vulnerabilities and threats, risk and impact, in order to preserve confidentiality, integrity and availability. Then, the protected data do not restrict the how of shared knowledge, they allow to prevent loss, and maintain confidentiality. While in the past, however, information traveled for long distances to be shared, now the real-time accelerates the transmission of data, shared knowledge and transmission of personal information. The information in the form of voice or data, written, stored or communicated, shared or processed, is a source of knowledge, and the power of knowledge should not be underestimated. Information is a resource that has acquired, increasingly, more value. Meanwhile, the simplicity with which all kinds of personal information are shared increases due to false illusion of an always possible repair. An ICT system sometimes gives a false perception to keep safe with a series of consecutive procedures, which are activated after a threat has exploited a vulnerability. Security is not a repair process but a strategic decision and preventive planning. Information Security is repair for any damage suffered, noting that risk is a prerogative of the system. If a failure occurs we run for cover, with heavy investments which sometimes cause interruption of processes and so on. This is an attitude, modus operandi, which belongs to a society that has lost the concept of value of information [9][10] . Today, the process towards the convergence of the triple play(voice, video,data) and quadruple play (voice, video, data and mobile communications) has been developed to take into account new realities and new perspectives in the world of telecommunications and to account for the need to converge networks and the existing fixed and mobile services in the NGN (Next Generation Network) [2]. For example, Voice over IP , VoIP, is the first step towards convergence and represents a big challenge already well under way. VoIP has become a valid alternative to the traditional telephone network. With VoIP, ensuring privacy is extremely difficult [3][4]. VoIP is economically viable and highly effective, and the widespread use of VoIP in recent years is growing, but it introduces many security challenges, and the benefits of this technology are as great as the security issues . The need for security is linked to the value of the information that is transmitted, and with the new development of converged networks, the information also acquires an important shared-communication value. Therefore, in an ICT system is important to understand, first of all,why we must protect the information, than figure out what kind of risks are there, and the safety requirements, suitable to the system, even before to design an analysis of the assets and evaluate what to do.

4 Understanding Risk: Why Is the Security Important?

There are three requirements that must be maintained for an ICT system:

1. Confidentiality or Privacy: is the restriction of access to resources only to a specified set of individuals or processes. Also for the recipients of informations are valid the same restrictions and is necessary identify them to avoid a spread of information.
2. Integrity: is the prohibition to manipulate informations by those persons or processes that may damage them. In this way in a communication process data losses are prevented because sender and recipients are known.
3. Availability: is the attempt to limit the delay in information delivery, in this way the recipient receives informations as soon as possible, with a finite delay.

These requirements are commonly called the CIA set of security objectives. Information Security means improving the system and protecting the assets from the many different kinds of threats. Threats can cause damage by intervening directly in the information exchanged or by interrupting the continuity of certain procedures, affecting the system, destroying a part of it, or violating the rules of privacy. The damage may be multiple, and its nature is not limited to a few unfortunate events. Technology and communication thus becomes a vehicle for both good and bad actions. The importance of security is directly proportional to the value of information, which is shared in any system. With a proper strategic planning and risk analysis we can estimate the expected benefits from each investment in safety. In general the information can be more or less sensitive, and each database can be more or less confidential. So there may theoretically be a multitude of degrees of confidentiality and secrecy. With the convergence of the technologies and services the threats, to the entire systems and processes, are increasing. The attacks are more probable and this probability increases with the value of the information. First of all, to design a security infrastructure, it is necessary to properly assess the risk to protect the data to assure confidentiality, availability and integrity, and so the importance of security for a general system [5]. To benefit from the investments, it is essential to understand where, how and when to apply them [11][13]. We need to know the system information, assets and processes involved in a communication, and we must assess threats and vulnerabilities of the system. The analysis of risk requires knowledge of the probability of failure and its distribution and the probability that an attack occurs, that is when a threat exploits a vulnerability, because of the lack of proper security measures. Thus, we can assess the degree of system security and analyse the existing countermeasures to try to decrease the risk, minimise the losses, and successfully manage the security. To do this, we consider in subsequent sections, a biological approach to evaluate each complex ICT system as a biological organism. In this way we can apply the analytical models of failure distribution and define the trend of the risk value, which is useful in decision making strategies of countermeasures.

5 Bio-Inspired Security

5.1 Introduction and Related Work

The similarity between biological processes and computer security problems has long been recognised and studied, over the years. To prove this, in 1987, Adelman introduced the term"computer virus", inspired by biological terms, such as Spafford with the term "form of artificial life", referring to the virus, and so on.The analogy between the protection mechanisms of living organism and the security could be indeed appealing. Many comparisons have been studied according to several point of view[12][18][19]. Whereas there have been a moltitude of studies based on a biology-computer analogy for defense methodologies,there have been several studies about similarities between computer worms and biological virus, pathogens. There are many biological terms such us "worms", virus, which have been borrowed to name camputer attacks. The term virus is widely used for malicious code affecting computer systems and we could use this term for different threats affecting the communication systems in general. Such usage suggests the comparison with biological diseases. We can describe the computer virus as a program that can affect other programs or entire networks, by modifying them, exploiting vulnerabilities and compromising the security requirements. Recognising that there is a real parallelism between biological systems and computer networking[18], we consider the future convergence of the networks and the evolution of the communication systems as a process highly complex, such as a biological entity. Our networks are increasingly facing new challages and they grow larger in size, and we want to continue to be able to achieve the same robustness, availability and safety. Biological systems have been evolving over billions of years, adapting to a continuous changing of environment. If we consider information systems and biological systems, they share several properties such as complexity and interactions between individuals of a population (asset of the information systems). There are many analogies between computer systems and biology, and many research studies support this idea [20]. The reasearch in this area has mostly focused on leveraging epidemiological studies of disease propagation to predict computer virus propagation[12][15][17]. The use of epidemiological model to predict virus is based on a wide range of mathematical models, which has been developed over the years. When considering the spread of an infection through a population many factors are likely to be important, such as the trasmissibility of the agent pathogen, the immune response the general behaviour of the population and the risk perception[14]. We discuss about this in the next section.

5.2 Epidemiological Models

In recent years there had been a growing interest in using biological, epidemiological models to gain insights into information and communication systems. These approachs developed through the years are based on several models[17]. Some of this looked at the effect of the network topology on the speed of virus propagation, some looked at virus spread on different network, and some the risk perception in epidemics. Different studies have also used several biological models for immunisation strategies. As a result, biologically inspired research in information and communication security is a quickly growing field to elucidate how biological or epidemilogical concepts in particular have been most successfully applied and how we can apply these to the safety strategies for risk analysis and management[15]. Through this study, we notified a close similarities between the biological diseases and what we define as communication risk.Among the biomedical disciplines, epidemiology there seemed to be more suitable for comparison. Epidemiology is a methodology, a technical approach to problems, a "philosophy". Epidemiology is a way "different" to study health and disease, and it is cross-science. Epidemiology is working with the clinic and preventive medicine. It is involved in analyzing the causes, the course and obviously, the consequences of diseases, by mathematical-statistical models that analyze the spread of disease in populations. Purposes of Epidemiology are:

- Determine the origin of a disease whose cause is not known.
- Investigate and control a disease whose cause is known.
- Acquire informations about the natural history of disease.
- Plans programs and activities of control and monitoring of disease.
- Assess the economic effects of disease and analyze the cost-benefit.

In this way, individuals of a population exposed to a particular virus, are like the information assets of the communications system, exposed because of their vulnerability to network threats. The threat can damage one or all of the assets of the information system, and it represents a potential cause of an accident or deliberate accidental, as a malicious code. The vulnerability is a weakness of the system for the security of informations. The attack occurs when the vulnerability is exploited by threat agents. Many epidemiological studies are designed to verify the existence of associations between certain events, in this case, we talk of the incidence or of the onset of a disease in a population of individuals. Epidemiology is a methodology of study. The same can be said for safety. Using these concepts and models, it can inform, guide, inspire information security, and understand, prevent, detect, interdict and counter threats to information assets and system. The risk, like a disease, is the result of three factors: threat, vulnerability and explanatory variables. In the case of a communication system, we talk of the impact of a malicious code or a threat of the network, and therefore, also of damage caused as a result. These results can be distorted because of other variables that somehow might confuse the results, and so called confounding variables (counfonders). These variables may be confounding or interacting variables, also called in econometrics control variables or explanatory variables, which are used into the Cox regression model [7][8]. The explanatory variable plays a key role in understanding the relationship between cause and effect of a general threat. The explanatory variable is a variable which is used to explain or to predict changes in a value of another variable. This is important to evaluate the relationship between the point events that we define in the next section. To assess the confounding, it is necessary an analysis which provides a collection of epidemiological data sets and informations. This is complex in the case of communications systems and networks. In this case, the analysis is multivariate, because it involved so many factors that influence and change network vulnerabilities. The Cox Model also considers the variable time and helps to assess the risk of exposure to a threat in the time. This analysis is embedded in a broader analysis, survival analysis and failure time analysis. Thus, realizing a link between biomedical disciplines and safety review, the models of failure time analysis are needed to asses the probability of the the attacks distributions, to estimate the extent of the risk and of the security investments.This approach depends on having knowledge of the probability distributions associated with successful attacks on information assets. About this, little real data is available, and to estimate the risk we need to use a simplistic model. We deal with this in the next section, giving an overview of the failure time distribution models.

5.3 Failure Time Distribution

A failure event in general, could be due to an attack by malevolent individuals or groups that want to damage the security system. A failure doesn't meaning the total distruction of the system, but even the impairment of the informations that it holds. In this section we want to give a brief overview of the study done about survivor analysis and failure time distributions models, which allows us to estimate the risk. Ryan and Ryan [11][13] models a general information infrastructure in number of finite information systems $\{S_i : i \in I\}$, where, $S_i \neq S_j$ if $i \neq j$, and the set I= {0,1,2,3,. . . ..}. Each system, which purpose is to preserve the information, can be thought as a finite collection of information assets S_i={α_k:k ∈ I }. Threats can destroy or only degrade information, we can practice information security at the system and network level, where it is easier for designers to act in different ways to reduce risk. Each system is also characterized by a vector X_i. called the decision vector, where each element is determined by the decisions and the strategies chosen to manage risk. Obviously threats that can affect the system are many, and then many variables should be introduced, but in this case they are reduced to a small number in order to analyze the model more accurately. The potential threats can attack the system in a single finite set {T_j:j ∈ I}, and it can damage each information asset {α_k}. The consequences of a successful attack of a threat T_j at time $t_i$is called *impact*, but if the attack failed *impact* is zero. The danger of each threat on each information asset, is the impact may have on each characteristic, and adding all these quantities to obtain the total probability. For each information system during the time interval of observation, it is necessary to do a distinction between *complete data* and *censored data*. Complete Data come from those systems whose failure causes are well-known and which occur during the observation period. Censored Data come from those systems that have no failure during the observation period, or if a failure occurs and the cause is unknown. For each individual system it is possible to define the following functions, and summarise the main functions involved as follow:

- Survivor Function S(t), which is the probability of being operational at time t :
$$S[t] = P_r[T \geq t] = 1 - F(t) \quad (1)$$
Where F(t) is the Failure function, which tell us the probability of having a failure at time t.
- Failure Density Function f(t), which is the probability density function:
$$f(t) = \frac{dF(t)}{dt} = -\frac{dS(t)}{dt} \quad (2)$$
- Hazard Function h(t), which is the probability that an individual fails at time t, given that the individual has survived to that time:
$$h(t) = \lim_{\delta \to 0^+} P_r(t \leq T < t + \delta \mid T \geq)/\delta \quad (3)$$
where $h(t)\delta t$ is the approximate probability that an individual will die in the interval $(t, t + \delta t)$, having survived up until t
- Cumulative hazard function H(t):
$$H(t) = -\log S(t) \quad (4)$$

In this regard, it is possible to introduce estimators of these functions S(t), F(t), f(t), h(t), referring to *complete data*, then neglecting *censored data*. Consider N systems and suppose that n(t) is the number of failure that occur before time t, the number of systems that most likely will have a failure in the interval $[t, t+\delta]$ is denoted by $n(t + \delta) - n(t)$ and N-n(t) is the number of systems still operating at time t. An empirical estimator for the function S(t) is :

$$S(t) = \frac{N - n(t)}{N} \quad (5)$$

An empirical estimator for the function F(t) is:

$$F(t) = \frac{n(t)}{N} \quad (6)$$

Instead for the function f(t) is :

$$f(t) = \frac{n(t + \delta) - n(t)}{\delta N} \quad (7)$$

For small value of , it is possible to calculate an empirical estimator for the function h(t), which is:

$$h(t)\delta = \frac{f(t)\delta}{S(t)} \quad (8)$$

Unfortunately, these definitions are not valid for data which are Censored, but it is possible to give alternative definitions to match the previous one to the case of interest. Using the first method, called Kaplan-Meier, we denote by:

$$t_{(1)} < t_{(2)} < t_{(3)} \ldots < t_{(m)}$$

the distinct ordered times of death(not considering the censored data). We define $d_{(i)}$ the number of failure at time $t_{(i)}$, and $n_{(i)}$ the number of surviving system just before the instant $t_{(i)}$. The estimator for the function S(t) can then be defined as follows:

$$S^{KM}(t) = \prod_i (\frac{n_i - d_i}{n_i}) \quad (9)$$

for $t_j \leq t < t_{j+1}, j = 1, 2, 3, \ldots, k-1$. The explanation of why this expression is valid even if there are Censored Data is very simple. To be alive at time t, a system must surely survive even in the moments before time $t_{(1)}$,$t_{(2)}$ .. because for sure in this interval $[t_{(i)}, t_{(i+1)}]$ there is no failure. So the probability of a failure at time $t_{(i)}$ is equal to $d_{(i)}/n_{(i)}$, taken for granted that the system survived in the previous interval. Obviously if there are not Censored Data, the estimator expression coincides with the previous case. The expression for other functions are:

$$F^{KM}(t) = 1 - S^{KM}(t) \quad (10)$$

$$H^{KM}(t) = -\log S^{KM}(t) \quad (11)$$

$$h^{KM}(t) = \frac{d_j}{n_j(t_{j+1} - t_j)} \quad (12)$$

The second method, called Nelson-Aalen, may be considered better than the Kaplan-Meier, as the latter can be considered an approximation when d_j is smaller than n_j. In this case the expressions become:

$$S^{NA}(t) = \prod_j \exp(-\frac{d_j}{n_j}) \quad (13)$$

$$F^{NA}(t) = 1 - S^{NA}(t) \quad (14)$$

$$H^{NA}(t) = -\log S^{NA}(t) = \sum_{j=1}^{r} \frac{d_j}{n_j} \quad (15)$$

$$h^{NA}(t) = \frac{d_j}{n_j(t_{j+1} - t_j)} \quad (16)$$

6 Risk Analysis Model

The risk analysis model, presented in this section, is based on a bio-inspired approach. Through this approach, epidemiological matching and failure time distribution models, we build the foundation for a complex risk analysis and for security descision-making strategies to manage the safety of the system. Each ICT system consists of a series of assets. Each asset is linked to each other and both exchange information and are actively involved in the processes within the system, and are interconnected and communicate with the external environment. A biological system, a human body for example, has many similarities with ICT systems:

- High Complexity.
- High Connectivity.
- Numerous Access Points.
- Communication, Cooperation and Coordination on micro and macro level
- Vulnerabilities to several threats
- Relation with other systems of the same nature
- Relation and Communication with external environment

Thus, we can define in both contexts:

- Biological Risk: the probability that a virus exploits a vulnerability. This can provide a disease of a single individual human body or it can spread causing an epidemic.
- ICT Risk: the probability that a threat exploits a vulnerability of an asset or of the system to cause an attack, compromising the security requirements of confidentiality, integrity and availability.
- Biological Failure: it is the event linked to an outbreak of a disease.
- ICT Failure: it is an event linked to the damage in the system, which is manifested, for example, with a denial of service.

Based on previous observations and definitions, we can define the risk of a system as the sum of two components:

$$R = R^* + R_r \tag{17}$$

the first term is a function of the failure time distribution $F(t)$, while the second represents the Residual Risk R_r. It is the minimum achievable risk threshold of each system. Below this threshold it is impossible to get off, because there are not systems with a risk threshold equal to zero. The risk analysis aims to estimate these values and contextualizing them in the test system, we can then determine the safety measures to be taken to minimize component dependent failure distribution. Thus, Risk R^* can assume three different values :

- R_{nt}= Not Tolerable Risk. It is the maximum risk threshold above which the system has serious security problems.
- R_u= Unprotected Risk. It is the risk threshold of a system where there are not investments of any kind.
- R_t=Tolerable Risk. It is a risk threshold that we want to achieve, decreasing the threshold R_u, applying a certain investment I.

Now we identify four ideal cases, based on the strategic choice of investment in the system. In each case we show the trend of the risk as a function of the time. We identify the four limits values described above. The risk will change within certain ranges, depending on the adopted security strategies, investments and preventive actions or shelter when damage occurs.

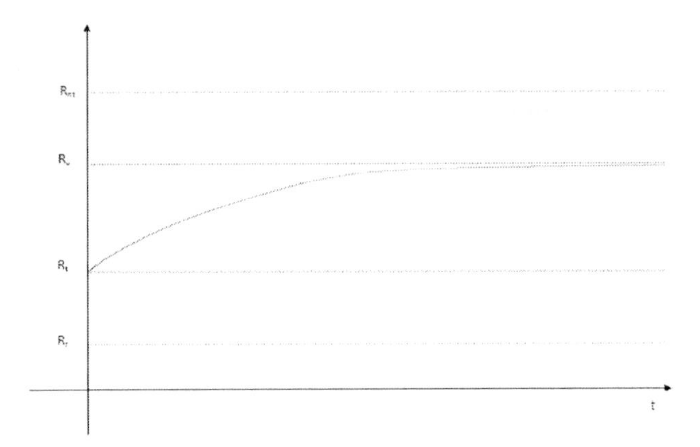

Fig. 1. Case 1

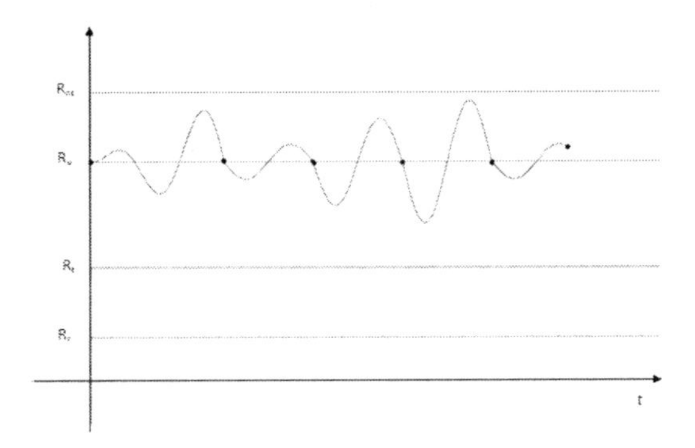

Fig. 2. Case 2

6.1 Case 1: Initial investment, Made in the Initial Planning Stage

The initial risk is R_t, because there is an initial investment I. The risk of the system is maintained within the range $R_t - R_u$, because the system is not totally exposed to threats. There can be intermediate investments but they are inappropriate or negligible and therefore they are not considered (for example an antivirus update unscheduled).

6.2 Case 2: Intermediate Investments (Maintenance, Protection/Shelter after an Attack of a Threat)

In this second case, the initial investments are negligible. Investments, in this case, are done for a scheduled maintenance of the system, or after a successful attack of a threat. The points indicated by the arrows are times where the investments are applied. The function oscillates aroud R_u.

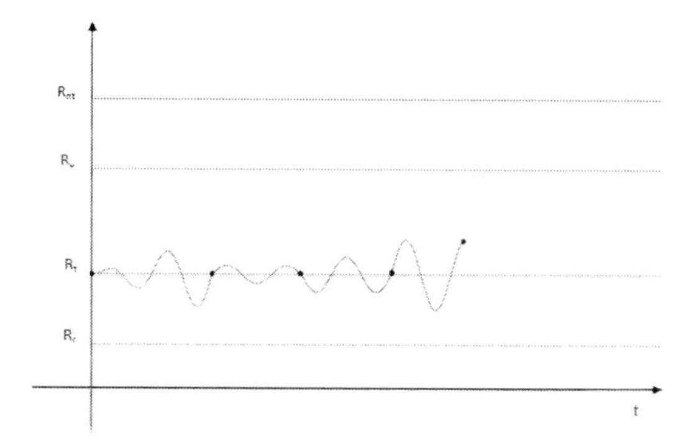

Fig. 3. Case 3

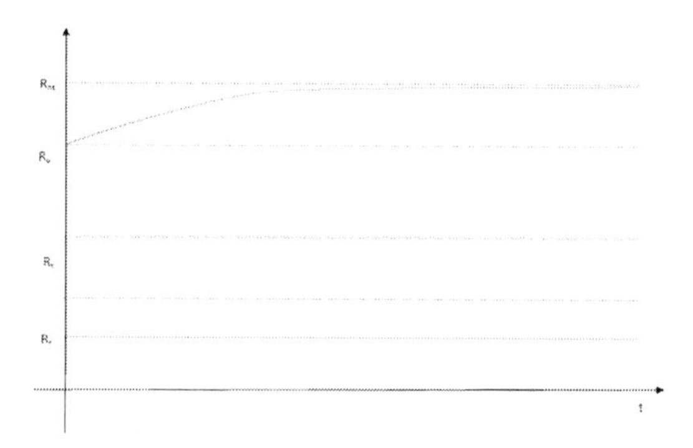

Fig. 4. Case 4

6.3 Case 3: Initial and Intermediate Investments

The risk, in this case, has small peaks over the threshold R_t, and oscillates around R_t, and sometimes tends to the minimum threshold of R_r, without touching it.

6.4 Case 4: No Investments

In this case the initial and intermediate investments are not necessary equal to zero, but they can be inappropriate or applied in non- strategic instants of time. There is the absence of a strategic security planning.The risk increases dramatically.

7 The Economics of Risk: Strategic Decisions and Business Investments

The ideal analysis should allow so great a risk estimate. The Case 3 is the ideal case where investments are allocated at the beginning and at intermediate time.

An optimal strategy also considers the economic tradeoff between the timing of investment, its value and cost-effective. To improve the security of a system it should make investments a way to protect the system from possible threats. It is obvious that this investment must be profitable for the system and prevent any attacks, or survive if there was one. The developments in technology have improved performance, in terms of security systems, but it has meant that the threats, that can attack a system, evolved. For this reason, before investing the money, it is necessary to make appropriate assessments. Resuming the previous definitions of hazard function, we can consider two different systems and compare them [11]. In the first system we suppose to make investments to improve security and define $h_1(t)$ its instantaneous failure rate, in the second system we decide not to invest in security and call $h_0(t)$ its hazard rate. It is possible to relate the two quantities through the following relatioship[11] :

$$h_1(t) = kh_0(t) \tag{18}$$

Assuming that the hazard function is continuous, although in reality it is not very likely, it is assumed that the processes of censored always occur after a failure. From this report it is possible to evaluate the benefit that an investment can result in increasing the security of a system. The parameter k is called Hazard Ratio, and through its value it is possible to make the following observations :

- If $k < 1$ the probability of succumbing to an attack is less in a system on which we decide to invest.
- If $k = 1$ there is no advantage on investing, because the two systems, when being attacked, would behave the same way.
- If $k > 1$ the system in which we have invested money succumb more easily to an attack and then invest money is not beneficial.

If we decide to invest for the security of the system, the benefits that we expect to obtain is given by the following formula [13]:

$$E_{NB}[i] = p_0 L - piL - i \tag{19}$$

where p_0 and p_i are the loss probabilities for the system on which we invested and the one without, respectively, and L is the loss. A positive value of E_{NB} means that the investment has made benefits. If we apply the investment, the Survivor Function is shifted to the right, this means that systems survive longer than others. Thus, we want to introduce and propose the loss variable definition for an ICT system, as:

$$L = L_{inf} + L_E \tag{20}$$

Where L_{inf}, is the information loss, reported to confidentiality, integrity and availability:

$$L_{inf} = L_{infC} + L_{infINT} + L_{infAVAIL} \tag{21}$$

And L_E is the economic loss due to the information loss. At each risk threshold it is possible to match a loss threshold.

- L_u is the information and economic loss in an unprotected system.
- L_t is a tolerable information and economic loss of system whit a tolerable risk threshold.
- L_r is the minimum information and economic loss (ideal case) of a system ideally with no risk.
- L_{nt} is a non tolerable loss associated to a non tolerable maximum risk.

8 Conclusion and Future Works

ICT networks are critical part of the infrastructure needed to operate a modern industrial society and facilitate efficient communication between people, efficient use of resource, and efficient way to ensure security of the systems fram threats that may exploit vulnerabilities and endanger security requirements. Most networks have emerged without clear global strategy to project information and communication technology. Historically, security decisions are taken outside of the context and after the damage occures, with the false perception to obtain high security efficiency at reasonable cost. A security management strategy implies a complex dynamic risk analysis of the system,of information exchanged and of communication and cooperation with other systems. Introducing robustness inevitable requires additional cost, computational and redundant resources on making network tollerant to failures, but this is not cost-effective in the short term. Drawing inspiration from biology has led to useful approaches to problem-solving, with this paper we want to propose and develop a risk analysis model based on bio-inspired approach for information and communication systems and for general future converged networks. There is a great opportunity to nd solutions in biology that can be applied to security issues of information systems. We intend to develop a biologically inspired model for dynamic, adaptive converged information and communication systems to estime a quantitative measure of risk based on this presented approach. This involves proportional hazard models, Cox regression models, angent based models, cross-correlation, life table of assets,adaptive networks, etc. The measure of risk will be useful to evaluate the most appropriate security countermeasures,expected benets of an investment and therefore, to manage the safety, balancing costs, benets and efficiency of a network.

References

1. Leveque, V.: Information Security: A Strategic Approach. IEEE Computer Society, J. Wiley and Sons (2006)
2. Ayoama, T.: A New Generation Network: Beyond the Internet and NGN, Keio University and National institute of Information and Communications Technology, ITU-T KALEIDOSCOPE, IEEE Communications Magazine (2009)
3. VoIP Security Alliance,VoIP Security and Privacy Threat Taxonomy (2010), `http://www.voipsa.org`

4. Keromytis, A.D.: Voice-over-IP Security: Research and Practice, IEEE Computer and Reliability Societies, Secure Systems (2010)
5. Shneier, B.: Architecture of privacy. IEEE Computer Society, Security and Privacy (2009)
6. Quittek, J., Niccolini, S., Tartarelli, S., Schlegel, R.: NEC Europe Ltd, 2008 On Spam over Internet Telephony (SPIT) Prevention IEEE Communications Magazine (2008)
7. Roxbee Cox, D., Oakes, D. (eds.): Analysis of Survival data. Chapman & Hall/CRC (1984)
8. Roxbee Cox, D.: Regression Models and life-tables. Journal of the Royal Society, Series B (Methodological) 34(2) (1972)
9. International Standard ISO/IEC 27002:2005, Information Technology Security tech- niques. Code of Practice for information security management
10. International Standard ISO/IEC 27005:2008, Information Technology Security techniques. Information Security Risk Management
11. Ryan, J.C.H., Ryan, D.J.: Performance Metrics for Information security Risk management. IEEE Computer Society, Security and Privacy (2008)
12. Ryan, J.C.H., Ryan, D.J.: Biological System and models in informa- tion Security. In: Proceedings of the 12th Colloquium for Information System Security Education, University of Texas, Dallas (2008)
13. Ryan, J.C.H., Ryan, D.J.: Expected benefits of information security investments, Computer and Security, ScienceDirect (2006), http://www.sciencedirect.com
14. Kitchovitch, S., Lió, P.: Risk perception and disease spread on social networks. In: International Conference on Computational Science (2010)
15. Lachin, J.M.: Biostatistical Methods: The Assessment of Relative Risks. John Wiley & Sons, NewYork (2000)
16. Kalbeish, J.D., Prentice, R.L.: The Statistical Analysis of Failure-Time Data, 2nd edn. Wiley, Chichester (2002)
17. Murray, W.H.: The application of epidemiology to computer viruses. Computer& Security 7(2) (1988)
18. Dressler, F., Akan, O.B.: A Survey on Bio-Inspired Networking. Elsevier Computer Networks 54(6) (2010)
19. Li, J., Knickerbocker, P.: Functional similarities between computer worms and biological pathogens. Elsevier Computer & Security (2007)
20. Meisel, M., Pappas, V., Zhang, L.: A taxonomy biologically inspired research in computer networking. Elsevier Computer Networks (2010)

Maximization of Network Survivability Considering Degree of Disconnectivity

Frank Yeong-Sung Lin[1], Hong-Hsu Yen[2], and Pei-Yu Chen[1,3,4,*]

[1] Department of Information Management, National Taiwan University
yslin@im.ntu.edu.tw
[2] Department of Information Management, Shih Hsin University
hhyen@cc.shu.edu.tw
[3] Information and Communication Security Technology Center
Taipei, Taiwan, R.O.C.
d96006@im.ntu.edu.tw
[4] Institute Information Industry Taipei, Taiwan, R.O.C.

Abstract. The issues of survivability of networks, especially to some open year round services have increased rapidly over the last few years. To address this topic, the effective survivability metric is mandatory for managerial responsibility. In this paper, we provide a survivability mechanism called Degree of Disconnectivity (DOD) for the network operator to detect risks. To evaluate and analyze the robustness of a network for network operators, this problem is modeled as a mathematical programming problem. An attacker applies his limited attack power intelligently to the targeted network. The objective of the attacker is to compromise nodes, which means to disable the connections of O-D pairs, to achieve the goal of reaching a given level of the proposed Degree of Disconnectivity metric. A Lagrangean Relaxation-based algorithm is adopted to solve the proposed problem.

Keywords: Information System Survivability, Degree of Disconnectivity, Lagrangean Relaxation, Mathematical Programming, Network Attack, Optimization Problem, Resource Allocation.

1 Introduction

With customers' expectations of open year round service, the complete shutdown of an attacked system is not an acceptable option anymore. Nonetheless, over the past decade, malicious and intentional attacks have undergone enormous growth. With the number of attacks on systems increasing, it is highly probable that sooner or later an intrusion into those systems will be successful. As several forms of attacks are aimed at achieving pinpointed destruction to systems, like DDoS, the relatively new paradigms of survivability are becoming crucial. Compared with other metrics, like reliability and average availability, that measure a network, survivability is a network's ability to perform its designated set of functions under all potentially damaging events, such as failure in a network infrastructure component [1].

* Correspondence should be sent to d96006@im.ntu.edu.tw.

B. Murgante et al. (Eds.): ICCSA 2011, Part I, LNCS 6782, pp. 667–676, 2011.

Survivability focuses on preserving essential services in unbounded environments, even when systems in such environments are penetrated and compromised.

A number of papers have studied various theoretical aspects of survivability against antagonistic attacks. In [2], due to the diversified definitions of network survivability, the author categorized the methods to evaluate network survivability into three subcategories: connectivity, performance, and function of other quality or cost measures. The least twenty recognized quality models indicate that the survivability is that users could receive the services that they need without interruption and in a timely manner.

When evaluating survivability, the connectivity of a network to achieve a service level agreement is another vital issue. The definition of network connectivity is the minimum number of links or nodes that must be removed to disconnect an O-D (Original and Destination) pair [3]. In general, the more numbers of links or nodes that must be removed to disconnect an O-D pair, the higher the survivability of the network will be. Thus, there are many researches adopting the network connectivity with the quantitative analysis of network survivability. In [4], the author proposed using the network connectivity to measure the network survivability under intentional attacks and random disasters. In addition, in [5], the author also adopted the network connectivity to do the quantitative analysis of network survivability and the survivability metric is called the Degree of Disconnectivity (DOD).

Many network scenarios have traditionally assumed that the defender, i.e. network operator, only confront a fixed and immutable threat. However, the September 11 attacks in 2001 demonstrated that major threats today involve strategic attackers that can launch a serial action of malicious and intentional attacks, and choose the strategy that maximizes their objective function. The attacker utilizes his knowledge about the target network to formulate his attacking strategy order to inflict maximum damage on a system, a network, or a service under malicious and intentional attacks. Consequently, it is critical for the defender to take into consideration the attacker's strategy when it decides how to allocate its resource among several defensive measures [6].

However, the conflict interaction between the attacker and the defender suggests a need to assume that both of them are fully strategic optimizing agents with their different objectives. A number of papers have studied various theoretical aspects of protecting potential targets against attacks. Some of these papers discuss the scenarios and solved the problem with game theory. Here, in this paper, a conflict network attack-defense scenario is described as a mathematical model to optimize the resource allocation strategies for network operators and is expressed for both attackers and defenders considering the DOD [5].

2 Problem Formulation and Notations

The network scenario discussed in this paper can be seen as a game with attackers and defender entities: attackers represent intelligent or rational entities of the network that may choose a computer and corrupt its database and systems, such as hackers. The defender represents the distributed database administrator, whose goal is to maintain the integrity of the data. Once the database is compromised, the DOD is affected, which is defined as (1). Based on [5], the proposed survivability metric called degree of disconnectivity (DOD) is defined as S, which assesses the average damage level of a network; it can also be called the degree of segmentation, degree of segregation, or degree of separation.

The DOD metric in this paper is defined as S, which is evaluated on the disconnected numbers of O-D pairs among all O-D pairs. DOD can be generated as the residual index of the networks. In this case, t_{wi} is 1, while node i on an O-D pair w is dysfunctional. The Original node here is source node, and the destination node is the database. The transmission cost of dysfunctional node is M, otherwise it is ε. The greater the value of S, the more the network is damaged.

$$S = \frac{\sum_{w \in W} \sum_{i \in V} t_{wi} c_i}{C_2^N \times M} . \tag{1}$$

Because the attacker's resources, i.e. time, money, and man power, are limited, only part of a network can be compromised. Therefore, the resources must be fully utilized so that the attacker can cause the maximum harm to the target network. In order to discuss the worst case scenario, the concept of [6], which assumes complete information and perfect perception on behalf of both attackers and defenders, is adopted. Hence, both the attacker and the defender have complete information about the targeted network topology and the budget allocation is assumed. The serial of attack actions considering the survivability of a network is then modeled as an optimization problem, in which the objective is to minimize the total attack cost from an attacker's perspective, such that the given critical O-D pair is disconnected and the survivability is over the given threshold resulting in the inability of the network to survive. Note that the network discussed here is at the AS level.

The above problem is formulated as a mathematical model as follows. For simplicity, since the targeted network is at the AS level, the attacker cannot simply attack any node directly. The notations used in this paper and problem formulation is defined in Table 1.

Table 1. Given Parameters and Decision Variables

Given parameter Notation	Description
V	Index set of nodes
W	Index set of OD pairs
P_w	Set of all candidate paths of an OD pair w, where $w \in W$
M	Large amount of processing cost that indicates a node has been compromised
ε	Small amount of cost processing cost that indicates a node is functional
δ_{pi}	Indicator function, 1 if node i is on path p, 0 otherwise, where $i \in V$ and $p \in P_w$
$\hat{a}_i$	The threshold of attack cost leading to a successful node attack
S	The threshold of a network crash, which is the average damage level of all O-D pairs
R_w	The weight of O-D pair w, where $w \in W$
$\hat{a}_i$	The threshold of attack cost leading to a successful node attack

Table 1. *(continued)*

Decision variable Notation	Description
x_p	1 if path p is chosen, 0 otherwise, where $p\in P_w$
y_i	1 if node i is compromised by attacker, 0 otherwise (where $i\in V$)
t_{wi}	1 if node i is used by OD pair w, 0 otherwise, where $i\in V$ and $w\in W$
c_i	Processing cost of node i, which is ε if i is functional, M if i is compromised by attacker, where $i\in V$

The problem is then formulated as the following minimization problem:

$$\min_{y_i} \sum_{i\in V} y_i \hat{a}_i \ , \qquad \text{(IP 1)}$$

Subject to:

$$c_i = y_i M + (1-y_i)\varepsilon \qquad \forall i\in V \qquad \text{(IP 1.1)}$$

$$\sum_{i\in V} t_{wi} c_i \le \sum_{i\in V} \delta_{pi} c_i \qquad \forall\, p\in P_w,\ w\in W \qquad \text{(IP 1.2)}$$

$$\sum_{p\in P_w} x_p \delta_{pi} = t_{wi} \qquad \forall i\in V,\ w\in W \qquad \text{(IP 1.3)}$$

$$S \le \frac{\sum_{w\in W} R_w \sum_{i\in V} t_{wi} c_i}{|W| \times M} \qquad \text{(IP 1.4)}$$

$$\sum_{p\in P_w} x_p = 1 \qquad \forall w\in W \qquad \text{(IP 1.5)}$$

$$x_p = 0 \text{ or } 1 \qquad \forall p\in P_w,\ w\in W \qquad \text{(IP 1.6)}$$

$$y_i = 0 \text{ or } 1 \qquad \forall i\in V \qquad \text{(IP 1.7)}$$

$$t_{wi} = 0 \text{ or } 1 \qquad \forall\, i\in V,\ w\in W. \qquad \text{(IP 1.8)}$$

The objective of the formulation is to minimize the total attack cost by of the attacker by deciding which node to compromise. Constraint (IP 1.1) describes the definition of the transmission cost of node i, which is ε if node i is functional, and M if node i is compromised. Constraint (IP 1.2) requires that the selected path for an O-D pair w should be the minimal cost path. Constraint (IP 1.3) denotes the relationship between t_{wi} and $x_p\delta_{pi}$. To simplify the problem-solving procedure, the auxiliary set of decision variables t_{wi} is replaced by the sum of all $x_p\delta_{pi}$. (IP 1.1) to (IP 1.3) jointly require that, when a node is chosen for attack, there must be exactly one path from the attacker's initial position, s, to that node, and each node on the path must have been compromised. These constraints are jointly described as the continuity constraints. And constraint (IP 1.4) determines that if a target network has been compromised, the DOD metrics must be larger than the given threshold. Constraints (IP 1.5) and (IP 1.6) jointly entail that only one of the candidate paths of an OD pair w can be selected. Lastly, constraints (IP 1.6) to (IP 1.8) impose binary restrictions on decision variables.

3 Solution Approach

3.1 Solution Approach for Solving the Problem of (IP 1)

3.1.1 Lagrangean Relaxation

Lagrangean Relaxation (LR) [7, 8] has been very useful in conjunction with branch and bound, which serves as the basis for the development of heuristics (dual ascent) and variable fixing. This approach is composed by sets of constraints that are relaxed and dualized by adding them to the objective function with penalty coefficients, the Lagrangian multipliers. The objective, in the relaxation, is to dualize, possibly after a certain amount of remodeling, which is then transformed into disconnected and easier to solve subproblems. In such a way, these subproblems obtain bounds on the actual integer optimal value, and separate solutions to the individual subproblems which, while not necessarily consistent because they may violate some of the linking constraints, might however suggest ways of constructing good globally feasible solutions.

By applying this method with a vector of Lagrangean multipliers u^1, u^2, u^3, and u^3, the model can be transformed into the following Lagrangean relaxation problem (LR 1). In this case, constraints (1-1) to (1-4) are relaxed, and dualized by adding them to the objective function with penalty coefficients, the Lagrangian multipliers, which are defined as u^1, u^2, u^3, and u^4 with the vectors of $\{u_i^1\}$, $\{u_{wp}^2\}$,$\{u_{wi}^3\}$,$\{u^4\}$ respectively. The objective, in this case, in the relaxation, is to dualize, possibly after a certain amount of remodeling, the constraints linking the component together in such a way that the original problem is transformed into disconnected and easier to solve subproblems. Here, (LR 1) is decomposed into three independent and easily solvable optimization subproblems with respect to decision variables x_p, y_i, and t_{wi}, c_i; the respective subproblems can thus be optimally solved.

Subproblem 1.1 (related to decision variable x_p):

$$Z_{Sub1}(u^3) = \min \sum_{w\in W}\sum_{i\in V}\sum_{p\in P_w} u_{wi}^3 \delta_{pi} x_p , \tag{Sub 1.1}$$

Subject to:

$$\sum_{p\in P_w} x_p = 1 \qquad \forall w\in W \tag{IP 1.5}$$

$$x_p = 0 \text{ or } 1 \qquad \forall p\in P_w,\ w\in W \tag{IP 1.6}$$

To reduce the complexity, subproblem 1.1 is decomposed into $|W|$ problems, which are all independent shortest path problems. The value of x_p for each O-D pair w is individually determined. Hence, u_{wi}^3 can be viewed as the cost of node i on O-D pair w. Dijkstra's algorithm is adopted to obtain x_p for each O-D pair w. The time complexity of Dijkstra's algorithm is $O(|V|^2)$, where $|V|$ is the number of nodes; therefore, the time complexity of subproblem 1 is $O(|W|\times|V|)$.

Subproblem 1.2 (related to decision variable y_i):

$$Z_{Sub2}(u^1) = \min \sum_{i\in V} y_i \hat{a}_i(b_i) + \sum_{i\in V} u_i^1 y_i \varepsilon + \sum_{i\in V} u_i^1 y_i(-M) + u_i^1 \varepsilon$$
$$= \min \sum_{i\in V} \left[\hat{a}_i(b_i) + u_i^1 \varepsilon + u_i^1(-M) \right] y_i + u_i^1 \varepsilon, \qquad \text{(Sub 1.2)}$$

Subject to:

$y_i = 0 \text{ or } 1$ $\forall i \in V$ (IP 1.7)

To solve subproblem 1.2 optimally, this problem can also be decomposed into $/V/$ individual problems. The value of decision variable y_i is determined by its coefficient, whose value is $\hat{a}_i(b_i) + u_i^1 \varepsilon + u_i^1(-M)$. In order to minimize subproblem 2, if this coefficient is positive, y_i is set as zero; otherwise it is one. The time complexity of subproblem 2 is $O(/V/)$.

Subproblem 1.3 (related to decision variable t_{wi} and c_i):

$$Z_{Sub3}(u^1, u^2, u^3, u^4)$$
$$= \min \sum_{i\in V} u_i^1 c_i + \sum_{w\in W} \sum_{p\in P_w} u_{wp}^2 \sum_{i\in V} t_{wi} c_i + \sum_{w\in W} \sum_{p\in P_w} u_{wp}^2 \sum_{i\in V} (-\delta_{pi} c_i)$$
$$+ \sum_{w\in W} \sum_{i\in V} u_{wi}^3 (-t_{wi}) + u^4 (-\sum_{w\in W} R_w \sum_{i\in V} t_{wi} c_i) + u^4 S|W|M \qquad \text{(Sub 1.3)}$$
$$= \min \sum_{i\in V} \left\{ \left[u_i^1 - \sum_{w\in W} \sum_{p\in P_w} u_{wp}^2 \delta_{pi} + \sum_{w\in W} \left((\sum_{p\in P_w} u_{wp}^2) - u^4 R_w \right) t_{wi} \right] c_i - \sum_{w\in W} u_{wi}^3 t_{wi} \right\}$$
$$+ u^4 S|W|M,$$

Subject to:

$t_{wi} = 0 \text{ or } 1$ $\forall i \in V, w \in W$ (IP 1.8)

$c_i = \varepsilon \text{ or } M$ $\forall i \in V.$ (Sub 1.3.2)

To optimally solve subproblem 1.3, it is further decomposed it into $/V/$ independent subproblems. However, since each decision variable t_{wi} and c_i in (LR 4) and (LR 5) have only two kinds of value, the exhaustive search is applied here to find the optimal objective function value among the four combinations of t_{wi} and c_i. The time complexity of subproblem 1.3 is $O(|V| \times |W|)$.

These relaxed problems are solved optimally to get a lower bound for the primal problem. After solving (LR 1), the resulting bounds are taken as the initial bounds in the next stage. Three stage heuristics are adopted to derive feasible solutions to the primal problem, and the subgradient method is used to update the Lagrangean multipliers to obtain a better bound.

3.1.2 Getting Primal Feasible Solutions

To obtain the primal feasible solutions of (IP 1), the solutions obtained from (LR) are considered. By using the Lagrangean relaxation method and the subgradient method, it is possible to get the tightest possible bound on the optimal value. The *Zn* auxiliary problem consisting in optimizing the bound over all possible values of the multipliers is thus solved. This theoretical lower bound on the primal objective function value, as well as ample hints for getting primal feasible solutions, is obtained. However, as some critical and difficult constraints are relaxed to obtain the (LR) problem, the solutions may not be valid for the primal problem. Thus, there is the need to develop heuristics to tune the values of the decision variables so that primal feasible solutions can be obtained. As a result, a heuristic is adopted to improve this situation. In this heuristic, each solution to (LR) is adjusted to a feasible solution to (IP 1).

The concept of this heuristic arises from the attacker's strategy. Given that the node was traversed several times, the attacker would have a higher possibility of attacking it. Hence, the compromised nodes are separated in the *Attack-Bucket*, while the rest nodes are in the *Safety-Bucket*. The nodes in both buckets are separately sorted in descending order by their attacked frequencies. First, select nodes with most frequently from the Safety-Bucket to transfer to the Attacked-Bucket. Then adjust the nodes transferred to the Attacked-Bucket from the Safety-Bucket. In this manner, a heuristic for getting a primal feasible solution is developed. The time complexity for this heuristics is $O(|V|)$.

4 Computational Experiments

4.1 Experiment Environment

The proposed algorithms for the DOD model are coded in Visual C++ and run on a PC with an INTEL™ Core2 CPU 6400 2.13 GHz CPU. Two types of network topology, grid and scale-free networks, as attack targets are demonstrated here. The network size here is under 9, 16, 25, 16 nodes. The parameters used in the experiments are detailed as below.

Table 2. Experiment Parameter Settings

Parameters	Value
Network Topology	Grid (square), Scale-free
Number of Nodes $\|N\|$	9, 16, 25, 36
Total Defense Budget	Equal to Number of Nodes
No. of O-D pairs $\|W\|$	72, 240, 600, 1260
Degree of Disconnectivity (S)	80%, 60%, 40%, 20%
Defense Capability $\hat{a}_i(b_i)$	$\hat{a}_i(b_i) = 0.5b_i + \varepsilon$, b_i

4.2 Computational Experiment of (IP 1)

To demonstrate the effectiveness of the proposed heuristics, we implement one algorithm, Degree-based Attack Algorithm (DAA) for comparison purposes. The details are described in Table 3. The concept of the DAA is derived from the heuristic of stage_1 of the Three-Stage Heuristic for getting a primal feasible solution.

Table 3. Degree-based Attack Algorithm (DAA)

```
1. //Initialization
2. SumOfRTC= $\sum_{w\in W} R_w \sum_{i\in V} t_{wi} c_i$
3. Threshold= (S×|W|×M);
// Stage 1: MAX(Node_ Degree) in Safety-Bucket
4. WHILE (SumOfRTC < Threshold AND unfinished==TRUE ){
5.     //Find the node i among the Safety-Bucket
6.     FIND node i, whose degree is maximal;
7.       SET node i to attack; // switch node i to Attack-
Bucket;
8.IF (all the nodes' is in the Attack-Bucket){
9.unfinished==FALSE;
10. }
11. ELSE{
12. unfinished==TRUE;
13. }
14. RUN Dijkstra then to calculate the SumOfRTC;
15.}//end of while
```

4.3 Experiment Result of (IP 1)

To compare attack behavior under different scenarios, we use the attackers' attack cost to evaluate the degree to which the attacker's objective is achieved. The greater the attack cost, the more robust the network. The LR value means the attack cost is calculated by the optimal feasible solution derived from the Lagrangean Relaxation process. The experiment results under different topology types [10], numbers of nodes, and damage distribution patterns are shown in Table 4-4.

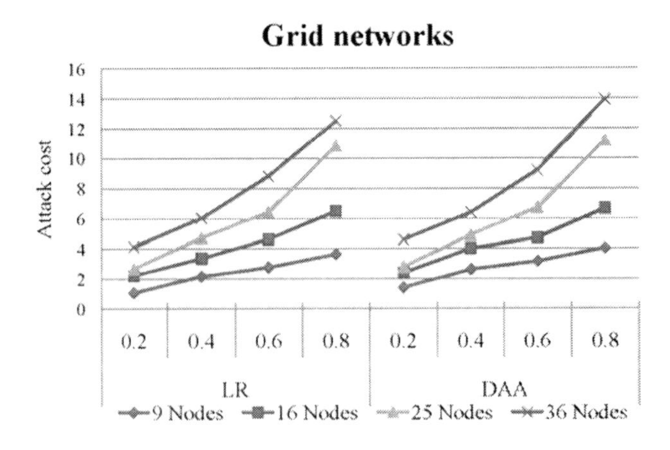

Fig. 1. The comparison of LR and DAA under grid networks

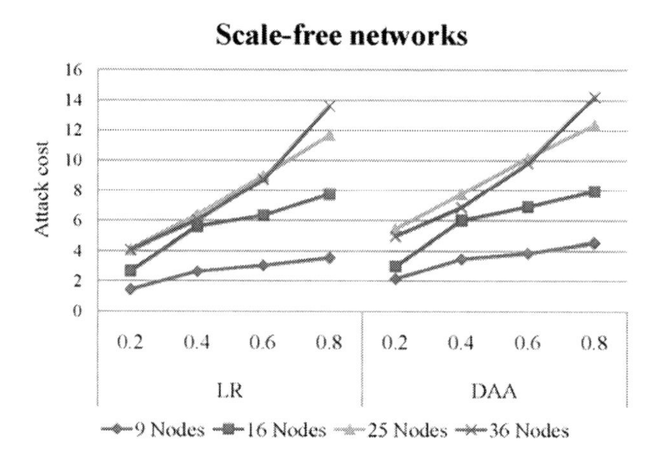

Fig. 2. The comparison of LR and DAA under scale-free networks

As Fig. 1 and Fig. 2 show, among these figures, the cross axle is the Attack cost. Each point on the chart represents the DOD value under degree-based initial budget allocation strategy. The vertical axle is the given threshold of DOD under the given network topologies and budget allocations. Compared to the solution quality of the proposed Lagrangean Relaxation-based algorithm (LR) with DAA, the LR obtained the lowest attack costs. The proposed heuristic outperforms in all cases, which always causes the lowest network survivability (the highest DOD value) in all network topologies and sizes. The attackers' resources are utilized and generalized by the solution derived from LR among various types of network topology. Meanwhile, the survivability of grid networks is lower than others using the DOD metric, since grid networks are more regular and more connected. As a result, some nodes are used more often by OD pairs. If these nodes are compromised by the attacker, the DOD value increases.

5 Conclusions

In this paper, the attack and defense scenarios consider the DOD metric to describe attacker and defender behavior of networks by simulating the role of the defender and the attacker. The attacker tries to maximize network damage by compromising nodes in the network, whereas the defender's goal is to minimize the impact by deploying defense resources to nodes and enhance their defense capability. In this context, the DOD is used to measure the damage of the network. The problem is solved by a Lagrangean Relaxation-based algorithm, and the solution to the problem is obtained from the subgradient-like heuristic and budget adjustment algorithm.

The main contribution of this research is the generic mathematical model for solving the network attack-defense problem, which is modeled as a generic mathematical model. The model is then optimally solved by the proposed heuristic. With this mathematical technique, the complex problems based on the optimized methodology is resolved.

The novel network DOD reflects the aim of an attacker to separate the target network into pieces. This metric enables the indication of the damage of the residual networks. Also, we have examined the survivability of networks with different topologies, sizes and budget allocation policies. Their survivability can be significantly improved by adjusting the defense budget allocation. From the outcomes of the experiments, we can conclude that the defense resources should be allocated according to the importance of nodes.

The current research considers a one-round scenario in which both actors deploy their best strategies under the given topology, but it would be more comprehensive if the scenario is demonstrated in multi-rounds. Both actors may not exhaustively distribute their resources in single round. Moreover, the defenders could deploy some false targets, i.e. honeypots, to attract attackers to waste their budgets. This more complex attacker behavior, therefore, should be considered in further research.

Acknowledgments. This research was supported by the National Science Council of Taiwan, Republic of China, under grant NSC-99-2221-E-002-132.

References

1. Snow, A.P., Varshney, U., Malloy, A.D.: Reliability and Survivability of Wireless and Mobile Networks. IEEE Computer 33(7), 449–454 (2000)
2. Ellison, R.J., Fisher, D.A., Linger, R.C., Lipson, H.F., Longstaff, T., Mead, N.R.: Survivable Network Systems: An Emerging Discipline, Technical Report CMU/SEI-97-TR-013 (November 1997)
3. Westmark, V.R.: A Definition for Information System Survivability. In: Proceedings of the 37th Hawaii International Conference on System Sciences (2004)
4. Al-Kofahi, O.M., Kamal, A.E.: Survivability Strategies in Multihop Wireless Networks. IEEE Wireless Communications (2010)
5. Bier, V.M., Oliveros, S., Samuelson, L.: Choosing What to Protect: Strategic Defense Allocation Against an Unknown Attacker. Journal of Public Economic Theory 9, 563–587 (2007)
6. Powell, R.: Defending Against Terrorist Attacks with Limited Resources. American Political Science Review 101(3), 527–541 (2007)
7. Fisher, M.L.: An Applications Oriented Guide to Lagrangian Relaxation. Interfaces 15(2), 10–21 (1985)
8. Ahuja, R.K., Magnanti, T.L., Orlin, J.B.: Network Flows: Theory, Algorithms, and Applications: Chapter 16 Lagrangian Relaxation and Network Optimization, pp. 598–639. Prentice-Hall, Englewood Cliffs (1993)

Application of the GFDM for Dynamic Analysis of Plates

Francisco Ureña*, Luis Gavete, Juan José Benito, and Eduardo Salete

francisco.urena@uclm.es,
lu.gavete@upm.es,
jbenito@ind.uned.es,
esalete@ind.uned.es

Abstract. This paper shows the application of generalized finite difference method (GFDM) to the problem of dynamic analysis of plates. We investigated stability and its relation with the irregularity of a cloud of nodes.

Keywords: meshless methods, generalized finite difference method, moving least squares, plates, stability.

1 Introduction

The rapid development of computer technology has allowed the use several methods to solve partial differential equations(12,13,14). The numerical method of lines (NML) discretizes the PDE with respect to only one variable (usually space) preserving the continuous differential with respect to time. This method can be applied to the control of the parabolic partial differential equation and the dynamic analysis of plates (15,16).

The Generalized finite difference method (GFDM) is evolved from classical finite difference method (FDM). GFDM can be applied over general or irregular clouds of points (6). The basic idea is to use moving least squares (MLS) approximation to obtain explicit difference formulae which can be included in the partial differential equations (8). Benito, Ureña and Gavete have made interesting contributions to the development of this method (1,2,4,5,7). The paper (3) shows the application of the GFDM in solving parabolic and hyperbolic equations.

This paper decribes how the GFDM can be applied for solving dynamic analysis problems of plates.

The paper is organized as follows. Section 1 is the introduction. Section 2 describes the explicit generalized finite difference schemes. In section 3 is studied the consistency and von Neumann stability. In Section 4 is analyzed the relation between stability and irregularity of a cloud of nodes. In Section 5 some applications of the GFDM for solving problems of dynamic analysis are included. Finally, in Section 6 some conclusions are given.

* Universidad de Castilla La Mancha, Spain.

B. Murgante et al. (Eds.): ICCSA 2011, Part I, LNCS 6782, pp. 677–689, 2011.

2 The Generalized Finite Difference Method

Let us to consider the problem governed by:

$$\frac{\partial^2 U(x,y,t)}{\partial t^2} + A^2[\frac{\partial^4 U(x,y,t)}{\partial x^4} + 2\frac{\partial^4 U(x,y,t)}{\partial x^2 \partial^2} + \frac{\partial^4 U(x,y,t)}{\partial y^4}] = F(x,y,t)$$
$$(x,y) \in (0,1) \times (0,1), \quad t > 0 . \quad (1)$$

with boundary conditions

$$\begin{cases} U(x,y,t)|_{\Gamma} = 0 \\ \frac{\partial^2 U(x,y,t)}{\partial y^2}|_{(0,y,t)} = \frac{\partial^2 U(x,y,t)}{\partial y^2}|_{(1,y,t)} = 0, \forall y \in [0,1] \\ \frac{\partial^2 U(x,y,t)}{\partial x^2}|_{(x,0,t)} = \frac{\partial^2 U(x,y,t)}{\partial x^2}|_{(x,1,t)} = 0, \forall x \in [0,1] \end{cases} . \quad (2)$$

and initial conditions

$$U(x,y,0) = 0; \quad \frac{\partial U(x,y,t)}{\partial t}|_{(x,y,0)} = G(x,y) . \quad (3)$$

where F, G are two known smoothed functions and the constant A depends of the material and geometry of the problem.

2.1 Explicit Generalized Differences Schemes

The intention is to obtain explicit linear expressions for the approximation of partial derivatives in the points of the domain. First of all, an irregular grid or cloud of points is generated in the domain. On defining the composition central node with a set of N points surrounding it (henceforth referred as nodes), the star then refers to the group of established nodes in relation to a central node. Each node in the domain have an associated star assigned.

If u_0 is an approximation of fourth-order for the value of the function at the central node (U_0) of the star, with coordinates (x_0, y_0) and u_j is an approximation of fourth-order for the value of the function at the rest of nodes, of coordinates (x_j, y_j) with $j = 1, \cdots, N$.

Firstly, we use the explicit difference formulae for the values of partial derivatives in the space variables. On including the explicit expressions for the values of the partial derivatives the star equation is obtained as

$$[\frac{\partial^4 U(x,y,t)}{\partial x^4} + 2\frac{\partial^4 U(x,y,t)}{\partial x^2 \partial^2} + \frac{\partial^4 U(x,y,t)}{\partial y^4}]_{(x_0,y_0,n)} = m_0 u_0 + \sum_{j=1}^{N} m_j u_j . \quad (4)$$

with

$$m_0 + \sum_{j=1}^{N} m_j = 0 . \quad (5)$$

Secondly, we shall use an explicit formula for the part of the equation 1 that depends on time. This explicit formula can be used to solve the Cauchy initial value problem. This method involves only one grid point at the advanced time level. The second derivative with respect to time is approached by

$$\frac{\partial^2 U}{\partial t^2}|(x_0, y_0, n) = \frac{u_0^{n+1} - 2u_0^n + u_0^{n-1}}{(\triangle t)^2} \ . \tag{6}$$

If the equations 5 and 6 are substituted in equation 1 the following recursive relationship is obtained

$$u_0^{n+1} = 2u_0^n - u_0^{n-1} + A^2(\triangle t)^2[m_0 u_0^n + \sum_{j=1}^{N} m_j u_j^n] + F(x_0, y_0, n\triangle t) \ . \tag{7}$$

The first derivative with respect to the time is approached by the central difference formula.

3 Stability Criterion

For the stability analysis the first idea is to make a harmonic decomposition of the approximated solution at grid points and at a given time level n. Then we can write the finite difference approximation in the nodes of the star at time n, as

$$u_0^n = \xi^n e^{i\boldsymbol{\nu}^T \boldsymbol{x_0}}; \quad u_j^n = \xi^n e^{i\boldsymbol{\nu}^T \boldsymbol{x_j}} \ . \tag{8}$$

where: ξ is the amplification factor,

$$\boldsymbol{x_j} = \boldsymbol{x_0} + \boldsymbol{h_j}; \qquad \xi = e^{-iw\triangle t}$$

$\boldsymbol{\nu}$ is the column vector of the wave numbers

$$\boldsymbol{\nu} = \left\{ \begin{array}{c} \nu_x \\ \nu_y \end{array} \right\} = \nu \left\{ \begin{array}{c} \cos\varphi \\ \sin\varphi \end{array} \right\}$$

then we can write the stability condition as: $\|\xi\| \leq 1$.

Including the equation 7 into the equation 6, cancelation of $\xi^n e^{i\boldsymbol{\nu}^T \boldsymbol{x_0}}$, leads to

$$\xi = 2 + \frac{1}{\xi} - (\triangle t)^2 A^2 (m_0 + \sum_{1}^{N} m_j e^{i\boldsymbol{\nu}^T \boldsymbol{h_j}}) \ . \tag{9}$$

Using the expression 5 and after some calculus we obtain the quadratic equation

$$\xi^2 - \xi[2 + A^2(\triangle t)^2(\sum_{1}^{N} m_j(1 - \cos\boldsymbol{\nu}^T\boldsymbol{h_j}) - i\sum_{1}^{N} m_j \sin\boldsymbol{\nu}^T\boldsymbol{h_j})] + 1 = 0 \tag{10}$$

Hence the values of are

$$\xi = b \pm \sqrt{b^2 - 1} \ . \tag{11}$$

where

$$b = 1 + \frac{A^2(\triangle t)^2}{2} \sum_1^N m_j (1 - \cos \boldsymbol{\nu}^T \boldsymbol{h_j}) - i \frac{A^2(\triangle t)^2}{2} \sum_1^N m_j \sin \boldsymbol{\nu}^T \boldsymbol{h_j} \ . \tag{12}$$

If we consider now the condition for stability, we obtain

$$\|b \pm \sqrt{b^2 - 1}\| \leq 1 \ . \tag{13}$$

Operating with the equations 12 and 13, cancelling with conservative criteria, the condition for stability of star is obtained

$$\triangle t \leq \frac{1}{4A\sqrt{m_0}} \ . \tag{14}$$

4 Irregularity of the Star (IIS) and Stability

In this section we are going to define the index of irregularity of a star (IIS) and also the index of irregularity of a cloud of nodes (IIC).

The coefficient m_0 is function of:

- The number of nodes in the star
- The coordinates of each star node referred to the central node of the star
- The weighting function (see references $(1, 4)$)

If the number of nodes by star and the weighting function are fixed, then the equation 14 is function of the coordinates of each node of star referred to its central node.

Denoting τ_0 as the average of the distances between of the nodes of the star and its central node with coordinates (x_0, y_0) and denoting τ the average of the τ_0 values in the stars of the cloud of nodes, then

$$\overline{m_0} = m_0 \tau^4 \ . \tag{15}$$

The stability criterion can be rewritten as

$$\triangle t < \frac{\tau^2}{4a\sqrt{|\overline{m_0}|}} \ . \tag{16}$$

For the regular mesh case, the inequality 14 is for the cases of one and two dimensions as follows

$$\triangle t < \frac{9\tau^2}{A\sqrt{13}[3(1+\sqrt{2}) + 2\sqrt{5}]^2} \quad if \quad N = 24 \ . \tag{17}$$

Multiplying the right-hand side of inequalities 17, respectively, by the factors

$$\frac{\sqrt{13}[3(1+\sqrt{2})+2\sqrt{5}]^2}{36\sqrt{|\overline{m_0}|}} \quad if \quad N=24\,. \tag{18}$$

the inequality 16 is obtained.

For each one of the stars of the cloud of nodes, we define the IIS for a star with central node in (x_0, y_0) as Eq. 18

$$IIS_{(x_0,y_0)} = \frac{\sqrt{13}[3(1+\sqrt{2})+2\sqrt{5}]^2}{36\sqrt{|\overline{m_0}|}} \quad if \quad N=24\,. \tag{19}$$

that takes the value of one in the case of a regular mesh and $0 < IIS \leq 1$.

If the index IIS decreases, then absolute values of $\overline{m_0}$ increases and then according with Eq. 14, $\triangle t$ decreases.

The irregularity index of a cloud of nodes (IIC) is defined as the minimum of all the IIS of the stars of a cloud of nodes.

5 Numerical Results

This section provides some of the numerical results of the applications of GFDM for dynamic analysis of plates (consider a rectangular plate. 1×1), using the weighting function

$$\Omega(h_j,k_j) = \frac{1}{(\sqrt{h_j^2+k_j^2})^3}\,. \tag{20}$$

The global exact error can be calculated as

$$Global \quad exact \quad error = \sqrt{\frac{\sum_{i=1}^{N} e_i^2}{N}}\,. \tag{21}$$

where N is the number of nodes in the domain, e_i is the exact error in the node i.

5.1 Free Vibrations of a Simply Supported Plate

The equation is

$$\frac{\partial^2 U(x,y,t)}{\partial t^2} + \frac{1}{4\pi^4}[\frac{\partial^4 U(x,y,t)}{\partial x^4} + 2\frac{\partial^4 U(x,y,t)}{\partial x^2 \partial^2} + \frac{\partial^4 U(x,y,t)}{\partial y^4}] = 0$$
$$(x,y) \in (0,1)\times(0,1), \quad t>0\,. \tag{22}$$

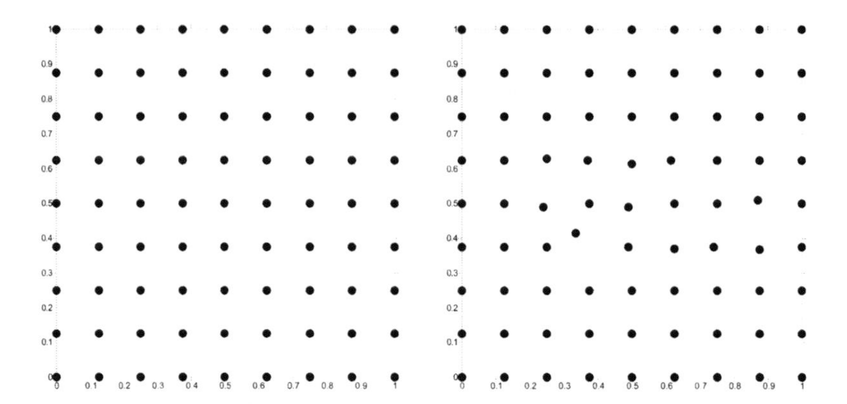

Fig. 1. Regular and irregular mesh

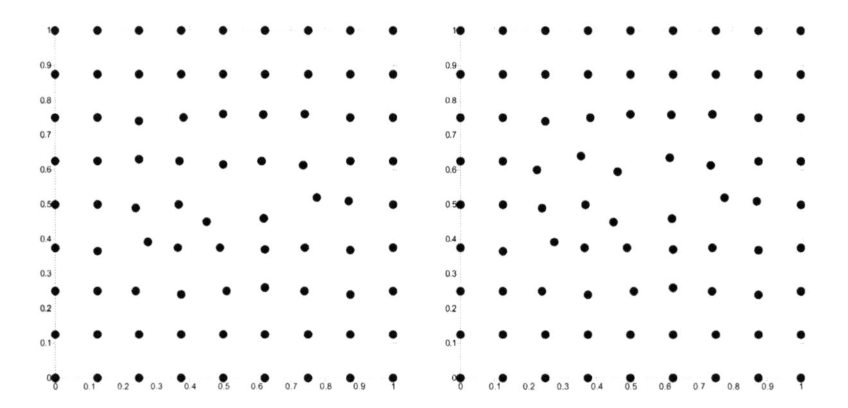

Fig. 2. Irregular meshes

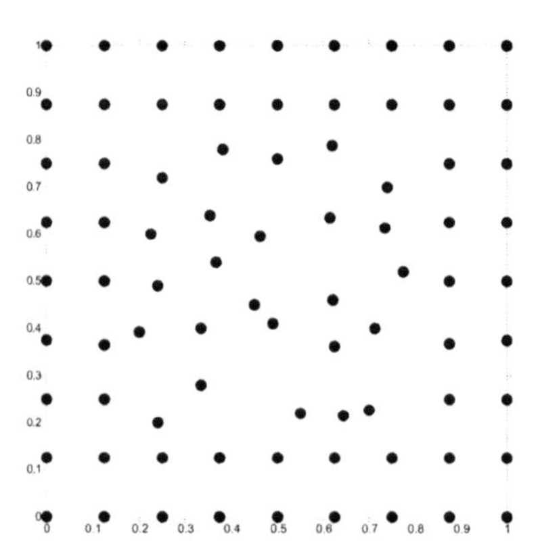

Fig. 3. Irregular mesh

with boundary conditions

$$\begin{cases} U(x,y,t)|_{\Gamma} = 0 \\ \frac{\partial^2 U(x,y,t)}{\partial y^2}|_{(0,y,t)} = \frac{\partial^2 U(x,y,t)}{\partial y^2}|_{(1,y,t)} = 0, \forall y \in [0,1] \\ \frac{\partial^2 U(x,y,t)}{\partial x^2}|_{(x,0,t)} = \frac{\partial^2 U(x,y,t)}{\partial x^2}|_{(x,1,t)} = 0, \forall x \in [0,1] \end{cases} . \quad (23)$$

and initial conditions

$$U(x,y,0) = 0; \qquad \frac{\partial U(x,y,t)}{\partial t}|_{(x,y,0)} = \sin(\pi x)\sin(\pi y) . \quad (24)$$

Table 1 shows the results of the global error, using a regular mesh of 81 nodes (figure 1), for several values of $\triangle t$.

Table 1. Influence of $\triangle t$ in the global error

$\triangle t$	% error
0.01	0.009298
0.005	0.002319
0.003	0.000834
0.001	0.000093

Table 2. Influence of irregularity of mesh in the global error

IIC	% error
0.92	0.002315
0.83	0.002547
0.76	0.002735
0.58	0.002770

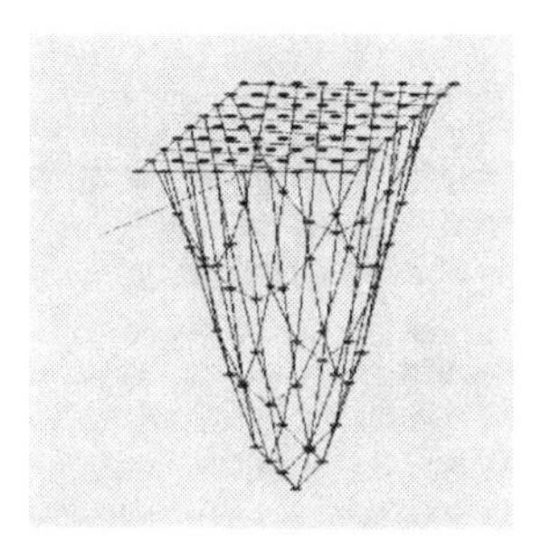

Fig. 4. Approximated solution in the last time step

Table 2 shows the results of global error with $\triangle t = 0.005$ for several irregular meshes of 81 nodes (figures 1, 2 and 3).

Figure 4 shows the solution of the equation 22 in the last time step ($n = 2000$). As new initial conditions let us assume that due to impact an initial velocity is given to a point ($x = y = 0.5$) of the plate, which give the conditions

$$U(x,y,0) = 0; \quad \begin{cases} \frac{\partial U(x,y,t)}{\partial t}|_{(x,y,0)} = 1 \; if \quad x = y = 0.5 \\ \frac{\partial U(x,y,t)}{\partial t}|_{(x,y,0)} = 0 \; if \; (x,y) \neq (0.5, 0.5) \end{cases} . \tag{25}$$

The exact solution is given by

$$U(x,y,t) = 2[\sin(\pi x)\sin(\pi y)\sin(t) - \frac{1}{9}\sin(3\pi x)\sin(3\pi y)\sin(9t) + \frac{1}{25}\sin(5\pi x)\sin(5\pi y)\sin(25t) - \cdots] . \tag{26}$$

Table 3 shows the results of the global error, using a regular mesh of 81 nodes (figure 1) and $\triangle t = 0.001$, versus the number of time steps (n).

Table 3. Variation of global error versus the number of time steps

n	Global error
100	0.01122
200	0.01858
600	0.02690
1200	0.03363

Figures 5 and 6 show the approximated solution of the equation 22 with the initial conditions 25 in the last time steps for the cases $n = 100$, $n = 200$, $n = 600$ and $n = 1200$ time steps respectively.

5.2 Forced Vibrations of a Simply Supported Plate

The equation is

$$\frac{\partial^2 U(x,y,t)}{\partial t^2} + \frac{1}{4\pi^4}[\frac{\partial^4 U(x,y,t)}{\partial x^4} + 2\frac{\partial^4 U(x,y,t)}{\partial x^2 \partial^2} + \frac{\partial^4 U(x,y,t)}{\partial y^4}] = 15 \sin t \sin(2\pi x)\sin(2\pi y)) \quad (x,y) \in (0,1) \times (0,1), \quad t > 0 . \tag{27}$$

with boundary conditions

$$\begin{cases} U(x,y,t)|_{\Gamma} = 0 \\ \frac{\partial^2 U(x,y,t)}{\partial y^2}|_{(0,y,t)} = \frac{\partial^2 U(x,y,t)}{\partial y^2}|_{(1,y,t)} = 0, \forall y \in [0,1] \\ \frac{\partial^2 U(x,y,t)}{\partial x^2}|_{(x,0,t)} = \frac{\partial^2 U(x,y,t)}{\partial x^2}|_{(x,1,t)} = 0, \forall x \in [0,1] \end{cases} . \tag{28}$$

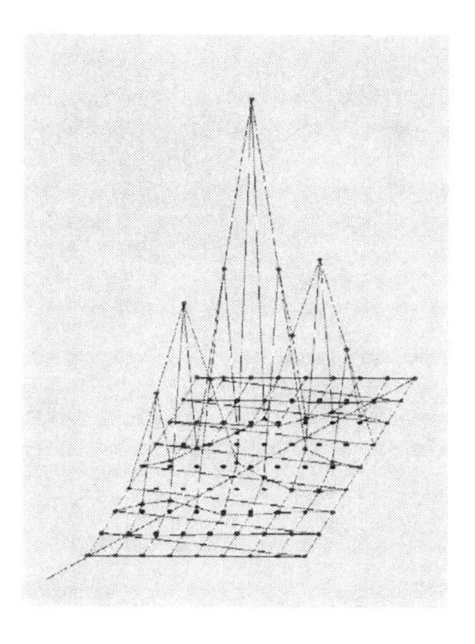

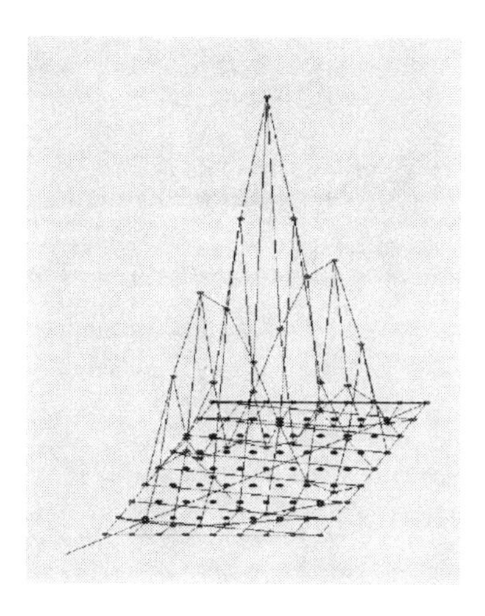

Fig. 5. Approximated solution with n=100 Approximated solution with n=200

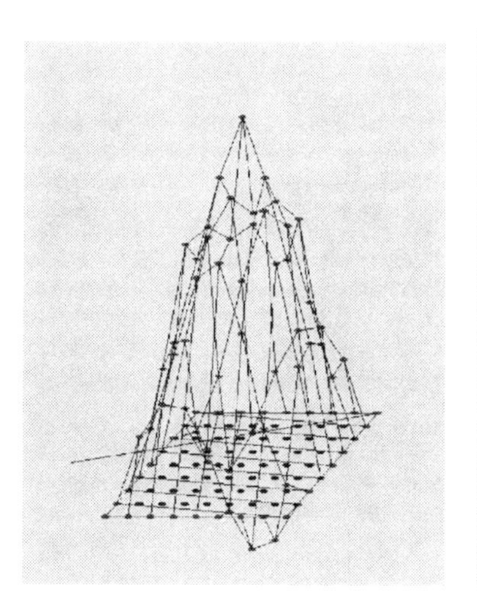

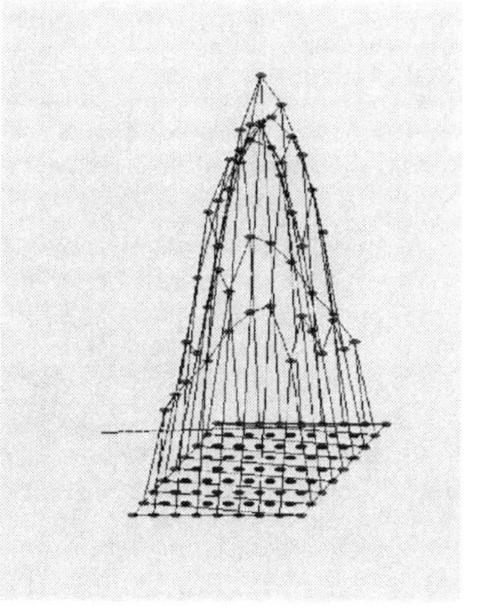

Fig. 6. Approximated solution with n=600 Approximated solution with n=1200

and initial conditions

$$U(x,y,0) = 0; \quad \frac{\partial U(x,y,t)}{\partial t}|_{(x,y,0)} = \sin(\pi x)\sin(\pi y) + \sin(2\pi x)\sin(2\pi y) \; . \quad (29)$$

Table 4 shows the results of the global error, using regular mesh of 81 nodes (figure 1), for several values of $\triangle t$.

Table 5 shows the results of global error with $\triangle t = 0.005$ for several irregular meshes of 81 nodes (figures 1, 2 and 3).

Table 4. Influence of $\triangle t$ in the global error

$\triangle t$	% error
0.01	0.332700
0.005	0.084030
0.003	0.030310
0.001	0.003372

Table 5. Influence of $\triangle t$ in the global error

IIC	% error
0.92	0.08403
0.83	0.08427
0.76	0.08565
0.58	0.08611

Figure 7 shows the solution of the equation 27 in the last time step ($n = 2000$).

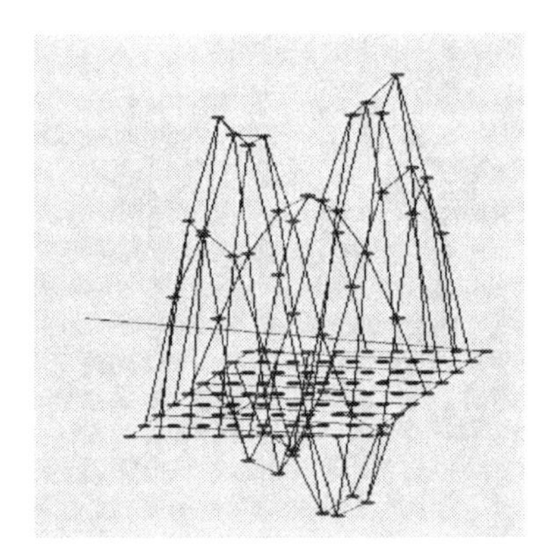

Fig. 7.

5.3 Free Vibrations of a Fixed Plate

In this section, the weighting function used is 20 and the global exact error can be calculated by 21.

The pde is

$$\frac{\partial^2 U(x,y,t)}{\partial t^2} + \frac{1}{4(4.73)^4}[\frac{\partial^4 U(x,y,t)}{\partial x^4} + 2\frac{\partial^4 U(x,y,t)}{\partial x^2 \partial^2} + \frac{\partial^4 U(x,y,t)}{\partial y^4}] =$$
$$15 \sin t \sin(2\pi x) \sin(2\pi y) \quad (x,y) \in (0,1) \times (0,1), \quad t > 0 \,. \quad (30)$$

with boundary conditions

$$\begin{cases} U(x,y,t)|_\Gamma = 0 \\ \frac{\partial U(x,y,t)}{\partial y}|_{(0,y,t)} = \frac{\partial U(x,y,t)}{\partial y}|_{(1,y,t)} = 0, \forall y \in [0,1] \\ \frac{\partial U(x,y,t)}{\partial x}|_{(x,0,t)} = \frac{\partial U(x,y,t)}{\partial x}|_{(x,1,t)} = 0, \forall x \in [0,1] \end{cases} . \quad (31)$$

and initial conditions

$$\begin{cases} U(x,y,0) = 0 \\ \frac{\partial U(x,y,t)}{\partial t}|_{(x,y,0)} = (\cos(4.73x) - \cosh(4.73x) - 0.982501[\sin(4.73x) - \\ \sinh(4.73x)])(\cos(4.73y) - \cosh(4.73y) - 0.982501[\sin(4.73y) - \sinh(4.73y)]) \end{cases} . \quad (32)$$

The exact solution is given by

$$U(x,y,t) = (\cos(4.73x) - \cosh(4.73x) - 0.982501[\sin(4.73x) - \sinh(4.73x)]) \\ (\cos(4.73y) - \cosh(4.73y) - 0.982501[\sin(4.73y) - \sinh(4.73y)]) \sin t \,. \quad (33)$$

Table 6 shows the results of the global error, using a regular mesh of 81 nodes (figure 1), for several values of $\triangle t$.

Table 6. Influence of $\triangle t$ in the global error

$\triangle t$	Global error
0.005	0.36490
0.002	0.03519
0.001	0.00492
0.0005	0.00064

Table 7. Influence of irregularity of mesh in the global error

IIC	Global error
0.92	0.00492
0.83	0.00494
0.76	0.00496
0.58	0.00504

Table 7 shows the results of global error with $\triangle t = 0.001$ for several irregular meshes of 81 nodes (figures 1, 2 and 3).

Figure 8 shows the approximated solution of the equations 30, 31 and 32 in the last time step ($n = 500$).

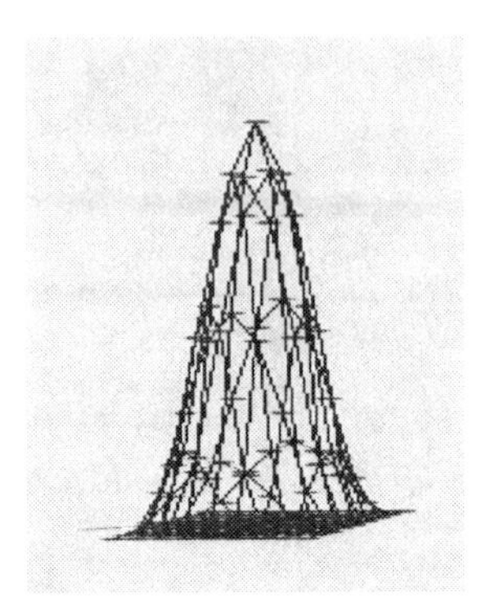

Fig. 8. Approximated solution in the last time step

6 Conclusions

The use of the generalized finite difference method using irregular clouds of points is an interesting way of solving partial differential equations. The extension of the generalized finite difference to the explicit solution problem of dynamic analysis has been developed.

The von Neumann stability criterion has been expressed in function of the coefficients of the star equation for irregular cloud of nodes.

As it is shown in the numerical results, a decrease in the value of the time step, always below the stability limits, leads to a decrease of the global error.

Acknowledgments. The authors acknowledge the support from Ministerio de Ciencia e Innovación of Spain, project CGL2008 – 01757/CLI.

References

1. Benito, J.J., Ureña, F., Gavete, L.: Influence several factors in the generalized finite difference method. Applied Mathematical Modeling 25, 1039–1053 (2001)
2. Benito, J.J., Ureña, F., Gavete, L., Alvarez, R.: An h-adaptive method in the generalized finite difference. Comput. Methods Appl. Mech. Eng. 192, 735–759 (2003)
3. Benito, J.J., Ureña, F., Gavete, L., Alonso, B.: Solving parabolic and hyperbolic equations by Generalized Finite Difference Method. Journal of Computational and Applied Mathematics 209(2), 208–233 (2007)
4. Benito, J.J., Ureña, F., Gavete, L., Alonso, B.: Application of the Generalized Finite Difference Method to improve the approximated solution of pdes. Computer Modelling in Engineering & Sciences 38, 39–58 (2009)
5. Gavete, L., Gavete, M.L., Benito, J.J.: Improvements of generalized finite difference method and comparison other meshless method. Applied Mathematical Modelling 27, 831–847 (2003)

6. Liszka, T., Orkisz, J.: The Finite Difference Method at Arbitrary Irregular Grids and its Application in Applied Mechanics. Computer & Structures 11, 83–95 (1980)
7. Benito, J.J., Ureña, F., Gavete, L.: Leading-Edge Applied Mathematical Modelling Research, ch. 7. Nova Science Publishers, New York (2008)
8. Orkisz, J.: Finite Difference Method (Part, III). In: Kleiber, M. (ed.) Handbook of Computational Solid Mechanics, Springer, Heidelberg (1998)
9. Timoshenko, S.P., Young, D.H.: Teoría de Estructuras. Urmo S.A. de Ediciones, Spain
10. Thomson, W.T.: Vibration Theory and Applications. Prentice-Hall, Englewood Cliffs (1965)
11. Vinson, J.R.: The Behavoir or Thin Walled Strutures: Beams, Plates ans Shells. Kluwer Academic Publishers, Boston
12. Evans, L.C.: Partial Differential Equations. American Mathematical Society. Graduate Studies in Mathematics 19 (2010)
13. Knabner, P., Angerman, L.: Numerical Methods for Elliptic and Para bolic Partial Differential Equations. Texts in Applied Mathematics, vol. 44. Springer, New York (2003)
14. Morton, K.W., Mayers, D.F.: Numerical solution of partial differential equations: An introduction. Cambridge University Press, Cambridge (1996)
15. Respondek, J.: Numerical Simulation in the Partial Differential Equations Controllability Analyssis with Physically Meaningful Constraints. Mathematics and Computers in Simulation 81(1), 120–132 (2010)
16. Respondek, J.: Approximate controllability of the n-th order infinite dimensional systems with controls delayed by the control devices. Int. J. Systems Sci. 39(8), 765–782 (2008)

A Viewer-Dependent Tensor Field Visualization Using Particle Tracing

Gildo de Almeida Leonel, João Paulo Peçanha, and Marcelo Bernardes Vieira

Universidade Federal de Juiz de Fora, DCC/ICE,
Cidade Universitária, CEP: 36036-330, Juiz de Fora, MG, Brazil
{gildo.leonel,joaopaulo,marcelo.bernardes}@ice.ufjf.br

Abstract. Tensor field visualization is a hard task due to the multivariate data contained in each local tensor. In this paper, we propose a particle-tracing strategy to let the observer understand the field singularities. Our method is a viewer-dependent approach that induces the human perceptual system to notice underlying structures of the tensor field. Particles move throughout the field in function of anisotropic features of local tensors. We propose a easy to compute, viewer-dependent, priority list representing the best locations in tensor field for creating new particles. Our results show that our method is suitable for positive semi-definite tensor fields representing distinct objects.

Keywords: Tensor Field, Particle Tracing, Dynamic Visualization, Scientific Visualization.

1 Introduction

Arbitrary tensor fields are very useful in a large number of knowledge areas like physics, medicine, engineering and biology. The main goal of the study of tensors in these areas is to investigate and seek for collinear and coplanar objects represented by tensors. These objects or artifacts are formed by subsets of arranged and structured tensors which capture some geometric continuity like, for example, fibers.

The best visualization methods must offer different features to allow the observer to see as many aspects of tensor multivariate data as possible. Therefore, it is very hard to combine in a single method all the expected functionalities. In this paper we introduce a visualization process suitable for many different positive semi-definite tensor fields. Our goal is to highlight most continuity information in a simple and adaptive fashion. An interesting approach may take into account not only the static data given by an ordinary tensor field. It can also use other information like the object's surrounding space and the observer (i.e. camera model) to generate and modify the visual data.

In this paper we present a dynamic method to visualize tensor fields. It takes into account the observer point of view and other attributes aiming to highlight collinear and coplanar information. The particle motion incites the human perceptual system to fuse and perceive salient features. The work of [1] also use

B. Murgante et al. (Eds.): ICCSA 2011, Part I, LNCS 6782, pp. 690–705, 2011.

particle tracing to extract visual information of a tensor field. However, the criterion to create particles is purely random, which may generate some confusing results. In this paper we defined a priority list to choose the best places in the space where particles should born to produce a superior viewing result. The priorities are computed by a linear combination of anisotropic measures of tensors and by the viewer camera parameters.

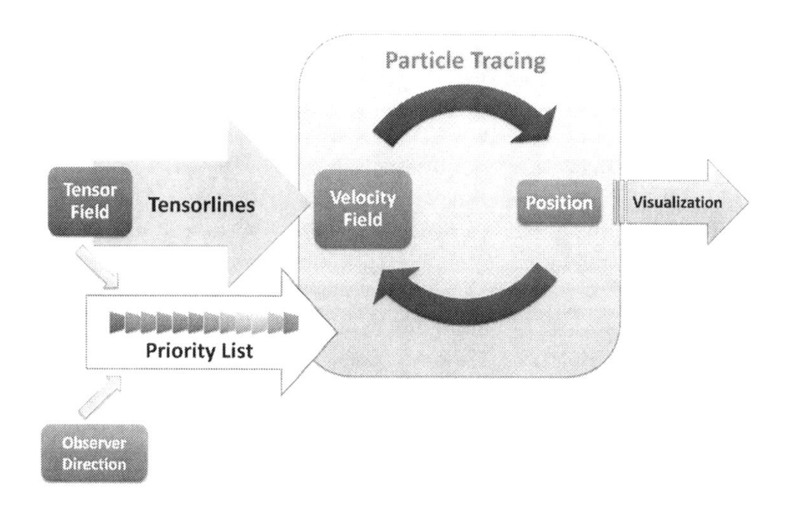

Fig. 1. Schematic representation of the proposed method

An overview of our method is depicted in Figure 1, having the following steps: extract the best velocity vector from the tensor field, use a priority list to define where new particles will appear, perform the advection of particles.

2 Related Works

In tensor field visualization we can adopt different approaches to represent information. The discrete approach is commonly used when punctual data is sufficient to obtain the required information. A superquadric glyph is an ordinary fashion to represent local information given by the field mapping into geometric primitives, like cubes, ellipses and cylinders this information. Using the concept of glyphs, Shaw et al [2] have developed their work for multidimensional generic data visualization. Their main contribution is to connect the advantages of the visual human being perception and superquadrics intrinsic interpolation feature. In a later extension [3], they propose to measure how many forms assumed by superquadric can be distinguished by human vision system. Westin et al [4] proposed an anisotropic metric to identify and compare a set of glyphs. Kindlmann [5] defines a linear mapping of shape coefficients in order to view anisotropic and isotropic tensors. They present the problem of ambiguous glyphs that can induce to wrong visual conclusions when the glyphs adopt planar or linear forms. In one hand, all problems that involves symmetry can be solved using ellipsoidal glyphs. In other hand, there are ambiguity situations in the visual identification of the tensor. If the point of view direction is aligned to the main eigenvector, ellipsoidal linear tensors can be identified as spheres. To overcome this problem, he presents a new parametrization of superquadric tensor glyphs to better represent shape and orientation.

There are also continuous methods for tensor visualization based upon the tensor interpolation of two distinct points in a multidimensional space. In [6] the concept of tensor field lines - extended from [7] - is generalized, and the concept of hyperstreamlines is introduced. In that work they represent all information of a tensor field taking into account not points, but the trajectory generated by the tensor using its eigenvectors. This approach is interesting to visualize symmetric tensor fields, where its eigenvectors are real and orthogonal. However, the field becomes hard to visualize for a large number of hypersetreamlines. Delmarcelle et al [8] have presented another problem with hypersetreamlines: the degeneration when a tensor has at least two equal eigenvectors. In [9] is presented a method to avoid degeneration due to planar and spherical tensors in input data. This method was applied in tensors fields obtained from magnetic resonance images.

Zheng and Pang [10] proposed a method to visualize tensor field using the concept of linear integral convolution. Their work is an extension of [11], which uses a white texture noise and hyperstreamlines to generate the visual information.

Dynamical particles walking through a tensor field is a powerful and recent method for visualization. The sensation of movement incites the human perceptual system making easier the understanding of some field properties. Kondratieva et al [1] has proposed a dynamical approach using particle tracing in GPU (Graphic Processing Unit). They argue that particle tracing gives an efficient and intuitive way to understand the tensor field dynamics. The advection of a set of particles in a continuous flow is used to induce particle motion. Through the tensor field, a direction vector field is generated - based on [12] - and then, the advection using this vector field is performed.

3 Fundamentals

3.1 Orientation Tensor

A local orientation tensor is a special case of non-negative symmetric rank 2 tensor. It was introduced by Westin [13] to estimate orientations in a field. This tensor is symmetric and can be saw as a pondered sum of projections:

$$\mathbf{T} = \sum_{i=1}^{n} \lambda_i e_i e_i^T, \tag{1}$$

where $\{e_1, e_2, ..., e_m\}$ is a base of $\mathbb{R}^n$. Therefore, it can be decomposed into:

$$\mathbf{T} = \lambda_n \mathbf{T}_n + \sum_{i=1}^{n-1} (\lambda_i - \lambda_i + 1)\, \mathbf{T}_i, \tag{2}$$

where λ_i are the eigenvalues corresponding to each eigenvector e_i. This is an interesting decomposition because of its geometric interpretation. In fact, in $\mathbb{R}^3$, an orientation tensor $\mathbf{T}$ decomposed using Equation 2 can be represented using the contribution of its linear, planar, and spherical intrinsic features:

$$\mathbf{T} = (\lambda_1 - \lambda_2)\,\mathbf{T}_l + (\lambda_2 - \lambda_3)\,\mathbf{T}_p + \lambda_3\mathbf{T}_s. \tag{3}$$

A $\mathbb{R}^3$ tensor decomposed by Equation 3, with eigenvalues $\lambda_1 \geq \lambda_2 \geq \lambda_3$, can be interpreted as following:

- $\lambda_1 \gg \lambda_2 \approx \lambda_3$ corresponds to an approximately linear tensor, with the spear component being dominant.
- $\lambda_1 \approx \lambda_2 \gg \lambda_3$ corresponds to an approximately planar tensor, with the plate component being dominant.
- $\lambda_1 \approx \lambda_2 \approx \lambda_3$ corresponds to an approximately isotropic tensor, with the ball component being dominant, and no main orientation present.

For many purposes only the main direction of the tensor is necessary. Furthermore, the shape of the tensor is generally more important than its magnitude. Using the sum of the tensor eigenvalues, one may obtain the linear, planar, and spherical coefficients of anisotropy:

$$c_l = \frac{\lambda_1 - \lambda_2}{\lambda_1 + \lambda_2 + \lambda_3}, \tag{4}$$

$$c_p = \frac{2\,(\lambda_2 - \lambda_3)}{\lambda_1 + \lambda_2 + \lambda_3}, \tag{5}$$

$$c_s = \frac{3\lambda_3}{\lambda_1 + \lambda_2 + \lambda_3}. \tag{6}$$

Note that coefficients in Equations 5 and 6 were scaled by 2 and 3, respectively, so that each of them independently lie in the range $\in [0, 1]$ with $c_l + c_p + c_s = 1$ [13].

3.2 Invariants towards Eigenvalues

The eigenvalues of a tensor $\mathbf{D}$ can be calculated solving:

$$\mathtt{det}(\lambda\mathbf{I} - \mathbf{D}) = 0.$$

Hence:

$$\mathtt{det}(\lambda\mathbf{I} - \mathbf{D}) = \begin{vmatrix} \lambda - D_{xx} & -D_{xy} & -D_{xz} \\ & \lambda - D_{yy} & -D_{yz} \\ & & \lambda - D_{zz} \end{vmatrix} = \lambda^3 - J_1\lambda^2 + J_2\lambda - J_3,$$

where,

$$\begin{aligned}
J_1 &= D_{xx} - D_{yy} - D_{zz} = \mathtt{tr}(\mathbf{D}),\\
J_2 &= D_{xx}D_{yy} + D_{xx}D_{zz} + D_{yy}D_{zz} - D_{xy}^2 - D_{xz}^2 - D_{yz}^2 = \frac{\mathtt{tr}(\mathbf{D})^2 - \mathtt{tr}(\mathbf{D}^2)}{2},\\
J_3 &= 2D_{xy}D_{xz}D_{yz} + D_{xx}D_{yy}D_{zz} - D_{xz}^2D_{yy} - D_{yz}^2D_{xx} - D_{xy}^2D_{zz} = \mathtt{det}(\mathbf{D}).
\end{aligned} \tag{7}$$

so that $\mathtt{tr}(\mathbf{D})$ and $\mathtt{det}(\mathbf{D})$ are the trace and the determinant of tensor $\mathbf{D}$, respectively.

The matrix determinant is invariant to basis changing and thus is classified as an algebraic invariant. Another useful invariant used to determine the eigenvalues of a tensor is the squared norm:

$$\begin{aligned}
J_4 = \|\mathbf{D}\|^2 &= J_1^2 - 2J_2\\
&= D_{xx}^2 + 2D_{xy}^2 + 2D_{xz}^2 + D_{yy}^2 + 2D_{yz}^2 + D_{zz}^2\\
&= \lambda_1^2 + \lambda_2^2 + \lambda_3^2.
\end{aligned} \tag{8}$$

3.3 Eigenvalue Wheel

Kindlmann [14] describes other three invariants used to solve a cubic polynomial:

$$Q = \frac{J_1^2 - 3J_2}{9} = \frac{J_4 - J_2}{9} = \frac{3J_4 - 3J_1^2}{18} \tag{9}$$

$$R = \frac{-9J_1J_2 + 27J_3 + 2J_1^3}{54} = \frac{-5J_1J_2 + 27J_3 + 2J_1J_4}{54} \tag{10}$$

$$\Theta = \frac{1}{3}\cos^{-1}\left(\frac{R}{\sqrt{Q^3}}\right). \tag{11}$$

The wheel eigenvalues can be defined as a wheel with three equally placed radii centered on the real number line at $J_3/3$. The radius of the wheel is $2\sqrt{Q}$, and Θ measures the orientation of the first radius [14].

The central moments of a tensor determines the geometric parameters of the eigenvalue wheel. The central moments are defined as:

$$\begin{aligned}
\mu_1 &= \frac{1}{3}\sum \lambda_i = \frac{\lambda_1 + \lambda_2 + \lambda_3}{3} = J_1/3\\
\mu_2 &= \frac{1}{3}\sum (\lambda_i - \mu_1)^2 = \frac{2(\lambda_1^2 + \lambda_2^2 + \lambda_3^2 - \lambda_1\lambda_2 - \lambda_1\lambda_3 - \lambda_2\lambda_3)}{9} = 2Q\\
\mu_3 &= \frac{1}{3}\sum (\lambda_i - \mu_1)^3 = 2R.
\end{aligned}$$

The second central moment μ_2 is the variance of the eigenvalues, and the standard deviation is $\sigma = \sqrt{\mu_2} = \sqrt{2Q}$. The asymmetry A_3 of the eigenvalues is defined as follows [15]:

$$A_3 = \frac{\mu_3}{\sigma^3} = \frac{\sum(\lambda_i - \mu_1)^3}{3\mu_2\sqrt{\mu_2}} = \frac{R}{\sqrt{2Q^3}}. \tag{12}$$

3.4 Anisotropy

In literature we can find many forms to measure the tensor anisotropy. The fractional anisotropy (FA), relative anisotropy (RA), volume ratio and others, can be computed using the tensor eigenvalues [16].

The FA [16] and RA [17] are defined as following:

$$FA = \frac{3}{\sqrt{2}}\sqrt{\frac{\mu_2}{J_4}} = 3\sqrt{\frac{Q}{J_4}} = \sqrt{\frac{J_4 - J_2}{J_4}}$$

$$RA = \frac{\sqrt{\mu_2}}{\sqrt{2}\mu_1} = \frac{3\sqrt{Q}}{J_1}. \tag{13}$$

3.5 Tensorlines

The tensorlines concept is an extension of the hyperstreamlines method proposed in [6]. Hyperstreamlines is obtained by a smooth path tracing. This is done by using the main tensor eigenvector to perform line integration. The degeneration problem in this method incited Weinstein *et al* [9] to develop an extension called tensorlines. The tensorlines method uses multiple tensor features to determine the correct path to follow. It stabilizes the propagation incorporating two additional terms $\mathbf{v}_{int}$ and $\mathbf{v}_{out}$ given by:

$$\mathbf{v}_{out} = \mathbf{T}\mathbf{v}_{in}, \tag{14}$$

so that $\mathbf{v}_{in}$ is the incoming direction, $\mathbf{v}_{out}$ the outgoing direction and $\mathbf{T}$ the local tensor. The $\mathbf{v}_{in}$ vector corresponds to the propagation direction in the previous step, and $\mathbf{v}_{out}$ is the input vector transformed by the tensor.

The propagation vector used in the integral is a linear combination of e_1, $\mathbf{v}_{in}$, and $\mathbf{v}_{out}$. The next propagation vector, $\mathbf{v}_{prop}$, depends on the shape of the tensor:

$$\mathbf{v}_{prop} = c_l e_1 + (1 - c_l)\left((1 - w_{punct})\,\mathbf{v}_{in} + w_{punct}\mathbf{v}_{out}\right), \tag{15}$$

where $w_{punct} \in [0,1]$ is a parameter defining the penetration into isotropic regions [9].

3.6 Particle Tracing

The tensorline method generates a vector field that can be visualized using many approaches. One of those is called particle tracing. In this method, massless particles are inserted into the field subspace and their movements are coordinated by its vectors.

It is necessary to compute the particle position $\vec{x}$ in time t over velocity $\vec{v}$ each time-step. The mathematical model for this problem is straightforward. A given particle p, is identified by your initial position $\vec{x}_{po}$ with velocity $\vec{v}_p\,(p,t)$. We must find $\vec{x} \in \mathbb{R}^n$:

$$\begin{cases} \frac{d\vec{x}_p}{dt} = \vec{v}_p\,(\vec{x}_p, t) & t \in [t_0, T_p] \\ \vec{x}_p|_{t=t_0} = \vec{x}_{p_0}\,. \end{cases}, \tag{16}$$

where Tp is the time for particle p walk through all domain Ω.

4 Proposed Method

One common problem in tensor field visualization is ambiguity. In glyph-based visualization, tensors with different forms may appear similar in a particular point of view. Tensors with linear anisotropy may be identified as an isotropic if the main eigenvector is aligned to the observer. To solve this problem we can adopt a metric to evaluate the tensor orientation in regard to the observer. This strategy can be efficient not only to treat the degeneration problem, but also to improve other visualization methods. We will apply the benefits of observer metrics to propose a visualization method based on particle tracing.

In our work, particles in motion will represent the features of the tensor field. One critical point in visualization using particle tracing is to define the particle starting point. The most intuitive approach used to insert particles into the domain is to compute new positions randomly. However, a fixed distribution function will generally not insert new particles in most interesting sites.

4.1 Priority Features

Let $\mathbb{T}_{x\times y\times z}$ being a discrete and finite tensor field with lattice given by $x, y, z \in \mathbb{N}$ so that $\mathbb{T} = \{\mathbf{t}_1, \mathbf{t}_2, \mathbf{t}_3 ... \mathbf{t}_n\}$, and composed by $|\mathbb{T}| = N$ tensors. For a given voxel (a, b, c) where $a, b, c \in \mathbb{N}$ and such that $a \leq x$, $b \leq y$ and $c \leq z$ we have the correspondent tensor $\mathbf{t}_i \in \mathbb{T}$.

To correct visualize the tensor field we need to define a criterion to generate and insert particles into the domain $\mathbb{T}$. This should be done in accordance with field variants and properties. In this work we propose a scalar $\Upsilon \in \mathbb{R}$, which defines the priority of a voxel to have a particle being created on it. The Υ is also used to define a color palette aiming to highlight the desired properties. This priority is calculated using tensors characteristics and geometric features of the scene (Fig. 2).

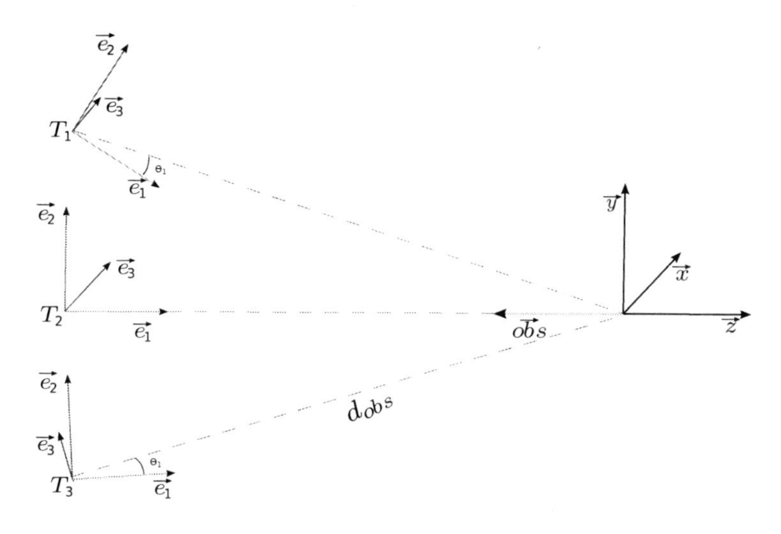

Fig. 2. Simulation space

To evaluate the tensor position in relation to the observer we propose the coefficients k_1, k_2 e k_3:

$$k_1 = 1 - |\boldsymbol{e}_1 \cdot \boldsymbol{obs}| \tag{17}$$

$$k_2 = 1 - |\boldsymbol{e}_2 \cdot \boldsymbol{obs}| \tag{18}$$

$$k_3 = |\boldsymbol{e}_3 \cdot \boldsymbol{obs}|, \tag{19}$$

where $\boldsymbol{e}_1$, $\boldsymbol{e}_2$ and $\boldsymbol{e}_3$ are the eigenvectors of the tensor and $\boldsymbol{obs}$ is the vector that corresponds to the camera view.

Another important coefficient is the Euclidean distance between the observer and the tensor d_{obs}:

$$d_{obs} = \frac{|\boldsymbol{x}_T - \boldsymbol{x}_{obs}|}{MAX(d_{obs})}, \tag{20}$$

this distance is normalized by the greatest distance in the field $MAX(d_{obs})$.

These coefficients are used together with tensor attributes to evaluate the priority of a voxel receive a particle. For this proposal, we will calculate the scalar Υ as the linear combination of the following terms:

- average of the eigenvalues of the tensor (μ_1): related to the tensor size;
- variance of the eigenvalues of the tensor (μ_2): a bigger variance indicates that the tensor will probably have a planar or linear anisotropy;
- asymmetry of the tensor eigenvalues (A_3): changes from negative to positive as the tensor vary from planar to linear;
- standard square of a tensor (J_4): related to amplification imposed by the tensor;
- coefficient of fractional anisotropy (FA) and relative anisotropy (RA): used to detect anisotropy and isotropic regions;

- coefficient of orthogonally between the observer and the first eigenvalue (k_1), with the second eigenvalue (k_2) and the third eigenvalue (k_3): quantify the relative position of the observer in relation to the tensor eigensystem, so we can prioritize tensors that are parallel or orthogonal to the observer;
- normalized distance to the observer (d_{obs}): reveal tensors closer to the screen;
- coefficient of linear anisotropy (c_l), coefficient of planar anisotropy (c_p): also allow to differentiate anisotropy.

Tensor fields may present multivariate information coming from many different applications. Aiming to generate appropriate results, the scalar Υ will be parameterized by the user in order to focus on the desired characteristics:

$$\begin{aligned}\Upsilon_{\mathbf{t}} =&\alpha_1\mu_1 + \alpha_2\mu_2 + \alpha_3 A_3 + \alpha_4 J_4 + \alpha_5 FA + \alpha_6 RA \\ &+ \alpha_7 k_1 + \alpha_8 k_2 + \alpha_9 k_3 + \alpha_{10} d_{obs} + \alpha_{11} c_l + \alpha_{12} c_p.\end{aligned} \tag{21}$$

where $\alpha_i \in [-1, 1]$ and $\mathbf{t} \in \mathbb{T}$. So, the $\Upsilon_{\mathbf{t}}$ ponders how much the tensor $\mathbf{t} \in \mathbb{T}$ presents the required information.

4.2 Priority List and Particle Insertion

In the application beginning, a number $N_p \in \mathbb{N}$ of particles will be established by the user. The program will allocate all the necessary memory and particles are initialized, but not immediately inserted into the space. In the next step, all tensors $\mathbf{t} \in \mathbb{T}$ will be sorted and ranked in a list with most important elements (higher $\Upsilon_{\mathbf{t}}$) positioned on the top (Fig. 3).

To insert a particle p_i into the domain, a random number $\kappa \in [0, 1]$ is generated using a standard normal distribution and then we select the correspondent z-th tensor, $z \in [0, N-1]$, in the priority list:

$$z = \frac{\kappa N}{\varsigma} \tag{22}$$

where N is the total number of tensors and ς defines a Gaussian distribution. A bigger ς implies a higher frequency of choice of the top tensors in the priority

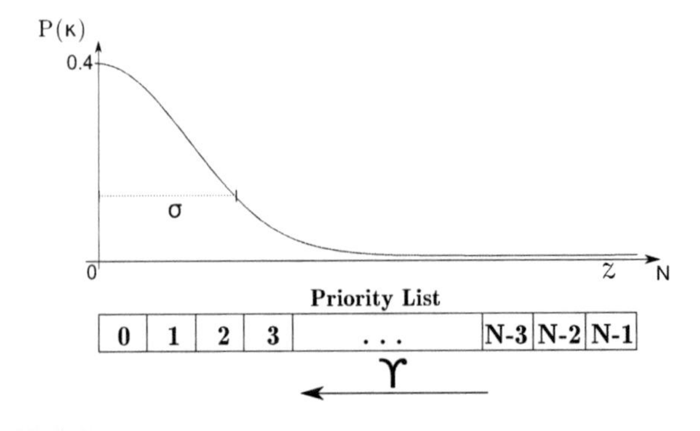

Fig. 3. Normal probability distribution on the priority list

list. The particle p_i will be created in the position of the z-th tensor spatial position. The process of particle insertion stops when the domain contains at least N_p particles.

The next algorithm step is to define a main direction $\vec{v} \in \mathbb{R}^3$ for the new particle. We may define a unique direction for each particle. This can be done using a gradient of a specific field attribute. However, the created particle initially will always move towards the same direction. A better solution is to invert the initial direction of half the number of creations.

The Υ scalar has viewer-dependent terms, so, it is necessary to reorder the priority list on every change of the camera position and orientation. This process can be computationally expensive, and impairs the visualization performance.

To deal with this problem we use the following implementation strategy: after the first iteration, the priority list stays partially ordered only if the camera changes are not abrupt. If a full reordering is needed, the quicksort algorithm with median-of-three partitioning [18] is performed. This algorithm has presented relatively good performance results, leading to a real time visualization.

4.3 Particle Removal

The particle is removed from the visualization space when it reaches one of the following situations: a) it is located at a bigger isotropic region b) get away from the visualization lattice, and c) when the absolute value of the dot product between the entry direction into a voxel and the current voxel propagation direction is equal to zero or smaller than a threshold $\gamma \in \mathbb{R}$. We have found empirically the value $\gamma = 0.3$ as a good parameter to avoid that a particle get stuck between two voxels with opposite directions. It implies that the angle among these two directions should be in the interval $(72.54°, 90.00°]$.

In our implementation, for performance reasons, no particle is deallocated until the application ends. When the stop criterion is reached for a determined particle, its computational resources are reused and it is recreated using the priority list.

In Kondratieva *et al.* [1] work, it is proposed that particles should be restarted in its original initial position. Later, Kondratieva [19] concludes that the previous approach needed modifications. They observed a flicker behavior in the display due to the presence of high frequencies in the field. This situation may distract the user and disturb the visualization. The authors proposed that particles should always restart at random in the tensor field.

In our method, a fixed number of particles is created, and during the visualization process they are inserted into the lattice taking into account the priority list and removed when it is necessary. When a particle is destroyed, it restarts using the creation criterion. Thus, particles may reborn and highlight different features in the same simulation. Further information will be highlighted if the user, at runtime, manually changes the parameters α_i presented in Equation 21, or changes the observer's point of view.

Our algorithm for tensor field visualization may be summarized as following:

1. select the tensor field to be visualized (domain $\mathbb{T}$) and the number of particles (N_p);
2. compute $\Upsilon_{\mathbf{t}}$ (Eq. 21) for each tensor;
3. for each tensor $\mathbf{t} \in \mathbb{T}$, sort and rank it in the priority list;
4. select an available particle and insert it into the visualization space using the priority list;
5. perform the particle advection loop;
6. verify what particles must be killed using the stop criterion;
7. if the viewer position or orientation changes too much, perform step 3.
8. if the number of particles in visualization space is smaller than N_p, go to step 4, otherwise go to step 5;

5 Results

In section, we present shots of different types of tensor fields. The particles were represented by a pointer glyph (otherwise specified) and the color gradient adopted flows from blue (minimum) to red (maximum) for a given Υ (Fig. 4).

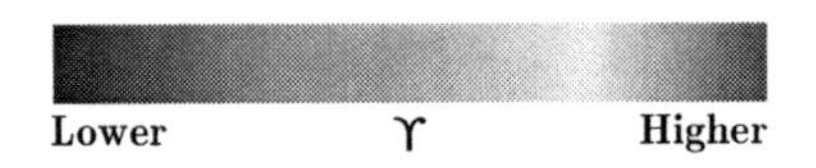

Fig. 4. Color palette for the Υ values

We have inserted into a 38x39x40 grid three spherical charges, located at (0,0,0), (38,39,40) and (38,0,40). For all voxels in the grid we use a formulation to ponder the influence of each charge in that space region and then compute a local tensor. The Figure 5(a) shows the obtained result using discrete glyphs and in Figure 5(b) we draw a few tensorlines.

An important tensor feature is the anisotropy A_3 (Eq. 12). Thus, if the user want to seek for regions of high anisotropy, the Υ function may be adjusted. It varies from positive to negative as the tensors changes it form from linear to planar.

We have defined $\Upsilon = -A_3 + FA$ and the results are shown in Figure 6. The anisotropy in this field can be seen using the proposed method (Fig. 6). We have used superquadric glyphs (Fig. 6(a)) and pointer glyphs (Fig. 6(b)) as particles to understand the field properties. In Figure 6(a) there are less particles than Figure 6(b). Note that the superquadric glyph particles, in Figure 6(a), are more stretched in regions near to the charges showing a linear behavior. This regions presents a high anisotropy. Using the Υ function in Figure 6(b) one may note: a) a large number of particles are inserted into that region, b) the particles flows smoothly making the field variation more understandable. We consider

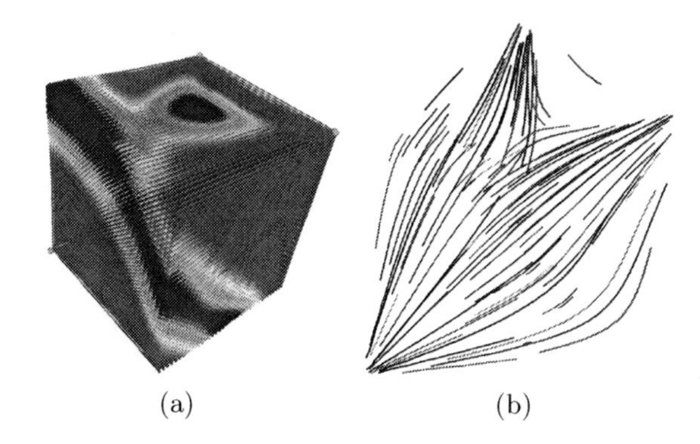

Fig. 5. A tensor field with three charges: (a) represented by superquadric glyphs and (b) by a few tensorlines

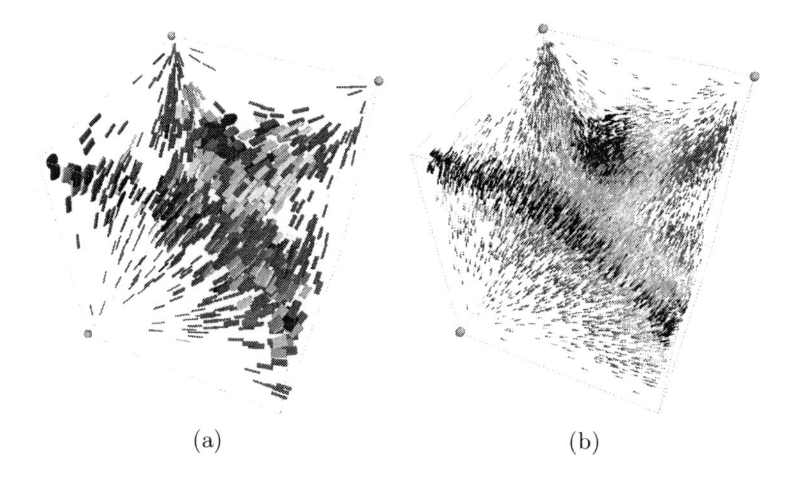

Fig. 6. Three charges field represented by our method: (a) particles assuming superquadric glyph shapes (particles near the charges are more stretched) and (b) pointer glyphs smoothly flowing through the domain

the pointer glyph the best way to represent the particles because it makes the visualization cleaner and allows a complete view inside the volume.

The next example is a helical tensor field (Fig. 7) with visualization depending on the observer. The tensors in this field suffer a torsion process along the z-axis. Using k_1 and having the z-axis orthogonal to the observer (Fig. 7(a)), one may see that the tensors in the internal regions tend to have high priority values (reddish colors) as they are orthogonal to the viewer. Using k_1 with the z-axis aligned with the observer (Fig. 7(b)), the now bluish sites (low priority values) represent tensors highly parallel to the viewer, in regard to the new camera orientation. The proposed method is highly efficient and suitable to extract volumetric information from tensor fields.

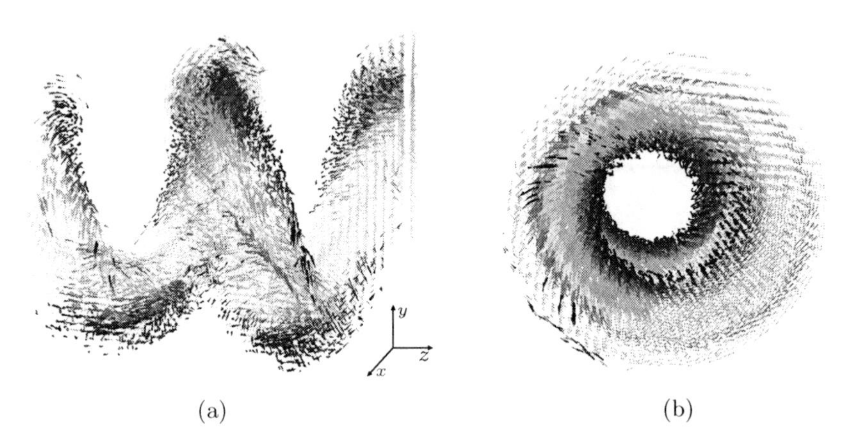

Fig. 7. Helical field: color palette given by k_1 in two different views

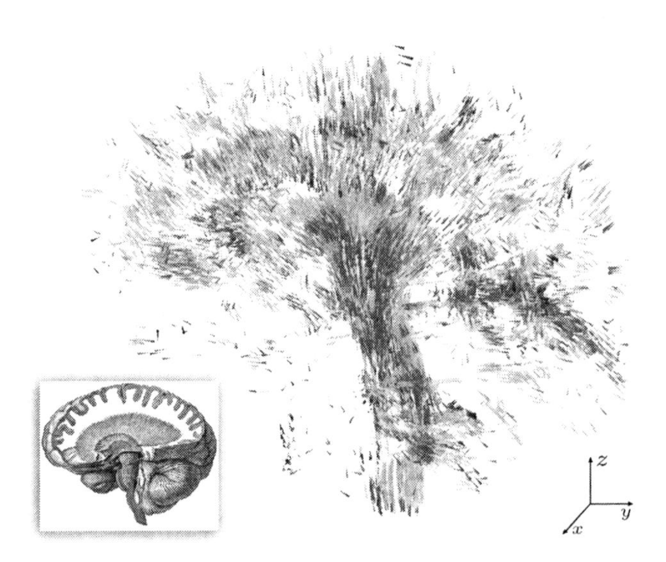

Fig. 8. Brain fiber visualization. Lower left corner: brain image from [19].

Diffusion tensor magnetic resonance imaging (DT-MRI) is generally used to detect fibrous structures of biological tissues. In this work we have used a diffusion tensor field of a brain available at [20] to test our method. The results are shown in Figures 8 and 9.

The branching and crossing of brain's white matter tracts generates local tensors with high planar anisotropy [5]. To find these brain regions, we adjust the priority and the colorization using $\Upsilon = \mu_2 + A_3 - FA$ (Fig. 8). So, we are searching for tensors with higher variance, amplitude and anisotropy - we are penalizing the isotropic regions with $-FA$. In the central regions of the

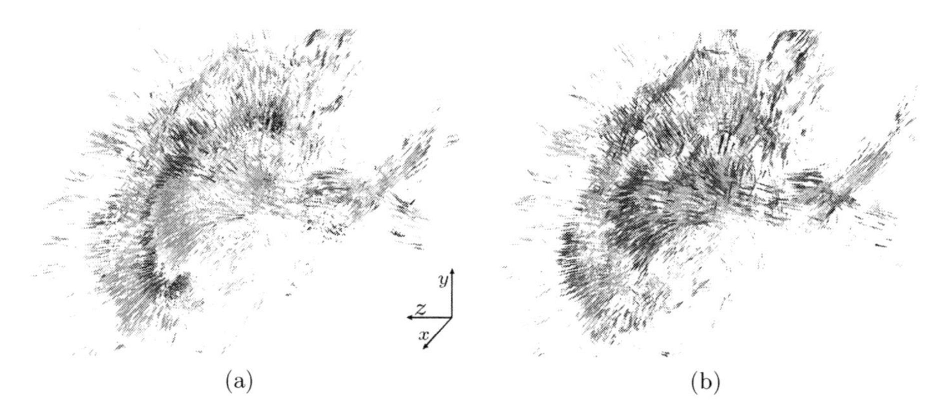

Fig. 9. Influence of the viewer-dependent terms: (a) highlighting tensors orthogonal to e_1 and (b) orthogonal to e_2.

brain, which is composed by a larger number of fiber and tissues, we can see a concentration of the required information, as expected.

The influence of viewer-dependent terms can be also observed in the brain field (Fig. 9). When we are searching for tensors with eigenvectors orthogonal to the observer, the k_1 and k_2 view-dependent terms could be used. The k_1 coefficient (Fig. 9(a)) highlights tensors which has main eigenvectors orthogonal to the observer. In a opposite fashion, the k_2 coefficient is emphasizing tensors that presents main direction non-orthogonal to the observer, for the same point of view (Fig. 9(b)).

6 Conclusions

In this paper we presented a tensor field visualization method based upon particle tracing using viewer-dependent terms. We proposed a priority list which defines where particles should born in the tensor field domain. This is done aiming to highlight regions of interest. We also present a set of observer-dependent coefficients that contributes to the final visualization, generating suitable results. In order to cover a wide range of different tensor fields, we developed a Υ scalar which can be adjusted to the user needs. The Υ quantifies the importance of a tensor in the visualization process for a given set of parameters (Eq. 21). Thus, a previous knowledge of the field is required to achieve a better visual interpretation.

We provide results using three different tensor fields. In each field, the anisotropy analysis showed correctly collinear and coplanar structures formed by the tensors throughout the domain. The view-dependent attributes contributed to the visualization process, highlighting orthogonality and proximity of tensors in relation of the observer (Fig. 7 and 9).

A flicker problem occurs when a new created particle reaches isotropic regions and are instantly destroyed by the removal criterion. This effect can be avoided by filtering the noise present in the tensor field and smoothing the transition between isotropic and anisotropic regions.

Acknowledgment

The authors thank to FAPEMIG (*Fundação de Amparo à Pesquisa do Estado de Minas Gerais*), CAPES (*Coordenação de Aperfeiçoamento de Pessoal de Ensino Superior*) and UFJF for funding this research.

References

1. Kondratieva, P., Krüger, J., Westermann, R.: The application of gpu particle tracing to diffusion tensor field visualization. In: Visualization (VIS 2005), pp. 73–78. IEEE, Los Alamitos (2005)
2. Shaw, C.D., Ebert, D.S., Kukla, J.M., Zwa, A., Soboroff, I., Roberts, D.A.: Data visualization using automatic, perceptually-motivated shapes. In: Proceeding of Visual Data Exploration and Analysis, SPIE (1998)
3. Shaw, C.D., Hall, J.A., Blahut, C., Ebert, D.S., Roberts, D.A.: Using shape to visualize multivariate data. In: NPIVM 1999: Proceedings of the 1999 workshop on new paradigms in information visualization and manipulation in conjunction with the eighth ACM internation conference on Information and knowledge management, pp. 17–20. ACM, New York (1999)
4. Westin, C.F., Peled, S., Gudbjartsson, H., Kikinis, R., Jolesz, F.A.: Geometrical diffusion measures for MRI from tensor basis analysis. In: ISMRM 1997, Vancouver Canada, April 1997, pp. 17–42 (1997)
5. Kindlmann, G.: Superquadric tensor glyphs. In: Proceedings of IEEE TVCG/EG Symposium on Visualization 2004, May 2004, pp. 147–154 (2004)
6. Delmarcelle, T., Hesselink, L.: Visualization of second order tensor fields and matrix data. In: VIS 1992: Proceedings of the 3rd conference on Visualization 1992, pp. 316–323. IEEE Computer Society Press, Los Alamitos (1992)
7. Dickinson, R.R.: A unified approach to the design of visualization software for the analysis of field problems. In: Three-Dimensional Visualization and Display Technologies, SPIE Proceedings, January 1989, vol. 1083 (1989)
8. Delmarcelle, T., Hesselink, L.: Visualizing second-order tensor fields with hyper streamlines. IEEE Computer Graphics and Applications 13(4), 25–33 (1993)
9. Weinstein, D., Kindlmann, G., Lundberg, E.: Tensorlines: advection-diffusion based propagation through diffusion tensor fields. In: VIS 1999: Proceedings of the conference on Visualization 1999, pp. 249–253. IEEE Computer Society Press, Los Alamitos (1999)
10. Zheng, X., Pang, A.: Hyperlic. In: VIS 2003: Proceedings of the 14th IEEE Visualization 2003 (VIS 2003), p. 33. IEEE Computer Society, Washington, DC (2003)
11. Cabral, B., Leedom, L.C.: Imaging vector fields using line integral convolution. In: SIGGRAPH 1993: Proceedings of the 20th annual conference on Computer graphics and interactive techniques, pp. 263–270. ACM, New York (1993)
12. Krüger, J., Kipfer, P., Kondratieva, P., Westermann, R.: A particle system for interactive visualization of 3D flows. IEEE Transactions on Visualization and Computer Graphics 11(6), 744–756 (2005)

13. Westin, C.F.: A Tensor Framework for Multidimensional Signal Processing. PhD thesis, Linköping University, Sweden, S-581 83 Linköping, Sweden (1994), Dissertation No 348, ISBN 91-7871-421-4
14. Kindlmann, G.: Visualization and Analysis of Diffusion Tensor Fields. PhD thesis (September 2004)
15. Bahn, M.: Invariant and Orthonormal Scalar Measures Derived from Magnetic Resonance Diffusion Tensor Imaging. Journal of Magnetic Resonance 141(1), 68–77 (1999)
16. Masutani, Y., Aoki, S., Abe, O., Hayashi, N., Otomo, K.: MR diffusion tensor imaging: recent advance and new techniques for diffusion tensor visualization. European Journal of Radiology 46, 53–66 (2003)
17. Basser, P.J., Pierpaoli, C.: Microstructural and physiological features of tissues elucidated by quantitative-diffusion-tensor mri. Journal of Magnetic Resonance, Series B 111(3), 209–219 (1996)
18. Mahmoud, H.M.: Sorting: a distribution theory. John Wiley & Sons, Chichester (2000)
19. Kondratieva, P.: Real-Time Approaches for Model-Based Reconstruction and Visualization of Flow Fields. PhD thesis, Institut fur Informatik Technische Universitat Munchen (2008)
20. Kindlmann, G.: Diffusion tensor mri datasets, `http://www.sci.utah.edu/~{}gk/DTI-data/`

Statistical Behaviour of Discrete-Time Rössler System with Time Varying Delay

Madalin Frunzete[1,2], Adrian Luca[1], Adriana Vlad[1,3], and Jean-Pierre Barbot[2,4]

[1] Faculty of Electronics, Telecommunications and Information Technology, Politehnica University of Bucharest, 1-3, Iuliu Maniu Bvd., Bucharest 6, Romania
[2] Electronique et Commande des Systmes Laboratoire, EA 3649 (ECS-Lab/ENSEA) ENSEA, Cergy-Pontoise, France
[3] The Research Institute for Artificial Intelligence, Romanian Academy, 13, Calea 13 Septembrie, Bucharest 5, Romania
[4] EPI Non-A INRIA, France
madalin.frunzete@upb.ro, adrian.luca@upb.ro, avlad@racai.ro, barbot@ensea.fr

Abstract. In this paper a modified discrete-time chaotic system is presented from the statistical point of view. This chaotic system is used in a cryptosystem and, for improving the presented method used in security data transmission, its structure is changed. The technique is implemented for a Rössler hyperchaotic system. The improvement consists in modifying the existing system in order to obtain a higher robustness for the cryptosystem; for this, a time varying delay is added in its structure.

Keywords: discrete chaotic systems, chaotic systems with delay, transient time, statistical independence.

1 Introduction

Improving the existing methods of secure data transmission for more secured communications is always a popular research domain. Using the chaotic maps in new ciphering methods is one of these directions.

When it is talking about information exchange it is understood the existence of, at least, one emitter and one receiver (the number is increasing in multicast or broadband communications). This paper deals with the case of "Single Input Single Output" (S.I.S.O.), one emitter and one receiver. A very important condition for establishing a communication is the synchronization of both subsystems; without that, the receiver could not "understand" what it is receiving. The possibility of synchronizing two chaotic systems (proved in [1]) and the control of synchronization (see [2]) permitted the existence of cryptography based on multidimensional chaotic systems.

The chaotic Rössler system referred in this paper is used in [3] for implementing a ciphering method generically named *inclusion method* (I.M.). In this type of method the message is embedded in the structure of the system so, roughly

B. Murgante et al. (Eds.): ICCSA 2011, Part I, LNCS 6782, pp. 706–720, 2011.

speaking, the evolution of the chaotic behaviour means the ciphering. In context of the addition method (A.M.), see [4], the original message is added to a sequence generated by a chaotic system.

The Rössler discrete system presents hyperchaotic behaviour; such systems have more than one unstable Lyapunov exponents, [3], [5], and their behaviour is more irregular than classical chaos.

In **section 2** it will be presented the weaknesses of the classical I.M., here based on Rössler map. By using a time delay technique, a new structure of Rössler system will be introduced.

In **section 3** the two structures of Rössler map (without delay and with delay) will be put into comparison from the statististical point of view. The statistical aspects concern: the transient time and the probability law (related by using Smironv test) and the statistical dependence/independence issue (treated by using on original test [6]).

2 Rössler Map without and with Delay

The identifiability and some parametric cryptanalysis are used to characterize, in general, a cryptosystem created by I.M., in this paper the case study is a cryptosystem based on Rössler map. The equations of this system are given in (1),

$$\begin{cases} z_1^+ = a_1 z_1(1-z_1) + a_2 z_2 \\ z_2^+ = b_1[(1-b_2 z_1)(z_2+b_3)-1](1-b_4 z_3) \\ z_3^+ = c_1 z_3(1-z_3) - c_2(1-b_2 z_1)(z_2+b_3) \end{cases} \tag{1}$$

where z_1, z_2 and z_3 are the state variables of the system; rewriting in a short form $z := (z_1, z_2, z_3)^T \in \Re^3$ represents the state vector evaluated at the step k (i.e. $z(k)$), so $z^+ := z(k+1)$. The vector $\theta \in \Re^8$ is the parameter vector of system (1), $\theta = (a_1, a_2, b_1, b_2, b_3, b_4, c_1, c_2)^T$. Equation (1) is a three-dimensional hyperchotic system, see [3] and its behaviour can be observed in Fig. 1 (computed for parameter vector $\theta = (3.78, 0.2, 0.1, 2, 0.35, 1.9, 3.8, 0.05)^T$ and initial condition $z(0) = (0.542, 0.087, 0.678)^T$).

In [3] a cryptosystem is created as follows:

$$\begin{cases} z_1^+ = a_1 z_1(1-z_1) + a_2 z_2 \\ z_2^+ = b_1[(1-b_2 z_1)(z_2+b_3)-1](1-b_4 z_3) \\ z_3^+ = c_1 z_3(1-z_3) - c_2(1-b_2 z_1)(z_2+b_3) + m \\ w = z_1 \end{cases} \tag{2}$$

where m is the original message (the input of the cryptosystem) and w is the cryptogram (the output of the cryptosystem). The secret key can be represented by a part or totally of hyperchaotic system parameters. Variable $m \in \Re$ is considered the confidential message to be transmitted; for not influencing the hyperchaotic behaviour the amplitude of m must be less than 10^{-2}. Hence, system (2) is considered a "Single Input Single Output" (S.I.S.O.) system which has m as input and w as output.

The structure of the cryptosystem is proposed in this way because, from the observability point of view, see [7] and [8], choosing the state variable z_1 as output is the most appropriate decision. The observability gives important information about the possibility to inverse the system, so by having a sufficient number of samples from the selected output w all the state vector $z := (z_1, z_2, z_3)^T \in \Re^3$ can be reconstructed.

For selecting the state variable where the input will be introduced (included) the singularity manifold is analyzed in [7]. The singularity manifold defines the states where the system cannot be inversed. As conclusion the third state variable is chosen for including the original message, m, and the first state variable as output, w.

By performing a known plain text attack it can be proved that the secret key can be found just using a few samples from the known message and the corresponding samples from the cryptogram. In the context of dynamical system the parameter identifiability defined in [9] is used for proving how the parameter vector can be found. This is proved in [9] for two discrete bi-dimensional cryptosystems build as I.M., Burger map and Hennon map.

For verifying the parameter identifiability, the system (2) is reiterated s times, where s is the observability index of the system. In general, s value is equal to the dimension of the system; for the case study the observability index is $s = 3$, see [3]. So the system (2) is reiterated three times:

$$\begin{cases} z_1^+ - a_1 z_1(1-z_1) - a_2 z_2 = 0 \\ z_1^{++} - a_1 z_1^+(1-z_1^+) - a_2 z_2^+ = 0 \\ z_1^{+++} - a_1 z_1^{++}(1-z_1^{++}) - a_2 z_2^{++} = 0 \\ z_2^+ - b_1[(1-b_2 z_1)(z_2+b_3)-1](1-b_4 z_3) = 0 \\ z_2^{++} - b_1[(1-b_2 z_1^+)(z_2^+ + b_3)-1](1-b_4 z_3^+) = 0 \\ z_2^{+++} - b_1[(1-b_2 z_1^{++})(z_2^{++}+b_3)-1](1-b_4 z_3^{++}) = 0 \\ z_3^+ - c_1 z_3(1-z_3) + c_2(1-b_2 z_1)(z_2+b_3) + m = 0 \\ z_3^{++} - c_1 z_3^+(1-z_3^+) + c_2(1-b_2 z_1^+)(z_2+b_3^+) + m^+ = 0 \\ z_3^{+++} - c_1 z_3^{++}(1-z_3^{++}) + c_2(1-b_2 z_1^{++})(z_2+b_3^{++}) + m^{++} = 0 \\ w - z_1 = 0 \\ w^+ - z_1^+ = 0 \\ w^{++} - z_1^{++} = 0 \\ w^{+++} - z_1^{+++} = 0 \end{cases} \tag{3}$$

An input/output relation is obtained from (3), as in [10], and shows that exists an equation only between the parameter vector θ, the output w and the message m; it can be summarized:

$$\mathscr{F}(\theta, w^{+++}, w^{++}, w^+, m) = 0 \tag{4}$$

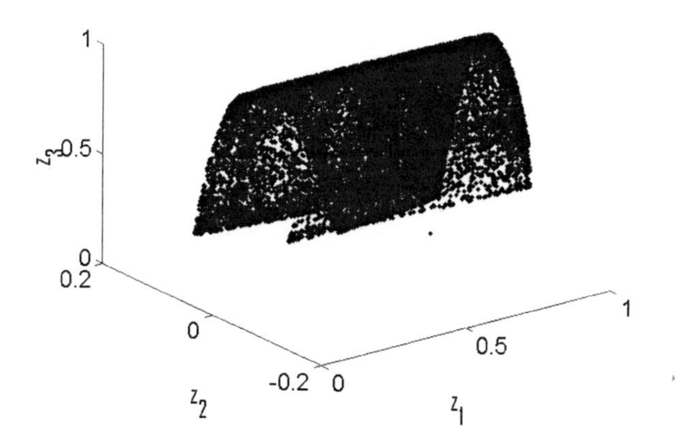

Fig. 1. Rössler map phase portrait, "Folded-Towel" (system (1))

Then, the input/output relation (4) is iterated 7 times and yields $\mathscr{F}_1$, $\mathscr{F}_2, \dots, \mathscr{F}_8$. In this case there are 8 equations for 8 parameters, where:

$$\mathscr{F}_1(a_1, w^{+++}, w^{++}, w^{+}, m) = 0 \tag{5}$$

So, performing plain text attack and having the parameter identifiability, see [11] and [12], it is proved that the robustness of the cryptosystem (2) seems to be weak.

For increasing the distance between the cryptograph and the cryptanalyst an improvement is proposed for the method presented. The structure of the resulted cryptosystem is based on the following system:

$$\begin{cases} z_1^+ = a_1 z_1(1 - z_1) + a_2 z_2^{d-} \\ z_2^+ = b_1[(1 - b_2 z_1)(z_2^{d-} + b_3) - 1](1 - b_4 z_3) \\ z_3^+ = c_1 z_3(1 - z_3) - c_2(1 - b_2 z_1)(z_2^{d-} + b_3) \end{cases} \tag{6}$$

where $z^{d-} := z(k - d(k))$ and d is a vector who contains the delay corresponding to the step k. At a first glance, the hyperchaotic behaviour seems to be conserved, the phase portrait corresponding to the system (6) looks like in Fig. 1. The aim is to build a new cryptosystem on this modified system. A proposed scheme can be similar with (2). The way to recover the ciphered message showed in [3] indicates that the delay can be added in the state variable z_2 and the simplicity of reconstructing the original message is not too much affected, but the robustness is improved. Trying to find an input/output relation as (4) will be more difficult (to find how many times the system (6) has to be reiterated). So the dimension of system (3) is increasing more and more if the delay is different from iteration to iteration.

The experiments from section 3 are performed for a variable delay with maximum value 3, delay randomly generated. For example, for the first 15 iterations the elements of delay vector were randomly generated as follows:

k	1	2	3	4	5	6	7	8	9	10	11	12	13	14	15
d	0	0	0	3	1	2	1	2	1	3	2	3	1	2	2

For exemplification of using the delay vector is presented its influence on the first equation of system (6) at the first iterations:

k	d	System (6), first equation
1	0	$z_1(1) = a_1 z_1(0)(1 - z_1(0)) + a_2 z_2(0 - 0)$
2	0	$z_1(2) = a_1 z_1(1)(1 - z_1(1)) + a_2 z_2(1 - 0)$
3	0	$z_1(3) = a_1 z_1(2)(1 - z_1(2)) + a_2 z_2(2 - 0)$
4	3	$z_1(4) = a_1 z_1(3)(1 - z_1(3)) + a_2 z_2(3 - 3)$
5	1	$z_1(5) = a_1 z_1(4)(1 - z_1(4)) + a_2 z_2(4 - 1)$
6	2	$z_1(6) = a_1 z_1(5)(1 - z_1(5)) + a_2 z_2(5 - 2)$
7	1	$z_1(7) = a_1 z_1(6)(1 - z_1(6)) + a_2 z_2(6 - 1)$

For future applications based on Rössler system with delay it is necessary to verify if the new system has the same statistical properties as the initial Rössler map. With respect to this aspect, the statistical behaviour of two type Rössler system, (1) and (6), are put into comparison in the next section.

3 Statistical Analysis on Systems Behaviour

In this section it will be verified if and how the statistical properties of Rössler map with delay (6) are different than those given in [13] for the initial system. The main experimental results obtained in [13] were focused on the followings:

- the computational measurements of the transient time - the time elapsed from the initial condition (initial state vector) of the system up to its entrance in stationarity.
- the evaluation of the minimum sampling distance which enables the statistical independence between two random variable extracted from the same state variable (z_1 or z_2 or z_3).
- the evaluation of the minimum sampling distance which enables the statistical independence between two different outputs of the system (*i.e.* the pair (z_1, z_3)).

Note that, three random processes assigned to the three state variables are obtained for a fixed set of parameters $\theta = (3.78, 0.2, 0.1, 2, 0.35, 1.9, 3.8, 0.05)^T$, but different initial conditions. In Fig. 3(a) are presented two trajectories for each of the three random processes assigned to the Rössler system without delay. The trajectories are specified by the following two sets of initial conditions: $(z_1(0), z_2(0), z_3(0)) = (0.25, 0.1, 0.75)$ and $(z_1(0), z_2(0), z_3(0)) = (0.8, -0.1, 0.27)$.

3.1 Measuring Transient Time for the Rössler System with Delay

Here the statistical analysis is based on applying Smirnov test in two ways: by measuring the transient time for each state variable of system (6) and by verifying if the probability law for z_i, with $i = \{1, 2, 3\}$, from system (6), is the same with the probability law corresponding to z_i from the system (1).

As for Rössler map the first order probability law appropriate to the random process (in the stationarity region) is unknown, the Smirnov test was selected for measurements of the transient time. The test is based on two independent experimental data sets, $(x_1, x_2, \ldots, x_n)$ and $(y_1, y_2, \ldots, y_m)$, considered as the observed values of two random variables X and Y, which comply with the *i.i.d.* model. The two hypothesis of the test are:

- H_0 : the two random variables X and Y have the same probability law
- H_1 : the two random variables X and Y do not have the same probability law

The test relies on the experimental cumulative distribution function (c.d.f.) $Fe_X(x)$ and $Fe_Y(y)$ for the two random variables X and Y. The aim of the test is to establish if the two random variables X and Y obey to the same probability law.

For measuring the transient time there were considered $n = m = N = 100000$ (the size of the experimental data sets) and the significance level $\alpha = 0.05$. For generating these data sets are needed $2N = 200000$ different initial conditions of the type $(z_1(0), z_2(0), z_3(0))$ which were generated according to the uniform law in the domain $[0.2; 0.8] \times [-0.1; 0.1] \times [0.2; 0.8]$. For the two random variables X and Y the experimental data sets are obtained as follows: by sampling the random process at the iteration k_1 results the set $(x_1, x_2, \ldots, x_N)$ and by sampling, the same state variable, at iteration k_2 results the set $(y_1, y_2, \ldots, y_N)$. The set $(x_1, x_2, \ldots, x_N)$ is obtained from the first N initial conditions and the other set $(y_1, y_2, \ldots, y_N)$ comes from the other N initial conditions (from the ensemble of $2N$). This way of choosing the two data sets ensures the independence between them. The Smirnov test was applied for different values of k_1 and each time k_2 value was kept the same; this k_2 value was selected in the stationarity region. Smirnov test is applied for each pair (k_1, k_2) and if the hypotheses H_0 is accepted, k_1 may indicate the entrance in the stationarity region.

For an accurate decision a Monte Carlo analysis is applied, which consists in resuming the Smirnov test by 500 times. Thus, for each pair (k_1, k_2) and keeping the same volume of experimental data sets $n = m = N = 100000$ the proportion of acceptance of H_0 null hypothesis is recorded.

In Table 1 are presented the experimental results referring to the three random processes assigned to the Rössler map with delay (6), in comparison with the experimental results obtained in [13] for Rössler map without delay (1). On the first row are indicated the iterations k_1 used to obtain the set corresponding to the random variable X; for Y was considered the iteration $k_2 = 200$, which is supposed to be in the stationarity region. For each and every state variable z_i are given the numerical results of the Monte Carlo analysis for the two Rössler maps (with and without delay).

Table 1. The proportion of accepting H_0 for Smirnov test (transient time evaluation)

k_1	10	15	20	25	30	35	40	45	50
z_1	0	33.6	90	**94.2**	**95.8**	**95**	**94**	**94.8**	**95**
	0.51	85.2	91.7	**93.6**	**95.6**	**95.1**	**94.1**	**94.8**	**94.8**
z_2	0	92.6	76	**95**	**95.2**	**96.4**	**93.8**	**95.8**	**94.2**
	0.16	76.7	92.7	**94.8**	**95**	**94.7**	**95.5**	**96.1**	**94.4**
z_3	0	10	90	**94**	**94.2**	**95.4**	**94**	**95**	**95**
	14.6	91.6	92.4	**94.2**	**94.4**	**95.9**	**95**	**95.4**	**95.1**

How to read the table: for example, considering $k_1 = 25$, the state variable z_1 and the Rössler system with delay, the proportion of accepting H_0 null hypothesis is 93.6% (from the total of 500 Smirnov tests applied for the pair $(k_1; k_2) = (25; 200)$). For the same $k_1 = 25$, the same state variable z_1, but the Rössler system without delay, the proportion of accepting H_0 null hypothesis is 94.2%. By analyzing the results, the value $k_1 = 25$ iteration may indicate the beginning of the stationarity region for all state variables of Rössler system without or with delay.

Note. According to the probability estimation theory, if the proportion of H_0 hypothesis acceptance lies in the [93%; 97%] interval, it can be said, with a 95% confidence, that k_1 and k_2 belong to the stationarity region.

Supplementary to the transient time measurement, there were used Smirnov tests for verifying if the probability law for z_i, with $i = \{1, 2, 3\}$, from system (6), is the same with the probability law corresponding to z_i from the system (1).

The size of the experimental data sets was $n = m = N = 100000$ and the significance level $\alpha = 0.05$. For generating these data sets are needed $2N = 200000$ different initial conditions of the type $(z_1(0); z_2(0); z_3(0))$ which were generated according to the uniform law in the domain $[0.2; 0.8] \times [-0.1; 0.1] \times [0.2; 0.8]$. Both systems were initialized by using the same set of initial conditions.

The two data sets required by Smirnov test are obtained as follows:

- for a fixed state variable z_i, the data set $(x_1, x_2, \ldots, x_N)$ is obtained from the system (6) by sampling the state variable at the iteration k_1. The values k_1 were chosen in the stationarity region, so more than 25 iterations;
- the data set $(y_1, y_2, \ldots, y_N)$ is obtained from the system (1) by sampling the same state variable z_i at the iteration $k_2 = 100$ (this iteration is considered in the stationarity region).

For each pair $(k_1; k_2)$ the Smirnov test was resumed 500 times for a more accurate decision. The results are presented in Table 2 and concerning it, the decision upon the probability law of the new system will be discussed. For example, taking into consideration state variable z_3 and iteration $k_1 = 40$, in 96.4% of times from the ensemble of 500, the Smirnov test indicates that the random variable X has the same probability law as the random variable Y. Note that the random variable X is obtained from the third state variable of system (6) at $k_1 = 40$ and the

Table 2. The proportion of accepting H_0 for Smirnov test (verifying probability law for each state variable of systems (1) and (6))

k_1	30	40	50
x_1	96.6	94.7	95
x_2	95.7	97.4	96.2
x_3	95.2	96.4	95.7

random variable Y is obtained from the system (1) at the iteration $k_2 = 100$ also from the third state variable.

Fig. 2 presents the experimental cumulative distribution functions of the random variables extracted from the random processes assigned to each state variable of the two systems. The random processes were sampled at $k_1 = 50$ iterations for the Rössler system without delay (1), respectively $k_2 = 50$ for Rössler system with delay (6).

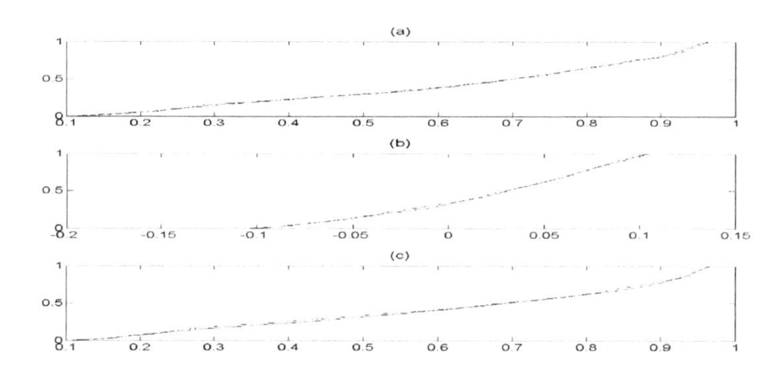

Fig. 2. Experimental Cumulative Distribution Function for Rössler map without delay (solid line) and with delay (dashed line): (a) z_1 state variable, (b) z_2 state variable, (c) z_3 state variable

Remark. The experimental results presented in this subsection sustained that the statistical behaviour of the systems (1) and (6) is quite the same, so the first visual observation concerning the phase portrait of both systems is confirmed. All the advantages of hyperchaotic behaviour could be taken into consideration in the context of future applications based on Rössler system with delay.

3.2 Statistical Independence

In what follows the minimum sampling distance which enables statistical independence between two jointly distributed random variables X and Y sampled from Rössler map is determinate. X and Y are obtained from the random processes at k_1 and $k_2 = k_1 + d$, respectively, where d is the sampling distance under investigation for statistical independence. The following two cases are considered:

(a) X and Y are sampled from a same random process. Each of the three random processes assigned to Rössler map (1), respectively modified Rössler map (6), see Fig. 3(a).
(b) X and Y are sampled from distinct random processes. For example, the two random processes assigned to z_1 and z_3 state variables in (1), or in (6), are investigated, see Fig. 3(b).

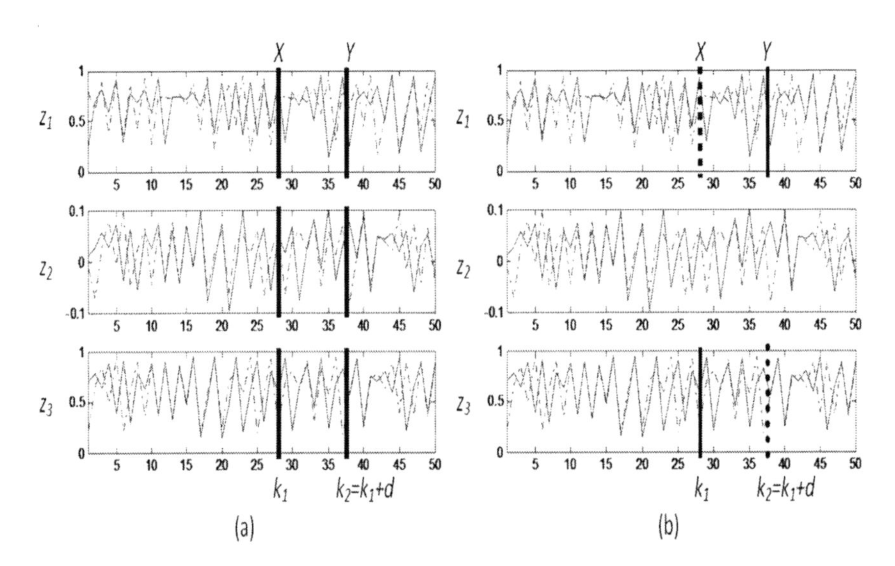

Fig. 3. Trajectories of the random processes; the random variables X and Y are sampled from the same process (a) or distinct random processes (b)

To evaluate the minimum sampling distance d which enables statistical independence between X and Y the test procedure given in [6] was applied. For the first case (a), the independence test was applied in a similar way as in [13] and [14].

The statistical independence in the context of Rössler map without delay was discussed in [13] and in this subsection some results are confirmed and a new analysis is presented. A complete investigation was realised here with respect to the analysis in pair, $(z_i; z_j)$, for all three state variable, where $i \neq j$ and $i, j = 1, 2, 3$; also the statistical independence was verified for the proposed system (6). The aim is to verify the behaviour of Rössler map with delay in order to see if this system can provide new information concerning statistical independence that may be useful in future applications.

The statistical independence procedure starts with the acquisition of two experimental data sets of N size, $(x_1, x_2, \ldots, x_N)$ and $(y_1, y_2, \ldots, y_N)$, measurements of the two random variables X and Y corresponding to the (a) or (b) mentioned case. Each of the two data sets has to comply with the *i.i.d.* statistical model. Here, the data set $(x_1, x_2, \ldots, x_N)$ consists of the values obtained at k_1 iteration, and $(y_1, y_2, \ldots, y_N)$ consists of the values at k_2 iteration. The N initial conditions of the type $(z_1(0), z_2(0), z_3(0))$ were randomly chosen in the domain $[0.2; 0.8] \times [-0.1; 0.1] \times [0.2; 0.8]$.

Note that when a single random process is concerned, as in the case (a), each pair of values (x_i, y_i), $i = \overline{1, N}$ is obtained on the same trajectory at k_1 and $k_2 = k_1 + d$ iterations. In case (b), when the statistical independence is investigated on two distinct random processes, the pair (x_i, y_i) is obtained on two trajectories (x_i on first trajectory at k_1 iteration and y_i on the second trajectory at k_2 iteration).

The two hypotheses of the independence test are: the null hypothesis, H_0, meaning that the random variables X and Y are statistically independent, and the alternative hypothesis, H_1, meaning that the random variables X and Y are dependent. By the test procedure, two successive transforms are applied on each of the two investigated random variables X and Y:

$$\begin{aligned} X &\longrightarrow X' \longrightarrow U \\ Y &\longrightarrow Y' \longrightarrow V \end{aligned}$$

According to the test theory, X' and Y' are uniformly distributed in the $(0; 1)$ interval, and U and V are standard normal random variables. By this approach, the entire independence analysis is shifted by the two transforms from the (x, y) coordinates to (x', y') coordinates and finally to the (u, v) coordinates. This results from: U is a function only of X and V is a function only of Y.

By the test theory, if U and V are dependent then X and Y are dependent; if U and V are independent then X and Y could be statistically independent. For an accurate decision upon statistical independence, a very large amount of data is needed. The sizing of the experiment is discussed in the test theory and is based on some results from Kolmogorov-Smirnov concordance test.

The independence test algorithm consists of the following main steps:

1. The experimental cumulative distribution functions $Fe_X(x)$ and $Fe_Y(y)$, for X and Y, respectively, is computed.
2. Applying $Fe_X(x)$ transform on the x_i values and $Fe_Y(y)$ on the y_i values, $i = \overline{1, N}$, the new data $x_i^{'} = Fe_X(x_i)$ and $y_i^{'} = Fe_Y(y_i)$ are obtained, which are uniformly distributed in $(0; 1)$.
3. Applying a new random variable transform, which is the inverse of the distribution function of the standard normal law (of mean 0 and variance 1) computed by numerical methods, on data $(x_1^{'}, x_2^{'}, \ldots, x_i^{'}, \ldots, x_N^{'})$ and $(y_1^{'}, y_2^{'}, \ldots, y_i^{'}, \ldots, y_N^{'})$, the new data of the type $(u_1, u_2, \ldots, u_i, \ldots, u_N)$ and $(v_1, v_2, \ldots, v_i, \ldots, v_N)$ are obtained. According to the known theory, both u_i and v_i values are standard normally distributed. As U and V are normal

random variables, one can now benefit from Pearson's independence test on Gaussian populations.

4. The correlation coefficient r between U and V and the t test value are computed:

$$r = \frac{\sum_{i=1}^{N}(u_i - \bar{u})(v_i - \bar{v})}{\sqrt{\sum_{i=1}^{N}(u_i - \bar{u})^2 \sum_{i=1}^{N}(v_i - \bar{v})^2}} \tag{7}$$

$$t = r\sqrt{\frac{N-2}{1-r^2}} \tag{8}$$

If the null hypothesis H_0 is correct (X and Y are independent), then the test value should obey the Student law with $N-2$ degrees of freedom. Based on the α significance level, the $t_{\alpha/2}$ point value of the Student law of $N-2$ degrees of freedom is computed. If $|t| \leq t_{\alpha/2}$, then the two random variables X and Y could be independent and in order to decide upon independence the test has to continue with step 5. Conversely, if $|t| \geq t_{\alpha/2}$, the random variables X and Y are dependent and no further investigation is required (the test stops).

5. If $|t| \leq t_{\alpha/2}$, then test if U and V are jointly normally distributed. If U and V are proven jointly normally distributed, the decision is that X and Y are statistically independent. If U and V are not jointly normally distributed, X and Y are dependent. In order to decide upon the normal bivariate law, the test uses the visual inspection of the (u, v) scatter diagram. Thus, to accept statistical independence between X and Y, the (u, v) scatter diagram should be similar to that shown in Fig. 4 - the (u, v) reference scatter diagram.

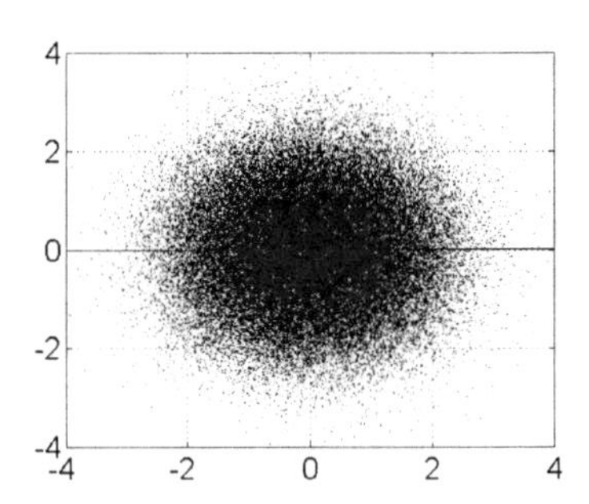

Fig. 4. The reference scatter diagram for two independent random variables in (u, v) coordinates

As it is known, the correlation of the two random variables is equivalent to the existence of a linear dependence between them: the correlation exists if there is a non-zero linear regression slope (an ascending or a descending data trend). In our illustrations the linear regression slope can be also graphically inspected on (u, v) coordinates; it is plotted on the respective scatter diagram and is equal to r correlation coefficient from (7).

The experiments were done for $N = 100000$ initial conditions following a uniform probability law in $[0.2; 0.8] \times [-0.1; 0.1] \times [0.2; 0.8]$ intervals. The significance level was chosen $\alpha = 0.05$, thus the $\alpha/2$ point value of the Student law of $N - 2$ degrees of freedom is $t_{\alpha/2} = 1.96$.

For exemple, let consider X and Y both sampled from the random process assigned to z_1 state variable of Rössler map without delay, with the sampling distance $d = 10$ iterations. The numerical results obtained by means of (7) and (8) are:

$$r = -0.0228 \Rightarrow t = 7.2241 \Rightarrow |t| \geq t_{\alpha/2}$$

Following the test procedure by the numerical results the conclusion is that the two random variable X and Y are statistically dependent for $d = 10$, so the test algorithm stops. The result is confirmed observing that the scater diagram in (u, v) coordinates, Fig. 5(a), differs from the respective reference scater diagram from Fig. 4. Also, the regression slope indicates correlated data.

For the same state variable z_1 of Rössler map without delay, but the sampling distance $d = 35$ iterations, the numerical results:

$$r = -0.0042 \Rightarrow t = 1.3145 \Rightarrow |t| \leq t_{\alpha/2}$$

The test procedure decides that the two random variables are statistically independent. The correlation coefficient r has a small value so it can results a t value lower than the given threshold $t_{\alpha/2}$ and the scatter diagram in (u, v) coordinates, Fig. 5(b), resembles at the respective reference scatter diagram in Fig. 4.

The final conclusions on the minimum sampling distance which enables statistical independence are supported by a Monte Carlo analysis. The numerical results of these analysis are presented in Table 3 for system (1) and in Table 4 for system (6). The test procedure was resumed 500 times. The experiments were computed for $k_1 = 100$ and $k_2 = k_1 + d$ (for different values k_1 in the stationarity region the results were similar).

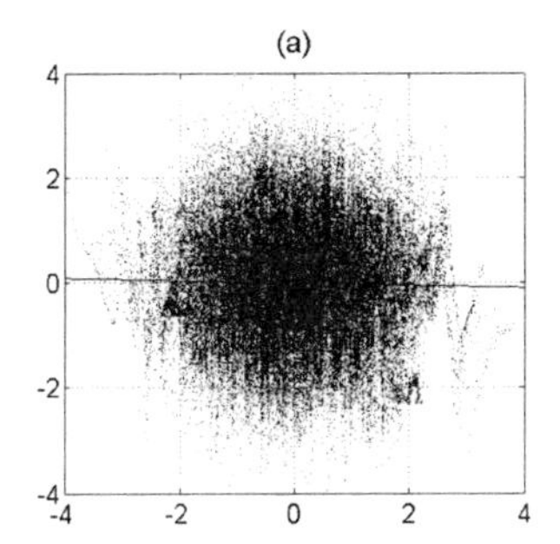

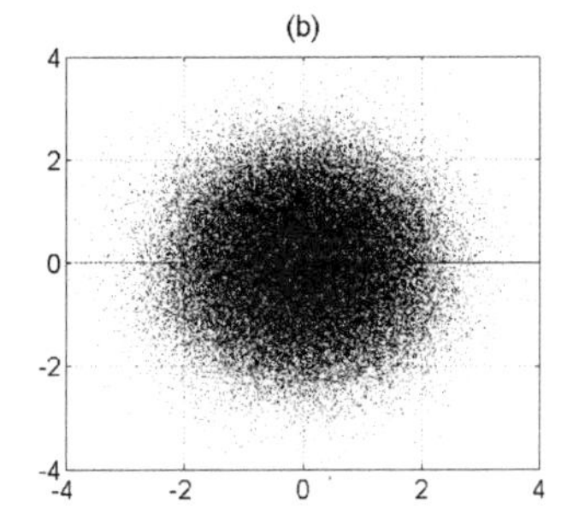

Fig. 5. Scater diagram for (u, v) coordinates and z_1 state variable from system (1): (a) distance $d = 10$; (b) distance $d = 35$

How to read the tables with the experimental results: for example, on the last column of Table 3, the random variable X is obtained at iteration k_1 from the random process assigned to state variable z_1 and the random variable Y is obtained at iteration k_2 from the random process assigned to state variable z_3, $X := z_1[k_1]$ and $Y := z_3[k_2]$. In this case the decision is that $d = 30$ is the minimum sampling distance which enables statistical independence, because for $d \geq 30$ the proportion of acceptance of H_0 null hypothesis remains in the range $[93\%; 97\%]$. For $d = 30$, the proportion is 95.6% (see last column of Table 3), so it is in the interval mentioned, as the other bold values from the table.

According to first three columns Table 3, for each state variable a value $d = 35$ iterations is a minimum sampling distance which enables statistical independence. The analysis on last six columns of this table presents the independence investigation between two different state variables. For example from the last two columns it can be noticed that z_3 becomes independente with respect to z_1 after 30 iterations, while z_1 becomes independent with respect to z_3 after 40 iterations.

The Table 4 presents the results by applying the independence test on system (6) - where the delay is adjusted on the system structure. It can be observed

Table 3. The proportion of acceptance for H_0 null hypothesis for independence test applied on system (1)

d	X / Y	$z_1[k_1]$ / $z_1[k_2]$	$z_2[k_1]$ / $z_2[k_2]$	$z_3[k_1]$ / $z_3[k_2]$	$z_1[k_1]$ / $z_2[k_2]$	$z_2[k_1]$ / $z_1[k_2]$	$z_2[k_1]$ / $z_3[k_2]$	$z_3[k_1]$ / $z_2[k_2]$	$z_3[k_1]$ / $z_1[k_2]$	$z_1[k_1]$ / $z_3[k_2]$
10		0	84.8	0.98	0	0	11.8	71	77.6	23.8
20		95	95.8	89.8	47.6	37.2	69.8	62.8	0	59
25		60.4	92.4	87	**94.6**	66	**95.6**	89	86.2	85
30		83.6	**95.2**	89.6	**93.2**	**95**	**93.4**	**94**	86	**95.6**
35		**96**	**94.4**	**96**	**94.4**	**93.8**	**95.8**	**94.8**	89.6	**93.4**
40		**95.6**	**94.6**	**95.6**	**95.2**	**95.4**	**96.6**	**94.8**	**96.6**	**94.8**
45		**95.2**	**95.8**	**95.2**	**96.2**	**94.6**	**95**	**95.4**	**95**	**95.6**
50		**93.2**	**97.2**	**93.2**	**94.4**	**93.4**	**95.8**	**94.2**	**95.8**	**94.2**

Table 4. The proportion of acceptance for H_0 null hypothesis for independence test applied on system (6) - the system with delay

d	X / Y	$z_1[k_1]$ / $z_1[k_2]$	$z_2[k_1]$ / $z_2[k_2]$	$z_3[k_1]$ / $z_3[k_2]$	$z_1[k_1]$ / $z_2[k_2]$	$z_2[k_1]$ / $z_1[k_2]$	$z_2[k_1]$ / $z_3[k_2]$	$z_3[k_1]$ / $z_2[k_2]$	$z_3[k_1]$ / $z_1[k_2]$	$z_1[k_1]$ / $z_3[k_2]$
10		16	26.6	27.2	0	8.20	11.6	0.80	3	27.4
20		84.6	**92.6**	90	95.4	91	89.2	64.6	80.2	79.2
25		**93.4**	**94.6**	**93.6**	**94.2**	**94.4**	**94.4**	**93.6**	**91.8**	**91.6**
30		**94.4**	**95.6**	**95.4**	**97.2**	**95**	**94.2**	**94.2**	**93.6**	**95.8**
35		**95.4**	**96.2**	**94.6**	**94.8**	**94.4**	**95.2**	**92.6**	**95.2**	**94.8**
40		**96**	**95.8**	**95.8**	**96**	**94.6**	**96.8**	**94.6**	**96.8**	**94.6**
45		**96**	**96**	**94.8**	**96**	**94.8**	**95.2**	**94.2**	**94.8**	**94.2**
50		**93.2**	**95.8**	**95.8**	**94.8**	**95.2**	**95.6**	**94.4**	**95.6**	**94.4**

that, in all cases, the minimum sampling distance which enables statistical independence is decreasing with respect of the same situation from Table 3, by comparing column by column the two tables.

4 Conclusions

The idea of chaotic discrete system with delay, introduced with the purpose of designing a more robust cryptosystem, was proved as being also useful from the statistical point of view, with respect to the "identification" attack. Consequently, the robustness will increase and the statistical behaviour is not changing if the delay is adjusted in the structure of the initial system; so, all the advantages of hyperchaotic dynamics can be preserved.

The results on the statistical independence show that the modified system gives a smaller distance with respect to the minimum sampling distance which enables the statistical independence. Thus, it can be said that the distance between the cryptograph and the cryptanalyst is increasing because the output considered variable becomes faster independed with respect to the dynamic variable where the original message is included.

Acknowledgement. The work has been funded by the Sectoral Operational Programme Human Resources Development 2007-2013 of the Romanian Ministry of Labour, Family and Social Protection through the Financial Agreement POSDRU/6/1.5/S/19.

References

1. Pecora, L.M., Carroll, T.L.: Synchronizing chaotic circuits. IEEE Trans. Circuits Systems 38, 453–456 (1991)
2. Nijmeijer, H., Marels, M.Y.: An observer looks at synchronization. IEEE Trans. Circuits and Syst. I 44, 307 (1997)
3. Belmouhoub, I., Djemaï, M., Barbot, J.-P.: Cryptography By Discrete-Time Hyperchaotic Systems. In: Proceedings of the 42nd IEEE Conference on Decision and Control Maui, Hawaii, USA, pp. 1902–1907 (2003)
4. Larger, L., Goedgebuer, J.-P.: Le chaos chiffrant, Pour La Science. N36 (2002)
5. Wolf, A., Swift, J.B., Swinney, H.L., Vastano, J.A.: Determining Lyapunov Exponents from a time series. Physica 16D, 285–317 (1985)
6. Badea, B., Vlad, A.: Revealing statistical independence of two experimental data sets: An improvement on spearman's algorithm. In: Gavrilova, M.L., Gervasi, O., Kumar, V., Tan, C.J.K., Taniar, D., Laganá, A., Mun, Y., Choo, H. (eds.) ICCSA 2006. LNCS, vol. 3980, pp. 1166–1176. Springer, Heidelberg (2006)
7. Perruquetti, W., Barbot, J.-P.: Chaos in automatic control. CRC Press, Boca Raton (2006), ISBN 0824726537
8. Letellier, C., Aguirre, L.A.: On the interplay among synchronization, observability and dynamics. Physical Review E 82(1), id. 016204 (2010)
9. Anstett, F., Millerioux, G., Bloch, G.: Chaotic Cryptosystems: Cryptanalysis and Identifiability. IEEE Transactions on Circuits and Systems I: Regular Papers, pp. 2673–2680 (2006), ISSN: 1549-8328

10. Fliess, M.: Automatique en temps discret et algèbre aux differences. Mathematicum 2, 213–232 (1990)
11. Ljung, L.: System Identification - Theory for the User, 2nd edn., pages 607. Prentice-Hall, Upper Saddle River (2002) (Chinese edn.), ISBN 0-13-656695-2
12. Nomm, S., Moog, C.H.: Identifiability of discrete-time nonlinear systems. In: Proc. of the 6th IFAC Symposium on Nonlinear Control Systems, NOLCOS, Stuttgart, Germany, pp. 477–489 (2004)
13. Frunzete, M., Luca, A., Vlad, A.: On the Statistical Independence in the Context of the Rössler Map. In: 3rd Chaotic modeling and Simulation International Conference (CHAOS 2010), Chania, Greece (2010)
14. Vlad, A., Luca, A., Frunzete, M.: Computational Measurements of the Transient Time and of the Sampling Distance That Enables Statistical Independence in the Logistic Map. In: Gervasi, O., Taniar, D., Murgante, B., Laganà, A., Mun, Y., Gavrilova, M.L. (eds.) ICCSA 2009. LNCS, vol. 5593, pp. 703–718. Springer, Heidelberg (2009)

Author Index

Abánades, Miguel A. IV-353
Abenavoli, R. Impero IV-258
Aberer, Karl III-566
Addesso, Paolo II-354
Agarwal, Suneeta V-398
Aguilar, José Alfonso V-421
Aguirre-Cervantes, José Luis IV-502
Ahn, Deukhyeon III-495
Ahn, Jin Woo II-463
Ahn, Minjoon IV-173
Ahn, Sung-Soo IV-225, IV-248
Akman, Ibrahim V-342
Alghathbar, Khaled V-458
Ali, Amjad IV-412
Ali, Falah H. I-573
Alizadeh, Hosein I-526
Almendros-Jiménez, Jesús M. I-177
Aloisio, Giovanni IV-562, IV-572
Alonso, César L. I-550
Amjad, Sameera V-383
Anjos, Eudisley V-270
Arabi Naree, Somaye II-610
Ardanza, Aitor IV-582
Arias, Enrique I-615
Arolchi, Agnese II-376
Aryal, Jagannath I-439
Asche, Hartmut I-329, I-492, II-366
Asif, Waqar IV-133
Astrakov, Sergey N. III-152
Azad, Md. Abul Kalam III-245
Azam, Farooque V-383

Bae, Doohwan V-326
Bae, Sueng Jae V-11, V-32
Bagci, Elife Zerrin V-521
Baldassarre, Maria Teresa V-370
Balucani, Nadia III-453
Bang, Young-Cheol IV-209
Baraglia, Ranieri III-412
Baranzelli, Claudia I-60
Barbot, Jean-Pierre I-706
Baresi, Umberto I-162
Barrientos, Antonio III-58
Bastianini, Riccardo III-466
Becerra-Terón, Antonio I-177
Bechtel, Benjamin I-381
Bélec, Carl I-356
Benedetti, Alberto I-162
Benito, Juan José IV-35
Bertolotto, Michela II-51
Bertot, Yves IV-368
Berzins, Raitis II-78
Bhardwaj, Shivam V-537
Bhowmik, Avit Kumar I-44
Bicho, Estela III-327
Bimonte, Sandro I-17
Blat, Josep IV-547
Blecic, Ivan I-423, I-477, II-277
Blondia, Chris III-594
Bocci, Enrico IV-316
Böhner, Jürgen I-381
Bollini, Letizia I-501
Borfecchia, Flavio II-109
Borg, Erik II-366
Borges, Cruz E. I-550
Borruso, Giuseppe I-454
Botana, Francisco IV-342, IV-353
Botón-Fernández, María II-475
Bouaziz, Rahma V-607
Bouroubi, Yacine I-356
Bravi, Malko III-412
Brennan, Michael I-119
Buccarella, Marco IV-270
Bugarín, Alberto IV-533
Burdalski, Maciej II-63
Butt, Wasi Haider V-383

Cabral, Pedro I-44, I-269
Cação, I. III-271, III-316
Caeiro-Rodriguez, Manuel II-506
Cafer, Ferid V-342
Caglioni, Matteo I-135
Caminero, A.C. III-582
Campobasso, Francesco I-342
Campos, Alessandra M. III-654
Capannini, Gabriele III-412
Carlini, Maurizio IV-277, IV-287
Carlucci, Angelo II-243

Carneiro, Tiago IV-75
Carrasco, Eduardo IV-582
Carretero, J. III-582
Casas, Giuseppe Las II-243
Casavecchia, Piergiorgio III-453
Castellucci, Sonia IV-277
Castrillo, Francisco Prieto II-475
Catasta, Michele III-566
Cattani, Carlo IV-287, IV-644
Cavinato, Gian Paolo I-92
Cazorla, Diego I-615
Cecchini, Arnaldo I-423, I-477, II-277
Cecchini, Massimo IV-296, IV-307
Cestra, Gabriele I-225
Cha, Myungsu V-193
Chacón, Jonathan IV-547
Chan, Weng Kong III-668
Charvat, Karel II-78
Chau, Ming II-648, II-664
Chaudhuri, Amartya IV-472
Chen, Chao IV-582
Chen, Gao V-458
Chen, Jianyong IV-604
Chen, Ming V-562
Chen, Pei-Yu I-667
Chen, Xin-Yi III-608
Chen, Yen Hung III-141
Chengrong, Li IV-50
Chiabai, Aline II-227
Chiarullo, Livio II-227
Cho, Hyung Wook V-32
Cho, Yongyun IV-452, IV-462
Choi, Bum-Gon V-11, V-22
Choi, Seong Gon V-205
Choo, Hyunseung IV-148, IV-173, V-32, V-181, V-193
Chua, Fang-Fang V-471
Chung, GyooPil V-133
Chung, Min Young V-11, V-22, V-32
Chung, Tai-Myoung I-537
Ciotoli, Giancarlo I-92
Cividino, Sirio IV-270
Clementini, Eliseo I-225
Colantoni, Andrea IV-270, IV-296, IV-307
Coll, Eloina I-152
Colorado, Julian III-58
Conte, Roberto II-354
Convery, Sheila I-119
Coors, Volker I-300
Corcoran, Padraig II-51
Costa, Lino III-343
Costa, M. Fernanda P. III-231, III-327
Costa e Silva, Eliana III-327
Costachioiu, Teodor II-293
Costantini, A. III-387
Crocchianti, Stefano III-453
Cruz, Carla III-358

Daneke, Christian I-381
Daneshpajouh, Shervin III-132
D'Angelo, Gianlorenzo II-578
Dantas, Sócrates de O. III-654
Dao, Manh Thuong Quan IV-148
Das, Sandip III-84
DasBit, Sipra IV-472
Dasgupta, Arindam I-643
de Almeida, Rafael B. III-654
de Almeida Leonel, Gildo I-690
de Castro, Juan Pablo I-76
De Cecco, Luigi II-109
Decker, Hendrik V-283
Deffuant, Guillaume I-17
De Florio, Vincenzo III-594
de la Dehesa, Javier I-550
del Cerro, Jaime III-58
dela Cruz, Pearl May I-269
Della Rocca, Antonio Bruno II-376
De Mauro, Alessandro IV-582
D'Emidio, Mattia II-578
De Paolis, Lucio Tommaso IV-562, IV-572
de Rezende, Pedro J. III-1
De Santis, Fortunato II-330
Desnos, Nicolas V-607
de Souza, Cid C. III-1
Dévai, F. III-17
Dias, Joana M. III-215
Diego, Vela II-624
Di Martino, Ferdinando II-15
Di Rosa, Carmelo II-151
Di Trani, Francesco I-410
do Carmo Lopes, Maria III-215
Domínguez, Humberto de Jesús Ochoa II-522
Doshi, Jagdeep B. II-695
Dragoni, Aldo F. IV-572
Drlik, Martin V-485
Duarte, José II-185
Dutta, Goutam II-695

Dzerve, Andris II-78
Dzik, Karol II-63

Ebrahimi Koopaei, Neda II-610
Elias, Grammatikogiannis II-210
El-Zawawy, Mohamed A. V-355
Engemaier, Rita I-329
Eom, Young Ik III-495, V-147, V-217
Erdönmez, Cengiz IV-103
Erlhagen, Wolfram III-327
Erzin, Adil I. III-152, V-44
Escribano, Jesús IV-353
e Silva, Filipe Batista I-60
Esnal, Julián Flórez IV-582
Espinosa, Roberto II-680
Ezzatti, P. V-643

Falcão, M.I. III-200, III-271, III-358
Falk, Matthias I-423
Fanizzi, Annarita I-342
Faria, Sergio IV-75
Fattoruso, Grazia II-376
Fazio, Salvatore Di I-284
Ferenc, Rudolf V-293
Fernandes, Edite M.G.P. III-174, III-185, III-231, III-245, III-287
Fernández, Juan J. II-303
Fernández-Sanz, Luis V-257
Ferreira, Brigida C. III-215
Ferreira, Manuel III-343
Fichera, Carmelo Riccardo I-237
Fichtelmann, Bernd II-366
Finat, Javier II-303
Fontenla-Gonzalez, Jorge II-506
Formosa, Saviour II-125
Fouladgar, Mohammadhani V-622
Freitag, Felix III-540
Frigioni, Daniele II-578
Fritz, Steffen II-39
Frunzete, Madalin I-706
Fuglsang, Morten I-207
Fusco, Giovanni I-135
Fúster-Sabater, Amparo I-563

Galli, Andrea I-369
Gámez, Manuel V-511
Garay, József V-511
Garcia, Ernesto III-453
García, Félix V-370
Garcia, Inma V-547
García, Ricardo I-76
Garcia, Thierry II-648, II-664
García-Castro, Raúl V-244
García-García, Francisco I-177
Garg, Sachin III-107
Garrigós, Irene V-421
Garzón, Mario III-58
Gavete, Luis I-677, III-676, IV-35
Gavete, M. Lucía IV-35
Gervasi, O. III-387
Ghedira, Khaled II-594
Ghodsi, Mohammad III-132
Gholamalifard, Mehdi I-32
Ghosal, Amrita IV-472
Ghosh, S.K. I-643
Giacchi, Evelina I-652
Giaoutzi, Maria II-210
Gil-Agudo, Ángel IV-582
Gilani, Syed Zulqarnain Ahmad II-534
Giorguli, Silvia I-192
Giuseppina, Menghini IV-270
Goličnik Marušić, Barbara II-136
Gomes, Carla Rocha I-60
Gomes, Jorge II-185
González, María José IV-384
González-Aguilera, Diego II-303
González-Vega, Laureano IV-384
Goswami, Partha P. III-84
Graj, Giorgio I-162
Gruber, Marion IV-518
Guillaume, Serge I-356
Gulinck, Hubert I-369
Guo, Cao IV-50
Gupta, Pankaj III-300
Gutiérrez, Edith I-192
Gyimóthy, Tibor V-293

Hailang, Pan IV-50
Halder, Subir IV-472
Hamid, Brahim V-607
Hammami, Moez II-594
Han, Chang-Min II-635
Han, Soonhee II-635
Handoyo, Sri I-315
Hansen, Henning Sten I-207
Hanzl, Małgorzata II-63
Hashim, Mazlan II-318
Hernández-Leo, Davinia IV-547
Hilferink, Maarten I-60
Hobza, Ladislav III-30

Hodorog, Mădălina III-121
Hong, Kwang-Seok V-58
Hong, Qingqi IV-592
Hong, Young-Ran III-506
Hou, Xianling L. IV-619, IV-633
Hreczany, David III-479
Hu, Shaoxiang X. IV-619, IV-633
Hur, Kunesook II-31

Ilieva, Sylvia V-232
İmrak, Cevat Erdem IV-103
Iqbal, Muddesar IV-412
Irshad, Azeem IV-412
Iyer, Ravishankar K. III-479

James, Valentina II-109, II-376
Janecka, Karel II-78
Jang, JiNyoung V-133
Jeon, Gwangil IV-185
Jeon, Jae Wook V-96, V-110
Jeong, EuiHoon IV-185, IV-209
Jeong, Jongpil IV-235
Jeong, Seungmyeong V-70
Jeong, Soon Mook V-96, V-110
Jeung, Hoyoung III-566
Jeung, Jaemin V-70
Jin, Seung Hun V-110
José, Jesús San II-303
José Benito, Juan I-677, III-676
Josselin, Didier I-439
Jung, Hyunhee V-593
Jung, Sung-Min I-537

Kanade, Gaurav III-107
Kang, Miyoung V-96
Karmakar, Arindam III-84
Kelle, Sebastian IV-518
Khan, Bilal Muhammad I-573
Khan, Muhammad Khurram V-458
Khan, Zeeshan Shafi V-447
Ki, Junghoon II-31
Kim, ByungChul IV-424
Kim, Cheol Hong II-463
Kim, Dae Sun V-167
Kim, Dong In V-157
Kim, Dong-Ju V-58
Kim, Dong Kyun V-110
Kim, Dongsoo III-506
Kim, Hongsuk V-181
Kim, Hyungmin V-96
Kim, Hyun Jung III-622
Kim, Hyun-Sung III-608, III-622, V-593
Kim, Inhyuk V-147, V-217
Kim, Jeehong III-495
Kim, Jong Myon II-463
Kim, Jung-Bin V-133
Kim, Junghan V-147, V-217
Kim, Junghoon III-495
Kim, Jun Suk V-22
Kim, Mihui IV-173, V-193
Kim, Minsoo IV-225
Kim, Moonseong V-193
Kim, Myung-Kyun IV-197
Kim, Nam-Uk I-537
Kim, SunHee IV-209
Kim, Taeseok III-528
Kim, Young-Hyuk V-83
Kim, Youngjoo III-528
Kinoshita, Tetsuo V-410
Kitatsuji, Yoshinori V-167
Klemke, Roland IV-518
Kloos, Carlos Delgado IV-488
Knauer, Christian III-44
Ko, Byeungkeun III-528
Kocsis, Ferenc V-293
Kodama, Toshio III-556
Kolingerová, Ivana III-30, III-163
Koomen, Eric I-60
Kovács, István V-293
Kowalczyk, Paulina II-63
Kriegel, Klaus III-44
Krings, Axel II-490
Kubota, Yuji II-547
Kujawski, Tomasz II-63
Kunigami, Guilherme III-1
Kunii, Tosiyasu L. III-556
Kuzuoglu, Mustafa IV-11
Kwak, Ho-Young V-1
Kwiecinski, Krystian II-63
Kwon, Key Ho V-96, V-110
Kwon, NamYeong V-181
Kwon, Young Min V-11

Lachance-Bernard, Nicolas II-136
La Corte, Aurelio I-652
Laganà, Antonio III-387, III-397, III-412, III-428, III-442, III-453, III-466
Lama, Manuel IV-533
Langkamp, Thomas I-381
Lanorte, Antonio II-330, II-344

Lanza, Viviana II-265
La Porta, Luigi II-376
Lasaponara, Rosa II-330, II-344, II-392, II-407
Lavalle, Carlo I-60
Lazarescu, Vasile II-293
Leal, José Paulo V-500
Lee, Byunghee V-437
Lee, Chien-Sing V-471
Lee, Dongyoung IV-225, IV-248
Lee, Jae-Joon V-133
Lee, Jae-Kwang V-83
Lee, JaeYong IV-424
Lee, Jongchan IV-452
Lee, Junghoon V-1
Lee, Kue-Bum V-58
Lee, Kwangwoo IV-123, V-437
Lee, MinWoo V-133
Lee, Sang-Woong II-635
Lee, Sook-Hyoun V-120
Lee, Tae-Jin V-120
Lei, Shi IV-50
Leng, Lu V-458
Leung, Ying Tat II-93
Li, Qingde IV-592
Li, Shangming IV-26
Li, Sikun V-577
Liao, Zhiwu W. IV-619, IV-633
Liguori, Gianluca I-225
Lim, Il-Kown V-83
Lim, JaeSung V-70, V-133
Lim, SeungOk IV-209
Lima, Tiago IV-75
Limiti, M. IV-258
Liu, Lei V-577
Llorente, I.M. III-582
Lobosco, Marcelo III-654
Lo Curzio, Sergio II-376
Longo, Maurizio II-354
López, Inmaculada V-511
López, Luis María II-436
López, Pablo I-76
López, Rosario II-436
Losada, R. IV-328
Luca, Adrian I-706
Lucas, Caro I-588
Luo, Jun III-74

Magri, Vincent II-125
Mahapatra, Priya Ranjan Sinha III-84
Mahboubi, Hadj I-17
Mahini, Reza I-588
Mahiny, Abdolrassoul Salman I-32
Maier, Georg IV-91
Malonek, H.R. III-261, III-271, III-316, III-358
Mancera-Taboada, Juan II-303
Mancini, Marco I-92
Manfredi, Gaetano II-109
Manfredini, Fabio II-151
Manso-Callejo, Miguel-Angel I-394
Manuali, C. III-397
Marcheggiani, Ernesto I-369
Marconi, Fabrizio I-92
Marghany, Maged II-318
Marras, Serena I-423
Marsal-Llacuna, Maria-Lluïsa II-93
Martínez, Brian David Cano II-522
Martínez, José II-303
Martínez, Rubén II-303
Martinez-Llario, Jose I-152
Martini, Sandro II-109
Martins, Tiago F.M.C. III-185
Marucci, Alvaro IV-307
Masi, Angelo I-410
Masini, Nicola II-392
Mateu, Jorge I-269
Maurizio, Vinicio II-578
Maynez, Leticia Ortega II-522
Mazón, Jose-Norberto II-680, V-421
McCallum, Ian II-39
Medina, Esunly III-540
Mendes, José I-1
Messeguer, Roc III-540
Miklós, Zoltán III-566
Milani, Alfredo V-537
Min, Sangyoon V-326
Minaei, Behrouz I-526
Minaei-Bidgoli, Behrouz V-622
Miranda, Fernando III-200
Mirmomeni, Masoud I-588
Misra, Sanjay V-257, V-342, V-398
Miszkurka, Michał II-63
Modica, Giuseppe I-237, I-284
Molina, Pedro IV-35
Monarca, Danilo IV-296, IV-307
Montaña, José L. I-550
Montenegro, Nuno II-185
Montesano, Tiziana II-330
Montrone, Silvestro I-342

Moon, Jongbae IV-452, IV-462
Mooney, Peter II-51
Moreira, Adriano I-1
Moreira, Fernando V-500
Moscatelli, Massimiliano I-92
Moura-Pires, João I-253
Mourrain, Bernard III-121
Mubareka, Sarah I-60
Münier, Bernd I-207
Munk, Michal V-485
Muñoz-Caro, Camelia I-630
Murgante, Beniamino I-410, II-255, II-265

Nagy, Csaba V-293
Nalli, Danilo III-428
Nam, Junghyun IV-123, V-437
Narboux, Julien IV-368
Nasim, Mehwish IV-159
Neuschmid, Julia II-125, II-162
Ngan, Fantine III-374
Ngo, Hoai Phong IV-197
Nguyen, Ngoc Duy IV-148
Nikšič, Matej II-136
Niño, Alfonso I-630
Nita, Iulian II-293
Niyogi, Rajdeep V-537
Nolè, Gabriele II-407
Ntoutsi, Irene II-562
Nuñez, A. III-582

Obersteiner, Michael II-39
Oh, Chang-Yeong V-120
Oh, DeockGil IV-424
Oh, Kyungrok V-157
Oh, Seung-Tak V-181
Oliveira, Lino V-500
Oliveira, Miguel III-343
Onaindia, Eva V-547
Opioła, Piotr IV-112
Ortigosa, David II-450
Oßenbrügge, Jürgen I-381
Oyarzun, David IV-582
Ozgun, Ozlem IV-11

Pacifici, Leonardo III-428
Paik, Juryon IV-123, V-437
Paik, Woojin IV-123
Pajares, Sergio V-547
Palazuelos, Camilo III-638
Pallottelli, Simonetta III-466
Palomino, Inmaculada IV-35
Pampanelli, Patrícia III-654
Panneton, Bernard I-356
Paolillo, Pier Luigi I-162
Parada G., Hugo A. IV-488
Pardo, Abelardo IV-488
Pardo, César V-370
Park, Gyung-Leen V-1
Park, Jae Hyung II-463
Park, Jeong-Seon II-635
Park, Kwangjin IV-185
Park, ManKyu IV-424
Park, Sangjoon IV-452
Park, Seunghun V-326
Park, Young Jin II-463
Parsa, Saeed II-610
Parvin, Hamid I-526, V-622
Pascale, Carmine II-109, II-376
Pascual, Abel IV-384
Patti, Daniela II-162
Pavlov, Valentin V-232
Peçanha, João Paulo I-690, III-654
Pech, Pavel IV-399
Perchinunno, Paola I-342
Pereira, Ana I. III-287
Perger, Christoph II-39
Pernin, Jean-Philippe IV-502
Petrov, Laura I-119
Petrova-Antonova, Dessislava V-232
Pham, Tuan-Minh IV-368
Piattini, Mario V-370
Pierri, Francesca II-422
Pino, Francisco V-370
Plaisant, Alessandro II-277
Plotnikov, Roman V. V-44
Pollino, Maurizio I-237, II-109, II-376
Pons, Josè Luis IV-582
Poplin, Alenka II-1
Poturak, Semir II-63
Prasad, Rajesh V-398
Produit, Timothée II-136
Prud'homme, Julie I-439
Pyles, David R. I-423

Qaisar, Saad IV-133, IV-159
Queirós, Ricardo V-500
Quintana-Ortí, E.S. V-643

Raba, N.O. V-633
Radliński, Łukasz V-310

Radulovic, Filip V-244
Rajasekharan, Shabs IV-582
Rambaldi, Lorenzo IV-316
Randrianarivony, Maharavo IV-59
Rao, Naveed Iqbal II-534
Rashid, Khalid V-447
Recio, Tomás IV-328, IV-384
Regueras, Luisa María I-76
Remón, A. V-643
Ren, Guang-Jie II-93
Restaino, Rocco II-354
Reyes, Sebastián I-630
Rezazadeh, Hassan I-588
Ricci, Paolo II-109
Ricciardi, Francesco IV-572
Ristoratore, Elisabetta II-109
Rocca, Lorena II-227
Rocha, Ana Maria A.C. III-185, III-343
Rocha, Humberto III-215
Rocha, Jorge Gustavo II-172
Rodríguez-Gonzálvez, Pablo II-303
Rolewicz, Ian III-566
Romero, Francisco Romero V-370
Rossi, Claudio III-58
Rotondo, Francesco II-199
Royo, Dolors III-540
Rubio, Julio IV-384
Ruiz-Lopez, Francisco I-152
Ruskin, Heather J. I-602
Ryu, Yeonseung III-518

Said, Nesrine II-594
Sajavičius, Svajūnas IV-1
Salete, Eduardo I-677, III-676
Sánchez, José L. I-615
Sánchez, Landy I-192
Sánchez, Vianey Guadalupe Cruz II-522
San-Juan, Juan Félix II-436, II-450
San-Martín, Montserrat II-450
Santiago, Manuel III-374
Santo, Isabel A.C.P. Espírito III-174
Santos, Cristina P. III-343
Sanz, David III-58
Saracibar, Amaia III-453
Sayikli, Cigdem V-521
Scatá, Marialisa I-652
Schicho, Josef III-121
Schill, Christian II-39
Schindler, Andreas IV-91
Schoier, Gabriella I-454
Schrenk, Manfred II-125, II-162
Scorza, Francesco II-243, II-255, II-265
Sebastia, Laura V-547
See, Linda II-39
Seki, Yoichi III-556
Selicato, Francesco II-199
Selmane, Schehrazad V-527
Sen, Jaydip IV-436
Seo, Dae-Young IV-185
Sessa, Salvatore II-15
Shafiq, Muhammad IV-412
Shahumyan, Harutyun I-119
Sharma, Anuj Kumar V-398
Shen, Jie II-624
Sher, Muhammad V-447
Shin, Eunhwan V-147, V-217
Shin, MinSu IV-424
Shon, Min Han V-193
Shu, Jian-Jun III-668
Siabato, Willington I-394
Silva, Ricardo I-253
Singh, Alok V-398
Singh, Sanjeet III-300
Skouteris, Dimitrios III-442
Skouteris, Dimitris III-428
Śliwka, Anna II-63
Smirnov, Arseny III-94
Sohn, Sung Won V-205
Son, Dong Oh II-463
Son, Zeehan V-193
Song, Tae Houn V-96, V-110
Spano, Donatella I-423
Spassov, Ivaylo V-232
Specht, Marcus IV-518
Spiliopoulou, Myra II-562
Spiteri, Pierre II-648, II-664
Stankiewicz, Ewa II-63
Stankova, E.N. V-633
Stankutė, Silvija I-492
Stehn, Fabian III-44
Stein, Ariel F. III-374
Stigliano, Francesco I-92
Sztajer, Szymon I-512

Tagliolato, Paolo II-151
Takahashi, Daisuke II-547
Tan, Li II-490
Tasso, Sergio III-466
Terlizzi, Luca I-162

Theodoridis, Yannis II-562
Tian, Jie IV-592
Tilio, Lucia I-410, II-265
Tomaz, G. III-261
Tominc, Biba II-136
Torre, Carmelo M. I-466
Torricelli, Diego IV-582
Trčka, Jan III-30
Tremblay, Nicolas I-356
Trunfio, Giuseppe A. I-423, I-477
Tucci, Andrea O.M. IV-287

Uchiya, Takahiro V-410
Ukil, Arijit IV-436
Urbano, Paulo II-185
Ureña, Francisco I-677, III-676, IV-35
Uribe-Paredes, Roberto I-615

Valcarce, José L. IV-328, IV-353
Valente, João III-58
Valero-Lara, Pedro I-615
Varga, Zoltán V-511
Vasic, Jelena I-602
Vázquez-Poletti, J.L. III-582
Vega, Davide III-540
Vega-Rodríguez, Miguel A. II-475
Vello, Michela IV-270
Verderame, Gerardo Mario II-109
Verdú, Elena I-76
Verdú, María Jesús I-76
Vidács, László V-293
Vidal, Juan C. IV-533
Vieira, Marcelo B. III-654
Vieira, Marcelo Bernardes I-690
Vigneault, Philippe I-356
Villarini, M. IV-258
Villegas, Osslan Osiris Vergara II-522
Vivanco, Marta G. III-374
Vivanco, Marta García IV-35
Vivone, Gemine II-354
Vizzari, Marco I-103
Vlach, Milan III-300
Vlad, Adriana I-706
Vona, Marco I-410
Vrábelová, Marta V-485
Vyatkina, Kira III-94

Wachowicz, Monica I-1
Walia, Sudeep Singh I-643
Walkowiak, Krzysztof I-512
Wang, Hao IV-173
Westrych, Katarzyna II-63
White, Roger I-119
Wierzbicka, Agata II-63
Williams, Brendan I-119
Winstanley, Adam II-51
Wójcicki, Mateusz II-63
Won, Dongho IV-123, V-437
Won, YunJae IV-209
Woźniak, Michał I-512
Wylie, Tim III-74

Xing, Changyou V-562
Xu, Zhao I-300

Yan, Ming V-577
Yang, Liu V-577
Yang, Soo-Hyeon III-518
Yang, Ziyu V-577
Yasmina Santos, Maribel I-1, I-253
Yeong-Sung Lin, Frank I-667
Yen, Hong-Hsu I-667
Yim, Keun Soo III-479
Yokota, Hidetoshi V-167
Yong, Kian Yan III-668
Yoon, David II-624
Yu, Jinkeun IV-185
Yu, Lidong V-562
Yu, Myoung Ju V-205
Yunes, Tallys III-1

Zalyubovskiy, Vyacheslaw IV-148
Zambonini, Edoardo III-412
Zemek, Michal III-163
Zenha-Rela, Mário V-270
Zhang, Jiashu V-458
Zhou, Junwei IV-604
Zhu, Binhai III-74
Zoccali, Paolo I-284
Zorrilla, Marta III-638
Zubcoff, José II-680